D1788168

BIBLIOGRAPHY ON EXPERIMENT AND TREATMENT DESIGN

PRE-1968

BIBLIOGRAPHY ON EXPERIMENT AND TREATMENT DESIGN PRE-1968

WALTER T. FEDERER

Professor of Biological Statistics
and in Charge of the Biometrics Unit
Cornell University

and

LESLIE N. BALAAM

Director of Biometrical Services
University of Sydney

Published for the International Statistical Institute by

OLIVER & BOYD · EDINBURGH

OLIVER & BOYD
Tweeddale Court
14 High Street
Edinburgh EH1 1YL
A Division of Longman Group Limited

First published 1972

ISBN 0 05 002670 4

Printed in Great Britain by T. & A. Constable Ltd., Edinburgh

CONTENTS

I. GENERAL INTRODUCTION

Publications on the various aspects of experiment design (the arrangement of treatments in an experiment) and of treatment design (the selection of the treatments included in an experiment) have become numerous, dispersed, and more frequent in recent years. Thus, searching the literature for papers on a specified topic has become a difficult task which can be considerably alleviated by a comprehensive bibliography, especially if a classification of the citations has been made. The general statistics bibliographies by Kendall and Doig [1962, 1965, 1968] are helpful in this connection; however, bibliographies on specific topics are even more useful. This was the reason for the preparation of a previous bibliography, which covered literature on the philosophy, design, conduct, and statistical analyses of experimental investigations prior to 1950 (see Federer [1964]); no classification of the literature citations was presented except that a partial classification was given in the textual part of the report.

The present bibliography covers literature citations on the philosophy, layout, conduct, and statistical analysis of experimental investigations; literature on the construction and use of experiment (experimental) and treatment designs and their associated statistical analyses received the major emphasis in this work which contains an estimated 8700 different literature citations. Most listings are single literature citations but a number are multiple references. The latter type of listing was used for references to abstracts, corrections, and addenda to an article from the same journal. Also, two or more consecutive papers in the same journal by the same author on related topics frequently were included together as a single listing in the bibliography. In some instances, additional references were discovered after the citations had been numbered. In order to maintain the alphabetical-chronological listing, these additional references were inserted between two numbered references and given the lower number with an a, b, c, etc. appended.

The bibliography is divided into five sub-bibliographies or parts. The first part is given in Section VI and covers the same area of literature covered by Federer [1964] for the pre-1950 era; it is a revised and considerably expanded version of the previous one. More than 1900 published works are included, with the citations being numbered from 1 to 1884. The second part (Section VII) consists of more than 1700 references on experiment design for the 1950 to 1967 period. The listings are numbered from 2001 to 3683. The third part of the bibliography (Section VIII) contains approximately 2000 references on the subject of treatment design. Some of the listings under treatment design were also listed under experiment design as it was desired to have a complete, self-contained bibliography on each of the subjects of experiment and treatment design. Part three on treatment design was the only section in which duplication of listings was allowed. These listings are numbered from 3684 to 5622 and cover the years 1950 to 1967. The fourth part of the bibliography (Section IX) pertains to published literature on the philosophy, conduct, application, analysis, and other aspects of the planning and conduct of experimental investigations. The approximately 2750 listings are numbered from 5623 through 8378 and cover the years from 1950 through 1967. Part five (Section X) is a listing of books on experiment and treatment designs.

In addition to subdividing the bibliography into sub-bibliographies as described above, a classification of listings was developed. A hierarchical classification system was used to classify papers. First a paper was classified as to whether it was chiefly on experiment (experimental) design as E, on treatment design as T, on analyses of experiments as A, on the general conduct of experiments or on special topics as C, on bibliographies as related to the design and analysis of experiments as B, and on sequential and fixed sample size procedures as S.

At the second level of classification each E, T, A, and S categories (but not C and B) were divided by numbers 1,2,... ≤ 15 into subclasses.

Finally, at the third level a paper in the E and T categories (but not in the A, B, C, and S categories) was classified as to whether it was on the general conduct, on the application, or on the analysis for an experiment or a treatment design, which was coded a; whether it was on the construction procedure for an experiment or a treatment design which was coded c; or whether the paper contained a numerical example of an experiment or treatment design which was coded e. The classification system is described in sections II, III, and IV for the E, T, and other categories, respectively.

As with all bibliographies some references have been omitted accidently and some publications containing articles on the design and analysis of experiments did not come to the authors' attention. Also, the searching was done mostly by title with some searching being done from summaries or abstracts. In the latter case the reference and the checking of the citation came from the abstract. The references at the end of an article were scanned to determine if any should be included. Quite often an entire volume was scanned if it contained one or more appropriate articles for the bibliography. The following journals were searched in their entirety for references to be included. The abbreviation listed on the right is used in the bibliography. The listing is in alphabetical order, omitting articles and prepositions from the title. (Several additional journals were partially or completely searched. Some of those completely scanned were omitted from the following list because they did not include papers on experiment and treatment design.)

Agronomy Journal, Vol. 42 (1950) to Vol. 59 (1967) (U.S.)	Agron. J.
American Society for Quality Control, Technical Conference Transactions, ninth (1955) to twenty-first (1967) (U.S.)	ASQC, Tech. Conf. Trans.
The American Statistician, Vol. 4 (1950) to Vol. 21 (1967) (U.S.)	Amer. Stat.
Annals of the Institute of Statistical Mathematics, Vol. 2 (1950) to Vol. 19 (1967) (Tokyo, Japan)	Ann. Inst. Stat. Math.
The Annals of Mathematical Statistics, Vol. 21 (1950) to Vol. 38 (1967) (U.S.)	AMS
Applied Statistics, Vol. 1 (1952) to Vol. 16 (1967) (Great Britain)	Appl. Stat.
Australian Journal of Agricultural Research, Vol. 1 (1950) to Vol. 18 (1967) (Australia)	Australian J. Agri. Res.
The Australian Journal of Statistics, Vol. 1 (1959) to Vol. 9 (1967) (Australia)	Australian J. Stat.
Biometrics, Vol. 6 (1950) to Vol. 23 (1967) (International)	Biometrics
Biométrie-Praximétrie, Vol. 1 (1960) to Vol. 8 (1967) (Belgium)	Biométrie-Praximétrie
Biometrika, Vol. 39 (1950) to Vol. 54 (1967) (Great Britain)	Biometrika
Biometrische Zeitschrift, Vol. 1 (1959) to Vol. 9 (1967) (Germany)	Biom. Zeit.
Bragantia, Vol. 10 (1950) to Vol. 26 (1967) (Brazil)	Bragantia
The British Journal of Statistical Psychology, Vol. 1 (1950) to Vol. 19 (1966) (Great Britain)	British J. Stat. Psychology
Bulletin of the Calcutta Mathematical Society, Vol. 42 (1950) to Vol. 57 (1965) (India)	Bull. Calcutta Math. Soc.
Bulletin de l'Institut International de Statistique, Vol. 32 (1950) to Vol. 41 (1966)	Bull. ISI
Bulletin of Mathematical Statistics, Vol. 4 (1950) to Vol. 12 (1967) (Fukuoka, Japan)	Bull. Math. Stat.

Calcutta Statistical Association Bulletin, Vol. 2 (1950) to Vol. 16 (1967) (India)	Calcutta Stat. Assoc. Bull.
Canadian Journal of Mathematics, Vol. 2 (1950) to Vol. 19 (1967) (Canada)	Canadian J. Math.
Comptes Rendus, Des Séances de l'Académie des Sciences, Paris, Vol. 230 (1950) to Vol. 265 (1967) (France)	Comptes Rendus
Crop Science, Vol. 1 (1961) to Vol. 7 (1967) (U.S.)	Crop Sci.
Current Science, Vol. 19 (1950) to Vol. 36 (1967) (India)	Current Sci.
Der Züchter, Zeitschrift für theoretische und angewandte Genetik, Vol. 20 (1950) to Vol. 37 (1967) (Germany)	Der Züchter
Dissertation Abstracts, Vol. 10 (1950) to Vol. 27 (1967) (U.S.)	Dissertation Abs.
Estadística, Journal of the Inter-American Statistical Institute, Vol. 8 (1950) to Vol. 25 (1967)	Estadística
Euphytica, Journal of Plant Breeding, Vol. 1 (1952) to Vol. 16 (1967) (Netherlands)	Euphytica
Genetical Research, Vol. 1 (1960) to Vol. 10 (1967) (Great Britain)	Genetical Res.
Heredity, Vol. 4 (1950) to Vol. 14 (1960) (Great Britain)	Heredity
Industrial and Engineering Chemistry, Vol. 42 (1950) to Vol. 59 (1967) (U.S.)	Ind. Eng. Chemistry
Industrial Quality Control, Vol. 6 (1950) to Vol. 24 (1967) (U.S.)	IQC
Journal of the American Statistical Association, Vol. 45 (1950) to Vol. 62 (1967) (U.S.)	JASA
Journal of Combinatorial Theory, Vol. 1 (1966) to Vol. 3 (1967) (U.S.)	J. Combinatorial Theory
Journal of the Indian Society for Agricultural Statistics, Vol. 2 (1950) to Vol. 19 (1967) (India)	JISAS
Journal of the Indian Statistical Association, Vol. 1 (1963) to Vol. 5 (1967) (India)	JISA

Journal of Research of the National Bureau of Standards, Mathematics and Mathematical Physics, Series B, Vol. 44 (1950) to Vol. 71 (1967) (U.S.)	J. Res. NBS
Journal of the Royal Statistical Society, Series A, (General), Vol. 113 (1950) to Vol. 130 (1967) (Great Britain)	JRSSA
Journal of the Royal Statistical Society, Series B, (Methodological), Vol. 12 (1950) to Vol. 29 (1967) (Great Britain)	JRSSB
Journal of Science of the Hiroshima University, Series A-1 (Mathematics), Vol. 17 (1953) to Vol. 31 (1967) (Japan)	J. Sci. Hiroshima Univ. Ser. A-1
Journal of the Society for Industrial and Applied Mathematics, Vol. 1 (1953) to Vol. 15 (1967) (U.S.)	J. Soc. Ind. Appl. Math.
Mathematics of Computation from 1960, formerly was Mathematical Tables and Aids to Computation, Vol. 4 (1950) to Vol. 21 (1967) (U.S.)	Math. Tables Aids Comp. or Math. Comp.
Melhoramento, Estudos da Estação de Melhoramento de Plantas, Vol. 4 (1950) to Vol. 16 (1963) (Portugal)	Melhoramento
Memoirs of the Faculty of Science, Series A (Mathematics), Kyūshū University, Vol. 6 (1951) to Vol. 21 (1967) (Fukuoka, Japan)	Mem. Faculty Sci.
Methods of Information in Medicine, Methodik der Information in der Medizin, Vol. 1 (1962) to Vol. 5 (1966) (Germany)	Meth. Inf. Med.
Metrika, International Journal for Theoretical and Applied Statistics, Vol. 1 (1958) to Vol. 12 (1967) (Germany)	Metrika
Metron, International Review of Statistics, Vol. 15 (1950) to Vol. 26 (1967) (France)	Metron
Mitteilungsblatt für Mathematische Statistik (became Metrika), Vol. 4 (1952) to Vol. 9 (1957) (Germany)	Mitteilungsblatt Math. Stat.
Plant Breeding Abstracts, Vol. 20 (1950) to Vol. 37 (1967) (U.S.)	Pl. Br. Abs.
Proceedings of the American Mathematical Society, Vol. 1 (1950) to Vol. 18 (1967) (U.S.)	Proc. Amer. Math. Soc.

Proceedings of the American Society for Horticultural Science, Vol. 55 (1950) to Vol. 91 (1967) (U.S.)	Proc. Amer. Soc. Hort. Sci.
Proceedings of the nth Conference on the Design of Experiments in Army Research Development and Testing, Vol. 1 (1957) to Vol. 12 (1967) (U.S.)	Proc. Conf. Design Expt. Army Res. Dev. Testing
Proceedings of the kth Berkeley Symposium on Mathematical Statistics and Probability	Proc. kth Berkeley Symp. Math. Stat. Prob.
Psychometrika, Vol. 15 (1950) to Vol. 32 (1967) (U.S.)	Psychometrika
Publications de l'Institut de Statistique de l'Universite de Paris, Vol. 1 (1952) to Vol. 16 (1967) (France)	Pub. Inst. Stat. Univ. Paris
Quality Control and Applied Statistics (Abstracts) Vol. 11 (1966) to Vol. 12 (1967) (U.S.)	QC Appl. Stat.
Reports of Statistical Application Research, Union of Japanese Scientists and Engineers, Vol. 1 (1951-2) to Vol. 14 (1967) (Japan)	Reports Stat. Appl. Res.
Revue de l'Institut International de Statistique, Vol. 18 (1950) to Vol. 35 (1967) (International)	Rev. ISI
Revue de Statistique Appliquée, Institut de Statistique Universite de Paris, Vol. 1 (1953) to Vol. 15 (1967) (France)	Rev. Stat. Appl.
Sankhyā, The Indian Journal of Statistics, Vol. 10 (1950) to Vol. 29 (1967) (India)	Sankhyā
Statistica Neerlandica (named Statistica prior to Vol. 9, 1955), Vol 4 (1950) to Vol. 21 (1967) (Netherlands)	Statistica or Stat. Neerlandica
Suid-Afrikaanse Tydskrif vir Landbouwetenskap, Vol. 1 (1958) to Vol. 10 (1967) (South Africa)	Suid-Afri. Tydskrif Landbouwetenskap
Technometrics, Vol. 1 (1959) to Vol. 9 (1967) (U.S.)	Technometrics
Trabajos de Estadística, Vol. 1 (1950) to Vol. 15 (1964)	Trabajos Estad.
Zeitschrift für Pflanzenzüchtung, Vol. 29 (1950) to Vol. 58 (1967) (Germany)	Z. Pflanzen-züchtung

In addition to the above journals the following books and special publications containing a series of papers represent some of the other sources scanned for papers on experiment design and treatment design:

Biometrical Genetics

Bulletin de l'Institut Agronomique et des Stations de Recherches de Gembloux (Hors Serie), 1960, Vol. 1.

Cold Spring Harbour Symposia on Quantitative Biology, Proceedings.

Colloques Internationaux du Centre National de la Recherche Scientifique, No. 10, Le Plan d'Experience.

Contributions to Probability and Statistics: Essays in Honor of Harold Hotelling.

Contributions to Statistics: Presented to Professor P. C. Mahalanobis on the Occasion of his 70th Birthday.

Economic and Technical Analysis of Fertilizer Innovations and Resource Use.

Experimental Designs in Industry.

Heterosis.

Mathematics and Computer Science in Biology and Medicine.

Statistical Genetics and Plant Breeding, NAS-NRC.

Statistical Theory of Reliability.

Statistics and Mathematics in Biology.

Stochastic Models in Medicine and Biology.

II. CLASSIFICATION FOR EXPERIMENT (EXPERIMENTAL) DESIGNS

In the published literature two terms, experimental design and experiment design, have been utilized to refer to the arrangement of treatments in a comparative experiment. Perhaps the term experiment design is a more appropriate term than experimental design when one compares it with the terms treatment design and survey design. Also published literature is highly confused on the use of the term experimental design in that this term has been used for almost all statistical aspects of experimentation from the selection of sample size and of treatments to the conduct and analyses of experiments. This would be one reason for discarding the term experimental design and for using the term experiment design.

In general, experiment designs are classified into designs with zero-way, one-way, two-way, etc. elimination of heterogeneity in experimentation. (As may be noted from published literature, e.g. reference 2194, one alternate form of classification is to denote these designs as one-dimensional, two-dimensional, three-dimensional, etc., respectively; other classifications have been devised.) This refers to the type of blocking in an experiment. The categorization can result in some confusion for such designs as the split plot (or the split block) experiment designs and the lattice square experiment design. The latter experiment design is classified as having three-way elimination of heterogeneity, i.e., complete blocks and rows and columns within each complete block. In the split-block design analyses using only one-way blocking are possible. Similarly in a resolvable balanced incomplete block design, an intrablock analysis on treatments and residual is possible without considering the fact that complete blocks were designed into the experiment.

The various subclassifications of experiment designs are listed below as El, E2, ···, El4.

1. Zero-way elimination of heterogeneity E1

   i) completely randomized design including experiments for which the treatments may be grouped in various ways in a hierarchical manner.

2. One-way elimination of heterogeneity E2

   i) balanced
      -randomized complete block
      -balanced incomplete block
      both resolvable and nonresolvable.

ii) partially balanced E3
- -circular
- -cyclic
- -group divisible
- -rectangular lattices
- -right angular
- -triangular
- -2,3,···,n associate class

iii) other incomplete block designs E4
- -augmented
- -calibration
- -chain block
- -direct product
- -extended incomplete block
- -linked block
- -reinforced
- -repetitions of designs with same treatments
- -slipped block
- -staircase
- -star
- -unequal block sizes
- -unequal replication on treatments

iv) others E5
- -blocking on a trend or gradient
- -constrained randomization
- -n-ary
- -rotation and other serially correlated
- -simple changeover
- -simple multiresponse
- -split plot (whole plots in a completely randomized design)
- -supercomplete
- -systematic

v) missing, mixed-up, or damaged experimental units (plots) E6

3. Two-way elimination of heterogeneity

i) row-column designs E7
- -gradients within blocks for designs with one-way elimination of heterogeneity
- -latin square and generalized latin square
- -simple change-over
- -tied-latin squares
- -Youden square and extended Youden square
- -other latin rectangle or row-column

ii) row-column designs to measure residual effects E8
- double change-over
- rotation
- tied double change-over

iii) others E9
- augmented
- balancing in two groups
- direct product
- multivariate hierarchical grouping
- nested balance
- plaid and half-plaid latin square
- quasi-latin square
- Trojan square

iv) missing, mixed-up, or damaged plots E10

4. Three-way elimination of heterogeneity

i) lattice square, lattice rectangle, and latin cube (including missing, mixed-up, and damaged plots) E11

ii) others E12
- augmented
- graeco-latin square
- latin cube
- magic latin square
- quasi-latin square
- split block (or criss-cross)
- split and split, split plot with one- or higher-way elimination of heterogeneity among whole plots
- U:VW and U:VW:XYZ plans for two sets of treatments

5. Four and higher-way elimination of heterogeneity E13
- hyper-graeco-latin square and cube
- lattice square with split plots
- ℓ-restrictional lattice
- multidimensional
- split, split, split ··· plot

6. Properties of experiment designs E14
- balance and partial balance
- blocking
- connectedness
- efficiency
- orthogonality
- randomization

-replication
-sensitivity
-others

Each of the above categories (except for E14) was further classified as a, c, and/or e. The a refers to application and/or analysis, the c refers to construction, and the e refers to a numerical example (mostly nonfictitious) for a particular experiment design.

III. CLASSIFICATION FOR TREATMENT DESIGN

Treatment design refers to the selection of treatments in an experiment. Many writers in statistical literature have used the term experimental design or design of experiments in their discussion of treatment designs. This has led to some confusion. This usage has been especially prevalent in connection with factorial and fractional factorial treatment designs.

The following classification of treatment designs has been utilized to classify statistical publications:

1. Selection of X values and of the X variables in regression studies. T1

2. Factorial T2
   - -complete factorial
   - -confounding in factorial experiments
   - -mixed and fixed effects excluding variance component analysis
   - -quantity x quality interaction

3. Fractional replication

   i) regular fractions (only complete confounding of effects in aliasing structure) T3

   ii) other fractions T4
   - -response surface, rotatable, central composite, steepest ascent, EVOP, response curves, PARTAN
   - -simplex, vertices, mixtures, constrained T5
   - -weighing T6
   - -others (additional or missing combinations, group testing, random balance, randomized factorials, complementation, screening X-variables) T7

4. Dosage response (sequential and fixed) T8

5. Genetic (diallel crossing systems, parents and crosses, other genetic treatment designs) T9

6. Properties T10
   - -aliasing
   - -balance
   - -invariance

-minimum variance, trace, determinant, etc.
-optimal
-orthogonality
-randomized
-rotatability
-unbiased

7. Paired, tripled, $\cdots$, k comparisons T11

As with experiment designs, the treatment design categories (except T10) were further subdivided into the a, c, and e where these symbols have the same meaning as for experiment designs.

The categories classified as T8 and T9 are considered to be incomplete. A bibliography could be constructed in these two areas alone. The coverage for classifications T8 and T9 is considered to be sufficient to allow the reader to obtain a good share of the pertinent literature on treatment design in these two areas. The references listed plus the references in the listed articles should cover most of the material published in these areas.

IV. CLASSIFICATION FOR TOPICS RELATED TO THE PLANNING AND CONDUCT OF EXPERIMENTS AND TO THE ANALYSIS OF EXPERIMENT AND TREATMENT DESIGNS

A complete coverage was attempted for statistical publications (including special mimeograph series and theses) on experiment and treatment design. The coverage of statistical literature for the A, B, C, and S categories was not complete; publications relating solely to properties of statistical procedures were omitted. The selection of topics was arbitrary, with variance component analyses, properties of statistical procedures, time series and regression analyses in economics, sample survey design, and chi-square analyses being excluded. The topics included were those considered to be of most importance in the analysis of experiments. The topics included and the classification symbols follow:

1. Analysis and some aspects of experimentation

i)	error rates	A1
ii)	ranking procedures (ordering of treatment means, tournaments, picking a winner, screening; see also S2)	A2
iii)	multiple range and simultaneous confidence interval procedures	A3
iv)	nonadditivity of effects	A4
v)	randomization and permutation tests	A5
vi)	outliers, extreme values, stragglers, runs, residual analyses	A6
vii)	covariance and multivariate analyses including growth curves and nonlinear regression	A7
viii)	calculus of factorials and direct products	A8
ix)	transformation of data	A9
x)	unequal numbers procedures	A10
xi)	heterogeneity, pooling, and combining variances Behrens-Fisher procedures, and random sample size	A11

xii) distribution free procedures A12

xiii) summarization of results from groups of experiments, allocation of resources and units to various categories, long term experiments A13

xiv) miscellaneous topics (plot size and shape, uniformity trials, Kronecker products in statistical analysis, cumulants and poly-kays, plant density, competition, border effect, experimental controls, subsampling, random versus systematic arrangements, vector analysis) A14

xv). computer programs for analyses of experiments A15

2. Sample size

i) fixed S1

ii) sequential sampling and sequential experimentation S2

3. Conduct and philosophy of experimentation and special procedures (general conduct, grouping experiments, clinical trials, survey of experimentation, measurement of tastes and threshholds, data analyses including jackknife procedures) C

4. Bibliography (main purpose of paper was to present a bibliography or if a good bibliography was included in the article) B

No further subdivision of the above categories was made. It should be noted that some categories have much in common; for example, many papers listed under A2 are concerned with determining sample size, some papers in A7 are concerned with long term experiments, and some papers in A13 are concerned with allocation of samples within various strata.

V. LISTING OF REFERENCE NUMBERS BY CLASSIFICATION

In order to facilitate use of the bibliography for special studies, all reference numbers pertaining to a given classification are listed in numerical order in the section following the books on experiment and treatment design. The classifications for which this listing was made are E1, E2, ···, E14, T1, T2, ···, T11, A1, A2, ···, A15, B, C, S1, and S2.

E1 (Unblocked designs)

7, 71, 134, 358, 407, 445, 449, 564, 569, 709, 711, 872, 994, 1027, 1061, 1071, 1204, 1401, 1491, 1497, 1751, 1752, 1753, 1754, 1768, 1824, 2034, 2043, 2052, 2053, 2063, 2071, 2072, 2074, 2087, 2095, 2101, 2106, 2109, 2136, 2195, 2208, 2209, 2236, 2291, 2324, 2374, 2376, 2416, 2424, 2428, 2441, 2444, 2454, 2459, 2484, 2509, 2512, 2513, 2539, 2602, 2617, 2683, 2698, 2699, 2700, 2701, 2703, 2704, 2735, 2747, 2771, 2772, 2774, 2810, 2816, 2823, 2825, 2842, 2843, 2844, 2865, 2891, 3087, 3091, 3099, 3115, 3117, 3132, 3164, 3193, 3223, 3233, 3242, 3243, 3245, 3281, 3309, 3420, 3433, 3434, 3463, 3466, 3532, 3533, 3575, 3587, 3588, 3592, 3593, 3681, 3682, 3814, 4016, 4074, 4246, 4354, 4356, 4602, 4603, 4604, 4811, 4869, 5075, 5100, 5142, 5187, 5365, 5369, 5531, 5537, 5621, 5622.

E2 (Balanced designs)

15, 20, 40, 41, 58, 73, 74, 76, 79, 82, 88, 94, 101, 102, 103, 106, 114, 115, 116, 117, 118, 119, 121, 125, 131, 132, 135, 141, 158, 166, 167, 168, 169, 170, 171, 172, 173, 176, 177, 180, 181, 184, 185, 186, 187, 188, 191, 193, 199, 201, 202, 213, 216, 242, 253, 254, 258, 261, 266, 267, 280, 281, 283, 299, 304, 305, 318, 320, 324, 332, 334, 342, 343, 349, 351, 354, 358, 361, 363, 364, 365, 366, 384, 387, 403, 412, 415, 424, 444, 445, 449, 453, 459, 460, 477, 511, 532, 542, 548, 549, 557, 559, 564, 570, 573, 581, 583, 607, 608, 611, 614, 634, 645, 651, 692, 698, 706, 720, 721, 757, 758, 765, 766, 767, 768, 769, 774, 778, 787, 788, 795, 800, 813, 840, 849, 868, 872, 878, 903, 904, 905, 909, 920, 921, 922, 930, 934, 939, 961, 972, 973, 974, 983, 988, 1000, 1008, 1013, 1022, 1027, 1041, 1044, 1058, 1061, 1063, 1071, 1074, 1084, 1089, 1091, 1092, 1093, 1099, 1101, 1106, 1116, 1121, 1128, 1130, 1132, 1139, 1140, 1141, 1145, 1151, 1156, 1159, 1160, 1162, 1164, 1166, 1173, 1174, 1179, 1180, 1181, 1182, 1187, 1189, 1192, 1195, 1196, 1209, 1214, 1228, 1241, 1242, 1243, 1245, 1250, 1253, 1255, 1271, 1273, 1276, 1279, 1283, 1296, 1297, 1304, 1307, 1316, 1325, 1332, 1333, 1334, 1336, 1337, 1340, 1343, 1351, 1373, 1376, 1385, 1386, 1387, 1388, 1389, 1390, 1391, 1392, 1393, 1396, 1400, 1402, 1419, 1420, 1421, 1422, 1423, 1435, 1439, 1454, 1455, 1468, 1470, 1487, 1490, 1491, 1494, 1499, 1502, 1518, 1551, 1573, 1584, 1585, 1592, 1596, 1597, 1607, 1616, 1620, 1621, 1625, 1641, 1644, 1649, 1658, 1660, 1672, 1684, 1685, 1686, 1692, 1694, 1695, 1712, 1722, 1724, 1728, 1729, 1734, 1751, 1752, 1754, 1757, 1760, 1761, 1775, 1776, 1778, 1782, 1785, 1791, 1795, 1799, 1806, 1807, 1808, 1817, 1819, 1820, 1821, 1822, 1823, 1824, 1825, 1827, 1828, 1829, 1833, 1834, 1835, 1836, 1843, 1844, 1847, 1849, 1850, 1852, 1854, 1855, 1869, 1875, 2008, 2011, 2015, 2021, 2025, 2027, 2028, 2033, 2035, 2036, 2039, 2040, 2041, 2043, 2050, 2051, 2052, 2053, 2054, 2057, 2059, 2060, 2061, 2062, 2063, 2065, 2069, 2070, 2073, 2075, 2078, 2079, 2082, 2083, 2084, 2085, 2086, 2094, 2098, 2099, 2100, 2101, 2102, 2105, 2106, 2108, 2116, 2117, 2118, 2119, 2127, 2133, 2134, 2136, 2137, 2138, 2140, 2141, 2142, 2147, 2148, 2153, 2155, 2156, 2164, 2174, 2181, 2186, 2188, 2190, 2196, 2197, 2198, 2199, 2200, 2202, 2203, 2204, 2205, 2206, 2207, 2209, 2216, 2217, 2218, 2219, 2220, 2221, 2222, 2223, 2224, 2225, 2226, 2227, 2228, 2236, 2239, 2244, 2245, 2246, 2247, 2248, 2248a, 2251, 2263, 2267, 2270, 2271, 2272, 2273, 2274, 2275, 2275a, 2278, 2283, 2288, 2291, 2300, 2307, 2314, 2316, 2321, 2324, 2325, 2328, 2331, 2332, 2333, 2334, 2335, 2337, 2338, 2339, 2346, 2347, 2352, 2359, 2360, 2361, 2372, 2374, 2375, 2391, 2392, 2393, 2400,

E2 (cont'd.)

2408, 2410, 2411, 2414, 2416, 2417, 2418, 2433, 2434, 2435, 2438, 2439, 2441, 2442, 2443, 2451, 2454, 2455, 2456, 2457, 2462, 2469, 2471, 2473, 2474, 2474b, 2477, 2481, 2484, 2485, 2486, 2492, 2493, 2494, 2495, 2497, 2499, 2507, 2509, 2512, 2513, 2519, 2520, 2534, 2539, 2540, 2543, 2546, 2548, 2549, 2550, 2551, 2564, 2566, 2567, 2572, 2575, 2576, 2577, 2581, 2582, 2583, 2585, 2595, 2596, 2597, 2598, 2600, 2602, 2604, 2605, 2607, 2608, 2609, 2611, 2617, 2618, 2619, 2620, 2622, 2623, 2624, 2635, 2638, 2639, 2640, 2643, 2644, 2645, 2646, 2647, 2649, 2650, 2651, 2652, 2654, 2658, 2659, 2660, 2661, 2662, 2663, 2665, 2666, 2676, 2677, 2678, 2681, 2683, 2684, 2686, 2687, 2692, 2693, 2702, 2710, 2718, 2719, 2721, 2722, 2723, 2725, 2726, 2727, 2728, 2729, 2730, 2731, 2734, 2735, 2736, 2738, 2739, 2740, 2742, 2747, 2750, 2751, 2752, 2754, 2757, 2758, 2760, 2766, 2768, 2773, 2777, 2780, 2784, 2786, 2789, 2790, 2791, 2793, 2794, 2795, 2796, 2797, 2806, 2807, 2809, 2813, 2816, 2821, 2824, 2828, 2833, 2836, 2838, 2842, 2843, 2844, 2856, 2858, 2863, 2864, 2868, 2870, 2871, 2872, 2873, 2875, 2876, 2878, 2882, 2892, 2893, 2895, 2896, 2897, 2899, 2900, 2901, 2902, 2903, 2904, 2908, 2913, 2914, 2915, 2917, 2918, 2919, 2920, 2921, 2922, 2924, 2927, 2928, 2932, 2944, 2945, 2946, 2951, 2952, 2954, 2955, 2959, 2961, 2969, 2971, 2982, 2983, 2985, 2986, 2989, 2990, 2995, 2996, 2997, 2999, 3000, 3003, 3004, 3005, 3006, 3009, 3011, 3017, 3018, 3019, 3022, 3023, 3036, 3048, 3051, 3055, 3059, 3060, 3061, 3064, 3070, 3074, 3076, 3077, 3082, 3088, 3089, 3090, 3095, 3096, 3099, 3100, 3113, 3114, 3115, 3117, 3129, 3132, 3137, 3140, 3149, 3157, 3158, 3161, 3162, 3164, 3166, 3168, 3173, 3177, 3181, 3182, 3183, 3184, 3185, 3191, 3192, 3194, 3195, 3197, 3199, 3200, 3201, 3202, 3207, 3215, 3221, 3222, 3223, 3227, 3229, 3232, 3233, 3235, 3239, 3244, 3246, 3251, 3252, 3254, 3259, 3260, 3261, 3262, 3264, 3265, 3266, 3269, 3277, 3281, 3282, 3283, 3285, 3286, 3287, 3289, 3293, 3298, 3300, 3301, 3308, 3310, 3312, 3322, 3328, 3331, 3332, 3334, 3335, 3336, 3341, 3342, 3343, 3344, 3345, 3346, 3355, 3357, 3358, 3359, 3361, 3363, 3364, 3365, 3369, 3373, 3374, 3382, 3383, 3388, 3391, 3396, 3398, 3399, 3401, 3403, 3404, 3405, 3407, 3408, 3409, 3410, 3411, 3413, 3418, 3419, 3420, 3421, 3428, 3430, 3431, 3432, 3436, 3437, 3439, 3440, 3443, 3445, 3453, 3456, 3459, 3460, 3461, 3462, 3464, 3465, 3468, 3472, 3474, 3475, 3476, 3479, 3480, 3482, 3483, 3486, 3488, 3489, 3491, 3492, 3494, 3496, 3498, 3512, 3515, 3519, 3520, 3521, 3525, 3529, 3532, 3535, 3537, 3541, 3542, 3549, 3550, 3551, 3552, 3553, 3554, 3556, 3563, 3565, 3566, 3567, 3568, 3570, 3573, 3577, 3578, 3579, 3580, 3584, 3586, 3588, 3589, 3590, 3591, 3593, 3596, 3597, 3598, 3601, 3605, 3607, 3608, 3609, 3612, 3613, 3614, 3616, 3625, 3627, 3628, 3634, 3635, 3639, 3641, 3648, 3651, 3652, 3656, 3663, 3665, 3666, 3669, 3673, 3675, 3676, 3677, 3678, 3679, 3680, 3681, 3682, 3683, 3716, 3727, 3764, 3772, 3780, 3814, 3859, 3861, 3869, 3874, 3884, 3940, 3963, 3964, 3965, 3966, 3984, 3985, 3986, 3987, 3988, 3989, 3990, 3991, 3992, 4016, 4052, 4063, 4064, 4074, 4094, 4098, 4111, 4211, 4212, 4299, 4304, 4337, 4342, 4344, 4345, 4346, 4351, 4354, 4356, 4448, 4472, 4519, 4553, 4556, 4570, 4578, 4784, 4800, 4811, 4818, 4828, 4833, 4839, 4851, 4855, 4866, 4876, 4885, 4889, 4890, 4891, 4903, 4904, 4932, 4956, 4968, 5002, 5013, 5027, 5068, 5075, 5100, 5123, 5142, 5155, 5156, 5162, 5180, 5184, 5187, 5273, 5350, 5382, 5390, 5391, 5392, 5399, 5426, 5427, 5450, 5471, 5492, 5494, 5531, 5557, 5558, 5572, 5607, 5608, 5609, 5615, 5621, 5622.

E3 (Partially balanced incomplete block designs)

15, 20, 39, 54, 119, 170, 171, 180, 181, 185, 188, 191, 203, 236, 241, 242, 245, 272, 273, 305, 308, 309, 312, 320, 339, 340, 341, 342, 343, 345, 349, 354, 356, 404, 415, 444, 488, 511, 542, 559, 610, 611, 614, 620, 645, 676, 677, 678, 679, 680, 698, 718, 733, 787, 828, 840, 868, 869, 905, 909, 974, 975, 976, 983, 1027, 1084, 1089, 1164, 1166, 1170, 1173, 1174, 1209, 1221, 1228, 1276, 1284, 1299, 1331, 1333, 1334, 1336, 1337, 1340, 1343, 1376, 1377, 1391, 1392, 1420, 1421, 1455, 1495, 1518, 1628, 1671, 1684, 1685, 1686, 1694, 1791, 1843, 1847, 1848, 1853, 1855, 1869, 1883, 1884, 2008, 2011, 2012, 2013, 2014, 2015, 2016, 2020, 2023, 2024, 2025, 2027, 2035 2038, 2047, 2061, 2107, 2110, 2111, 2115, 2117, 2119, 2122, 2126, 2133, 2134, 2142, 2149, 2150, 2151, 2152, 2157, 2158, 2159, 2161, 2169, 2170, 2171, 2172, 2173, 2174, 2175, 2176, 2177, 2178, 2179, 2180, 2182, 2191, 2194, 2203, 2210, 2211, 2214, 2219, 2220, 2221, 2227, 2228, 2229, 2230, 2231, 2232, 2238, 2239, 2245, 2246, 2247, 2250, 2251, 2262, 2275a, 2278, 2285, 2286, 2287, 2289, 2294, 2295, 2296, 2297, 2298, 2306, 2309, 2310, 2311, 2312, 2314, 2315, 2316, 2317, 2318, 2321, 2329, 2330, 2332, 2333, 2336, 2339, 2340, 2343, 2344, 2345, 2352, 2355, 2356, 2357, 2373, 2375, 2381, 2394, 2396, 2402, 2405, 2406, 2408, 2412, 2413, 2417, 2418, 2419, 2420, 2422, 2427, 2460, 2475, 2478, 2479, 2481, 2483, 2498, 2500, 2509, 2521, 2522, 2523, 2526, 2533, 2544, 2545, 2547, 2552, 2557, 2564, 2570, 2589, 2591, 2604, 2605, 2620, 2625, 2626, 2630, 2635, 2640, 2644, 2646, 2649, 2667, 2669, 2672, 2683, 2686, 2695, 2696, 2697, 2706, 2708, 2709, 2710, 2711, 2716, 2717, 2719, 2743, 2744, 2745, 2746, 2757, 2758, 2759, 2767, 2769, 2771, 2777, 2778, 2779, 2780, 2781, 2782, 2783, 2785, 2787, 2788, 2790, 2806, 2808, 2809, 2813, 2817, 2818, 2819, 2820, 2824, 2828, 2829, 2831, 2832, 2837, 2838, 2839, 2840, 2847, 2848, 2850, 2851, 2852, 2853, 2854, 2866, 2867, 2869, 2871, 2873, 2876, 2879, 2880, 2881, 2893, 2896, 2897, 2904, 2907, 2912, 2915, 2918, 2921, 2922, 2923, 2924, 2925, 2926, 2927, 2928, 2929, 2930, 2931, 2932, 2933, 2934, 2935, 2936, 2937, 2938, 2939, 2940, 2941, 2947, 2948, 2949, 2950, 2951, 2952, 2959, 2960, 2961, 2962, 2963, 2964, 2965, 2969, 2973, 2974, 2975, 2976, 2977, 2978, 2979, 2980, 2981, 3015, 3016, 3017, 3020, 3024, 3025, 3027, 3028, 3031, 3032, 3033, 3034, 3035, 3036, 3037, 3038, 3039, 3040, 3041, 3044, 3048, 3053, 3060, 3061, 3069, 3071, 3072, 3073, 3075, 3078, 3079, 3080, 3083, 3085, 3095, 3096, 3098, 3099, 3116, 3132, 3140, 3143, 3149, 3150, 3158, 3167, 3169, 3170, 3171, 3173, 3174, 3178, 3179, 3180, 3185, 3188, 3189, 3190, 3191, 3195, 3196, 3197, 3203, 3206, 3208, 3209, 3210, 3211, 3212, 3213, 3214, 3219, 3224, 3225, 3226, 3228, 3229, 3247, 3249, 3251, 3253, 3254, 3255, 3256, 3259, 3261, 3262, 3263, 3265, 3266, 3267, 3268, 3269, 3270, 3271, 3272, 3273, 3274, 3276, 3284, 3294, 3296, 3297, 3302, 3307, 3311, 3315, 3320, 3321, 3325, 3326, 3327, 3328, 3329, 3330, 3331, 3333, 3337, 3338, 3339, 3340, 3346, 3347, 3349, 3351, 3352, 3353, 3358, 3359, 3361, 3362, 3365, 3366, 3367, 3368, 3369, 3372, 3373, 3375, 3381, 3384, 3385, 3386, 3389, 3390, 3391, 3392, 3393, 3395, 3396, 3400, 3401, 3402, 3403, 3404, 3406, 3416, 3417, 3435, 3438, 3441, 3442, 3444, 3445, 3446, 3448, 3451, 3454, 3455, 3472, 3475, 3482, 3483, 3484, 3486, 3488, 3489, 3493, 3495, 3498, 3504, 3505, 3508, 3512, 3513, 3514, 3516, 3525, 3539, 3545, 3564, 3568, 3569, 3577, 3579, 3583, 3586, 3609, 3610, 3611, 3623, 3624, 3625, 3626, 3627, 3640, 3650, 3654, 3658, 3659, 3660, 3661, 3662, 3663, 3664, 3665, 3666, 3667, 3668, 3672, 3676, 3874, 3901, 3963, 3969, 3982, 3986, 3987, 3991, 3992, 3993, 3997,

E3 (cont'd.)

3998, 3999, 4098, 4199, 4200, 4211, 4212, 4213, 4354, 4364, 4365, 4609, 4621, 4624, 4713, 4784, 4800, 4814, 4815, 4816, 4817, 4838, 4839, 4840, 4885, 4889, 4904, 4932, 4956, 5011, 5012, 5013, 5068, 5074, 5075, 5100, 5123, 5156, 5171, 5180, 5285, 5291, 5294, 5295, 5301, 5350, 5351, 5390, 5391, 5392, 5399, 5493, 5558, 5596, 5606, 5607, 5608, 5609, 5610.

E4 (Additional incomplete block designs)

16, 17, 311, 645, 909, 1166, 1175, 2009, 2010, 2015, 2016, 2022, 2032, 2042, 2064, 2093, 2096, 2104, 2123, 2124, 2162, 2163, 2174, 2258, 2259, 2260, 2265, 2306, 2342, 2348, 2354, 2358, 2390, 2397, 2398, 2399, 2400, 2401, 2403, 2409, 2421, 2466, 2489, 2496, 2501, 2503, 2508, 2509, 2521, 2522, 2523, 2550, 2586, 2590, 2610, 2612, 2621, 2733, 2741, 2745, 2756, 2761, 2762, 2765, 2766, 2775, 2776, 2806, 2831, 2832, 2835, 2839, 2840, 2841, 2939, 3029, 3042, 3043, 3045, 3046, 3047, 3048, 3128, 3140, 3141, 3142, 3175, 3176, 3195, 3198, 3204, 3205, 3217, 3218, 3219, 3220, 3254, 3256, 3295, 3323, 3350, 3352, 3356, 3462, 3467, 3482, 3484, 3497, 3510, 3522, 3523, 3531, 3536, 3543, 3544, 3545, 3546, 3547, 3548, 3571, 3583, 3623, 3624, 3625, 3627, 3629, 3637, 3638, 3653, 4041, 4116, 4182, 4217, 4354, 4364, 4365, 4784, 4840, 5013, 5171, 5294, 5397, 5474.

E5 (Miscellaneous designs)

16, 17, 63, 66, 155, 157, 159, 164, 200, 209, 214, 238, 245, 253, 254, 255, 278, 289, 303, 315, 316, 336, 374, 422, 433, 474, 495, 562, 564, 574, 575, 577, 634, 651, 691, 693, 694, 736, 820, 857, 902, 905, 918, 919, 940, 947, 948, 949, 950, 951, 952, 953, 954, 955, 959, 980, 998, 1024, 1062, 1117, 1192, 1196, 1197, 1203, 1211, 1216, 1255, 1256, 1261, 1262, 1270, 1271, 1322, 1345, 1350, 1356, 1371, 1388, 1390, 1391, 1392, 1410, 1436, 1437, 1442, 1445, 1462, 1474, 1482, 1502, 1547, 1554, 1570, 1584, 1587, 1589, 1594, 1610, 1624, 1631, 1669, 1691, 1729, 1746, 1809, 1810, 1835, 1847, 1852, 1861, 2003, 2005, 2017, 2018, 2019, 2044, 2046, 2048, 2055, 2056, 2058, 2069, 2102, 2105, 2108, 2128, 2131, 2142, 2233, 2235, 2250, 2255, 2256, 2257, 2258, 2259, 2266, 2292, 2301, 2308, 2319, 2322, 2323, 2325, 2326, 2327, 2367, 2368, 2369, 2370, 2377, 2382, 2385, 2386, 2415, 2432, 2436, 2437, 2445, 2446, 2452, 2453, 2458, 2465, 2468, 2471, 2472, 2473, 2480, 2509, 2512, 2518, 2537, 2538, 2558, 2562, 2568, 2593, 2641, 2655, 2664, 2674, 2675, 2681, 2688, 2689, 2691, 2705, 2707, 2723, 2755, 2786, 2803, 2805, 2814, 2815, 2824, 2859, 2860, 2882, 2883, 2884, 2898, 2916, 2953, 2967, 2972, 2987, 2988, 3002, 3008, 3010, 3014, 3056, 3058, 3060, 3089, 3097, 3099, 3121, 3122, 3123, 3124, 3127, 3131, 3132, 3133, 3134, 3135, 3136, 3138, 3139, 3140, 3144, 3145, 3146, 3165, 3186, 3187, 3231, 3275, 3279, 3280, 3371, 3426, 3433, 3447, 3449, 3450, 3452, 3469, 3470, 3475, 3477, 3478, 3481, 3485, 3486, 3487, 3490, 3499, 3507, 3525, 3530, 3531, 3532, 3547, 3571, 3582, 3593, 3599, 3603, 3604, 3608, 3615, 3629, 3630, 3631, 3633, 3642, 3643, 3644, 3645, 3647, 3649, 3678, 3874, 4179, 4259, 4260, 4290, 4328, 4354, 4356, 4556, 4608, 4773, 4891, 5005, 5075, 5095, 5100, 5101, 5389, 5390, 5557, 5566.

E6 (Missing or mixed up observations for blocked designs)

8, 70, 88, 185, 199, 203, 228, 340, 341, 343, 345, 423, 487, 512, 698, 778, 934, 936, 937, 1071, 1159, 1160, 1162, 1166, 1253, 1583, 1592, 1661, 1712, 1836, 2004, 2005, 2073, 2094, 2127, 2128, 2140, 2236, 2248a, 2269, 2291, 2349, 2350, 2351, 2352, 2353, 2359, 2392, 2394, 2396, 2407, 2408, 2410, 2411, 2426, 2429, 2460, 2477, 2509, 2520, 2540, 2548, 2549, 2584, 2589, 2595, 2596, 2597, 2600, 2631, 2657, 2680, 2686, 2690, 2720, 2749, 2750, 2834, 2866, 2867, 2868, 2895, 2943, 2985, 3007, 3057, 3067, 3081, 3082, 3132, 3133, 3148, 3202, 3236, 3300, 3313, 3345, 3425, 3428, 3430, 3431, 3474, 3509, 3525, 3574, 3596, 3597, 3598, 3601, 3607, 3658, 4016, 4074, 4337, 4354, 4876, 4903, 5100, 5101.

E7 (Row-column designs)

5, 20, 29, 41, 82, 89, 90, 104, 125, 132, 139, 158, 164, 165, 178, 185, 189, 192, 193, 199, 202, 207, 208, 215, 222, 231, 232, 237, 238, 252, 253, 254, 268, 275, 276, 299, 304, 320, 321, 334, 343, 358, 410, 432, 433, 434, 454, 455, 460, 474, 478, 481, 492, 494, 506, 509, 513, 516, 532, 540, 544, 546, 548, 549, 551, 555, 557, 558, 559, 564, 571, 582, 583, 597, 599, 600, 608, 617, 618, 628, 644, 645, 646, 647, 648, 653, 654, 660, 674, 675, 682, 686, 699, 724, 757, 792, 793, 795, 800, 814, 825, 831, 873, 877, 879, 882, 901, 907, 911, 915, 935, 939, 961, 986, 988, 989, 1000, 1008, 1013, 1019, 1020, 1027, 1044, 1053, 1054, 1057, 1063, 1074, 1075, 1078, 1079, 1080, 1081, 1084, 1085, 1086, 1091, 1093, 1099, 1125, 1128, 1145, 1169, 1195, 1197, 1205, 1214, 1218, 1242, 1244, 1245, 1250, 1260, 1276, 1279, 1282, 1297, 1301, 1325, 1335, 1385, 1386, 1387, 1388, 1389, 1390, 1391, 1392, 1398, 1400, 1419, 1420, 1441, 1454, 1461, 1462, 1463, 1473, 1483, 1487, 1491, 1501, 1522, 1528, 1538, 1548, 1550, 1551, 1552, 1553, 1554, 1555, 1578, 1579, 1585, 1587, 1589, 1592, 1604, 1614, 1615, 1616, 1639, 1642, 1643, 1645, 1646, 1647, 1672, 1676, 1689, 1700, 1702, 1703, 1704, 1705, 1706, 1707, 1708, 1724, 1732, 1757, 1760, 1772, 1775, 1776, 1785, 1791, 1795, 1807, 1830, 1836, 1837, 1843, 1845, 1847, 1849, 1852, 1864, 1866, 1868, 1869, 1870, 1871, 1874, 1875, 1878, 2001, 2002, 2006, 2007, 2029, 2030, 2031, 2035, 2045, 2046, 2064, 2069, 2076, 2077, 2078, 2079, 2080, 2089, 2090, 2091, 2092, 2102, 2103, 2106, 2108, 2120, 2121, 2125, 2127, 2129, 2136, 2139, 2143, 2144, 2145, 2146, 2147, 2148, 2154, 2161, 2165, 2166, 2167, 2168, 2183, 2184, 2185, 2187, 2189, 2192, 2193, 2194, 2234, 2236, 2237, 2238, 2239, 2241, 2242, 2243, 2248a, 2252, 2253, 2254, 2261, 2264, 2277, 2278, 2279, 2280, 2290, 2291, 2293, 2299, 2302, 2313, 2315, 2320, 2321, 2341, 2352, 2362, 2363, 2364, 2365, 2366, 2369, 2371, 2379, 2384, 2395, 2404, 2408, 2423, 2440, 2443, 2449, 2461, 2461a, 2470, 2474a, 2474b, 2484, 2487, 2488, 2495, 2504, 2505, 2506, 2509, 2511, 2512, 2513, 2527, 2528, 2530, 2531, 2536, 2541, 2553, 2563, 2565, 2569, 2573, 2574, 2577, 2578, 2579, 2580, 2585, 2587, 2588, 2592, 2599, 2601, 2603, 2608, 2613, 2615, 2616, 2629, 2632, 2633, 2634, 2636, 2637, 2642, 2648, 2649, 2653, 2655, 2679, 2681, 2682, 2683, 2685, 2687, 2712, 2713, 2714, 2715, 2723, 2724, 2726, 2728, 2729, 2730, 2732, 2750, 2753, 2770, 2811, 2816, 2822, 2827, 2830, 2833, 2845, 2846, 2849, 2855, 2862, 2863, 2875, 2878, 2882, 2889, 2890, 2893, 2905, 2909, 2910, 2942, 2947, 2948, 2949, 2950, 2956, 2957, 2958, 2968, 2970, 2985, 2990, 2991, 2992, 2994, 2997, 3004, 3012, 3013, 3048, 3052, 3054, 3059, 3060, 3061, 3063, 3065, 3066, 3068, 3084, 3089, 3090, 3092, 3093, 3094, 3099, 3101, 3102, 3103, 3104,

E7 (cont'd.)

3105, 3106, 3107, 3108, 3109, 3110, 3111, 3112, 3119, 3130, 3132, 3140, 3150, 3151, 3153, 3156, 3163, 3196, 3199, 3221, 3222, 3230, 3233, 3234, 3237, 3238, 3248, 3250, 3251, 3257, 3258, 3276, 3281, 3283, 3288, 3289, 3290, 3291, 3292, 3304, 3306, 3312, 3357, 3360, 3361, 3373, 3376, 3379, 3380, 3387, 3394, 3397, 3399a, 3415, 3421, 3423, 3427, 3429, 3433, 3457, 3467, 3471, 3492, 3501, 3502, 3503, 3506, 3508, 3511, 3525, 3527, 3534, 3562, 3572, 3581, 3586, 3590, 3593, 3594, 3597, 3598, 3600, 3606, 3607, 3608, 3617, 3618, 3619, 3620, 3621, 3622, 3625, 3632, 3634, 3636, 3646, 3655, 3656, 3669, 3670, 3671, 3673, 3675, 3707, 3814, 3847, 3901, 4016, 4019, 4020, 4059, 4074, 4098, 4137, 4183, 4349, 4354, 4356, 4483, 4484, 4517, 4556, 4575, 4578, 4630, 4811, 4828, 4845, 4865, 4866, 4890, 4891, 5013, 5027, 5075, 5094, 5100, 5124, 5187, 5242, 5433, 5483, 5554, 5557, 5579, 5615.

E8 (Row-column designs to measure residual effects)

200, 209, 212, 214, 237, 303, 319, 336, 374, 493, 495, 562, 574, 575, 577, 617, 618, 634, 698, 716, 736, 825, 919, 998, 1007, 1025, 1027, 1062, 1117, 1256, 1270, 1345, 1350, 1371, 1388, 1390, 1391, 1392, 1437, 1445, 1461, 1462, 1474, 1482, 1594, 1711, 1746, 1759, 1809, 1810, 1850, 1861, 2017, 2018, 2019, 2037, 2066, 2067, 2112, 2113, 2114, 2131, 2132, 2201, 2240, 2241, 2256, 2257, 2282, 2301, 2308, 2325, 2364, 2379, 2446, 2461a, 2490, 2509, 2514, 2517, 2535, 2542, 2568, 2580, 2606, 2614, 2627, 2628, 2691, 2694, 2753, 2805, 2859, 2877, 2883, 2884, 2885, 2886, 2887, 2888, 2889, 2890, 2898, 2916, 2953, 2972, 3026, 3030, 3097, 3118, 3119, 3120, 3121, 3122, 3123, 3124, 3125, 3126, 3127, 3132, 3136, 3216, 3299, 3341, 3370, 3371, 3377, 3378, 3424, 3469, 3470, 3500, 3503, 3517, 3558, 3595, 3602, 3630, 3631, 3633, 3856, 4019, 4354, 5094, 5095, 5100.

E9 (Miscellaneous row-column designs)

1047, 1335, 1356, 1676, 1772, 1849, 2026, 2130, 2389, 2502, 2509, 2532, 2555, 2556, 2590, 2748, 2809, 2861, 2862, 3159, 3160, 3240, 3275, 3279, 3280, 3360, 3414, 3449, 3452, 3484, 3548, 3670, 3671, 4354, 4696, 4800, 4865, 5316, 5474.

E10 (Missing and mixed up observations for row-column designs)

8, 199, 228, 343, 410, 512, 1583, 1592, 1702, 1705, 1836, 1845, 1864, 1870, 2030, 2088, 2127, 2248a, 2349, 2351, 2395, 2408, 2423, 2429, 2509, 2510, 2565, 2592, 2631, 2657, 2750, 2809, 2822, 2985, 3236, 3238, 3415, 3423, 3525, 3597, 3598, 4354, 4800.

E11 (Lattice square and rectangle, latin cube, etc.)

138, 206, 236, 239, 308, 312, 313, 320, 341, 342, 344, 346, 354, 490, 718, 733, 828, 864, 869, 1027, 1209, 1331, 1338, 1376, 1387, 1388, 1389, 1390, 1391, 1392, 1421, 1673, 1722, 1847, 1848, 1855, 1856, 2081, 2249, 2262, 2316, 2350, 2476, 2482, 2491, 2509, 2515, 2524, 2525, 2529, 2631, 2668, 2670, 2671, 2673, 2683, 2792, 2863, 2951, 2952, 3049, 3060, 3061, 3182, 3183, 3184, 3185, 3251, 3365, 3518, 3657, 3674, 4354, 4359, 4550, 4551, 4760, 4866, 5155, 5156.

E12 (Other designs for three-way elimination of heterogeneity)

14, 18, 20, 29, 42, 43, 51, 52, 56, 73, 76, 85, 89, 90, 112, 126, 159, 236, 297, 375, 376, 389, 390, 406, 443, 445, 451, 477, 481, 483, 506, 509, 513, 514, 516, 547, 552, 558, 559, 616, 617, 618, 622, 645, 652, 698, 707, 786, 788, 824, 840, 853, 864, 872, 888, 906, 920, 938, 973, 1047, 1053, 1054, 1079, 1080, 1081, 1084, 1085, 1086, 1099, 1132, 1165, 1169, 1175, 1183, 1205, 1218, 1241, 1242, 1244, 1245, 1246, 1282, 1283, 1297, 1301, 1334, 1386, 1387, 1388, 1389, 1390, 1391, 1392, 1398, 1402, 1416, 1435, 1443, 1453, 1463, 1490, 1491, 1502, 1522, 1528, 1538, 1578, 1579, 1597, 1614, 1615, 1671, 1672, 1678, 1689, 1713, 1730, 1732, 1757, 1760, 1761, 1785, 1791, 1792, 1811, 1815, 1816, 1817, 1818, 1824, 1830, 1838, 1849, 1850, 1862, 1874, 1877, 2044, 2049, 2068, 2089, 2090, 2091, 2092, 2103, 2108, 2125, 2130, 2136, 2143, 2144, 2145, 2154, 2160, 2165, 2166, 2167, 2183, 2184, 2185, 2187, 2189, 2192, 2193, 2234, 2236, 2237, 2238, 2252, 2253, 2254, 2264, 2280, 2282, 2291, 2299, 2303, 2304, 2305, 2326, 2327, 2377, 2378, 2380, 2383, 2430, 2447, 2448, 2449, 2450, 2467, 2474a, 2480, 2502, 2509, 2510, 2512, 2518, 2554, 2555, 2556, 2558, 2559, 2560, 2561, 2571, 2585, 2599, 2601, 2633, 2634, 2636, 2637, 2642, 2644, 2646, 2647, 2648, 2649, 2653, 2656, 2681, 2713, 2714, 2723, 2748, 2763, 2764, 2770, 2893, 2905, 2918, 2956, 2957, 2966, 2970, 2984, 3006, 3008, 3012, 3013, 3021, 3048, 3052, 3054, 3056, 3059, 3062, 3065, 3086, 3088, 3094, 3099, 3101, 3102, 3103, 3104, 3105, 3106, 3107, 3108, 3109, 3110, 3111, 3112, 3132, 3139, 3147, 3149, 3150, 3152, 3154, 3155, 3157, 3230, 3233, 3241, 3276, 3289, 3292, 3305, 3307, 3314, 3360, 3394, 3397, 3399a, 3411, 3412, 3414, 3457, 3458, 3490, 3499, 3524, 3528, 3532, 3555, 3559, 3560, 3561, 3572, 3590, 3593, 3608, 3619, 3620, 3621, 3622, 3656, 3673, 3678, 3762, 3890, 4016, 4074, 4159, 4354, 4356, 4556, 4696, 4721, 4722, 5002, 5013, 5027, 5038, 5075, 5100, 5123, 5187, 5316, 5477, 5479, 5481, 5482, 5557.

E13 (Designs with four and higher way elimination of heterogeneity)

18, 76, 159, 165, 192, 231, 232, 490, 506, 509, 513, 514, 516, 544, 546, 547, 548, 551, 558, 559, 597, 645, 647, 911, 915, 1079, 1080, 1081, 1084, 1169, 1205, 1301, 1398, 1463, 1522, 1528, 1614, 1615, 1689, 1759, 1830, 2042, 2091, 2092, 2125, 2130, 2143, 2144, 2145, 2165, 2166, 2167, 2194, 2234, 2239, 2280, 2282, 2299, 2304, 2449, 2502, 2516, 2518, 2599, 2608, 2633, 2634, 2636, 2637, 2646, 2647, 2648, 2649, 2770, 2855, 2951, 2952, 2954, 3013, 3054, 3094, 3102, 3104, 3107, 3108, 3110, 3111, 3182, 3183, 3184, 3185, 3289, 3292, 3365, 3397, 3399a, 3411, 3412, 3448, 3586, 3622, 3673, 3901, 4360, 5155, 5156, 5351.

E14 (Properties of experiment designs)

54, 70, 146, 203, 241, 242, 308, 309, 354, 415, 610, 828, 1046, 1089, 1209, 1421, 1455, 1470, 1628, 1689, 1728, 1843, 1847, 1850, 1883, 1884, 2030, 2034, 2038, 2062, 2081, 2097, 2135, 2212, 2215, 2249, 2268, 2276, 2288, 2373, 2387, 2388, 2425, 2431, 2463, 2464, 2474, 2512, 2547, 2566, 2578, 2594, 2649, 2751, 2786, 2788, 2790, 2798, 2799, 2800, 2801, 2802, 2804, 2812, 2824, 2826, 2857, 2869, 2874, 2904, 2906, 2911, 2913, 2919, 2942, 2967, 2993, 2998, 3036, 3050, 3059, 3070, 3115, 3140, 3143, 3158, 3213, 3215, 3252, 3258, 3278, 3296, 3311, 3316, 3317, 3318,

E14 (cont'd.)

3319, 3320, 3324, 3341, 3348, 3354, 3355, 3422, 3459, 3473, 3475, 3493, 3495, 3538, 3540, 3549, 3551, 3557, 3576, 3585, 3605, 3613, 3975, 4242, 4314, 4356, 4768, 4770, 4771, 4775, 5027, 5323, 5478.

T1 (Selection of X variables and values in regression)

301, 746, 2425, 2801, 2803, 3422, 3686, 3692, 3714, 3715, 3723, 3738, 3746, 3747, 3774, 3804, 3807, 3808, 3835, 3870, 3921, 3952, 4042, 4043, 4070, 4071, 4072, 4073, 4090, 4091, 4092, 4109, 4120, 4160, 4177, 4214, 4215, 4225, 4239, 4242, 4243, 4278, 4297, 4317, 4327, 4329, 4330, 4336, 4436, 4439, 4460, 4482, 4518, 4529, 4559, 4626, 4629, 4631, 4632, 4633, 4634, 4635, 4728, 4745, 4747, 4757, 4764, 4769, 4770, 4772, 4773, 4774, 4776, 4777, 4778, 4779, 4780, 4781, 4813, 4852, 4854, 4861, 4877, 4878, 4887, 4888, 4897, 4898, 4907, 4908, 4909, 4915, 4916, 4949, 4961, 4988, 5029, 5049, 5060, 5064, 5097, 5165, 5194, 5195, 5243, 5246, 5247, 5254, 5278, 5323, 5324, 5335, 5384, 5385, 5487, 5490, 5509, 5512, 5547, 5597.

T2 (Factorial treatment design)

40, 42, 56, 69, 91, 123, 131, 135, 159, 174, 175, 179, 182, 183, 190, 207, 208, 211, 258, 269, 300, 318, 320, 338, 339, 352, 372, 377, 378, 402, 403, 405, 418, 424, 457, 459, 460, 489, 497, 511, 518, 521, 532, 545, 547, 551, 557, 581, 602, 603, 612, 623, 627, 638, 639, 749, 753, 754, 764, 785, 817, 829, 830, 863, 864, 865, 872, 907, 908, 910, 922, 935, 943, 990, 994, 997, 1021, 1048, 1049, 1058, 1073, 1084, 1093, 1099, 1120, 1121, 1138, 1157, 1158, 1161, 1171, 1172, 1173, 1176, 1195, 1206, 1212, 1222, 1245, 1271, 1279, 1289, 1290, 1291, 1298, 1323, 1330, 1338, 1341, 1396, 1400, 1419, 1420, 1441, 1468, 1487, 1488, 1490, 1491, 1495, 1498, 1499, 1508, 1518, 1530, 1556, 1566, 1586, 1588, 1595, 1596, 1597, 1601, 1616, 1626, 1627, 1646, 1663, 1664, 1682, 1699, 1720, 1750, 1754, 1756, 1757, 1758, 1766, 1783, 1784, 1785, 1791, 1793, 1795, 1807, 1814, 1818, 1836, 1838, 1839, 1843, 1849, 1865, 1873, 1874, 2033, 2039, 2068, 2070, 2083, 2106, 2129, 2138, 2141, 2142, 2210, 2211, 2214, 2229, 2230, 2231, 2232, 2236, 2263, 2275a, 2283, 2291, 2321, 2342, 2365, 2383, 2404, 2428, 2436, 2437, 2472, 2485, 2493, 2494, 2497, 2504, 2507, 2509, 2512, 2516, 2573, 2607, 2613, 2615, 2632, 2638, 2671, 2676, 2681, 2684, 2685, 2687, 2699, 2700, 2701, 2706, 2711, 2715, 2748, 2764, 2806, 2809, 2816, 2817, 2818, 2819, 2820, 2821, 2833, 2837, 2838, 2839, 2849, 2865, 2868, 2873, 2878, 2882, 2895, 2921, 2982, 3006, 3039, 3040, 3048, 3059, 3062, 3095, 3099, 3119, 3124, 3132, 3133, 3149, 3164, 3184, 3185, 3232, 3233, 3325, 3347, 3352, 3353, 3362, 3414, 3463, 3466, 3476, 3485, 3486, 3488, 3489, 3519, 3521, 3527, 3555, 3557, 3559, 3560, 3561, 3563, 3564, 3566, 3588, 3592, 3606, 3608, 3609, 3632, 3661, 3666, 3667, 3668, 3669, 3681, 3682, 3684, 3685, 3688, 3689, 3690, 3691, 3716, 3727, 3728, 3729, 3731, 3732, 3733, 3734, 3735, 3736, 3737, 3745, 3759, 3760, 3762, 3763, 3764, 3765, 3768, 3772, 3773, 3795, 3797, 3799, 3800, 3801, 3802, 3805, 3809, 3810, 3813, 3814, 3815, 3831, 3832, 3838, 3839, 3845, 3846, 3847, 3849, 3850, 3851, 3852, 3853, 3854, 3857, 3861, 3864, 3867, 3869, 3874, 3875, 3876, 3877, 3881, 3882, 3888, 3893, 3898, 3900, 3903, 3909, 3911, 3920, 3921, 3933, 3951, 3959, 3969, 3972, 3980, 3981, 3982, 3993, 3996, 3997, 3998, 3999, 4006, 4007, 4012, 4014,

T2 (cont'd.)

4015, 4016, 4032, 4039, 4040, 4044, 4048, 4052, 4054, 4055, 4057, 4063, 4064, 4068, 4074, 4075, 4084, 4097, 4098, 4103, 4110, 4112, 4113, 4114, 4116, 4119, 4134, 4135, 4137, 4143, 4146, 4147, 4148, 4151, 4152, 4159, 4161, 4162, 4167, 4168, 4169, 4171, 4176, 4180, 4181, 4183, 4184, 4186, 4189, 4191, 4192, 4193, 4196, 4204, 4205, 4220, 4224, 4245, 4246, 4250, 4251, 4252, 4259, 4260, 4277, 4298, 4301, 4310, 4311, 4320, 4323, 4325, 4326, 4328, 4332, 4341, 4342, 4344, 4345, 4346, 4347, 4348, 4349, 4350, 4351, 4354, 4356, 4360, 4363, 4368, 4369, 4370, 4371, 4372, 4373, 4380, 4381, 4382, 4401, 4410, 4422, 4423, 4433, 4434, 4435, 4437, 4443, 4444, 4445, 4446, 4454, 4462, 4463, 4469, 4471, 4472, 4473, 4475, 4476, 4477, 4483, 4484, 4487, 4492, 4511, 4512, 4513, 4516, 4517, 4519, 4523, 4524, 4531, 4533, 4541, 4542, 4544, 4549, 4551, 4552, 4553, 4556, 4557, 4558, 4570, 4572, 4575, 4577, 4578, 4579, 4602, 4603, 4604, 4605, 4606, 4609, 4624, 4625, 4630, 4639, 4641, 4644, 4649, 4652, 4654, 4667, 4668, 4670, 4671, 4672, 4679, 4680, 4681, 4684, 4685, 4686, 4687, 4688, 4691, 4696, 4697, 4697a, 4702, 4705, 4721, 4730, 4733, 4734, 4749, 4754, 4763, 4783, 4784, 4785, 4786, 4787, 4788, 4789, 4790, 4791, 4792, 4793, 4794, 4795, 4800, 4806, 4807, 4811, 4814, 4815, 4816, 4817, 4818, 4819, 4820, 4822, 4823, 4824, 4828, 4829, 4830, 4831, 4832, 4834, 4835, 4838, 4839, 4840, 4841, 4842, 4845, 4850, 4869, 4871, 4872, 4873, 4876, 4882, 4885, 4890, 4891, 4899, 4902, 4903, 4906, 4912, 4913, 4918, 4920, 4924, 4925, 4928, 4930, 4931, 4932, 4934, 4935, 4943, 4944, 4945, 4950, 4951, 4955, 4962, 4963, 4968, 4975, 4976, 4980, 4981, 4982, 4983, 4984, 4985, 4986, 4987, 5001, 5002, 5006, 5007, 5008, 5011, 5012, 5013, 5018, 5019, 5024, 5027, 5030, 5032, 5038, 5041, 5043, 5047, 5052, 5053, 5054, 5055, 5056, 5063, 5068, 5069, 5071, 5075, 5080, 5094, 5095, 5098, 5100, 5101, 5109, 5115, 5120, 5123, 5136, 5137, 5138, 5139, 5141a, 5142, 5144, 5154, 5155, 5156, 5157, 5158, 5159, 5164, 5167, 5176, 5177, 5178, 5184, 5187, 5193, 5207, 5208, 5211, 5213, 5223, 5228, 5231, 5232, 5234, 5235, 5236, 5237, 5238, 5241, 5249, 5250, 5251, 5252, 5255, 5262, 5265, 5266, 5267, 5277, 5280, 5282, 5283, 5285, 5286, 5287, 5289, 5291, 5292, 5293, 5294, 5295, 5300, 5301, 5304, 5305, 5311, 5316, 5317, 5326, 5343, 5344, 5345, 5346, 5347, 5349, 5358, 5365, 5369, 5382, 5383, 5387, 5388, 5389, 5390, 5391, 5392, 5403, 5408, 5412, 5415, 5416, 5426, 5427, 5431, 5433, 5443, 5460, 5461, 5464, 5475, 5476, 5477, 5478, 5479, 5480, 5481, 5482, 5484, 5492, 5493, 5494, 5502, 5508, 5510, 5514, 5515, 5522, 5523, 5531, 5533, 5534, 5535, 5536, 5537, 5538, 5539, 5543, 5544, 5545, 5546, 5552, 5554, 5557, 5558, 5563, 5565, 5567, 5568, 5577, 5578, 5579, 5580, 5581, 5582, 5586, 5587, 5602, 5604, 5606, 5607, 5608, 5609, 5610, 5611, 5612, 5614, 5615, 5616, 5617, 5618, 5620, 5621, 5622.

T3 (Fractional replication - regular)

60, 62, 230, 507, 515, 518, 859, 863, 913, 914, 1289, 1290, 1291, 1330, 1341, 1344, 1758, 1843, 2241, 2277, 2386, 2445, 2455, 2509, 2512, 2806, 2873, 2921, 3048, 3098, 3445, 3535, 3694, 3701, 3703, 3704, 3705, 3706, 3708, 3709, 3710, 3713, 3733, 3736, 3763, 3775, 3789, 3790, 3791, 3792, 3834, 3850, 3851, 3882, 3885, 3886, 3887, 3888, 3889, 3891, 3894, 3903, 3912, 3921, 3948, 3949, 3972, 3994, 4014, 4015, 4019, 4036, 4037, 4038, 4049, 4056, 4059, 4076, 4110, 4113, 4117, 4118, 4125, 4126, 4127, 4129, 4163, 4164, 4165, 4167, 4171, 4172, 4178, 4179, 4184, 4185, 4198, 4206, 4223, 4224, 4287, 4290, 4291, 4299, 4315, 4318, 4319, 4320, 4321, 4323, 4331, 4340, 4341, 4354, 4356, 4358, 4367, 4385, 4400, 4403, 4404, 4419, 4420,

T3 (cont'd.)

4454, 4533, 4536, 4580, 4639, 4640, 4648, 4650, 4666, 4672, 4697, 4703, 4726, 4727, 4744, 4748, 4759, 4765, 4784, 4885, 4932, 4943, 4950, 4978, 4979, 4996, 5013, 5025, 5074, 5083, 5084, 5085, 5120, 5140, 5163, 5227, 5253, 5298, 5350, 5366, 5393, 5394, 5395, 5430, 5450, 5480, 5500, 5502, 5507, 5522, 5530, 5571, 5589, 5590, 5601, 5613, 5616.

T4 (Response surface treatment designs)

148, 382, 467, 747, 1509, 2160, 2197, 2509, 2512, 2684, 2705, 2803, 2806, 3048, 3194, 3219, 3632, 3687, 3688, 3703, 3720, 3722, 3731, 3734, 3735, 3761, 3766, 3767, 3796, 3806, 3811, 3812, 3827, 3840, 3841, 3848, 3868, 3890, 3892, 3893, 3895, 3896, 3897, 3904, 3908, 3910, 3912, 3913, 3914, 3916, 3917, 3918, 3919, 3920, 3923, 3924, 3926, 3927, 3928, 3929, 3930, 3931, 3932, 3934, 3935, 3936, 3937, 3939, 3940, 3942, 3943, 3944, 3945, 3946, 3947, 3950, 3952, 3953, 3954, 3955, 3956, 3957, 3958, 3968, 3972, 4004, 4006, 4007, 4008, 4012, 4026, 4027, 4028, 4029, 4030, 4031, 4046, 4047, 4048, 4051, 4053, 4065, 4066, 4067, 4078, 4079, 4080, 4090, 4097, 4124, 4136, 4144, 4149, 4155, 4162, 4187, 4188, 4190, 4195, 4196, 4197, 4201, 4202, 4203, 4224, 4230, 4232, 4233, 4234, 4235, 4236, 4238, 4240, 4241, 4244, 4254, 4261, 4262, 4264, 4265, 4267, 4268, 4269, 4270, 4272, 4273, 4274, 4275, 4276, 4279, 4280, 4283, 4284, 4285, 4292, 4293, 4303, 4322, 4335, 4354, 4356, 4377, 4405, 4406, 4407, 4408, 4413, 4414, 4421, 4427, 4428, 4429, 4430, 4431, 4440, 4441, 4465, 4474, 4475, 4496, 4515, 4532, 4533, 4534, 4535, 4543, 4549, 4555, 4568, 4569, 4570, 4571, 4573, 4585, 4586, 4587, 4588, 4589, 4590, 4591, 4592, 4593, 4601, 4606, 4608, 4612, 4613, 4614, 4615, 4629, 4636, 4637, 4638, 4643, 4653, 4654, 4655, 4656, 4657, 4658, 4659, 4660, 4661, 4662, 4663, 4669, 4670, 4671, 4672, 4673, 4677, 4678, 4679, 4682, 4683, 4689, 4704, 4729, 4731, 4732, 4734, 4736, 4737, 4739, 4742, 4746, 4749, 4750, 4757, 4762, 4763, 4773, 4784, 4796, 4797, 4798, 4799, 4801, 4805, 4808, 4809, 4810, 4813, 4844, 4861, 4871, 4874, 4880, 4882, 4883, 4884, 4896, 4906, 4907, 4917, 4924, 4926, 4933, 4946, 4960, 4965, 4966, 4969, 4972, 4973, 4989, 5010, 5013, 5017, 5026, 5028, 5031, 5034, 5035, 5036, 5037, 5048, 5051, 5057, 5077, 5079, 5106, 5107, 5110, 5113, 5114, 5150, 5160, 5162, 5170, 5171, 5207, 5213, 5228, 5230, 5245, 5296, 5297, 5308, 5309, 5322, 5325, 5333, 5337, 5367, 5368, 5375, 5376, 5379, 5388, 5401, 5404, 5405, 5406, 5407, 5425, 5428, 5429, 5436, 5437, 5438, 5444, 5451, 5452, 5454, 5458, 5465, 5466, 5467, 5468, 5469, 5490, 5491, 5515, 5517, 5520, 5521, 5526, 5527, 5532, 5533, 5540, 5551, 5562, 5565, 5569, 5570, 5579, 5585, 5586, 5588, 5591, 5605.

T5 (Simplex, vertices, mixtures, constrained fractional replicates)

2509, 3718, 3828, 3929, 3939, 4083, 4124, 4255, 4263, 4281, 4282, 4354, 4481, 4496, 4543, 4660, 4843, 4848, 4905, 4953, 4954, 5004, 5009, 5145, 5268, 5269, 5270, 5271, 5272, 5502.

T6 (Weighing designs)

57, 58, 59, 60, 61, 62, 639, 748, 866, 912, 1137, 1339, 1580, 1843, 2085, 2260, 2275a, 2509, 3014, 3776, 3777, 3778, 3779, 3780, 3786, 3787, 3788, 4041, 4354, 4543, 4802, 5005, 5148, 5149, 5151, 5166, 5169, 5459, 5575, 5592.

T7 (Other fractional replicates)

580, 821, 1256, 1431, 1572, 2007, 2188, 2194, 2241, 2242, 2277, 2509, 2515, 2792, 2858, 2921, 3153, 3229, 3288, 3448, 3548, 3693, 3694, 3695, 3696, 3697, 3698, 3699, 3700, 3701, 3702, 3703, 3704, 3705, 3707, 3708, 3711, 3712, 3713, 3720, 3722, 3742, 3743, 3744, 3775, 3781, 3782, 3783, 3784, 3785, 3789, 3790, 3791, 3792, 3793, 3829, 3830, 3833, 3840, 3880, 3887, 3889, 3899, 3901, 3902, 3903, 3915, 3922, 3925, 3938, 3954, 4019, 4020, 4021, 4022, 4023, 4024, 4025, 4027, 4030, 4031, 4036, 4037, 4050, 4057, 4058, 4059, 4077, 4114, 4115, 4121, 4122, 4123, 4141, 4142, 4145, 4170, 4173, 4174, 4175, 4201, 4237, 4247, 4248, 4258, 4266, 4271, 4286, 4288, 4289, 4295, 4302, 4305, 4312, 4315, 4318, 4319, 4320, 4321, 4338, 4340, 4341, 4347, 4352, 4354, 4355, 4358, 4359, 4367, 4400, 4418, 4420, 4422, 4449, 4455, 4456, 4482, 4529, 4530, 4536, 4539, 4577, 4616, 4617, 4664, 4666, 4672, 4674, 4675, 4676, 4677, 4702, 4714, 4715, 4716, 4717, 4718, 4719, 4720, 4723, 4724, 4725, 4760, 4836, 4837, 4848, 4854, 4855, 4881, 4892, 4919, 4932, 4964, 5020, 5070, 5072, 5073, 5081, 5082, 5083, 5084, 5086, 5087, 5088, 5089, 5090, 5091, 5121, 5124, 5137, 5138, 5139, 5140, 5141, 5152, 5179, 5180, 5191, 5233, 5242, 5248, 5253, 5256, 5257, 5258, 5281, 5284, 5286, 5287, 5292, 5296, 5299, 5311, 5327, 5328, 5329, 5330, 5331, 5332, 5336, 5343, 5348, 5351, 5352, 5353, 5354, 5355, 5356, 5357, 5359, 5361, 5371, 5372, 5381, 5393, 5394, 5395, 5398, 5412, 5421, 5437, 5438, 5447, 5453, 5455, 5474, 5487, 5498, 5500, 5501, 5502, 5503, 5504, 5505, 5506, 5507, 5510, 5528, 5550, 5567, 5568, 5573, 5574, 5592, 5593, 5594, 5598, 5602, 5603, 5619, 5620.

T8 (Dosage response treatment designs)

429, 492, 493, 494, 499, 500, 502, 503, 504, 505, 508, 517, 519, 522, 523, 525, 552, 584, 624, 625, 716, 743, 801, 804, 805, 806, 809, 810, 924, 925, 926, 927, 1011, 1094, 1095, 1096, 1097, 1100, 1127, 1141, 1292, 1308, 1318, 1349, 1433, 1483, 1507, 1517, 1602, 1623, 1804, 2132, 2137, 2307, 2398, 2412, 2413, 2512, 2836, 2861, 2862, 2863, 3562, 3606, 3717, 3719, 3721, 3724, 3740, 3749, 3750, 3751, 3753, 3754, 3755, 3756, 3757, 3758, 3798, 3803, 3817, 3818, 3819, 3820, 3821, 3822, 3824, 3825, 3826, 3836, 3837, 3842, 3843, 3844, 3855, 3856, 3858, 3859, 3860, 3862, 3863, 3866, 3872, 3873, 3941, 4002, 4003, 4005, 4009, 4010, 4011, 4062, 4069, 4073, 4082, 4085, 4086, 4087, 4088, 4089, 4094, 4095, 4099, 4100, 4101, 4108, 4130, 4131, 4132, 4133, 4138, 4139, 4140, 4150, 4166, 4182, 4194, 4199, 4200, 4221, 4222, 4226, 4227, 4228, 4229, 4253, 4296, 4312, 4316, 4333, 4356, 4376, 4384, 4386, 4387, 4388, 4389, 4390, 4391, 4396, 4397, 4398, 4399, 4415, 4417, 4425, 4426, 4438, 4464, 4468, 4470, 4494, 4495, 4497, 4514, 4525, 4526, 4527, 4528, 4538, 4547, 4559, 4560, 4574, 4576, 4594, 4595, 4596, 4597, 4598, 4599, 4600, 4607, 4627, 4642, 4665, 4678, 4690, 4691, 4692, 4693, 4738, 4743, 4782, 4803, 4804, 4821, 4825, 4833, 4849, 4862, 4863, 4864, 4865, 4866, 4868, 4881, 4886, 4895, 4901, 4910, 4921, 4929, 4952, 4967, 4970, 4971, 4974, 4990, 4991, 4992, 5023, 5033, 5039, 5040, 5058, 5059, 5061, 5062, 5096, 5105, 5111, 5116, 5117, 5118, 5182, 5188, 5189, 5190, 5192, 5226, 5240, 5275, 5306, 5307, 5312, 5373, 5400, 5414, 5424, 5445, 5446, 5448, 5449, 5470, 5473, 5483, 5485, 5486, 5489, 5497, 5511, 5518, 5519, 5529, 5554, 5555, 5561, 5564, 5576.

T9 (Genetic treatment designs)

221, 265, 333, 417, 452, 487, 491, 529, 703, 708, 854, 956, 1031, 1033, 1192, 1280, 1430, 1510, 1511, 1512, 1577, 1590, 1674, 1855, 1858, 2477, 2512, 2522, 2523, 2709, 2763, 2856, 3497, 3615, 3725, 3726, 3739, 3741, 3752, 3769, 3770, 3771, 3794, 3816, 3878, 3879, 3905, 3906, 3907, 3960, 3961, 3962, 3995, 4000, 4001, 4013, 4035, 4045, 4061, 4093, 4096, 4102, 4104, 4105, 4106, 4107, 4128, 4153, 4154, 4156, 4157, 4158, 4207, 4249, 4256, 4257, 4294, 4306, 4307, 4308, 4309, 4334, 4337, 4339, 4343, 4348, 4353, 4356, 4357, 4358, 4361, 4362, 4364, 4365, 4366 4374, 4375, 4378, 4379, 4383, 4392, 4393, 4394, 4395, 4412, 4416, 4424, 4432, 4447, 4449, 4450, 4451, 4452, 4453, 4457, 4458, 4459, 4461, 4478, 4479, 4480, 4485, 4486, 4488, 4489, 4490, 4491, 4502, 4503, 4504, 4505, 4506, 4507, 4508, 4509, 4510, 4537, 4540, 4545, 4546, 4548, 4558, 4562, 4563, 4564, 4565, 4566, 4567, 4581, 4582, 4583, 4584, 4610, 4611, 4616, 4618, 4619, 4620, 4621, 4622, 4623, 4645, 4646, 4647, 4698, 4699, 4700, 4701, 4706, 4707, 4708, 4709, 4710, 4711, 4712, 4722, 4735, 4751, 4752, 4753, 4755, 4756, 4758, 4766, 4767, 4812, 4826, 4827, 4846, 4847, 4851, 4853, 4856, 4857, 4858, 4859, 4860, 4867, 4870, 4875, 4879, 4893, 4894, 4900, 4911, 4914, 4922, 4923, 4927, 4936, 4937, 4938, 4939, 4940, 4941, 4947, 4994, 4998, 5003, 5015, 5016, 5021, 5022, 5044, 5045, 5046, 5066, 5067, 5076, 5092, 5093, 5099, 5112, 5119, 5122, 5125, 5126, 5127, 5128, 5129, 5130, 5131, 5132, 5133, 5134, 5135, 5136, 5143, 5153, 5161, 5172, 5173, 5174, 5175, 5181, 5183, 5186, 5196, 5197, 5198, 5199, 5200, 5201, 5202, 5203, 5204, 5205, 5206, 5209, 5210, 5212, 5214, 5215, 5216, 5217, 5218, 5219, 5220, 5221, 5222, 5224, 5225, 5229, 5239, 5244, 5259, 5260, 5261, 5276, 5279, 5288, 5302, 5303, 5313, 5315, 5320, 5321, 5334, 5338, 5339, 5340, 5341, 5342, 5377, 5378, 5380, 5386, 5397, 5409, 5410, 5411, 5413, 5417, 5418, 5419, 5420, 5422, 5434, 5435, 5463, 5488, 5495, 5496, 5499, 5513, 5516, 5524, 5525, 5548, 5549, 5553, 5556, 5559, 5566, 5583, 5584, 5599, 5600.

T10 (Properties of treatment designs)

2464, 2799, 2801, 2802, 2803, 2804, 3945, 4006, 4007, 4008, 4226, 4314, 4406, 4411, 4529, 4589, 4731, 4768, 4770, 4771, 4773, 4775, 4854, 4993, 5352, 5374, 5432, 5500, 5507, 5520, 5599.

T11 (Paired, tripled, ... k comparisons)

629, 632, 633, 692, 876, 1144, 1384, 1472, 1609, 1881, 2153, 2203, 2205, 2206, 2207, 2212, 2216, 2217, 2219, 2220, 2222, 2223, 2226, 2227, 2228, 2338, 2417, 2418, 2419, 2421, 2455, 2457, 2583, 2670, 2759, 2876, 2897, 2959, 3308, 3541, 3616, 3654, 3684, 3685, 3730, 3748, 3823, 3865, 3871, 3883, 3884, 3963, 3964, 3965, 3966, 3967, 3970, 3971, 3973, 3974, 3975, 3976, 3977, 3978, 3979, 3983, 3984, 3985, 3986, 3987, 3988, 3989, 3990, 3991, 3992, 4017, 4018, 4033, 4034, 4081, 4111, 4208, 4209, 4210, 4211, 4212, 4213, 4216, 4217, 4218, 4219, 4231, 4299, 4300, 4301, 4304, 4324, 4402, 4409, 4442, 4448, 4449, 4466, 4467, 4493, 4498, 4499, 4500, 4501, 4520, 4521, 4522, 4550, 4554, 4561, 4628, 4651, 4694, 4695, 4713, 4740, 4741, 4761, 4889, 4904, 4942, 4948, 4956, 4957, 4958, 4959, 4977, 4995, 4997, 4999, 5000, 5014, 5042, 5050, 5065, 5078, 5102, 5103, 5104, 5108,

T11 (cont'd.)

5146, 5147, 5168, 5185, 5263, 5264, 5273, 5274, 5290, 5310, 5314, 5318, 5319, 5360, 5362, 5363, 5364, 5370, 5396, 5402, 5423, 5423a, 5439, 5440, 5441, 5442, 5456, 5457, 5462, 5471, 5472, 5541, 5542, 5560, 5572, 5595, 5596.

A1 (Error rates)

535, 6382, 6383, 6734, 6735, 6736, 7964, 8071.

A2 (Ranking procedures)

1457, 2162, 2163, 2216, 2417, 2503, 2508, 2538, 2594, 3852, 3984, 4001, 4002, 4003, 4208, 4209, 4211, 4222, 4392, 4409, 4457, 4458, 4459, 4504, 4505, 4506, 4507, 4554, 4641, 4651, 4698, 4699, 4922, 4942, 5039, 5122, 5424, 5439, 5442, 5593, 5595, 5637, 5638, 5640, 5641, 5642, 5718, 5735, 5736, 5737, 5749, 5750, 5752, 5753, 5754, 5755, 5762, 5763, 5779, 5780, 5781, 5782, 5783, 5784, 5785, 5786, 5787, 5788, 5789, 5790, 5791, 5792, 5794, 5795, 5848, 5851, 5852, 5884, 5895, 5901, 5907, 5923, 5944, 5967, 5989, 6029, 6030, 6036, 6042, 6043, 6096, 6099, 6101, 6102, 6209, 6254, 6270, 6272, 6273, 6274, 6296, 6297, 6298, 6299, 6313, 6314, 6315, 6335, 6364, 6377, 6434, 6454, 6548, 6549, 6600, 6601, 6647, 6648, 6649, 6651, 6652, 6653, 6654, 6655, 6656, 6657, 6658, 6659, 6660, 6661, 6667, 6668, 6674, 6693, 6694, 6830, 6920, 6932, 6936, 6959, 7006, 7013, 7093, 7095, 7102, 7103, 7135, 7137, 7180, 7186, 7218, 7219, 7220, 7221, 7227, 7278, 7289, 7290, 7374, 7403, 7443, 7447, 7492, 7509, 7511, 7512, 7514, 7515, 7542, 7576, 7578, 7580, 7603, 7612, 7613, 7626, 7669, 7670, 7671, 7674, 7677, 7703, 7711, 7758, 7790, 7802, 7832, 7834, 7835, 7836, 7838, 7839, 7892, 7901, 7927, 7928, 7930, 7933, 7934, 7938, 7946, 7948, 7949, 7986, 7991, 8043, 8065, 8104, 8239, 8342, 8348.

A3 (Simultaneous confidence intervals)

449, 497, 535, 978, 1189, 1250, 1380, 1637, 2217, 2534, 2794, 2823, 2824, 2842, 2843, 2844, 3164, 3239, 3244, 3301, 3434, 3465, 3468, 3533, 3985, 5142, 5659, 5728, 5729, 5801, 5889, 5899, 5916, 5922, 5945, 6047, 6051, 6078, 6087, 6095, 6106, 6107, 6126, 6144, 6169, 6188, 6189, 6190, 6191, 6237, 6285, 6286, 6287, 6288, 6289, 6290, 6291, 6292, 6294, 6295, 6296, 6297, 6298, 6299, 6300, 6302, 6303, 6306, 6307, 6308, 6309, 6310, 6311, 6312, 6316, 6317, 6325, 6326, 6358, 6373, 6477, 6478, 6480, 6524, 6525, 6550, 6553, 6556, 6560, 6566, 6567, 6700, 6734, 6735, 6736, 6737, 6738, 6739, 6740, 6741, 6746, 6747, 6748, 6833, 6874, 6995, 6996, 7008, 7047, 7049, 7058, 7065, 7068, 7077, 7078, 7079, 7080, 7081, 7083, 7084, 7085, 7093, 7104, 7135, 7142, 7147, 7162, 7241, 7294, 7295, 7296, 7297, 7301, 7302, 7308, 7316, 7404, 7412, 7421, 7426, 7427, 7428, 7455, 7471, 7510, 7531, 7543, 7547, 7548, 7549, 7561, 7562, 7597, 7598, 7599, 7600, 7601, 7646, 7654, 7655, 7656, 7657, 7672, 7675, 7676, 7677, 7687, 7703, 7704, 7713, 7714, 7715, 7717, 7718, 7719, 7720, 7721, 7724, 7747, 7759, 7760, 7761, 7762, 7805, 7806, 7808, 7834, 7850, 7870, 7882, 7947, 7954, 7959, 7960, 7961, 7962, 7963, 7965, 7966, 7967, 7988, 8026, 8043, 8044, 8061, 8062, 8071, 8072, 8076, 8143, 8144, 8158, 8159, 8172, 8174, 8186, 8269, 8270, 8271, 8272.

A4 (Nonadditivity)

1638, 2002, 2263, 2371, 2499, 2527, 2676, 3245, 3534, 3570, 3573, 3594, 4052, 4176, 4348, 4553, 5246, 5712, 6363, 6528, 6529, 6742, 6829, 6872, 7021, 7023, 7202, 7243, 7358, 8075, 8197, 8198, 8200, 8377.

A5 (Randomization and permutation tests)

2063, 2327, 2328, 2369, 2547, 2548, 2549, 2591, 2791, 2792, 3059, 3076, 3077, 3197, 3518, 3583, 3590, 3591, 3592, 3593, 3678, 3679, 3680, 3683, 4760, 5027, 5115, 5267, 5298, 5536, 5537, 6089, 6134, 6137, 6140, 6173, 6214, 6973, 6980, 7368, 7543, 8377.

A6 (Outliers, runs, and residual analysis)

318, 1263, 1272, 1361, 1362, 1539, 1723, 2045, 2076, 2598, 3300, 3310, 3774, 4165, 4168, 4169, 5670, 5671, 5673, 5676, 5679, 5765, 5769, 5826, 5829, 5895, 5897, 5906, 5925, 5935, 5975, 6145, 6186, 6192, 6193, 6194, 6206, 6211, 6255, 6256, 6258, 6259, 6271, 6272, 6273, 6283, 6391, 6392, 6414, 6462, 6490, 6500, 6523, 6534, 6600, 6625, 6640, 6641, 6642, 6643, 6644, 6645, 6670, 6671, 6672, 6673, 6692, 6701, 6743, 6798, 6821, 6889, 6923, 6931, 6945, 6997, 7015, 7075, 7088, 7089, 7090, 7092, 7094, 7120, 7127, 7128, 7153, 7164, 7165, 7166, 7198, 7199, 7268, 7311, 7360, 7371, 7380, 7386, 7389, 7392, 7489, 7569, 7573, 7586, 7587, 7588, 7589, 7702, 7860, 7880, 7911, 7945, 8030, 8033, 8047, 8048, 8053, 8081, 8162, 8169, 8180, 8249, 8265, 8266, 8378.

A7 (Covariance and multivariate analysis)

7, 10, 47, 48, 53, 64, 72, 76, 100, 131, 138, 163, 197, 204, 205, 207, 208, 229, 263, 264, 294, 295, 300, 310, 317, 342, 357, 391, 399, 400, 407, 409, 412, 424a, 428, 454, 465, 466, 467, 469, 475, 476, 483, 492, 511, 529, 530, 531, 550, 554, 560, 581, 599, 604, 626, 702, 709, 734, 737, 738, 739, 740, 750, 751, 778, 796, 798, 799, 808, 811, 812, 830, 845, 846, 848, 849, 851, 852, 867, 888, 909, 932, 936, 942, 967, 989, 1010, 1019, 1020, 1030, 1052, 1073, 1074, 1125, 1152, 1159, 1160, 1163, 1168, 1178, 1183, 1190, 1193, 1212, 1228, 1236, 1243, 1254, 1279, 1288, 1321, 1325, 1342, 1346, 1369, 1395, 1416, 1437, 1457, 1487, 1491, 1500, 1509, 1526, 1527, 1575, 1594, 1612, 1618, 1620, 1630, 1636, 1658, 1660, 1665, 1666, 1670, 1697, 1710, 1712, 1719, 1757, 1770, 1774, 1776, 1778, 1781, 1784, 1788, 1789, 1790, 1796, 1797, 1801, 1831, 1851, 1872, 2095, 2117, 2269, 2322, 2323, 2324, 2330, 2349, 2352, 2366, 2370, 2379, 2385, 2396, 2408, 2471, 2473, 2485, 2486, 2487, 2493, 2494, 2497, 2528, 2529, 2539, 2550, 2579, 2602, 2603, 2606, 2611, 2618, 2688, 2698, 2703, 2824, 2908, 2990, 2991, 3051, 3087, 3088, 3093, 3116, 3132, 3138, 3186, 3187, 3245, 3277, 3279, 3449, 3468, 3472, 3476, 3491, 3492, 3519, 3524, 3528, 3530, 3532, 3537, 3550, 3554, 3607, 3634, 3662, 3664, 3688, 3715, 3732, 3745, 3804, 3807, 3808, 3810, 3853, 3904, 3919, 3920, 3953, 3956, 4053, 4068, 4086, 4087, 4109, 4152, 4162, 4252, 4322, 4327, 4342, 4344, 4345, 4346, 4373, 4446, 4471, 4473, 4490, 4582, 4584, 4605, 4636, 4638, 4669, 4682, 4683, 4684, 4699, 4705, 4796, 4803, 4824, 4830, 4835, 4914, 4926, 5010, 5063, 5097, 5098, 5100, 5183, 5266, 5278, 5317, 5336, 5382, 5394, 5395, 5426, 5431, 5484, 5508, 5544, 5555, 5623,

A7 (cont'd.)

5625, 5628, 5629, 5631, 5632, 5633, 5641, 5642, 5655, 5656, 5658, 5661, 5695, 5696, 5700, 5701, 5705, 5708, 5709, 5717, 5731, 5734, 5742, 5743, 5745, 5751, 5757, 5758, 5759, 5766, 5770, 5774, 5777, 5796, 5803, 5805, 5806, 5818, 5820, 5821, 5822, 5823, 5824, 5825, 5827, 5835, 5837, 5846, 5847, 5849, 5865, 5873, 5874, 5875, 5883, 5891, 5893, 5902, 5903, 5905, 5919, 5927, 5931, 5932, 5934, 5936, 5937, 5955, 5960, 5962, 5976, 5981, 5987, 5997, 5999, 6010, 6012, 6013, 6014, 6015, 6016, 6017, 6018, 6023, 6024, 6025, 6026, 6032, 6039, 6044, 6045, 6046, 6048, 6061, 6071, 6074, 6077, 6079, 6089, 6092, 6115, 6120, 6123, 6125, 6127, 6129, 6130, 6131, 6132, 6137, 6138, 6139, 6143, 6153, 6155, 6156, 6161, 6162, 6164, 6165, 6166, 6170, 6171, 6172, 6178, 6180, 6183, 6204, 6208, 6210, 6214, 6217, 6223, 6233, 6234, 6235, 6236, 6246, 6249, 6251, 6252, 6275, 6276, 6279, 6282, 6301, 6304, 6305, 6318, 6319, 6320, 6321, 6324, 6333, 6334, 6338, 6343, 6344, 6346, 6351, 6361, 6364, 6365, 6366, 6369, 6371, 6373, 6376, 6378, 6404, 6405, 6412, 6415, 6416, 6430, 6438, 6440, 6441, 6442, 6446, 6461, 6467, 6469, 6474, 6484, 6492, 6498, 6499, 6501, 6513, 6526, 6527, 6532, 6533, 6542, 6543, 6546, 6550, 6552, 6553, 6555, 6557, 6558, 6559, 6562, 6563, 6574, 6582, 6583, 6585, 6593, 6598, 6603, 6628, 6629, 6630, 6631, 6632, 6633, 6634, 6635, 6636, 6637, 6638, 6639, 6652, 6669, 6675, 6676, 6678, 6686, 6697, 6698, 6699, 6700, 6702, 6703, 6708, 6712, 6725, 6726, 6728, 6732, 6745, 6750, 6751, 6754, 6755, 6756, 6764, 6767, 6774, 6777, 6781, 6794, 6795, 6796, 6798, 6801, 6807, 6809, 6815, 6817, 6820, 6826, 6827, 6840, 6846, 6847, 6851, 6860, 6861, 6862, 6879, 6881, 6882, 6883, 6885, 6886, 6888, 6895, 6896, 6898, 6900, 6905, 6907, 6910, 6911, 6916, 6917, 6918, 6919, 6921, 6923, 6933, 6939, 6946, 6948, 6950, 6962, 6963, 6964, 6965, 6966, 6977, 6978, 6980, 6981, 6982, 6989, 6990, 6993, 6994, 7001, 7002, 7003, 7004, 7011, 7016, 7028, 7029, 7039, 7048, 7050, 7051, 7052, 7067, 7069, 7073, 7074, 7080, 7081, 7083, 7084, 7085, 7086, 7089, 7094, 7095, 7096, 7097, 7100, 7106, 7108, 7122, 7127, 7132, 7133, 7148, 7155, 7156, 7162, 7163, 7169, 7172, 7174, 7189, 7207, 7211, 7212, 7213, 7214, 7228, 7229, 7230, 7240, 7244, 7253, 7254, 7255, 7257, 7258, 7259, 7262, 7264, 7265, 7269, 7284, 7285, 7286, 7291, 7292, 7314, 7318, 7320, 7324, 7345, 7356, 7361, 7400, 7402, 7406, 7411, 7418, 7419, 7429, 7430, 7433, 7440, 7441, 7447, 7448, 7453, 7459, 7460, 7493, 7495, 7501, 7502, 7503, 7504, 7522, 7541, 7556, 7562, 7563, 7564, 7575, 7579, 7581, 7582, 7583, 7585, 7592, 7595, 7610, 7611, 7613, 7614, 7615, 7616, 7617, 7618, 7619, 7621, 7623, 7624, 7625, 7626, 7627, 7632, 7634, 7635, 7644, 7645, 7650, 7659, 7663, 7667, 7685, 7698, 7707, 7709, 7710, 7712, 7716, 7717, 7718, 7719, 7721, 7722, 7723, 7725, 7726, 7727, 7753, 7792, 7793, 7798, 7807, 7812, 7813, 7815, 7825, 7829, 7833, 7843, 7854, 7855, 7856, 7863, 7864, 7865, 7866, 7867, 7879, 7880, 7881, 7882, 7895, 7896, 7901, 7903, 7905, 7909, 7916, 7917, 7918, 7919, 7931, 7931, 7939, 7940, 7941, 7942, 7944, 7945, 7946, 7948, 7958, 7968, 7974, 7993, 8001, 8011, 8013, 8018, 8027, 8028, 8031, 8034, 8035, 8036, 8051, 8054, 8055, 8058, 8066, 8067, 8069, 8073, 8090, 8092, 8093, 8094, 8095, 8096, 8099, 8100, 8103, 8116, 8118, 8122, 8129, 8132, 8133, 8136, 8141, 8142, 8145, 8150, 8160, 8181, 8183, 8184, 8190, 8191, 8192, 8193, 8194, 8201, 8206, 8218, 8222, 8225, 8233, 8245, 8249, 8250, 8251, 8252, 8253, 8254, 8257, 8258, 8259, 8260, 8261, 8262, 8264, 8267, 8275, 8278, 8281, 8284, 8289, 8290, 8296, 8334, 8363, 8365, 8369.

A8 (Calculus of factorials and direct products)

2831, 2838, 2840, 2929, 2938, 3546, 3670, 3671, 3763, 4114, 4369, 4370, 4371, 4372, 4839, 4841, 4842, 5617, 5618.

A9 (Transformation of data)

2, 23, 51, 73, 74, 75, 79, 82, 94, 129, 130, 132, 136, 246, 247, 248, 249, 251, 261, 291, 299, 304, 314, 383, 462, 466, 467, 598, 621, 652, 803, 925, 1200, 1432, 1504, 1507, 1652, 1683, 1734, 1767, 2138, 2248, 2331, 2551, 3133, 3553, 3570, 3717, 3842, 3861, 3864, 3865, 3933, 3980, 3981, 4088, 4277, 4298, 4348, 4738, 4822, 5024, 5101, 5125, 5576, 5678, 5725, 5817, 5819, 5836, 5898, 5900, 5936, 5972, 6001, 6168, 6176, 6181, 6262, 6357, 6419, 6435, 6450, 6455, 6482, 6496, 6534, 6618, 6757, 6768, 6769, 6770, 6778, 6803, 6872, 6985, 7006, 7125, 7170, 7326, 7337, 7338, 7351, 7354, 7358, 7376, 7434, 7435, 7437, 7554, 7606, 7607, 7608, 7631, 7649, 7668, 7710, 7734, 7859, 7950, 7975, 8050, 8078, 8176, 8177, 8192, 8197, 8244, 8303.

A10 (Unequal numbers procedures)

210, 224, 331, 407, 616, 702, 708, 757, 867, 934, 1162, 1247, 1259, 1488, 1491, 1498, 1499, 1530, 1619, 1632, 1633, 1634, 1661, 1687, 1688, 1756, 1757, 1768, 1838, 1840, 1859, 2082, 2094, 2104, 2199, 2244, 2398, 2432, 2493, 2494, 2497, 2507, 2565, 2615, 2740, 2754, 2766, 2943, 2944, 2945, 3081, 3082, 3230, 3324, 3428, 4103, 4182, 4220, 4344, 4345, 4346, 4351, 4369, 4370, 4371, 4372, 4484, 4487, 4557, 4558, 5071, 5098, 5262, 5280, 5617, 5619, 5630, 5950, 5952, 6180, 6346, 6357, 6395, 6447, 6476, 6485, 6570, 6571, 6602, 6762, 6804, 6871, 7064, 7070, 7071, 7136, 7271, 7283, 7293, 7328, 7528, 7541, 7550, 7584, 7615, 7630, 7633, 7668, 7764, 7886, 7915, 7924, 8256, 8303.

A11 (Unequal variances and pooling procedures)

82, 105, 462, 466, 681, 725, 865, 1200, 1259, 1287, 1417, 1418, 1427, 1428, 1477, 1485, 1598, 1725, 1726, 1783, 1786, 1787, 2052, 2099, 2100, 2195, 2196, 2200, 2377, 2416, 2443, 2444, 2461, 2527, 2546, 2581, 2618, 2619, 2954, 2955, 2960, 3285, 3286, 3309, 3485, 3486, 3672, 3773, 3774, 4384, 4421, 4524, 4899, 5053, 5211, 5237, 5277, 5317, 5389, 5390, 5698, 5699, 5704, 5727, 5732, 5733, 5734, 5746, 5748, 5756, 5760, 5778, 5800, 5804, 5815, 5816, 5828, 5841, 5842, 5869, 5928, 5937, 5939, 5940, 5961, 6004, 6008, 6031, 6033, 6035, 6066, 6073, 6184, 6185, 6188, 6235, 6243, 6270, 6295, 6323, 6336, 6356, 6378, 6400, 6415, 6416, 6422, 6423, 6425, 6426, 6429, 6494, 6495, 6499, 6503, 6521, 6526, 6547, 6551, 6555, 6581, 6622, 6623, 6662, 6663, 6664, 6665, 6744, 6758, 6799, 6845, 6849, 6867, 6868, 6869, 6872, 6875, 6883, 6899, 6900, 6901, 6902, 6924, 6928, 6947, 6950, 6979, 7082, 7146, 7149, 7154, 7176, 7177, 7178, 7299, 7300, 7306, 7391, 7409, 7455, 7498, 7499, 7500, 7507, 7527, 7560, 7561, 7562, 7567, 7573, 7577, 7600, 7602, 7641, 7682, 7692, 7701, 7712, 7729, 7731, 7732, 7733, 7734, 7739, 7755, 7794, 7893, 7951, 8025, 8045, 8052, 8121, 8147, 8173, 8175, 8199, 8203, 8208, 8209, 8296, 8302.

A12 (Distribution-free procedures)

1082, 1083, 1138, 1287, 1751, 1752, 1753, 1754, 1755, 1802, 2098, 2108, 2118, 2141, 2202, 2204, 2205, 2206, 2208, 2209, 2218, 2221, 2222, 2223, 2224, 2225, 2226, 2374, 2435, 2442, 2451, 2455, 2456, 2457, 2469, 2511, 2617, 2624, 2665, 2678, 2692, 2721, 2722, 2725, 2825, 2864, 2892, 3018, 3064, 3166, 3168, 3232, 3310, 3533, 3565, 3585, 3587, 3588, 3684, 3685, 3714, 3721, 3815, 3865, 3869, 3964, 3965, 3980, 3981, 3983, 3988, 3989, 3990, 4017, 4081, 4299, 4300, 4304, 4467, 4499, 4695, 4761, 4807, 4957, 4958, 4959, 5078, 5104, 5168, 5184, 5235, 5289, 5305, 5314, 5387, 5423, 5531, 5552, 5631, 5645, 5707, 5752, 5753, 5762, 5764, 5792, 5802, 5807, 5813, 5821, 5822, 5823, 5824, 5827, 5861, 5864, 5871, 5902, 5903, 5943, 5945, 5946, 5947, 5976, 5980, 6029, 6048, 6049, 6050, 6088, 6089, 6158, 6175, 6182, 6187, 6195, 6214, 6239, 6240, 6241, 6242, 6261, 6271, 6308, 6327, 6335, 6337, 6370, 6394, 6433, 6488, 6490, 6491, 6502, 6505, 6531, 6555, 6565, 6578, 6608, 6646, 6681, 6723, 6784, 6785, 6786, 6787, 6788, 6790, 6807, 6816, 6819, 6820, 6821, 6822, 6823, 6825, 6827, 6833, 6859, 6861, 6862, 6863, 6864, 6884, 6894, 6904, 6936, 6955, 6980, 6991, 7030, 7035, 7045, 7046, 7051, 7058, 7076, 7087, 7091, 7106, 7134, 7136, 7137, 7138, 7175, 7210, 7231, 7253, 7274, 7275, 7298, 7301, 7302, 7312, 7335, 7359, 7367, 7368, 7369, 7370, 7417, 7426, 7427, 7428, 7454, 7488, 7493, 7532, 7565, 7576, 7577, 7578, 7579, 7580, 7581, 7582, 7583, 7591, 7594, 7638, 7647, 7654, 7655, 7656, 7657, 7666, 7671, 7672, 7686, 7693, 7699, 7700, 7704, 7712, 7725, 7727, 7747, 7797, 7799, 7800, 7846, 7847, 7848, 7849, 7850, 7851, 7852, 7853, 7854, 7855, 7856, 7857, 7870, 7884, 7885, 7894, 7908, 7923, 7928, 7929, 7956, 7959, 7960, 7961, 7962, 7963, 7966, 7967, 7983, 7990, 7994, 8015, 8029, 8032, 8034, 8038, 8047, 8048, 8061, 8070, 8080, 8097, 8103, 8111, 8114, 8117, 8120, 8126, 8127, 8128, 8130, 8131, 8139, 8161, 8162, 8164, 8165, 8167, 8168, 8169, 8170, 8171, 8202, 8220, 8221, 8223, 8225, 8236, 8237, 8238, 8239, 8265, 8315, 8316, 8320, 8360.

A13 (Combining groups of experiments and allocation of resources)

297, 487, 491, 498, 530, 531, 582, 583, 707, 749, 777, 854, 862, 865, 946, 986, 1046, 1087, 1129, 1191, 1224, 1254, 1379, 1381, 1450, 1826, 1862, 1867, 1879, 2017, 2018, 2019, 2242, 2284, 2429, 2477, 2568, 2590, 2609, 2610, 2918, 3097, 3123, 3128, 3135, 3480, 3483, 3557, 3631, 3688, 4020, 4044, 4064, 4337, 4339, 4363, 4368, 4445, 4738, 4936, 4941, 5224, 5225, 5279, 5342, 5478, 5577, 5578, 5581, 5710, 5722, 5730, 5773, 5794, 5814, 5938, 6004, 6069, 6076, 6080, 6081, 6082, 6083, 6084, 6085, 6086, 6238, 6356, 6410, 6445, 6481, 6507, 6561, 6852, 7150, 7151, 7159, 7160, 7161, 7242, 7250, 7252, 7365, 7383, 7408, 7410, 7439, 7455, 7458, 7505, 7506, 7553, 7605, 7690, 7691, 7692, 7749, 7750, 7779, 7787, 7788, 7899, 7900, 7914, 7936, 8005, 8007, 8009, 8042, 8138, 8211, 8214, 8343, 8356.

A14 (Miscellaneous topics)

4, 13, 22, 27, 28, 30, 31, 32, 33, 34, 35, 36, 44, 45, 46, 49, 50, 53, 55, 63, 67, 71, 75, 77, 80, 81, 83, 86, 87, 95, 96, 97, 98, 99, 107, 110, 113, 120, 127, 132, 133, 142, 145, 146, 147, 149, 150, 151, 152, 153, 154, 160, 162, 194, 195, 196, 198, 201, 202, 217, 218, 219, 220, 223, 225, 234, 235, 236, 243, 262, 270, 274,

A14 (cont'd.)

277, 278, 279, 282, 283, 284, 285, 286, 287, 289, 290, 292, 296, 298, 302, 306, 307, 308, 311, 316, 322, 323, 325, 326, 327, 328, 330, 334, 335, 337, 347, 348, 350, 360, 362, 369, 379, 380, 381, 384, 385, 386, 395, 396, 397, 398, 401, 408, 411, 413, 414, 419, 420, 430, 434, 435, 436, 437, 438, 439, 441, 450, 454, 455, 456, 457, 461, 463, 464, 468, 470, 471, 472, 473, 474, 479, 480, 486, 496, 498, 501, 510, 524, 528, 532, 536, 556, 560a, 563, 564, 565, 566, 567, 576, 577, 578, 579, 583, 585, 586, 587, 588, 589, 590, 591, 592, 593, 594, 595, 596, 599, 600, 601, 608, 619, 621, 626, 636, 640, 641, 644, 650, 655, 656, 657, 658, 660, 663, 664, 668, 669, 670, 671, 672, 683, 687, 688, 689, 690, 693, 695, 696, 698, 699, 704, 705, 713, 717, 722, 727, 729, 732, 745, 752, 755, 756, 759, 760, 761, 762, 763, 770, 771, 773, 776, 780, 781, 782, 784, 789, 790, 812, 815, 816, 822, 824, 826, 832, 833, 835, 836, 837, 838, 842, 843, 844, 847, 848, 849, 850, 852, 855, 856, 858, 870, 881, 883, 884, 885, 886, 887, 889, 891, 892, 893, 894, 895, 896, 898, 899, 901, 902, 916, 917, 918, 923, 939, 940, 941, 947, 948, 949, 950, 951, 952, 953, 954, 955, 957, 959, 963, 967, 968, 971, 972, 973, 974, 976, 977, 980, 982, 984, 985, 988, 989, 991, 992, 993, 995, 999, 1001, 1005, 1013, 1014, 1015, 1018, 1019, 1020, 1021, 1024, 1030, 1033, 1036, 1037, 1039, 1040, 1042, 1043, 1050, 1055, 1063, 1064, 1067, 1068, 1069, 1070, 1072, 1076, 1087, 1088, 1090, 1093, 1104, 1109, 1110, 1111, 1112, 1114, 1118, 1119, 1122, 1123, 1124, 1129, 1131, 1133, 1142, 1143, 1148, 1152, 1153, 1154, 1155, 1177, 1184, 1185, 1186, 1188, 1193, 1198, 1199, 1203, 1207, 1208, 1210, 1211, 1213, 1216, 1217, 1220, 1223, 1224, 1225, 1226, 1227, 1229, 1230, 1231, 1232, 1233, 1234, 1235, 1236, 1237, 1238, 1239, 1248, 1249, 1252, 1266, 1267, 1268, 1269, 1274, 1275, 1277, 1286, 1293, 1294, 1300, 1302, 1309, 1310, 1311, 1314, 1315, 1319, 1324, 1326, 1327, 1328, 1348, 1353, 1354, 1355, 1356, 1357, 1358, 1360, 1366, 1367, 1370, 1372, 1375, 1376, 1386, 1403, 1404, 1415, 1424, 1425, 1435, 1436, 1444, 1448, 1449, 1450, 1451, 1452, 1454, 1455, 1460, 1464, 1465, 1466, 1467, 1469, 1470, 1475, 1476, 1478, 1479, 1480, 1482, 1484, 1485, 1503, 1504, 1505, 1513, 1514, 1523, 1524, 1533, 1534, 1541, 1542, 1543, 1544, 1548, 1554, 1555, 1557, 1558, 1559, 1560, 1561, 1564, 1565, 1567, 1568, 1569, 1570, 1581, 1582, 1587, 1595, 1599, 1606, 1610, 1611, 1624, 1628, 1631, 1650, 1653, 1654, 1658, 1659, 1662, 1663, 1664, 1667, 1668, 1669, 1675, 1677, 1679, 1691, 1695, 1696, 1709, 1714, 1715, 1716, 1717, 1718, 1731, 1735, 1736, 1737, 1739, 1742, 1743, 1745, 1748, 1749, 1763, 1764, 1792, 1800, 1803, 1813, 1831, 1832, 1841, 1842, 1846, 1857, 1863, 1867, 1876, 2045, 2065, 2077, 2078, 2079, 2102, 2341, 2366, 2373, 2431, 2458, 2459, 2462, 2474, 2544, 2723, 2734, 2812, 2869, 2871, 2893, 3089, 3090, 3132, 3193, 3242, 3243, 3281, 3298, 3421, 3475, 3554, 3567, 3608, 3745, 4061, 4251, 4488, 4498, 4706, 4951, 5046, 5100, 5193, 5206, 5223, 5315, 5380, 5557, 5574, 5626, 5634, 5636, 5639, 5644, 5649, 5663, 5665, 5681, 5682, 5684, 5685, 5702, 5721, 5725, 5726, 5767, 5776, 5830, 5840, 5876, 5877, 5878, 5879, 5880, 5885, 5886, 5887, 5888, 5917, 5918, 5924, 5926, 5964, 5966, 5977, 5978, 5979, 5985, 6001, 6005, 6006, 6008, 6015, 6022, 6027, 6028, 6034, 6037, 6040, 6056, 6057, 6060, 6094, 6103, 6104, 6113, 6114, 6115, 6117, 6118, 6125, 6157, 6174, 6202, 6205, 6224, 6247, 6248, 6250, 6264, 6265, 6266, 6267, 6268, 6269, 6278, 6280, 6328, 6329, 6341, 6342, 6374, 6375, 6379, 6390, 6396, 6411, 6432, 6436, 6443, 6448, 6449, 6453, 6457, 6458, 6459, 6464, 6465, 6471, 6473, 6483, 6497, 6508, 6509, 6512, 6515, 6516, 6517, 6518, 6519, 6520, 6535, 6544, 6554, 6564, 6568, 6572, 6580, 6595, 6624, 6688, 6691, 6709, 6711, 6714, 6715, 6716, 6717, 6718, 6719, 6721, 6759, 6765, 6766, 6771, 6772, 6779, 6783, 6796, 6797, 6814, 6824, 6831, 6832, 6834, 6855, 6856, 6857, 6865, 6873, 6876, 6890, 6891, 6897, 6913, 6914, 6915, 6922, 6934, 6940, 6943, 6944, 6949, 6954, 6956, 6957,

A14 (cont'd.)

6958, 6968, 6980, 6999, 7000, 7014, 7017, 7018, 7019, 7020, 7031, 7032, 7037, 7040, 7056, 7061, 7062, 7089, 7105, 7109, 7114, 7115, 7131, 7152, 7157, 7158, 7183, 7208, 7217, 7247, 7251, 7276, 7277, 7279, 7281, 7282, 7287, 7303, 7304, 7305, 7313, 7327, 7329, 7330, 7331, 7333, 7334, 7336, 7344, 7348, 7349, 7352, 7353, 7379, 7381, 7384, 7388, 7394, 7397, 7422, 7424, 7444, 7446, 7452, 7456, 7457, 7461, 7463, 7464, 7473, 7474, 7475, 7480, 7482, 7483, 7485, 7486, 7490, 7497, 7505, 7508, 7516, 7517, 7518, 7520, 7521, 7529, 7530, 7538, 7551, 7552, 7555, 7557, 7558, 7571, 7574, 7604, 7637, 7653, 7658, 7664, 7665, 7679, 7684, 7695, 7696, 7705, 7706, 7708, 7767, 7768, 7769, 7770, 7771, 7772, 7773, 7774, 7775, 7776, 7777, 7782, 7789, 7795, 7796, 7809, 7823, 7824, 7826, 7827, 7830, 7842, 7844, 7845, 7858, 7868, 7869, 7871, 7872, 7873, 7875, 7876, 7877, 7889, 7890, 7906, 7907, 7912, 7970, 7972, 7978, 7979, 7980, 7981, 7987, 7989, 7997, 7998, 8002, 8020, 8021, 8023, 8039, 8040, 8057, 8063, 8064, 8068, 8074, 8079, 8088, 8105, 8106, 8113, 8115, 8123, 8124, 8137, 8156, 8178, 8182, 8187, 8188, 8189, 8195, 8196, 8210, 8215, 8226, 8230, 8231, 8232, 8246, 8255, 8263, 8283, 8287, 8288, 8291, 8292, 8293, 8294, 8295, 8312, 8313, 8314, 8317, 8319, 8337, 8345, 8352, 8361, 8362.

A15 (Computer programs for statistical analyses)

717, 862, 2096, 2107, 2262, 2565, 2687, 2690, 2764, 2907, 2982, 3059, 3284, 3765, 3932, 4044, 4068, 4135, 4536, 4578, 4579, 4643, 4721, 4930, 4931, 4968, 5027, 5030, 5055, 5056, 5164, 5232, 5281, 5282, 5383, 5502, 5503, 5507, 5577, 5578, 5580, 5680, 5712, 5713, 5796, 5905, 6007, 6013, 6014, 6016, 6063, 6072, 6153, 6438, 6510, 6579, 6583, 6593, 6703, 6749, 6753, 6793, 6909, 7107, 7144, 7211, 7233, 7448, 7451, 7620, 7763, 7764, 7973, 8031, 8193, 8213, 8305, 8306, 8307, 8308, 8309, 8311, 8312.

B (Bibliographies)

41, 137, 143, 453, 631, 635, 717, 791, 798, 799, 802, 808, 811, 819, 895, 1237, 1262, 1363, 1371, 1419, 1420, 1733, 1738, 1739, 1740, 1741, 1779, 1882, 2051, 2090, 2275a, 2291, 2416, 2417, 2508, 2635, 2637, 2806, 2816, 2966, 3048, 3132, 3749, 3858, 4074, 4211, 4426, 4612, 4673, 4757, 4784, 4811, 4970, 5013, 5100, 5315, 5413, 5662, 5744, 5860, 5920, 5941, 5988, 6000, 6020, 6024, 6055, 6186, 6191, 6201, 6225, 6226, 6227, 6228, 6229, 6230, 6260, 6349, 6350, 6380, 6384, 6434, 6576, 6577, 6618, 6644, 6650, 6679, 6684, 6685, 6775, 6776, 6791, 6832, 6849, 6887, 6890, 6894, 6927, 6935, 6960, 6986, 6987, 6988, 6998, 7026, 7091, 7111, 7192, 7245, 7268, 7280, 7317, 7364, 7401, 7413, 7414, 7465, 7494, 7496, 7525, 7618, 7619, 7625, 7629, 7697, 7735, 7736, 7737, 7738, 7797, 7833, 7878, 7918, 7952, 7977, 8014, 8037, 8050, 8085, 8282, 8310.

C (Conduct of experiments and special procedures)

1, 2, 3, 4, 6, 9, 11, 12, 20, 24, 25, 27, 37, 38, 41, 46, 49, 50, 65, 68, 78, 83, 84, 86, 87, 92, 93, 108, 109, 110, 121, 122, 124, 129, 133, 134, 137, 139, 140, 141, 143, 144, 154, 157, 163, 181, 194, 226, 227, 233, 240, 244, 246, 247, 248, 249, 250, 251, 256, 257, 259, 263, 264, 265, 271, 274, 288, 293, 307, 312, 320, 323,

C (cont'd.)

328, 329, 330, 333, 350, 353, 354, 355, 359, 361, 367, 368, 370, 371, 373, 388, 392, 393, 394, 405, 416, 420, 424a, 425, 427, 428, 429, 430, 431, 432, 440, 442, 445, 446, 447, 448, 451, 456, 458, 467, 474, 482, 484, 485, 489, 494, 499, 519, 526, 527, 532, 533, 534, 537, 538, 539, 541, 543, 556, 561, 564, 568, 572, 603, 605, 606, 607, 609, 612, 613, 615, 642, 643, 649, 661, 662, 665, 666, 667, 668, 669, 670, 684, 685, 696, 697, 698, 700, 701, 710, 711, 712, 714, 719, 723, 726, 728, 729, 735, 741, 742, 743, 744, 757, 758, 759, 772, 774, 775, 778, 779, 783, 785, 787, 794, 797, 801, 807, 812, 814, 818, 821, 823, 827, 832, 834, 839, 841, 853, 857, 861, 867, 871, 874, 875, 880, 881, 883, 885, 890, 895, 900, 901, 902, 918, 928, 929, 940, 944, 945, 946, 947, 948, 949, 950, 951, 952, 953, 954, 955, 958, 959, 960, 962, 964, 965, 966, 967, 970, 972, 979, 981, 985, 987, 995, 996, 1000, 1002, 1003, 1006, 1009, 1010, 1011, 1012, 1015, 1016, 1017, 1020, 1021, 1023, 1024, 1026, 1027, 1028, 1029, 1032, 1034, 1035, 1036, 1037, 1038, 1043, 1045, 1051, 1055, 1056, 1059, 1060, 1061, 1063, 1065, 1066, 1068, 1088, 1090, 1092, 1098, 1102, 1103, 1105, 1107, 1108, 1113, 1114, 1115, 1126, 1128, 1130, 1131, 1133, 1134, 1135, 1136, 1145, 1146, 1147, 1148, 1149, 1150, 1167, 1190, 1191, 1192, 1193, 1194, 1195, 1200, 1201, 1203, 1211, 1214, 1215, 1216, 1219, 1228, 1229, 1230, 1231, 1232, 1233, 1234, 1237, 1239, 1245, 1251, 1255, 1257, 1258, 1260, 1261, 1262, 1264, 1265, 1266, 1267, 1268, 1269, 1276, 1278, 1279, 1281, 1285, 1296, 1303, 1304, 1305, 1306, 1308, 1311, 1312, 1313, 1317, 1318, 1319, 1320, 1321, 1322, 1323, 1329, 1347, 1348, 1352, 1354, 1359, 1361, 1364, 1365, 1366, 1368, 1370, 1373, 1374, 1378, 1379, 1381, 1382, 1383, 1394, 1395, 1397, 1399, 1404, 1405, 1406, 1407, 1408, 1409, 1410, 1411, 1412, 1413, 1419, 1420, 1426, 1434, 1436, 1438, 1440, 1442, 1445, 1446, 1447, 1451, 1452, 1453, 1454, 1458, 1459, 1460, 1461, 1469, 1471, 1474, 1481, 1482, 1484, 1486, 1489, 1490, 1492, 1493, 1496, 1506, 1515, 1516, 1518, 1519, 1525, 1526, 1527, 1529, 1531, 1532, 1535, 1536, 1537, 1540, 1545, 1546, 1547, 1549, 1550, 1551, 1552, 1553, 1554, 1557, 1563, 1564, 1565, 1566, 1568, 1570, 1574, 1577, 1588, 1589, 1590, 1591, 1593, 1596, 1602, 1603, 1605, 1606, 1608, 1610, 1613, 1616, 1617, 1618, 1621, 1622, 1624, 1629, 1635, 1640, 1648, 1651, 1655, 1656, 1657, 1661, 1667, 1669, 1670, 1672, 1679, 1680, 1681, 1682, 1691, 1692, 1693, 1695, 1698, 1701, 1710, 1714, 1715, 1721, 1727, 1729, 1733, 1737, 1739, 1740, 1741, 1744, 1745, 1747, 1750, 1760, 1762, 1765, 1769, 1771, 1773, 1775, 1776, 1777, 1778, 1780, 1784, 1785, 1790, 1791, 1793, 1794, 1795, 1797, 1798, 1805, 1812, 1813, 1826, 1831, 1833, 1834, 1835, 1839, 1843, 1849, 1860, 1880, 2069, 2079, 2105, 2106, 2139, 2197, 2203, 2210, 2211, 2212, 2213, 2215, 2219, 2220, 2227, 2228, 2235, 2236, 2248, 2266, 2283, 2284, 2321, 2352, 2369, 2428, 2462, 2484, 2509, 2512, 2519, 2585, 2608, 2654, 2655, 2683, 2693, 2713, 2714, 2724, 2734, 2735, 2755, 2793, 2801, 2806, 2856, 2884, 2888, 2889, 2890, 2893, 2917, 2965, 2987, 2988, 2996, 3001, 3003, 3009, 3011, 3095, 3098, 3129, 3132, 3134, 3164, 3165, 3275, 3281, 3310, 3420, 3427, 3477, 3478, 3582, 3584, 3586, 3606, 3608, 3609, 3632, 3639, 3645, 3646, 3647, 3649, 3669, 3681, 3682, 3731, 3733, 3741, 3749, 3774, 3801, 3823, 3838, 3839, 3841, 3854, 3909, 3911, 3912, 3940, 3951, 3963, 3969, 3970, 3971, 3975, 3976, 3978, 3979, 3986, 3987, 3991, 3992, 4010, 4016, 4032, 4049, 4054, 4063, 4064, 4066, 4098, 4131, 4132, 4144, 4162, 4165, 4224, 4231, 4232, 4245, 4246, 4300, 4325, 4339, 4347, 4350, 4354, 4356, 4396, 4397, 4415, 4426, 4454, 4498, 4501, 4541, 4542, 4544, 4576, 4671, 4692, 4693, 4703, 4733, 4734, 4750, 4754, 4810, 4851, 4946, 4977, 5014, 5065, 5068, 5074, 5079, 5100,

C (cont'd.)

5108, 5111, 5142, 5146, 5147, 5188, 5207, 5229, 5238, 5275, 5402, 5429, 5435, 5457, 5554, 5557, 5558, 5561, 5565, 5579, 5615, 5621, 5622, 5624, 5649, 5650, 5651, 5652, 5653, 5654, 5664, 5674, 5683, 5703, 5711, 5715, 5716, 5717, 5719, 5720, 5723, 5724, 5738, 5739, 5744, 5756, 5760, 5761, 5768, 5771, 5772, 5773, 5775, 5793, 5798, 5799, 5809, 5811, 5812, 5829, 5830, 5831, 5832, 5833, 5834, 5847, 5853, 5854, 5855, 5856, 5857, 5858, 5859, 5866, 5867, 5868, 5870, 5872, 5881, 5882, 5890, 5892, 5894, 5896, 5904, 5908, 5909, 5912, 5920, 5921, 5929, 5930, 5933, 5941, 5942, 5951, 5953, 5954, 5957, 5958, 5959, 5963, 5965, 5967, 5970, 5973, 5974, 5983, 5984, 5986, 5990, 5991, 5992, 5993, 5994, 5995, 5996, 5998, 6002, 6003, 6009, 6011, 6021, 6024, 6054, 6062, 6064, 6065, 6068, 6070, 6075, 6091, 6093, 6100, 6105, 6109, 6110, 6111, 6112, 6116, 6119, 6124, 6128, 6141, 6146, 6147, 6148, 6149, 6150, 6151, 6152, 6154, 6159, 6160, 6163, 6173, 6177, 6179, 6196, 6197, 6198, 6200, 6201, 6203, 6207, 6213, 6215, 6216, 6218, 6219, 6220, 6221, 6222, 6231, 6231, 6237, 6244, 6245, 6253, 6257, 6263, 6277, 6281, 6284, 6293, 6322, 6330, 6331, 6332, 6339, 6340, 6345, 6347, 6348, 6349, 6350, 6351, 6352, 6353, 6354, 6355, 6367, 6368, 6370, 6380, 6385, 6388, 6389, 6393, 6397, 6398, 6399, 6401, 6402, 6403, 6406, 6407, 6408, 6409, 6413, 6418, 6420, 6421, 6422, 6423, 6424, 6427, 6428, 6434, 6437, 6444, 6451, 6452, 6453, 6456, 6460, 6468, 6470, 6475, 6479, 6486, 6487, 6489, 6493, 6497, 6504, 6506, 6507, 6511, 6514, 6537, 6538, 6539, 6540, 6541, 6564, 6569, 6573, 6584, 6587, 6594, 6597, 6599, 6605, 6606, 6607, 6609, 6610, 6611, 6612, 6613, 6614, 6615, 6616, 6617, 6620, 6666, 6677, 6679, 6680, 6689, 6690, 6691, 6704, 6705, 6706, 6707, 6710, 6713, 6716, 6722, 6727, 6729, 6730, 6731, 6733, 6752, 6760, 6761, 6763, 6780, 6789, 6792, 6796, 6800, 6802, 6805, 6806, 6808, 6810, 6811, 6812, 6813, 6828, 6835, 6836, 6837, 6838, 6839, 6841, 6842, 6843, 6844, 6848, 6850, 6853, 6854, 6858, 6866, 6870, 6877, 6878, 6880, 6892, 6893, 6908, 6912, 6935, 6938, 6941, 6942, 6951, 6952, 6953, 6954, 6961, 6967, 6969, 6970, 6971, 6974, 6975, 6976, 6979, 6983, 6984, 6992, 7009, 7010, 7022, 7024, 7025, 7026, 7036, 7038, 7039, 7041, 7043, 7044, 7053, 7054, 7055, 7057, 7060, 7063, 7066, 7098, 7099, 7101, 7102, 7110, 7113, 7116, 7117, 7118, 7119, 7120, 7121, 7123, 7124, 7126, 7129, 7130, 7139, 7140, 7141, 7143, 7145, 7167, 7168, 7171, 7179, 7181, 7182, 7184, 7187, 7188, 7193, 7194, 7195, 7196, 7197, 7200, 7201, 7203, 7204, 7205, 7206, 7209, 7216, 7222, 7223, 7224, 7225, 7226, 7232, 7234, 7238, 7239, 7245, 7246, 7248, 7256, 7260, 7261, 7262, 7263, 7266, 7267, 7270, 7272, 7273, 7280, 7288, 7315, 7319, 7322, 7323, 7325, 7332, 7333, 7339, 7340, 7341, 7342, 7343, 7347, 7355, 7357, 7362, 7363, 7366, 7373, 7377, 7378, 7382, 7385, 7387, 7393, 7395, 7398, 7401, 7405, 7407, 7415, 7416, 7420, 7423, 7425, 7431, 7432, 7436, 7438, 7442, 7449, 7450, 7462, 7465, 7466, 7467, 7469, 7470, 7472, 7476, 7477, 7478, 7479, 7481, 7484, 7487, 7491, 7519, 7520, 7523, 7524, 7526, 7533, 7534, 7535, 7536, 7537, 7540, 7545, 7546, 7554, 7559, 7568, 7570, 7572, 7593, 7594, 7622, 7628, 7636, 7639, 7642, 7643, 7647, 7651, 7678, 7683, 7688, 7689, 7728, 7740, 7741, 7742, 7745, 7746, 7748, 7751, 7757, 7766, 7778, 7780, 7781, 7782, 7783, 7784, 7791, 7803, 7804, 7810, 7811, 7814, 7816, 7817, 7818, 7821, 7822, 7831, 7833, 7840, 7849, 7861, 7862, 7876, 7877, 7883, 7887, 7888, 7889, 7899, 7902, 7904, 7910, 7913, 7921, 7922, 7925, 7932, 7935, 7936, 7937, 7943, 7953, 7955, 7969, 7971, 7976, 7977, 7995, 7996, 7999, 8003, 8004, 8006, 8007, 8008, 8009, 8012, 8016, 8017, 8019, 8045, 8049, 8056, 8059, 8060, 8074, 8077, 8082, 8083, 8084, 8086, 8087, 8089, 8091, 8098, 8101, 8102, 8107, 8108, 8109,

C (cont'd.)

8110, 8112, 8119, 8134, 8135, 8140, 8146, 8151, 8152, 8153, 8154, 8179, 8204, 8209, 8211, 8212, 8215, 8216, 8219, 8224, 8234, 8235, 8240, 8241, 8242, 8243, 8244, 8247, 8248, 8255, 8268, 8276, 8277, 8280, 8286, 8290, 8297, 8298, 8299, 8301, 8304, 8306, 8310, 8318, 8321, 8322, 8323, 8324, 8325, 8326, 8327, 8328, 8329, 8330, 8331, 8332, 8333, 8335, 8336, 8338, 8339, 8340, 8341, 8344, 8346, 8347, 8349, 8350, 8351, 8353, 8354, 8355, 8356, 8357, 8358, 8359, 8364, 8366, 8367, 8368, 8370, 8371, 8373, 8374, 8375, 8376

S1 (Fixed sample size)

19, 21, 26, 30, 95, 111, 128, 150, 152, 153, 156, 161, 221, 275, 277, 333, 350, 367, 381, 408, 421, 426, 446, 464, 482, 486, 536, 539, 553, 630, 637, 641, 659, 673, 695, 696, 713, 715, 722, 730, 731, 756, 772, 773, 780, 784, 789, 836, 856, 860, 867, 883, 897, 923, 931, 933, 969, 992, 993, 1004, 1012, 1017, 1024, 1025, 1031, 1032, 1038, 1050, 1077, 1126, 1133, 1155, 1202, 1208, 1210, 1224, 1225, 1240, 1277, 1286, 1295, 1310, 1313, 1360, 1405, 1414, 1424, 1425, 1429, 1434, 1438, 1450, 1456, 1511, 1514, 1520, 1548, 1562, 1563, 1564, 1566, 1569, 1571, 1576, 1599, 1608, 1631, 1655, 1656, 1657, 1681, 1696, 1748, 1755, 2034, 2372, 2534, 2544, 2608, 2796, 2884, 3132, 4232, 4316, 4350, 4393, 4394, 4537, 4545, 5044, 5100, 5206, 5244, 5320, 5371, 5372, 5421, 5625, 5626, 5647, 5648, 5649, 5654, 5666, 5669, 5710, 5741, 5770, 5795, 5797, 5810, 5913, 5914, 5915, 5926, 5956, 5968, 5983, 6001, 6019, 6027, 6038, 6058, 6059, 6067, 6097, 6101, 6108, 6167, 6199, 6281, 6359, 6360, 6362, 6372, 6381, 6385, 6387, 6409, 6439, 6463, 6466, 6512, 6515, 6556, 6575, 6588, 6589, 6590, 6591, 6592, 6596, 6604, 6621, 6626, 6627, 6720, 6734, 6736, 6773, 6903, 6906, 6915, 6930, 6937, 6956, 7012, 7027, 7042, 7059, 7072, 7185, 7190, 7191, 7249, 7282, 7289, 7307, 7309, 7310, 7350, 7353, 7372, 7375, 7383, 7396, 7415, 7445, 7461, 7468, 7475, 7521, 7539, 7553, 7554, 7574, 7590, 7652, 7662, 7664, 7665, 7680, 7681, 7692, 7694, 7750, 7767, 7786, 7795, 7817, 7827, 7828, 7837, 7871, 7872, 7874, 7897, 7929, 7982, 8041, 8046, 8157, 8163, 8166, 8185, 8274, 8300.

S2 (Sequential sample size and procedures)

260, 467, 1520, 1521, 1600, 1690, 2292, 2293, 2513, 2733, 2772, 2956, 2957, 3223, 3542, 3704, 3721, 3840, 4099, 4100, 4101, 4221, 4226, 4227, 4228, 4229, 4232, 4262, 4284, 4297, 4393, 4394, 4443, 4444, 4492, 4548, 4559, 4560, 4627, 4643, 4665, 4666, 4676, 4769, 4796, 4906, 4921, 5048, 5081, 5082, 5086, 5091, 5195, 5309, 5445, 5446, 5518, 5519, 5564, 5627, 5635, 5643, 5646, 5657, 5660, 5667, 5668, 5672, 5675, 5677, 5686, 5687, 5688, 5689, 5690, 5691, 5692, 5693, 5694, 5697, 5706, 5714, 5735, 5736, 5737, 5740, 5747, 5756, 5782, 5783, 5785, 5786, 5789, 5808, 5838, 5839, 5843, 5844, 5845, 5850, 5862, 5863, 5910, 5911, 5943, 5944, 5946, 5947, 5948, 5949, 5968, 5969, 5971, 5973, 5982, 6036, 6041, 6044, 6045, 6046, 6052, 6053, 6054, 6078, 6090, 6098, 6101, 6102, 6121, 6122, 6124, 6133, 6135, 6136, 6140, 6142, 6209, 6212, 6239, 6362, 6386, 6394, 6417, 6431, 6472, 6488, 6521, 6522, 6530, 6536, 6545, 6569, 6586, 6619, 6626, 6667, 6668, 6682, 6683, 6687, 6695, 6696, 6724, 6773, 6818, 6886, 6888, 6925, 6926, 6927, 6929, 6932, 6972, 7005, 7006, 7007, 7013, 7033, 7034, 7035, 7112, 7173, 7180, 7215, 7230,

S2 (cont'd.)

7235, 7236, 7237, 7289, 7321, 7346, 7372, 7390, 7399, 7443, 7510, 7511, 7513, 7514, 7515, 7544, 7566, 7596, 7609, 7612, 7640, 7648, 7660, 7661, 7673, 7674, 7675, 7676, 7693, 7730, 7731, 7732, 7733, 7743, 7744, 7752, 7754, 7755, 7756, 7765, 7785, 7801, 7819, 7820, 7838, 7839, 7841, 7891, 7892, 7898, 7920, 7926, 7938, 7947, 7949, 7957, 7984, 7985, 7992, 8000, 8010, 8022, 8024, 8041, 8062, 8114, 8125, 8131, 8148, 8149, 8155, 8205, 8206, 8207, 8217, 8227, 8228, 8229, 8237, 8239, 8273, 8279, 8285, 8300, 8315.

VI. THE PRE-1950 BIBLIOGRAPHY WITH SOME COMMENTS ON ASSOCIATED DEVELOPMENTS

The contribution by L. Euler (481)* in 1782 has had and is having a profound effect upon research on the construction and properties of experiment and treatment designs as defined above. In 1832 the paper by Hassler (692) is important to modern experiment design, especially in the area of standardization of weights and measures. In agriculture, modern field experimentation began when Jean Baptiste Boussingault started a series of tests on his farm near Bechelbronne, Alsace in 1834; he was the first man to undertake field experiments on a practical scale (see Leonard and Clark (985)). In 1841, John Bennet Lawes established the Rothamsted Experimental Station on his farm in England. He was joined by J. H. Gilbert in 1843. These two men worked together for the next 57 years and their field lay-outs were models of carefully planned and well-executed experiments. In Denmark, P. Nielson began to duplicate his pasture experiments about 1870 and by 1872 (see R. A. Fisher, "The Design of Experiments") some of the Danish workers began to use the 5 x 5 square known as the Knut Vik square (238, 433, 1197, 1538, 1676). In the 1880's the systematic arrangement ABCDDCBA was used in England while other systematic schemes were devised and used in the U.S. about 1888. The next experiment designs of importance appear to be the chessboard (1547, 1551) design used at Warminster, England, since 1909, and the half-drill-strip methods proposed by Beavan and Student (1550 - 1554). The latter method was devised to test two varieties of drilled crops; it was extended to deal with more than two varieties but the mechanical convenience of the half-drill-strip system was lost.

Although many workers in the first quarter of the twentieth century kept insisting that the results from any given experiment must be considered as a random sample from a specified population and although the concept of random samples dates back to biblical times at least, randomization of experimental lay-outs was not suggested and seldom if ever used. Hall and Russell (644) in 1911 used the term "scattered" plots, and Hubback (756) and Engledow and Wadham (473) used random sampling procedures in 1924. It remained for Fisher (532) to set forth the theoretical need for randomization and to stress its importance in designing experiments.

It might be stated that experiment design as known today, had its beginning with Sir Ronald A. Fisher and began in the second quarter of the twentieth century. During this quarter the writings of Fisher on experiment and treatment design are of prime importance to researchers in these areas. This is especially true of his book entitled "The Design of Experiments", first published in 1935. Two other

---

* Numbers in parentheses refer to reference numbers associated with a particular listing in the bibliography.

greats in this era are F. Yates and R. C. Bose. Certainly the "The Design and Analysis of Factorial Experiments" by Yates (1849) is a classic and several others are important (e.g. 1837, 1838, 1843, 1844, 1845, 1847 - 1856). Likewise, one should consider Bose's 1938 paper on hyper-graeco-latin squares (165) and the Bose and Nair paper in 1939 on "Partially Balanced Incomplete Block Designs" (191) as classics; Bose is associated with several important papers from 1938 to 1949 (see 166 - 190, 192). Another person to make significant contributions in this era is W. G. Cochran (see 297 - 299, 303, 305, 307 - 309, 313, 319, 320, 1862) whose works on combining results from different experiments with the same set of treatments, on long term experiments, on incomplete block designs, and on experiments with residual effects represent noteworthy contributions to the design and analysis of experiments.

Two other people to make noteworthy contributions to the theory of experiment and treatment designs in the 1940's were D. J. Finney and C. R. Rao. The former made contributions to the theory and application of fractional replication (507, 515), confounding in factorial experiments (514, 518, 521), latin square designs (506, 509, 513, 516), and biological assays (499, 500, 502 - 505, 508, 517, 519, 522, 523, 525). C. R. Rao's contributions were in the areas of incomplete block designs (1174, 1175, 1331 - 1333, 1336, 1337, 1340, 1343), confounding in factorials (1171 - 1173, 1176, 1331, 1335, 1338, 1341), fractional replication (1338, 1341), and weighing designs (1339).

A number of other researchers contributed significantly to experiment and treatment design construction, lay-out, and analyses during the last half of the second quarter; one finds several important papers on the construction and analysis of experiment and treatment designs by such people as Banerjee (weighing designs (57 - 62)), Bliss (biological assays (131, 133 - 135, 137, 139)), Bruck (orthogonality of latin squares and balanced incomplete block designs (232)), Chowla (balanced incomplete block designs (187, 188)), Federer (combining experiments with different treatments (487, 491), prime-power lattice designs (488, 490, 868, 869)), Hall (latin square designs (645 - 647)), Hussain (incomplete block designs (764 - 769)), Kempthorne (fractional replication and confounding in factorials (863, 864), weighing designs (866), prime-power lattice designs (868, 869)), Kishen (latin and hyper-graeco-latin cubes and hypercubes (911, 915), confounding and analysis in factorials (190, 907, 908, 910), fractional replication (913, 914), weighing designs (912), MacNeish (orthogonality of latin squares (1054)), Mann (orthogonality of latin squares (1079 - 1081, 1084)), Nair (incomplete block designs (191, 1164, 1166), latin square designs (192, 1169), confounding in factorials (1157, 1158, 1161, 1171 - 1173, 1176)), Nandi (incomplete block designs (1180 - 1183)), Norton (orthogonality of latin squares (1205)), Plackett (fractional replication (1289 - 1291)), Ryser (orthogonality of latin squares and balanced incomplete block designs (232)), Youden (Youden "square" design (1868 - 1871, 1875)), and perhaps others. With regards to other aspects of planning experiments the papers by Smith (1479) and by Christidis (282, 286) on the size and shape of the experimental unit and the paper by Yates and Zacopanay (1867) on subsampling and competition represent important contributions to statistical literature. There are many worthwhile papers on analyses of data, some of which have already been mentioned; one such paper is the one by Tukey on a single degree of freedom for nonadditivity (1638).

One of the major factors affecting the quality and quantity of research in any area is the leadership and sympathetic nature of an administrator. Certainly Sir E. John Russell was a driving force at the Rothamsted Experimental Station in obtaining first rate individuals to work on the design and analysis of experiments. Likewise, H. H. Love was instrumental in motivating agriculturists to consider the various aspects of designing and analyzing experiments in the U.S. A short while later, P. C. Mahalanobis was performing these same functions in India. The efforts of such men as these had a significant effect on the acceptance and use of statistical procedures. In the U.S. Government area Charles F. Sarle was instrumental in obtaining supporting monies for studying the various aspects of the design and analysis of experiments at Iowa State University and at North Carolina State. Arnold J. King and George W. Snedecor at Iowa State University and Gertrude M. Cox at North Carolina State were driving forces in building statistical departments and in providing the scientific climate for conducting research on the many aspects of experiment and treatment designs. These three individuals made significant contributions to Statistics in the U.S. through their administrative and educational efforts. Another person making similar contributions was E. A. Cornish in Australia. The role of administrators and educators is frequently omitted in a discussion of contributions to a field of endeavor. This omission may be occasioned by the fact that a list of publications on this subject cannot be ascribed to them. This is considered unfortunate and is the reason for mentioning some outstanding educators and administrators contributing to this bibliography through their educational and administrative efforts.

We have attempted to highlight some of the important papers and developments related to experiment and treatment designs. No doubt some important contributors were overlooked, and for this the authors apologize. It is suggested that the reader simply insert additions where appropriate and delete those authors whose work he considers of lesser importance than the authors did. One of the more important developments (mostly by R. A. Fisher) in the second quarter of the twentieth century has not been discussed as yet. Probably one of the most important aspects of design theory is to develop principles and the properties, and criteria whereby one is able to compare designs with respect to each other. The following is similar to a diagram which is reported to have hung on the wall of R. A. Fisher's office at Rothamsted:

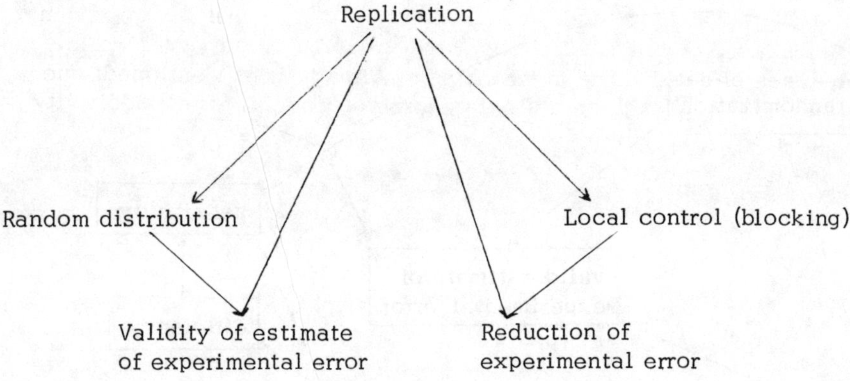

The principles of replication, randomization, and local control (blocking) form the basis for the construction and use of all experiment designs. The principles of confounding, orthogonality, and balance have been enunciated by F. Yates and the partially balanced incomplete design properties has been rigorously defined by R. C. Bose. The Fisherian concepts of efficiency and sensitivity are important design properties. We may relate all these properties and principles in a diagram similar to the above. One such diagram is given in Figure 1.

A treatment design property defined by Yates (1849) states that a factorial design is balanced if all effects of the same order, say all two-factor interactions, are confounded with blocks an equal number of times. This means that all effects of the same order for m factors each at k levels will have the same variance. Other treatment design properties appear not to have been discussed in the 1925 - 1949 period.

The pre-1950 bibliography represents a revised and enlarged version of the MRC Technical Summary Report #405 by Federer [1964]; this part of the bibliography was not subdivided since there are less than 2000 listings. It was not integrated into the 1950 - 1967 bibliographies partly because of the preparation of Report #405 and partly because it was felt that the research in the third quarter of the twentieth century involved many more people, was less of a pioneering and more of a specific nature, and was pursued at a more active pace than the research in the second quarter.

The references in the pre-1950 part as well as in other parts are listed first in alphabetical order and then in chronological order and each reference carries a code, such as for example E3ace; T2ae; A12, which classifies the article with regard to its main content. Publications with no code or with an incomplete code

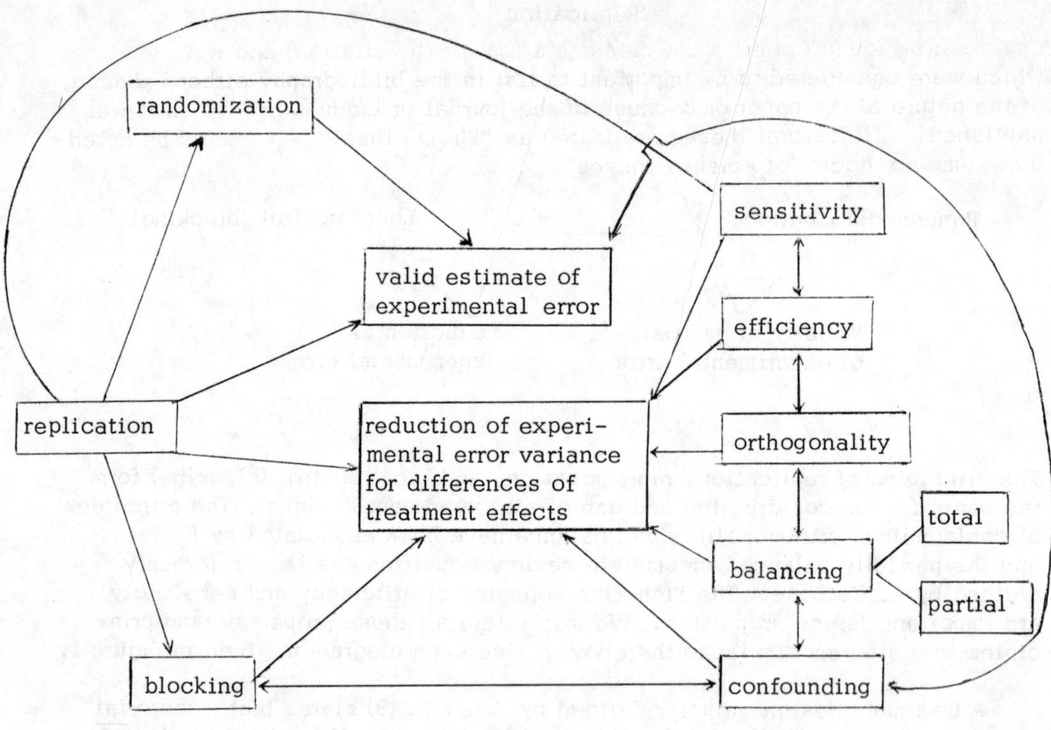

Figure 1. Some relationships among the properties of experiment designs.

were those cited in another paper but for which the original paper or an abstract could not be found to verify the classification (or the citation) and were those which were considered to be important to list in the bibliography either because of the nature of the paper or because of the journal or language in which it was published. All doctoral theses are listed as "Ph.D. Thesis"; it should be noted that some are doctor of science theses.

Advisory Entomologists Conference, Report [1944], Wireworms and food production - a wireworm survey of England and Wales (1939-1942). Ministry Agri. Bull. 128. C (1)

Afzal, M., Nanda, D. N., and Abbas, M. [1943], Studies on the cotton jassid (Empoasca devastans Distant) in the Punjab, IV. A note on the statistical study of jassid population. Indian J. Agri. Sci. 13:634-638. A9;C (2)

Agricultural Experimental Stations of France [1943], L'expérimentation au champ dans les stations de recherches agronomiques Françaises. Ann. Agron. Paris 13:295-331. C (3)

Ahmad, N. [1938], A brief note on work of statistical interest done at the cotton technological laboratory, Matunga. Sankhyā 3:295-298. A14;C (4)

Aitken, A. C. [1945], Statistical Mathematics. Oliver and Boyd 4th ed. (1st ed. 1939), London, pp. 139-142. E7ae (5)

Alexander, W. P. [1920], How to get reliable results from field experiments. Hawaiian Sugar Planters' Assoc. Bull. No. 13. C (6)

Allan, F. E. [1936], Some principles of statistics and their application to agricultural experiments III. Australian Inst. Agri. Sci. J. 1:109-116. E1ae;A7 (7)

Allan, F. E. and Wishart, J. [1930], A method of estimating yield of a missing plot in field experimental work. J. Agri. Sci. 20:399-406. E6,10ae (8)

Allen, E. W. [1930], Initiating and executing agronomic research. J. American Soc. Agron. 22:341-348. C (9)

Alumnus, [1932], A comparison of the effect of rain fall on spring- and autumn-dressed wheat at Rothamsted Experimental Station, Harpenden. J. Agri. Sci. 22:101-114. A7 (10)

Alwood, W. B. and Price, R. H. [1890], Variety tests with potatoes. Va. Agri. Expt. Bull. 6:1-20. C (11)

Anderson, J. A. [1945], The role of statistics in technical papers. Trans. Amer. Cereal Chemists Assoc. 3:69-73. C (12)

Anderson K. L. [1942], A comparison of line transects and permanent quadrats in evaluating composition and density of pasture vegetation of the tall prairie grass type. J. American Soc. Agron. 34:805-822. A14 (13)

Anderson, R. L. [1946], Missing-plot techniques. Biometrics 2:41-47. E12ae (14)

Anderson, R. L. [1946], The analysis of orthogonal square lattice experiments with d duplications of the basic design (abstract). Biometrics 2:58. E2,3a (15)

Anderson, R. L. [1949], Query 64. Biometrics 5:68-71. E4,5a (16)

Anderson, R. L. and Manning, H. L. [1948], An experimental design used to estimate the optimum planting date for cotton. Biometrics 4:171-196. E4,5ae (17)

Andrews, W. S. [1908], Magic Squares and Cubes. The Open Court Publishing Co., Chicago. E12,13c (18)

Angelo, E. and Potter, G. F. [1939], The error of sampling in studying distribution of the root systems of tung trees by means of the Veihmeyer soil tube. Proc. American Soc. Hort. Sci. 37:518-520. S1 (19)

Anonymous, [1949], Handleiding voor Veldproeven. (Manual for field experiments.) Meded. Landbvoorlicht Dienst. Wageningen, No. 59, 139 pp. E2,3,7,12ae;C (20)

Anos, A. [1945], Un metodo grafico para hallar el numero necesario de repeticiones en las experiencias comparativas. (A graphical method for determining the number of replications necessary in comparative experiments.) Bol. Inst. Nac. Invest. Agron. Madr. 13:1-16. S1 (21)

Ansari, M. A. A. and Sant, G. K. [1943], A study of soil heterogeneity in relation to size and shape of plots in wheat field at Raya (Muttra District). Indian J. Agri. Sci. 13:652-658. A14 (22)

Anscombe, F. J. [1948], The transformation of Poisson, binomial and negative-binomial data. Biometrika 35(3/4):246-254. A9 (23)

Anscombe, F. L. [1948], The validity of comparative experiments. JRSSA 111:181-200. C (24)

Anthony, R. D. and Waring, J. H. [1922], Methods of interpreting yield records in apple fertilization experiments. Pennsylvania Agri. Expt. Sta. Bull. 173: 1-42. C (25)

Arceneaux, G. [1935], Determining theoretical yields of sugar in connection with sugar cane variety tests on the basis of field sampling and laboratory-scale milling tests. Proc. Fifth Congress Int. Soc. Sugar Cane Tech. Brisbane, pp. 568-578. S1 (26)

Arceneaux, G. [1939], Report of the Committee on Technique of Field Experiments. Proc. Sixth Congress Int. Soc. Sugar Cane Tech. pp. 387-390. A14;C (27)

Arceneaux, G. [1939], Studies of some problems of experimental technique in testing sugarcane varieties. Ph. D. Thesis, Cornell Univ. A14 (28)

Arceneaux, G. and Herbert, L. P. [1943], A statistical analysis of varietal yields of sugarcane obtained over a period of years. J. American Soc. Agron. 35: 148-160. E7,12ae (29)

Arceneaux, G., Stokes, I. E., Belcher, B. A., and Gibbens, R. T. [1939], A study of the relation between size of field sample and experiment errors of juice analysis and sugar-yield determination in connection with sugarcane variety tests. Proc. Sixth Congress Int. Soc. Sugar Cane Tech., La pp. 744-759. A14;S1 (30)

Arny, A. C. [1921], Further experiments in field technic in plot tests. J. Agri. Res. 21:483-499. A14 (31)

Arny, A. C. [1922], Border effect and ways of avoiding it. J. American. Soc. Agron. 14:266-278. A14 (32)

Arny, A. C. and Garber, R. J. [1919], Field technic in determining yields of plots of grain by the rod-row method. J. American Soc. Agron. 11:33-47. A14 (33)

Arny, A. C. and Hayes, H. K. [1918], Experiments in field technic in plot tests. J. Agri. Res. 15:251-262. A14 (34)

Arny, A. C. and Schmid, A. R. [1942], A study of the inclined point quadrat method of botanical analysis of pasture mixtures. J. American Soc. Agron. 34:238-247. A14 (35)

Arny, A. C. and Steinmetz, F. H. [1919], Field technic in determining yields of experimental plots by the square yard method. J. American Soc. Agron. 11: 81-106. A14 (36)

Artemova, P. K. and Povolotskoi, E. E. (Editors), [1935], Varietal testing in the main agricultural crops. Part I. Main principles and methods of judging varieties in competitive variety tests. Part II. General methods and technique of variety tests. Lenin Academy Agri. Sci. Inst. Plant Ind. Variety Testing Service, Part I pp. 101, Part II. pp. 106. C (37)

Atkins, I. M. [1945], The use of nursery plots vs. field plots for varietal evaluation. Report Fifth Hard Red Winter Wheat Improvement Conf., Manhattan, Kansas, Feb. 12, 13 and 14, 1945. Div. Cereal Crops Dis., Plant Ind. Sta., Beltsville, Md.(Mimeo.). C (38)

Atkins, R. E. [1948], An evaluation of certain agronomic and disease characters in advanced generations of bulk hybrid oat populations. Ph. D. Thesis, Iowa State College. E3ae (39)

Autrey, K. M., Knodt, C. B., and Williams, P. S. [1947], Grass and legume silage studies using 2-quart glass jars as miniature silos. J. Dairy Sci. 30:775-785. E2ae;T2ae (40)

Bachér, I. [1933], Moderna synpunkter på fältförsökets metodik och den statistika analysen av försöksresultatet. Nordisk Jordbrugsforskning 15:220-239. E2,7ae;B;C (41)

Bachér, I. [1935], Planläggning av komplicerade försök. (The laying out of complicated experiments.) Beretn. Nordisk JordbrForsker. Foren. 5th Kongr. København Juli 4-7, Hefte:329-342. E12a;T2a (42)

Bachet, C. C. [1612], Problèmes Plaisants et Délectables (second edition, 1624). Lyons. E12c (43)

Bailey, B. [1941], A study of sampling in cross-section measurements of wool fiber. J. Agri. Res. 63:407-415. A14 (44)

Bair, R. A. [1942], Climatological measurements for use in the prediction of maize yield. Ecology 23:79-88. A14 (45)

Baker, G. A. [1941], Fundamental distribution of errors for agricultural field trials. Nat. Math. Magazine 16:1-13. A14;C (46)

Baker, G. A. [1943], Length-growth curves for the razor clam. Growth 7:439-443. A7 (47)

Baker, G. A. [1945], Graduation of human growth curves. Growth 9:299-301. A7 (48)

Baker, G. A. and Briggs, F. N. [1945], Wheat bunt field trials. J. American Soc. Agron. 37:127-133. A14;C (49)

Baker, G. A. and Briggs, F. N. [1949], Wheat bunt field trials. II. Proc. Berkeley Symp. Math. Stat. Prob. pp. 485-491. A14;C (50)

Baker, G. A. and Hanna, G. C. [1949], Transformation of split-plot yield data to improve analysis of variance. Proc. American Soc. Hort. Sci. 53:273-275. E12a;A9 (51)

Ball, W. W. R. [1939], Mathematical Recreations and Essays, 11th edition (1st edition 1922). Macmillan, London. E12c (52)

Balmukand, B. [1928], Studies in crop variation V. The relation between yield and soil nutrients. J. Agri. Sci. 18:602-627. A7,14 (53)

Bancroft, T. A. and Smith, A. L. [1949], Efficiency of the simple lattice design relative to randomized complete blocks design in cotton variety and strain testing. Agron. J. 41:157-160. E3,14a (54)

Banerjee, K. C., Bose, S. S., and Mahalanobis, P. C. [1936], Statistical notes for agricultural workers No. 20. Studies in tiller variation. Indian J. Agri. Sci. 6:1122-1133. A14 (55)

Banerjee, K. S. [1941], Split plot arrangements and confounding higher order interactions. Indian Sci. Congress. E12a;T2a (56)

Banerjee, K. S. [1948], On the design of experiments for weighing and making other types of measurements. Sci. Culture 13:344. T6c (57)

Banerjee, K. S. [1948], Weighing designs and balanced incomplete blocks. AMS 19:394-399. E2c;T6c (58)

Banerjee, K. S. [1948], On Hotelling's weighing problem. Current Sci. 17:356. T6c (59)

Banerjee, K. S. [1949], A note on weighing design. AMS 20:300-304. T3,6c (60)

Banerjee, K. S. [1949], On the variance factors of weighing designs in between two Hadamard matrices. Calcutta Stat. Assoc. Bull. 2:38-42. T6ac (61)

Banerjee, K. S. [1949], On certain aspects of spring balance designs. Sankhyā 9:367-376. T3,6c (62)

Bär, A. L. S. [1939], Interpretatie van proefveld-resultaten. (Interpretation of the results of field experiments.) Landbouwk. Tijdschr. Wageningen 51:229-246. E5a;A14 (63)

Barankin, E. W. [1949], Extension of the Romanovsky-Bartlett-Scheffé test. Proc. Berkeley Symp. Math. Stat. Prob. pp. 433-449. A7 (64)

Barbacki, S. [1935], Ogólna metodyka doświadczeń polowych w zarysie. (An outline of the common methods of field investigation.) Bibl. Pulawska 12:119. C (65)

Barbacki, S. and Fisher, R. A. [1936], A test of the supposed precision of systematic arrangements. Ann. Eugenics 7:189-193. E5a (66)

Barber, C. W. [1914], Note on the influence of shape and size of plots in tests of varieties of grain. Maine Agri. Expt. Sta. Bull. 226:76-84. A14 (67)

Barclay, C. and Grantham, J. [1933], Statistical methods in field experiments with rubber. Archief voor de Rubbercultuur in Nederlandsch-Indië 17:213-237. C (68)

Barnard, M. M. [1936], An enumeration of the confounded arrangements in the 2x2x2... factorial designs. JRSSB 3:195-202. T2a (69)

Bartlett, M. S. [1934], Recent work on the analysis of variance. JRSSB 1:252-255. E6,14a (70)

Bartlett, M. S. [1935], An examination of the value covariance in dairy cow nutrition experiments. J. Agri. Sci. 25:238-244. Elae;A14 (71)

Bartlett, M. S. [1936], A note on the analysis of covariance. J. Agri. Sci. 26:488-491. A7 (72)

Bartlett, M. S. [1936], The square root transformation in analysis of variance. JRSSB 3:68-78. E2,12ae;A9 (73)

Bartlett, M. S. [1936], Some notes on insecticide tests in the laboratory and in the field. JRSSB 3:185-194. E2ae;A9 (74)

Bartlett, M. S. [1937], Sub-sampling for attributes. JRSSB 4:131-135. A9,14 (75)

Bartlett, M. S. [1937], Some examples of statistical methods of research in agriculture and applied biology. JRSSB 4:137-183. E2,12,13ae;A7 (76)

Bartlett, M. S. [1938], The approximate recovery of information from replicated field experiments with large blocks. J. Agri. Sci. 28:418-427. A14 (77)

Bartlett, M. S. [1946], A modified probit technique for small probabilities. JRSSB 8:113-117. C (78)

Bartlett, M. S. [1947], The use of transformations. Biometrics 3:39-52. E2ae;A9 (79)

Bartlett, M. S. [1948], Determination of plant densities. Nature 162(4120):621. A14 (80)

Bartlett, M. S. and Greenhill, A. W. [1936], The relative importance of plot variation and of field and laboratory sampling errors in small plot pasture productivity experiments. J. Agri. Sci. 26:258-262. A14 (81)

Bartlett, M. S. and Kendall, D. G. [1946], The statistical analysis of variance-heterogeneity and the logarithmic transformation. JRSSB 8:128-138. E2,7a;A9,11 (82)

Bartlett, S. [1926], Variations in the live weight of dairy cows. J. Agri. Sci. 16:383-391. A14;C (83)

Bartlett, S. [1929], Normal day to day variability of yield of milk and fat of individual cows. J. Agri. Sci. 19:438-451. C (84)

Bartlett, S. and Blaxter, K. L. [1947], The value of urea as a substitute for protein in the rations of dairy cattle I. Field trials with dairy cows. J. Agri. Sci. 37:32-44. E12a (85)

Batchelor, L. D., Parker, E. R., and McBride, R. [1928], Studies preliminary to the establishment of a series of fertilizer trials in a bearing citrus grove. Univ. California Agri. Expt. Sta. Berkeley Bull. 451. A14;C (86)

Batchelor, L. D. and Reed, H. S. [1918], Relation of the variability of yields of fruit trees to the accuracy of field trials. J. Agri. Res. 12:245-283. A14;C (87)

Baten, W. D. [1939], Formulas for finding estimates for two and three missing plots in randomized layouts. Tech. Bull. Michigan Agri. Expt. Sta. 165:1-16. E2,6a (88)

Baten, W. D., Northam, J. I., and Yeager, A. F. [1941], Grouping of strains or varieties by use of a latin square. J. American Soc. Agron. 33:616-622. E7,12ae (89)

Bates, G. H. [1935], A study of the factors influencing size of potato tubers. J. Agri. Sci. 25:297-313. E7,12a (90)

Baxter, B. [1940], The application of factorial design to a psychological problem. Psychological Rev. 47:494-500. T2a (91)

Baxter, B. [1941], Problems in the planning of psychological experiments. American J. Psychology 54:270-280. C (92)

Beall, G. [1940], The technique of randomization in field work. Canadian Ent. 72:45-48. C (93)

Beall, G. [1942], The transformation of data from entomological field experiments so that the analysis of variance becomes applicable. Biometrika 32:243-262. E2ae;A9 (94)

Bean, H. W. [1946], Single weight versus a three-day average weight for swine. J. Animal Sci. 5:30-35. A14;S1 (95)

Bean, H. W. [1948], Single weight versus three-day average weight for sheep. J. Animal Sci. 7:50-54. A14 (96)

Bear, F. E. and McClure, G. M. [1920], Sampling soil plots. Soil Sci. 9:65-75. A14 (97)

Beattie, J. H. and Boswell, V. R. [1936], Statistical studies of apparently uniform fields of carrots and onions on peat soils. Die Gartenbauwissenshaft 10:279-288. A14 (98)

Beattie, J. H., Boswell, V. R., and Batten, E. T. [1936], Plot and plant variation in Virginia peanuts. Proc. American Soc. Hort. Sci. 34:586-589. A14 (99)

Beaumont, J. H. [1939], An analysis of growth and yield relationships of coffee-trees in the Kona district, Hawaii. J. Agri. Res. 59:223-235. A7 (100)

Becker, E. R. and Derbyshire, R. C. [1937], Biological assay of feeding stuffs in a basal ration for coccidium-growth-promoting substance I. Iowa State College J. Sci. 11:311-322. E2ae (101)

Becker, E. R. and Derbyshire, R. C. [1937], Biological assay of feeding stuffs in a basal ration for coccidium-growth-promoting substance II. Iowa State College J. Sci. 12:211-215. E2ae (102)

Becker, E. R. and Waters, P. C. [1939], Biological assay of feeding stuffs in a basal ration for coccidium-growth-promoting substance III. Iowa State College J. Sci. 13:243-247. E2ae (103)

Beckett, W. H. [1932], Field experiments. A note on residual variance in field experiments. Tropical Agri. 9:7-10. E7ae (104)

Behrens, W. V. [1929], Ein Beitrag zur Fehlerberechnung bei wenigen Beobachtungen. Landwirtschaftliche Jahrbücher 68:807-837. A11 (105)

Bell, G. D. H. et al. [1942], Experiments on cutting potato tubers. J. Agri. Sci. 32:255-273. E2ae (106)

Bergh, O. I. [1922], Missing hills in potatoes. Minn. Agri. Ext. Service Reports pp. 136-139. A14 (107)

Berkson, J. [1949], Minimum χ^2 and maximum likelihood solution in terms of a linear transform, with particular reference to bio-assay. JASA 44:273-278. C (108)

Berkson, J., Rulon, P. J., and Moulton, F. R. [1936], Significant figures in statistical constants. Science 84:437, 483-484, 574-575. C (109)

Berry, R. A. and O'Brien, D. G. [1921], Errors in feeding experiments with cross-bred pigs. J. Agri. Sci. 11:275-286. A14;C (110)

Berry, R. A. and O'Brien, D. G. [1924], Field experiments. J. Agri. Sci. 14: 407-412. S1 (111)

Bescoby, H. B. [1932], The technique of a barley manuring trial. J. South-East Agri. College, Wye, Kent 30:215-230. E12ae (112)

Bever, W. M. and Slife, F. W. [1948], Sample size and plot size for testing resistance of strawberry varieties to Verticillium wilt. Phytopathology 38: 1034-1038. A14 (113)

Bhattacharya, K. N. [1943], A note on two-fold triple systems. Sankhyā 6: 313-314. E2c (114)

Bhattacharya, K. N. [1944], On a new symmetrical balanced incomplete block design. Bull. Calcutta Math. Soc. 36:91-96. E2c (115)

Bhattacharya, K. N. [1944], A new balanced incomplete block design. Sci. Culture 9:508. E2c (116)

Bhattacharya, K. N. [1945], The balanced incomplete block design v=b=31, r=k=10, λ=3. Proc. Thirty-Second Indian Sci. Congress, Sec XII, No. 37, E2c (117)

Bhattacharya, K. N. [1946], A new solution in symmetrical balanced incomplete block designs (v=b=31, r=k=10, λ=3). Sankhyā 7:423-424. E2c (118)

Bhattacharyya, A. [1942], A note on Ramamurti's problem of maximal sets. Sankhyā 6:189-192. E2,3c (119)

Bickford, C. A. [1935], A simple, accurate method of computing basal area of forest stands. J. Agri. Res. 51:425-433. A14 (120)

Bigot, A. [1934], De toepassing von "Analysis of variance" bij proefvelden. (The application of "Analysis of variance" in field experiments.) Landbouwk. Tijdschr. Wageningen 46:640-652, 705-717. E2ae;C (121)

Bigot, A. [1941], Grafische voorstelling van proefveldresultaten. (Graphic representation of experiment plot results.) Archief voor de Suikerindustrie in Nederlanden Nederlandsch-Indië 2:44-48. C (122)

Bigot, A. [1941], Factorieele vakkenproeven in de Java-suikerindustrie. (Factorial plot experiments in the Java Sugar industry.) Archief voor de Suikerindustrie in Nederlanden Nederlandsch-Indië 2:361-364. T2a (123)

Biot, J. B. [1816], <u>Traité de Physique Expérimentale et Mathematique</u>, Paris. C (124)

Bjerke, B. [1937], Sammenregning av observasjonsrekker. (Notes on the mathematical treatment of observations.) Meld. Norg. LandbrHøisk 17:305-330. E2,7a (125)

Bjerke, B. [1938], Sammenregning av observasjoner II. Det tredimensjonale tilfelle. (Notes on the mathematical treatment of observations II. The three dimensional case.) Meld. Norg. LandbrHøisk 18:98-111. E12a (126)

Blackman, G. E. and Bartlett, M. S. [1935], A study by statistical methods of the distribution of species in grassland associations. Ann. Botany 49:749-777. A14 (127)

Blaney, J. E. and Smith, J. B. [1931], Sampling market garden soils for nitrates. Soil Sci. 31:281-290. S1 (128)

Bliss, C. I. [1937], The analysis of field experimental data expressed in percentages. Plant Protection, Leningrad 12:67-77. A9;C (129)

Bliss, C. I. [1938], The transformation of percentages for use in the analysis of variance. Ohio J. Sci. 38:9-12. A9 (130)

Bliss, C. I. [1940], Factorial design and covariance in the biological assay of vitamin D. JASA 35:498-506. E2ae;T2ae;A7 (131)

Bliss, C. I. [1941], Statistical problems in estimating populations of Japanese beetle larvae. J. Economic Entomology 34:221-232. E2,7a;A9,14 (132)

Bliss, C. I. [1941], A simplified calculation of the potency of penicillin and other drugs assayed biologically with a graded response. JASA 39:479-487. A14;C (133)

Bliss, C. I. [1946], An experimental design for slope-ratio assays. AMS 17: 232-237. Ela;C (134)

Bliss, C. I. [1947], 2×2 factorial experiments in incomplete groups for use in biological assays. Biometrics 3:69-88. E2ae;T2ae (135)

Bliss, C. I., Anderson, E. O., and Marland, R. E. [1943], A technique for testing consumer preferences, with special reference to the constituents of ice cream. Storrs Agri. Expt. Sta. Bull. 251. A9 (136)

Bliss, C. I. and Cattell, M. K. [1943], Biological assay. Annual Review Physiology 5:479-539. B;C (137)

Bliss, C. I. and Dearborn, R. B. [1942], The efficiency of lattice squares in corn selection tests in New England and Pennsylvania. Proc. American Soc. Hort. Sci. 41:324-342. Ellae;A7 (138)

Bliss, C. I. and Marks, H. P. [1939], The biological assay of insulin I. II. The estimation of drug potency from a graded response. Qtly. J. Pharmacy Pharmacology 12:82-110, 182-205. E7ae;C (139)

Bloomers, P. and Linquist, E. F. [1942], Experimental and statistical studies: application of newer statistical techniques. Rev. Educational Res. 12:501-520. C (140)

Blosser, T.H. and Smith, V. R. [1947], A comparison of total milk and fat production and per cent fat of a 26- and 24-hour test day. J. Dairy Sci. 30:951-956. E2ae;C (141)

Blunn, C. T. and Grandstaff, J. O. [1945], Comparison of the yields of side samples from weanling and yearling sheep. J. Animal Sci. 4:122-127. A14 (142)

Boggs, M. M. and Hanson, H. L. [1949], Analysis of foods by sensory difference tests. Advances Food Res. 2:219-258. B;C (143)

Bonazzi, A. [1933], Errors in field experimentation with ratoon cane. Acos Tec. Azucareros Cuba Proc. 7:32-40. C (144)

Bond, T. E. T. [1947], Some Ceylon examples of the logarithmic series and the index of diversity of plant and animal populations. Ceylon J. Sci. Sect. A. Bot. 12(4):195-202. A14 (145)

Bondorff, K. A. [1942], Arbejdsfeljlens Størrelse ved lokale Forsøg. (The magnitude of experimental error in field experimrnts.) Nord. JordbrForskn. 24:151-152. E14;A14 (146)

Bonnett, R. K. and Burkar, F. L. [1923], Rate of seeding ---A factor in variety tests. J. American Soc. Agron. 15:161-171. A14 (147)

Booth, A. D. [1949], An application of the method of steepest descents to the solution of systems on non-linear simultaneous equations. Quart. J. Mechanics Appl. Math. 2:460-468. T4a (148)

Borden, R. J. [1930], Experimental technique. <u>Facts About Sugar</u> 25:834-835. A14 (149)

Borden, R. J. [1930], Replication of plot treatments in field experiments. Hawaiian Planters' Record 34:151-155. A14;S1 (150)

Borden, R. J. [1930], A study of the probable error of a single plot from field tests with sugar cane in Hawaii. Hawaiian Planters' Record 34:467-476. A14 (151)

Borden, R. J. [1931], Field experiment technique. Hawaiian Planters' Record 35:245-277. A14;S1 (152)

Borden, R. J. [1931], Studies in experimental technique - shape, size and replication. Hawaiian Planters' Record 35:295-304. A14;S1 (153)

Borden, R. J. [1932], Improving field experiments. Hawaiian Planters' Record 36:369-373. A14;C (154)

Borden, R. J. [1936], Studies in experimental technique - the balanced block arrangement of treatments. Hawaiian Planters' Record 40:113-116. E5a (155)

Borden, R. J. [1937], Better planning for field experiments with fertilizers. Hawaiian Planters' Record 41:99-109. S1 (156)

Borden, R. J. [1939], Studies in experimental technique: Plot arrangement. Proc. Sixth Congress Int. Soc. Sugar Cane Tech., La (1938): 733-744.
E5a;C (157)

Borden, R. J. [1939], Studies in experimental technique. Selection of layout: Blocks versus Latin Squares. Hawaiian Planters' Record 43:7-10.
E2,7a (158)

Borden, R. J. [1939], A modern statistical analysis for field experiments. The analysis of variance for simple factorial experiments. Hawaiian Planters' Record 43:73-114. E5,12,13ae;T2a (159)

Borden, R. J. [1940], Border effect in field experiments that are concerned with fertilizer practices. Hawaiian Planters' Record 44:11-14. A14 (160)

Borden, R. J. [1943], Replication: the safeguard for uncontrolled variation. Hawaiian Planters' Record 47:135-153. S1 (161)

Borden, R. J. [1943], Yield variations with special reference to border effects in field tests. Hawaiian Planters' Record 47:195-203. A14 (162)

Borden, R. J. [1946], Modified Mitscherlich method for soil cultures. Soil Sci. 62:51-60. A7;C (163)

Borden, R. J. and Kutsunai, Y. [1934], For better field experiments with fertilizers. Hawaiian Planters' Record 38:215-231. E5,7a (164)

Bose, R. C. [1938], On the application of the properties of Galois fields to the problem of construction of Hyper-Greco-Latin squares. Sankhyā 3:323-338.
E7,13c (165)

Bose, R. C. [1939], On the construction of balanced incomplete block designs. Ann. Eugenics 9:353-399. E2c (166)

Bose, R. C. [1941], Discussion on the mathematical theory of design of experiments. Sankhyā 5:170-174. E2c (167)

Bose, R. C. [1942], The affine analogue of Singer's theorem. J. Indian Math. Soc. 6:1-15. E2c (168)

Bose, R. C. [1942], On some new series of balanced incomplete block designs. Bull. Calcutta Math. Soc. 34:17-31. E2c (169)

Bose, R. C. [1942], A note on the resolvability of balanced incomplete block designs. Sankhyā 6:105-110. E2,3c (170)

Bose, R. C. [1942], A note on two combinatorial problems having applications in the theory of design of experiments. Sci. Culture 8:192-193.
E2,3c (171)

Bose, R. C. [1943], A note on two series of balanced incomplete block designs. Bull. Calcutta Math. Soc. 34:129-130. E2c (172)

Bose, R. C. [1944], The fundamental theorem of linear estimation (abstract). Proc. Thirty-First Indian Sci. Congress 3:2-3. E2c (173)

Bose, R. C. [1944], On the problem of balancing in symmetrical factorial designs. Proc. Thirty-First Indian Sci. Congress 3:5-6. T2a (174)

Bose, R. C. [1945], The symmetrical factorial designs of the class s^5 with s^3 plot blocks. Proc. Thirty-Second Indian Sci. Congress 3:168. T2a (175)

Bose, R. C. [1945], Minimum functions in Galois fields. Proc. Nat. Academy Sci. 14:191. E2c (176)

Bose, R. C. [1945], On the roots of a well known congruence. Proc. Nat. Academy Sci. 14:193. E2c (177)

Bose, R. C. [1945], A note on Youden's squares. Proc. Thirty-Second Indian Sci. Congress, Sec. 12:43. E7c (178)

Bose, R. C. [1946], A theorem on balancing. Proc. Thirty-Third Indian Sci. Congress 3:168. T2a (179)

Bose, R. C. [1947], Recent work on "Incomplete block designs" in India. Biometrics 3:176-178. E2,3c (180)

Bose, R. C. [1947], The design of experiments. Presidential address to the Section of Statistics. Proc. Thirty-Fourth Indian Sci. Congress. Part II. E2,3c;C (181)

Bose, R. C. [1947], The maximum number of factors, that can be accommodated in a symmetrical factorial experiment, when the interactions up to a given order are left unconfounded. Proc. Thirty-Fourth Indian Sci. Congress 3:6-7. T2a (182)

Bose, R. C. [1947], Mathematical theory of the symmetrical factorial design. Sankhyā 8:107-166. T2a (183)

Bose, R. C. [1947], On a resolvable series of balanced incomplete block designs. Sankhyā 8:249-256. E2c (184)

Bose, R. C. [1949], Least squares aspects of analysis of variance (first dittoed by VPI, 1947). Mimeo., Univ. N. Carolina. E2,3,6,7ac (185)

Bose, R. C. [1949], A note on Fisher's inequality for balanced incomplete block designs. AMS 20:619-620. E2c (186)

Bose, R. C. and Chowla, S. [1945], A new series of balanced incomplete block designs. Proc. Thirty-Second Indian Sci. Congress Assoc. Sec. 12:5. E2c (187)

Bose, R. C., Chowla, S., and Rao, C. R. [1944], On the integral order (mod p.) of quadratics x^2+ax+b, with applications to the construction of minimum functions for GF(p^2), and to some number theory results. Bull. Calcutta Math. Assoc. 36:153-174. E2,3c (188)

Bose, R. C. and Kishen, K. [1939], On partially balanced Youden squares. Sci. Culture 4:136-137. E7c (189)

Bose, R. C. and Kishen, K. [1940], On the problem of confounding in the general symmetrical factorial design. Sankhyā 5:21-36. T2a (190)

Bose, R. C. and Nair, K. R. [1939], Partially balanced incomplete block designs. Sankhyā 4:337-372. E2,3c (191)

Bose, R. C. and Nair, K. R. [1941], On complete sets of Latin squares. Sankhyā 5:361-382. E7,13c (192)

Bose, R. D. [1932], Another yield trial with Pusa barleys: The method of analysis of variance. Agri. Livestock Ind. 2:603-618. E2,7ae (193)

Bose, R. D. [1933], Application of modern statistical methods to field trials. Agri. Livestock Ind. 3:330-340. A14;C (194)

Bose, R. D. [1935], Some soil-heterogeneity trials at Pusa and the size and shape of experimental plots. Indian J. Agri. Sci. 5:579-608. A14 (195)

Bose, S. S., Ganguli, P. M., and Mahalanobis, P. C. [1936], Statistical notes for agricultural workers No. 19. The frequency distribution of plot yields and the optimum size of plots in a uniformity trial with rice in Assam. Indian J. Agri. Sci. 6:1107-1121. A14 (196)

Bose, S. S. and Gupta, S. C. S. [1935], A study in covariance with fodder crops. Twenty-Second Indian Sci. Congress Proc. p. 347. A7 (197)

Bose, S. S., Khanna, K. L., and Mahalanobis, P. C. [1939], Statistical notes for agricultural workers No. 24. Note on the optimum shape and size of plots for sugarcane experiments in Bihar. Indian J. Agri. Sci. 9:807-816. A14 (198)

Bose, S. S. and Mahalanobis, P. C. [1938], On estimating individual yields in the case of mixed-up yields of two or more plots in field experiments. Sankhyā 4:103-120. E2,6,7,10ace (199)

Bose, S. S. and Menon, T. V. G. [1937], Statistical notes for agricultural workers No. 23. A statistical study of the deterioration of yields of the permanent manurials, Pusa. Indian J. Agri. Sci. 7:205-213. E5,8a (200)

Bourne, J. B. [1938], The importance and use of appropriate assumed means in collating field experimental results statistically. Tropical Agri. 15:247-258. E2a;A14 (201)

Bowen, M. F. [1947], Population distribution of the beet leafhopper in relation to experimental field-plot lay-out. J. Agri. Res. 75:259-278. E2,7a;A14 (202)

Boyce, S. W. [1945], Statistical studies on New Zealand wheat trials. I. The efficiency of lattice design. II. The analysis of lattice trials with incomplete data. New Zealand J. Sci. Tech. 27A:270-280. E3,6,14a (203)

Boyd, D. A. [1939], The estimation of Rothamsted temperature from the temperature of Oxford and Greenwich. Ann. Eugenics 9:341-352. A7 (204)

Boyd, D. A. [1940], The influence of meteorological factors on arable crop yields, based on an examination of the results of the Saxmundham rotation experiment. Ph. D. Thesis, Univ. London. A7 (205)

Boyd, L. L. [1948], The effect of moisture content of wood on the withdrawal resistance of roofing nails. M. S. Thesis, Iowa State College. E11ae (206)

Brady, J. [1934], Some factors influencing lodging in cereals. J. Agri. Sci. 24:209-232. E7a;T2a;A7 (207)

Brady, J. [1935], A biological application of the analysis of co-variance. JRSSB 2:99-106. E7ae;T2ae;A7 (208)

Brandon, J. F. [1925], Crop rotation and cultural methods at the Akron (Colorado) Field Station. Dept. Bull. 1304, U. S. Dept. Agri. E5,8a (209)

Brandt, A. E. [1933], The analysis of variance in a "2×s" table with disproportionate frequencies. JASA 28:164-173. A10 (210)

Brandt, A. E. [1937], Factorial design. J. American Soc. Agron. 29:658-667. T2ae (211)

Brandt, A. E. [1938], Tests of significance in reversal or switchback trials. Iowa Agri. Expt. Sta. Res. Bull. 234:60-87. E8ae (212)

Brandt, A. E. [1941], The relation between the design of an experiment and the analysis of variance. JASA 36:283-292. E2ae (213)

Brandt, A. E. [1945], Principles of experimental design applied to long-time rotations. Soil Sci. Soc. Amer. Proc. 10:306-315. E5,8a (214)

Braude, R. [1947], The effect of feeding iodinated casein to pigs. J. Agri. Sci. 37:45-50. E7ae (215)

Braude, R. and Foote, A. S. [1942], Pig-feeding experiment using dried Clostridium residue. J. Agri. Sci. 32:324-329. E2ae (216)

Brewbaker, H. E. and Bush, H. L. [1942], Pre-harvest estimate of yield and sugar percentage based on random-sampling technique. Amer. Soc. Sugar Beet Tech., Proc. 3:184-196. A14 (217)

Brewbaker, H. E. and Bush, H. L. [1946], Four years' results of pre-harvest sampling in estimating yield and sugar percentage. Amer. Soc. Sugar Beet Tech., Proc. 4:141-153. A14 (218)

Brewbaker, H. E. and Deming, G. W. [1935], Effect of variations in stand on yield and quality of sugar beets grown under irrigation. J. Agri. Res. 50: 195-210. A14 (219)

Brewbaker, H. E. and Immer, F. R. [1931], Variations in stand as sources of experimental error in yield tests with corn. J. American Soc. Agron. 23: 469-480. A14 (220)

Brieger, F. G. [1947], A determinação dos números de individuos mínimos necessários na experimentação genética. (The determination of the minimum number of individuals necessary in genetical experimentation.) Anais da Escola Superior de Agricultura "Luiz de Queiroz", Piracicaba 4:217-262. T9a;S1 (221)

Brouwer, E. and Dijkstra, N. D. [1938], On alimentary acetonuria and ketonuria in dairy cattle induced by feeding grass silage of the butyric acid type. J. Agri. Sci. 28:695-700. E7ae (222)

Brown, B. [1932], The evaluation of a statistical technique for analysis of rat feeding data, based on uniformity trials. M. S. Thesis, Iowa State College. A14 (223)

Brown, B. [1932], A sampling test of the technique of analyzing variance in a 2×N table with disappropriate frequencies. Proc. Iowa Academy Sci. 39:205-207. A10 (224)

Brown, B. A. [1922], Plot competition with potatoes. J. American Soc. Agron. 14:257-258. A14 (225)

Brown, B. A. [1937], Technic in pasture research. J. American Soc. Agron. 29:468-476. C (226)

Brown, G. W. [1947], Experimental design principles in industrial research. Int. Stat. Conf. Proc. 3A:314-318. C (227)

Brown, G. W. and Flood, M. M. [1947], Tumbler mortality. JASA 42:562-574. E6,10a (228)

Brown, R. M. [1934], Statistical analyses for finding a simple method for estimating the percentage heart rot in Minnesota aspen. J. Agri. Res. 49:929-942. A7 (229)

Brownlee, K. A., Kelly, B. K., and Loraine, P. K. [1948], Fractional replication arrangements for factorial experiments with factors at two levels. Biometrika 35:268-276. T3ac (230)

Brownlee, K. A. and Loraine, P. K. [1948], The relationship between finite groups and completely orthogonal squares, cubes and hyper-cubes. Biometrika 35: 277-282. E7,13c (231)

Bruck, R. H. and Ryser, H. J. [1949], The nonexistence of certain finite projective planes. Canadian J. Math. 1:88-99. E7,13c (232)

Brunswik, E. [1949], Systematic and representative design of psychological experiments. Proc. Berkeley Symp. Math. Stat. Prob. pp. 143-202. C (233)

Bryan, A. A. [1931], A statistical study in the relation of size and shape of plot and number of replications to precision in yield comparisons with corn. Iowa Agri. Expt. Sta. Report No. 67. A14 (234)

Bryan, A. A. [1933], Factors affecting experimental error in field plot tests with corn. Iowa Agri. Expt. Sta. Bull 163:243-280. A14 (235)

Bryan, A. A., Eckhardt, R. C., and Sprague, G. F. [1940], Spacing experiments with corn. J. American Soc. Agron. 32:707-714. E3,11,12a;A14 (236)

Bugelski, B. R. [1949], A note on Grant's discussion of the Latin square principle in the design of experiments. Psychological Bull. 46:49-50. E7,8a (237)

Burdet, W. [1935], Het uitschakelen der systematische variatie oorzaken bij het proefveldonderzoek volgens Knut Vik. (The elimination of the causes of systematic variation in field experimentation, by the method of Knut Vik.) Landbouwk. Tijdschr. Wageningen 46 sic. 47 :88-94. E5,7a (238)

Burkett, A. L., Gordon, E. D. et al. [1949], Effects of age of plant and retting procedures on kenaf, Hibiscus cannabinus L., fiber. J. American Soc. Agron. 41:255-260. E11ae (239)

Burn, J. H. [1930], The errors of biological assay. Physiol. Rev. 10:146-169. C (240)

Bush, H. L. [1940], The three dimensional quasi-factorial experiment with three groups of sets for testing sugar beet breeding strains. American Soc. Sugar Beet Tech., Proc. 2:113-116. E3,14a (241)

Bush, H. L. [1942], Further studies in newer designs for large-scale variety tests. American Soc. Sugar Beet Tech. Proc. 3:365-372. E2,3,14a (242)

Cain, S. A., Friesner, R. C., and Patzger, J. E. [1930], A comparison of strip and quadrat analyses of the woody plants on a central Indiana river bluff. Butler Univ. Bot. Studies 1:157-171. A14 (243)

Call, L. E. [1922], Increasing the efficiency of agronomic research. J. American Soc. Agron. 14:329-338. C (244)

Calvet, R. P., and de Zulueta, M. M. [1946], Métodos estadísticos para la comparación de gran número de variedades. (Statistical methods for comparing a large number of varieties.) Bol. Inst. Nac. Invest. Agron. Madr. 14:29-62. E3,5a (245)

Campbell, F. L., Snedecor, G. W., and Simanton, W. A. [1939], Biostatistical problems involved in the standardization of liquid household insecticides. JASA 34:62-70. A9;C (246)

Campbell, J. A. and Emslie, A. R. G. [1945], Studies on the chick assay for vitamin D. III and IV. Poultry Sci. 24:296-304. A9;C (247)

Campbell, J. A. and Emslie, A. R. G. [1947], Studies on the chick assay for vitamin D. V. A comparison of A.O.A.C. and B.S.I. diets and feeding periods. Poultry Sci. 26:255-261. A9;C (248)

Campbell, J. A. and Emslie, A. R. G. [1947], Studies on the chick assay for vitamin D. VI. Sources of variation in the response of replicate groups with time. Poultry Sci. 26:568-572. A9;C (249)

Campbell, J. A. and Emslie, A. R. G. [1947], Variability in chick growth data. Poultry Sci. 26:573-575. C (250)

Campbell, J. A., Migicovsky, B. B., and Emslie, A. R. G. [1945], Studies on the chick assay for vitamin D. I and II. Poultry Sci. 24:3-7, 72-80. A9;C (251)

Cannon, C. Y., Hansen, E. N., and O'Neil, J. R. [1932], The use of water bowls in the dairy barn. Iowa Agri Expt. Sta. Bull. 292:101-114. E7ae (252)

Capó, B. G. [1937], A comparative study of the statistical methods most commonly used in agricultural research. J. Agri. Univ. Puerto Rico 21:201-224. E2,5,7ae (253)

Capó, B. G. [1944], A method of interpreting the results of field trials. J. Agri. Univ. Puerto Rico 28:7-21. E2,5,7ae (254)

Capó, B. G. [1944], A new method of performing field trials. J. Agri. Univ. Puerto Rico 28:22-34. E5ae (255)

Carey, T. M. [1945], Fertiliser requirements of sugar cane. Ph. D. Thesis, Univ. London. C (256)

Carleton, M. A. [1909], Limitations in field experiments. Soc. Promotion Agri. Sci. Proc. 30:55-61. C (257)

Carmichael, R. D. [1937], Introduction to the Theory of Groups of Finite Order. Ginn. & Co. Boston. E2c;T2c (258)

Caroll, W. E. [1930], Group feeding as a method of livestock experimentation. Amer. Soc. Animal Prod. Proc. 23:34-44. C (259)

Carpenter, O. [1949], Sequential tests of the linear hypotheses. Ph. D. Thesis, Iowa State College. S2 (260)

Carruth, L. A. [1936], Experiments for the control of larvae of Heliothis obsoleta Fabr. on western Long Island during 1935. J. Economic Ent. 29:205-209. E2ae;A9 (261)

Cartwright, W. B. and Larrimer, W. H. [1926], Determination by the 5-square-yard method of the yield of wheat plots used in studies of the hessian fly, Phytophaga destructor Say. J. Agri. Res. 32:1045-1048. A14 (262)

Cashen, R. O. [1944], The effect of rainfall and associated weather on manurially treated permanent grassland as demonstrated by manurial experiments carried out since 1856 on the grass plots on Rothamsted Park. Ph. D. Thesis, Univ. London. A7;C (263)

Cashen, R. O. [1947], The influence of rainfall on the yield and botanical composition of permanent grass at Rothamsted. J. Agri. Sci. 37:1-10. A7;C (264)

Cavalli, L. L. and Magni, C. [1947], Methods of analysing the virulence of bacteria and viruses for genetical purposes. Heredity 1:127-132. T9a;C (265)

Cayley, A. [1850], On the triadic arrangements of seven and fifteen things. Philos. Mag. 37:50-53 (Collected mathematical Papers, I, 481-484). E2c (266)

Cayley, A. [1863], On a tactical theorem relating to the triads of fifteen things. London, Edinburgh, and Dublin Philos. Mag. and J. 25 (Ser. 4):59-61 (Collected mathematical Papers, V. 95-97). E2c (267)

Cayley, A. [1890], On Latin squares. Messenger Math. 19:135-137. E7c (268)

Chakravertti, S. C., Bose, S. S., and Mahalanobis, P. C. [1936], Statistical notes for agricultural workers No. 16. A complex experiment on rice at the Chinsurah Farm, Bengal, 1933-34. Indian J. Agri. Sci. 6:34-51. T2ae (269)

Chakravertti, S. C., Bose, S. S , and Mahalanobis, P C. [1936], Statistical notes for agricultural workers No. 21. The effect of different methods of harvest on the estimated error of field experiments on rice. Agri. Livestock India 6:814-825. A14 (270)

Champlin, M [1921], The technique of field husbandry experimentation. Sci. Agri. 2:14-18. C (271)

Chang-Choong, P. A. [1939], Testing large numbers of rice varieties by the quasi-factorial method. Agri. J. Brit. Guiana. 10:78-88. E3a (272)

Chang, C. C. [1937], (A new method of randomized field arrangement involving a large number of varieties of crops) J. Agri. Assoc. China 158:48-56. E3a (273)

Chardon, F. [1932], Limitations of the theoretical check yield method for laying out and interpreting results of field experiments with sugar cane. Proc. Fourth Congress Int. Soc. Sugar-Cane Tech. Bull. 82. A14;C (274)

Cheeseman, E. E. and Pound, F. J. [1932], Uniformity trials on cacao. Tropical Agri. 9:277-288. E7a;S1 (275)

Chen, K. K., Bliss, C. I., and Robbins, E. B. [1942], The digitalis-like principles of calotropis compared with other cardiac substances. J. Pharmacology Expt. Therapeutics 74:223-234. E7ae (276)

Chenfong, K and Lee, S. [1935], (A study on the number of replication in pot experiments.) J. Agri. Res. Nat. Cent. Univ. Nanking 2:23-28. A14;S1 (277)

Chiang, Ti-Chiu. [1937], A statistical analysis of field technique used in cotton improvement. M S. Thesis, Cornell Univ. E5a;A14 (278)

Chittenden, F. J. [1915], On the influence of planting distance on the yield of crops. J. Royal Hort Soc. 41:88-93. A14 (279)

Chowdhury, S. [1944], Rhizoctonia root-rot of pan (Piper betle) in relation to manuring. Indian J. Agri. Sci. 14:391-394. E2ae (280)

Chowla, S. [1949], On difference sets. Proc. Nat. Academy Sci. 35:92-94. E2c (281)

Christidis, B G [1931], The importance of shape of plots in field experimentation. J. Agri. Sci. 21:14-37. A14 (282)

Christidis, B. G. [1935], Intervarietal competition in yield trials with cotton. J. Agri. Sci. 25:231-237. E2ae;A14 (283)

Christidis, B. G. [1937], Competition between cotton varieties: A reply. J. American Soc. Agron. 29:703-705. Al4 (284)

Christidis, B. G. [1939], Further studies on competition in yield-trials with cotton. Empire J. Exptl. Agri. 7:111-120. Al4 (285)

Christidis, B. G. [1939], Variability of plots of various shapes as affected by plot orientation. Empire J. Exptl. Agri. 7:330-342. Al4 (286)

Chu, T. L. [1949], A study on the technique of field experimentation with sugar cane. I. Suitable size and shape of plot and number of replications. Reports Taiwan Sugar Expt. Sta. 5:77-90. Al4 (287)

Churchman, C. W. [1948], Theory of Experimental Inference. Macmillan, New York. C (288)

Clapham, A. R. [1929], The estimation of yield in cereal crops by sampling methods. J. Agri. Sci. 19:214-235. E5a;Al4 (289)

Clapham, A. R. [1931], Studies in sampling technique: Cereal experiments. J. Agri. Sci. 21:366-371, 376-390. Al4 (290)

Clark, A. and Leonard, W. H. [1939], The analysis of variance with special reference to data expressed as percentages. J. American Soc. Agron. 31: 55-66. A9 (291)

Clements, F. E., Weaver, J. E., and Hanson, H. C. [1929], Plant Competition: An Analysis of Community Functions. Carnegie Inst. Washington. Al4 (292)

Cline, M. G. [1944], Principles of soil sampling. Soil Sci. 58:275-288. C (293)

Cochran, W. G. [1934], The distribution of quadratic forms in a normal system, with applications to the analysis of covariance. Proc. Cambridge Philosophical Soc. 36:178-191. A7 (294)

Cochran, W. G. [1935], A note on the influence of rainfall on the yield of cereals in relation to manurial treatment. J. Agri. Sci. 25:510-522. A7 (295)

Cochran, W. G. [1936], Statistical analysis of field counts of diseased plants. JRSSB 3:49-67. Al4 (296)

Cochran, W. G. [1937], Problems arising in the analysis of a series of similar experiments. JRSSB 4:102-118. El2a;Al3 (297)

Cochran, W. G. [1937], A catalogue of uniformity trial data. JRSSB 4:233-253. A14 (298)

Cochran, W. G. [1938], Some difficulties in the statistical analysis of replicated experiments. Empire J. Exptl. Agri. 6:157-175. E2,7ae;A9 (299)

Cochran, W. G. [1938], Recent advances in mathematical statistics. Recent work on the analysis of variance. JRSSA 101·434-449. E2,3,7,11,14a; T2a;A7 (300)

Cochran, W. G. [1938], The omission or addition of an independent variate in multiple linear regression. JRSSB 5:171-176. T1a (301)

Cochran, W. G. [1938], The information supplied by the sampling results. Ann. Appl. Biology 25:383-389. A14 (302)

Cochran, W G. [1939], Long-term agricultural experiments. JRSSB 6:104-148. E5,8a (303)

Cochran, W. G. [1940], The analysis of variance when experimental errors follow the Poisson or binomial laws. AMS 11:335-347. E2,7ae;A9 (304)

Cochran, W. G. [1940], The analysis of lattice and triple lattice experiments in corn varietal tests II. Mathematical theory. Iowa Agri. Expt. Sta. Res. Bull. 281:45-66. E2,3a (305)

Cochran, W. G. [1940], The estimation of the yields of cereal experiments by sampling for the ratio of grain to total produce. J. Agri. Sci. 30:262-275. A14 (306)

Cochran, W. G. [1940], A survey of experimental design. Agri. Marketing Service Mimeo., U.S.D.A., 38 pp. E2,5,7a;T2a; A14;C (307)

Cochran, W. G. [1941], An examination of the accuracy of lattice and lattice square experiments on corn. Iowa Agri. Expt. Sta. Res. Bull. 289:400-415. E3,11,14a;A14 (308)

Cochran, W. G. [1941], Lattice designs for wheat variety trials. J. American Soc. Agron. 33:351-360. E3,14a (309)

Cochran, W. G. [1942], The theory of linear hypotheses, (or of linear regression). Mimeo. Iowa State College. A7 (310)

Cochran, W. G. [1942], Sampling theory when the sampling units are of unequal sizes. JASA 37:199-212. E4a;A14 (311)

Cochran, W. G. [1943], Some developments in statistics. Chronica Botanica 7:383-386. E3,11a;C (312)

Cochran, W. G. [1943], Some additional lattice square designs. Iowa Agri. Expt. Sta. Res. Bull. 318:731-748. Ellace (313)

Cochran, W. G. [1943], The comparison of different scales of measurement for experimental results. AMS 14:205-216. A9 (314)

Cochran, W. G. [1944], "Student's" collected papers. AMS 15:435-438. E5a (315)

Cochran, W. C. [1946], Relative accuracy of systematic and stratified random samples for a certain class of populations. AMS 17(2):164-177. E5a;Al4 (316)

Cochran, W. G. [1946], Analysis of covariance. Inst. Stat. Mimeo. Ser. No. 6, Univ. N. Carolina. A7 (317)

Cochran, W. G. [1947], Some consequences when the assumptions for the analysis of variance are not satisfied. Biometrics 3:22-38. E2ae;T2ae;A6 (318)

Cochran, W. G., Autrey, K. M., and Cannon, C. Y. [1941], A double change over design for dairy cattle feeding experiments. J. Dairy Sci. 24:937-951. E8ace (319)

Cochran, W. G. and Cox, G. M. [1944], Experimental designs. Mimeo., Iowa State College Univ. N. Carolina. E2,3,7,llae;T2ae;C (320)

Cochran, W. G. and Watson, D. J. [1936], An experiment on observer's bias in the selection of shoot-heights. Empire J. Exptl. Agri. 4:69-76. E7ae (321)

Coene, R. de [1948], Méthodes statistiques pour l'etude des essais de rendements cotonniers à Bambesa. (Statistical methods for the study of yield trials of cottons at Bambesa.) Bull. Agri. Congo Belge 39:803-818. Al4 (322)

Coggins, P. P. [1928], Some general results of elementary sampling theory for engineering use. Bell System Tech. J. 7:26-69. Al4;C (323)

Cole, F. N. [1922], Kirkman parades. Bull. Amer. Math. Soc. 28:435-437. E2c (324)

Cole, J. S. and Hallsted, A. L. [1926], The effect of outside rows on the yields of plots of kafir and milo at Hays, Kans. J. Agri. Res. 32:991-1002. Al4 (325)

Cole, L. C. [1946], A study of the cryptozoa of an Illinois woodland. Ecol. Monographs 16:49-86. Al4 (326)

Cole, L. C. [1946], A theory for analyzing contagiously distributed populations. Ecology 27:329-341. A14 (327)

Collins, G. N. [1929], The application of statistical methods to seed testing. Circular 79, U.S. Dept. Agri. A14;C (328)

Collins, S. H. [1911], The application of the theory of errors to investigations of milk. J. Board Agri. (London), Suppl. 7:48-55. C (329)

Comin, D. [1935], Plot technique in field celery experiments. Amer. Soc. Hort. Sci. Proc. 33:524-527. A14;C (330)

Comstock, R. E. [1943], Overestimation of mean squares by the method of expected numbers. JASA 38:335-340. A10 (331)

Comstock, R. E., Peterson, W. J., and Stewart, H. A. [1948], An application of the balanced lattice design in a feeding trial with swine. J. Animal Sci. 7:320-331. E2ae (332)

Comstock, R. E. and Winters, L. M. [1942], Design of experimental comparisons between lines of breeding in livestock. J. Agri. Res. 64:523-532. T9a;C;S1 (333)

Conagin, A. [1949], Eficiência na experimentaçao com canteiros. (Efficiency of plot experimentation). Rev. Agri. Sao Paulo 24:123-128. E2,7a;A14 (334)

Conrad, J. P. [1937], Distribution of residual soil moisture and nitrates in relation to border effect of corn and sorgo. J. American Soc. Agron. 29:367-378. A14 (335)

Cook, R. L., Millar, C. E., and Robertson, L. S. [1945], A crop rotation field layout with an illustration of the statistics involved in combining several years data. Soil Sci. Soc. Amer. Proc. 10:213-218. E5,8ae (336)

Coombs, A. V. [1934], The border effect in plot experiments. Empire J. Exptl. Agri. 2:315-323. A14 (337)

Cornish, E. A. [1936], Non-replicated factorial experiments. J. Australian Inst. Agri. Sci. 2:79-82. T2a (338)

Cornish, E. A. [1938], Factorial treatments in incomplete randomized blocks. J. Australian Inst. Agri. Sci. 4:199-203. E3ae;T2ae (339)

Cornish, E. A. [1940], The estimation of missing values in incomplete randomized blocks experiments. Ann. Eugenics 10:112-118. E3,6ae (340)

Cornish, E. A. [1940], The estimation of missing values in quasi-factorial designs. Ann. Eugenics 10:137-143. E3,6,11ae (341)

Cornish, E. A. [1940], The analysis of covariance in quasi-factorial designs. Ann. Eugenics 10:269-279. E2,3,11ae;A7 (342)

Cornish, E. A. [1940], The analysis of quasi-factorial designs with incomplete data I. Incomplete randomized blocks. J. Australian Inst. Agri. Sci. 6:31-39. E2,3,6,7,10ae (343)

Cornish, E A. [1941], The analysis of quasi-factorial designs with incomplete data 2. Lattice squares. J. Australian Inst. Agri. Sci. 7:19-26. E11ae (344)

Cornish, E. A. [1943], The recovery of inter-block information in quasi-factorial designs with incomplete data I. Square, triple, and cubic lattices. Australian Coun. Scientific Ind. Res. Bull. 158. E3,6a (345)

Cornish, E. A. [1944], The recovery of inter-block information in quasi-factorial designs with incomplete data II. Lattice squares. Australian Coun. Scientific Ind. Res. Bull. 175. E11a (346)

Costello, D. F. and Klipple, G. E. [1939], Sampling intensity in vegetation surveys made by the square-foot density method. J. American Soc. Agron. 31:800-810. A14 (347)

Covas, G. and Christensen, J. R. [1945], Determinación del tamaño de parcelas para ensayos comparativos de rendimientos en la vid. (Determination of the size of plots for comparative yield tests in vine.) Rev. Argent. Agron. 12:26-29. A14 (348)

Covas, G. and Sivori, E. M. [1939], Eficiencia en ensayos comparativos de rendimientos planeados según el método de "blocks incompletos simétricos." (Efficiency of comparative yield tests based on the incomplete symmetrical block method.) Rev. Argent. Agron. 6:126-130. E2,3a (349)

Cowan, F. T. [1934], Application of the variance method to the comparison of grasshopper baits. J. Economic Ent. 27:705-713. A14;C;S1 (350)

Cox, G. M. [1940], Enumertaion and construction of balanced incomplete block configurations. AMS 11:72-85. E2ac (351)

Cox, G. M. [1943], Modernized field designs at Rothamsted. Soil Sci. Soc. Amer. Proc. 8:20-22. E2,3,7,8,12a; T2a (352)

Cox, G. M. [1944], Statistics as a tool for research. J. Home Economics 36:575. C (353)

Cox, G. M. [1949], A survey of types of experimental designs. Unpublished paper presented at Biometric Society meetings, Geneva, Switzerland. E2,3,11,14a;C (354)

Cox, G. M. and Cochran, W. G. [1946], Designs of greenhouse experiments for statistical analysis. Soil Sci. 62:87-98. C (355)

Cox, G. M. and Eckhardt, R. C. [1940], The analysis of lattice and triple lattice experiments in corn varietal tests. I. Construction and numerical analysis. Iowa Agri. Expt. Sta. Res. Bull. 281:1-44. E3ace (356)

Cox, G. M. and Snedecor, G. W. [1936], Covariance used to analyze the relation between corn yield and acreage. J. Farm Economics 18:597-607. A7 (357)

Cramér, H. [1946], Mathematical Methods of Statistics. Princeton Univ. Press, pp. 536-556. E1,2,7a (358)

Crampton, E. W. [1926], The interpretation of the feeding trial. Sci. Agri. 7:41-50. C (359)

Crampton, E. W. [1929], Competition at the feed trough - a factor in group feeding of pigs. Sci. Agri. 10:268-271. A14 (360)

Crampton, E. W. [1930], Individual feeding for the comparative feeding trial. American Soc. Animal Production Proc. 23:56-63. E2ae;C (361)

Crampton, E. W. [1930], Cod liver oil for growing pigs. Sci. Agri. 10:523-529. A14 (362)

Crampton, E. W. [1931], Statistical analysis of comparative feeding trial data. Sci. Agri. 11:281-285. E2ae (363)

Crampton, E. W. [1931], Powdered skimmilk for weanling pigs. Sci. Agri. 11: 347-350. E2ae (364)

Crampton, E. W. [1931], Soaked beet pulp versus malt sprout-barley screening-molasses mixture as a succulent feed for cows in milk. Sci. Agri. 12:174-177. E2ae (365)

Crampton, E. W. [1932], Provendeine for market pigs. Sci. Agri. 12:553-563. E2ae (366)

Crampton, E. W. [1932], Estimating statistically the significance of differences in comparative feeding trials. Sci. Agri. 13:16-25. C;S1 (367)

Crampton, E. W. [1942], The design of animal husbandry experiments. J. Animal Sci. 1:263-276. C (368)

Crombie, A. C. [1947], Interspecific competition. J. Animal Ecology 16:44-73. A14 (369)

Crowther, C. [1915], Probable error in pig-feeding trials. J. Agri. Sci. 7:137-141. C (370)

Crowther, E. M. [1936], The technique of modern field experiments. J. Royal Agri. Soc. 97:54-81. C (371)

Crowther, F. [1937], Multiple-factor experiments on the manuring of cotton in Egypt. Empire J. Exptl. Agri. 5:169-179. T2a (372)

Crowther, F. and Bartlett, M. S. [1938], Experimental and statistical technique of some complex cotton experiments in Egypt. Empire J. Exptl. Agri. 6:53-68. C (373)

Crowther, F. and Cochran, W. G. [1942], Rotation experiments with cotton in the Sudan Gezira. J. Agri. Sci. 32:390-405. E5,8a (374)

Crowther, F. and Mahmoud, A. [1935], Experiments in Egypt on the interaction of factors in crop growth. 1. A preliminary investigation of the interrelation of variety, spacing, nitrogen, and water supply, with reference to yields of cotton. Royal Agri. Soc. Egypt Tech. Section, Bull. 22. El2a (375)

Crowther, F., Tomforde, A., and Mahmoud, A. [1936], Experiments in Egypt on the interaction of factors in crop growth 4. Nitrogenous and phosphatic manuring of cotton and their relation to variety and spacing. Royal Agri. Soc. Egypt Tech. Section, Bull. 26. El2a (376)

Crutchfield, R. S. [1938], Efficient factorial design. J. Psychology 5:339-346. T2a (377)

Crutchfield, R. S. and Tolman, E. C. [1940], Multiple-variable design for experiments involving interaction of behavior. Psychological Rev. 47:38-42. T2a (378)

Currence, T. M. [1935], Relation of plot size and shape to variability of carrot yields on peat soils. Proc. American Soc. Hort. Sci. 33:484-488. Al4 (379)

Currence, T. M. [1947], Studies related to field plot technique with tomatoes. Proc. American Soc. Hort. Sci. 50:290-296. Al4 (380)

Currence, T, M. and Krantz, F. A. [1936], The relation of plot size and shape to potato yield variations. American Potato J. 13:310-313. Al4;Sl (381)

Curry, H. B. [1944], The method of steepest descent for non-linear minimization problems. Quarterly Appl. Math. 2:258-261. T4c (382)

Curtiss, J. H. [1943], On transformations used in the analysis of variance. AMS 14:107-122. A9 (383)

Dabral, B.M. and Chiney, S. S. [1938], A note on border effects in manurial experiments on cotton at Sakrand during 1935-36. Indian J. Agri. Sci. 8:629-635. E2ae;A14 (384)

Darroch, J.G. [1948], Sampling as a method of estimating experimental plot yields in Iowa. Unpublished manuscript, Iowa State College. A14 (385)

Das, A.C. [1949], Determination of the best shape of a sample cut (abstract). Proc. Thirty-Sixth Indian Sci. Congress, Part III, pp. 7. A14 (386)

Das, S. [1945], Availability of superphosphate with depth of its placement in calcareous soils. Indian J. Agri. Sci. 15:47-49. E2ae (387)

Dash, J.S., Turner, P.E., et al. [1930], Report of sub-committee on standardization of field experiments for the West Indian Conference of Agricultural Officers (Report published in full in the Proceedings of the Conference). Field experiments on sugar cane. Tropical Agri. 7:101-104, 125-131. C (388)

Dastur, R.H. and Singh, M. [1942], Studies in the periodic partial failures of the Punjab-American cottons in the Punjab VII. Amelioration of tirak on soils with saline subsoils (sandy loams). Indian J. Agri. Sci. 12:679-696. E12ae (389)

Dastur, R.H. and Singh, M. [1943], Studies in the periodic partial failures of the Punjab-American cottons in the Punjab IX. The interrelation of manurial factors and water-supply on the growth and yield of 4-F cotton on light sandy soils. Indian J. Agri. Sci. 13:610-630. E12ae (390)

Dastur, R.H. and Tashna, U.C. [1943], Studies in the periodic partial failures of the Punjab-American cottons in the Punjab VIII. The relation of weather factors with the spread of tirak in American cottons. Indian J. Agri. Sci. 13:449-467. A7 (391)

Davenport, E. and Frazer, W.J. [1896], Experiments with wheat, 1888-1895. Ill. Agri. Expt. Sta. Bull. 41:147-155. C (392)

David, F.N. [1934], On the P_{λ_n} test for randomness. Biometrika 26:1-11. C (393)

David, F.N. [1947], A power function for tests of randomness in a sequence of alternatives. Biometrika 34:335-339. C (394)

Davies, J.G. [1930], The sampling of cane in the field. Tropical Agri., Sugar Tech. Ser. No. 3. A14 (395)

Davies, J.G. [1931], The experimental error of the yield from small plots of "natural" pasture. Council Sci. Ind. Res. Australia, Bull. 48, 22 pp. A14 (396)

Davies, J. G. [1931], The sub-sampling of bagasse. Tropical Agri. 8:298-299. A14 (397)

Davies, W. M. [1931], Cultivation of the sugar beet crop. Three years' investigation of the effects of spacing. J. Ministry Agri. (London), 37:973-985. A14 (398)

Davis, F. E. and Harrell, G. D. [1942], Relation of weather and its distribution to corn yields. U.S.D.A. Tech. Bull. 806. A7 (399)

Davis, F. E. and Pallesen, J. E. [1940], Effect of the amount and distribution of rainfall and evaporation during the growing season on yields of corn and spring wheat. J. Agri. Res. 60:1-23. A7 (400)

Davis, J. F. [1940], The relationship between leaf area and yield of the field bean with a statistical study of methods for determining leaf area. J. American Soc. Agron. 32:323-329. A14 (401)

Davis, J. F., Baten, W. D., and Cook, R. L. [1946], The effect of time of application and levels of nitrogen, phosphorous and potash on the growth of sugar beets with a detailed statistical procedure of confounding in a 3×3×3 design. Michigan State College Agri. Expt. Sta. Tech. Bull. 203:1-40. T2ace (402)

Davis, J. F., Cook, R. L., and Baten, W. D. [1942], A method of statistical analysis of a factorial experiment involving influence of fertilizer analyses and placement of fertilizer on stand and yield of cannery peas. J. American Soc. Agron. 34:521-532. E2ae;T2ae (403)

Dawson, C. D. R. [1939], An example of the quasi-factorial design applied to a corn breeding experiment. Ann. Eugenics 9:157-173. E3ae (404)

Day, B. B. [1949], The statistical part in welding investigations. Welding J. Supplement, October. T2ae;C (405)

Day, B. B. and Austin, L. [1939], A three-dimensional lattice design for studies in forest genetics. J Agri. Res. 59:101-119. E12ac (406)

Day, B. B. and Fisher, R. A. [1937], The comparison of variability in populations having unequal means. An example of the analysis of covariance with multiple dependent and independent variates. Ann. Eugenics 7:333-348. E1ae;A7,10 (407)

Day, J. W. [1920], The relation of size, shape, and number of replications of plats to probable error in field experimentation. J. American Soc. Agron. 12: 100-105. A14;S1 (408)

Dean, L. A. [1939], Relationships between rainfall and coffee yields in the Kona district, Hawaii. J. Agri. Res. 59:217-222. A7 (409)

DeLury, D. B. [1946], The analysis of latin squares when some observations are missing. JASA 41:370-389. E7,10a (410)

DeLury, D. B. [1947], On the estimation of biological populations. Biometrics 3:145-167. A14 (411)

DeLury, D. B. [1948], The analysis of covariance. Biometrics 4:153-170. E2ae;A7 (412)

Delwiche, E. J. [1928], A new system for variety test plats. J. American Soc. Agron. 20:771-773. A14 (413)

Deming, G. W. and Brewbaker, H. E. [1934], Border effect in sugar beets. J. American Soc. Agron. 26:615-619. A14 (414)

Deming, G. W. and Coleman, O. H. [1942], Comparative efficiency of lattice and random-block designs for a sugar-beet variety test. American Soc. Sugar Beet Tech., Proc. 3:181-183. E2,3,14a (415)

Deming, W. E. [1941], Some thoughts on statistical inference. J. Wash. Academy Sci. 31:85-92. C (416)

Dempster, E. R. and Lerner, I. M. [1949], Selection problems in animal breeding. Proc. Berkeley Symp. Math. Stat. Prob. pp. 481-483. T9a (417)

den Doop, J. E. A. [1939], Factorial diagrams. Soil Sci. 48:497-504. T2a (418)

Dendrinos, A. D. [1931], The question of the shape of plots in field experimentation. Bull. Assoc. Int. Sélect. Plantes 4:106-113. A14 (419)

Dennett, J. H. [1934], The layout of field experiments. Malayan Agri. J. 22: 276-283. A14;C (420)

Denny, F. E. [1922], Formula for calculating the number of fruits required for an adequate sample for analysis. Botanical Gaz. 73:44-57. S1 (421)

Derevitzky, N. F. [1933], The investigation of the accuracy of comparing means in the experiment with a scattered situation of replications (in Russian with English summary). Lenin Academy Agri. Sci. Inst. Plant Ind. Appl. Botany, Genetics, Plant Breeding, Ser. II, Bull. 3:137-187. E5a (422)

Derewitzky, N. [1934], Bearbeitung der Ergebnisse von Versuchen mit ausfallenden Parzellendaten. Arch. PflBau. 10·573-585. E6a (423)

deVerteuil, J. [1934], Manurial experiments on coconuts on Perseverance Estate, Couva. Tropical Agri. 11:313-315. E2ae;T2ae (424)

de Vries, O. [1938], Het serieprincipe bij veldproeven. (The "scale-principle" for field trials . Landbouwk. Tijdschr. Wageningen 50:340-358. A7;C (424a)

Dewey, J. [1929], The Quest for Certainty: A study of the relation of knowledge and action. Minton Balch & Co., New York. C (425)

Dickerson, G. E. [1942], Experimental design for testing inbred lines of swine. J. Animal Sci. 1:326-341. S1 (426)

Dijkveld Stol, J. J. [1942], Het uitschakelen van systematische fouten bij proefvelden in vierkantsvorm. (Elimination of systematic errors in square field experiments.) Landbouwk. Tijdschr. Wageningen 54:185-202. C (427)

Dijkveld Stol, J. J. [1942], Oogstformuleering. (Formulae for yield.) Landbouwk. Tijdschr. Wageningen 54:726-738, 798-817. A7;C (428)

Dixon, W. J. and Mood, A. M. [1948], A method for obtaining and analyzing sensitivity data. JASA 43:109-126. T8a;C (429)

Dodge, H. F. and Roming, H. G. [1929], Method of sampling inspection. Bell System Tech. J. 8:613-631. A14;C (430)

Dojes, R. P. [1936], Een modern rassenproefveld. (A modern experimental field for testing varieties.) Meded. Nat. Comité voor Brouwgerst, Wageningen No. 1, pp. 20. C (431)

Donald, H. P. [1939], The relative importance of sow and litter during the growth of suckling pigs: A comparison of fostered with normally reared pigs. Empire J. Expt. Agri. 7:32-42. E7a;C (432)

Dorph-Petersen, K. [1942], Fejlberegning paa Forsøg med systematisk Parcelfordeling. (Estimation of standard error in experiments with systematic plot distribution.) Nordisk Jordbrugsforskning 24:140-150. E5,7a (433)

Dorph-Petersen, K. [1949], Parcelfordeling i markforsøg. (Plot arrangement in field experiments.) Tidsskr. Planteavl. (Copenhagen) 52 (1):111-175. E7a;A14 (434)

Dorsey, M. J. and McMunn, R. L. [1931], A comparison of different methods of taking samples of apples in experimental plots. Proc. American Soc. Hort. Sci. 28:619-626. A14 (435)

Doughty, L. R. and Engledow, F. L. [1928], Investigations on yield in the cereals. V. A study of four fields of wheat: The limiting effect of population-density on yield and an analytical comparison of yields. J. Agri. Sci. 18:317-345. A14 (436)

Down, E. E. [1942], Plot technic studies with small grains. J. American Soc. Agron. 34:472-481. A14 (437)

Down, E. E. and Thayer, J. W. [1942], Plot technic studies with navy beans. J. American Soc. Agron. 34:919-922. A14 (438)

Doxtator, C. W. [1940], Harvest sampling studies with five varieties of sugar beets. American Soc. Sugar Beet Tech. Proc. 2:103-106. A14 (439)

Doxtator, C. W., Tolman, B., Cormany, C. E., Bush, H. L., and Jensen, V. [1942], Standardization of experimental methods. American Soc. Sugar Beet Tech. Proc. 3:595-599. C (440)

Drew, W. B. [1944], Studies on the use of the point-quadrat method of botanical analysis of mixed pasture vegetation. J. Agri. Res. 69:289-297. A14 (441)

Ducomet, V. [1932], Essais comparatifs de rendement-simplification dans le calcul des erreurs et méthode simple d'interprétation des résultats d'après la comparison des produits parcellaires. Sélectionneur Français 1:22-44. C (442)

Dudeney, H. E. [1917], Amusements in Mathematics. Thomas Nelson and Sons, London, Edinburgh, New York, Toronto, and Paris. E12c (443)

Dudeney, H. E. [1919], The Canterbury Puzzles. Thomas Nelson and Sons, London. E2,3c (444)

Dudley, F. J. and Read, D. R. [1949], The design of experiments in egg production in poultry. Harper Adams Utility Poultry J. 34:65-82. E1,2,12ae;C (445)

Dudok van Heel, J. P. [1939], Iets over proefvelden. (Some comments on experiment plots.) Van Zaad tot Suiker 9:136-138. C;S1 (446)

Dufrenoy, I. [1932], Tabac. Bull. Tutunului, 21:291-306. C (447)

Dufrénoy, J. [1936], Méthodes statistiques appliquées à la pathologie végétale. Ann. Épiphyt. Phytogénét. (1934/35) 1(N.S.):147-256. C (448)

Duncan, D. B. [1947], Significance tests for differences between ranked variates drawn from normal populations. Ph. D. Thesis, Iowa State College. E1,2a;A3 (449)

Dungan, G. H. [1946], Distribution of corn plants in the field. J. American Soc. Agron. 38:318-324. A14 (450)

Dunlop, G. [1933], Methods of experimentation in animal nutrition. J. Agri. Sci. 23:580-614. E12a;C (451)

East, E. M [1910], The role of selection in plant breeding. Pop. Sci. Mo. 77·190-203. T9a (452)

Eckenstein, O. [1911], Bibliography of Kirkman's school girl problem. Messenger Math. 41:33-36. E2c;B (453)

Eden, T. [1931], Studies in the yield of tea. I. The experimental errors of field experiments with tea. J. Agri. Sci. 21:547-573. E7a;A7,14 (454)

Eden, T. [1931], The experimental errors of field experiments with tea. Tea Res. Inst. Ceylon, Bull. 6. E7a;A14 (455)

Eden, T. [1932], Studies in the yield of tea II. The seasonal and sampling variation in yield and mineral composition of the tea leaf. J. Agri. Sci. 22:386-395. A14;C (456)

Eden, T. [1935], Studies in the yield of tea. III. Field experiments with potash and nitrogen in relation to pruning cycle. Empire J. Exptl. Agri. 3:105-118. T2a;A14 (457)

Eden, T. [1935], The development of field experiments in agricultural research. Tropical Agriculturist (Ceylon) 84:63-69, 131-149, 188-195. C (458)

Eden, T. and Fisher, R. A. [1927], Studies in crop variation. IV. The experimental determination of the value of top dressings with cereals. J. Agri. Sci. 17:548-562. E2ae;T2ae (459)

Eden, T. and Fisher, R. A. [1929], Studies in crop variation. VI. Experiments on the response of the potato to potash and nitrogen. J. Agri. Sci. 19:201-213. E2,7ae;T2ae (460)

Eden, T. and Maskell, E. J. [1928], The influence of soil heterogeneity on the growth and yield of successive crops. J. Agri. Sci. 18:163-185. A14 (461)

Eden, T. and Yates, F. [1933], On the validity of Fisher's z test when applied to an actual example of non-normal data. J. Agri. Sci. 23:6-17. A9,11 (462)

Edgar, J. L. [1938], Strawberry cultivation studies. II. Variability in individual plant size and cropping, with special reference to area and shape of plots for field experiments. J. Pomology 16:91-100. A14 (463)

Edwards, J. [1932], The progeny test as a method of evaluating the dairy sire. J. Agri. Sci. 22:811-837. A14;S1 (464)

Eisenhart, C. [1939], The interpretation of certain regression methods and their use in biological and industrial research. AMS 10:162-186. A7 (465)

Eisenhart, C. [1947], The assumptions underlying the analysis of variance. Biometrics 3:1-21. A7,9,11 (466)

Eisenhart, C., Hastay, M. W., and Wallis, W. A. [1947], Selected Techniques of Statistical Analysis. McGraw-Hill, N.Y. T4a;A7,9;C;S2 (467)

Ellison, L. [1942], A comparison of methods of quadratting short-grass vegetation. J. Agri. Res. 64:595-614. A14 (468)

Emmert, E. M. [1940], Partial elimination of experimental error from data by the use of significance tests. Proc. American Soc. Hort. Sci. 37:272-278. A7 (469)

Engledow, F. L. [1926], A census of an acre of corn (by sampling). J. Agri. Sci. 16:166-195. A14 (470)

Engledow, F. L. and Ramiah, K. [1930], Investigation on yield in cereals VII. A study of development and yield of wheat based upon varietal comparison. J. Agri. Sci. 20:265-344. A14 (471)

Engledow, F. L. and Wadham, S. M. [1923], Investigations on the yields of cereals. Part I. J. Agri. Sci. 13:390-439. A14 (472)

Engledow, F. L. and Wadham, S. M. [1924], Investigations on the yields of cereals. Part II. J. Agri. Sci. 14:66-98, 287-345. A14 (473)

Engledow, F. L. and Yule, G. U. [1926], The principles and practices of yield trials. Empire Cotton Growing Corp. Vol.3 (2&3). E5,7a;A14;C (474)

Englehart, M. D. [1936], The technique of path coefficients. Psychometrika 1: 287-293. A7 (475)

Englehart, M. D. [1941], The analysis of variance and covariance techniques in relation to the conventional formulas for the standard error of a difference. Psychometrika 6:221-233. A7 (476)

Ensign, R. D., Tucker, R. et al. [1947], 1947 Pinto bean variety experiments and demonstrations in cooperation with Agronomy, Extension, and Extension Pathology. Mimeo. Report, Colorado A&M, Fort Collins. E2,12ae (477)

Erdös, P. and Kaplansky, I. [1946], The asymptotic number of Latin rectangles. American J. Math. 68:230-236. E7c (478)

Erickson, R. O. and Stehn, J. R. [1945], A technique for analysis of population density data. Amer. Midland Nat. 33(3):781-787. A14 (479)

Erlee, T. J. D. [1943], Twee problemen op het gebied van de variatie-statistiek, bestudeerd op basis van de uitkomsten van de 203 in het jaar 1931 op Java geoogste blanco suikerrietvakkenproeven. (Two problems in statistical variation based on 203 blank experiments, harvested in 1931, on sugar cane in Java,) Landbouwk. Tijdschr. Wageningen 55:660-668. A14 (480)

Euler, L. [1782], Recherches sur une nouvelle espèce des quarrés magiques. Verhandlingen Zeeuwach Genootschap Wetenschappen, Vlissigen 9:85-239 (also in Commentationes Arithmeticae 2:302-361, 1849 and in Euleri Opera Omnia 7:291-392, 1923). E7,12c (481)

Evvard, J. M., Culbertson, C. C., and Snedecor, G. W. [1928], How many animals per experimental lot? Proc. American Soc. Animal Production 18: 161-176. C;S1 (482)

Fan, F. R. and Koo, W. F. [1941], (Application of analysis of covariance to quasi-factorial experiment.) Kwangsi Agri. 2·14-20. E12a;A7 (483)

Farrell, F. D. [1913], Interpreting the variation of plat yields. U.S. Dept. Agri. Circular 109:27-32. C (484)

Farrell, F. D. [1914], Basing alfalfa yields on green weights. J. American Soc. Agron. 6:42-45. C (485)

Federer, W. T. [1945], Studies on sample size and number of replicates for guayule investigations. J. American Soc. Agron. 37:469-478. A14;S1 (486)

Federer, W. T. [1948], Evaluation of variance components from a group of experiments with multiple classifications. Ph. D. Thesis, Iowa State College. E6a;T9a;A13 (487)

Federer, W. T. [1949], The general theory of prime power lattice designs. III. The analysis for p^3 varieties in blocks of p plots for more than 3 replicates. Biometrics 5:144-161. E3ace (488)

Federer, W. T. [1949], Designs for research and observational experiments. Mimeo. Cornell Univ. E2,3,7,8,11,12ae; T2ae;C (489)

Federer, W. T. [1949], The general theory of prime power lattice designs. VI-A. The design and analysis for a k×k lattice square with split plots of k plots in three and six replicates, illustrated for k=2 and 5. Unpublished manuscript, Cornell Univ. E11,13ace (490)

Federer, W. T. and Sprague, G. F. [1947], A comparison of variance components in corn yield trials. I. Error, tester × line, and line components in top-cross experiments. J. American Soc. Agron. 39:453-463. T9a;A13 (491)

Fieller, E. C. [1940], The biological standardization of insulin. JRSSB 7:1-64. E7ae;T8a;A7 (492)

Fieller, E. C. [1944], A fundamental formula in the statistics of biological assay and some applications. Quarterly J. Pharmacology 17:117-123. E8a;T8a (493)

Fieller, E. C., Irwin, J. O., Marks, H. P., and Shrimpton, E. A. G. [1939], The dosage-response relation in the cross-over rabbit test for insulin. Quarterly J. Pharmacy Pharmacology 12:206-211, 724-742. E7a;T8a;C (494)

Finney, D. J. [1940], The little Hoos field experiment on the residual values of certain manures. Empire J. Exptl. Agri. 8:111-125. E5,8ae (495)

Finney, D. J. [1941], Wireworm populations and their effect on crops. Ann. Appl. Biology 28:282-295. A14 (496)

Finney, D. J. [1941], The joint distribution of various ratios based on a common error mean square. Ann. Eugenics 11:136-140. T2a;A3 (497)

Finney, D. J. [1941], The relationship of plant number and yield in sugar-beet and mangolds. Empire J. Exptl. Agri. 9:57-64. A13,14 (498)

Finney, D. J. [1942], The analysis of toxicity tests on mixtures of poisons. Ann. Appl. Biology 29:82-94. T8a;C (499)

Finney, D. J. [1942], Examples of the planning and interpretation of toxicity tests involving more than one factor. Ann. Appl. Biology 29:330-332. T8a (500)

Finney, D. J. [1943], Recent developments in the wireworm survey. Agri. Prog. 8:36-38. A14 (501)

Finney, D. J. [1943], The statistical treatment of toxicological data relating to more than one dosage factor. Ann. Appl. Biology 30:71-79. T8a (502)

Finney, D. J. [1944], The application of the probit method to toxicity test data adjusted for mortality in the controls. Ann. Appl. Biology 31:68-74. T8a (503)

Finney, D. J. [1944], Mathematics of biological assay. Nature 53:284. T8a (504)

Finney, D. J. [1944], The application of probit analysis to the results of mental tests. Psychometrika 9:31-39. T8a (505)

Finney, D. J. [1945], Some orthogonal properties of the 4×4 and 6×6 Latin squares. Ann. Eugenics 12:213-219. E7,12,13c (506)

Finney, D. J. [1945], The fractional replication of factorial experiments. Ann. Eugenics 12:291-301 (correction 15:276). T3ace (507)

Finney, D. J. [1945], The microbiological assay of vitamins: The estimate and its precision. Quarterly J. Pharmacology 18:77-82. T8a (508)

Finney, D. J. [1946], Orthogonal partitions of the 5×5 Latin squares. Ann. Eugenics 13:1-3. E7,12,13c (509)

Finney, D. J. [1946], Field sampling for the estimation of wireworm populations. Biometrics 2:1-7. A14 (510)

Finney, D. J. [1946], Standard errors of yields adjusted for regression on an independent measurement. Biometrics 2:53-55. E2,3a;T2a;A7 (511)

Finney, D. J. [1946], A note on "missing-plot techniques". Biometrics 2:94. E6,10a (512)

Finney, D. J. [1946], Latin squares of the sixth order. Experientia, Basel 2: 404-405. E7,12,13c (513)

Finney, D. J. [1946], Recent developments in the design of field experiments I. Split-plot confounding. II. Unbalanced split-plot confounding. J. Agri. Sci. 36:56-68. E12,13ace (514)

Finney, D. J. [1946], Recent developments in the design of field experiments III. Fractional replication. J. Agri. Sci. 36:184-191. T3ace (515)

Finney, D. J. [1947], Orthogonal partitions of the 6x6 Latin squares. Ann. Eugenics 13:184-196. E7,12,13c (516)

Finney, D. J. [1947], The estimation from individual records of the relationship between dose and quantal response. Biometrika 34:320-334. T8a (517)

Finney, D. J. [1947], The construction of confounding arrangements. Empire J. Exptl. Agri. 15:107-112. T2,3ac (518)

Finney, D. J. [1947], The principles of biological assay. JRSSB 9:46-91. T8a;C (519)

Finney, D. J. [1947], The significance of associations in a square point lattice. JRSSB 9:99-103. A14 (520)

Finney, D. J. [1948], Main effects and interactions. JASA 43:566-571. T2a (521)

Finney, D. J. [1949], The choice of a response metameter in bioassay. Biometrics 5:261-272. T8a (522)

Finney, D. J. [1949], The adjustment for a natural response rate in probit analysis. Ann. Appl. Biology 36:187-195. T8a (523)

Finney, D. J. and Palca, H. [1949], The elimination of bias due to edge effects in forest sampling. Forestry 23:31-47. A14 (524)

Finney, D. J. and Stevens, W. L. [1948], A table for the calculation of working probits and weights in probit analyses. Biometrika 35:191-201. T8a (525)

Fippin, E. O. [1941], The objectives and methods of field plot fertilizer tests and a proposed improvement of methods. Proc. American Soil Sci. Soc. 5:274-280. C (526)

Fischer, G. J. [1937], Interpretación estadistica de experiencias biológicas. (Statistical interpretation of biological experiments.) Arch. Fitotécn. Uruguay 2:85-106. C (527)

Fisher, R. A. [1911], Appendix to Mercer and Hall's paper on "The experimental error of field trials." J. Agri. Sci. 4:128-132. A14 (528)

Fisher, R. A. [1918], The correlation between relatives on the supposition of Mendelian inheritance. Royal Soc. Edinburgh Trans. 52:399-433. T9a;A7 (529)

Fisher, R. A. [1921], Studies in crop variation. I. An examination of the yield of dressed grain from Broadbalk. J. Agri. Sci. 11:107-135. A7,13 (530)

Fisher, R. A. [1924], The influence of rainfall on the yield of wheat at Rothamsted. Philosophical Trans. Royal Soc. 213(B):89-142, A7;13 (531)

Fisher, R. A. [1926], The arrangement of field experiments. J. Ministry Agri. 33:503-513. E2,7a;T2a;A14;C (532)

Fisher, R. A. [1929], The statistical method in psychical research. Proc. Soc. Psychical Res. 39:189-192. C (533)

Fisher, R. A. [1929], Statistics and biological research. Nature 124:266-267. C (534)

Fisher, R.A. [1929], Tests of significance in harmonic analysis. Proc. Royal Soc. London 125(A):54-59. A1,3 (535)

Fisher, R. A. [1931], The technique of field experiments. Principles of plot experimentation in relation to the statistical interpretation of the results. Rothamsted Conf. 13:11-13. A14;S1 (536)

Fisher, R. A. [1933], The contributions of Rothamsted to the development of statistics. Report of 1933, Rothamsted Exptl. Sta. pp. 43-50. C (537)

Fisher, R. A. [1934], Randomization, and an old enigma of card play. Math. Gazette 18:294-297. C (538)

Fisher, R. A. [1935], The independence of experimental evidence in agricultural research. Trans. Third Int. Congress Soil Sci. 2:112-119. C;S1 (539)

Fisher, R. A. [1936], The half-drill strip system agricultural experiments. Nature 138:1101. E7a (540)

Fisher, R. A [1938], The mathematics of experimentation. Nature 142:442. C (541)

Fisher, R. A. [1940], An examination of the different possible solutions of a problem in incomplete blocks. Ann. Eugenics 10:52-75. E2,3c (542)

Fisher, R. A. [1941], The interpretation of experimental four-fold tables. Science 94:210-211. C (543)

Fisher, R. A. [1942], New cyclic solutions to problems in incomplete blocks. Ann. Eugenics 11:290-299 (correction 11:402-403). E7,13c (544)

Fisher, R. A. [1942], The theory of confounding in factorial experiments in relation to the theory of groups. Ann. Eugenics 11:341-353. T2c (545)

Fisher, R. A. [1942], Some combinatorial theorems and enumerations connected with the numbers of diagonal types of a Latin square. Ann. Eugenics 11:395-401. E7,13c (546)

Fisher, R. A. [1945], A system of confounding for factors with more than two alternatives, giving completely orthogonal cubes and higher powers. Ann. Eugenics 12:283-290. E12,13c;T2c (547)

Fisher, R. A. [1945], Recent progress in experimental design. L'application du calcul des probabilités. (Colloquium held at Geneva 12-15 July, 1939) (Inst. Int. Coopération Intellectuelle). E2,7,13c (548)

Fisher, R. A. [1946], Statistical Methods for Research Workers. 10th edition (1st ed., 1925) Oliver & Boyd, London. E2,7ace (549)

Fisher, R. A. [1947], The analysis of covariance method for the relation between a part and the whole. Biometrics 3:65-68. A7 (550)

Fisher, R. A. [1947], Development of the theory of experimental design. Proc. Int. Stat. Conf. 3:434-439. E7,13c;T2c (551)

Fisher, R. A. [1949], A biological assay of tuberculins. Biometrics 5:300-316. E12ae;T8ae (552)

Fisher, R. A., Corbet, A. S., and Williams, C. B. [1943], The relation between the number of species and the number of individuals in a random sample of an animal population. J. Animal Ecology 12:42-58. S1 (553)

Fisher, R. A. and MacKenzie, W. A. [1922], The correlation of weekly rainfall. Quarterly J. Royal Meteorological Soc. 48:234-245. A7 (554)

Fisher, R. A. and MacKenzie, W. A. [1923], Studies in crop variation II. The manurial response of different potato varieties. J. Agri. Sci. 13:311-320. E7ae (555)

Fisher, R. A., Thornton, H. G., and MacKenzie, W. A. [1922], The accuracy of the plating method of estimating the density of bacterial populations. Ann. Appl. Biology 9:325-359. A14;C (556)

Fisher, R.A. and Wishart, J. [1930], The arrangement of field experiments and the statistical reduction of the results. Imperial Bureau Soil Sci. Tech. Comm. No. 10. E2,7ae;T2ae (557)

Fisher, R.A. and Yates, F. [1934], The 6×6 Latin squares. Proc. Cambridge Philosophical Soc. 30:492-507. E7,12,13c (558)

Fisher, R.A. and Yates, F. [1948], Statistical Tables for Biological, Agricultural and Medical Research. Oliver & Boyd, Edinburgh, 3rd Edition (1st ed., 1938). E2,3,7,12,13c (559)

Fix, E. [1949], Distributions which lead to linear regressions. Proc. Berkeley Symp. Math. Stat. Prob. pp. 79-91. A7 (560)

Fleming, W.E. and Baker, F.E. [1936], A method for estimating populations of larvae of the Japanese beetle in the field. J. Agri. Res. 53:319-331. A14 (560a)

Forsaith, C.C. [1943], Statistics for foresters. New York State College Forestry Tech. Bull. 16(4):1-69. C (561)

Forester, H.C. [1937], Design of agronomic experiments for plots differentiated in fertility by past treatments. Iowa State College Res. Bull. 226:139-172. E5,8a (562)

Forbes, E.B., Elliot, R.F. et al. [1946], Variation in determinations of digestive capacity of sheep. J. Animal Sci. 5:298-305. A14 (563)

Forster, H.C. and Vasey, A.J. [1932], The development of accuracy in agricultural experiments. Victoria J. Agri. 30:35-52. E1,2,5,7a;A14;C (564)

Fracker, S.B. and Brischle, H.A. [1944], Measuring the local distribution of Ribes. Ecology 25(3):283-303. A14 (565)

Franke, A. [1946], Veredelingsonderzoekingen met rogge III: Blanco proeven. (Rye breeding research III: Uniformity trials.) Landbouwk. Tijdschr. Wageningen 58:77-87. A14 (566)

Frankena, H. J. [1935], Over blanco--of blinde proeven. (On blank tests.) Versl. RijkslandbProefst. 's-Grav. 41(A):173-209. A14 (567)

Frankena, H. J. and Both, M. P. [1939], Enkele opmerkingen over proefveldtechniek. (Some observations on the technique of field experiments.) Versl. RijkslandbProefst. 's-Grav. 45(16)A:427-437. C (568)

Fraser, A.H.H. and Robertson, D. [1933], The influence of nutritional state of the sheep on its susceptibility to infestation with the stomach worm, Haemonchus contortus. Empire J. Expt. Agri. 1:17-21. E1a (569)

Frost, A. [1871], General solution and extension of the problem of fifteen schoolgirls. Quarterly J. Appl. Math. 11:26-37. E2c (570)

Gaines, J. C. [1937], Cotton flea hopper control tests using the Latin square plat arrangement and analysis of variance. J. Economic Entomology 30: 119-125. E7ae (571)

Gaines, J. C. [1938], Analysis of data in plat designs for cotton insect control. J. Economic Entomology 31:656-659. C (572)

Galinat, W. C. and Everett, H. L. [1949], A technique for testing flavor in sweet corn. Agron. J. 41:443-445. E2ae (573)

Garber, R. J. and Hoover, M. M. [1930], Persistence of soil differences with respect to productivity. J. American Soc. Agron. 22:883-890. E5,8a (574)

Garber, R. J. and McIlvaine, T. C. [1935], Analysis of variance of corn yields obtained in crop rotation experiments. J. American Soc. Agron. 27:480-485. E5,8a (575)

Garber, R. J., McIlvaine, T. C., and Hoover, M. M. [1926], A study of soil heterogeneity in experiment plots. J. Agri. Res. 33:255-268. A14 (576)

Garber, R. J., McIlvaine, T. C., and Hoover, M. M. [1931], A method of laying out experiment plats. J. American Soc. Agron. 23:286-298. E5,8a;A14 (577)

Garber, R. J. and Odland, T. E. [1926], Influence of adjacent rows of soybeans on one another. J. American Soc. Agron. 18:605-607. A14 (578)

Garber, R. J. and Pierre, W. H. [1933], Variation of yields obtained in small artificially constructed field plats. J. American Soc. Agron. 25:98-105. A14 (579)

Garcia, R. F. [1932], The triangle system of fertilizer experiments as applied to sugar-cane. Proc. Fourth Congress Int. Soc. Sugar-Cane Tech. Bull. 89. T7a (580)

Garner, F. H., Grantham, J. and Sanders H. G. [1934], The value of covariance in analyzing field experimental data. J. Agri. Sci. 24:250-259. E2a;T2a;A7 (581)

Garner, F. H. and Sanders, H. G. [1938], A study of the effect of feeding oils to dairy cows and of the value of the latin square lay-out in animal experimentation. J. Agri. Sci. 28:541-555. E7a;A13 (582)

Garner, H. V. and Weil, J. W. [1939], The standard errors of field plots at Rothamsted and outside centres. Empire J. Exptl. Agri. 7:369-379. E2,7a;A13,14 (583)

Garwood, F. [1941], The application of maximum likelihood to dosage-mortality curves. Biometrika 32:46-58. T8ae (584)

Ghose, R. L. M. and Sanyal, A. T. [1945], Technique of field experiments in jute. Agri. Res. Mem. Indian Cent. Jute Cttee. 4:1-25. A14 (585)

Ghosh, B. [1943], On the distribution of random distances in a rectangle. Sci. Culture 8:388. A14 (586)

Ghosh, B. [1943], On the random distances between two rectangles. Sci. Culture 8:464. A14 (587)

Ghosh, B. [1943], On sampling in unknown fields. Sci. Culture 9:129-130. A14 (588)

Ghosh, B. [1943], On the construction of some natural fields. Sci. Culture 9:213-214. A14 (589)

Ghosh, B. [1945], Efficiency of rectangular "plots" of different shapes and sizes in field experiments or sample surveys. Proc. Thirty-Second Indian Sci. Congress, Sec. XII, No. 48. A14 (590)

Ghosh, B. [1946], Measures of heterogeneity in agricultural and similar fields, and their interrelations. Sci. Culture 11:382-383. A14 (591)

Ghosh, B. [1949], Topographic variation in statistical fields. Calcutta Stat. Assoc. Bull. 2:11-28. A14 (592)

Ghosh, B [1949], Topographic randomness and its statistical tests. Calcutta Stat. Assoc. Bull. 2:65-70. A14 (593)

Ghosh, B. [1949], On a particular type of natural field (abstract). Variance and covariance of rectangular and "linear" sample units for a type of natural field. (abstract). Proc. Thirty-Sixth Indian Sci. Congress, Part III, pp. 7-8. A14 (594)

Ghosh, M. N. [1948], Tests of randomness. Calcutta Stat, Assoc. Bull. 1:135-137. A14 (595)

Ghosh, M. N. [1948], A test for field uniformity based on the space correlation method. Sankhyā 9:39-46. A14 (596)

Ghurye, S. G. [1948], A characteristic of species of 7×7 Latin squares. Ann. Eugenics 14:133. E7,13c (597)

Ghurye, S. G. [1949], Transformations of a binomial variate for the analysis of variance. JISAS 2:94-109. A9 (598)

Gilbert, S. M. [1938], Variability in yield of Coffea arabica. E. Africa Agri. J. 4:131-139. E7a;A7,14 (599)

Gilbert, S. M. [1938], Planning field experiments on Coffea arabica. Tropical Agri. 15:16-18. E7a;A14 (600)

Gilbert, S. M. [1938], Plot size in field experiments with Coffea arabica. Tropical Agri. 15:52-55. A14 (601)

Giles, P. L. [1914], On the plans of fertilizer experiments. J. American Soc. Agron. 6:36-41. T2a (602)

Gobeil, R. [1940], Importance of statistics in forestry work (in French). La Forêt Québec 2 (2):31-48. T2ae;C (603)

Gomes, F. P. and Malavolta, E. [1949], Aspectos matemáticos e estatísticos da lei de Mitcherlich. An. Ex. Sup. Agri. "Luiz de Queiroz" 6:193-229. A7 (604)

Goodsell, W. D. [1942], Paired and grouped observations contrasted. Illinois Univ. Dept. Agri. Econ. Agri. Expt. Sta. Report AE-1837. C (605)

Goulden, C. H. [1927], Some applications of biometry to agronomic experiments. Sci. Agri. 7:365-376. C (606)

Goulden, C. H. [1929], Statistical methods in agronomic research. Plant Breeders' Series Publications No. 2, Canadian Seed Growers Assoc., Ottawa, Ont. E2ae;C (607)

Goulden, C. H. [1931], Modern methods of field experimentation. Sci. Agri. 11:681-701. E2,7ae;A14 (608)

Goulden, C. H. [1932], Application of the variance analysis to experiments in cereal chemistry. Cereal Chemistry 9:239-260. C (609)

Goulden, C. H. [1937], Efficiency in field trials of pseudo-factorial and incomplete randomized block methods. Canadian J. Res. C. 15:231-241. E3,14a (610)

Goulden, C. H. [1937], Modern methods for testing a large number of varieties. Dominion Canada, Dept. Agri. Tech. Bull. 9:1-36. E2,3a (611)

Goulden, C. H. [1939], Methods of Statistical Analysis (2nd ed., 1952). John Wiley & Sons, Inc. New York. E2,3,6,7,11,12ae; T2ae;C (612)

Goulden, C. H. [1942], Fundamentals of experimentation. Spragg Memorial Lectures Plant Breeding Michigan State College (1941) pp. 16-27. C (613)

Goulden, C. H. [1944], A uniform method of analysis for square lattice experiments. Sci. Agri. 25:115-136. E2,3a (614)

Goulden, C. H. [1944], Experimental design for cereal chemists. Cereal Chemistry 21:159-171. C (615)

Gowe, R. S. [1949], Studies of the physiological basis for a genetic type of infertility in the domestic fowl. Ph. D. Thesis, Cornell Univ.
E12a;A10 (616)

Grant, D. A. [1948], The latin square principle in the design and analysis of psychological experiments. Psychological Bull. 45:427-442.
E7,8,12ac (617)

Grant, D. A. [1949], The statistical analysis of a frequent experimental design. American J. Psychology 62:119-122. E7,8,12a (618)

Grantham, A. E. [1914], The effect of rate of seeding on competition in wheat varieties. J. American Soc. Agron. 6:124-128. A14 (619)

Green, J. M. [1947], The inheritance of combining ability in maize hybrids and the relative value of two testers for estimating topcross performance. Ph. D. Thesis, Iowa State College. E3ae (620)

Greenslade, R. M. and Pearce, S. C. [1940], Field sampling for the comparison of infestations of strawberry crops by the aphis Capitophorus fragariae Theob. J. Pomology Hort. Sci. 17:308-317. A9,14 (621)

Gregory, F. G., Growther, F., and Lambert, A. R. [1932], The interrelation of factors controlling the production of cotton under irrigation in the Sudan. J. Agri. Sci. 22:617-638. E12ae (622)

Gregory, F. G. and Richards, F. J. [1929], Physiological studies in plant nutrition I. The effect of manurial deficiency on the respiration and assimilation rate in barley. Ann. Botany 43:119-161. T2a (623)

Gridgeman, N. T. [1944], Mathematics of biological assay. Nature 153:461-462.
T8a (624)

Gridgeman, N. T. [1945], Special designs for vitamin D assays. Quarterly J. Pharmacy Pharmacology 18:15-23. T8a (625)

Griffee, F. [1928], Correcting yields in rod-row trials with the aid of the regression equation. J. American Soc. Agron. 20:569-581. A7,14 (626)

Griswold, R.M. and Blakeslee, L.H. [1939], The effect of different wrappings, temperatures, and length of storage on keeping qualities of frozen pork chops. Proc. American Soc. Animal Production 32:305-314. T2ae (627)

Grootenhuis, J.A. and Post, J.J. [1946], The suitability of the Latin square for trials of a fairly simple character. Meded. Dir. Tuinb. pp. 173-175.
E7a (628)

Grossnickle, L.T. [1942], The scaling of test scores by the method of paired comparisons. Psychometrika 7:43-64. T11a (629)

Grubbs, F. E. [1949], On designing single sampling inspection plans. AMS 20:242-256. S1 (630)

Guha Roy, K. K. and Mahalanobis, P. C. [1936], Statistical methods and their application to agronomy. A bibliography. Misc. Bull. Imp. Coun. Agri. Res. Delhi 9:1-120. B (631)

Gulliksen, H. [1946], Paired comparisons and the logic of measurements. Psychological Rev. 53:199-213. T11a (632)

Guttman, L. [1946], An approach for quantifying paired comparisons and rank order. AMS 17:144-163. T11a (633)

Haber, E. S. and Snedecor, G. W. [1946], Forecasts from an incomplete experiment on asparagus. Proc. American Soc. Hort. Sci. 48:481-487. E2,5,8ae (634)

Haemer, K. [1949], Question 23. Amer. Stat. 2(5):10. B (635)

Haig, I. T. [1929], Accuracy of quadrat sampling in studying forest reproduction on cut-over areas. Ecology 10:374-381. A14 (636)

Haines, W. B. [1938], Manuring hevea II. Revision of experimental results by means of a sampling method for yield. Empire J. Exptl. Agri. 6:11-19. S1 (637)

Haines, W. B. and Crowther, E. M. [1940], Manuring hevea III. Results on young buddings in British Malaya. Empire J. Exptl. Agri. 8:169-184. T2a (638)

Hajek, J. [1949], Užití komplexní methody a intervalu spolehlivosti při važeni. (An application of factorial design and confidence intervals to weighing.) (Czech. English and Russian summaries.) Statis. Obzor. 29:258-273. T2,6a (639)

Haldane, J. B. S. [1930], A mathematical theory of natural and artificial selection. Part VII. Selection intensity as a function of mortality rate. Proc. Cambridge Philosophical Soc. 27:131-136. A14 (640)

Hale, R. W. [1931], Experimental errors in chicken-rearing experiments. J. Agri. Sci. 21:716-725. A14;S1 (641)

Hall, A. D. [1905], The Book of the Rothamsted Experiments. E. P. Dutton Co. New York. C (642)

Hall, A. D. [1909], The experimental error in field trials. J. Board Agri. (London) 16:365-370. C (643)

Hall, A. D. and Russell, E. J. [1911], Field trials and their interpretation. J. Board Agri. (London), Suppl. 7:5-14. E7a;A14 (644)

Hall, M. [1943], Projective planes. Trans. Amer. Math. Soc. 54:229-277. E2,3,4,7,12,13c (645)

Hall, M. [1945], An existence theorem for latin squares. Bull. American Math. Soc. 51:387-388. E7c (646)

Hall, M. [1947], Cyclic projective planes. Duke Math. J. 14:1079-1090. E7,13c (647)

Hamming, G. [1947], Grafische verwerking van een Fisher-proef. (Graphic correction of a Fisher experiment.) Landbouwk. Tijdschr. Wageningen 59: 496-504. E7a (648)

Hamming, G. [1949], Het samenvatten van rassenproeven en het toepassen van vruchtbaarheidscorrecties mit niet-orthogonale methoden. Versl. Landbouwk. Onderzoek.'s - Gravenhage, pp. 197. C (649)

Hancock, N. I. [1936], Row competition and its relation to cotton varieties of unlike plant growth. J. American Soc. Agron. 28:948-957. A14 (650)

Hancock, N. I. [1947], Variations in length, strength, and fineness of cotton fibres from bolls of known flowering dates, locks, and nodes. J. American Soc. Agron. 39:122-134. E2,5ae (651)

Hanna, G. C. and Baker, G. A. [1949], Transformation of split-plot yield trial data to improve analysis of variance. Proc. American Soc. Hort. Sci. 53:273-275. E12a;A9 (652)

Hansberry, T. R. and Richardson, C. H. [1935], A design for testing technique in codling moth spray experiments. Iowa State College J. Sci. 10:27-35. E7ae (653)

Hansford, C. G., Hosking, H. R., Stoughton, R. H., and Yates, F. [1933], An experiment on the incidence and spread of angular leaf-spot disease of cotton in Uganda. Ann. Appl. Biology 20:404-420. E7ae (654)

Hanson, H. C. [1934], A comparison of methods of botanical analysis of the native prairie in western North Dakota. J. Agri. Res. 49:815-842. A14 (655)

Hanson, H. C. and Love, L. D. [1930], Comparison of methods of quadratting. Ecology 11:734-748. A14 (656)

Hanson, H. C. and Love, L. D. [1930], Size of list quadrat for use in determining effects of different systems of grazing upon <u>Agropyron smithii</u> mixed prairie. J. Agri. Res. 41:549-560. A14 (657)

Harper, H. J. [1946], Effect of row spacing on the yield of small grain nurse crops. J. American Soc. Agron. 38:785-794. A14 (658)

Harrington, J. B. [1939], The number of replicated small plat tests required in regional variety trials. J. American Soc. Agron. 31:287-299. S1 (659)

Harrington, J. B. [1941], The effect of having rows different distances apart in rod row plot tests of wheat, oats and barley. Sci. Agri. 21:589-606. E7ae;A14 (660)

Harris, F. S. and Butt, N. I. [1920], The unreliability of short-time experiments. J. American Soc. Agron. 12:158-167. C (661)

Harris, J. A. [1913], On the calculation of intra-class and inter-class coefficients of correlation from class moments when the number of possible combinations is large. Biometrika 9:446-472. C (662)

Harris, J. A. [1915], On a criterion of substratum homogeneity (or heterogeneity) in field experiments. American Naturalist 49:430-454. Al4 (663)

Harris, J. A. [1920], Practical universality of field heterogeneity as a factor influencing plot yields. J. Agri. Res. 19:279-314. Al4 (664)

Harris, J. A. [1926], The service of statistical formulae in the analysis of plat yields. J. American Soc. Agron. 18:247-273. C (665)

Harris, J. A. [1928], Mathematics in the service of agronomy. J. American Soc. Agron. 20:443-454. C (666)

Harris, J. A. [1930], Criticism of the limitations of the statistical method. J. American Soc. Agron. 22:263-269. C (667)

Harris, J. A., Harrison, G. J., and Lockwood, E. K. [1929], A criterion of the differentiation of varieties or of experimental areas with respect to their capacity to produce seedling stands of cotton. J. Agri. Res. 38:601-621. Al4;C (668)

Harris, J. A., Harrison, G. J., and Wadley, F. M. [1928], Illustrations of the application of a criterion of the deviation of an observed from a random distribution to the problem of seedling stand in sea-island, Egyptian, and upland cotton. J. Agri. Res. 36:603-614. Al4;C (669)

Harris, J. A. and Ness, M. M. [1928], Applicability of Pearson's equivalent probability r method to the problem of seedling mortality in sea-island, Egyptian, and upland cotton. J. Agri. Res. 36:615-623. Al4;C (670)

Harris, J. A. and Scofield, C. S. [1921], Permanence of differences in the plots of an experimental field. J. Agri. Res. 20:335-356. Al4 (671)

Harris, J. A. and Scofield, C. S. [1928], Further studies on the permanence of differences in the plots of an experimental field. J. Agri. Res. 36:15-40. Al4 (672)

Harris, M., Horvitz, D. G., and Mood, A. M. [1948], On the determination of sample sizes in designing experiments. JASA 43:391-402. Sl (673)

Harrison, C. J. and Bose, S. S. [1942], Studies of the variations in value of tea and its chemical constituents. Sankhyā 6:151-166. E7ae (674)

Harrison, C. J., Bose, S. S., and Mahalanobis, P. C. [1935], The effect of manurial dressings, weather conditions, and manufacturing processes on the quality of tea at Tocklai Experimental Station, Assam. Sankhyā 2:33-42. E7ae (675)

Harshbarger, B. [1944], On the analysis of a certain six-by-six four-group lattice design. AMS 15:307-320. E3ae (676)

Harshbarger, B. [1945], On the analysis of a certain six-by-six four-group lattice design using the recovery of interblock information. AMS 16:387-390. E3ae (677)

Harshbarger, B. [1946], Preliminary report on the rectangular lattices. Biometrics 2:115-119. E3a (678)

Harshbarger, B. [1947], Rectangular lattices. Va. Agri. Expt. Sta. Memoir 1:1-26. E3ace (679)

Harshbarger, B. [1949], Triple rectangular lattices. Biometrics 5:1-13. E3ace (680)

Hartley, H. O. [1940], Testing the homogeneity of a set of variances. Biometrika 31:249-255. A11 (681)

Hartley, H. O. and Smith, C. A. B. [1948], The construction of Youden squares. JRSSB 10:262-263. E7c (682)

Hartman, J. D. and Stair, E. C. [1942], Field plot technique studies with tomatoes. Proc. American Soc. Hort. Sci. 41:315-320. A14 (683)

Hartman, J. D. and Stair, E. C. [1946], Correlation of means and standard deviations in tomato field experiments. Proc. American Soc. Hort. Sci. 48:337-340. C (684)

Hartzell, F. Z. [1924], The use of biometrical methods in the interpretation of codling moth experiments. J. Economic Entomology 17:183-192. C (685)

Hartzell, F. Z. [1930], The Latin square arrangement of experimental plats. J. Economic Entomology 23:747-753. E7a (686)

Hartzell, F. Z. [1946], Methods of estimating foliage area injured by grape leafhoppers. New York State Agri. Expt. Sta., Geneva, Tech. Bull. 277:1-49. A14 (687)

Hasel, A. A. [1938], Sampling error in timber surveys. J. Agri. Res. 57:713-736. A14 (688)

Hasel, A. A. [1942], Estimation of volume in timber stands by strip sampling. AMS 13:179-206. A14 (689)

Hasel, A. A. [1942], Sampling error of cruises in the California pine region. J. Forestry 40(3):211-217. A14 (690)

Hasel, A. A. [1949], Long-term silvicultural experiment on methods of cutting. Proc. Berkeley Symp. Math. Stat. Prob. pp. 477-479. E5a (691)

Hassler, F. R. [1832], An account of the means and methods employed in comparison of weights and measures, ordered by the Senate of the United States, under direction of the Treasury Department, in 1831. Document No. 299, House of Representatives, 22nd Congress, 1st Session, pp. 37-122. E2ace;T11ace (692)

Hayes, H. K. [1923], Controlling experimental error in nursery trials. J. American Soc. Agron. 15:177-192. E5ae;A14 (693)

Hayes, H. K. [1925], Control of soil heterogeneity and use of the probable error concept in plant breeding studies. Minnesota Agri. Expt. Sta. Tech. Bull. 30:1-21. E5a (694)

Hayes, H. K. and Arny, A. C. [1917], Experiments in field technic in rod row tests. J. Agri. Res. 11:399-419. A14;S1 (695)

Hayes, H. K. and Garber, R. J. [1927], Breeding Crop Plants. McGraw-Hill, New York. A14;C;S1 (696)

Hayes, H. K and Immer, F. R. [1928], A study of probable error methods in field experiments. Sci. Agri. 8:345-352. C (697)

Hayes, H. K. and Immer, F. R. [1942], Methods of Plant Breeding. McGraw-Hill, New York. E2,3,6,8,12ae;A14;C (698)

Hayes, H. K., Wilson, H. K., and Ausemus, E. R. [1932], An experimental study of the rod-row method with spring wheat. J. American Soc. Agron. 24:950-960. E7a;A14 (699)

Hayford, J. F. [1893], On the least square adjustment of weighings. Appendix 10, (U.S. Coast and Geodetic Survey) Report for 1892. Government Printing Office. C (700)

Hays, F. A. [1932], The significance of differences in means in repetition experiments. Poultry Sci. 11:14-17. C (701)

Hazel, L. N. [1946], The covariance analysis of multiple classification tables with unequal subclass numbers. Biometrics 2:21-25. A7,10 (702)

Hazel, L. N. and Lush, J. L. [1942], The efficiency of three methods of selection. J. Heredity 33:393-399. T9a (703)

Hazel, L. N. and Terrill, C. E. [1946], Effects of some environmental factors on weanling traits of range Columbia, Corriedale, and Targhee lambs. J. Animal Sci. 5:318-325. A14 (704)

Hazel, L. N. and Terrill, C. E. [1946], Effects of some environmental factors on fleece and body characteristics of range Rambouillet yearling ewes. J. Animal Sci. 5:382-388. A14 (705)

Heath, O. V. S. [1934], A simple apparatus for measuring the compactness of soil in the field, and some results obtained in a cultivation experiment. Empire J. Exptl. Agri. 2:205-212. E2ae (706)

Hedayetullah, S., Sen, S., and Nair, K. R. [1944], Statistical notes for agricultural workers No. 26. Influence of dates of planting and spacings on some winter varieties of rice. Indian J. Agri. Sci. 14:248-259. E12ace;A13 (707)

Henderson, C. R. [1948], Estimation of general, specific, and maternal combining abilities in crosses among inbred lines of swine. Ph. D. Thesis, Iowa State College. T9a;A10 (708)

Hendricks, W. A. [1935], The use of "differential regression" in analysis of variance. J. Agri. Sci. 25:258-263. E1ae;A7 (709)

Hendricks, W. A. [1935], Effects of mineral supplements on the length of the tail and wing feathers in white leghorns. Poultry Sci. 14:221-227. C (710)

Hendricks, W. A. [1935], The statistical treatment of hatchability data. Poultry Sci. 14:365-372. E1ae;C (711)

Hendricks, W. A. [1936], Some elementary concepts in the theory of sampling. Poultry Sci. 15:462-465. C (712)

Henry, G. F., Down, E. E., and Baten, W. D. [1942], An adequate sample of corn plots with reference to moisture and shelling percentages. J. American Soc. Agron. 34:777-781. A14;S1 (713)

Herchenroder, M. V. M. [1943], A review of modern practical methods of analysis in statistics. Rev. Agri. Maurice 22:51-68. C (714)

Hetzer, H. O. and Brier, G. W. [1939], Estimates of mean differences necessary for significance between pigs in group feeding experiments. Amer. Soc. Animal Production Proc. 32:157-161. S1 (715)

Heubner, W. and Stühlmann, M. [1940], Zur Methodique biologischer Auswendungen. Arch. Int. Pharmacodyn. 64:485- E8a;T8a (716)

Hey, G. B. [1938], A new method of experimental sampling illustrated on certain non-normal populations. Biometrika 30:68-80. A14,15;B (717)

Heyne, E. G. [1945], Use of new experimental designs. Report Fifth Hard Red Winter Wheat Improvement Conf., Manhattan, Kansas, Feb. 12, 13 and 14, 1945. Div. Cereal Crops Dis., Plant Ind. Sta., Beltsville, p. 17 (Mimeo.). E3,11a (718)

Hoblyn, T. N. [1929], The relationships between the experimental and the demonstration plot and their respective value to the investigator. County Officer, Fruit Grower. Annual Report East Malling Res. Sta. 17:40-45. C (719)

Hoblyn, T. N. [1931], Field experiments in horticulture. Imperial Bur. Fruit Production, Tech. Comm. No. 2. E2ae (720)

Hoblyn, T. N. [1931], The lay-out and conduct of two manurial trials with rasperries: Together with deductions that can validly be drawn. J. Pomology Hort. Sci. 9:303-330. E2a (721)

Hoblyn, T. N. [1938], A study of the variation in keeping quality of apples in store: As illustrated by the behaviour of the variety McIntosh Red from an Ontario apple orchard. JRSSB 5:129-170. A14;S1 (722)

Hoblyn, T. N. [1945], The design of field experiments with cacao. Report Cacao Res. Conf. London, May-June, 1945, pp.164-168. C (723)

Hoblyn, T. N. and Edgar, J. L. [1938], A study of the technique of variety trials, as illustrated by the comparative yields of four black currant varieties grown in three different localities. J. Pomology 15:326-337. E7a (724)

Hochberg, M. [1946], The effect of non-normality on tests of significance. M. S. Thesis, Iowa State College. A11 (725)

Hoffman, M. B. [1929], The use of performance records in laying out a raspberry fertilizer experiment. American Soc. Hort. Sci. Proc. 26:203-207. C (726)

Hollowell, E. A. and Heusinkveld, D. [1933], Border effect studies of red clover and alfalfa. J. American Soc. Agron. 25:779-789. A14 (727)

Holman, L. J. [1938], Simplified Statistics. Sir Isaac Pitman & Sons, Ltd., London, pp. xii+142+32. C (728)

Holme, R. V. [1941], Corn microplots and their interpretation. An interim discussion. J.A.S.T. Quarterly 5(2):3-8. A14;C (729)

Holmes, M. C. [1935], Sampling analysis and sample size. Franklin Inst. J. 219(4):483-486. S1 (730)

Holmes, R. L. and O'Neal, A. M. [1939], A study of the number of stalks of cane required for accuracy in sampling experimental plots. Proc. Sixth Congress Int. Soc. Sugar Cane Tech., La. (1938):760-764. S1 (731)

Homeyer, P. G. and Black, C. A. [1946], Sampling replicated field experiments on oats for yield determinations. Soil Sci. Soc. America Proc. 11:341-344. A14 (732)

Homeyer, P. G., Clem, M. A., and Federer, W. T. [1947], Punched card and calculating machine methods for analyzing lattice experiments including lattice squares and the cubic lattice. Iowa Agri. Expt. Sta. Res. Bull. 347:27-171. E3,11ae (733)

Hoogland, J. J. [1941], Note on the numerical calculation of the orthogonal polynomials. Ann. Eugenics 11:77-79. A7 (734)

Hoogland, J. J. and Bär, A. L. S. [1941], Gedetailleerde varietie-analyse van proefveldresultaten. (Detailed analysis of variance of the results of plot trials.) Arch. Suikerind. Nederland. Ned.-Ind. 2:402-407. C (735)

Hopkins, E. S. [1945], Long-time crop and culture rotations. Soil Sci. Soc. Amer. Proc. 10:295-299. E5,8a (736)

Hopkins, J. W. [1935], Weather and wheat yield in western Canada I. Influence of rainfall and temperature during the growing season on plot yields. Canadian J. Res. 12:306-334. A7 (737)

Hopkins, J. W. [1941], Agricultural meteorology: Monthly sequence of summer precipitation at Winnipeg, Swift Current and Edmonton. Canadian J. Res. 19 (C):85-94. A7 (738)

Hopkins, J. W. [1941], Agricultural meteorology: Seasonal incidence of rainless and rainy periods at Winnipeg, Swift Current, and Edmonton. Canadian J. Res. 19 (C):267-277. A7 (739)

Hopkins, J W. [1941], Agricultural meteorology: Summer sequence of monthly mean temperature at Winnipeg, Swift Current, and Edmonton. Canadian J. Res. 19 (C):485-492. A7 (740)

Hopkins, J. W [1946], Precision of assessment of palatability of foodstuffs by laboratory panels. Canadian J. Res. 24 (F):203-214. C (741)

Hopkins, J. W. [1946], Statistical design of experiments. Canadian Chemistry Process Ind. 30 (5):113-116. C (742)

Horsfall, J. G. and Rich, S. [1949], Spirally arranged plots in a design for field assay of fungicides. Connecticut Agri. Expt. Sta. Bull. 530:1-19. T8a;C (743)

Horton, H. B. [1948], A method for obtaining random numbers. AMS 19:81-85. C (744)

Horton, J. S. [1941], The sample plot as a method of quantitative analysis of chaparral vegetation in southern California. Ecology 22 (4):457-468. A14 (745)

Hotelling, H. [1940], The selection of variates for use in prediction with some comments on the general problem of nuisance parameters. AMS 11:271-283. T1ac (746)

Hotelling, H. [1941], Experimental determination of the maximum of a function. AMS 12:20-45. T4ac (747)

Hotelling, H. [1944], Some improvements in weighing and other experimental techniques. AMS 15:297-306. T6c (748)

Houghland, G. V. C. and Strong, W. O. [1941], Results of a 5-year factorial experiment with potato fertilizer. J. American Soc. Agron. 33:189-199. T2a;A13 (749)

Houseman, E. E. [1942], Methods of computing a regression of yield on weather. Iowa Agri. Expt. Sta. Res. Bull. 302. A7 (750)

Houseman, E. E. and Davis, F. E. [1942], Influence of distribution of rainfall and temperature on corn yields in western Iowa. J. Agri. Res. 65:533-545. A7 (751)

Houseman, E. E., Weber, C. R., and Federer, W. T. [1946], Pre-harvest sampling of soybeans for yield and quality. Iowa Agri. Expt. Sta. Res. Bull. 341:808-826. A14 (752)

Hsü, E. H. and Sherman, M. [1946], The factorial analysis of the electro-encephalogram. J. Psychology 22:89-92. T2a (753)

Hsu, P. L. [1943], Some simple facts about the separation of degrees of freedom in factorial experiments. Sankhyā 6:253-254. T2a (754)

Hsu, T. S. [1940], (An experiment on intervarietal competition in rice.) Kwangsi Agri. 1:14-20. A14 (755)

Hubback, J. A. [1946], Sampling for rice yield in Bihar and Orissa. Sankhyā 7:281-294. A14;S1 (756)

Huber, L. L. and Sleesman, J. P. [1934], Technique of field experimentation in entomology. I. Some principles involved in a well-planned experiment. J. Economic Entomology 27:1166-1170. E2,7ae; A10;C (757)

Huber, L. L. and Sleesman, J. P. [1935], Technique of field experimentation in entomology. II. The reduction of data by the method of analysis of variance. J. Economic Entomology 28:70-76. E2ae;C (758)

Hudson, H. G. [1939], Population studies with wheat I. Sampling. J. Agri. Sci. 29:76-110. A14;C (759)

Hudson, H. G. [1941], Population studies with wheat II. Propinquity; III. Seed rates in nursery trials and field plots. J. Agri. Sci. 31:116-144.
A14 (760)

Hulbert, H. W. and Remsberg, J. D. [1927], Influence of border rows in variety tests of small grains. J. American Soc. Agron. 19:585-589.
A14 (761)

Hulbert, H. W. et al. [1931], Border effect in variety tests of small grains. Idaho Agri. Expt. Sta. Tech. Bull. 9. A14 (762)

Hull, A. C. [1948], Depth, season, and row spacing for planting grasses on southern Idaho range lands. J. American Soc. Agron. 40:960-969.
A14 (763)

Hussain, Q. M. [1943], A note on interaction. Sankhyā 6:321-322.
T2a (764)

Hussain, Q. M. [1945], Symmetrical incomplete block designs with λ = 2, k = 8 or 9. Bull. Calcutta Math. Soc. 37:115-123. E2c (765)

Hussain, Q. M. [1945], On the totality of the solutions for the symmetrical incomplete block designs: λ = 2, k = 5 or 6. Sankhyā 7:204-208.
E2c (766)

Hussain, Q. M. [1946], Impossibility of symmetrical incomplete block design with λ = 2, k = 7. Sankhyā 7:317-322. E2c (767)

Hussain, Q. M. [1948], Structure of some incomplete block designs. Sankhyā 8:381-383. E2c (768)

Hussain, Q. M. [1948], Alternate proof of the impossibility of the symmetrical design with λ = 2, k = 7. Sankhyā 8:384. E2c (769)

Hutchinson, A. H. and Knapp, F. M. [1947], Random sampling, planned sampling, and selective sampling as applied to forest ecology and silviculture. Trans. Roy. Soc. Canada (3rd Ser.) (1946) 40(5):77-79. A14 (770)

Hutchinson, G. E. [1947], A note on the competition between two social species. Ecology 28:319-321. A14 (771)

Hutchinson, J. B. and Kubersingh, [1936], Studies in plant breeding technique I Indian J. Agri. Sci. 6:672-683. C;S1 (772)

Hutchinson, J B. and Panse, V. G [1935], Studies in the technique of field experiments I. Size, shape and arrangement of plots in cotton trials. Indian J. Agri. Sci. 5:523-538. A14;S1 (773)

Hutchinson, J. B. and Panse, V. G. [1935], Studies in the technique of field experiments II. Sampling for staple-length determinations in cotton trials. Indian J. Agri. Sci. 5:545-553. E2a;C (774)

Hutchinson, J. B. and Panse, V. G. [1935], Studies in the technique of field experiments III. An application of the method of covariance to selection for disease resistance in cotton. Indian J. Agri. Sci. 5:554-558.
C (775)

Hutchinson, J. B. and Panse, V. G. [1935], Studies in the technique of field experiments IV. A study of margin effect in variety trials with cotton and wheat. Indian J. Agri. Sci. 5:671-692. A14 (776)

Hutchinson, J. B. and Panse, V. G. [1937], An examination of an analysis of a serial experiment. Agri. Livestock India 7:332-338. A13 (777)

Hutchinson, J. B. and Panse, V. G. [1937], Studies in plant breeding technique II. The design of field tests of plant breeding material. Indian J. Agri. Sci. 7:531-564.
E2,6ae;A7;C (778)

Hutchinson, J. B., Panse, V. G., Apte, N. S., and Pugh, B. M. [1938], Studies in plant breeding technique III. Crop analysis and varietal improvement in Malvi _jowar_ (_Andropogon sorghum_). Indian J. Agri. Sci. 8:131-152.
C (779)

Immer, F. R. [1932], A study of sampling technic with sugar beets. J. Agri. Res. 44:633-647. A14;S1 (780)

Immer, F. R. [1932], Size and shape of plot in relation to field experiments with sugar beets. J. Agri. Res. 44:649-668. A14 (781)

Immer, F. R. [1934], Varietal competition as a factor in yield trials with sugar beets. J. American Soc. Agron. 26:259-261. A14 (782)

Immer, F. R. [1936], A study of the association between mean yields and standard deviations of varieties tested in replicated yield trials. J. American Soc. Agron. 28:24-27. C (783)

Immer, F. R. [1936], Sampling technic. Mimeo. Univ. Minnesota. A14;S1 (784)

Immer, F. R. [1936], Manual of Applied Statistics. Mimeo. Univ. Minnesota. (Published by Burgess in 1950). E1,2,3,7,12ae; T2ae;C (785)

Immer, F. R. [1942], Distribution of yields of single plants of varieties and F_2 crosses of barley. J. American Soc. Agron. 34:844-850. E12ae (786)

Immer, F. R. [1945], Some uses of statistical methods in plant breeding. Biometrics 1:13-15. E2,3a;C (787)

Immer, F. R., Hayes, H. K., and Powers, L. [1934], Statistical determination of barley varietal adaption. J. American Soc. Agron. 26:403-419. E2,12ae (788)

Immer, F. R. and LeClerg, E. L. [1936], Errors of routine analysis for percentage of sucrose and apparent purity coefficient with sugar beets taken from field experiments. J. Agri. Res. 52:505-515. A14;S1 (789)

Immer, F. R. and Raleigh, S. M. [1933], Further studies of size and shape of plot in relation to field experiments with sugar beets. J. Agri. Res. 47:591-598. A14 (790)

Immer, F. R., Tysdal, H. M., and Steece, H. M. [1939], Bibliography of field experiments. J. American Soc. Agron. 31:1049-1052. B (791)

Inman, W. R. [1941], Digestibility studies with foxes II. Sci. Agri. 22:33-39. E7ae (792)

Inman, W. R. and Smith, G. E. [1941], Digestibility studies with foxes I. Sci. Agri. 22:18-32. E7ae (793)

Innes, R. F. [1939], Simple methods of field experimentation for the investigation of sugar cane problems. Jamaican Assoc. Sugar Tech. 3:30-46. C (794)

Irwin, J. O [1931], Mathematical theorems involved in the analysis of variance. JRSSA 94:284-300. E2,7a (795)

Irwin, J. O. [1931], Recent advances in mathematical statistics. JRSSA 94:568-578. A7 (796)

Irwin, J. O. [1931], Precision records in horticulture. J. Pom. Hort. Sci. 9: 149-194. C (797)

Irwin, J. O. [1932], Recent advances in mathematical statistics (1931). JRSSA 95:498-530. A7;B (798)

Irwin, J. O. [1934], Recent advances in mathematical statistics (1932). JRSSA 97:114-154. A7;B (799)

Irwin, J. O. [1934], On the independence of the constituent items in the analysis of variance. JRSSB 1:236-251. E2,7a (800)

Irwin, J. O. [1937], Statistical method applied to biological assays. JRSSB 4:1-60. T8a;C (801)

Irwin, J. O. [1938], Recent advances in mathematical statistics. Bibliography of mathematical statistics (1935, 1936, and first half of 1937). JRSSA 101: 394-433. B (802)

Irwin, J. O [1943], A table of the variance of $\sqrt{x}$ when x has a Poisson distribution. JRSSA 106:143-144. A9 (803)

Irwin, J. O. [1943], On the calculation of the error of biological assays. J. Hygiene 43:121-128. T8a (804)

Irwin, J. O [1943], The error of the biological assay of insulin by the mouse-convulsion test. Quarterly J. Pharm. 16:352-362. T8a (805)

Irwin, J. O. [1944], A statistical examination of the accuracy of vitamin A assays. An analysis of three co-operative experiments designed to ascertain the value of the conversion factor for transforming spectrophotometric values into international units. J. Hygiene 43:291-314. T8a (806)

Irwin, J. O. [1946], On the interpretation of within and between class analysis of variance when the intra class correlation is negative. JRSSA 109:157-158. C (807)

Irwin, J. O., Bartlett, M. S., Cochran, W. G , and Fieller, E. C. [1935], Recent advances in mathematical statistics (1933). JRSSA 98:83-127. A7;B (808)

Irwin, J. O and Cheeseman, E.A. [1939], On the maximum-likelihood method of determining dosage-response curves and approximations to the median-effective dose, in cases of a quantal response. JRSSB 6:174-185.
T8a (809)

Irwin, J. O. and Cheeseman, E. A. [1939], On an approximate method of determining the median effective dose and its error in the case of a quantal response. J. Hygiene 39:574-580. T8a (810)

Irwin, J O., Cochran, W. G., Fieller, E. C., and Stevens, W. L. [1936], Recent advances in mathematical statistics (1934). JRSSA 99:714-769.
A7;B (811)

Irwin, J O., Cochran, W. G., and Wishart, J. [1938], Crop-estimation and its relation to agricultural meteorology. JRSSB 5:1-45. A7,14;C (812)

Jablonski, E. [1892], Théorie des permutations et des arrangements circulaires complets. J. Math. Pures Appl. 8:331-349. E2c (813)

Jackson, F. K. and Wad, Y. D. [1933], The design and conduct of field experiments. Agri. Livestock India 3:211-233. E7ae;C (814)

Jacob, W. C. [1939], The importance of border effect in certain kinds of field experiments with potatoes. Proc. American Soc. Hort. Sci. 37:866-870. A14 (815)

Jacob, W. C. [1942], Statistical analysis of border effect in field experiments with potatoes and cauliflower. Ph. D. Thesis, Cornell Univ. A14 (816)

Jacob, W. C. [1944], Análise estatistica de experimentos factoriais. (Statistical analysis of factorial experiments.) Brazilian Ministr. Agri. Rio de Janeiro. 33:41-56. T2ae (817)

James, N. [1939], The accuracy of the plating method for estimating the number of bacteria, actinomyces and fungi in a laboratory sample of soil. Iowa State College J. Sci. 14:50-52. C (818)

Jardine, W. M., Lyon, T. L., and Wiancko, A. T. [1917], Report of the committee on standardization of field experiments. J. American Soc. Agron. 9:402-419. B (819)

Jeffreys, H. [1939], Random and systematic arrangements. Biometrika 31:1-8. E5a (820)

Jellinek, E M. [1946], Clinical tests on comparative effectiveness of analgesic drugs. Biometrics 2:87-91. T7a;C (821)

Jessen, R. J. [1942], Statistical investigation of a sample survey for obtaining farm facts. Iowa Agri. Expt. Sta. Res. Bull. 304:1-104. A14 (822)

Jevons, W. S. [1889], Elementary Lessons in Logic: deductive and inductive. (1st ed. in 1870) (reprinted 1928). MacMillan, N. Y. C (823)

Jodon, N. E. and Beachell, H. M. [1938], Row spacing and rate of seeding for rice nursery plats. J. American Soc. Agron. 30:212-219. E12ae;A14 (824)

Johnson, B. C., Hamilton, T. S., Mitchell, H. H., and Robinson, W. B. [1942], The relative efficiency of urea as a protein substitute in the ration of ruminants. J. Animal Sci. 1:236-245. E7,8ae (825)

Johnson, F. A. [1943], A statistical study of sampling methods for tree nursery inventories. J. Forestry 41:674-679. A14 (826)

Johnson, H. M. [1942], Pre-experimental assumptions as determiners of experimental results. Psychological Rev. 47:338-346. C (827)

Johnson, I. J. and Murphy, H. C. [1943], Lattice and lattice square designs with oat uniformity data and in variety trials. J. American Soc. Agron. 35:291-305. E3,11,14a (828)

Johnson, P. O. and Tsao, F. [1944], Factorial design in the determination of differential limen values. Psychometrika 9:107-144. T2a (829)

Johnson, P. O and Tsao, F. [1945], Factorial design and covariance in the study of individual educational development. Psychometrika 10:133-162. T2a;A7 (830)

Johnson, S. T. [1929], A note on the sampling of sugar beet. J. Agri. Sci. 19:311-314. E7ae (831)

Joint Committee of A. S. Agron., A. Dairy Sci. Assoc., and A. S. An. Prod. [1943], Preliminary report on pasture investigations technique. J. Dairy Sci. 26:353-369. A14;C (832)

Jolly, A. L. [1942], Uniformity trials on estate cacao fields in Grenada, B.W.I. Tropical Agri. 19:167-174. A14 (833)

Jones, E. W, [1937], Practical field methods of sampling soils for wireworms. J. Agri. Res. 54:123-134. C (834)

Jones, J. P. [1927], Some values of the statistical method in plat work. J. American Soc. Agron. 19:675. A14 (835)

Jones, J. P. [1928], The checker-board method of laying out plats. J. American Soc. Agron. 20:400-402. A14;S1 (836)

Joret, G. [1933], Evaluation du rendement par l'échantillonage des récoltes sur pied. Annales Agronomiques 3:430-452. A14 (837)

Justesen, S. H. [1932], Influence of size and shape of plots on the precision of field experiments with potatoes. J, Agri. Sci. 22:366-372. A14 (838)

Justesen, S. H. [1939], Toepassing van de strooiingsanalyse bij vakkenproeven. (Application of the analysis of variance in field experiments.) Landbouw. 15:346-363. C (839)

Justesen, S. H. [1948], Quasi-factoriele opzet van veldproeven met een groot aantal objecten. (Quasi-factorial design for field experiments having a large number of treatments.) Meded. Alg. Proefst. Landb., Java 81:31. E2,3,12a (840)

Kadam, B. S. [1942], How to conduct yield trials of improved strains of crops in districts. Poona Agri. College Mag. 34:117-123. C (841)

Kadam, B. S. and Patel, S. M. [1937], Studies in field-plot technique with P. typhoideum Rich. Empire J. Expt. Agri. 5:219-230. A14 (842)

Kalamkar, R. J. [1932], Experimental error and field-plot technique with potatoes. J. Agri. Sci. 22:373-385. A14 (843)

Kalamkar, R. J. [1932], A study in sampling technique with wheat. J. Agri. Sci. 22:783-796. A14 (844)

Kalamkar, R. J. [1933], A statistical examination of the yield of mangolds from Barnfield at Rothamsted. J. Agri. Sci. 23:161-175. A7 (845)

Kalamkar, R. J. [1933], The influence of rainfall on the yields of mangolds at Rothamsted. J. Agri. Sci. 23:571-579. A7 (846)

Kalamkar, R. J. [1941], A note on crop cutting experiments on cotton. Sankhyā 5:345-348. A14 (847)

Kalamkar, R. J. and Dhannalal, L. A. [1939], Variability of plant density and the estimation of yield of cotton by sampling (abstract). Proc. Twenty-Sixth Indian Sci. Congress, Lahore, Part III, p. 198. A7,14 (848)

Kalamkar, R. J. and Dhannalal, L. A. [1940], Sampling studies in a cotton variety trial. Sankhyā 4:567-576. E2ae;A7,14 (849)

Kalamkar, R. J., Kadam, B. S., et. al. [1943], Studies on precision observations on rice at Karjat. Indian J. Agri. Sci. 13:204-231. A14 (850)

Kalamkar, R. J. and Satakopan, V. [1940], The influence of the rainfall distribution on the cotton yields at the government experimental farms at Akola and Jalgon. Indian J. Agri. Sci. 10:960-974. A7 (851)

Kalamkar, R. J. and Singh, S. [1935], A statistical examination of the yield of wheat at the Cawnpore Agricultural College Farm, Part I. Indian J. Agri. Sci. 5:346-354. A7,14 (852)

Keen, B. A. [1933], Experimental methods for the study of soil cultivation. Empire J. Expt. Agri. 1:97-102. E12a;C (853)

Keller, K. R. [1949], A comparison involving number of, and relationship between, testers in evaluating inbred lines of maize. Agron. J 41:323-331. T9a;A13 (854)

Keller, K. R. [1949], Uniformity trial on hops, Humulus lupulus L., for increasing the precision of field experiments. Agron. J. 41:389-392. A14 (855)

Keller, K. R. and Li, J. C. R. [1949], The relationship between number of vines per hill and yield in hops. Agron. J. 41:569-573. A14;S1 (856)

Keller, W. [1946], Designs and technic for the adaption of controlled competition to forage plant breeding. J. American Soc. Agron. 38:580-588. E5a;C (857)

Kelley, O. J., Haise, H. R., Markham, L. C., and Hunter, A. S. [1946], Increased rubber production from thickly seeded guayule. J. American Soc. Agron. 38:589-613. A14 (858)

Kelly, B. K. et. al. [1948], Fractional replication arrangements for factorial experiments with factors at two levels. Biometrika 35:268-276. T3ac (859)

Kempthorne, O. [1944], Comments on the note "On a theorem concerning sampling". JRSSA 107:58. S1 (860)

Kempthorne, O. [1945], Statistics in biology. Biology Human Affairs 10(3):1-8. C (861)

Kempthorne, O. [1946], Analysis of a series of experiments by the use of punched cards. JRSSB 8:118-127. A13,15 (862)

Kempthorne, O. [1947], A simple approach to confounding and fractional replication in factorial experiments. Biometrika 34:255-272. T2,3c (863)

Kempthorne, O. [1947], Recent developments in the design of field experiments IV. Lattice squares with split plots. J. Agri. Sci. 37:156-162. E11,12ae;T2ae (864)

Kempthorne, O. [1947], A note on differential response in blocks. J. Agri. Sci. 37:245-248. T2a;A11,13 (865)

Kempthorne, O. [1948], The factorial approach to the weighing problem. AMS 19:238-245. T6a (866)

Kempthorne, O. [1949], Design of Experiments. Mimeo. Iowa State College. E1-3,6-13ace;T1-3ace; A7,10;C;S1 (867)

Kempthorne, O. and Federer, W. T. [1948], The general theory of prime-power lattice designs. I. Introduction and designs for p^n varieties in blocks of p plots. Biometrics 4:54-79. E2,3ac (868)

Kempthorne, O. and Federer, W. T. [1948], The general theory of prime-power lattice designs. II. Designs for p^n varieties in blocks of p^s plots, and in squares. Biometrics 4:109-121. E3,11ac (869)

Kempthorne, O., Schmidt, J. L. and Snedecor, G. W. [1948], The estimation of yield of corn of standard moisture content in hybrid seed corn production. J. American Soc. Agron. 40:645-654. A14 (870)

Kendall, M. G. [1941], A theory of randomness. Biometrika 32:1-15. C (871)

Kendall, M. G. [1948], The Advanced Theory of Statistics, Vol. II. Charles Griffin & Co., Ltd., London. E1,2,12ae;T2ae (872)

Kendall, M. G. [1948], Who discovered the Latin square? Amer. Stat. 2(4):13. E7a (873)

Kendall, M. G. and Smith, B. B. [1938], Randomness and random sampling numbers. JRSSA 101:147-166. C (874)

Kendall, M. G. and Smith, B. B. [1939], Second paper on random sampling numbers. JRSSB 6:51-61. C (875)

Kendall, M. G. and Smith, B. B. [1940], On the method of paired comparisons. Biometrika 31:324. T11a (876)

Kerawala, S. M. [1941], The enumeration of the Latin rectangle of depth three by means of a difference equation. Bull. Calcutta Math. Soc. 33:119-127. E7c (877)

Kerawala, S. M. [1946], Note on symmetrical incomplete block designs: $\lambda = 2$, $k = 6$ or 7. Bull. Calcutta Math. Soc. 38:190-192. E2c (878)

Kerawala, S. M. [1947], A note on self-conjugate Latin squares of prime degree. Math. Student 15:16. E7c (879)

Kermack, W. O. and McKendrick, A. G. [1940], The design and interpretation of experiments based on a fourfold table: the statistical assessment of the effect of treatment. Proc. Royal Soc. Edinburgh 60:362-375. C (880)

Kerr, H. W. [1932], 1. Plot technique 2. Field sampling. Proc. Fourth Congress Int. Soc. Sugar-Cane Tech. Bull. 53 and 55. A14;C (881)

Kerr, H. W. [1934], Recent developments in the agriculture of sugar-cane in Queensland. Empire J. Expt. Agri. 2:20-28. E7ae (882)

Kerr, H. W. [1935], Report of the committee on field experimentation. Proc. Fifth Congress Int. Soc. Sugar Cane Tech. Brisbane pp. 244-279. A14;C;S1 (883)

Kerr, H. W. [1939], Notes on plot technique. Proc. Sixth Congress Int. Soc. Sugar Cane Tech., La.(1938), pp. 764-778. A14 (884)

Kerr, H. W. [1939], Field experimentation with sugar cane. Queensland Bur, Sugar Expt. Sta. Tech. Comm. 11. A14;C (885)

Kerr, H. W. and von Stieglitz, C. R. [1938], Some studies in soil sampling technique. Queensland, Bur. Sugar Expt. Sta. Tech. Comm. 10:205-217. A14 (886)

Khan, A. R. and Dalal, J. R. [1943], Optimum size and shape of plots for brassica experiments in the Punjab. Sankhyā 6:317-320. A14 (887)

Khargonkar, S. A. [1948], The estimation of missing plot value in split-plot and strip trials. J. Indian Soc. Agri. Stat. 1:147-161. E12ae;A7 (888)

Kiesselbach, T. A. [1918], Studies concerning the elimination of experimental error in comparative crop tests. Nebraska Agri. Expt. Sta. Res. Bull. 13:1-95. A14 (889)

Kiesselbach, T. A. [1919], Experimental error in field trials. J. American Soc. Agron. 11:235-241. C (890)

Kiesselbach, T. A. [1919], Plat competition as a source of error in crop tests. J. American Soc. Agron. 11:242-247. A14 (891)

Kiesselbach, T. A. [1922], Corn investigations. Nebraska Agri. Expt. Sta. Res. Bull. 20:1-151. A14 (892)

Kiesselbach, T. A. [1923], Competition as a source of error in comparative corn yields. J. American Soc. Agron. 15:199-215. A14 (893)

Kiesselbach, T. A. [1928], The mechanical procedure of field experimentation. J. American Soc. Agron. 20:433-442. A14 (894)

Kiesselbach, T. A., Garber, R. J. et. al. [1933], Standardization of field experiments. J. American Soc. Agron. 25:803-828. A14;B;C (895)

Kiesselbach, T. A. and Weihing, R. M. [1933], Effect of stand irregularities upon the acre yield and plant variability of corn. J. Agri. Res. 47:399-416. A14 (896)

King, A. J. and Houseman, E. E. [1942], Determination of sample sizes. Illinois Univ. Dept. Agri. Econ. Agri. Expt. Sta. Report AE-1837. S1 (897)

King, A. J. and Jebe, E. H. [1940], An experiment in pre-harvest sampling of wheat fields. Iowa Agri. Expt. Sta. Res. Bull. 273:624-649. A14 (898)

King, A. J., McCarty, D. E., and McPeek, M. [1942], An objective method of sampling wheat fields to estimate production and quality of wheat. U.S. Dept. Agri. Tech. Bull. 814:1-87. A14 (899)

King, F. G., Gramlich, H. J., Evvard, J. M., Morrison, F. B., and Good, E. S. [1922], Report of the committee on uniform methods of livestock experimental work. American Soc. Animal Production Proc. 15:100-105. C (900)

Kirk, L. E. [1929], Field plot technique with potatoes with special reference to the latin square. Sci. Agri. 9:719-729. E7ae;A14;C (901)

Kirk, L. E. and Goulden, C. H. [1925], Some statistical observations on a yield test of potato varieties. Sci. Agri. 6:89-97. E5a;A14;C (902)

Kirkman, T. P. [1847], On a problem in combinations. Cambridge Dublin Math. J. 2:191-204. E2c (903)

Kirkman, T. P. [1850], Note on an unanswered prize question. Cambridge Dublin Math. J. 5:255-262. E2c (904)

Kishen, K. [1939], Symmetrical incomplete block arrangements with blocks of unequal size. Sci. Culture 5:133-134. E2,3,5c (905)

Kishen, K. [1939], Split plot technique in field experimentation. Proc. Twenty-Sixth Indian Sci. Congress, Lahore, Part III, Section IX, No. 47. E12a (906)

Kishen, K. [1939], On a general method of constructing the appropriate compounds in terms of the twelve 3×3 Latin squares for any of the eight components $N_1K_1P_1$, $N_1K_1P_2$, $N_1K_2P_1$, etc., of the second order interaction in a 3×3 factorial arrangement (abstract). Proc. Twenty-Sixth Indian Sci. Congress, Lahore, Part III, p. 199. E7c;T2c (907)

Kishen, K. [1940], On a simplified method of expressing components of the second order interaction in a 3^3 factorial design. Sankhyā 4:577-580. T2a (908)

Kishen, K. [1941], Symmetrical unequal block arrangements. Sankhyā 5:329-344. E2,3,4ac;A7 (909)

Kishen, K. [1942], On expressing any single degree of freedom for treatments in an s^m factorial arrangement in terms of its sets for main effects and interactions. Sankhyā 6:133-140. T2a (910)

Kishen, K. [1942], On Latin and Hyper-Graeco-Latin cubes and hyper-cubes. Current Sci. 11(3):98-99. E7,13c (911)

Kishen, K. [1945], On the design of experiments for weighing and making other types of measurements. AMS 16:294-300 (also in Current Sci. 14:194-195) T6c (912)

Kishen, K. [1947], On fractional replication of the general symmetrical factorial design. Current Sci. 16:138-139. T3c (913)

Kishen, K. [1948], On fractional replication of the general symmetrical factorial design. JISAS 1:91-106. T3c (914)

Kishen, K. [1949], On the construction of Latin and Hyper-Graeco-Latin cubes and hyper-cubes. JISAS 2:20-48. E7,13c (915)

Klages, K. H. [1928], Yields of adjacent rows of sorghums in variety and spacing tests. J. American Soc. Agron. 20:582-598. A14 (916)

Klages, K. H. [1931], A modification of Delwiche's system of laying out cereal variety test plats. J American Soc. Agron. 23:186-189. A14 (917)

Klages, K. H. [1933], The reliability of nursery tests as shown by correlated yields from nursery rows and field plats. J. American Soc. Agron. 25:464-472. E5a;A14;C (918)

Klages, K. H. [1938], Trend studies in relation to the analysis of yield data from rotation experiments. J. American Soc. Agron. 30:624-631. E5,8a (919)

Klingebiel, A. A. and Brown, P. E. [1937], Effect of applications of fine limestone: I. J. American Soc. Agron. 29:944-959. E2,12ae (920)

Klingebiel, A. A. and Brown, P. E. [1937], Effect of applications of fine limestone: II. J. American Soc. Agron. 29:978-989. E2ae (921)

Klingebiel, A. A. and Brown, P. E. [1938], Effect of applications of fine limestone: III. J. American Soc. Agron. 30:1-9. E2ae;T2ae (922)

Knapp, B., Phillips, R. W., Black, W. H., and Clark, R. T. [1942], Length of feeding period and number of animals required to measure economy of gain in progeny tests of beef bulls. J. Animal Sci. 1:285-292. A14;S1 (923)

Knudsen, L. F. [1945], The use of statistics in biological experimentation and assay. J. Assoc. Off. Agri. Chem. 28:806-813. T8a (924)

Knudsen, L. F. and Curtis, J. M. [1947], The use of the angular transformation in biological assays. JASA 42:282-296. T8a;A9 (925)

Knudsen, L. F. and Randall, W. A. [1945], Penicillin assay and its control chart analysis. J. Bacteriology 50:187-200. T8a (926)

Knudsen, L. F., Smith, R. B., Vos, B. J., and McClosky, W. T. [1946], The biological assay of epinephrine. J. Pharmacology 86:339-343. T8a (927)

Kogan, L. S. [1948], Analysis of variance-repeated measurements. Psychological Bull. 45:131-143. C (928)

Kosambi, D. D. [1942], A test of significance for multiple observations. Current Sci. 11:271-274. C (929)

Koshal, R. S., Gulati, A. N., and Ahmad, N. [1940], The inheritance of mean fibre-length, fibre-weight per unit length and fibre-maturity of cotton. Indian J. Agri. Sci. 10:975-989. E2ae (930)

Koshal, R. S. and Turner, A. J. [1930], Studies in the sampling of cotton for the determination of fibre-properties III. The size and reliability of a satisfactory sample. Tech. Bull. Indian Central Cotton Comm. Ser. B, 10:1-39. S1 (931)

Kossack, C. F. [1949], Some techniques for simple classification. Proc. Berkeley Symp. Math. Stat. Prob. pp. 345-352. A7 (932)

Kramer, H. H. and Albrecht, H. R. [1948], The adaptation to small samples of the pearling test for kernel hardness in wheat. J. American Soc. Agron. 40:422-431. S1 (933)

Krishna Iyer, P. V. [1940], The analaysis of simple non-symmetrical experiments. Indian J. Agri. Sci. 10:686-690. E2,6ae;A10 (934)

Krishna Iyer, P. V. [1940], Symmetrical incomplete randomized blocks. Proc. Twenty-Seventh Indian Sci. Congress, Madras, Part III, Sect. Agri. Abst. 59:231. E7a;T2ac (935)

Krishna Iyer, P. V. [1940], The analysis of simple non-symmetrical experiments. Proc. Twenty-Seventh Indian Sci. Congress, Madras, pp. 231-232. E6ae;A7 (936)

Krishna Iyer, P. V. [1941], Standard error of the difference between two estimates for incomplete block experiments. Current Sci. 10:165. E6a (937)

Krishna Iyer, P. V. [1941], The analysis of incomplete split plot designs. Sci. Culture 6:487. E12a (938)

Krishna Iyer, P. V [1942], Studies with wheat uniformity trial data. I. Size and shape of experimental plots and the relative efficiency of different lay-outs. Indian J. Agri. Sci. 12:240-262. E2,7a;A14 (939)

Krishna Iyer, P. V. [1942], Studies with wheat uniformity trial data. II. Balanced versus randomized arrangements. Indian J. Agri. Sci. 12:263-273. E5a;A14;C (940)

Krishna Iyer, P. V. [1942], Studies with wheat uniformity trial data. III. Distributions of variances and ratio of variances. Indian J. Agri. Sci. 12: 274-280. A14 (941)

Krishna Iyer, P. V. [1945], Tests of significance by analysis of covariance in multivariate populations. Current Sci. 14:297. A7 (942)

Krishnaswami Ayyangar, A. A. [1945], Interaction formulae in analysis of variance. Current Sci. 14:35. T2a (943)

Kristensen, R. K. [1911], Om anvendelse af fejlloven på afgrøder af roer. (On the application of the theory of errors in the yield of turnips.) Tidsskr. Planteavl 18:744-765. C (944)

Kristensen, R. K. [1915], Om bestemmelse af middelfejlen ved markforsøg. (On the determination of the mean error in agricultural experiments.) Tidsskr. Planteavl 22:349-364. C (945)

Kristensen, R. K. [1922], Nøjagticheden ved varietets-og stammeforsøg med roer, udførte paa Statens Forsøgsstatione i aarene 1886-1919. (The accuracy of experiments with turnips at State Experimental Stations during the period 1886-1919.) Tidsskr. Planteavl 28:95-118. A13;C (946)

Kristensen, R. K. [1929], Bestemmelse af middelfejlen ved markforsøg med forskellige parcelfordelinger. (Determination of mean errors in experimental designs with different plot arrangements.) Tidsskr. Planteavl 35:615-645. E5a;A14;C (947)

Kristensen, R. K. [1931-1932], Ausgleichung und Fehlerberechnung beim Feldversuch. Eine kurzgefasste Darstellung. Arch. Pflanzenbau 8:436-455. E5a;A14;C (948)

Kristensen, R. K. [1932], Bestemmelse af middelfejlen ved forsøgsresultater med uregelmaessige ensidige afvigelser. (Determination of mean errors for experimental results containing the systematic components.) Tidsskr. Planteavl 38:161-164. E5a;A14;C (949)

Kristensen, R. K. [1933], Bestemmelse af middelfejlen ven hjaelf af differensdannelser. (Determination of mean error by means of the taking of differences.) Tidsskr. Planteavl 39:349-353. E5a;A14;C (950)

Kristensen, R. K. [1933], Raekkemetoden og den spredte parcelfordeling. (The row method and other arrangements of plots in experimental designs.) Tidsskr. Planteavl 39:699-709. E5a;A14;C (951)

Kristensen, R. K. [1933, 1935], Om variationsanalyse. I, II and III. (On the analysis of variance. I, II and III.) Tidsskr. Planteavl 39:535-539, 40:521-527, 835-839. E5a;A14;C (952)

Kristensen, R. K. [1934], Fejlberegning ved markforsøg. Erstaningstal. (Calculation of error in field experiments. Exchange values.) Tidsskr. Planteavl 40:161-168. E5a;A14;C (953)

Kristensen, R. K. [1936], Om variationsanalyse. IV. (On the analysis of variance IV.) Tidsskr. Planteavl 41:360-367. E5a;A14;C (954)

Kristensen, R. K. [1936], Middelfejlen paa de udlignede forsøgsresultater ved markforsøg. (The mean error in the adjusted experimental results in field experiments.) Nord. JordbrForskn. Nos,3-4:286-291. E5a;Al4;C (955)

Kudrjawzew, P. N. [1934], Polyallele Kreuzung als Prüfungsmethode für die Leistungsfähigkeit von Zuchtebern. Züchtungskunde 9:444-452. T9a (956)

Kulkarni, R. K., Bose, S. S., and Mahalanobis, P. C. [1936], Statistical notes for agricultural workers No. 17. On the influence of shape and size of plots on the effective precision of field experiments with Juar (Andropogon sorghum). Indian J. Agri. Sci. 6:460-474. Al4 (957)

Kurtz, A. K. and Edgerton, H. A. [1939], Statistical Dictionary of Terms and Symbols. Wiley, N.Y. C (958)

Kutsunai, Y. [1931], Notes on interpreting experimental results. Hawaiian Planters' Record 35:279-294. E5a;Al4;C (959)

Lachman, W. H. [1938], A statistical analysis of form variations in specific strains of tomatoes. Proc. American Soc. Hort. Sci. 35:559-561. C (960)

Ladell, W. R. S. (Appendix by Cochran, W. G.) [1938], Field experiments on the control of wireworms. Appendix: The information supplied by the sampling results. Ann. Appl. Biology 25:341-389. E2,7ae (961)

Lamas, P. J. A. [1944], The number of frequencies and accuracy in field trials. Granos 8:3-16. C (962)

Lambert, E. B [1934], Size and arrangement of plots for yield tests with cultivated mushrooms. J. Agri. Res. 48:971-980. A14 (963)

Lamm, R. [1939], Årsskrift fran Alnarps Lantbruks-, Majeri- och Trädgårdsinstitut 1938. Redogörelse för stamförsök och statskontroll av köksväxtstammar vid statens trädgårdsförsök år 1937. (Annual Report of the Alnarp Agricultural, Dairying and Horticultural Institute 1938--Report on trials of strains and state control of vegetable strains in the state horticultural trials in the year 1937.) Malmö, pp. 166. C (964)

Landahl, H. D. [1938-39], A contribution to the mathematical biophysics of psychophysical discriminations I-III. Psychometrika 3:107-125, Bull. Math. Biophysics 1:159-176. C (965)

Landhal, H. D. [1941], Studies in mathematical biophysics of discrimination and conditioning. II. Special case. Errors, trials, and number of possible responses. Bull. Math. Biophysics 3:71-77. C (966)

Lander, P. E., Narain, R., and Singh, A. [1938], Soil uniformity trials in the Punjab, I. Indian J. Agri. Sci. 8:271-307. A7,14;C (967)

Lander, P. E., Narain, R., and Singh, A. [1941], Soil uniformity trials in the Punjab, II. Indian J. Agri. Sci. 11:338-355. A14 (968)

Laude, H. H. [1945], Field vs. nursery plots for varietal evaluation. Report Fifth Hard Red Winter Wheat Improvement Conf., Manhattan, Kansas, Feb. 12, 13 and 14, 1945. Div. Cereal Crops Dis., Plant Ind. Sta., Beltsville, Md. pp. 16-17 (Mimeo.). S1 (969)

Lawley, D. N. [1943], On problems connected with item selection and test construction. Proc. Royal Soc. Edinburgh 62:74-82. C (970)

Leasure, J. K. [1949], Determining the species composition of swards. Agron. J. 41:204-206. A14 (971)

LeClerg, E. L. [1935], Factors affecting experimental error in greenhouse pot tests with sugar beets. Phytopathology 25:1019-1025. E2a;A14;C (972)

LeClerg, E. L. [1937], Relative efficiency of randomized-block and split-plot designs of experiments concerned with damping-off data for sugar beets. Phytopathology 27:942-945. E2,12a;A14 (973)

LeClerg, E. L. [1939], Relative efficiency of quasi-factorial and randomized-block designs of experiments concerned with damping-off of sugar beets. Phytopathology 29:637-641. E2,3ae;A14 (974)

LeClerg, E. L. [1942], Relation of field plot design to seed source tests of Irish potatoes in the south. American Potato J. 19:75-79. E3a (975)

LeClerg, E. L. and Henderson, M. T. [1940], Relative efficiency of the two-dimensional quasi-factorial design as compared with a randomized-block arrangement when concerned with yields of Irish potatoes. American Potato J. 17:279-282. E3ae;A14 (976)

Legatt, C. W. [1941], A study of the relative efficiency of seed sampling methods. Canadian J. Res. Sec. C-D 19:156-162. A14 (977)

Lehmer, E. [1944], Inverse tables of probabilities of errors of the second kind. AMS 15:388-398. A3 (978)

Lein, A. [1949], Zur Anlage und Auswertung van Sortenversuchen. Z. Pflanzenz. 28:113-143. C (979)

Lennox, C. G. and Mangelsdorf, A. J. [1935], Plot technique in the testing of sugar-cane varieties. Proc. Fifth Congress Int. Soc. Sugar Cane Tech. Brisbane pp. 280-285 (Facts about Sugar 21:150, 1936). E5a;A14 (980)

Leonard, H. [1932], Ueber die Genauigkeit und Zuverlassigkeit der quantitativ-botanischen Untersuchung bei Wiesenversuchen. Arch. Pflanzenbau 8:650-682. C (981)

Leonard, O. A. and Anderson, W. S. [1947], Some problems in sampling the sweet-potato plant. Proc. American Soc. Hort. Sci. 50:299-302. A14 (982)

Levi, F. W. [1942], <u>Finite Geometrical Systems</u>. Univ. Calcutta. E2,3c (983)

Leonard, W. H. and Clark, A. G. [1936], Protein content of corn as influenced by laboratory analyses and field replication. Colorado Expt. Sta. Tech. Bull. 19:1-9. A14 (984)

Leonard, W. H. and Clark, A. G. [1939], <u>Field Plot Technique</u>. Burgess Pub. Co. Minneapolis, Minn. E2,7,12ae;T2ae; A14;C (985)

Lewis, A. H. and Trevains, D. [1934], Investigations of the manurial effectiveness of ammonium phosphate II. Empire J. Exptl. Agri. 2:244-250. E7ae;A13 (986)

L'Heritier, P. [1949], Les méthodes statistiques dans l'expérimentation biologique. Cen. Nat. Rech. Sci. Paris 93:25. C (987)

Li, C. -C. [1940], The competition effect, size and shape of plat and the use of check plats in cotton experiments. Ph. D. Thesis, Cornell Univ. E2,7ae;A14 (988)

Li, H. W., Meng, C. J., and Liu, T. N. [1936], Field results in a millet breeding experiment. J. American Soc. Agron. 28:1-15. E7ae;A7,14 (989)

Li, J. C. R. [1944], Design and statistical analysis of some confounded factorial experiments. Iowa Agri. Expt. Sta. Res. Bull. 333:452-492. T2ac (990)

Li, J. C. R. and Keller, K. R. [1949], The relationship between the number of vines per hill and yield in hops. Agron. J. 41:569-573. A14 (991)

Li, J. C. R. and Overman, A. [1949], Use of replications in deep-fat frying experiments. Food Res. 14:278-282. A14;S1 (992)

Ligon, L. L. [1930], Size of plat and number of replications in field experiments with cotton. J. American Soc. Agron. 22:689-699. A14;S1 (993)

Linquist, E. F. [1940], Statistical Analysis in Educational Research. Houghton Mifflin Co., Boston, Mass. E1ae;T2ae (994)

Lindquist, E. F. [1940], Sampling in educational research. J. Educ. Psychology 31:561-574. A14;C (995)

Lindquist, E. F. and Cook, W. W. [1933], Experimental procedures in test evaluation. J. Exptl. Education 1:163-185. C (996)

Lins Martin, O.A. [1946], O método factorial de investigaçao das faculdades mentais; análise de resultados experimentais obtidos em Sao Paulo em 1944. Rev. Brasil. Estatist. 8:303-338. T2a (997)

Lister, P. B. and Schumacher, F. X. [1937], The influence of rainfall upon tuft area and height growth of three semidesert range grasses in southern Arizona. J. Agri. Res. 54:109-121. E5,8a (998)

Livermore, J. R. [1927], A critical study of some of the factors concerned in measuring the effect of selection in the potato. J. American Soc. Agron. 19:857-896. A14 (999)

Livermore, J. R. [1932], Laboratory Exercises in Statistical Methods of Analysis. Mimeo. Cornell Univ. (latest mimeo. copy, 1950). E2,7ae;C (1000)

Livermore, J. R. [1932], Plot technique for field experiments with the potato. Proc. Potato Assoc. America 18:7-19. A14 (1001)

Livermore. J. R. [1934], The interrelations of various probability tables and a modification of student's probability table for the argument "t". J. American Soc. Agron. 26:665-673. C (1002)

Livermore, J. R. [1935], The potato field trial. American Potato J. 12:142-150. C (1003)

Livermore, J. R. and Neely, W. [1933], The determination of the number of samples necessary to measure differences with varying degrees of precision. J. American Soc. Agron. 25:573-577. S1 (1004)

Loesell, C. M. [1936], Size of plat and number of replications necessary for varietal trials with white pea beans. J. American Soc. Agron. 28:534-547. A14 (1005)

Löfvenmark, H. [1941], Ritning av försöksplaner med skrivmaskin. (Delineation of plans of experiments with the typewriter.) Lantmannen 25:796-798. C (1006)

Loosli, J. K. and Lucas, H. L. [1943], The effect of thiamin feeding upon milk and fat production. J. Dairy Sci. 26:291-294. E8ae (1007)

Loosli, J. K., Maynard, L. A., and Lucas, H. L. [1944], IV. Further studies of the influence of different levels of fat intake upon milk secretion. Cornell Univ. Agri. Expt. Sta. Memoir 265:1-32. E2,7ae (1008)

Loevinger, J. [1947], A systematic approach to the construction and evaluation of tests of ability. Psychological Monogr. 61(285):1-49. C (1009)

Loevinger, J. [1948], The technic of homogeneous tests compared with some aspects of "scale analysis" and factor analysis. Psychological Bull. 45: 507-529. A7;C (1010)

Loewe, S. [1947], Bio-assay by direct potency estimation. Science 106:89-91. T8a;C (1011)

Lord, F. M. [1944], Reliability of multiple-choice tests as a function of number of choices per item. J. Educational Psychology 35:175-180. C;S1 (1012)

Lord, L. [1931], A uniformity trial with irrigated broadcast rice. J. Agri. Sci. 21:178-188. E2,7ae;A14 (1013)

Lotka, A. J. [1932], The growth of mixed populations: Two species competing for a common food supply. J. Washington Academy Sci. 22:461-469. A14 (1014)

Love, H. H. [1919], The experimental error in field trials. J. American Soc. Agron. 11:212-216. A14;C (1015)

Love, H. H. [1923], The importance of the probable error concept in the interpretation of experimental results. J. American Soc. Agron. 15:217-224.
C (1016)

Love, H. H. [1924], The rôle of statistics inagronomic experimentation. Sci. Agri. 5:84-92. C;S1 (1017)

Love, H. H. [1928], Planning the plat experiment. J. American Soc. Agron. 20:426-432. A14 (1018)

Love, H. H. [1936], Are uniformity trials useful? J. American Soc. Agron. 28:234-245. E7ae;A7,14 (1019)

Love, H. H. [1936], Application of Statistical Methods to Agricultural Research. The Commercial Press, Ltd., Shanghai. E7ae;A7,14;C (1020)

Love, H. H. [1943], Experimental Methods in Agricultural Research. Univ. Puerto Rico, Agri. Expt. Sta. E2,3,7,11,12ae;
T2ae;A14;C (1021)

Love, H. H. and Brunson, A. M. [1924], Student's method for interpreting paired experiments. J. American Soc. Agron. 16:60-68. E2a (1022)

Love, H. H. and Craig, W. T. [1918], Methods used and results obtained in cereal investigations at the Cornell Station. J. American Soc. Agron. 10:145-157.
C (1023)

Love, H. H. and Craig, W. T. [1938], Investigations in plot technic with small grains. Cornell Univ. Agri. Expt. Sta. Memoir 214:1-26. E5a;A14;C;S1 (1024)

Lucas, H. L. [1943], A method of equalized feeding for studies with dairy cows. J. Dairy Sci. 26:1011-1022. E8a;S1 (1025)

Lucas, H. L. [1948], Techniques in animal science research. Proc. Auburn Conf. Stat. Appl. Soc. Sci., Plant Sci., Animal Sci. pp. 62-73. C (1026)

Lucas, H. L. [1948], Designs in animal science research. Proc. Auburn Conf. Stat. Appl. Soc. Sci., Plant Sci., Animal Sci. pp. 77-86. E1,2,3,7,8,11a;C
(1027)

Lucas, H. L. [1949], Design and analyses of feeding experiments with milking dairy cattle. Inst. Stat. Mimeo. Ser. No. 18, Univ. N. Carolina.
C (1028)

Lush, J. L. [1930], Interpreting the results of group feeding experiments. Proc. American Soc. Animal Production 23:44-55. C (1029)

Lush, J. L. [1931], Predicting gains in feeder cattle and pigs. J. Agri. Res. 42: 853-881. A7,14 (1030)

Lush, J. L. [1931], The number of daughters necessary to prove a sire. J. Dairy Sci. 14:209-220. T9a;S1 (1031)

Lush, J. L. [1932], The use of statistical methods in animal husbandry. Proc. American Soc. Animal Production 25:15-19. C;S1 (1032)

Lush, J. L. [1932], An empirical test of the approximate method of calculating coefficients of inbreeding and relationship from livestock pedigrees. J. Agri. Res. 45:565-569. T9a;A14 (1033)

Lush, J. L. and Black, W. H. [1926], How much accuracy is gained by weighing cattle three days instead of one at the beginning and end of feeding experiments. Proc. American Soc. Animal Production pp. 206-210. C (1034)

Lush, J. L., Black, W. H., and Semple, A. T. [1929], The use of dressed-beef appraisals in measuring the market desirability of beef cattle. J. Agri. Res. 39:147-162. C (1035)

Lush, J. L., Christensen, F. W., Wilson, C. V. and Black, W. H. [1928], The accuracy of cattle weights. J. Agri. Res. 36:551-580. A14;C (1036)

Lush, J. L. and Copeland, O. C. [1930], A study of the accuracy of measurements of dairy cattle. J. Agri. Res. 41:37-50. A14;C (1037)

Lush, J. L., Gramlich, H. J., et. al. [1932], Summary of recommendations by the committee on methods of investigation. Proc. American Soc. Animal Production 25:380-393. C;S1 (1038)

Lynch, P. B. [1940], Sampling methods for the estimation of grain yields in cereal trials. New Zealand J. Sci. Tech. 22:151A-157A. A14 (1039)

Lynch, P. B. [1942], New harvesting technique with cereal trials. Sampling methods necessitated by use of header-harvester. New Zealand J. Agri. 64:173-175. A14 (1040)

Lynes, F. F. [1935], Statistical analyses applied to research in weed eradication. J. American Soc. Agron. 27:980-987. E2ae (1041)

Lyon, T. L. [1910], A comparison of the error in yield of wheat from plats and from single rows in multiple series. Proc. American Soc. Agron. 2:38-39. A14 (1042)

Lyon, T. L. [1911], Some experiments to estimate errors in field plat tests. Proc. American Soc. Agron. 3:89-114. A14;C (1043)

Ma, P. C. [1937], (New conception of field experimentation.) J. Agri. Assoc. China 158:13-20. E2,7ae (1044)

Ma, P. C. and Fan, F. C. [1936], The new arrangement of field experiments according to randomized methods (in Chinese). J. Agri. Assoc. China 149:14-30. C (1045)

Ma, R. H. and Harrington, J. B. [1948], The standard errors of different designs of field experiments at the University of Saskatchewan. Sci. Agri. 28:461-474. E14a;A13 (1046)

Ma, R. H. and Harrington, J. B. [1949], A study on field experiments of semi-latin square design. Sci. Agri. 29:241-251. E9,12a (1047)

Ma, R. H. and Kao, L. M. [1940], A factorial experiment on rice culture. Empire J. Expt. Agri. 8:23-33. T2ae (1048)

Ma, R. H. and Yieh, H. T. [1935], Studies on technics of confounding I. The chess-board method of analyzing higher order interactions of 2^n factorial experiment and its method of confounding. Nanking J. 7: 65-68. T2a (1049)

MacDonald, D., Fielding, W. L., and Ruston, D. F. [1939], Experimental methods with cotton. I. The design of plots for variety trials. J. Agri. Sci. 29:35-47. A14;S1 (1050)

Mack, W. B. [1935], Methods in variety trials. Proc. American Soc. Hort. Sci. 32:491-492. C (1051)

Mackenzie, W. A. [1924], Studies in crop variation. III. An examination of the yield of dressed grain from Hoos field. J. Agri. Sci. 14:434-460. A7 (1052)

MacNeish, H. F. [1921], Das problem der 36 offiziere. Jahresbericht Deutschen Mathematiker-Vereinigung 30:151-153. E7,12c (1053)

MacNeish, H. F. [1922], Euler's squares. Ann. Math. 23:221-227. E7,12c (1054)

Magistad, O. C. and Farden, C. A. [1934], Experimental error in field experiments with pineapples. J. American Soc. Agron. 26:631-644. A14;C (1055)

Mahalanobis, P. C. [1928], On the need for standardization in measurements on the living. Biometrika 20A:1-31. C (1056)

Mahalanobis, P. C. [1932], A statistical note on certain rice-breeding experiments in the central provinces. Indian J. Agri. Sci. 2:157-169. E7ae (1057)

Mahalanobis, P. C. [1932], Statistical notes for agricultural workers No. 4. Rice and potato experiments at Sriniketan (1931). Indian J. Agri. Sci. 2:694-703. E2ae;T2ae (1058)

Mahalanobis, P. C. [1933], The use of the method of paired differences for estimating the significance of field trials. Indian J. Agri. Sci. 3:349-352. C (1059)

Mahalanobis, P. C. [1933], On the need for randomization of plots in field trials. Indian J. Agri. Sci. 3:549-551. C (1060)

Mahalanobis, P. C. [1933], Statistical notes for agricultural workers No. 6, 11, 13, 14. Indian J. Agri. Sci. 3:131-138, 349-352, 549-551, 1108-1115. E1,2ae;C (1061)

Mahalanobis, P. C. [1934], Statistical notes for agricultural workers No. 15. Indian J. Agri. Sci. 4:361-385. E5,8ae (1062)

Mahalanobis, P. C. [1940], A review of the application of statistical theory to agricultural field experiments in India. Indian J. Agri. Sci. 10:192-212. E2,7a;A14;C (1063)

Mahalanobis, P. C. [1940], A sample survey of the acreage under jute in Bengal. Sankhyā 4:511-530. A14 (1064)

Mahalanobis, P. C. [1940], Discussion on planning of experiments. Sankhyā 4:530-531. C (1065)

Mahalanobis, P. C. [1941], A note on random fields. Sci. Culture 7:54. C (1066)

Mahalanobis, P. C. [1944], On large-scale sample surveys. Philosophical Trans. Royal Soc. B, 231:329-451. A14 (1067)

Mahalanobis, P. C. [1945], Report on the Bihar crop survey: Rabi season, 1943-44. Sankhyā 7:29-106. A14;C (1068)

Mahalanobis, P. C. [1946], Use of small-size plots in sample surveys for crop yields. Nature, London 158:798-799. A14 (1069)

Mahalanobis, P. C. [1946], Sample surveys of crop yields in India. Sankhyā 7:269-280. A14 (1070)

Mahalanobis, P. C. and Bose, S. S. [1933], Statistical notes for agricultural workers No. 7, 8, 9, 10, 12. Indian J. Agri. Sci. 3:139-154, 339-348, 544-548. E1,2,6ae (1071)

Mahalanobis, P. C., Chakravertti, S. C., and Banerjee, E. A. R. [1934], Influence of shape and size of plots on the accuracy of field experiments with rice, Chinsurah, Bengal (abstract). Twenty-First Ann. Meet. Ind. Sci. Congress Bombay, Sect. Agri. 8:3. Al4 (1072)

Mahalanobis, P. C. and Ray, S. [1936], School marks and intellegence test scores. Sankhyā 2:397-402. T2a;A7 (1073)

Mahoney, C. H. and Baten, W. D. [1939], The use of the analysis of covariance and its limitation in the adjustment of yields based upon stand irregularities. J. Agri. Res. 58:317-328. E2,7ae;A7 (1074)

Main, W. R. and Tippett, L. H. C. [1941], The design of weaving experiments. Shirley Inst. Memoirs 18:109-120. E7ac (1075)

Mallik, A. K., Satakopan, V., and Gopal Rao, S. [1945], A study on the estimation of the yield of wheat by sampling. Indian J. Agri. Sci. 15:219-225. Al4 (1076)

Mandelson, J. [1946], Estimation of optimum sample size in destructive testing by attributes. IQC 3(3):24-26. Sl (1077)

Manis, H. C. and Leffert, I. [1939], Preliminary studies on the comparative value of some sprays and dusts in potato insect control. Iowa State Coll. J. Sci. 14:155-161. E7ae (1078)

Mann, H. B. [1942], The construction of orthogonal Latin squares. AMS 13: 418-423. E7,12,13c (1079)

Mann, H. B. [1943], On the construction of sets of orthogonal Latin squares. AMS 14:401-414. E7,12,13c (1080)

Mann, H. B. [1944], On orthogonal Latin squares. Bull. American Math. Soc. 50:249-257. E7,12,13c (1081)

Mann, H. B. [1945], Nonparametric tests against trend. Econometrica 13:245-259. Al2 (1082)

Mann, H. B. [1945], On a test for randomness based on signs of differences. AMS 16:193-199. Al2 (1083)

Mann, H. B. [1949], <u>Analysis and Design of Experiments</u>. Dover, N.Y. E2,3,7,12,13ac;T2ac (1084)

Margossian, A. [1931], Carrés Latins et carrés d'Euler (modules impairs). Enseign. Math. 30:41-49. E7,12c (1085)

Margossian, A. [1935], Carrés Latins semi-diagonaux. Enseign. Math. 34: 365-377. E7,12c (1086)

Marcuse, S. [1949], Optimum allocation and variance components in nested sampling with an application to chemical analysis. Biometrics 5:189-205. A13;14 (1087)

Marino, A. E. [1940], Efficiencia de ensayos planeados por el metodo de las parcelas divididas. (Efficiency (or accuracy) of trials calculated by the divided-plot method.) Rev. Argentina Agron. 7:69-88. A14;C (1088)

Marino, A. E. and Luna, J. T. [1941], Planeos de ensayos en blocks incompletos (Lattice y lattice balanceado). Análisis de resultados en sayos con maíces. (Designs for trials laid down in incomplete blocks (lattice and balanced lattice). Analysis of results of trials with maize.) Rev. Argentina Agron. 8:281-316. E2,3,14a (1089)

Martin, W. H., Fay, A. C., and Renner, K. M. [1930], The limits of error of the Babcock test for cream. J. Agri. Res. 41:147-159. A14;C (1090)

Martins, R. G. [1945], Experimentação agrícola. (Agricultural experimentation.) Brazilian Ministry Agri. Rio de Janeiro 34:1-14. E2,7a (1091)

Maskell, E. J. [1928], Experimental error. A survey of recent advances in statistical method. Tropical Agri. 5:306-309. E2a;C (1092)

Maskell, E J. [1929], Experimental error. A survey of recent advances in statistical method. Tropical Agri. 6:5-11, 45-48, 97-99. E2,7ae;T2ae;A14 (1093)

Masuyama, M, [1946], A formula for assaying Penicillin. Res. Memoirs Inst. Stat. Math., Tokyo 2:466. T8a (1094)

Masuyama, M. [1947], Superposition method of assaying antibiotic substances and its fundamental formula. Res. Memoirs Inst. Stat. Math., Tokyo 3:48. T8a (1095)

Masuyama, M. [1947], Method of estimating the minimum effective dose pooling the results of replicated experiments. Res. Memoirs Inst. Stat. Math., Tokyo 3:129. T8a (1096)

Masuyama, M. [1949], On a one-dimensional diffusion method of assaying antibiotic substances and its fundamental formulas. Biometrics 5:317-329. T8a (1097)

Mather, K. [1940], The design and significance of synergic action tests. J. Hygiene 40:513-531. C (1098)

Mather, K. [1946], Statistical Analysis in Biology. 2nd Edition, Interscience Publishers, Inc., N.Y. E2,7,12ae;T2ae (1099)

Mather, K. [1949], The analysis of extinction time data in bioassay. Biometrics 5:127-143. T8a (1100)

Mathew, N. T. [1941], The influence of seasons on human reproduction. Sankhyā 5:261-268. E2ae (1101)

Maynard, L. A. [1930], Experimental feeding methods with dairy cattle and the interpretation of the results they yield. Proc. American Soc. Animal Production 23:74-81. C (1102)

Maynard, L. A. and Mc Cay, C. M. [1932], The use of simple statistical methods in nutrition investigations. Proc. American Soc. Animal Production 25:20-24. C (1103)

McCall, A. G. [1917], A new method for harvesting small grain and grass plots. J. American Soc. Agron. 9:138-140. A14 (1104)

McCallan, S. E. A. and Wellman, R. H. [1943], A greenhouse method of evaluating fungicides by means of tomato foliage diseases. Contr. Boyce Thompson Inst. 13:93-134. C (1105)

McCarthy, M. D. [1939], On the application of the z-test to randomized blocks. AMS 10:337-359. E2a (1106)

McCleery, F. C. [1935], Experimental method. Applications of the analysis of variance in experimental arrangement. J. Australian Inst. Agri. Sci. 1:96-105. C (1107)

McClelland, C. K. [1926], New methods with check plats. J. American Soc. Agron. 18:566-575. C (1108)

McClelland, C. K. [1926], Some determinations of plat variability. J. American Soc. Agron. 18:819-823. A14 (1109)

McClelland, C. K. [1929], The effect of narrow alleys on small grain yields. J. American Soc. Agron. 21:524-532. A14 (1110)

McClelland, C. K. [1934], Border rows of oat plats as affecting yields and variability. J. American Soc. Agron. 26:491-496. A14 (1111)

McHatton, T. H. [1947], The comparison of plot size in a peach experiment. Proc. American Soc. Hort. Sci. 49:18-20. A14 (1112)

McHatton, T. H. [1947], Variations in peach tree yields. Proc. American Soc. Hort. Sci. 49:121-124. C (1113)

McKee, R. [1914], Moisture as a factor of error in determining forage yields. J. American Soc. Agron. 6:113-117. A14;C (1114)

Meded. LandbvoorlichtDienst. Wageningen [1949], Handleiding voor veldproeven. (Manual for field experiments.) No. 59, pp. 139. C (1115)

Mendonca, P. de Varennese, [1942], Ortogonalidade e análise de variância. (Orthogonality and analysis of variance.) An. Inst. Sup. Agron. Lisboa 13:15-37 and Portug. Math. 3:234-252. E2a (1116)

Menon, T. V. G. and Bose, S. S. [1937], Statistical notes for agricultural workers No. 22. A statistical study of the yields of the permanent manurials, Pusa. Indian J. Agri. Sci. 7:193-204. E5,8ae (1117)

Manusan, H. [1935], Size of plot and its relationship to field spraying experiments with potatoes. J. Economic Entomology 28:190-192. A14 (1118)

Mercer, W. B. and Hall, A. D. [1911], The experimental error of field trials. J. Agri. Sci. 4:107-127. A14 (1119)

Merrill, S., Kilby, W. W., and Greer, S. R. [1942], Fertilization of tung seedlings in the nursery. Proc. American Soc. Hort. Sci. 41:167-170. T2ae (1120)

Merrill, T. A. [1944], Effects of soil treatments on the growth of the highbush blueberry. J. Agri. Res. 69:9-20. E2ae;T2ae (1121)

Meyers, C. H. and Perry, F. R. [1923], Analysis and interpretation of data obtained in comparative tests of potatoes. J. American Soc. Agron. 15:239-253. A14 (1122)

Meyers, M. T. and Patch, L. H. [1937], A statistical study of sampling in field surveys of the fall population of the European corn borer. J. Agri. Res. 55: 849-871. A14 (1123)

Michels, C. A. and Schwenderman, J. [1934], Determining yields on experimental plats by the square yard method. J. American Soc. Agron. 26:993-1001. A14 (1124)

Miles, L. G. and Bryan, W. W. [1937], The analysis of covariance and its use in correcting for irregularities of stand in agricultural trials for yield. Proc. Royal Soc. Queensland 48(4):30-34. E7a;A7 (1125)

Mill. Baking Mtg. [1944], Agronomic aspects of hard spring wheat breeding. 8. Newer technics, equipment, and experimental designs. Rep. Mill. Baking Mtg., Seventh Hard Spring Wheat Conf., Minneapolis, Minn., Feb. 28 and 29--March 1, pp. 50-51. (Mimeo.). C;S1 (1126)

Miller, L. C. and Tainter, M. L. [1944], Estimation of the ED_{50} and its error by means of logarithmic-probit graph paper. Proc. Soc. Expt. Biology 57:261-264. T8a (1127)

Ming, W. C. [1937], Comparison between some methods of estimating field experimental errors (in Japanese). Proc. Crop. Sci. Soc. Japan 9:97-112. E2,7ae;C (1128)

Miranda, R. M., Culbertson, C. C., and Lush, J. L. [1946], Factors affecting rate of gain and their relation to allotment of pigs for feeding trials. J. Animal Sci. 5:243-250. A13,14 (1129)

Mitchell, H. H. [1930], The paired-feeding method: Its value and limitations in livestock experimentation. Proc. American Soc. Animal Production 23:63-73. E2ae;C (1130)

Mitchell, H. H. and Grindley, H. S. [1913], Element of uncertainty in the interpretation of feeding experiments. Illinois Agri. Expt. Sta. Bull. 165:459-579. A14;C (1131)

Mitra, M. [1937], Studies on the stinking smut or bunt of wheat in India. Indian J. Agri. Sci. 7:459-478. E2,12ae (1132)

Mitra, S. K. and Ganguli, P. M. [1934], Uniformity trial in rice (abstract). Twenty-First Ann. Meet. Ind. Sci. Cong. Bombay, Sect. Agri. 9:3-4. A14;C;S1 (1133)

Mitscherlich, E. A. [1934], Die Verarbeitung landwirtschaftlicher und anderer biologischer Versuchsergebnisse. Schr. Koenigsb. gelehrt Ges. naturw. Kl. 10:199-229. C (1134)

Mitscherlich, E. A. [1947], A method for eliminating the systematic error due to soil heterogeneity in drainage and irrigation experiments. Zeit. Pfl. Ernähr. Düng. 37:259-264. C (1135)

Moffatt, J. R. [1939], Agricultural methods adopted in the Rothamsted classical and modern field experiments. Empire J. Exptl. Agri. 7:251-260. C (1136)

Mood, A. M. [1946], On Hotelling's weighing problem. AMS 17:432-446. T6c (1137)

Mood, A. M. and Brown, G. W. [1948], Non-parametric test for factorial experiments. Unpublished paper, Iowa State College. T2a;A12 (1138)

Moore, E. H. [1893], Concerning triple systems. Mathematische Annalen 43: 271-285. E2c (1139)

Moore, E. H. [1896], Factorial memoranda. Amer. J Math. 18:264-303. E2c (1140)

Moore, W. and Bliss, C. I. [1942], A method for determining insecticidal effectiveness using Aphis rumicis and certain organic compounds. J. Economic Entomology 35:544-553. E2ae;T8a (1141)

Morgan, A. and Beruldsen, E. T. [1931], Sampling technique as applied to irrigated pasture in regard to botanical composition and carrying capacity under different grazing systems. J. Dept. Agri. Victoria (Australia) 29:36-45. A14 (1142)

Morgan, J. O. [1908], Some experiments to determine the uniformity of certain plats for field tests. Proc. American Soc. Agron. 1:58-67. A14 (1143)

Mosteller, F. [1940], Some miscellaneous contributions to scale theory: Remarks on the method of paired comparisons. Harvard Univ. Lab. Soc. Relations Report No. 10, chapter III. T11a (1144)

Mottley, C. M. [1942], Experimental designs for developing and testing a stocking policy. North American Wildlife Conf. Transactions 7:224-238. E2,7a;C (1145)

Mottley, C. M. [1942], Modern methods of studying fish populations. North American Wildlife Conf. Transactions 7:356-360. C (1146)

Mottley, C. M. and Embody, D. R. [1942], The effect of the full moon on trout fishing. JASA 37:41-47. C (1147)

Mottley, C. M., Embody, D. R., and Hess, A.D. [1939], The design of an observation. Lab. Limnology Fisheries, Cornell Univ. (Mimeo.). A14;C (1148)

Muench, H. [1945], Experiments and the statistician. Nutr. Rev. 3:321-322. C (1149)

Muench, H. [1947], Statistics in the planning and evaluation of health practices. American J. Public Health 37:1273-1276. C (1150)

Mulder, P. [1917], Kirkman-Systemen (Groningen Dissertation). Leiden. E2c (1151)

Murray, R. K. S. [1934], The value of a uniformity trial in field experimentation with rubber. J. Agri. Sci. 24:177-184. A7,14 (1152)

Musgrave, G. W. [1924], Depression of check-row yields by adjoining high yielding plot rows in potatoes. J. American Soc. Agron. 16:633-635. A14 (1153)

Musgrave, G. W. [1926], A further note on competition in potatoes. J. American Soc. Agron. 18:166-171. A14 (1154)

Myers, C. H. and Perry, F. R. [1923], Analysis and interpretation of data obtained in comparative tests of potatoes. J. American Soc. Agron. 15:239-253. A14;S1 (1155)

Naik, K. C. [1939], Some citrus nursery technique trials at the Fruit Research Station, Anantarajupet, Madras Presidency. Indian J. Agri. Sci. 9:651-673. E2ae (1156)

Nair, K. R. [1938], On a method of getting confounded arrangements in the general symmetrical type of experiment. Sankhyā 4:121-138. T2ac (1157)

Nair, K. R. [1939], Some balanced confounded arrangements for the 5^n type of experiment (abstract). Proc. Twenty-Sixth Indian Sci. Congress, Lahore, Part III, p. 200. T2ac (1158)

Nair, K. R. [1939], The application of co-variance technique to field experiments with mixed-up yields. Sci. Culture 4:474. E2,6a;A7 (1159)

Nair, K. R. [1940], The application of the technique of analysis of covariance to field experiments with several missing or mixed-up plots. Sankhyā 4:581-588. E2,6a;A7 (1160)

Nair, K. R. [1940], Balanced confounded arrangements for the 5^n type of experiment. Sankhyā 5:57-70. T2c (1161)

Nair, K. R. [1941], A note on the method of "fitting of constants" for analysis of non-orthogonal data arranged in a double classification. Sankhyā 5:317-328. E2,6a;A10 (1162)

Nair, K. R. [1942], Efficiency of the adjustment for concomitant characters in biological experiments. Sankhyā 6:167-174. A7 (1163)

Nair, K. R. [1943], Certain inequality of relationships among the combinatorial parameters of incomplete block designs. Sankhyā 6:255-259. E2,3ac (1164)

Nair, K. R. [1944], Statistical notes for agricultural workers No. 27. Calculation of the standard errors and tests of significance of different types of treatment comparisons in split-plot and strip arrangements of field experiments. Indian J. Agri. Sci. 14:315-319. E12a (1165)

Nair, K. R. [1944], The recovery of inter-block information in incomplete block designs. Sankhyā 6:383-390. E2,3,4,6a (1166)

Nair, K. R. [1948], Statistical methods and experimental design. Indian Forester 74:247-250. C (1167)

Nair, K. R. [1948], Certain symmetrical properties of unbiased estimates of variance and covariance. JISAS 1:162-172. A7 (1168)

Nair, K. R. and Bose, R. C. [1939], On completely orthogonalised sets of Latin squares. Sci. Culture 4:666-667. E7,12,13c (1169)

Nair, K. R. and Mahalanobis, P. C. [1940], Statistical notes for agricultural workers No. 25. A simplified method of analysis of quasi-factorial experiments in square lattice with a preliminary note on the joint analysis of yield of paddy and straw. Indian J. Agri. Sci. 10:663-685. E3ae (1170)

Nair, K. R. and Rao, C. R. [1941], Confounded designs for asymmetrical factorial experiments. Sci. Culture 7:313-314. T2a (1171)

Nair, K. R. and Rao, C. R. [1942], Confounded designs for $k \times p^m \times q^n$ type of factorial experiments. Sci. Culture 7:361-362. T2a (1172)

Nair, K. R. and Rao, C. R. [1942], A general class of quasi-factorial designs leading to confounded designs for factorial experiments. Sci. Culture 7: 457-458. E2,3ac;T2a (1173)

Nair, K. R. and Rao, C. R. [1942], A note on partially balanced incomplete block designs. Sci. Culture 7:568-569. E2,3ac (1174)

Nair, K. R. and Rao, C. R. [1942], Incomplete block designs for experiments involving several groups of varieties. Sci. Culture 7:615-616. E4,12ac (1175)

Nair, K. R. and Rao, C. R. [1948], Confounding in asymmetrical factorial experiments. JRSSB 10:109-131. T2ac (1176)

Nair, K. R., Hedayetullah, S., and Sen, S. [1944], Influence of dates of planting and spacings on some winter varieties of rice. Indian J. Agri. Sci. 14:248-259. A14 (1177)

Nanda, D. N. [1949], Efficiency of the application of discriminant function in plant-selection. JISAS 2·8-19. A7 (1178)

Nandi, H. K. [1945], Note on the enumeration of all possible non-isomorphic solutions of balanced incomplete block designs. Proc. Thirty-Second Indian Sci. Congress, Sec. XII, No. 54. E2c (1179)

Nandi, H. K. [1945], On the relation between certain types of tactical configurations. Bull. Calcutta Math. Soc. 37:92-94. E2c (1180)

Nandi, H. K. [1946], Enumeration of non-isomorphic solutions of balanced incomplete block designs. Sankhyā 7:305-312. E2c (1181)

Nandi, H. K. [1946], A further note on non-isomorphic solutions of incomplete block designs. Sankhyā 7:313-316. E2c (1182)

Nandi, H. K. [1947], A mathematical set-up leading to analysis of a class of designs. Sankhyā 8:172-176. E12ac;A7 (1183)

Narain, R. and Singh, A. [1937], Sampling of sugarcane for chemical analysis, I. Indian J. Agri. Sci. 7:601-625. A14 (1184)

Narasinga Rao, M. B. V. [1937], A note on a few experimental observations in the Rice Research Station, Berhampur (Madras). Indian J. Agri. Sci. 7:733-744. A14 (1185)

Neely, J. W. [1934], Plat technique studies with the fieldbean (Phaseolous vulgaris). Ph. D. Thesis, Cornell Univ. A14 (1186)

Netto, E. [1927], Lehrbuch der Combinatorik. Zeweite Auglage, Verlag Druck B.G. Teubner Leipzig and Berlin. E2c (1187)

Nevens, W. B. [1945], A comparison of sampling procedures in making pasture yield determinations. J. Dairy Sci. 28:171-185. A14 (1188)

Newman, D. [1939], The distribution of range in samples from a normal population, expressed in terms of an independent estimate of standard deviation. Biometrika 31:20-30. E2ae;A3 (1189)

Neyman, J. [1929], The theoretical basis of different methods of testing cereals. Part II. The method of parabolic curves. Sci. Pub. K. Buszczynski and Sons, Warzawa, Wiadom. Mat. pp. 1-35. A7;C (1190)

Neyman, J. [1932], O metodach opracowywonia doświadczeń wielokrotnych. (On the methods of interpreting results of multiple agricultural trials.) Rocz. Nauk Rol. (Polish Agri. and For. Ann.) 32:154-210. A13;C (1191)

Neyman, J. [1938], Lectures and conferences on mathematical statistics. Pub. Graduate School, U.S. Dept. Agri., Washington, pp. 160. E2,5a;T9a;C (1192)

Neyman, J. [1939], On a new class of "contagious" distributions, applicable in entomology and bacteriology. AMS 10:35-57. A7,14;C (1193)

Neyman, J. [1939], On the hypotheses underlying the applications of statistical methods to routine laboratory analyses (abstract). AMS 10:87. C (1194)

Neyman, J. in cooperation with K. Iwaszkiewicz and St. Kolodziejczyk. [1935], Statistical problems in agricultural experimentation. JRSSB 2:107-180. E2,7ae;T2a;C (1195)

Neyman, J. and Pearson, E. S. [1937], Note on some points in "Student's" paper on "Comparison between balanced and random arrangements of field plots" Biometrika 29:380-388. E2,5a (1196)

Nissen, Ø. [1942], Feilbetegning på forsøk med systematisk parsellfordeling. (The calculation of error in experiments with a systematic distribution of plots.) Nord. JordbrForskn. 7:357-358. E5,7a (1197)

Nissen, Ø. [1947], The interaction between neighbouring plots in root experiments. Meldinger Norges Landbrukshøgskole 27:155-164. A14 (1198)

Nissen, Ø. [1949], The use of visual estimates in place of separation analysis in experiments with hay crops. Meldinger Norges Landbrukshøgskole 29:225-256. A14 (1199)

Nissen, Ø. and Ottestad, P. [1943], On the analysis of variance and the effect of non-normality. Meldinger Norges Landbrukshøgskole 23:475-496. A9,11;C (1200)

Noll, C. F. [1928], The type of problem adapted to field plat experimentation. J. American Soc. Agron. 20:421-425. C (1201)

Nordskog, A. W., Clark, R. T., and Van Horn, L. [1945], Sampling wool clips for clean yield by the core boring method. J. Animal Sci. 4:113-121. S1 (1202)

Nordskog, A. W. and Crump. S. L. [1948], Systematic and random sampling for estimating egg production in poultry. Biometrics 4:223-233. E5a;A14;C (1203)

Norton, C. L., Eaton, H. D., Loosli, J. K., and Spielman, A. A. [1946], Controlled experiments on the value of supplementary vitamins for young dairy calves. J. Dairy Sci. 29:231-238. E1ae (1204)

Norton, H. W. [1939], The 7×7 squares. Ann. Eugenics 9:269-307 (errata 10:1). E7,12,13c (1205)

Norton, H. W. [1945], Calculation of chi-square for complex contingency tables. JASA 40:251-258. T2a (1206)

Nuckols, S. B. [1936], The use of actual and competitive yield data from sugar beet experiments. J. American Soc. Agron. 28:924-934. A14 (1207)

Odland, T. E. and Garber, R. J. [1928], Size of plat and number of replications in field experiments with soybeans. J. American Soc. Agron. 20:93-108. A14;S1 (1208)

Oliveira, A. J. de, [1946], Estudos de estatística agrónomica. III. Efficiência relativa dos diversos delineamentos estatísticos usados na comparaçao de grande numero de variedades. (Studies in agronomic statistics. III. Relative efficiency of different statistical layouts used in the comparison of a large number of varieties.) Agron. Lusitana 8:315-340. E2,3,11,14a (1209)

Olmstead, L. B. [1914], Some applications of the method of least squares to agricultural experiments. J. American Soc. Agron. 6:190-204. A14;S1 (1210)

Olsen, H. K. [1935], Raekkemetoden i dansk lokal forsøgsvirksomhed. (The row method in Danish local field trials.) Beretn. Nordisk JordbrForsker. Foren. Fifth Kongr. Kobenhavn Juli 4-7, Hefte 318-328. E5a;A14;C (1211)

O'Neil, J. B. [1942], The analysis of covariance by the method of individual comparisons. Sci. Agri. 22:721-724. T2ae;A7 (1212)

Opsomer, J. E. [1937], Recherches sur la "methodique" de l'amélioration du riz à Yangambi. I. La technique des essais. Publ. Inst. Agron. Congo Belge Sér. Sci. 12:1-25. A14 (1213)

Opsomer, J. E. [1937], Notes techniques sur la conduite des essais avec plantes annuelles et l'analyse des résultats. Publ. Inst. Agron. Congo Belge Sér. Sci. 14:1-79. E2,7ae;C (1214)

Osborne, J. G. [1939], A design for experiments in thinning forest stands. J. Forestry 37:296-304. C (1215)

Osborne, J. G. [1942], Sampling errors of systematic and random surveys of cover-type areas. JASA 37:256-264. E5a;A14;C (1216)

Osborne, J. G. and Schumacher, F. X. [1935], The construction of normal-yield and stand tables for even-aged timber stands. J. Agri. Res. 51:547-564. A14 (1217)

Ostle, B. [1949], Query 65. Biometrics 5:71-75. E7,12ae (1218)

Overman, A. and Li, J. C. R. [1948], Dependability of food judges as indicated by analysis of scores of food testing panels. Food Res. 13:441-449. C (1219)

Ovesnov, A. M. [1937], (On the size of the sample square in the study of phytocoenoses on continental grasslands.) Uchen. Zap. Perm. Gos. Univ. 2(4):137-143. A14 (1220)

Palacios, A. R. [1947], Eficiencia del diseño experimental lattice. Chapingo 15:5-16. E3a (1221)

Paley, R. E. A. C. [1933], On orthogonal matrices. J. Math. Physics 12:311-320. T2c (1222)

Pan, C. L. [1935], Uniformity trials with rice. J. American Soc. Agron. 27: 279-285. A14 (1223)

Panse, V. G. [1938], Preliminary studies on sampling in field experiments. Sankhyā 4:139-148. A13,14;S1 (1224)

Panse, V. G. [1941], Studies in the technique of field experiments. V. Size and shape of blocks and arrangement of plots in cotton trials. Indian J. Agri. Sci. 11:850-865. A14;S1 (1225)

Panse, V. G. [1946], Plot size in yield surveys on cotton. Current Sci. 15:218-219. A14 (1226)

Panse, V. G. [1947], Plot-size in yield surveys. Nature 159:820. A14 (1227)

Panse, V. G., Sukhatme, P. V., Kishen, K., and Sahasrabudhe, V. B. [1947], Symposium on statistical methods in plant and animal breeding. Proc. Indian Academy Sci. 25(B): 126-154. E2,3a;A7;C (1228)

Papadakis, J. [1931], Some considerations on the technique of field experiments. Bull. l'Assoc. Int. Sélectioneurs Plantes 4:59-66. A14;C (1229)

Papadakis, J. [1931], Varieties experiments in countries with great variability in ecological conditions from year to year. Bull. Assoc. Int. Sélectioneurs Plantes 4:71-75. A14;C (1230)

Papadakis, J. S. [1937], (A statistical method for field experiments.) Bull. Sci. Inst. Amélior. Plantes Salonique 23:1-30. A14;C (1231)

Papadakis, J. S. [1937], (On the breadth and length of plots in field experiments). Bull. Sci. Inst. Amélior. Plantes Salonique 24:1-24. A14;C (1232)

Papadakis, J. S. [1937], (Experiments and improvements in the method of pockets for variety trials.) Bull. Sci. Inst. Amélior. Plantes Salonique 27:1-32. A14; C (1233)

Papadakis, J. S. [1940], Comparison de différentes méthodes d'expérimentation phytotechnique. Revista Argentina Agronomia 7:297-362. A14;C (1234)

Park, T. [1948], Experimental studies of interspecies competition. I. Ecol. Monogr. 18:265-308. A14 (1235)

Parker, E. R. [1942], Adjustment of yields in an experiment with orange trees. Proc. American Soc. Hort. Sci. 41:23-33. A7,14 (1236)

Parker, E. R. and Batchelor, L. D. [1932], Variation in the yields of fruit trees in relation to the planning of future experiments. Hilgardia 7:81-161. A14;B;C (1237)

Parker, K. W. and Savage, D. A. [1944], Reliability of the line interception method in measuring vegetation on the southern great plains. J. American Soc. Agron. 36:97-110. A14 (1238)

Parker, W. H. [1931], Methods employed in variety trials by the National Institute of Agricultural Botany. J. Nat. Inst. Agri. Botany 3:5-22. A14;C (1239)

Parsons, R. H. [1942], Size of samples. Engineering, London 154:294-295. S1 (1240)

Paterson, D. D. [1933], The influence of time of cutting on the growth, yield and composition of tropical fodder grasses. I. J. Agri. Sci. 23:615-641. E2;12ae (1241)

Paterson, D. D. [1933], Experimentation and applied statistics for the practical agriculurist. Tropical Agri. 10:267-276, 303-311, 346-351. E2,7,12ae (1242)

Paterson, D. D. [1934], A note on the value of correlation and regression in statistical analysis. Tropical Agri. 11:160-169, 220-229. E2ae;A7 (1243)

Paterson, D. D. [1935], The growth, yield and composition of certain tropical fodders. J. Agri. Sci. 25:369-396. E7,12ae (1244)

Paterson, D. D. [1939], Statistical Technique in Agricultural Research. McGraw-Hill, New York. E2,7,12ae;T2ae;C (1245)

Paterson, D. D. and Hanschell, D. M. [1938], The comparison of four sugar-cane varieties in Trinidad. Tropical Agri. 15:199-201. E12ae (1246)

Patterson, R. E. [1946], The use of adjusting factors in the analysis of data with disproportionate subclass numbers. JASA 41:334-346. A10 (1247)

Patterson, R.E. [1947], The comparative efficiency of single versus three-day weight of steers. J. Animal Sci. 6:237-246. A14 (1248)

Paul, W. R. C. and Fernando, M. [1939], Field-plot technique with chillies (Capsicum annuum L.). Tropical Agri. 93:270-275. A14 (1249)

Paulson, E. [1949], A multiple decision procedure for certain problems in the analysis of variance. AMS 20:95-98. E2,7a;A3 (1250)

Peach, P. [1946], The use of statistics in the design of experiments. IQC 3(3): 15-17. C (1251)

Pearce, S. C. [1944], Sampling methods for the measurement of fruit crops. JRSSA 107:117-126. A14 (1252)

Pearce, S. C. [1948], Randomized blocks with interchanged and substituted plots. JRSSB 10:252-256. E2,6ae (1253)

Pearce, S.C. [1949], The variability of apple trees I. The extent of crop variation and its minimization by statistical methods. J. Hort. Sci. 25:3-9. A7,13 (1254)

Pearce, S. C. and Hoblyn, T. N. [1948], A review of experimental design at East Malling, 1919-1947. Annual Report, E. Malling Res. Sta., 1947, pp. 88-100. E2,5a;C (1255)

Pearce, S. C. and Taylor, J. [1948], The changing of treatments in a long-term trial. J. Agri. Sci. 38:402-410. E5,8ae;T7ae (1256)

Pearl, R. [1923], The interrealtion of the biometric and experimental methods of acquiring knowledge; with special reference to the problem of the duration of life. Metron 2(4):697-721. C (1257)

Pearson, E. S. [1929], Statistics and biological research. Nature 124:615. C (1258)

Pearson, E. S. [1931], The analysis of variance in cases of non-normal variation. Biometrika 23:114-133. A10,11 (1259)

Pearson, E. S. [1937], Some aspects of the problem of randomization. Biometrika 29:53-64. E7a;C (1260)

Pearson, E. S. [1938], Some aspects of the problem of randomization. II. An illustration of "Student's" inquiry into the effect of "Balancing" in agricultural experiments. Biometrika 30:159-179. E5a;C (1261)

Pearson, E. S. [1938], William Sealy Gosset 1876-1937. "Student" as statistician. Biometrika 30:210-250. E5a;B;C (1262)

Pearson, E. S. and Chandra Sekar, C. [1936], The efficiency of statistical tools and a criterion for the rejection of outlying observations (with an appendix by J.M.C. Scott). Biometrika 28:308-320. A6 (1263)

Pearson, K. [1911], The Grammar of Science, 3rd ed. A. and C. Black, London. C (1264)

Pearson, K. [1933], On a method of determining whether a sample of size n supposed to have been drawn from a parent population having a known probability integral has probably been drawn at random. Biometrika 25: 379-410. C (1265)

Pechanec, J. F. [1941], Sampling error in range surveys of sagebrush-grass vegetation. J. Forestry 39(1):52-54. A14;C (1266)

Pechanec, J. F. and Pickford, G. D. [1937], A comparison of some methods used in determining percentage utilization of range grasses. J. Agri. Res. 54:753-765. A14;C (1267)

Pechanec, J. F. and Stewart, G. [1940], Sagebrush-grass range sampling studies: Size and structure of sampling unit. J. American Soc. Agron. 32:669-682. A14;C (1268)

Pechanec, J. F. and Stewart, G. [1941], Sagebrush-grass range sampling studies: variability of native vegetation and sampling error. J. American Soc. Agron. 33:1057-1071. A14;C (1269)

Peevy, W. J., Smith, F. B., and Brown, P. E. [1940], Effects of rotational and manurial treatments for twenty years on the organic matter, nitrogen, and phosphorous contents of Clarion and Webster soils. J. American Soc. Agron. 32:739-753. E5,8a (1270)

Peh, S. C. [1937], Studies in yield comparisons of rice. J. American Soc. Agron. 29:167-185. E2,5ae;T2ae (1271)

Peirce, B. [1852], Criterion for the rejection of doubtful observations. Gould's Astr. J. (Mass.) 2:161-163. A6 (1272)

Peirce, B. [1860], Cyclic solutions of the school-girls' puzzles. Astr. J. 6: 169-174. E2c (1273)

Pepper, J. [1929], Studies in the theory of sampling. Biometrika 21:231-258. A14 (1274)

Peregudov, V. [1931], Mathematical study of the field experiment method as per data of the fertilizers station (in Russian with English summary). Proc. All-Union Scientific Res. Inst. Cotton Culture and Industry (NIHI) Issue 35:1-59. A14 (1275)

Pereira, H. C. [1948], Mathematical statistics in agricultural research. Proc. Nairobi Sci. Phil. Soc. 2:28-32. E2,3,7a;C (1276)

Pérez Calvet, R. [1945], Un estudio sobre las experiencias de uniformidad y su empleo la elección de parcela de repetición. (A study on uniformity experiments and their use un the choice of replication plots.) Bol. Inst. Invest. Exper. Agron. For. Madrid 12:329-348. A14;S1 (1277)

Pérez Calvet, R. and de Zulueta, M. M. [1946], (Statistical methods for comparing a large number of varieties.) Bol. Inst. Invest. Exper. Agron. For., Madrid 14:29-62. C (1278)

Pérez Calvet, R., de Zulueta, M. M., and Anós, A. [1943], Experimentación agricola. Fundamentos estadísticos y métodos operatorios. (Agricultural experimentation. Statistical principles and methods of operation.) Instituto Nacional de Investigaciones Agronómicas, Madrid pp. vii+272. E2,7ae;T2ae;A7;C (1279)

Perotti, J. M. [1943], Mean improvement in a normal variate under direct and indirect selection. M. S. Thesis, Iowa State Univ. T9a (1280)

Peters, C. C., Toronsend, A., and Traxler, A. E. [1945], Research methods and designs. Rev. Educ. Res. 15:377-393. C (1281)

Petersen, J. [1902], Les 36 officiers. Annu. Math. pp. 413-427. E7,12c (1282)

Peterson, M. L. [1947], Effect of method of grazing unimproved Kentucky bluegrass on beef production, botanical composition, and herbage yields. J. American Soc. Agron. 39:412-422. E2,12ae (1283)

Phipps, I. F., Pugsley, A. T., Hockley, S. R., and Cornish, E. A. [1944], The analysis of cubic lattice designs in varietal trials. Council Sci. Ind. Res. Australian Bull. 176:1-41. E3ae (1284)

Pickering, S. U. [1911], Experimental error in horticultural work. J. Board Agri. (London), Suppl. 7:38-47. C (1285)

Piper, C. V. and Stevenson, W. H. [1910], Standardization of field experimental methods in agronomy. Proc. American Soc. Agron. 2:70-76. A14;S1 (1286)

Pitman, E. J. G. [1937], Significance tests which may be applied to samples from any populations III. The analysis of variance test. Biometrika 29:322-335. A11,12 (1287)

Pitman, E. J. G. [1939], A note on normal correlation. Biometrika 31:9-12. A7 (1288)

Plackett, R. L. [1946], Some generalizations in the multifactorial design. Biometrika 33:328-332. T2,3c (1289)

Plackett, R. L. [1947], Cyclic intrablock subgroups and allied designs. Sankhyā 8:275-276. T2,3c (1290)

Plackett, R. L. and Burman, J. P. [1946], The design of optimum multifactorial experiments. Biometrika 33:305-325. T2,3a (1291)

Plackett, R. L. and Hewlett, P. S. [1948], Statistical aspects of the independent joint action of poisons, particularly insecticides, I. The toxicity of a mixture of poisons. Ann. Appl. Biology 35:347-358. T8a (1292)

Pohle, E. M. and Hazel, L. N. [1944], Clean-wool yields in small samples from eight body regions as related to whole-fleece yields in four breeds of sheep. J. Animal Sci. 3:159-165. A14 (1293)

Pohle, E. M., Hazel, L. N., and Keller, H. R. [1945], The infleunce of location and size of sample in predicting whole-fleece clean yields. J. Animal Sci. 4:104-112. A14 (1294)

Pope, O. A. [1931], The determination of sample size for diameter measurements in cotton fiber studies. J. Agri. Res. 43:957-984. S1 (1295)

Pope, O. A. [1935], Effects of certain soil types, seasonal conditions, and fertilizer treatments on length and strength of cotton fiber. Arkansas Agri. Expt. Sta. Bull. 319:1-98. E2a;C (1296)

Pope, O. A. [1936], Efficiency of single and double restrictions in randomized field trials with cotton when treated by the analysis of variance. Arkansas Agri. Expt. Sta. Bull. 326:1-28. E2,7,12ac (1297)

Pope, O. A. [1938], The applicability of certain recent experimental designs to cotton research (abstract). Proc. Thirty-Ninth Annual Convent. Assoc. S. Agri. Workers, Atlanta, Ga. pp. 66-67. T2a (1298)

Pope, O. A. [1944], The use of a cubic lattice design in cotton strain studies. Mimeo. Arkansas Expt. Sta. E3ae (1299)

Pope, O. A. [1947], Effect of skips, or missing row segments, on yield of seed cotton in field experiments. J. Agri. Res. 74:1-13. A14 (1300)

Poritsky, H. [1948], Method of Graeco-Latin squares. Elec. Eng. 67:1061. E7,12,13c (1301)

Post, J. J. [1947], Blanco proeven. (Uniformity trials.) Statistica 1:317-321. A14 (1302)

Potter, G. F. [1930], Significance of the probable error as applied to field experiments with apple trees. Proc. American Soc. Hort. Sci. 27:534-535. C (1303)

Potter, G. F. [1947], Research on problems of tung production and improvement, 1938-1946 (Presidential address). Proc. American Soc. Hort. Sci. 50:443-457. E2ae;C (1304)

Potter, G. F., Angelo, E., Painter, J. H., and Brown, R. T [1939], A statistical study of variation in tung fruits. Proc. American Soc. Hort. Sci. 37:515-517. C (1305)

Pound, F. J. [1932], The fruitfulness of cacao. The significance of the relation between total variance and mean annual yield. Tropical Agri. 9:288-290. C (1306)

Power, J. [1867], On the problem of the fifteen school girls. Quarterly J. Pure Appl. Math. 8:236-251. E2c (1307)

Powers, L. [1942], The nature of the series of environmental variances and the estimation of the genetic variances and the geometric means in crosses involving species of Lycopersicon. Genetics 27:561-575. T8a;C (1308)

Pratt, A. D. [1932], A study of methods used in conducting a silage feeding experiment. J. Dairy Sci. 15:303-311. Al4 (1309)

Pratt, A. J. [1933], Statistical methods in the study of tomato varieties and strains. Ph. D. Thesis, Cornell Univ. Al4;Sl (1310)

Preston, F. W. [1948], The commonness, and rarity, of species. Ecology 29: 254-283. Al4;C (1311)

Pridham, A. M. S. [1933], Application of probable error concept in floriculture. Proc. American Soc. Hort. Sci. 30:594-595. C (1312)

Pritchard, F. J. [1916], The use of checks and repeated plantings in varietal tests. J. American Soc. Agron. 8:65-81. C;Sl (1313)

Probst, A. H. [1943], Border effect in soybean nursery plots. J. American Soc. Agron. 35:662-666. Al4 (1314)

Proebsting, E. L. [1942], The relative yields of border fruit trees. Proc. American Soc. Hort. Sci. 41:34-36. Al4 (1315)

Pruthi, H. S. and Narayanan, E. S. [1939], A statistical study of the loss caused by borers and termites to mature sugarcane at Pusa in 1935 and 1936. Indian J. Agri. Sci. 9:15-37. E2ae (1316)

Przyborowski, J. [1933], O metodzie wykorzystania dawniejszych analiz laboratoryjnych w celu dokladniejszego wyznaczenia średniego bledu średniej arytmetycznej nielicznej serji równoleglych analiz. (On the method of estimating the standard error of a mean of a small series of routine analysis data, using the results of analysis made previously.) Rocz. Nauk Rol. 30:303-332. C (1317)

Przyborowski, J. and Ruebenbauer, T. [1936], Doświadczenia z odmianami pszenicy ozimej przeprowadzone w latach 1926-1934. (Varietal tests of winter wheat, carried out from 1926 to 1934 in southwest Poland.) Wydaw. Sekcji.Nasiennej Przy M.T. R. w Krakowie z Udziałem Zakładu Hodowli Roślin i Doświadczalnictwa U.J. Kraków 13:1-56. T8a;C (1318)

Przyborowski, J. and Wilénski, H. [1935], Statistical principles of routine work in testing clover seed for dodder. Biometrika 27:273-292. A14;C (1319)

Przyborowski, J. and Wilénski, H. [1935], Wyzyskanie korelacji między plonem I innemi mierzalnemi cechami poletek przy opracowywaniu doswiadczeń polowych. (On the use of correlation between yield and other measurable characters of single plots in reducing results of field experiments.) Rocz. Nauk. Roi. 34: 273-286. C (1320)

Przyborowski, J. and Wilénski, H. [1937], Zastosowanie Korelacji w niektórych przypadkach opracowywania wyników doświadczeń polowych. (On the use of correlation in some cases of practice of field experiments.) Roczn. Nauk.Rol. 38:205-208. A7;C (1321)

Przyborowski, J. and Wilénski, H. [1937], Metoda przeprowadzania doświadczeń z zastosowaniem poletek wzorcowych. (The check-parcel method.) Wydaw. Sekcji Nasiennej Przy M.T.R. w Karkowie i Zakładu Hodowli Roślin i Doświadczalnictwa U.J. Kraków 16:1-24. E5a;C (1322)

Przyborowski, J. and Wilénski, H. [1938], Analiza zmienności wyników doświadczén wielokrotnych. (Analysis of variance of results of multiple agricultural trials.) Wydawnictwa Sekcji Nasiennej Przy M.T.R. w Krakowie i Zakładu Hodowli Roślin i Doświadczalnictwa U.J. 20:1-19; supplement to Roczn. Nauk. Rol. 44. T2a;C (1323)

Przyborowski, J. and Wilénski, H. [1940], Homogeneity of results in testing samples from Poisson series. With an application to testing clover seed for dodder. Biometrika 31:313-323. A14 (1324)

Quenouille, M. H. [1948], The analysis of covariance and non-orthogonal comparisons. Biometrics 4:240-246. E2,7ae;A7 (1325)

Quenouille, M. H. [1949], Problems in plane sampling. AMS 20:355-375. A14 (1326)

Quinby, L. R., Killough, D. T., and Stansel, R. H. [1937], Competition between cotton varieties in adjacent rows. J. American Soc. Agron. 29:269-279. A14 (1327)

Rahman, K. A. and Singh, D. [1944], Estimation of sugarcane top-borer (Scirpophaga nivella F.) infestation. Indian J. Agri. Sci. 14:233-239. A14 (1328)

Rajabhooshanam, D. S. [1935], Application of modern statistical methods to yield trials (Woodhouse Memorial Prize Essay, 1933). Agri. Livestock India 5:145-155. C (1329)

Ramamurti, B. and Sitaraman, B. [1942], On maximal sets of confounded interactions in a $(2^n, 2^k)$ confounded design. Sankhyā 6:183-188. T2,3c (1330)

Rao, C. R [1943], Quasi-Latin squares in experimental arrangements. Current Sci. 12:322-323. E3,11c (1331)

Rao, C. R. [1944], Extension of the difference sets of Singer and Bose. Sci. Culture 10:57. E2c (1332)

Rao, C. R. [1944], On the linear set-up leading to intra- and inter-block informations. Sci. Culture 10:259-260. E2,3a (1333)

Rao, C. R. [1946], On the linear combination of observations and the general theory of least squares. Sankhyā 7:237-256. E2,3,12ace (1334)

Rao, C. R. [1946], Confounded designs in quasi-Latin squares. Sankhyā 7: 295-304. E7,9ac (1335)

Rao, C. R. [1945], Finite geometries and certain derived results in theory of numbers. Proc. Nat. Inst. Sci. India 11:136-149. E2,3c (1336)

Rao, C. R. [1946], Difference sets and combinatorial arrangements derivable from finite geometries. Proc. Nat. Inst. Sci. India 12:123-135. E2,3c (1337)

Rao, C. R. [1946], Hypercubes of strength "d" leading to confounded designs in factorial experiments. Bull. Calcutta Math. Soc. 38:67-78. E11c;T2c (1338)

Rao, C. R. [1946], On the most efficient designs in weighing. Sankhyā 7:440. T6a (1339)

Rao, C. R. [1947], General methods of analysis for incomplete block designs. JASA 42:541-561. E2,3ae (1340)

Rao, C. R. [1947], Factorial experiments derivable from combinatorial arrangements of arrays. JRSSB 9:128-139. T2,3c (1341)

Rao, C. R. [1948], Tests of significance in multivariate analysis. Biometrika 35:58-79. A7 (1342)

Rao, C. R. [1949], On a class of arrangements. Proc. Edinburgh Math. Soc. 8:119-125. E2,3c (1343)

Rao, C. R. [1949], The general theory of fractional replication in factorial experiments (abstract) Proc. Thirty-Sixth Indian Sci. Congress Part III, p. 9. T3c (1344)

Rau, A. A. [1948], A statistical examination of some experiments on coffee in India. M. S. Thesis, Iowa State College. E5,8a (1345)

Rau, A. A. [1949], Harmonic analysis and experimental data. Current Sci. 18:119-120. A7 (1346)

Read, D. R. [1949], A study of the accuracy of simple sampling methods for the estimation of egg production and mean egg weight. J. Agri. Sci. 39:259-264. C (1347)

Reed, J. F. and Rigney, J. A. [1947], Soil sampling from fields of uniform and nonuniform appearance and soil types. J. American Soc. Agron. 39:26-40. A14;C (1348)

Reed, L. J. and Muench, H. [1938], A simple method of estimating fifty per cent endpoints. American J. Hygiene 27:493-497. T8a (1349)

Reinhardt, F. [1932], Der Jahresfehler bei Felversuchen. Arch. Pflanzenbau 8:187-275. E5,8a (1350)

Reiss, M. [1859], Über eine Steinersche combinatorische Aufgabe. J. Reine Angewandte Math. 56:326-344. E2c (1351)

Remussi, C. [1938], Aplicación del análisis de la variancia a los ensayos comparativos de rendimiento. (Application of the analysis of variance to comparative yield tests.) Rev. Argent. Agron. 5:254-283. C (1352)

Reynolds, E. B., Killough, D. T., and Vantine, J. T. [1934], Size, shape, and replication of plats for field experiments with cotton. J. American Soc. Agron. 26:725-734. A14 (1353)

Richer, W. E. [1938], On adequate quantitative sampling of the pelagic net plankton of a lake. J. Fish. Res. Board Canada 4:19-32. A14;C (1354)

Richey, F. D. [1924], Adjusting yields to their regression on a moving average, as a means of correcting for soil heterogeneity. J. Agri. Res. 27:79-90. A14 (1355)

Richey, F. D. [1926], The moving average as a basis for measuring correlated variation in agronomic experiments. J. Agri. Res. 32:1161-1175. E5,9a;A14 (1356)

Richey, F. D. [1930], Some applications of statistical methods to agronomic experiments. JASA 25:269-283. A14 (1357)

Richmond, T. R. [1943], Competition in cotton variety tests. J. American Soc. Agron. 35:606-612. A14 (1358)

Richmond, T. R. and Fulton, H. J. [1936], Variability of fiber length in a relatively uniform strain of cotton. J. Agri. Res. 53:749-763. C (1359)

Ricker, W. E. [1948], Methods of Estimating Vital Statistics of Fish Populations. Indiana Univ. Pub. Sci. Ser. No. 15, 101 pp. A14;S1 (1360)

Riddle, O. C. and Baker, G. A. [1944], Biases encountered in large-scale yield tests. Hilgardia 16:1-14. A6;C (1361)

Rider, P. R. [1933], Criteria for rejection of observations. Washington Univ. Studies, New Ser., Sci. Tech. 8. A6 (1362)

Rigg, F. A. [1946], Recent advances in mathematical statistics. Bibliography of mathematical statistics (1940-42). JRSSA 109:395-450. B (1363)

Rigney, J. A. [1946], Good experiments are not accidental. Res. Farming 5, Progress Reports Nos,1,10,12. C (1364)

Rigney, J, A. [1946], Some statistical problems confronting horticultural investigators. Proc. American Soc. Hort. Sci. 48:351-357. C (1365)

Rigney, J. A. [1948], Techniques in field plot experimentation. Proc. Auburn Conf. Stat. Appl. to Soc. Sci., Plant Sci., and Animal Sci., pp. 39-44. A14;C (1366)

Rigney, J. A. and Blaser, R. E. [1948], Sampling alyce clover for chemical analyses. Biometrics 4:234-239. A14 (1367)

Rigney, J. A., Miles, S. R., and Andrews, W. B. [1948], The choice of suitable experimental units and experimental designs. Proc. Nat. Joint Comm. Fert. Applications (1947) 23:228-234. C (1368)

Rigney, J. A., Morrow, E. B., and Lott, W. L. [1949], A method of controlling experimental error for perennial horticultural crops. Proc. American Soc. Hort. Sci. 54:209-212. A7 (1369)

Rigney, J. A. and Reed, J. F. [1945], Some factors affecting the accuracy of soil sampling. Proc. Soil Sci. Soc. Amer. 10:257-259. A14;C (1370)

Ripley, P. O. [1941], The influence of crops upon those which follow. Sci. Agri. 21:522-583. E5,8a;B (1371)

Robertson, D. W. and Koonce, D. [1934], Border effect in irrigated plots of Marquis wheat receiving water at different times. J. Agri. Res. 48:157-166. A14 (1372)

Robertson, H. F. [1932], Fisher's analysis of variance with paddy on a field-scale. Indian J. Agri. Sci. 2:53-58. E2ae;C (1373)

Robinson, G. W. and Halnan, E. T. [1912], Probable error in pig feeding trials. J. Agri. Sci. 5:48-51. C (1374)

Robinson, G. W. and Lloyd, W. E. [1915], On the probable error of sampling in soil surveys. J. Agri. Sci. 7:144-153. A14 (1375)

Robinson, H. F., Rigney, J. A., and Harvey, P. H. [1948], Investigations in plot technique with peanuts. N. Carolina Agri. Expt. Sta. Tech. Bull. 86: 1-19. E2,3,11a;A14 (1376)

Robinson, H. F. and Watson, G. S. [1949], An analysis of simple and triple rectangular lattice designs. N. Carolina Agri. Expt. Sta. Tech. Bull. 88:1-56. E3ae (1377)

Roessler, E. B. [1936], Significant figures in statistical constants. Science 84:289-290. C (1378)

Roessler, E. B. [1943], Valid estimates of variance in the analysis of pooled data. Proc. American Soc. Hort. Sci. 42:481-483. A13;C (1379)

Roessler, E. B. [1946], Testing the significance of observations compared with a control. Proc. American Soc. Hort. Sci. 47:249-251. A3 (1380)

Roessler, E. B. and Leach, L. D. [1944], Analysis of combined data for identical replicated experiments. Proc. American Soc. Hort. Sci. 44:323-328. A13;C (1381)

Romanovskii, V. I. [1934], On new methods of mathematical statistics which apply to field trials. Mathematical statistics and industry (Russian). Soc. Nauk Tekh., Tashkent pp. 3-4, 9. C (1382)

Roop, W. P. [1947], Transitional evaluations and treatment of experimental data subject to very wide scatter. Bull. American Soc. Test. Math. 147:73-76. C (1383)

Ross, R. T. [1934], Optimum orders for the presentation of pairs in the method of paired comparisons. J. Educ. Psychology 25:375-382. T11a (1384)

Rothamsted Experimental Station Report, 1925-26. E2,7ae (1385)

Rothamsted Experimental Station Report, 1927-28. E2,7,12ae;A14 (1386)

Rothamsted Experimental Station Report 1929. E2,7,11,12ae (1387)

Rothamsted Experimental Station Report, 1930. E2,5,7,8,11,12ae (1388)

Rothamsted Experimental Station Report, 1931. E2,7,11,12ae (1389)

Rothamsted Experimental Station Report, 1932. E2,5,7,8,11,12ae (1390)

Rothamsted Experimental Station Report, 1933. E2,3,5,7,8,11,12ae (1391)

Rothamsted Experimental Station Report, 1934-36. E2,3,5,7,8,11,12ae (1392)

Russell, C. S. [1945], Errors in the routine daily measurement of the puerperal uterus. Biometrika 33:213-221. E2ae (1393)

Russell, E. J. [1931], The technique of field experiments (foreward). Rothamsted Conf. 13:5-8. C (1394)

Russell, E. J., Voelcker, J. A., and Cochran, W. G. [1936], Fifty Years of Field Experiments at the Woburn Experimental Station with a statistical report. Longmans, Green and Co., London. A7;C (1395)

Russell, G. A. and Little, V. A. [1946], Response of rotenone-bearing devil's shoestring, Tephrosia virginiana (L.) Pers., to fertilizer applications. J. American Soc. Agron. 38:646-650. E2ae;T2ae (1396)

Russell, J. [1926], Field experiments. How they are made and what they are. J. Ministry Agri. 32:989-1001. C (1397)

Sade, A. [1948], Enumeration des carres latins. Application au 7^{e} ordre. Conjecture pour les ordres superieurs. Chez l'auteur, 14Bd. du Jardin Zoologique, Marseille. Imprimerie du Pharo, 31 Rue Charras, Marseille. E7,12,13c (1398)

Sachs, W. H. [1926], Student's method as applied to field data covering a period of years. J. American Soc. Agron. 18:1064-1067. C (1399)

Sakamoto, H. [1949], On the criteria of the independence and the degrees of freedom of statistics and their applications to the analysis of variance. Ann. Inst. Stat. Math. 1:109-122. E2,7a;T2a (1400)

Salisbury, G. W. [1944], A controlled experiment in feeding wheat germ oil as a supplement to the normal ration of bulls used for artificial insemination. J. Dairy Sci. 27:551-562. Elae (1401)

Salisbury, G. W., Willett, E. L., and Seligman, J. [1942], The effect of the method of making semen smears upon the number of morphologically abnormal spermatozoa. J. Animal Sci. 1:199-205. E2,12ae (1402)

Salmon, C. [1913], A practical method of reducing experimental error in varietal tests. J. American Soc. Agron. 5:182-184. A14 (1403)

Salmon, C. [1914], Check plats - a source of error in varietal tests. J. American Soc. Agron. 6:128-131. A14;C (1404)

Salmon, S. C. [1923], Some limitations in the application of the method of least squares to field experiments. J. American Soc. Agron. 15:225-239. S1;C (1405)

Salmon, S. C. [1924], Some misapplications and limitations in using Student's method to interpret field experiments. J. American Soc. Agron. 16:717-721. C (1406)

Salmon, S. C. [1929], Why we believe. J. American Soc. Agron. 21:854-859. C (1407)

Salmon, S. C. [1929], <u>Principles of Agronomic Experimentation.</u> Unpublished lectures, Kansas State College. C (1408)

Salmon, S. C. [1930], The statistical method: A reply. J. American Soc. Agron. 22:270-271. C (1409)

Salmon, S. C. [1938], Unbalanced arrangements of plats in latin squares. J. American Soc. Agron. 30:947-950. E5a;C (1410)

Salmon, S. C. [1938], Generalized standard errors for evaluating bunt experiments with wheat. J. American Soc. Agron. 30:647-663. C (1411)

Salmon, S. C. [1940], The use of modern statistical methods in field experiments. J. American Soc. Agron. 32:308-320. C (1412)

Salmon, S. C. [1945], Techniques and progress in breeding wheat. Report Fifth Hard Red Winter Wheat Improvement Conf., Manhattan, Kansas, Feb. 12,13 and 14, 1945. Div. Cereal Crops Dis., Plant Ind. Sta., Beltsville, Md. pp. 14-15. (Mimeo.) C (1413)

Salvekar, P. M. [1945], Minimum size of sample required for experimental work to estimate the population mean with a specified degree of accuracy for a specified level of reliability. J. Univ. Bombay A 13(5):2-6. S1 (1414)

Sanders, H. G. [1930], A note on the value of uniformity trials for subsequent experiments. J. Agri. Sci. 20:63-73. A14 (1415)

Sant, G. K. [1940], Some features in the analysis of covariance with split-plot designs. Proc. Twenty-Seventh Indian Sci. Congress, p. 231. E12a;A7 (1416)

Satterthwaite, F. E. [1941], Synthesis of variance. Psychometrica 6:309-316. A11 (1417)

Satterthwaite, F. E. [1946], An approximate distribution of estimates of variance components. Biometrics 2:110-114. A11 (1418)

Saunders, A. R. [1935], Statistical methods with special reference to field experiments. Sci. Bull. Dept. Agri. S. Africa 147:1-76. E2,7a;T2a;B;C (1419)

Saunders, A. R. [1939], Statistical methods with special reference to field experiments. Sci. Bull. Dept. Agri. S. Africa 2nd ed. 200:1-112. E2,3,7a;T2a;B;C (1420)

Saunders, A. R. [1944], Efficiency of design in field experiment at Potchefstroom, South Africa. Empire J. Expt. Agri. 12:157-162. E2,3,11,14a (1421)

Sauvage, P. J. [1938], A field trial of three popular green manure crops in Trinidad. Tropical Agri. 15:114-115. E2ae (1422)

Savur, S. R. [1939], A note on the arrangement of incomplete blocks, when k=3 and λ=1. Ann. Eugenics 9:45-49. E2c (1423)

Sayer, W. and Krishna Iyer, P. V. [1936], On some of the factors that influence the error of field experiments with special reference to sugarcane. Indian J. Agri. Sci. 6:917-929. A14;S1 (1424)

Sayer, W., Vaidyanathan, M., and Subramonia Iyer, S. [1936], Ideal size and shape of sugarcane experimental plots based upon tonnage experiments with Co.205 and Co.213 conducted in Pusa. Indian J. Agri. Sci. 6:684-714. A14;S1 (1425)

Schad, C., Meneret, G., and Mayer, R. [1941], Application des méthodes Fisher et Student à la comparison des variétés en petites parcelles. Ann.Phytogénét. Paris 7:91-107. C (1426)

Scheffé, H. [1943], On solutions of the Behrens-Fisher problem, based on the t-distribution. AMS 14:35-44. A11 (1427)

Scheffé, H. [1944], A note on the Behrens-Fisher problem. AMS 15:430-432. A11 (1428)

Scheffé, H. and Tukey, J. W. [1944], A formula for sample sizes for population tolerance limits. AMS 15:217. S1 (1429)

Schmidt, J. [1919], La valeur de l'individu à titre de générateur appréciée suivant la méthode du croisement diallèle. Comp. Rend. Lab. Carlsberg 14, 6:1-33. T9ac (1430)

Schreiner, O. and Skinner, J. J. [1918], The triangle system of fertilizer experiments. J. American Soc. Agron. 10:225-246. T7ac (1431)

Schrek, R. and Lipson, H. I. [1941], Logarithmic frequency distributions. Human Biology 13:1-22. A9 (1432)

Schild, H. D. [1942], A method of conducting a biological assay on a preparation giving repeated graded responses. J. Physiology 101:115-130. T8a (1433)

Schroeder, C. H. and Lawrence, H. B. [1932], The number of chicks required to demonstrate the significance of growth differences. Poultry Sci. 11:208-218. C;S1 (1434)

Schumacher, F. X. [1945], Statistical method in forestry. Biometrics 1:29-32. E2,12a;A14 (1435)

Schumacher, F. X. and Bull, H. [1932], Determination of the errors of estimate of a forest survey, with special reference to the bottom-land hardwood forest region. J. Agri. Res. 45:741-756. E5a;A14;C (1436)

Schumacher, F. X. and Meyer, H. A. [1937], Effect of climate on timber-growth fluctuations. J. Agri. Res. 54:79-107. E5, 8a;A7 (1437)

Schuster, G. L. [1929], Methods of research in pasture investigations. J. American Soc. Agron. 21:666-673. C;S1 (1438)

Schützenberger, M. P. [1949], A non-existence theorem for an infinite family of symmetrical block designs. Ann. Eugenics 14:286-287. E2c (1439)

Scossiroli, R. [1949], Experimental designs and statistical methods for agricultural tests (Italian). Ann. Sper. Agron. N.S. 3:547-584. C (1440)

Seath, D. M. [1944], A 2×2 factorial design for double reversal feeding experiments. J. Dairy Sci. 27:159-164. E7ae;T2ae (1441)

Sebelien, J. [1920], Modern methods for experiments with fertilizers and manures. J. Agri. Sci. 10:415-419. E5a;C (1442)

Sen, H. D. [1940], Conversion of cane molasses into manure by the biological method and the results of the cropping tests with the manures prepared (1938-39). Indian J. Agri. Sci. 10:172-191. E12ae (1443)

Sen Gupta, J. M. [1948], Crop cutting experiment on sugarcane in a farm cultivation. Sankhyā 9:47-50. A14 (1444)

Shaw, B. T. [1945], Long-time crop and fertilizer rotations. Soil Sci. Soc. American Proc. 10:300-305. E5,8a;C (1445)

Shaw, F. J. F. [1936], A handbook of statistics for use in plant breeding and agricultural problems. Imp. Coun. Agri. Res. Delhi. C (1446)

Shen, E. [1940], Experimental design and statistical treatment in educational research. J. Exptl. Educ. 8:346-353. C (1447)

Shen, Li-Ying, [1934], Statistical analysis of a blank test of rice with suggestions for field technique. Agricultura Sinica 1:107-138. A14 (1448)

Shen, T. H. [1930], Field technic for determining comparative yields in wheat under different environmental conditions in China. J. American Soc. Agron. 22:193-215. A14 (1449)

Shewhart, W. A. [1926], Correction of data for errors of measurement. Bell System Tech. J. 5:11-26. A13,14;S1 (1450)

Shewhart, W. A. [1926], Correction of data for errors of averages obtained from small samples. Bell System Tech. J. 5:308-319. A14;C (1451)

Shewhart, W. A. and Winters, F. W. [1928], Small samples--new experimental results. JASA 23:144-153. A14;C (1452)

Shorrock, R. W. [1940], Co-ordinated experiments in pig husbandry. Empire J. Exptl. Agri. 8:159-167. E12a;C (1453)

Siao, F. [1935], Uniformity trials with cotton. J. American Soc. Agron. 27: 974-979. E2,7a;A14;C (1454)

Siao, F. and Cheng, C. C. [1942], A study on the efficiency of quasifactorial design (Chinese). Kwangsi Agri. 3:371-384. E2,3,14a;A14 (1455)

Sillitto, G. P. [1948], The numbers of observations needed in experiments leading to a t-test of significance of a mean or the difference of two means. Research 1:520-525. Sl (1456)

Simlote, K. M. [1947], An application of discriminant function for selection in durum wheats. Indian Agri. Sci. 17:269-280. A2,7 (1457)

Simon, J. [1932], Principles and methods for the treatment of the results of field trials (Czech.). Proc. Inst. Agron. Res. Rep. Czech. 84:1-114. C (1458)

Simpson, G. G. [1945], Note on graphic biometric comparison of samples. American Nat. 79:95-96. C (1459)

Simpson, T. W. [1931], Studies in sampling technique: Cereal experiments. II. A small scale threshing and winnowing machine. J. Agri. Sci. 21:372-375. Al4;C (1460)

Simpson, T. W [1938], Experimental methods and human nutrition. JRSSB 5:46-69. E7,8a;C (1461)

Simpson, T. W. and Wood, E. C. [1935], The value of certain supplements to the diet of children. Medical Officer 53:125-127, 135-138. E5,7,8a (1462)

Singer, J. [1938], A theorem in finite projective geometry and some applications to number theory. Trans. American Math. Soc. 43:377-385. E7,12,13c (1463)

Singh, A. [1943], Sampling of sugarcane for chemical analyses, III. Indian J. Agri. Sci. 13:547-561. Al4 (1464)

Singh, B. N. and Chalam, G. V. [1937], A quantitative analysis of the weed flora on arable land. J. Ecology 25:213-221. Al4 (1465)

Singh, B. N. and Das, K. [1938], Distribution of weed species on arable land. J. Ecology 26:455-466. Al4 (1466)

Singh, B. N. and Das, K. [1939], Percentage frequency and quadrat size in analytical studies of weed flora. J. Ecology 27:66-77. Al4 (1467)

Singh, R. N. [1942], Control of the wooly aphis (Eriosoma lanigerum Hausmann) by spraying and other methods. Indian J. Agri. Sci. 12:588-602. E2ae;T2ae (1468)

Sites, J. W. and Reitz, H. J. [1949], The variation in individual Valencia oranges from different locations of the tree as a guide to sampling methods and spot-picking for quality I. Soluble solids in the juice. Proc. American Soc. Hort. Sci. 54:1-10. Al4;C (1469)

Skuderna, A. W. and Doxtator, C. W. [1940], Comparison of quasi-factorial and randomized block designs for testing sugar beet varieties. Proc. American Soc. Sugar Beet Tech. 2:116-118. E2,14a;A14 (1470)

Smith, A.D.B. and Donald, H.P. [1937], Weaning weight of pigs and litter sampling with reference to litter size. J. Agri. Sci. 27:485-502. C (1471)

Smith, B. B. and Kendall, M. G. [1940], On the method of paired comparisons. Biometrika 31:324-345. T11ac (1472)

Smith, C. A. B. and Hartley, H. O. [1948], Construction of Youden squares. JRSSB 10:262-263. E7c (1473)

Smith, G. E. [1942], Sanborn field. Fifty years of field experiments with crop rotations, manure and fertilizer. Mo. Agri. Expt. Bull. 458:1-61. E5,8a;C (1474)

Smith, H. F. [1928], A biometrical study of rod row plats of timothy. M. S. Thesis, Cornell Univ. A14 (1475)

Smith, H. F. [1936], Comparison of agricultural and nursery plots in variety experiments. J. Australian Council Sci. Ind. Res. 9:207-210. A14 (1476)

Smith, H. F. [1936], The problem of comparing the results of two experiments with unequal errors. J. Australian Council Sci. Ind. Res. 9:211-212. A11 (1477)

Smith, H. F. [1937], The variability of plant density in fields of wheat and its effect on yield. Council Australian Sci. Ind. Res. Bull. 109. A14 (1478)

Smith, H. F. [1938], An empirical law describing heterogeneity in the yields of agricultural crops. J. Agri. Sci. 28:1-23. A14 (1479)

Smith, H. F. [1939], The effect of spacing and time of sowing on yield and yield components of wheat varieties. Australian Council Sci. Ind. Res. Pamphlet No. 91. A14 (1480)

Smith, H. F. [1949], Replicated experiments. Forestry 23:56-58. C (1481)

Smith, H. F. and Myers, C. H. [1934] A biometrical analysis of yield trials with timothy varieties using rod rows. J. American Soc, Agron. 26:117-128. E5,8a;A14;C (1482)

Smith, K. W., Marks, H. P., Fieller, E. C., and Broom, W. A. [1944], An extended cross-over design and its use in insulin assay. Quarterly J. Pharmacology 17:108-117. E7a;T8a (1483)

Smith, L. H. [1909], Plot arrangement for variety experiments with corn. Proc. American Soc. Agron. 1:84-89. A14;C (1484)

Snedecor, G. W. [1925], Experiments with small numbers of observations. Proc. Iowa Academy Sci. 32:367-369. A11,14 (1485)

Snedecor, G. W. [1930] , A statistical test of experimental technique. Proc. Iowa Academy Sci. 37:279-287. C (1486)

Snedecor, G. W. [1934] , Calculation and Interpretation of Analysis of Variance and Covariance. Collegiate Press, Inc., Ames, Iowa. E2,7ae;T2ae;A7 (1487)

Snedecor, G. W. [1934] , The method of expected numbers for tables of multiple classification with disproportionate subclass numbers. JASA 29:389-393. T2ae;A10 (1488)

Snedecor, G. W. [1934] , Biological variation versus errors in measurement. Science 80:246-247. C (1489)

Snedecor, G. W. [1936] , The improvement of statistical techniques in biology. JASA 31:690-701. E2,12ae;T2a;C (1490)

Snedecor, G. W. [1946] , Statistical Methods. 4th Edition (first ed., 1937), Iowa State College Press, Ames, Iowa. E1,2,7,12ae;T2ae; A7,10 (1491)

Snedecor, G. W. [1947], An experiment in the collection of morbidity and mortality data on farm animals. Proc. U.S. Livestock Sanitary Assoc., 51st Annual Meeting. C (1492)

Snedecor, G. W. [1947], The utilization of experimental data for improving design. Proc. Internat'l Stat. Inst. C (1493)

Snedecor, G. W. [1948], Query 57. Biometrics 4:132-134. E2ae (1494)

Snedecor, G. W. [1948], Query 59. Biometrics 4:211-213. E3ae;T2ae (1495)

Snedecor, G. W. [1948], Some principles of experimental design. Proc. Auburn Conf. Stat. Applied to Soc. Sci., Plant Sci., Animal Sci. pp. 47-51. C (1496)

Snedecor, G. W. [1949], Query 70. Biometrics 5:250-251. E1ae (1497)

Snedecor, G. W. and Breneman, W. R. [1945], A factorial experiment to learn the effects of four androgens injected into male chicks. Iowa State College J. Sci. 19:333-342. T2a;A10 (1498)

Snedecor, G. W. and Cox, G. M. [1935], Disproportionate subclass numbers in tables of multiple classification. Iowa Agri. Expt. Sta. Res. Bull. 180: 233-272. E2ae;T2ae;A10 (1499)

Snedecor, G. W. and Cox, G. M. [1937], Analysis of covariance of yield and time to first silks in maize. J. Agri. Res. 54:449-459. A7 (1500)

Snedecor, G. W. and Culbertson, C. C. [1932], An improved design for experiments with groups of animals whose outcome may be estimated. Proc. American Soc. Animal Production 25:25-31. E7a (1501)

Snedecor, G. W. and Haber, E. S. [1946], Statistical methods for an incomplete experiment on a perennial crop. Biometrics 2:61-67. E2,5,12ae (1502)

Snedecor, G. W. and King, A. J. [1942], Recent developments in sampling for agricultural statistics. JASA 37:95-102. A14 (1503)

Sophister, [1928], Discussion of small samples drawn from an infinite skew population. Biometrika 20A:389-423. A9,14 (1504)

Sosnin, A. V. [1946], (Assessing methods of determining yield on breeding plots. Selekcija i Semenovodstva (Breeding and Seed Growing) 13: Nos. 11-12:37-39. A14 (1505)

Soucek, J. [1931], Calculation of field experiments. Zeitschr. Zuckerind. Czech. Rep. 55:351-357. C (1506)

Spiller, D. [1948], Truncated log-normal and root-normal frequency distributions of insect populations. Nature 162:530-531. T8a;A9 (1507)

Spillman, W. J. [1922], A plan for the conduct of fertilizer experiments. J. American Soc. Agron. 13:304-310. T2ce (1508)

Spillman, W. J [1933], Use of the exponential yield curve in fertilizer experiments. USDA Tech. Bull. 348:1-66. T4a;A7 (1509)

Sprague, G. F. [1935], Random sampling and the distribution of phenotypes on ears of backcrossed maize. J. Agri. Res. 51:751-758. T9a (1510)

Sprague, G. F. [1939], An estimation of the number of top-crossed plants required for adequate representation of a corn variety. J. American Soc. Agron. 31:11-16. T9a;S1 (1511)

Sprague, G. F. and Tatum, L. A. [1942], General vs. specific combining ability in single crosses of corn. J. American Soc. Agron. 34:923-932. T9a (1512)

Sreenivasan, P. S. [1943], Studies on the estimation of growth and yield of <u>jowar</u> by sampling. Indian J. Agri. Sci. 13:399-412. A14 (1513)

Stadler, L. J. [1921], Experiments in field plot technique for the preliminary determination of comparative yields in small grains. Mo. Agri. Expt. Sta. Res. Bull. 49. A14;S1 (1514)

Stadler, L. J. [1929], Experimental error in field plot tests. Proc. Int. Congress Plant Sci. 1:107-127. C (1515)

Stapledon, R. G. [1931], The technique of grassland experiments. Rothamsted Conf. 13:22-28. C (1516)

Starr, D. F. [1944], The theory of probits at high mortalities. J. Economic Entomology 37:850. T8a (1517)

State Agricultural Experimental Institute of Sweden. [1939], Handledning i försöksteknik. (The technique of field experiments.) Medd. Lantbrukshögskolan Jordbruksförsöksanstalten, Norrtälje 1:1-207. E2,3a;T2a;C (1518)

Steece, H. M. [1940], Agronomic research projects. J. American Soc. Agron. 32:135-140. C (1519)

Stein, C. [1945], A two-sample test for a linear hypothesis whose power is independent of the variance. AMS 16:243-258. S1,2 (1520)

Stein, C. [1948], On sequences of experiments (abstract). AMS 19:117-118. S2 (1521)

Steiner, J. [1853], Combinatorische Aufgabe. J. Reine Angewandte Math. 45:181-182. E7,12,13c (1522)

Stephens, J. C. and Vinall, H. N. [1928], Experimental methods and the probable error in field experiments with sorghum. J. Agri. Res. 37:629-646. A14 (1523)

Stephens, S. G. [1942], Yield characters of selected oat varieties in relation to cereal breeding technique. J. Agri. Sci. 32:217-254. A14 (1524)

Stevens, O. A. [1918], Variations in seed tests resulting from errors in sampling. J. American Soc. Agron. 10:1-19. C (1525)

Stevens, W. L. [1937], Significance of grouping. Ann. Eugenics 8:57-69. A7;C (1526)

Stevens, W. L. [1937], A test of uniovular twins in mice. Ann. Eugenics 8:70-73. A7;C (1527)

Stevens, W. L. [1939], The completely orthogonalized Latin square. Ann. Eugenics 9:82-93. E7,12,13c (1528)

Stevens, W. L. [1940], On the interpretation of the data of certain experiments in paranormal cognition. Proc. Soc. Psych. Res. 46:256-260. C (1529)

Stevens, W. L. [1948], Statistical analysis of a non-orthogonal tri-factorial experiment. Biometrika 35:346-367. T2ae;A10 (1530)

Stevenson, T. M. [1923], Technique of field husbandry experimentation. Sci. Agri. 4: 41-54. C (1531)

Stewart, A. B. [1947], Memoranda on colonial fertilizer experiments. II. Planning and conduct of fertilizer experiments. Colon. Office Memo. 214:3-9. C (1532)

Stewart, F. C. [1919], Missing hills in potato fields: Their effect upon yield. N. Y. State Agri. Expt. Bull. 459:45-69. A14 (1533)

Stewart, F. C. [1921], Further studies on the effect of missing hills in potato fields and on the variation in the yield of potato plants from halves of the same seed tuber. N.Y. (Geneva) Agri. Expt. Sta. Bull. 489:1-52. A14 (1534)

Stewart, G. and Hutchings, S. S. [1936], The point-observation-plot (square-foot density) method of vegetation survey. J. American Soc. Agron. 28:714-722. C (1535)

Stewart, G., Talbot, M. W., and Hurtt, L. C. [1936], A tentative recommendation of technic for grazing experiments on range pastures in arid or semi-arid regions. J. American Soc. Agron. 28:81-83. C (1536)

Stoffels, A. [1940], De berekening van middelbare fouten bij niet-homogene proefvelden. (The calculation of the mean errors on non-homogeneous field plots.) Landbouwk. Tijdschr., Wageningen 52:165-174. C (1537)

Stoffels, A. [1947], The treatment of field experiments by Knut Vik's method (Dutch). Statistica 1:209-218. E7,12a (1538)

Stone, E. J, [1868, 1873], On the rejection of discordant observations. Monthly Notices Royal Astronomical Soc. 28:165-168, 34:9-15. A6 (1539)

Strickland, A. G. [1935], Error in horticultural experiments. J. Dept. Agri. Victoria 33:408-416. C (1540)

Strickland, A. G., Forster, H. C., and Vasey, A. J. [1932], A vine uniformity trial. J. Dept. Agri. Victoria 30:584-593. A14 (1541)

Stringfield, G. H. [1927], Intervarietal competition among small grains. J. American Soc. Agron. 19:971-983. A14 (1542)

Stringfield, G. H. [1928], Types of field and plat in crop tests. J. American Soc. Agron. 20:1073-1096. A14 (1543)

Stringfield, G. H. and Thatcher, L. E. [1947], Stands and methods of planting for corn hybrids. J. American Soc. Agron. 39:995-1010. A14 (1544)

Student, [1908], The probable error of a mean. Biometrika 6:1-25. C (1545)

Student, [1911], Note on the method of arranging plots so as to utilize a given area of land to the best advantage in testing two varieties. J. Agri. Sci. 4:128-132. C (1546)

Student, [1923], On testing varieties of cereals. Biometrika 15:271-293. E5a;C (1547)

Student, [1926], Mathematics and agronomy. J. American Soc. Agron. 18:703-719. E7ae;A14;S1 (1548)

Student, [1927], Errors of routine analysis. Biometrika 19:151-164. C (1549)

Student, [1930], Agricultural field experiments. Nature 126:843. E7a;C (1550)

Student, [1931], Yield trials. Baillière's Encyclopaedia Sci. Agri., edited by H. Hunter, London, p. 1342. E2,7a;C (1551)

Student, [1936], The half-drill strip system agricultural experiments. Nature 138:971-972. E7a;C (1552)

Student (W.S. Gosset) [1936], Co-operation in large-scale experiments. JRSSB 3:115-136. E7a;C (1553)

Student, [1937], Comparison between balanced and random arrangements of field plots. Biometrika 29:363-379. E5,7a;A14;C (1554)

Subbaiya, M. [1939], Sampling in sugar cane experimental work (abstract). Proc. Twenty-Sixth Indian Sci. Congress, Lahore, Part III, pp. 197-198. E7a;A14 (1555)

Subramonia Iyer, S. [1940], A supplementary note on the analysis of 3^3 and 3^4 designs (with three-factor interactions confounded) in field experiments in agriculture. Indian J. Agri. Sci. 10:691-692. T2a (1556)

Sukhatme, B. V. [1949], Random association of points on a lattice. JISAS 2:60-85. A14;C (1557)

Sukhatme, P. V [1946], Bias in the use of small-size plots in sample surveys for yield. Current Sci. 15:119-120. A14 (1558)

Sukhatme, P. V. [1946], Bias in the use of small-size plots in sample surveys. Nature 157:630. A14 (1559)

Sukhatme, P. V. [1947], The problem of plot size in large-scale yield surveys. JASA 42:297-310. A14 (1560)

Sukhatme, P. V. [1947], Use of small-size plots in yield surveys. Nature 160:542. A14 (1561)

Sukhatme, P. V. and Panse, V. G. [1943], Size of experiments for testing sera or vaccines. Indian J. Vet. Sci. Animal Husbandry 13:75-82. S1 (1562)

Summerby, R. [1923], Replication in relation to accuracy in comparative crop tests. J. American Soc. Agron. 15:192-199. C;S1 (1563)

Summerby, R. [1925], A study of size of plats, numbers of replications, and the frequency and methods of using check plats, in relation to accuracy in field experiments. J. American Soc. Agron. 17:140-150. A14;C;S1 (1564)

Summerby, R. [1934], The value of preliminary uniformity trials in increasing the precision of field experiments. Macdonald Coll. Tech. Bull. 15:1-64. A14;C (1565)

Summerby, R. [1937], The use of the analysis of variance in soil and fertilizer experiments with a particular reference to interactions. Sci. Agri. 17:302-311. T2a;C;S1 (1566)

Suneson, C. A. [1949], Survival of four barley varieties in a mixture. Agron. J. 41:459-461. A14 (1567)

Surface, F. M. and Pearl, R. [1916], A method of correcting for soil heterogeneity in variety tests. J. Agri. Res. 5:1039-1050. A14;C (1568)

Swanson, A. F. [1930], Variability of grain sorghum yields as influenced by size shape, and number of plats. J. American Soc. Agron. 22:833-838. A14;S1 (1569)

Swed, F. S. and Eisenhart, C. [1943], Tables for testing randomness of grouping in a sequence of alternatives. AMS 14:66-87. E5a;A14;C (1570)

Swineford, F. [1946], Graphical and tabular aids for determining sample size when planning experiments which involve comparisons of percentages. Psychometrika 11:43-49. S1 (1571)

Sylvester, J. J. [1867], Thoughts on inverse orthogonal matrices, simultaneous sign successions and tesselated pavements in two or more colours, with applications to Newton's rule, ornamental tile work and the theory of numbers. Phil. Mag. 34:461-475. T7c (1572)

Sylvester, J. J. [1893], Note on a nine school-girls' problem. Mess. Math. 22:159-160 (correction 192). E2c (1573)

Syzrantsev, P. I. [1949], (The mathematical treatment of a small number of experimental data.) Socialistic Grain Farming, Saratov 2:163-182. C (1574)

Szegö, G. [1939], Orthogonal polynomials. American Math. Soc. Colloquium Pub. 23:101-104. A7 (1575)

Tang, P. C. [1938], The power function of the analysis of variance tests with tables and illustrations of their use. Stat. Res. Mem. 2:126-149. S1 (1576)

Tang, Y. [1938], Certain statistical problems arising in plant breeding. Biometrika 30:29-56. T9a;C (1577)

Tarry, G. [1899], Sur le probléme d'Euler des n^2 officiers. L'Intermédiaire des Mathématiciens 6:251-252. E7,12c (1578)

Tarry, G. [1900-1901], Le probléme des 36 officiers. C. R Assoc. Franç. Avance. Sci. Nat. 1:122-123 and 2:170-203. E7,12c (1579)

Tartaglia, [1556], Trattato de Numeri e Misure. Venice. T6c (1580)

Taylor, F. W. [1908], The size of experiment plots for field crops. Proc. American Soc. Agron. 1:56-58. A14 (1581)

Taylor, H. L [1948], An examination of the effect of plot shape on experimental error. M. S. Thesis, Iowa State College. A14 (1582)

Taylor, J. [1948], Errors of treatment comparisons when observations are missing. Nature 162:262-263. E6,10a (1583)

Taylor, J. [1949], A valid restriction of randomization for certain field experiments. J. Agri. Sci. 39:303-308. E2,5a (1584)

Taylor, S. A. [1949], Soil air-plant growth relationships with emphasis on means of charaterizing soil aeration. Ph. D. Thesis, Cornell Univ. E2,7ae (1585)

Taylor, W. S. [1942], The use of interactions in analysing psychological data. British J. Psychology 32:248-258. T2a (1586)

Tedin, O. [1931], The influence of systematic plot arrangement upon the estimate of error in field experiments. J. Agri. Sci. 21:191-208. E5,7a;A14 (1587)

Tedin, O. [1936], Einige Gesichtspunkte zur Benutzung der Statistik im Versuchswesen. Der Züchter 8:185-187. T2a;C (1588)

Tedin, O. [1938], Systematisk eller slumpmässig parcellfördelning vid fältförsök. (Systematic or random arrangement of plots in field experiments.) Nord. JordbrForskn. 8:137-148. E5,7a;C (1589)

Tedin, O. [1948], Intryck från amerikansk växförädling. (An impression of American plant breeding.) Sverig. Utsädesfören. Tidskr. 58:277-287. T9a;C (1590)

Telegdy Kováts, L. [1935], A szabadföldi kísérletezés modern módszerei Angliában. (The modern methods of field experiment in Great Britain.) Mezögazdas. Kutatás. 8:361-374. C (1591)

Telegdy Kováts, L. [1937], Matematikai módszerek a tudományos kutatás szolgálatában. 2. Egyes, hiányzó parcellák valószínü termésének kiszámítása a modern szabadföldi kísérleteknél. (Mathematical methods in the service of scientific research. 2. Estimation of the probable yield of single missing plots in modern field experiments.) Mezögazdas. Kutatás. 10:1-7. E2,6,7,10ae (1592)

Telegdy Kováts, L. [1943], The principles of modern field experimentation technique. Rep. Hung. Agri. Expt. Sta. 46:183-195. C (1593)

Terman, G. L. and Freeman, J. F. [1948], Interpretation of yield data from a long-time soil fertility experiment. J. American Soc. Agron. 40:874-884. E5,8a;A7 (1594)

Terrill, C. E., Pohle, E. M., Emik, L. O., and Hazel, L. N. [1945], Estimation of clean-fleece weight from grease-fleece weight and staple length. J. Agri. Res. 70:1-10. T2a;A14 (1595)

Tharp, W. H., Wadleigh, C. H., and Barker, H. D. [1941], Some problems in handling and interpreting plant disease data in complex factorial designs. Phytopathology 31:26-48. E2ae;T2ae;C (1596)

Thomas, W. D., Robertson, D. W., et al. [1949], General report on pinto bean investigation, 1948. Colorado Agri. Expt. Sta. Misc. Series. Art. No. 435. E2,12ae;T2ae (1597)

Thompson, C. M. and Merrington, M. [1946], Tables for testing the homogeneity of a set of estimated variances. Biometrika 33:296-304. A11 (1598)

Thompson, R. C. [1934], Size, shape, and orientation of plots and number of replications required in sweet potato field-plot experiments. J. Agri. Res. 48:379-399. A14;S1 (1599)

Thompson, W. R. [1933], On the likelihood that one unknown probability exceeds another in view of the evidence of two samples. Biometrika 25:285-294. S2 (1600)

Thompson, W. R. [1939], Biological control and the theories of the interactions of populations. Parasitology 31:299-388. T2a (1601)

Thompson, W. R. [1948], On the use of parallel or non-parallel systems of transformed curves in bio-assay: illustration in the quantitative complement-fixation test. Biometrics 4:197-208. T8a;C (1602)

Thompson, W. R. [1949], Statistical methods for evaluation of diagnostic and other procedures: An objective weeding-out process applicable to material used in surveys of diagnostic variability. Human Biology 21:17-34. C (1603)

Thomson, G. H. [1941], The use of the Latin square in designing educational experiments. British J. Educational Psychology 11:135-137. E7ae (1604)

Thorne, C. E. [1908], The interpretation of field experiments. Proc. American Soc. Agron. 1:45-55. C (1605)

Thorne, C. E. [1909], Essentials of successful field experimentation. Ohio Agri. Expt. Sta. Circular 96:1-38. A14;C (1606)

Thornton, H. G. [1929], The influence of the number of nodule bacteria applied to the seed upon nodule formation in legumes. J. Agri. Sci. 19:373-381. E2ae (1607)

Thorold, C. A. [1946], A study of yields, preparation out-turns, and quality in arabica coffee. Part I. Yields. Empire J. Exptl. Agri. 15:96-106. C;S1 (1608)

Thurstone, L. L. [1927], The method of paired comparisons for social values. J. Abnormal Soc. Psychology 21:384-400. T11ac (1609)

Ting, Y. and Sie, H. T. [1935], The field technique of the experimental method and its error in rice (in Chinese). J. Agri. Assoc. China 142-143:13-48. E5a;A14;C (1610)

Tinney, F. W., Aamodt, O. S., and Ahlgren, H. L. [1937], Preliminary report of a study on methods used in botanical analyses of pasture swards. J. American Soc. Agron. 29:835-840. A14 (1611)

Tippett, L. H. C. [1926], On the effect of sunshine on wheat yield at Rothamsted. J. Agri. Sci. 16:159-165. A7 (1612)

Tippett, L. H. C. [1931], The Methods of Statistics. An Introduction Mainly for Workers in the Biological Sciences. Williams and Norgate, Ltd., London. C (1613)

Tippett, L. H. C. [1935], Some applications of statistical methods to the study of variation of quality of cotton yarn. JRSSB 2:27-62. E7,12,13a (1614)

Tippett, L. H. C. [1936], Applications of statistical methods to the control of quality in industrial production. Trans. Manchester Stat. Soc. pp. 1-32. E7,12,13a (1615)

Tippett, L. H. C. [1937], The Methods of Statistics. An Introduction Mainly for Experimentalists. Williams and Norgate Ltd., 2nd Edition, London (1st ed. 1931). E2,7a;T2a;C (1616)

Titus, H. W. [1934], Statistical method in planning and interpreting animal nutrition experiments. Poultry Sci. 13:358-359. C (1617)

Titus, H. W. [1935], The z-test in covariance analysis. Poultry Sci. 14:291-293. A7;C (1618)

Titus, H. W. and Hammond, J. C. [1935], A method of analyzing the data of chick nutrition experiments. Poultry Sci. 14:164-173. A10 (1619)

Titus, H. W. and Harshaw, H. M. [1935], An adaption of the "randomized block" to nutrition experiments. Poultry Sci. 14:3-15. E2ae;A7 (1620)

Titus, H. W. and Nestler, R. B. [1935], Effect of vitamin D on production and some properties of eggs. Poultry Sci. 14:90-98. E2a;C (1621)

Tocher, J. F. [1928], An investigation of the milk yield of dairy cows. Biometrika 20B:105-244. C (1622)

Tocher, K. D. [1949], A note on the analysis of grouped probit data. Biometrika 36:9-17. T8a (1623)

Todd, H. [1940], A note on random associations in a square point lattice. JRSSB 7:78-82. E5a;A14;C (1624)

Todd, J. A. [1933], A combinatorial problem. J. Math. Physics 12:321-333. E2c (1625)

Tolman, B. [1942], Multiple versus single-factor experiments. Proc. American Soc. Sugar Beet Tech. pp. 170-180. T2a (1626)

Torrie, J. H. and Dickson, J. G. [1943], The use of statistical methods in quality evaluation of barley and malt data. Cereal Chemistry 20:579-594. T2a (1627)

Torrie, J. H., Shands, H. L., and Leith, B. D. [1943], Efficiency studies of types of design with small grain yield trials. J. American Soc. Agron. 35: 645-661. E3,14a;A14 (1628)

True, A. C. [1937], A history of agricultural experimentation and research in the United States, 1607-1925. U.S.D.A. Miscel. Pub. 251. C (1629)

Trumble, H. C. and Cornish, E. A. [1936], The influence of rainfall on the yield of a natural pasture. J. Australian Council Sci. Ind. Res. 9:19-28. A7 (1630)

Tsai, H. and Chow, C. Y. [1943], Studies on field plot technique in wheat. Chinese J. Sci. Agri. 1:117-118. E5a;A14;S1 (1631)

Tsao, F. [1942], Tests of statistical hypotheses in the case of unequal or disproportionate numbers of observations in the subclasses. Psychometrika 7:195-212. A10 (1632)

Tsao, F. [1946], General solution of the analysis of variance and covariance in the case of unequal or disproportionate numbers of observations in the subclasses. Psychometrika 11:107-128. A10 (1633)

Tsao, F. and Johnson, P. O. [1946], A note on solutions of analysis of variance for the problem of unequal or disproportionate subclass numbers. J. Exptl. Education 14:253-262. A10 (1634)

Tsoumis, J. [1948], The natural conditions and the environment of agricultural experimentation in Greece (in Greek). Georgikon Deltion (Agri. Bull.) Athens pp. 113-120. C (1635)

Tukey, J. W. [1949], Dyadic anova, an analysis of variance for vectors. Human Biology 21:65-110. A7 (1636)

Tukey, J. W. [1949], Comparing individual means in the analysis of variance. Biometrics 5:99-114. A3 (1637)

Tukey, J. W. [1949], One degree of freedom for non-additivity. Biometrics 5: 232-242. A4 (1638)

Tukey, J. W. [1949], Interaction in a row-by-column design. Statistical Res. Group Memorandum Report 18, Princeton University pp. 1-14. E7a (1639)

Turner, F. C. [1944], A model for quantitative statistical experiments. Engineer 178:26-27. C (1640)

Turner, P. E. [1932], Cultivation experiments with sugar-cane. Tropical Agri. 9:120-126. E2ae (1641)

Turner, P. E. [1932], Manurial experiments with sugar-cane. I. Tropical Agri. 9:153-155. E7ae (1642)

Turner, P. E. [1932], Manurial experiments with sugar-cane. II. Tropical Agri. 9:177-184. E7ae (1643)

Turner, P. E. [1932], Manurial experiments with sugar-cane. III. Tropical Agri. 9:206-210. E2ae (1644)

Turner, P. E. [1933], Manurial experiments on sugar-cane. Tropical Agri. 10:60-67. E7ae (1645)

Turner, P. E. [1935], Recent investigations on sugar-cane and sugar-cane soils in Trinidad II. Tropical Agri. 12:293-302, 320-332. E7ae;T2ae (1646)

Turner, P. E. and Potter, J. A. [1932], Manurial experiments with sugar cane IV. Tropical Agri. 9:44-53. E7ae (1647)

Turner, W. D. [1940], A method for the analysis and interpretation of intragroup changes in measurements. J. General Psychology 23:343-365. C (1648)

Tysdal, H. M. [1935], An analysis of soil and seasonal effects in alfalfa variety tests. J. American Soc. Agron. 27:384-391. E2ae (1649)

Tysdal, H. M. and Kiesselbach, T. A. [1939], Alfalfa nursery technic. J. American Soc. Agron. 31:83-98. A14 (1650)

Uchytil, J. [1946], Zjistováni a adrzováni nejvyhodnejsich podmínek ve vyrobním procesu. (Investigation and control of the most advantageous conditions in manufacturing processes.) Stronjnicky Obzor 26:267-272. C (1651)

Upholt, W. M. [1942], The use of the square root transformation and analysis of variance with contagious distributions. J. Econ. Ent. 35:536-543. A9 (1652)

Vagholkar, B. P., Apte, V. N., and Subramonia Iyer, S. [1940], A study of plot size and shape technique for field experiments on sugarcane. Ind. J. Agri. Sci. 10:388-403. A14 (1653)

Vagholkar, B. P., Patwardhan, N. B., and Subramonia Iyer, S. [1940], Sampling of sugarcane for chemical analysis. Ind. J. Agri. Sci. 10:45-61. A14 (1654)

Vaidyanathan, M. [1933], Analyses of manurial experiments in India. Vol. II. Results of experiments conducted in the past in the several provincial agricultural farms (with statistical tables). Imp. Counc. Agri. Res. Simla, pp. 662. C;S1 (1655)

Vaidyanathan, M. [1933], Analyses of manurial experiments in India. Vol. III. Results of experiments conducted in the past in Pusa, Mysore, and Tocklai Tea Experimental Station (Cinnamara, Assam) (with statistical tables). Imp. Counc. Agri. Res. Simla, pp. 101. C;S1 (1656)

Vaidyanathan, M. [1934], Analysis of manurial experiments in India. Vol. I. A general review of the results of past experiments with fertilisers in India. Imp. Counc. Agri. Res. Simla, pp. ii+121+18. C;S1 ((1657)

Vaidyanathan, M. [1934], The method of "covariance" applicable to the utilization of previous crop records for judging the improved precision of experiments. Indian J. Agri. Sci. 4:327-342. E2a;A7,14 (1658)

Vaidyanathan, M. [1934], A statistical study of "ideal size and shape of plot for sugarcane" based upon tonnage experiment with Co. 213, conducted at Meghoul (Pusa) in 1933 (abstract). 21st Ann. Meet. Indian Sci. Congress, Bombay, Sect. Agri. 1:1. A14 (1659)

Vaidyanathan, M. [1934], "Covariance" applied to experimental plots at Tocklai Tea Experimental Station (abstract). 21st Ann. Meet. Indian Sci. Congress Bombay, Sect. Agri. 2:1. E2a;A7 (1660)

Vaidyanathan, M. [1936], Application of statistics to field technique in agriculture. Current Sci. 4:457-468. E6a;A10;C (1661)

Vaidyanathan, M. [1938], Discussion of sampling methods in developmental studies on cotton and the statistical treatment of results. Indian Cent. Cotton Comm., 1st Conf. Sci. Res. Workers Cotton India, Bombay (1937), pp. 424-436. A14 (1662)

Vaidyanathan, M. and Subramonia Iyer, S. [1940], A note on the analysis of 3^3 and 3^4 designs (with three-factor interactions confounded) in field experiments in agriculture. Indian J. Agri. Sci. 10:213-236. T2ae;A14 (1663)

Vajda, S. [1947], Technique of the analysis of variance. Nature 160:27. T2a;A14 (1664)

Van de Sande-Bachuyzen, H. L. [1926], Growth and growth formulae in plants. Science 64:653-654. A7 (1665)

Van de Sande-Bachuyzen, H. L. and Alsberg, C. [1927], Growth curves in annual plants. Physiol. Rev. 7:151-187. A7 (1666)

van der Laan, E. [1943], Mededeelingen van den Studiekring voor Proeftechniek. (Proceedings of the Study Circle for Experimental Methods.) Landbouwk. Tijdschr. Wageningen 55:621. A14;C (1667)

van der Meulen, J. G. J. [1931], Over den invloed van randplanten op de opbrengst van sawakpadi in kleine proefvakken. Tijdschrift Vereeniging Landbouwconsulenten Nederlandsch-Indië 7:85-103. A14 (1668)

Van der Plank, J. E. [1947], A method for estimating the number of random groups of adjacent diseased plants in a homogeneous field. Trans. Royal Soc. South Africa 31:269-278. E5a;A14;C (1669)

van der Reyden, D. [1943], Curve fitting by the orthogonal polynomials of least squares. Onderstepoort J. Vet. Sci. Anim. Indust. 18:355-404. A7;C (1670)

Van Rest, E. D. [1937], Examples of statistical methods in forest products research. JRSSB 4:184-209. E3,12a (1671)

van Uven, M. J. [1946], Mathematical Treatment of the Results of Agricultural and Other Experiments. P. Noordhoff N. V., Groningen and Batavia, pp. vi+310, 2nd Edition (1st ed. 1935). E2,7,12a;C (1672)

Vervelde, G. J. [1948], Lattice squares - mathematically applied (in Dutch). Neder. Landb.-dienst. Maandbl. 5:117-121. E11a (1673)

Vervelde, G. J. [1948], Kweken op resistentie of op opbrengst. (Breeding for resistance or breeding for yield.) Studiekring voor Plantenveredeling (Plant Breeding Study Circle) 4 March, Wageningen pp. 216-218 (mimeo.). T9a (1674)

Vigor, H. D. [1928], Crop estimates in England. JRSSA 91:1-48. A14 (1675)

Vik, K. [1924], Bedømmelse av feilene på forsøksfelter med og uten målestokk. Norges Landbrukshøgskole Meldinger 4:129-181. E7,9a (1676)

Vinall, H. N. and McKee, R. [1916], Moisture content and shrinkage of forage and the relation of these factors to the accuracy of experimental data. U.S.D.A. Bull. 353. A14 (1677)

Violle, B. [1838], Traité Complet des Carrés Magiques (3 vols.). Paris. E12c (1678)

Visser, W.C. [1937], De Ongelijkmatigheid van den grond en de nauwkeurigheid bij proefvelden. (Soil heterogeneity and accuracy of trial plots.) Versl. RijkslandbProefst., 's Grav. 43(7)A:225-270. A14;C (1679)

Visser, W.C. [1942], Over de bruikbaarheid van de grafisch-statistische bewerkingstechniek. (On the applicability of the graphical and statistical method of treatment.) Landbouwk. Tijdschr., Wageningen 54:403-416. C (1680)

Vollema, J.S. [1931], Over de methodiek van veldproeven in de rubbercultuur. (On the methods of field experiments in rubber cultivation.) Arch. Rubberc. Ned.-Ind. 15:391-422. C;S1 (1681)

Wadleigh, C. and Tharp, W. H. [1940], Factorial design in plant nutrition experiments in the greenhouse. Arkansas Expt. Sta. Bull. 401:1-67. T2a;C (1682)

Wadley, F. M. [1943], Statistical treatment of percentage counts. Science 98:536-538. A9 (1683)

Wadley, F. M. [1945], Incomplete block experimental designs in insect population problems. J. Economic Entomology 38:651-654. E2,3a (1684)

Wadley, F. M. [1946], Incomplete-block design adapted to paired tests of mosquito repellents. Biometrics 2:30-31. E2,3a (1685)

Wadley, F. M. [1948], Experimental design in comparison of allergens on cattle. Biometrics 4:100-108. E2,3a (1686)

Wald, A. [1940], A note on the analysis of variance with unequal class frequencies. AMS 11:96-100. A10 (1687)

Wald, A. [1941], On the analysis of variance in case of multiple classifications with unequal class frequencies. AMS 12:346-350. A10 (1688)

Wald, A. [1943], On the efficient design of statistical investigations. AMS 14:134-140. E7,12,13,14a (1689)

Wald, A. [1947], Sequential Analysis. Wiley, N.Y. S2 (1690)

Wald, A. and Wolfowitz, J. [1943], An exact test for randomness in the non-parametric case based on serial correlation. AMS 14:378-388. E5a;A14;C (1691)

Walker, R. H. [1937], The need for statistical control in soils experiments. J. American Soc. Agron. 29:650-657. E2a;C (1692)

Walker, R. H. and Brown, P. E. [1936], Chemical analyses of Iowa soils for phosphorus, nitrogen and carbon: A statistical study. Iowa Agri. Expt. Sta. Res. Bull. 203:59-104. C (1693)

Wang, C.-M. [1947], Methods of determining the degrees of freedom for errors. (in Chinese). Taiwan Tahsueh Nanguan Yenchiu Paokao/Memoirs College Agri. Nat. Taiwan Univ. 1(5):1-20. E2,3a (1694)

Wang, C. M., Chen, L. T., and Yang, C. C. [1936], (Some results of field experiments in rice for comparison of the moving average method and the variance method.) J. Agri. Assoc. China 145:1-25. E2a;A14;C (1695)

Wang, S. [1935], Effect of the length of row, number of replications, and frequency of checks on the accuracy of soybean experiments under the environmental conditions of Nanking, China (in Chinese). J. Agri. Assoc. China 132:49-62. A14;S1 (1696)

Wang, S. [1937], The value of covariance in analyzing the yield trial of kaoling (sorghum) (in Chinese). Nanking J. 7:93-102. A7 (1697)

Wanner, H. [1941], Streuungszerlegung, ein Hilfsmittel der modernen Statistik zur Beurteilung von Sortenanbauversuchen. Landwirtschaftliches Jahrbuch Schweiz 55:773-782. C (1698)

Ward, H. M. and Bailey, B. [1943], Wearing tests on fabric blends of new and reclaimed wool fiber. J. Agri. Res. 67:485-500. T2a (1699)

Ward, G. M. and Smith, V. R. [1949], Total milk production as affected by time of milking after application of a conditioned stimulus. J. Dairy Sci. 32:17-21. E7ae (1700)

Ware, L. M. and Johnson, W. A. [1948], Use of field bins for experimental studies with vegetable crops. Proc. Auburn Conf. Stat. Applied Soc. Sci., Plant Sci. Animal Sci. pp. 54-60. C (1701)

Watson, C. J., Campbell, J. A. et al. [1939], Digestibility studies with ruminants VI. Sci. Agri. 20:238-253. E7,10ae (1702)

Watson, C. J., Campbell, J. A. et al. [1940], Digestibility studies with ruminants VII. Sci. Agri. 20:458-469. E7ae (1703)

Watson, C. J., Campbell, J. A. et al. [1942], Digestibility studies with ruminants IX. Sci. Agri. 22:561-570. E7ae (1704)

Watson, C. J., Campbell, J. A. et al. [1943], Digestibility studies with swine I. Sci. Agri. 23:708-724. E7,10ae (1705)

Watson, C. J., Davidson, W. M. et al. [1939], Digestibility studies with ruminants V. Sci. Agri. 20:175-204. E7ae (1706)

Watson, C. J., Davidson, W. M. et al. [1949], Digestibility studies with ruminants XIV. Sci. Agri. 29:400-407. E7ae (1707)

Watson, C. J., Kennedy, J. W. et al. [1949], Digestibility studies with ruminants XIII. Sci. Agri. 29:263-272. E7ae (1708)

Watson, D. J. [1937], The estimation of leaf area in field crops. J. Agri. Sci. 27:474-483. A14 (1709)

Watson, D. J. and Russell, E. J. [1945], The Rothamsted experiments on mangolds, 1876-1940. Part III. Causes of variation of yield. Empire J. Expt. Agri. 13: 61-79. A7;C (1710)

Watson, S. J. and Ferguson, W. S. [1936], The nutritive value of artificially dried grass and its effect on the quality of milk produced by cows of the main dairy breeds. J. Agri. Sci. 26:189-211. E8ae (1711)

Watson, S. J. and Ferguson, W. S. [1936], The value of artificially dried grass, silage made with added molasses and A.I.V. fodder in the diet of the dairy cow and their effect on the quality of milk, with special reference to the value of the non-protein nitrogen. J. Agri. Sci. 26:337-367. E2,6ae;A7 (1712)

Watson, S. J. and Ferguson, W. S. [1937], The losses of dry matter and digestible nutrients in low-temperature silage, with and without added molasses or mineral acids. J. Agri. Sci. 27:67-107. E12ae (1713)

Waynick, D. D. [1918], Variability in soils and its significance to past and future soil investigations: I. A statistical study of nitrification in soils. California Univ. Pub. Agri. Sci. 3:243-270. A14;C (1714)

Waynick, D. D. and Sharp, L. T. [1919], Variability in soils and its significance to past and future soil investigations. II. Variations in nitrogen and carbon in field soils and their relation to the accuracy of field trials. California Univ. Pub. Agri. Sci. 4:121-139. A14;C (1715)

Weaver, J. E. and Clements, F. E. [1929], Plant Ecology. McGraw-Hill, N.Y. A14 (1716)

Webster, C. C. [1939], A note on a uniformity trial with oil palms. Tropical Agri. 16:15-19. A14 (1717)

Weihing, R. M. and Robertson, D. W. [1941], Forage yields of five varieties of alfalfa grown in nursery rows and field plots. J. American Soc. Agron. 33:156-163. A14 (1718)

Weil, C. S. [1947], Statistical evaluation of growth curves. Proc. Soc. Exptl. Biology 64:468-470. A7 (1719)

Weinberg, D. [1945], Une expérience de contrôle des méthodes d'analyse factorielle. C.R. Academy Sci., Paris 220:214-216. T2a (1720)

Weir, W. W. [1925], Limitations of Student's method when applied to fertilizer experiments. J. Agri. Res. 31:949-956. C (1721)

Weiss, M. G. and Cox, G. M. [1939], Balanced incomplete block and lattice square designs for testing yield differences among large numbers of soybean varieties. Iowa Agri. Expt. Sta. Res. Bull. 257:291-316. E2,11ace (1722)

Welch, B. L. [1936], Specification of rules for rejecting too variable a product, with particular reference to an electric lamp problem. JRSSB 3:29-48. A6 (1723)

Welch, B. L. [1937], On the z-test in randomized blocks and latin squares. Biometrika 29:21-52. E2,7ae (1724)

Welch, B. L. [1938], The significance of the difference between two means when the population variances are unequal. Biometrika 29:350-361. All (1725)

Welch, B. L. [1947], The generalization of 'Student's' problem when several different population variances are involved. Biometrika 34:28-35. All (1726)

Wellensiek, S. J. [1941], Een studiekring voor proeftechniek. (A study circle for the technique of field experiments.) Landbouwk. Tijdschr., Wageningen 53:743. C (1727)

Wellhausen, E. J. [1943], The accuracy of incomplete block designs in varietal trials in West Virginia. J. American Soc. Agron. 35:66-76. E2,14ae (1728)

Wellman, R. H., Thurston, H. W., and Whaley, F. R. [1948], A method for correcting for geographic variation in field experiments. Contr. Boyce Thompson Inst. 15:153-163. E2,5ae;C (1729)

Wenger, L. E. [1941], Soaking buffalo grass (Buchloe dactyloides) seed to improve its germination. J. American Soc. Agron. 33:135-141. E12ae (1730)

Werner, H. O. and Kisselbach, T. A. [1929-30], The effects of vacant hills and of plot competition upon the yield of potatoes in field experiments. Ann. Proc. Potato Assoc. America 16:109-120. A14 (1731)

Wernicke, P. [1910], Das Problem der 36 Offiziere. Iber. Deutch. Math. Verein. 19:264-267. E7,12c (1732)

Wernimont, G. [1949], Statistics applied to analysis. Analytical Chemistry 21: 115-120. B;C (1733)

Weston, D. W. A. R. and Taylor, R. E. [1945], Seed disinfection VII. J. Agri. Sci. 35:239-242. E2ae;A9 (1734)

Westover, K. C. [1924], The influence of plot size and replication on experimental error in field trials with potatoes. West Virginia Agri. Expt. Sta. Bull. 189:1-32. A14 (1735)

Wheeler, H. J. [1908], Some desirable precautions in plat experimentation. Proc. American Soc. Agron. 1:39-44. A14 (1736)

Wiancko, A. T., Arny, A. C., and Salmon, S. C. [1922], Report of Committee on Standardization of Field Experiments. J. American Soc. Agron. 13:368-374. A14;C (1737)

Wiancko, A. T., Arny, A. C., and Salmon, S. C. [1923], Bibliography of standardization of field experiments. J. American Soc. Agron. 15:33-40. B (1738)

Wiancko, A. T., Harris, F. S., and Salmon, S. C. [1919], Report of the Committee on Standardization of Field Experiments. J. American Soc. Agron. 10:345-354. A14;B;C (1739)

Wiancko, A. T., Love, H. H., Mooers, C. A., Salmon, S. C., and Arny, A. C. [1924], Report of Committee on Standardization of Field Experiments. J. American Soc. Agron. 16:1-16. B;C (1740)

Wiancko, A. T., Mooers, C. A. et al. [1924], Report of Committee on Standardization of Field Experiments. J. American Soc. Agron. 16:804-805. B;C (1741)

Wiebe, G. A. [1935], Variation and correlation in grain yield among 1,500 wheat nursery plots. J. Agri. Res. 50:331-357. A14 (1742)

Wiebe, G. A. [1937], The error in grain yield attending misspaced wheat nursery rows and the extent of the misspacing effect. J. American Soc. Agron. 29:713-716. A14 (1743)

Wiener, W. T. and Broadfoot, R. [1925], The amount of variability which may be expected to occur in a determination of comparative yields in small grain. Sci. Agri. (Ottawa) 5:305-309. C (1744)

Wiener, W. T. and Broadfoot, R. [1925], The effect of fallow borders on the variability of plot yields. Sci. Agri. (Ottawa) 5:310-312. A14;C (1745)

Wiggans, R. G. [1924], Experiments in crop rotation and fertilization. Cornell Univ. Agri. Expt. Sta. Bull. 434. E5,8a (1746)

Wiggans, R. G. [1926], Pasture studies. Cornell Univ. Agri. Expt. Memoir 104. C (1747)

Wilcox, A. N. [1927], A study of field plot technique with strawberries. Sci. Agri. 8:171-174. A14;S1 (1748)

Wilcox, J. C. [1948], Soil sampling technique in orchards. Sci. Agri. 28:321-332. A14 (1749)

Wilcox, O. W. [1941], A critique of field experiments with plant nutrients. American Fertilizer 95:8-11, 24, 25. T2a;C (1750)

Wilcoxon, F. [1945], Individual comparisons by ranking methods. Biometrics 1:80-83. E1,2ae;A12 (1751)

Wilcoxon, F. [1946], Individual comparisons of grouped data by ranking methods. J. Economic Entomology 39:269-270. E1,2ae;A12 (1752)

Wilcoxon, F. [1947], Probability tables for individual comparisons by ranking methods. Biometrics 3:119-122. E1a;A12 (1753)

Wilcoxon, F. [1949], Some Rapid Approximate Statistical Procedures (1st ed., 1946) revised. American Cyanamid Co., Stamford, Connecticut. E1,2ae;T2ae;A12 (1754)

Wilkins, F. S. and Hyland, H. L. [1938], The significance and technique of dry matter determinations in yield tests of alfalfa and red clover. Iowa Agri. Expt. Sta. Res. Bull. 240:315-351. A12;S1 (1755)

Wilks, S. S. [1938], The analysis of variance and covariance in non-orthogonal data. Metron 13(2):141-154. T2a;A10 (1756)

Wilks, S. S. [1947], Mathematical Statistics (1st ed. 1944). Princeton Univ. Press, Princeton, N.Y. E2,7,12a;T2a;A7,10 (1757)

Williams, E. J. [1949], Confounding and fractional replication in factorial experiments. Australian J. Agri. Sci. 15:145-153. T2,3a (1758)

Williams, E. J. [1949], Experimental designs balanced for the estimation of residual effects of treatments. Australian J. Sci. Res. 2(A):149-168. E8,13ac (1759)

Williams, R. O. [1932], Citrus experiments. Tropical Agri. 9:301-306. E2,7,12a;C (1760)

Wilm, H. G. [1945], Notes on analysis of experiments replicated in time. Biometrics 1:16-20. E2,12ae (1761)

Wilm, H. G. [1946], The design and analysis of methods for sampling micro-climatic factors. JASA 41:221-232. C (1762)

Wilm, H. G., Costello, D. F., and Klipple, G. E. [1944], Estimating forage yield by the double-sampling method. J. American Soc. Agron. 36:194-203. A14 (1763)

Wilson, E. [1941], A simple technique for laying out sample plots. J. Forestry 39:650-651. A14 (1764)

Wilson, E. B. [1941], The controlled experiment and the four-fold table. Science 93:557-560. C (1765)

Wing, H. D. [1948], A factorial study of musical tests. British J. Psychology 31:341-355. T2a (1766)

Winsor, C. P. [1948], Question 6: Logarithmic transformation. Amer. Stat. 2(1):18. A9 (1767)

Winsor, C. P. and Clarke, G. L. [1940], A statistical study of variation in the catch of plankton nets. J. Marine Res. 3:1-34. E1ae;A10 (1768)

Winter, F. L. [1929], The mean and variability as affected by continuous selection for composition in corn. J. Agri. Res. 39:451-476. C (1769)

Wishart, J. [1930], On the secular variation of rainfall at Rothamsted. Mem. Royal Met. Soc. 3:127-137. A7 (1770)

Wishart, J. [1931], Methods of field experimentation and the statistical analysis of the results. Rothamsted Conf. 12:13-21. C (1771)

Wishart, J. [1931], The analysis of variance illustrated in its application to a complex agricultural experiment on sugar beet. Archiv Pflanzenbau 5:561-584. E7,9ae (1772)

Wishart, J. [1933], The place of statistics in the experimental sciences. Cambridge Univ. Agri. Soc. Mag. 49. C (1773)

Wishart, J. [1933], The theory of orthogonal polynomial fitting. JRSSA 96:487-491. A7 (1774)

Wishart, J. [1934], Field experimentation -- the modern technique. Agri. Progress 11:149-156. E2,7a;C (1775)

Wishart, J. [1934], Analysis of variance and analysis of co-variance, their meaning, and their application in crop experimentation. Empire Cotton Growing Corp., Conf. Report, pp. 83-96. E2,7a;A7;C (1776)

Wishart, J. [1934-35], The evaluation of the field experiment. Proc. First Plant Breeding Conf. Nanking, China, pp. 25-30. C (1777)

Wishart, J. [1934], Statistics in agricultural research (discussion). JRSSB 1:26-61. E2ae;A7;C (1778)

Wishart, J. [1934], Bibliography of agricultural statistics, 1931-1933. JRSSB 1:94-106. B ((1779)

Wishart, J. [1936], Statistics in Chinese agricultural research. JASA 31:127-128. C (1780)

Wishart, J. [1936], Tests of significance in analysis of covariance. JRSSB 3:79-82. A7 (1781)

Wishart, J. [1938], Growth-rate determinations in nutrition studies with the bacon pig, and their analysis. Biometrika 30:16-28. E2ae (1782)

Wishart, J. [1938], Field experiments of factorial design. J. Agri. Sci. 28: 299-306. T2a;A11 (1783)

Wishart, J. [1939], Statistical treatment of animal experiments. JRSSB 6:1-22. T2a;A7;C (1784)

Wishart, J. [1940], Field trials: Their lay-out and statistical analysis. Imperial Bur. Plant Breeding Genetics, Tech. Comm. 7:1-36. E2,7,12ae;T2ae;C (1785)

Wishart, J. [1947], The variance ratio test in statistics. J. Inst. Actuar. Students' Soc. 6:172-184. A11 (1786)

Wishart, J. [1947], Proof of the distributions of χ^2, of the estimate of variance, and of the variance ratio. J. Inst. Actuar. Students' Soc. 7:98-103. A11 (1787)

Wishart, J. [1948], Tests of significance in the simple regression problem. J. Inst. Actuar. Students' Soc. 8:38-43. A7 (1788)

Wishart, J. [1949], Test of homogeneity of regression coefficients, and its application in the analysis of covariance. Le calcul des probabilités et ses applications. Colloque Int. Centre Nat. Rech. Sci., Paris 13:93-99. A7 (1789)

Wishart, J. [1949], Análisis de la varianza y covarianza con referencia especial a la experimentacion agrícola. (Analysis of variance and covariance with special reference to agricultural experimentation.) Inst. Nac. Invest. Agron., Madrid pp. 108. A7;C (1790)

Wishart, J. [1949], Design of experiments. Inst. Stat. Mimeo. Univ. N. Carolina. E2,3,7,12ae;T2ae; C (1791)

Wishart, J. and Clapman, A. R. [1929], A study of sampling technique: The effect of artificial fertilizers on the yield of potatoes. J. Agri. Sci. 19:600-618. E12ae;A14 (1792)

Wishart, J. and Garner, H. V. [1930], Fertilizer trials in 1929. J. Ministry Agri. 37:793-802. T2a;C (1793)

Wishart, J. and Hammond, J. [1933], A statistical analysis of the inter-relations of litter size and duration of pregnancy on the birth weight of rabbits. J. Agri. Sci. 23:463-472. C (1794)

Wishart, J. and Hines, H. J. G. [1929], Fertilizer trials on the ordinary farm. J. Ministry Agri., London 36:524-532. E2,7ae;T2ae;C (1795)

Wishart, J. and Mackenzie, W. A. [1930], Studies in crop variation. VII. The influence of rainfall on the yield of barley at Rothamsted. J. Agri. Sci. 20: 417-439. A7 (1796)

Wishart, J. and Sanders, H. G. [1936], Principles and practice of field experimentation. Empire Cotton Growing Corporation, London. E1,2,7,12ae;T2ae; A7;C (1797)

Wishart, J., Woodman, H. E., et. al. [1936], The nutrition of the bacon pig. I. The influence of high levels of protein intake on growth, conformation and quality in the bacon pig. J. Agri. Sci. 26:546-619. C (1798)

Witt, E. [1938], Ueber Steinersche Systeme. Abh. Math. Sem. Hamburg 12: 265-275. E2c (1799)

Wolf, H. W., Dawson, W. M., and Pohle, E. M. [1943], Fiber density and some methods of its measurement in the fleeces of Rambouillet sheep. J. Animal Sci. 2:188-196. A14 (1800)

Wolfe, T. K. [1924], A mathematical inquiry into the influence of the amount and distribution of rainfall on the yield of corn. J. American Soc. Agron. 17:356-362. A7 (1801)

Wolfowitz, J. [1949], Non-parametric statistical inference. Proc. Berkeley Symp. Math. Stat. Prob. pp. 93-113. A12 (1802)

Wolters, W. [1931], Determination of the ripeness of cane by means of pre-harvest sampling. Hawaiian Planters' Record 35:411-418. A14 (1803)

Wood, E. C. and Finney, D. J. [1946], The design and statistical analysis of microbiological assays. Quarterly J. Pharm. 19:112-127. T8a (1804)

Wood, E. C. and Simpson, T. W. [1937], The supplementary feeding of school children. Medical Press Circular 195:8 (October). C (1805)

Wood, R. C. [1932], A variety test on sugar-cane. Tropical Agri. 9:290-293. E2ae (1806)

Wood, R. C. [1933], Experiments with yams in Trinidad, 1931-3. Empire J. Exptl. Agri. 1:316-324. E2,7ae;T2ae (1807)

Wood, R. C. [1936], A manurial experiment on bananas. Empire J. Exptl. Agri. 4:365-367. E2ae (1808)

Wood, R. C. [1938], Settled holdings in the tropics. Tropical Agri. 15:147-153. E5,8ae (1809)

Wood, R. C. and Hardy, F. [1941], The college permanent manurial experiment. Tropical Agri. 18:48-61. E5,8a (1810)

Wood, R. C. and Paterson, D. D. [1932], A variety test of sugar-cane. Tropical Agri. 9:14-21. E12ae (1811)

Wood, T. B. [1911], The interpretation of experimental results. J. Board Agri. (London) Suppl. 7:15-37. C (1812)

Wood, T. B. and Stratton, F. J. M. [1910], The interpretation of experimental results. J. Agri. Sci. 3:417-440. A14;C (1813)

Woodford, E. K. and Cooper, H. R. [1941], A factorial experiment on the manuring of young tea. Empire J. Exptl. Agri. 9:12-22. T2a (1814)

Woodford, E. K. and Gregory, F. G. [1948], Preliminary results obtained with an apparatus for the study of salt uptake and root respiration of whole plants. Ann. Botany 12:335-370. E12ae (1815)

Woodman, H. E. and Evans, R. E. [1939], Nutrition of the bacon pig IV. J. Agri. Sci. 29:502-525. E12ae (1816)

Woodman, H. E. and Evans, R. E. [1940], The nutrition of the bacon pig V. J. Agri. Sci. 30:83-97. E2,12ae (1817)

Woodman, H. E. and Evans, R. E. [1941], The nutrition of the bacon pig VI. J. Agri. Sci. 31:231-245. E12ae;T2ae (1818)

Woodman, H. E. and Evans, R. E. [1942], The nutrition of the bacon pig VII. J. Agri. Sci. 32:85-107. E2ae (1819)

Woodman, H. E. and Evans, R. E. [1943], Further investigations of the feeding value of artificially dried potatoes. J. Agri. Sci. 33:1-14. E2ae (1820)

Woodman, H. E. and Evans, R. E. [1943], The nutrition of the bacon pig IX. J. Agri. Sci. 33:155-168. E2ae (1821)

Woodman, H. E. and Evans, R. E. [1945], The nutrition of the bacon pig X. J. Agri. Sci. 35:44-55. E2ae (1822)

Woodman, H. E. and Evans, R. E. [1945], Nutrition of the bacon pig XI. J. Agri. Sci. 35:133-149. E2ae (1823)

Woodman, H. E., Evans, R. E., Callow, E. H., and Wishart, J. [1936], The nutrition of the bacon pig I. J. Agri. Sci. 26:546-619. E1,2,12ae (1824)

Woodman, R. M. and Johnson, D. A. [1946], Plant growth with nutrient solutions III. J. Agri. Sci. 36:87-94. E2ae (1825)

Woodworth, R. S. [1912], Combining the results of several tests: A study in the statistical method. Psychol. Rev. 19:97-123. A13;C (1826)

Woolhouse, W. S. B. [1861], On Rev. T. P. Kirkman's problem representing certain triadic arrangements of fifteen symbols. Phil. Mag. 21(4):510-515. E2c (1827)

Woolhouse, W. S. B. [1862], On the triadic combinations of 15 symbols. Ladys' and Gentlemen's Diary 1862:84-88 (also London Assurance Mag. 10:275-281.). E2c (1828)

Woolhouse, W. S. B. [1863], On triadic combinations. Ladys' and Gentlemen's Diary 1863:79-90. E2c (1829)

Worzella, W. W. [1943], Response of wheat varieties to different levels of soil productivity: I. J. American Soc. Agron. 35:114-124. E7,12,13a (1830)

Wray, W. D. [1941], Some applications of uniformity trials. Ph. D. Thesis, Cornell Univ. A7,14;C (1831)

Wyatt, F. A. [1927], Variation in plot yields due to soil heterogeneity. Sci. Agri. 7:248-256. A14 (1832)

Yang, Y. K. [1935], Suggestion of a new formula for the deviation of the mean method. Hopei Agri. For. p. 20. E2a;C (1833)

Yang, Y. K. and Tsuei, C. Y. [1935], A study of the application of some methods for calculating experimental errors. Hopei Agri. For. p. 24. E2a;C (1834)

Yardi, N. R. and Poornapregna, V. N. [1946], Studies in plant breeding technique. The use of control plots in varietal trials. Indian J. Genet. Plant Breeding 6:74-80. E2,5a;C (1835)

Yates, F. [1933], The analysis of replicated experiments when the field results are incomplete. Empire J. Exptl. Agri. 1:129-142. E2,6,7,10ae;T2ae (1836)

Yates, F. [1933], The formation of Latin squares for use in field experiments. Empire J. Exptl. Agri. 1:235-244. E7ac (1837)

Yates, F. [1933], The principles of orthogonality and confounding in replicated experiments. J. Agri. Sci. 23:108-145. E12ae;T2ae;A10 (1838)

Yates, F. [1934], A complex pig-feeding experiment. J. Agri. Sci. 24:511-531. T2ac;C (1839)

Yates, F. [1934], The analysis of multiple classifications with unequal numbers in the different classes. JASA 29:51-66. A10 (1840)

Yates, F. [1935], Some examples of biased sampling. Ann. Eugenics 6:202-213. A14 (1841)

Yates, F. [1935], The place of quantitative measurements on plant growth in agricultural meteorology and crop forecasting. Report Conf. Empire Meteorologists, London, Meteorological Office Pub. No. 393. A14 (1842)

Yates, F. [1935], Complex experiments. JRSSB 2:181-247. E2,3,7,14a;T2,3,6ae; C (1843)

Yates, F. [1936], Incomplete randomized blocks. Ann. Eugenics 7:121-140. E2ace (1844)

Yates, F. [1936], Incomplete Latin squares. J. Agri. Sci. 26:301-315. E7,10ae (1845)

Yates, F. [1936], Applications of the sampling technique to crop estimation and forecasting. Trans. Manchester Stat. Soc. 103 ff. A14 (1846)

Yates, F. [1936], A new method of arranging variety trials involving a large number of varieties. J. Agri. Sci. 26:424-455. E2,3,5,7,11,14ace (1847)

Yates, F. [1937], A further note on the arrangement of variety trials: Quasi-latin squares. Ann. Eugenics 7:319-332. E3,11ace (1848)

Yates, F. [1937], The design and analysis of factorial experiments. Imp. Bur. Soil Sci. Tech. Comm. 35:1-95. E2,7,9,12ae;T2ae; C (1849)

Yates, F. [1938], The gain in efficiency resulting from the use of balanced designs. JRSSB 5:70-74. E2,8,12,14a (1850)

Yates, F. [1938], Orthogonal functions and tests of significance in the analysis of variance. JRSSB 5:177-180. A7 (1851)

Yates, F. [1939], The comparative advantages of systematic and randomized arrangements in the design of agricultural and biological experiments. Biometrika 30:440-466. E2,5,7a (1852)

Yates, F. [1939], The recovery of inter-block information in variety trials arranged in three-dimensional lattices. Ann. Eugenics 9:136-156. E3ace (1853)

Yates, F. [1940], The recovery of inter-block information in balanced incomplete block designs. Ann. Eugenics 10:317-325. E2ace (1854)

Yates, F. [1940], Modern experimental design and its function in plant selection. Empire J. Exptl. Agri. 8:223-230. E2,3,11a;T9a (1855)

Yates, F. [1940], Lattice squares. J. Agri. Sci. 30:672-687. E11ace (1856)

Yates, F. [1946], A review of recent statistical developments in sampling and sampling surveys. JRSSA 109:12-43. A14 (1857)

Yates, F. [1947], Analysis of data from all possible reciprocal crosses between a set of parental lines. Heredity 1:287-301. T9ae (1858)

Yates, F. [1947], Technique of the analysis of variance. Nature 160:472-473. A10 (1859)

Yates, F. [1947], Recent developments in the design of experiments. U.S.D.A. Mimeo. C (1860)

Yates, F. [1949], Design of rotation experiments. Commonwealth Bureau Soil Sci. Tech. Comm. No. 46. E5,8a (1861)

Yates, F. and Cochran, W. G. [1938], The analysis of groups of experiments. J. Agri. Sci. 28:556-580. E12a;A13 (1862)

Yates, F. and Finney, D. J. [1942], Statistical problems in field sampling for wireworms. Ann. Appl. Biology 29:156-167. A14 (1863)

Yates, F. and Hale, R. W. [1939], The analysis of latin squares when two or more rows, columns, or treatments are missing. JRSSB 6:67-79. E7,10ae (1864)

Yates, F. and Neyman, J. [1936], Correspondence on "Complex experiments". JRSSB 3:83-85. T2a (1865)

Yates, F. and Watson, D. J. [1934], Observer's bias in sampling-observations on wheat. Empire J. Exptl. Agri. 2:174-177. E7ae (1866)

Yates, F. and Zacopanay, I. [1935], The estimation of the efficiency of sampling, with special reference to sampling for yield in cereal experiments. J. Agri. Sci. 25:545-577. A13,14 (1867)

Youden, W. J. [1937], Use of incomplete block replications in estimating tobacco-mosaic virus. Contr. Boyce Thompson Inst. 9:41-48. E7ace (1868)

Youden, W. J. [1937], Dilution curve of tobacco-mosaic virus. Contr. Boyce Thompson Inst. 9:49-58. E2,3,7a (1869)

Youden, W. J. [1940], Seed-treatments with talc and root-inducing substances. Contr. Boyce Thompson Inst. 11:207-218. E7,10ae (1870)

Youden, W. J. [1940], Experimental designs to increase accuracy of greenhouse studies. Contr. Boyce Thompson Inst. 11:219-228. E7ace (1871)

Youden, W. J. [1941], Fluctuations of atmospheric sulphur dioxide. Contr. Boyce Thompson Inst. 11:473-484. A7 (1872)

Youden, W. J. [1948], Multiple factor experiments in analytical chemistry. Analytical Chemistry 20:1136-1140. T2a (1873)

Youden, W. J. [1949], How statistics improves physical, chemical, and engineering measurements. U.S.D.A. Lecture. E7,12a;T2a (1874)

Youden, W. J. and Beale, H. P. [1934], A statistical study of the local lesion method for estimating tobacco-mosaic virus. Contr. Boyce Thompson Inst. 6:437-454. E2,7ae (1875)

Youden, W. J. and Mehlich, A. [1937], Selection of efficient methods for soil sampling. Contr. Boyce Thompson Inst. 9:59-70. A14 (1876)

Youden, W. J. and Zimmerman, P. W. [1936], Field trials with fibre pots. Contr. Boyce Thompson Inst. 8:317-331. E12ae (1877)

Young, D. M. and Romans, R. G. [1948], Assays of insulin with one blood sample per rabbit per test day. Biometrics 4:122-131. E7ae (1878)

Zaleski, M. E. [1930], Expériences multiannuelles et multilocales. Bull. Assoc. Int. Sélect. Pl. Grande Culture 3:204-216. A13 (1879)

Zavitz, C. A. [1912], Care and management of land used for experiments with farm crops. Proc. American Soc. Agron. 4:122-126. C (1880)

Zermelo, E. [1929], Die Berechnung der Turnier-Ergebnisse als ein Maximumproblem der Wahrscheinlichkeitsrechnung. Mathematische Zeitschrift 29:436-460. T11a (1881)

Zinzadze, C. and Yates, F. [1933], Bibliography of statistical methods, chiefly on the application of the analysis of variance. Stat. Dept. Rothamsted Expt. Sta. Harpenden, Herts, p. 27. B (1882)

Zuber, M. S. [1940], A comparison of the relative efficiency of various experimental designs for corn yield tests. M. S. Thesis, Iowa State College. E3,14ae (1883)

Zuber, M. S. [1942], Relative efficiency of incomplete block designs using corn uniformity trial data. J. American Soc. Agron. 34:30-47. E3,14ae (1884)

Co-authors list for category "Pre-'50"

Name	Ref. No.	Name	Ref. No.
Aamodt, O.S.	1611	Black, W.H.	1034
Abbas, M.	2		1035
Ahlgren, H.L.	1611		1036
Ahmad, N.	930	Blakeslee, L.H.	627
Albrecht, H.R.	933	Blaser, R.E.	1367
Alsberg, C.	1666	Blaxter, K.L.	85
Anderson, E.O.	136	Bliss, C.I.	276
Anderson, W.S.	982		1141
Andrews, W.B.	1368	Bose, R.C.	1169
Angelo, E.	1305	Bose, S.S.	55
Anós, A.	1279		269
Apte, N.S.	779		270
Apte, V.N.	1653		674
Arny, A.C.	695		675
	1737		957
	1738		1071
	1740		1117
Ausemus, E.R.	699	Bosewell, V.R.	98
Austin, L.	406		99
Autrey, K.M.	319	Both, M.P.	568
		Breneman, W.R.	1498
Bailey, B.	1699	Brewbaker, H.E.	414
Baker, F.E.	560a	Brier, G.W.	715
Baker, G.A.	652	Briggs, F.N.	49
	1361		50
Banerjee, E.A.R.	1072	Brischle, H.A.	565
Bär, A.L.S.	735	Broadfoot, R.	1744
Barker, H.D.	1596		1745
Bartlett, M.S.	127	Broom, W.A.	1483
	373	Brown, G.W.	1138
	808	Brown, P.E.	920
Batchelor, L.D.	1237		921
Baten, W.D.	402		922
	403		1270
	713		1693
	1074	Brown, R.T.	1305
Batten, E.T.	99	Brunson, A.M.	1022
Beachell, H.M.	824	Bryan, W.W.	1125
Beale, H.P.	1875	Bull, H.	1436
Belcher, B.A.	30	Burkar, F.L.	147
Beruldsen, E.T.	1142	Burman, J.P.	1291
Black, C.A.	732	Bush, H.L.	217
Black, W.H.	923		218

VII. BIBLIOGRAPHY FOR EXPERIMENT DESIGN

Most of the references listed in the second part of the bibliography are the same as those given by Federer [1970]. From this list it may be observed that there has been considerable research activity in the area of experiment design in the 1950 to 1967 period. This activity appears to be accelerating as the solution of an existing problem often opens up several unsolved problems. In particular, R. C. Bose, who has been able to interest a talented group of students in experiment design research, and his co-workers have made many and varied contributions to the theory of experiment design in this period. Perhaps the most startling result in this era was the Bose-Parker-Shrikhande result demonstrating the falsity of L. Euler's (481) conjecture that no latin square of order $n = 4t + 2$ possesses an orthogonal mate. It would appear that the main consequence of the work of the "Euler-Spoilers" (as these three men have been called) was to rekindle interest in many related problems. However, some of the more useful results by this group would appear to be the methodic and thorough investigation of the various aspects of partially balanced incomplete block designs and related concepts and in extending the related mathematical theory.

Several papers on experiment design with two and higher-way elimination of heterogeneity have appeared. Designs with two-way elimination of heterogeneity appear under the E7 and E9 classifications while those with more than two-way elimination of experimental heterogeneity are categorized under E11, E12, or E13. Also, active research on designs for estimating residual or carry-over effects has been carried on; these papers are classified under E8.

It is interesting to note that research in experiment design after its exportation to the U.S. and India by the British, flourished rapidly in these countries. Also with the unrestricted exchange of ideas and individuals between India and U.S., these two countries have come to dominate the work in experiment design in the 1950 to 1967 period. However, there is research being done in other countries, e.g. France, Canada, and Japan. It appears that many workers in experiment design are unaware of some of the results of the French (Barra, Guérin, Lafon, Sade) and especially of the Japanese (Yamamoto, Masuyama, Ikeda, Kitagawa, Mitome, Ogawa, Ogasawara, Taguchi, Takeuchi, and several others). Literature citations to research by the French and Japanese are seldom made. Perhaps this bibliography will rectify this.

Abifarah, A. [1966], Carrés latins incomplet. Ph. D. Thesis, Univ. Paris. E7c (2001)

Abraham, J. K. [1960], On an alternative method of computing Tukey's statistic for the latin square model. Biometrics 16:686-691 (correction 17:669). E7a;A4 (2002)

Adair, C. R., Cralley, E. M., and Johnston, T. H. [1958], Long-time trends in grain yields of five rice varieties. Agron. J. 50:233-235. E5e (2003)

Adams, R. W. [1963], Estimating missing values by a regression method in time-trend experiments. M. S. Thesis, Iowa State Univ. E6a (2004)

Adams, R. W. [1963], Estimating missing values by a regression method in time-trend experiments (abstract). Biometrics 19:656. E5,6a (2005)

Addelman, S. [1965], Proportional frequency squares (abstract). Technometrics 7:267. E7c (2006)

Addelman, S. [1967], Equal and proportional frequency squares. JASA 62:226-240. E7ace;T7ace (2007)

Addelman, S. and Bush, S. [1964], A procedure for constructing incomplete block designs. Technometrics 6:389-403. E2,3c (2008)

Adhikary, B. [1965], On the properties and construction of balanced block designs with variable replications. Calcutta Stat. Assoc. Bull. 14:36-64. E4ac (2009)

Adhikary, B. [1965], A difference theorem for the construction of balanced block designs with variable replications. Calcutta Stat. Assoc. Bull. 14:167-170. E4c (2010)

Adhikary, B. [1966], Contribution to incomplete block designs. Ph. D. Thesis, Calcutta Univ. E2,3c (2011)

Adhikary, B. [1966], Some types of m-associate P.B.I.B. association schemes. Calcutta Stat. Assoc. Bull. 15:47-74. E3c (2012)

Adhikary, B. [1966], Various types of m-associate P.B.I.B. association schemes (abstract). JISAS 18(2):105-106. E3c (2013)

Adhikary, B. [1967], A new type of higher associate cyclical association schemes. Calcutta Stat. Assoc. Bull. 16:40-44. E3c (2014)

Adhikary, B. [1967], On the symmetric differences of pairs of blocks of incomplete block designs. Calcutta Stat. Assoc. Bull. 16:45-48. E2,3,4c (2015)

Adhikary, B. [1967], Group divisible designs with variable replications. Calcutta Stat. Assoc. Bull. 16:73-92. E3,4c (2016)

Agarwal, K. N. [1961], Analysis of crop rotation experiments with unequal periods (abstract). JISAS 13:241-242. E5,8a;A13 (2017)

Agarwal, K. N. [1965], Analysis of crop rotation - comparison of a three-crop rotation with and without legume (abstract). JISAS 17:125. E5,8a;A13 (2018)

Agarwal, K. N. [1966], Analysis of crop rotations when cycle is incomplete and design augmented (abstract). JISAS 18(2):102. E5,8a;A13 (2019)

Agrawal, H. [1963], On the dual of BIB designs. Calcutta Stat. Assoc. Bull. 12: 104-105. E3c (2020)

Agrawal, H. [1963], Solution of problem 3. JISA 1:115-116. E2a (2021)

Agrawal, H. [1963], On balanced block designs with two different number of replications. JISA 1:145-151. E4ac (2022)

Agrawal, H. [1964], On the bounds of the number of common treatments between blocks of certain two associate P.B.I.B. designs. Calcutta Stat. Assoc. Bull. 13:76-79 (errata 14:88). E3c (2023)

Agrawal, H. [1964], On the bounds of the number of common treatments between blocks of semi-regular group divisible designs. JASA 59:867-871. E3c (2024)

Agrawal, H. [1965], A note on incomplete block designs. Calcutta Stat. Assoc. Bull. 14:80-83. E2,3c (2025)

Agrawal, H. [1966], A method of construction of three factor balanced designs. JISA 4:10-13. E9c (2026)

Agrawal, H. [1966], Some generalizations of distinct representatives with applications to statistical designs. AMS 37:525-528. E2,3c (2027)

Agrawal, H. [1966], Comparison of the bounds of the number of common treatments between blocks of certain partially balanced incomplete block designs. AMS 37:739-740. E2c (2028)

Agrawal, H. [1966], Some methods of construction of designs for two-way elimination of heterogeneity-1. JASA 61:1153-1171. E7ac (2029)

Agrawal, H. [1966], Two-way elimination of heterogeneity. Calcutta Stat. Assoc. Bull. 15:32-38. E7,10,14a (2030)

Agrawal, H. [1966], Some systematic methods of construction of designs for two-way elimination of heterogeneity. Calcutta Stat. Assoc. Bull. 15:93-108. E7ac (2031)

Agrawal, H. and Raghavachari, R. [1964], On balanced block designs with three different numbers of replications. Calcutta Stat. Assoc. Bull. 13:80-86.
E4ac (2032)

Aird, P. L. [1956], Fertilizers in forestry and their use in hardwood population establishment. Pulp Paper Mag. Canada, Convention Issue. E2ae;T2a (2033)

Aitchison, J. [1961], The construction of optimal designs for the one-way classification analysis of variance. JRSSB 23:352-367. E1,14c;S1 (2034)

Alanen, J. D. [1961], Tables of minimum functions for generating Galois fields GF (p^n) (abstract). AMS 32:621. E2,3,7c (2035)

Aliev, J. S. O. [1967], Symmetric algebras and Steiner systems (in Russian). Doklady Akademiia Nauk, S.S.S.R. 174:511-513. E2c (2036)

Alimena, B. S. [1962], A method of determining unbiased distribution in the Latin square. Psycometrika 27:315-317. E8c (2037)

Allard, R. W. [1952], The precision of lattice designs with a small number of entries in lima bean yield trials. Agron. J. 44:200-202. E3,14a (2038)

Allen, R. S., Homeyer, P. G., and Jacobson, N. L. [1955], Use of activated glycerol dichlorohydrin for estimating vitamin A in dairy calf blood plasma. Iowa State College J. Sci. 29:721-734. E2ae;T2ae (2039)

Alltop, W. O. [1966], On the construction of block designs. J. Combinatorial Theory 1:501-502. E2c (2040)

Amaral, A. Z., Verdade, F. C., Schmidt, N. C., Wutke, A. C. P., and Igue, K. [1965], Parcelamento e intervalo da apliçacão de calcário. (Application of parcelle lime in different intervals.) Bragantia 24:83-96. E2e (2041)

Anderson, D. A. [1966], Multidimensional partially balanced designs (abstract). AMS 37:1858. E4,13c (2042)

Anderson, K. E. [1961], Computation of t by analysis of variance. Amer. Stat. 15(2): 18-19. E1,2a (2043)

Anderson, R. L. [1953], Query 105. Biometrics 9:533-534. E5,12a (2044)

Anscombe, F. J. [1961], Examination of residuals. Proc. Sixth Conf. Design Expt. Army Res. Dev. Testing. pp.7-19. E7ae;A6,14 (2045)

Archbold, J. W. [1960], A combinatorial problem of T. G. Room. Mathematika 7: 50-55. E5,7c (2046)

Archbold, J. W. and Johnson, N. L. [1956], A method of constructing partially balance incomplete block designs. AMS 27:624-632. E3c (2047)

Archbold, J. W. and Johnson, N. L. [1958], A construction for Room's squares and an application in experimental design. AMS 29:219-225. E5ac (2048)

Archer, E. J. [1952], Some Greco-latin analysis of variance designs for learning studies. Psychological Bull. 49:521-537. E12ae (2049)

Assmus, E. F. and Mattson, H. F. [1966], On the number of inequivalent Steiner triple systems. J. Combinatorial Theory 1:301-305. E2c (2050)

Assmus, E. F. and Mattson, H. F. [1967], On tactical configurations and error-correcting codes. J. Combinatorial Theory 2:243-257. E2c;B (2051)

Astrachan, M. [1958], Significance tests II - Small samples. ASQC Nat. Conv. Trans. 12:491-501. E1,2ae;A11 (2052)

Astrachan, M. [1959], Experimental designs and analysis of variance. II. Analysis of the data. ASQC, Nat. Conv. Trans. 13:523-540. E1,2ae (2053)

Atanasiu, N. [1951], Die Auswertung der landwirtschaftlichen Feldversuche. Z. Pflanzenzüchtung 30:112-121. E2e (2054)

Atanasiu, N. [1953], Über die statistischen Auswertungsmethoden der Feldversuche. Z. Pflanzenernährung Düngung Bodenkunde 61:229-235. E5a (2055)

Atanasiu, N. [1954], Zum Aufsatz von K. H. Müller Kristische Bemerkungen zu der Arbeit von N. Atanasiu "Über....". Z. Pflanzenernahrung Düngung Bodenkunde 64:135-138. E5a (2056)

Atanasiu, N. [1954], Zur Frage der Anlage und Auswertung der Feldversuche. Z. Pflanzenzuchtung 33:151-156. E2ae (2057)

Atanasiu, N. [1955], Statistische Auswertung von Versuchsergebnissen. Z. Landw. Vers.- Untersuch Wes. 1:88-94. E5a (2058)

Atanasiu, N. [1955], Stellungnahme zu den „Bermerkungen zu der Arbeit von N. Atanasiu: Zur Frage der Anlage und Auswertung der Feldversuche". Z. Pflanzenzüchtung 34:206-208. E2a (2059)

Atiqullah, M. [1958], Some new solutions of the symmetrical balanced incomplete block design with λ=2 and k=9. Bull. Calcutta Math. Soc. 50:23-28. E2c (2060)

Atiqullah, M. [1958], On configurations and non-isomorphism of some incomplete block designs. Sankhyā 20:227-248. E2,3c (2061)

Atiqullah, M. [1961], On a property of balanced designs. Biometrika 48:215-218. E2,14c (2062)

Atiqullah, M. [1963], On the randomization distribution and power of the variance ratio test. JRSSB 25:334-347. E1,2a;A5 (2063)

Atiqullah, M. and Cox, D. R. [1962], The use of control observations as an alternative to incomplete block designs. JRSSB 24:464-471. E4,7a (2064)

Atkinson, G. F. [1958], Components of the third cumulant in a two-way classification with n observations per cell. Biometrics Unit Mimeo. Ser. BU-98-M, Cornell Univ. E2a;A14 (2065)

Atkinson, G. F. [1963], Designs for sequences of treatments with carry-over effects. Ph. D. Thesis, Cornell Univ. E8ace (2066)

Atkinson, G. F. [1966], Designs for sequences of treatments with carry-over effects. Biometrics 22:292-309. E8ace (2067)

Bahn, E. [1959], Betrachtungen zum faktoriellen Versuch vom Typ 2^2. Z. Landw. Vers. Untersuch Wes. 5:232-254. E12a;T2a (2068)

Bahn, E., Banneick, A., Möbius, H., Müller, K.-H., and Ortlepp, H. [1957], Anlageschemata für Feldversuche (2-16 Prüfglieder) erarbeitet nach den Prinzipien einer gelenkten, gerechten Verteilung der Teilstücke. Z. Landw. Vers. Untersuch Wes. 3:151-205. E2,5,7ac;C (2069)

Bainbridge, J. R. [1951], Factorial experiments in pilot plant studies. Ind. Eng. Chemistry 43:1300-1306. E2e;T2ae (2070)

Bainbridge, T. R. [1963], Staggered, nested designs for estimating variance components. ASQC, Ann. Conv. Trans. 17:93-103. E1ace (2071)

Bainbridge, T. R. [1965], Staggered, nested designs for estimating variance components. IQC 22:12-20. E1ae (2072)

Baird, H. R. and Kramer, C. Y. [1960], Analysis of variance of a balanced incomplete block design with missing observations. Appl. Stat. 9:189-198. E2,6ae (2073)

Baker, F. B. and Collier, R. O. [1966], Some empirical results on variance ratios under permutation in the completely randomized design. JASA 61:813-820. E1a (2074)

Baker, G. A. [1952], Uniformity field trials when differences in fertility levels of subplots are not included in experimental error. AMS 23:289-293. E2a (2075)

Baker, G. A. and Briggs, F. N. [1950], Yield trials with backcross derived lines of wheat. Ann. Inst. Stat. Math. 2:61-67. E7a;A6 (2076)

Baker, G. A. and Hoyle, B. J. [1964], Significant differences on the basis of stable rankings by the SD technique. Hilgardia 35:627-646. E7a;A14 (2077)

Baker, G. A. and Johnson, J. P. [1964], Uniformity field trials and Monte Carlo simulations. Hilgardia 35:615-625. E2,7a;A14 (2078)

Baker, G. A. and Roessler, E. B. [1957], Implications of a uniformity trial with small plots of wheat. Hilgardia 27:183-188. E2,7a;A14;C (2079)

Balaam, L. N. [1965], A two-period design for the estimation of treatment × periods interaction. Biometrics Unit Mimeo. Ser. BU-177-M, Cornell Univ. E7ac (2080)

Balasubramanyan, R., Kannian, K., and Ramachandran, C. K. [1953], The efficiency of lattice square design in cotton. Indian Cotton Growing Rev. 7:143-148. E11,14a (2081)

Ballas, J. A. and Webster, J. T. [1966], On dependent tests from a non-orthogonal design. JASA 61:803-812 (corrigenda 61:1246). E2a;A10 (2082)

Balmer, E. et. al [1965], Contribuição ao estudo da influência dos fatôres físicos do solo, sobre a incidência da murcha do algodoeiro, causada por Fusarium oxysporum f. vasinfectum (Atk.) Snyder and Hansen Anais Escola Superior Agri. "Luiz de Queiroz" 22:247-258. E2ae;T2ae (2083)

Banerjee, K. S. [1951], Incomplete block designs. Current Sci. 20:229. E2c (2084)

Banerjee, K. S. [1951], Some observations on the practical aspects of weighing designs. Biometrika 38:248-251. E2ace;T6ace (2085)

Bang, T. [1957], On the sequence $[n\alpha]$, $n = 1, 2 \ldots$ Supplementary note to the preceding paper by Th. Skolem. Mathematica Scandinavica 5:69-76. E2c (2086)

Bangdiwala, I. S. [1957], A general formula for the t test. Amer. Stat. 11(4):21. E1a (2087)

Baptista, J. G. [1954], Estudos de estadistica aplicada I Quadrado Latino com dois talhões falhados. Melhoramento 7:5-27. E10ae (2088)

Barra, J. R. [1963], A propos d'un théorème de R. C. Bose. Comptes Rendus (First Part) 256:5502-5507. E7,12c (2089)

Barra, J. R. [1965], Carres latins et Euleriens (with discussion). Rev. ISI 33:16-23. E7,12,12c;B (2090)

Barra, J. R. J. and Guérin, R. [1963], Extension des carres gréco-latins cycliques. Publ. l'Inst. Stat. Univ. Paris 12:67-82. E7,12,13c (2091)

Barra, J. R. and Guérin, R. [1963], Utilisation pratique de la méthode de Yamamoto pour la construction systématique de carrés gréco-latins. Publ. l'Inst. Stat. Univ. Paris 12:131-136. E7,12,13c (2092)

Basson, R. P. [1959], Incomplete block designs augmented with a repeated control. M. S. Thesis. Iowa State Univ. E4ac (2093)

Baten, W. D. [1952], Variances of differences between means when there are two missing values in randomized block designs. Biometrics 8:42-50. E2,6ae;A10 (2094)

Baten, W. D., Tack, P. I. and Baeder, H. A. [1958], Testing for differences between methods of preparing fish by use of a discriminant function. IQC 14(7):7-10. E1e;A7 (2095)

Beal, S. R., Carmer, S. G. and Seif, R. D. [1965], Computer program for analyzing augmented randomized complete block designs. Agron. J. 57:316.
E4a;A15 (2096)

Becker, W. A. and Bearse, G. E. [1964], All-or-none traits and the sensitivity of experiments (abstract). Biometrics 20:378. E14 (2097)

Beeler, F. A. [1957], Some useful nonparametric significance tests. ASQC, Nat. Conv. Trans. 11:343-350. E2e;A12 (2098)

Behari, V. [1961], Analysis of some experimental designs when groups of treatments have different variances (abstract). JISAS 13:238. E2a;A11 (2099)

Behari, V. [1963], Testing for equality of treatment means in a R.B.D. trial with heterogenous error variances (abstract). JISAS 15:265-266. E2a;A11 (2100)

Behrens, W. U. [1955], Der Durchschnitt der Prufgliedmittelwerte als Bezugsgrosse. Z. Acker.- Pflanzenbau, 99:397-402. E1,2ae (2101)

Behrens, W. U. [1956], Die Eignung verschiedener Feldversuchs-Anordnungen zum Ausgleich der Bodenunterschiede. Z. Acker- Pflanzenbau, 101:243-278.
E2,5,7a;A14 (2102)

Behrens, W. U. [1956], Feldversuchsanordnungen mit verbessertem Ausgleich der Bodenunterschiede. Zeitschrift landw. Vers.- Unter. 2:176-193, illus.
E7,12ac (2103)

Behrens, W. U. [1958], Die Auswertung von nichtorthogonalen Versuchsserien, insbesondere im Sortenprüfwesen. Z. Acker- Pflanzenbau 105:1-18.
E4ae;A10 (2104)

Behrens, W. U. [1959], Die Auswertung von Feldversuchen nach Mitscherlich. Z. Acker.- Pflanzenbau 109:255-290. E2,5a;C (2105)

Bejar, J. [1957], Diseno de experimentos. Trabajos Estad. 8:91-108.
E,1,2,7ae;T2a;C (2106)

Beni, G. [1963], Elaborazione dei dati ottenuti in experimenti impostati con piani a reticolato, con l'ausilio di calcolatrici scriventi a più totalizzatori e della mecconografia a schede perforate. (Calculation of data obtained in lattice experiments, by the use of tabulating calculating machines with several totalizers and of mechanical reproduction from punched cards.) Genetica Agraria, Roma 16:97-110. E3a;A15 (2107)

Bennett, B. M. [1966], Use and estimating efficiency of rank-order tests in randomized block, latin square, and split-plot designs. Biométrie - Praximétrie, 7:79-87.
E2,5,7,12ae;A12 (2108)

Bennett, C. A. [1954], Effect of measurement error on chemical process control. IQC 10(4):17-20. E1e (2109)

Bennett, L. T. [1966], On a method of constructing partially balanced incomplete block design. Ph. D. Thesis, Oklahoma State Univ. E3c (2110)

Benson, C. T. [1966], A partial geometry $(q^3+1, q^2+1, 1)$ and corresponding PBIB design. Proc. Amer. Math Soc. 17:747-749. E3c (2111)

Berenblut, I. I. [1964], Change-over designs with complete balance for first residual effects. Biometrics 20:707-712. E8ac (2112)

Berenblut, I. I. [1967], The analysis of changeover designs with complete balance for first residual effects. Biometrics 23:578-580. E8ac (2113)

Berenblut, I. I. [1967], A change-over design for testing a treatment factor at four equally spaced levels. JRSSB 29:370-373 (corrigendum 29:586). E8ac (2114)

Berman, G. [1955], A three parameter family of partially balanced incomplete block designs with two associate classes. Proc. Amer. Math. Soc. 6:490-493. E3c (2115)

Bhagwandas, [1965], A note on balanced incomplete block designs. JASA 3:41-45. E2c (2116)

Bhapkar, V. P. [1959], A note on confidence bounds connected with ANOVA and MANOVA for balanced and partially balanced incomplete block designs (abstract). AMS 30:618 (Inst. Stat. Mimeo. Ser. No. 216, N. Carolina). E2,3a;A7 (2117)

Bhapkar, V. P. [1959], Contributions to the statistical analysis of experiments with one or more responses (not necessarily normal). Ph. D. Thesis, Univ. N. Carolina (Inst. Stat. Mimeo. Ser. No. 229, N. Carolina). E2a;A12 (2118)

Bhapkar, V. P. [1960], Confidence bounds connected with ANOVA and MANOVA for balanced and partially balanced incomplete block designs. AMS 31:741-748. E2,3a (2119)

Bhargava, P. N. [1959], Latin Square with orthogonal partitioning (abstract). Proc. Forty-Sixth Indian Sci. Congress, Part III, p. 19. E7ac (2120)

Bhargava, P. N. [1960], Analysis of the (r^2-r, r) orthogonal portion of a Latin Square. JISAS 12:95-99. E7a (2121)

Bhargava, P. N. and Das, M. N. [1961], On Rectangular Lattice Designs (abstract). Proc. Forty-Eighth Indian Sci. Congress, Part III, p.24. E3a (2122)

Bhargava, R. P. [1967], Analaysis of variance for Scheffe's mixed model for stair-case design. Sankhyā A 29:391-398. E4a (2123)

Bhaskar Rao, M. [1967], Partially balanced designs with unequal block sizes. JISA 5:90-103. E4ac (2124)

Bhattacharya, N. [1957], A note on the construction of orthogonal latin squares. Sankhyā 17:351-352. E7,12,13c (2125)

Bhattacharyya, K. N. [1950], Problems in partially balanced incomplete block designs. Calcutta Stat. Assoc. Bull. 2:177-182. E3c (2126)

Biggers, J. D. [1959], The estimation of missing and mixed-up observations in several experimental designs. Biometrika 46:91-105. E2,6,7,10ae (2127)

Biggers, J. D. [1961], Estimation of missing observations in split-plot experiments where whole-plots are missing or mixed-up. Biometrika 48:468-472. E5,6a (2128)

Bingham, R. S. [1959], Design of experiments from a statistical viewpoint, part II. IQC 15(12):12-15. E7ae;T2ae (2129)

Birtwistle, R. [1955], The Hyper-Graeco-Latin square. C.S.I.R.O. Div. of Math. Stat. Technical paper No. 2, pp. 1-12. E9,12,13ace (2130)

Bishop, R. F., MacLeod, L. B., Jackson, L. P., MacEachern, C. R., and Goring, E. T. [1962], A long-term field experiment with commercial fertilizers and manure. II. Fertility levels and crop yields in a rotation of potatoes, oats and hay. Canadian J. Soil Sci. 42:49-60. E5,8a (2131)

Blackith, R. E. [1950], Bio-assay systems for the pyrethrins. III. Application of the twin cross-over design to crawling insect assays. Ann. Appl. Biol., 37:508-515. E8ace;T8ae (2132)

Blackwelder, W. C. [1966], Construction of balanced incomplete block designs from association matrices. M. S. Thesis, Univ. N. Carolina. E2,3ac (2133)

Blackwelder, W. C. [1966], On constructing balanced incomplete block designs from association matrices with special reference to association schemes of two and three classes. Inst. Stat. Mimeo. Ser. No. 496, Univ. N. Carolina, 15 pp. E2,3c (2134)

Blackwell, D. [1951], Comparison of experiments. Proc. Second Berkeley Symp. Math. Stat. Prob. pp. 93-102. E14 (2135)

Blenk, H. [1951], Grundsätzliche Betrachtungen zur Varianzanalyse. Z. Pflanzenzüchtung 30:122-142. E1,2,7,12ae (2136)

Bliss, C.I. [1952], The clinical assay of diuretic agents II. Estimation of the error in a clinical assay. Biometrics 8:237-248. E2ae;T8ae (2137)

Bliss, C. I. [1956], The analysis of insect counts as negative binomial distributions. (with discussion). Proc. Tenth Int'l. Congress Entomology 2:1015-1032. E2ae;T2ae;A9 (2138)

Bliss, C. I., Greenwood, M. L., and McKendrick, M. H. [1953], A comparison of scoring methods for taste tests with mealiness of potatoes. Food Tech. 7:491-495. E7ac;C (2139)

Bloch, D. [1966], A Bayesian approach to some missing value problems in ANOVA and contingency tables. Dept. Stat., Johns Hopkins Univ., Tech. Report No. 62, pp. 1-19. E2,6a (2140)

Blomquist, M. O. [1966], Rank analysis of complete block factorial designs. Ph. D. Thesis, Univ. Minnesota. E2a;T2a;A12 (2141)

Boguslawski, E. v. [1958], Der Feldversuch mit polyfaktorieller Fragestellung. Z. Landw. Vers.- Untersuch Wes. 4:316-342. E2,3,5ac;T2a (2142)

Bohun-Chudyniv, V. [1954], On a general method for constructing completely orthogonal $2^k \times 2^k$ squares ($k \geq 1$) by using closed orthogonal systems of K-nions (abstract). Proc. Int'l. Congress Math. 2:281-282. E7,12,13c (2143)

Bohun-Chudyniv, V. [1955], On determining all possible orthogonalizable sets of Latin $2^3 \times 2^3$ squares and pertaining to them completely orthogonal Latin squares (abstract). Bull. Amer. Math. Soc. 61:175. E7,12,13c (2144)

Bohun-Chudyniv, V. [1955], On a new method of constructing canonical sets of Latin $2^k \times 2^k$ squares ($k \geq 2$) (abstract). Bull. Amer. Math. Soc. 61:174-175. E7,12,13c (2145)

Boogaerdt, J. [1959], De oorsprong van de naam "Latijns vierkant". (Origin of the name "Latin Square".) Stat. Neerlandica 13:465-466. E7a (2146)

Bose, R. C. [1950], A note on orthogonal arrays (abstract). Biometrics 6:179-180. E2,7c (2147)

Bose, R. C. [1950], A note on orthogonal arrays (abstract). AMS 21:304-305. E2,7c (2148)

Bose, R. C. [1951], Partially balanced incomplete block designs with two associate classes involving only two replications. Calcutta Stat. Assoc. Bull. 3:120-125. E3c (2149)

Bose, R. C. [1951], Group divisible incomplete block designs (abstract). AMS 22:311-312. E3c (2150)

Bose, R. C. [1952], A note on Nair's condition for partially balanced incomplete block designs with $k > r$. Calcutta Stat. Assoc. Bull. 4:123-126. E3c (2151)

Bose, R. C. [1952], Partially balanced designs with two plots per block (abstract). AMS 23:639. E3c (2152)

Bose, R. C. [1956], Paired comparison designs for testing concordance between judges. Biometrika 43:113-121. E2c;T11c (2153)

Bose, R. C. [1959], Note on Parker's method of constructing pairwise orthogonal sets of Latin squares (preliminary report) (abstract 557-11). Notices Amer. Math. Soc. 6:179. E7,12c (2154)

Bose, R. C. [1959], On the application of finite projective geometry for deriving a certain series of balanced Kirkman arrangements. Calcutta Math. Soc. Golden Jubilee Commemoration Vol. (1958-1959), pp. 341-354. E2c (2155)

Bose, R. C. [1960], On a method of comparing Steiner's triple systems. Contributions to Probability and Statistics, pp. 133-141. E2c (2156)

Bose, R. C. [1962], Partial geometries and partially balanced designs (abstract). AMS 33:1206. E3c (2157)

Bose, R. C. [1963], Combinatorial properties of partially balanced designs and association schemes. Sankhyā A 25:109-136 (also published in Contributions to Statistics, pp. 21-48). E3c (2158)

Bose, R. C. [1963], Strongly regular graphs, partial geometries and partially balanced designs. Pacific J. Math. 13:389-419 (also, Inst. Stat. Mimeo. Ser. No. 358, Univ. N. Carolina). E3c (2159)

Bose, R. C. [1965], Session chairman's prepared remarks. Proc. of the IBM Scientific Symp. Stat. pp. 55-57. E12c;T4c (2160)

Bose, R. C, and Bush, K. A. [1952], Orthogonal arrays of strength two and three. AMS 23:508-524 (abstract 22:312). E3,7c (2161)

Bose, R. C. and Cameron, J. M. [1965], The bridge tournament problem and calibration designs for comparing pairs of objects. J. Res. NBS 69B:323-332. E4ac;A2 (2162)

Bose, R. C. and Cameron, J. M. [1967], Calibration designs based on solutions to tournament problem. J. Res. NBS 71B:149-160. E4ace;A2 (2163)

Bose, R. C. and Chakravarti, I. M. [1966], Hermitian varieties in a finite projective space PG(N,q^2). Canadian J. Math. 18:1161-1182. E2c (2164)

Bose, R. C., Chakravarti, I. M., and Knuth, D. E. [1960], On methods of constructing sets of mutually orthogonal Latin squares using a computer (abstract). AMS 31:813-814. E7,12,13c (2165)

Bose, R. C., Chakravarti, I. M., and Knuth, D. E. [1960], On methods of constructing sets of mutually orthogonal latin squares using a computer. I. Technometrics 2:507-516. E7,12,13c (2166)

Bose, R. C., Chakravarti, I. M., and Knuth, D. E. [1961], On methods of constructing sets of mutually orthogonal Latin squares using a computer. II. Technometrics 3:111-117. E7,12,13c (2167)

Bose, R. C., Chakravarti, I. M., and Knuth, D. E. [1961], On methods of constructing sets of mutually orthogonal Latin squares using a computer (summary). JASA 56:393. E7c (2168)

Bose, R. C. and Clatworthy, W. H. [1952], Partially balanced designs with $k>r=3$, $\lambda_1=1$, $\lambda_2=0$ (abstract). AMS 23:138. E3c (2169)

Bose, R. C. and Clathworthy, W. H. [1955], Some classes of partially balanced designs. AMS 26:212-232. E3c (2170)

Bose, R. C., Clatworthy, W. H., and Shrikhande, S. S. [1954], Tables of partially balanced designs with two associate classes. N. Carolina Agri. Expt. Sta. Tech. Bull. No. 107, pp. i-iv, 1-255. E3c (2171)

Bose, R. C. and Connor, W. S. [1951], Necessary conditions for the existence of a symmetrical group divisible design (abstract). AMS 22:611.E3c (2172)

Bose, R. C. and Connor, W. S. [1952], Combinatorial properties of group divisible incomplete block designs. AMS 23:367-383. E3c (2173)

Bose, R. C. and Kuebler, R. R. [1958], On the construction of a class of error-correcti binary signalling codes. Inst. Stat. Mimeo. Ser. No. 199, Univ. N. Carolina. E2,3,4c (2174)

Bose, R. C. and Laskar, R. [1967], A characterization of tetrahedral graphs. J. Combinatorial Theory 3:366-385. E3c (2175)

Bose, R. C. and Mesner, D. M. [1959], On linear associative algebras corresponding to association schemes of partially balanced designs. AMS 30:21-38 (also Inst. Stat. Mimeo. Ser. No. 188, Univ. N. Carolina). E3c (2176)

Bose, R. C. and Nair, K. R. [1952], Resolvable incomplete block designs with two replications (abstract). Biometrics 8:94-95. E3ac (2177)

Bose, R. C. and Nair, K. R. [1952], Resolvable incomplete block designs with two replications (abstract). AMS 23:299. E3ac (2178)

Bose, R. C. and Nair, K. R. [1962], Resolvable incomplete block designs with two replications. Sankhyā A 24:9-24 (also Inst. Stat. Mimeo. Ser. No. 69, Univ. N. Carolina). E3ac (2179)

Bose, R. C. and Ray-Chaudhuri, D. K. [1959], Application of the geometry of quadrics in finite projective space to the construction of PBIB designs (abstract). AMS 30:614. E3c (2180)

Bose, R. C. and Ray-Chaudhuri, D. K. [1960], On a class of error correcting binary group codes. Information Control 3:68-79 (also, Inst. Stat. Mimeo. Ser. No. 240, Univ. N. Carolina). E2c (2181)

Bose, R. C. and Shimamoto, T. [1952], Classification and analysis of partially balanced incomplete block designs with two associate classes. JASA 47:151-184. E3ace (2182)

Bose, R. C. and Skrikhande, S. S. [1959], A theorem on the construction of orthogonal Latin squares and the falsity of Euler's conjecture (preliminary report) (abstract 558-26). Notices Amer. Math. Soc. 6:378-379. E7,12c (2183)

Bose, R. C. and Shrikhande, S. S. [1959], On the falsity of Euler's conjecture about the non-existence of two orthogonal Latin squares of order 4t+2. Proc. Nat. Academy Sci. (USA) 45:734-737. E7,12c (2184)

Bose, R. C. and Shrikhande, S. S. [1959], On the falsity of Euler's conjecture for all orders exceeding 26 (abstract 558-27). Notices Amer. Math. Soc. 6:379. E7,12c (2185)

Bose, R. C. and Shrikhande, S. S. [1959], On the composition of balanced incomplete block designs of index unity (abstract 559-18). Notices Amer. Math. Soc. 6:404. E2c (2186)

Bose, R. C. and Shrikhande, S. S. [1959], On the construction of pairwise orthogonal sets of latin squares and the falsity of the conjecture of Euler: II. Inst. Stat. Mimeo. Ser. No. 225, Univ. N. Carolina. E7,12c (2187)

Bose, R. C. and Shrikhande, S. S. [1959], A note on a result in the theory of code construction. Information Control 2:183-194. E2c;T7c (2188)

Bose, R. C. and Shrikhande, S. S. [1960], On the construction of sets of mutually orthogonal Latin squares and the falsity of a conjecture of Euler. Trans. Amer. Math. Soc. 95:191-209. E7,12c (2189)

Bose, R. C. and Shrikhande, S. S. [1960], On the composition of balanced incomplete block designs. Canadian J. Math. 12:177-188 (also Inst. Stat. Mimeo. Ser. No. 239, Univ. N. Carolina). E2c (2190)

Bose, R. C., Shrikhande, S. S., and Bhattacharya, K. N. [1953], On the construction of group divisible incomplete block designs. AMS 24:167-195 (also Inst. Stat. Mimeo. Ser. No. 60, Univ. N. Carolina). E3c (2191)

Bose, R. C., Shrikhande, S. S., and Parker, E. T. [1960], Further results on the construction of mutually orthogonal Latin squares and the falsity of Euler's conjecture. Canadian J. Math. 12:189-203. E7,12c (2192)

Bose, R. C., Shrikhande, S. S., and Parker, E. T. [1963], Orthogonal Latin squares and Euler's conjecture (discussion). Colloque Internationaux du Centre National de la Recherche Scientifique, No. 110 Le Plan d'Experiences, pp. 233-255 (also Inst. Stat. Mimeo. Ser. No. 245, Univ. N. Carolina). E7,12c (2193)

Bose, R. C. and Srivastava, J. N. [1964], Multidimensional partially balanced designs and their analysis, with applications to partially balanced factorial fractions. Sankhyā A 26:145-168 (also Inst. Stat. Mimeo. Ser. No. 376, Univ. N. Carolina). E3,7,13a;T7ac (2194)

Box, G. E. P. [1954], Some theorems on quadratic forms applied in the study of analysis of variance problems. I. Effect of inequality of variance in the one-way classification. AMS 25:290-302.. E1a;A11 (2195)

Box, G. E. P. [1954], Some theorems on quadratic forms applied in the study of analysis of variance problems. II. Effects of inequality of variance and of correlation between errors in the two-way classification. AMS 25:484-498. E2a;A11 (2196)

Box, G. E. P. and Guttman, I. [1966], Some aspects of randomization. JRSSB 28:543-558 (also Dept. Stat. Tech. Report No. 64, Univ. Wisconsin). E2a;T4a;C (2197)

Box, G. E. P. and Mueller, M. E. [1959], Randomization and least squares estimates (summary). JASA 54:489. E2a (2198)

Bowles, R. [1951], Accuracy of certain approximate solutions in predicting the correct model for experiments with unequal frequencies in subclasses. M. S. Thesis, Iowa State Univ. E2a;A10 (2199)

Bozivich, H., Bancroft, T. A., and Hartley, H. O. [1956], Analysis of variance: Preliminary tests, pooling and linear models. Preliminary tests of significance and pooling procedures for certain incompletely specified models. Iowa State Univ. WADC Tech. Report 55-244, 1:1-118. E2a;A11 (2200)

Bradley, J. V. [1958], Complete counterbalancing of immediate sequential effects in a Latin Square design. JASA 53:525-528 (corrigenda, 53:1030-1031). E8c (2201)

Bradley, R. A. [1951], Rank analysis of incomplete block designs I (Preliminary report) (abstract). Biometrics 7:125. E2a;A12 (2202)

Bradley, R. A. [1952], Statistical methods for sensory difference tests of food quality. Va. Agri. Expt. Sta., Blacksburg, Va. Bi-Annual Report No. 5, 119 pp. (Mimeo.). E2,3a;T11a;C (2203)

Bradley, R. A. [1953], Some further extensions of the rank analysis of incomplete block designs (abstract). Virginia J. Sci. 4:289-290. E2a;A12 (2204)

Bradley, R. A. [1954], Incomplete block rank analysis: On the appropriateness of the m[illegible] for a method of paired comparisons. Biometrics 10:375-390. E2ae;T11ae;A12 (2205)

Bradley, R. A. [1954], Rank analysis of incomplete block designs, II. Additional tables for the method of paired comparisons. Biometrika 41:502-537 (corrigendum 51:288). E2a;T11a;A12 (2206)

Bradley, R. A. [1955], Rank analysis od incomplete block designs III. Some large-sample results on estimation and power for a method of paired comparisons. Biometrika 42:450-470. E2a;T11a (2207)

Bradley, R. A. [1955], Some notes on the theory and application of rank order statistics. Part I. IQC 11(5):12-16. E1ae;A12 (2208)

Bradley, R. A. [1955], Some notes on the theory and application of rank order statistics. Part II. IQC 11(6):5-9. E1,2ae;A12 (2209)

Bradley, R. A. [1958], Recent research on statistical problems in subjective testing. ASQC, Nat. Conv. Trans. 12:143-152. E3ae;T2ae;C (2210)

Bradley, R. A. [1958], Recent research in statistical problems in subjective testing. Proc. Second Conf. Design Expt. Army Res. Dev. Testing. pp. 5-38. E3ae;T2ae;C (2211)

Bradley, R. A. [1963], Some relationships among sensory difference tests. Biometrics 19:385-397. E14;T11a;C (2212)

Bradley, R. A. and Duncan, D. B. [1951], Statistical methods for sensory difference tests of food quality. Va. Agri. Expt. Sta., Blacksburg, Va. Bi-Annual Report No. 1, (Mimeo.). C (2213)

Bradley, R. A. and Kramer, C. Y. [1958], Addenda to "Intra block analysis for factorials in two-associate class group divisible designs". AMS 29:933-935. E3a;T2a (2214)

Bradley, R. A. and Schumann, D. E. W. [1957], The comparison of the sensitivities of similar experiments: Applications. Biometrics 13:496-510. E14;C (2215)

Bradley, R. A and Somerville, P. N. [1953], Statistical methods for sensory difference tests of food quality. Va. Agri. Expt. Sta., Blacksburg, Va. Bi-Annual Report No. 7 (Mimeo). E2a;T11a;A2 (2216)

Bradley, R. A. and Somerville, P. N. [1954], Statistical methods for sensory difference tests of food quality. Va. Agri. Expt. Sta., Blacksburg, Va., Bi-Annual Report No. 8 (Mimeo.). E2a;T11a;A3 (2217)

Bradley, R. A. and Terry, M. E. [1951], Rank analysis of incomplete block designs. Virginia J. Sci. 2:379. E2a;A12 (2218)

Bradley, R. A. and Terry, M. E. [1951], Statistical methods for sensory difference tests of food quality. Va. Agri. Expt. Sta., Blacksburg, Va., Bi-Annual Report No. 2 (Mimeo.). E2,3a;T11a;C (2219)

Bradley, R. A. and Terry, M. E. [1951], Statistical methods for sensory difference tests of food quality. Va. Agri. Expt. Sta., Blacksburg, Va., Bi-Annual Report No. 3 (Mimeo.). E2,3a;T11a;C (2220)

Bradley, R. A. and Terry, M. E. [1952], New designs and techniques for organoleptic testing - the rank analysis of incomplete block designs - I. (abstract). Virginia J. Sci. 3:360. E2,3a;A12 (2221)

Bradley, R. A. and Terry, M. E. [1952], Rank analysis of incomplete block designs I - The method paired comparisons (abstract). Biometrics 8:95. E2a;T11a;A12 (2222)

Bradley, R. A. and Terry, M. E. [1952], Rank analysis of incomplete block designs. I. The method of paired comparisons (abstract). AMS 23:299-300.
E2a;T11a;A12 (2223)

Bradley, R. A. and Terry, M. E. [1952], Rank analysis of incomplete block designs II. The method for blocks of three (Preliminary report) (abstract). Biometrics 8:96.
E2a;A12 (2224)

Bradley, R. A. and Terry, M. E. [1952], Rank analysis of incomplete block designs. II. The method for blocks of three (Preliminary report) (abstract). AMS 23:300.
E2a;A12 (2225)

Bradley, R. A. and Terry, M. E. [1952], Rank analysis of incomplete block designs. I. The method of paired comparisons. Biometrika 39:324-345.
E2ae;T11ae;A12 (2226)

Bradley, R. A. and Terry, M. E. [1952], Statistical methods for sensory difference tests of food quality. Va. Agri. Expt. Sta., Blacksburg, Va., Bi-Annual Report No. 4 (Mimeo.).
E2,3a;T11a;C (2227)

Bradley, R. A. and Terry, M. E. [1953], Statistical methods for sensory difference tests of food quality. Va. Agri. Expt. Sta., Blacksburg, Va., Bi-Annual Report No. 6 (Mimeo).
E2,3a;T11a;C (2228)

Bradley, R. A., Walpole, R. E., and Kramer, C. Y. [1960], Intra- and inter-block analysis for factorials in incomplete block designs. Biometrics 16:566-581.
E3a;T2a (2229)

Brenna, L. S. [1958], Factorial treatments in lattice designs. Ph. D. Thesis, Virginia Polytechnic Inst.
E3a;T2a (2230)

Brenna, L. S. and Kramer, C. Y. [1959], Factorial treatment combinations in lattices (abstract). 15:633-634.
E3a;T2a (2231)

Brenna, L. S. and Kramer, C. Y. [1961], Factorial treatments in rectangular lattice designs. JASA 56:368-378.
E3ac;T2ac (2232)

Brougham, R. W. [1955], A study in rate of pasture growth. Australian J. Agri. Res. 6:804-812.
E5e (2233)

Brown, J. W. [1966], Enumeration of latin squares and isomorphism detection in finite planes. Ph. D. Thesis, Univ. California, Los Angeles.
E7,12,13c (2234)

Brown, W. A. and Krane, S. A. [1963], Design and analysis of entomological field experiments. Proc. Eight Conf. Design Expt. Army Res. Dev. Testing. pp. 99-118.
E5a;C (2235)

Brownlee, K. A. [1957], The principles of experimental design. IQC 13(8):12-20. E1,2,6,7,12ae; T2ae;C (2236)

Bruck, R. H. [1951], Finite nets, I. Numerical invariants. Canadian J. Math. 3:94-107. E7,12c (2237)

Bruck, R. H. [1963], Finite notes II. Uniqueness and imbedding. Pacific J. Math. 13:421-457. E3,7,12c (2238)

Bruck, R. H. [1963], What is a loop? Studies in Modern Algebra, pp. 59-99. E2,3,7,13c (2239)

Brunk, M. E. [1955], Sample surveys and experimental designs (abstract). Econometrica 23:346. E8a (2240)

Brunk, M. E. and Federer, W. T. [1953], How marketing problems of the apple industry were attacked and the research results applied. Biometrics Unit Mimeo. Ser. BU-40-M. Cornell Univ. E7,8a;T3,7c (2241)

Brunk, M. E. and Federer, W. T. [1953], Experimental designs and probability sampling in marketing research. JASA 48:440-452. E7a;T7a;A13 (2242)

Bruno, O. P. [1959], Effects of ballistics and meteorological variables on accuracy of artillery fire. Proc. Fourth Conf. Design Expt. Army Res. Dev. Testing. pp. 127-135. E7a (2243)

Bryant, E. C. [1955], An analysis of some two-way stratifications. Ph. D. Thesis, Iowa State Univ. E2a;A10 (2244)

Bucher, B. [1957], The recovery of intervariety information (abstract). AMS 28:530. E2,3a (2245)

Bucher, B. D. [1957], The recovery of intervariety information in incomplete block designs. Ph. D. Thesis, Princeton Univ. E2,3a (2246)

Bucher, B. [1957], The recovery of intervariety information (abstract). Biometrics 13:412. E2,3a (2247)

Burghausen, R. [1963], Die Bedeutung der Vegetationsheobachtungen in Kartoffelversuchen. European Potato J. 6:168-177. E2ae;A9;C (2248)

Burrows, P. M. [1966], Estimation of multiple missing values in standard experimental designs. The Rhodesia, Zambia and Malawi J. Agri. Res. 4:45-47. E2,6,7,10a (2248a)

Burton, G. W. and Fortson, J. C. [1965], Lattice-square designs increase precision of pearl millet forage yield trials. Crop Sci. 5:595. E11,14a (2249)

Bush, H. L. and Oldemeyer, R. K. [1961], Comparison of statistical designs for a large number of entries of sugar beet strains. J. Amer. Soc. Sugar Beet Tech. 11:306-308. E3,5a (2250)

Bush, K. A. [1950], Orthogonal arrays. Ph. D Thesis, Univ. N. Carolina. E2,3c (2251)

Bush, K. A. [1952], Orthogonal arrays of index unity. AMS 23:426-434. E7,12c (2252)

Bush, K. A. [1952], A generalization of a theorem due to MacNeish (note). AMS 23:293-295. E7,12c (2253)

Bush, K. A [1960], Euler squares. <u>Contributions to Probability and Statistics</u>, pp. 150-152. E7,12c (2254)

Butcher, J. C. [1956], Treatment variances for experimental designs with serially correlated observations. Biometrika 43:208-212. E5a (2255)

Cady, F. B. and Mason, D. D. [1961], Statistical analysis of a short-term crop rotation (abstract). Agron. Abstracts p. 67. E5,8a (2256)

Cady, F. B. and Mason, D. D. [1964], Comparison of fertility treatments in a crop rotation experiment. Agron. J. 56:476-479. E5,8ace (2257)

Calvin, L. D. [1953], Doubly balanced incomplete block designs for experiments in which the treatment effects are correlated. Ph. D. Thesis, Univ. N. Carolina. E,4,5ac (2258)

Calvin, L. D. [1954], Doubly balanced incomplete block designs for experiments in which the treatment effects are correlated. Biometrics 10:61-88. E4,5ace (2259)

Cameron, J. M. [1965], Calibration designs (abstract). Technometrics 7:269. E4c;T6c (2260)

Carlitz, L. [1953], Congruences connected with three-line Latin rectangles. Proc. Amer. Math. Soc. 4:9-11. E7c (2261)

Carmer, S. G. [1965], Computer programs for the construction of experimental layouts. Agron. J. 57:312-313. E3,11c;A15 (2262)

Cassady, J. C. [1962], Individual degrees of freedom for non-additivity in two- and higher-way designs. M. S. Thesis, Cornell Univ. (also Biometrics Unit Mimeo. Ser. BU-136-M, Cornell Univ.). E2a;T2a;A4 (2263)

Castellano, V. [1963], Sur l'emploi des méthodes d'élimination et d'autres procédés discutables dans l'élaboration probabiliste des données expérimentales. Colloques Internationoux de Centre National de la Recherche Scientifique No. 110, Le Plan d'Experiences pp. 165-179. E7,12ae (2264)

Chacko, V. J. [1951], Generalized incomplete block designs (abstract). Proc. Thirty-Eighth Indian Sci. Congress, Part III, p. 10. E4ac (2265)

Chacko, V. J. [1966], Designs for forestry spacing and thinning experiments. International Advisory Group of Forest Statisticians, Second Conf. Institutionen För Skoglig Matematisk Statistik pp. 23-35. E5ac;C (2266)

Chakrabarti, M. C. [1950], A note on balanced incomplete block designs. Bull. Calcutta Math. Soc. 42:14-16. E2c (2267)

Chakrabarti, M. C. [1963], On the C matrix in design of experiments. JISA 1:8-23. E14 (2268)

Chakrabarti, M. C. [1963], Query. JISA 1:50-52. E6a;A7 (2269)

Chakrabarti, M. C. [1963], Solution of problem 7. JISA 1:180. E2c (2270)

Chakrabarti, M. C. [1963], Query. JISA 1:230-234. E2c (2271)

Chakrabarti, M. C. [1963], Solution to problem 14. JISA 1:237-238. E2a (2272)

Chakrabarti, M. C. [1964], Query 2. JISA 2:184-186. E2c (2273)

Chakrabarti, M. C. [1964], Solution of problem 25. JISA 2:191. E2c (2274)

Chakrabarti, M. C. [1965], Solution to problem 39. JISA 3:61. E2c (2275)

Chakrabarti, M. C. [1965], Some aspects of design and analysis of experiments (Presidential address). Proc. Fifty-First/Fifty-Second Indian Sci. Congress, Part II, pp. 23-36. E2,3a;T2,6a;B (2275a)

Chakravarti, I. M. [1958], Simplified proofs of some results in the theory of optimal designs. Sankhyā 19:189-194. E14 (2276)

Chakravarti, I. M. [1963], Orthogonal and partially balanced arrays and their application in design of experiments. Metrika 7:231-243. E7c;T3,7c (2277)

Chakravarti, I. M. [1964], Combinatorial problems in design of experiments. Aerospace Res. Lab. ARL 64-102, U.S. Air Force, pp. 1-23. E2,3,7c (2278)

Chakravarti, I. M. [1965], Difference sets, orthogonal mappings and orthogonal Latin squares (abstract). AMS 36:353-354. E7c (2279)

Chakravarti, I. M. [1966], On the construction of difference sets and their use in the search for orthogonal Latin squares and error correcting codes (with discussion) Bull. ISI 41:957-958 (also Inst. Stat. Mimeo No. 427, Univ. N. Carolina.). E7,12,13c (2280)

Chakravarty, M. C. [1964], Design of experiments. Res. Training School Publication No. SM 64-3, Indian Stat. Inst. E;T (2281)

Chan, M. W. [1964], A multivariate analysis of an experimental design involving a complete set of Latin squares. Psychometrika 29:233-240. E8,12,13a (2282)

Chand, U. [1953], Experiments on tree crops (abstract). Biometrics 9:429-430. E2a;T2a;C (2283)

Chand, U. and Abraham, T. P. [1957], Some considerations in the planning and analysis of fertilizer experiments in cultivators' fields. JISAS 9:101-134. A13;C (2284)

Chandrasekhararao, K. [1964], On further constructions of cubic designs. Calcutta Stat. Assoc. Bull. 13:71-75. E3c (2285)

Chang, L. C. [1959], The uniqueness and nonuniqueness of the triangular association schemes. Sci. Record New Series 3:604-613. E3c (2286)

Chang, L. C. [1960], Association schemes of partially balanced designs with parameters v=28, n_1=12, n_2=15, p^2_{11}=4. Sci. Record, New Series 4:12-18. E3c (2287)

Chang, L.-C. [1961], On the estimator of relative efficiency of the randomized complete block design (in Chinese). Taiwan Tahsueh Nunghsuehyuan Yenchiu Paokao/Mem. College Agri. Nat. Taiwan Univ. 6(1):31-42. E2,14a (2288)

Chang, L.-C. and Liu, W. R. [1964], Incomplete block designs with square parameters for which k≤10 and r≤10. Scientia Sinica 13:1493-1495. E3c (2289)

Chassan, J. B. [1964], On the analysis of simple cross-overs with unequal numbers of replicates. Biometrics 20:206-208. E7a (2290)

Chew, V. [1958], Basic experimental designs. Experimental Designs in Industry. pp. 3-57. E1,2,6,7,12ae; T2a;B (2291)

Chilton, N. W., Fertig, J. W., and Kutscher, A. H. [1961], Studies in the design and analysis of dental experiments. III. Sequential analysis (double dichotomy). J. Dental Res. 40:331-340. E5ac;S2 (2292)

Chilton, N. W., Fertig, J. W. and Varma, A. A. O. [1964], Studies in the design and analysis of dental experiments. 10. Sequential analysis (t-test). J. Oral Therapeutics Pharmacology 1:175-182. E7ae;S2 (2293)

Chopra, A. S. [1960], On circular designs (abstract). JISAS 12:211-212. E3a (2294)

Chopra, A. S. [1961], Circular designs and designs developable from initial blocks (abstract). Proc. Forty-Eighth Indian Sci. Congress, Part III, p. 25. E3c (2295)

Chopra, A. S. [1964], Circular designs and designs developable from initial blocks. Ph. D. Thesis, I.A.R.S., New Delhi. E3c (2296)

Chopra, A. S. and Das, M. N. [1962], A note on fractional replicate of balanced incomplete block designs. JISAS 14:145-150. E3c (2297)

Chopra, A. S. and Das, M. N. [1964], On circular design with 3 and 4 plot blocks. JISAS 16:212-237. E3ac (2298)

Chowla, S., Erdös, P., Straus, E. G. [1960], On the maximal number of pairwise orthogonal Latin squares. Canadian J. Math. 12:204-208. E7,12,13c (2299)

Chowla, S. and Ryser, H. J. [1950], Combinatorial problems. Canadian J. Math. 2:93-99. E2c (2300)

Christidis, B. G. [1955], Rotation experiments with cotton. Empire J. Expt. Agri. 23:49-54. E5,8a (2301)

Ciminera, J. L. and Wolfe, E. K. [1953], An example of the use of extended cross-over designs in the comparison of NPH insulin mixtures. Biometrics 9:431-446. E7ae (2302)

Clarke, G. M. [1955], A design for testing several treatments under controlled environmental conditions. Appl. Stat. 4:199-206. E12ac (2303)

Clarke, G. M. [1963], A second set of treatments in a Youden square design. Biometrics 19:98-104. E12,13ace (2304)

Clarke, G. M. [1967], Four-way balanced designs based on Youden squares with 5,6,or 7 treatments. Biometrics 23:803-812. E12c (2305)

Clarke, P. M. [1951], An analysis of rectangular lattices with unequal block sizes, using inter-block information. Biometrics 7:287-294. E3,4a (2306)

Clarke, P. M. [1955], A note on the estimation of error in slope-ratio assays arranged in randomized blocks. Analyst 80:396-397. E2a;T8a (2307)

Clarke, R. T. and Smith, C. W. R. [1964], The residual effects of potassium fertilizers on yields of arable crops: preliminary results of five rotation experiments. Expt. Husbandry No. 12, pp. 15-23. E5,8a (2308)

Clatworthy, W. H. [1952], Partially balanced incomplete block designs with two associate classes and three replications. Ph. D. Thesis, Univ. N. Carolina (also Inst. Stat. Mimeo Ser. No. 54, Univ. N. Carolina). E3ac (2309)

Clatworthy, W. H. [1954], A geometrical configuration which is a partially balanced incomplete block design. Proc. Amer. Math. Soc. 5:47-55. E3c (2310)

Clatworthy, W. H. [1955], Partially balanced incomplete block designs with two associate classes and two treatments per block. J. Res. NBS 54:177-190. E3c (2311)

Clatworthy, W. H. [1956], Contributions on partially balanced incomplete block desig with two associate classes. NBS Appl. Math. Ser. 47 (U.S. Government Printing Office, Washington 25, D. C.) E3ac (2312)

Clatworthy, W. H. [1965], Some new families of partially balanced designs of the Latin square type (abstract). AMS 36:1080-1081. E7c (2313)

Clatworthy, W. H. [1967], On John's cyclic incomplete block designs. JRSSB 29: 243-247. E2,3c (2314)

Clatworthy, W. H. [1967], Some new families of partially balanced designs of the Latin square type and related designs. Technometrics 9:229-244 (abstract 8: 204). E3,7c (2315)

Clem, M. A. and Federer, W. T. [1950], Random arrangements for lattice designs. Agri. Expt. Sta. Special Report No. 5, Iowa State Univ. E2,3,11ac (2316)

Cléroux, R. [1965], On the F-test in the intrablock analysis of two associate partially balanced incomplete block designs (Preliminary report) (abstract). AMS 36:1611. E3a (2317)

Cléroux, R. [1966], Sur la robustesse du Test F dans l'analyse intrabloc du schéma à blocs incomplets partiellement équilibré et à deux classes associées. Ph. D. Thesis, Univ. Montreal. E3a (2318)

Clysters, H., Grossmann, B., Gilbert, P., and Lenger, A. [1962], Etude biométrique des effets secondaires de quelques fongicides sur pommier au cours de trois années consecutives. Biométrie-Praximétrie 3:194-218. E5e (2319)

Cochran, W. G. [1954], Query 106. Biometrics 10:155-157. E7a (2320)

Cochran, W. G. [1960], The design of experiments. Operations Research and Systems Engineering, pp. 508-553. E2,3,7a;T2a;C (2321)

Cole, J. W. L. and Grizzle, J. E. [1965], Applications of multivariate analysis of variance to repeated measurements experiments (abstract). Biometrics 21:1022-1023. E5ac;A7 (2322)

Cole, J. W. L. and Grizzle, J. E. [1966], Applications of multivariate analysis of variance to repeated measurements experiments. Biometrics 22:810-828. E5ac;A7 (2323)

Cole, L. C. [1959], On simplified computations. Amer. Stat. 13(1):20. E1,2a;A7 (2324)

Collier, R. O. [1956], Experimental designs in which the observations are assumed to be correlated. Ph. D. Thesis, Univ. Minnesota. E2,5,8a (2325)

Collier, R. O. [1958], Analysis of variance for correlated observations. Psychometrika 23:223-236. E5,12a (2326)

Collier, R. O. and Baker, F. B. [1963], The randomization distribution of F-ratios for the split-plot design - an empirical investigation. Biometrika 50:431-438. E5,12a;A5 (2327)

Collier, R. O. and Baker, F. B. [1966], Some Monte Carlo results on the power of the F-test under permutation in the simple randomized block design. Biometrika 53:199-203. E2a;A5 (2328)

Conagin, A. [1954], Látices retangulares. (Rectangular lattices.) Bragantia 13: 187-197. E3ace (2329)

Conagin, A. [1954], Análise da covariância em um látice retangular simples. (Covariance analysis in rectangular lattices.) Bragantia 14:35-50. E3ae;A7 (2330)

Conagin, A. [1955], Transformações dos dados experimentais. (Transformations of experimental data.) Bragantia 14:141-147. E2e;A9 (2331)

Connor, W. S. [1951], On the structure of incomplete block designs and the impossibility of certain unsymmetrical cases. Ph. D. Thesis, Univ. N. Carolina. E2,3ac (2332)

Connor, W. S. [1951], On the structure of incomplete block designs. Inst. Stat. Mime Ser. No. 44, Univ. N. Carolina. E2,3c (2333)

Connor, W. S. [1951], The structure of balanced incomplete block designs, and the impossibility of certain unsymmetrical cases (abstract). AMS 22:312-313. E2c (2334)

Connor, W. S. [1952], On the structure of balanced incomplete block designs. AMS 23:57-71 (correction, 24:135). E2c (2335)

Connor, W. S. [1952], Some relations among the blocks of symmetrical group divisible designs. AMS 23:602-609. E3c (2336)

Connor, W. S. [1953], The correspondence between two classes of balanced incomplet block designs (abstract). AMS 24:490. E2c (2337)

Connor, W. S. [1954], New experimental designs for paired observations (summary of paper). JASA 49:360. E2a;T11a (2338)

Connor, W. S. [1958], Experiences with incomplete block designs. Experimental Designs in Industry pp. 193-206. E2,3ace (2339)

Connor, W. S. [1958], The uniqueness of the triangular association scheme. AMS 29:262-266. E3c (2340)

Connor, W. S. [1960], Precision in interlaboratory tests. Ind. Eng. Chemistry 52(10): 77A-78A & 80A. E7c;A14 (2341)

Connor, W. S. [1960], Factorial experiments in incomplete block designs. Ind. Eng. Chemistry 52(12):83A-84A. E4ace;T2ace (2342)

Connor, W. S. and Clatworthy, W. H. [1953], Necessary conditions for the existence of partially balanced incomplete block designs with two associate classes (abstrac Biometrics 9:261. E3c (2343)

Connor, W. S. and Clatworthy, W. H. [1953], Necessary conditions for the existence of partially balanced incomplete block designs with two associate classes (abstract). AMS 24:497. E3c (2344)

Connor, W. S. and Clatworthy, W. H. [1954], Some theorems for partially balanced designs. AMS 25:100-112. E3c (2345)

Connor, W. S. and Rupp, M. K. [1952], The complete solutions of the balanced incomplete block designs with ten or fewer replications. Nat. Bur. Standards NBS Report 1896. E2c (2346)

Connor, W. S. and Savage, I. R. [1952], Question 34: Randomized blocks versus balanced incomplete blocks. Amer. Stat. 6(4):28. E2a (2347)

Connor, W. S. and Youden, W. J. [1952], Analysis of chain block designs (Preliminary report) (abstract). AMS 23:136. E4a (2348)

Coons, I. [1957], The analysis of covariance as a missing plot technique. Biometrics 13:387-405. E6,10a;A7 (2349)

Cornish, E. A. [1956], The recovery of interblock information in quasi-factorial designs with incomplete data. 3. Balanced incomplete blocks. C.S.I.R.O. Div. Math. Stat. Tech. Paper No. 3, pp. 1-21. E6,11ac (2350)

Corsten, L. C. A. [1957], Een algemene methode tot leemtevulling. (General missing plot technique.) Stat. Neerlandica 11:141-151. E6,10a (2351)

Corsten, L. C. A. [1958], Vectors, a tool in statistical regression theory. Mededelingen Landbouwhogeschool 58(1):1-92. E2,3,6,7ac;A7;C (2352)

Corsten, L. C. A. [1958], General missing plot technique. (In Dutch with Eng. summary.) Wageningen Rijksinst. v. Rassenonderz. van Landbgewassen. Meded. 34:141-151. E6 (2353)

Corsten, L. C. A. [1959], Incomplete block designs in which the number of replicates is not the same for all treatments. Inst. Stat. Mimeo. Ser. No. 226, Univ. N. Carolina. E4ac (2354)

Corsten, L. C. A. [1959], Note on triangular partially balanced incomplete block designs. Inst. Stat. Mimeo. Ser. No. 242. Univ. N. Carolina. E3c (2355)

Corsten, L. C. A. [1959], On triangular partially balanced incomplete block designs (abstract). Biometrics 15:635-636. E3ac (2356)

Corsten, L. C. A. [1960], Proper spaces related to triangular partially balanced incomplete block designs. AMS 31:498-501. E3c (2357)

Corsten, L. C. A. [1962], Balanced block designs with two different numbers of replicates. Biometrics 18:499-519. E4ace (2358)

Costa Pires, A. J. and Baptista, J. G. [1955], Análise do delineamento em blocos casualizados quando existem dois e três talhões perdidos. Melhoramento 8: 61-77. E2,6ae (2359)

Coury, T. et al [1953], Estudos sôbre o "vermelhão" do algodoeiro (III). Anaiz Escola Superior Agri. "Luiz de Queiroz" 10:83-94. E2ae (2360)

Coury, T., Torres, A. P., and Ranzani, G. [1955-1956], Experiências de adubação mineral e orgânica com Capim Kikuyu (Pennisetum clandestinum Hochst.) Anais Escola Superior Agri. "Luiz de Queiroz" 12-13:19-35. E2ae (2361)

Cox, C. P. [1956], Latin-square designs with individual gradients in one direction. Nature 177:1092. E7a (2362)

Cox, C. P. [1958], The analysis of Latin square designs with individual curvatures in one direction. JRSSB 20:193-204. E7ae (2363)

Cox, C. P. [1958], Experiments with two treatments per experimental unit in the presence of an individual covariate. Biometrics 14:499-512. E7,8ae (2364)

Cox, C. P. [1961], A practical application of a theoretically inefficient design. Biometrics 17:646-649. E7ac;T2ac (2365)

Cox, C. P. [1962], The relation between covariance and individual curvature analyses of experiments with background trends. Biometrics 18:12-21. E7a;A7,14 (2366)

Cox, D. R. [1951], Some systematic experimental designs. Biometrika 38:312-323. E5ac (2367)

Cox, D. R. [1952], Some recent work on systematic experimental designs. JRSSB 14:211-219. E5ace (2368)

Cox, D. R. [1956], A note on weighted randomization. AMS 27:1144-1151. E5,7a;A5;C (2369)

Cox, D. R. [1957], The use of a concomitant variable in selecting an experimental design. Biometrika 44:150-158 (correction 44:534). E5ac;A7 (2370)

Cox, D. R. [1958], The interpretation of the effects of non-additivity in the Latin square. Biometrika 45:69-73. E7a;A4 (2371)

Cox, G. M. [1951], The value and usefulness of statistics in research. Committee Expt. Design Agri. Res. Administration Mimeo, U.S. Dept. Agri. pp. 1-11. E2a;S1 (2372)

Crews, J. W., Jones, G. L., and Mason, D. D. [1964], Field plot technique studies with flue-cured tobacco. II. Experimental designs and replications. Agron. J. 56:435-438. E3,14a;A14 (2373)

Crouse, C. F. [1967], A class of distribution-free analysis of variance tests. S. African Stat. J. 1:75-80. E1,2a;A12 (2374)

Cunningham, E. P. and Henderson, C. R. [1966], Analytical techniques for incomplete block experiments. Biometrics 22:829-842. E2,3ae (2375)

Cureton, E. E. [1957], Further note on the two-group t test. Amer. Stat. 11(4):21. E1a (2376)

Curnow, R. N. [1957], Heterogeneous error variances in split-plot experiments. Biometrika 44:378-383. E5,12a;A11 (2377)

Curnow, R. N. [1959], The analysis of a two phase experiment. Biometrics 15:60-73. E12ae (2378)

Curnow, R. N. and Sharpe, E. [1962], The analysis of covariance as a means of reducing standard errors in certain experiments involving sequences of treatments. Biometrics 18:410-413. E7,8ae;A7 (2379)

Dagnelie, P. [1959], Le carré latin magique: Technique d'analyse de la variance. Rev. l'Agriculture 12(3):1-12. E12ace (2380)

Dai, Z. D. and Feng, X. N. [1964], Notes on finite geometries and the construction of PBIB designs. IV. Some "Anzahl" Theorems in orthogonal geometry over finite fields of characteristic not 2. Scientia Sinca 13:2001-2004. E3c (2381)

Dalebroux, M. [1960], Blocs randomisés et interactions. Biométrie-Praximétrie 1(2):35-44. E5a (2382)

Dalebroux, M. [1961], Exemple d'analyse d'un plan factoriel $3^2 \times (2)$ en parcelles découpées (split-plot). Biométrie-Praximétrie 2:121-129. E12ae;T2ae (2383)

Dall 'Aglio, G. [1963], Blocs incomplets équilibrés orthogonoux (discussion). Colloques Internatioux du Centre de la Recherche Scientifique, No. 110 Le Plan d'Expériences, pp. 195-214. E7ac (2384)

Danford, M. B., Hughes, H. M., and McNee, R. C. [1960], On the analysis of repeated-measurements experiments. Biometrics 16:547-565 (abstract 15:149). E5a;A7 (2385)

Daniel, C. and Wilcoxon, F. [1966], Factorial 2^{p-q} plans robust against linear and quadratic trends. Technometrics 8:259-278. E5ace;T3ace (2386)

Dar, S. N. [1962], On the comparison of the sensitivities of experiments. JRSSB 24:447-453. E14 (2387)

Dar, S. N. [1964], Comparison of the sensitivities of dependent experiments. Biometrics 20:209-212. E14 (2388)

Darby, L. A. and Gilbert, N. [1958], The Trojan square. Euphytica 7:183-188. E9ace (2389)

Darroch, J. G. [1960], Statistical methods in the selection and evaluation of sugar can Proc. Tenth Congress Int. Soc. Sugar Cane Tech. pp. 836-841. E4a (2390)

Das, B. [1967], Tactical configurations and graph theory. Calcutta Stat. Assoc. Bull. 16:136-138. E2c (2391)

Das, M. N. [1954], Missing plots and a randomised block design with balanced incompleteness. JISAS 6:58-76. E2,6ae (2392)

Das, M. N. [1954], On parametric relations in a balanced incomplete block design. JISAS 6:147-152. E2c (2393)

Das, M. N. [1955], Missing plots in partially balanced and other incomplete block designs. JISAS 7:111-126. E3,6ae (2394)

Das, M. N. [1955], Latin Squares with several missing plots. JISAS 7:46-56. E7,10ae (2395)

Das, M. N. [1956], Analysis of covariance in incomplete block designs with or without missing plots. JISAS 8:76-83. E3,6ae;A7 (2396)

Das, M. N. [1957], A generalised balanced design. JISAS 9:18-30. E4ace (2397)

Das, M. N. [1957], Bio-assays with non-orthogonal data. JISAS 9:67-81. E4ace;T8a;A10 (2398)

Das, M. N. [1958], On reinforced incomplete block designs (abstract). Proc. Forty-Fifth Indian Sci. Congress, Part III, p. 19. E4c (2399)

Das, M. N. [1958], A balanced design with unequal replications within block (abstract). JISAS 10:160. E2,4ac (2400)

Das, M. N. [1958], On reinforced incomplete block designs. JISAS 10:73-77. E4ac (2401)

Das, M. N. [1959], Circular designs (abstract): Proc. Forty-Sixth Indian Sci. Congress, part III, p. 18-19. E3ac (2402)

Das, M. N. [1959], A new type of design (abstract). JISAS 11:195. E4c (2403)

Das, M. N. [1960], Augmented fields and factorial designs in general (abstract). JISAS 12:152. E7c;T2c (2404)

Das, M. N. [1960], Circular designs. JISAS 12:45-56. E3ac (2405)

Das, M. N. [1960], A class of P.B.I.B. designs (abstract). JISAS 12:151. E3c (2406)

Das, M. N. and Bhargava, P. N. [1958], On problems of analysis of experiments having missing or mixed up plots (abstract). Proc. Forty-Fifth Indian Sci. Congress, part III, p. 19. E6a (2407)

Das, M. N. and Bhargava, P. N. [1958], On missing and mixed up plot techniques. JISAS 10:78-82. E2,3,6,7,10a;A7 (2408)

Das, M. N. and Bhargava, P. N. [1960], Applications of the type of designs based on constant frequency differences (abstract). JISAS 12:210. E4c (2409)

Das, M. N. and Kulkarni, G. A. [1959], On the accuracy of current approximate variances of treatment differences in randomized block designs with missing observations (abstract). Proc. Forty-Sixth Indian Sci. Congress, part III, p. 18. E2,6a (2410)

Das, M. N. and Kulkarni, G. A. [1958], On the accuracy of current approximate variances of treatment differences in randomised block designs with missing observations. JISAS 10:83-89. E2,6a (2411)

Das, M. N. and Kulkarni, G. A. [1960], A series of designs for parallel line bio-assays with two or more preparations (abstract). JISAS 12:209-210. E3a;T8a (2412)

Das, M. N. and Kulkarni, G. A. [1966], Incomplete block designs for bio-assays. Biometrics 22:706-729. E3ace;T8a (2413)

Das, M. N. and Kulshreshtha, A. C. [1966], Construction of balanced incomplete block designs by using confounded factorial designs (abstract). JISAS 18(2): 104-105. E2c (2414)

Das, M. N. and Rao, S. V. S. P. [1966], Balanced n-ary designs and their alternative uses (abstract). JISAS 18(2):105 E5c (2415)

David, H. A. [1962], Special problems in testing hypotheses. <u>Contributions to Order Statistics</u>, Chapter 7A, pp. 94-128. E1,2ae;A11;B (2416)

David, H. A. [1963], <u>The Method of Paired Comparisons</u>. No. 12 Griffins Statistical Monographs and Courses. Chas. Griffin & Co. Ltd., London, and Hafner Publishing Co., New York. E2,3ac;T11ac,A2;B (2417)

David, H. A. [1963], The structure of cyclic paired-comparison designs. J. Australian Math. Soc. 3:117-127. E2,3c;T11c (2418)

David, H. A. [1965], Enumeration of cyclic paired-comparison designs. Amer. Math. Monthly 72:241-248. E3c;T11c (2419)

David, H. A. [1967], Resolvable cyclic designs. Sankhyā A 29:191-198. E3ac (2420)

David, H. A. and Wolock, F. W. [1965], Cyclic designs. Proc. Tenth Conf. Design Expt. Army Res. Dev. Testing. pp. 283-297. E4ac;T11ac (2421)

David, H. A. and Wolock, F. W. [1965], Cyclic designs. AMS 36:1526-1534. E3c (2422)

Day, B. B. and Del Priore, F. R. [1953], The statistics in a gear-test program. IQC 9(5):16-20. E7,10ae (2423)

de Almeida Leme, H. [1955-1956], Contribuição para o estudo da influência da profundide de trabalho do arado na produção agricola. Anais Escola Superior Agri. "Luiz de Queiroz" 12-13:95-111. E1ae (2424)

DeGroot, M. H. [1966], Optimal allocation of observations. Ann. Inst. Stat. Math. 18:13-28. E14;T1c (2425) de la Puente, J. [1962], Analysis of a randomized complete block design in the event that data is missing. Estadística 20:307-314. E6ae (2426)

Delate, E. J. and Walz, R. N. [1958], Evaluating alternative ingredients in the manufacture of a product. Ind. Eng. Chemistry 50:309-312. E3ace (2427)

DeLury, D. B. [1954], On the design of experiments. IQC 10(4):24-29. E1a;T2a;C (2428)

de Oliveira, A. J. [1958], Analysis of a group of experiments on oats. II. Incomplete data. Agronomia Lusitana 20:155-176. E6,10a;A13 (2429)

d'Herbemont, G. [1960], Quelques remarques à propos d'un plan d'expérience. Rev. Stat. Appl. 8(2):41-52. E12ae (2430)

d'Herbemont, G. [1962], Considérations géométriques sur les plans d'expérimentation. Bull. ISI 39(3):145-154. E14;A14 (2431)

Dick, I. D. and Whittle, P. [1951], Contributions to the statistical design of identical twin experiments. New Zealand J. Sci. Tech. B 33:145-172. E5ac;A10 (2432)

Di Paola, J. W. [1966], Block designs and graph theory. J. Combinatorial Theory 1:132-148. E2c (2433)

Di Paola, J. W. [1966], On a restricted class of block design games. Canadian J. Math. 18:225-236. E2c (2434)

Divers, C. K. [1961], Statistical aids to visual inspection. ASQC Ann. Conv. Trans. 15:525-531. E2ae;A12 (2435)

Djokoto, R. K. and Stephens, D. [1961], Thirty long-term fertilizer experiments under continuous cropping in Ghana. I. Crop yields and responses to fertilizers and manures. Empire J. Expt. Agri. 29:181-195. E5a;T2a (2436)

Djokoto, R. K. and Stephens, D. [1961], Thirty long-term fertilizer experiments under continuous cropping in Ghana. II. Soil studies in relation to the effects of fertilizers and manures on crop yields. Empire J. Expt. Agri. 29:245-258. E5a;T2a (2437)

Doehlert, D. H. [1965], Balanced sets of balanced incomplete block designs of block size three. Technometrics 7:561-577. E2c (2438)

Doksum, K. [1967], Robust procedures for some linear models with one observation per cell. AMS 38:878-883. E2a (2439)

Dominick, B. A. [1952], Methods of research in marketing. An illustration of the use of the latin square in measuring the effectiveness of retail merchandising practices. Dept. Agri. Economics Paper No. 2, Cornell Univ.
E7e (2440)

Dossett, W. F. [1957], A further note on the t test. Amer. Stat. 11(4):21.
E1,2a (2441)

Douglas, A. W. [1961], Some multiple comparison tests for two-way classifications. M. S. Thesis, Cornell Univ. E2a;A12 (2442)

Douglas, A. W. [1964], Tests of homogeneity of variance for the latin square and a class of balanced incomplete block designs. Ph. D. Thesis, Cornell Univ.
E2,7a;A11 (2443)

Draper, N. R. and Guttman, I. [1966], Unequal group variances in the fixed-effects one-way analysis of variance: a Bayesian sidelight. Biometrika 53:27-35.
E1a;A11 (2444)

Draper, N. R. and Stoneman, D. M. [1966], Factor changes and linear trends in eight-run two-level factorial designs. Dept. Stat. Tech. Report No. 82, Univ. Wisconsin pp. 1-14. E5c;T3c (2445)

Dubetz, S., Russell, G. C., and Hill, K. W. [1955], Crop sequence studies on irrigated land in southern Alberta. Canadian J. Agri. Sci. 35:564-567.
E5,8a (2446)

Dudeney, H. E. and Bose, R. C. [1965], Magic squares. Encyclopaedia Britannica 14:573-575. E12c (2447)

Dugue, D. [1964], Un mélange d'algebre et de statistique: le plan d'expériences (with discussion). Rev. Stat. Appl. 12(1):7-15. E12ac (2448)

Dulmage, A. L., Johnson, D., and Mendelsohn, N. S. [1959], Orthogonal latin squares. Canadian Math. Bull. 2:211-216. E7,12,13c (2449)

Duncan, D. B. and Walser, M. [1966], Multiple regression combining within- and between-plot information. Biometrics 22:26-43. E12a (2450)

Durbin, J. [1951], Incomplete blocks in ranking experiments. British J. Psychology, Stat. Sec. 4:85-90. E2a;A12 (2451)

Dutton, A. M. [1951], Statistical analysis of long-term agricultural experiments. Ph. D. Thesis, Iowa State Univ. E5a (2452)

Dutton, A. M. [1952], Statistical analysis of long-term agricultural experiments (abstract). Iowa State College J. Sci. 26:198-199. E5a (2453)

Dwyer, P. S. [1958], Computational formulas for t^2. Amer. Stat. 12(3):18-19.
E1,2a (2454)

Dykstra, O. [1956], Incomplete block rank analysis: 2^{p-q} fractional factorials using a method of paired comparisons (abstract). AMS 27:871. E2a;T3,11a;A12 (2455)

Dykstra, O. [1956], A note on the rank analysis of incomplete block designs - Applications beyond the scope of existing tables. Biometrics 12:301-306. E2a;A12 (2456)

Dykstra, O. [1960], Rank analysis of incomplete block designs: A method of paired comparisons employing unequal repetitions on pairs. Biometrics 16:176-188. E2a;T11a;A12 (2457)

Eaton, H. D., Gosslee, D. G., and Lucas, H. L. [1959], Effect of duration of experiment on experimental errors in calf nutrition growth studies. J. Dairy Sci. 42:1398-1400. E5a;A14 (2458)

Ebner, A. R. [1959], Cumulant component estimation in the balanced one-way nested classification. M. S. Thesis, Cornell Univ. E1a;A14 (2459)

Ebner, A. R. and Federer, W. T. [1958], An incomplete block design in a nutrition experiment. Biometrics Unit Mimeo. Ser. BU-100-M, Cornell Univ. E3,6ace (2460)

Edwards, A. L. [1950], Homogeneity of variance and the Latin square design. Psychological Bull. 47:118-129. E7ae;A11 (2461)

Edwards, A. L. [1951], Balanced Latin-square designs in psychological research. American J. Psychology 64:598-603. E7,8c (2461a)

Edwards, M. V. [1956], The design, layout and control of provenance experiments. Z. Forstgentik Forstpflanzenzüchtung 5:169-180. E2a;A14;C (2462)

Ehrenfeld, S. [1955], On the efficiency of experimental designs. AMS 26:247-255. E14 (2463)

Ehrenfeld, S. [1956], Complete class theorems in experimental design. Proc. Third Berkeley Symp. Math. Stat. Prob. 1:57-67. E14;T10 (2464)

Ehrenfeld, S. [1961], Some experimental design problems in attribute life testing. Stat. Eng. Group Tech. Report No. 15, Columbia Univ. E5a (2465)

Eisenhart, C. [1951], On the statistical analysis of linked-block experiments (abstract). Biometrics 7:125. E4c (2466)

Eisler, S. L. [1957], Determining the effectiveness of cutting oils in reducing machine tool wear. Proc. First Conf. Design Expt. Army Res. Dev. Testing. pp. 235-239. E12c (2467)

Eissner, R. M. [1958], The use of a special systematic design for surveillance testing. Proc. Second Conf. Design Expt. Army Res. Dev. Testing. pp. 103-109. E5a (2468)

Elandt, R. [1957], O mozliwościach stosowania testów nieparametrycznych w zagadnieniach dóswiadczalnictwa rolniczego. (The possibility of applying nonparametric tests to problems of agricultural experimentation.). Roczniki Nauk Rolniczych 75(A-2):161-187. E2a;A12 (2469)

Elandt, R.C. [1962], Eliminacja dwukierunkowej zmiennosci glebowej za pomoca krzywych parabolicznych w układzie dlugich pasow. (Elimination of two-directional soil variability by the method of parabolic curves in long strip experiment.) Roczniki Nauk Rolniczych 87(A-1):117-134. E7ac (2470)

Elandt, R. C. [1964], Applicability of the extended method of parabolic curves in the analysis of agricultural data. Sankhyā B 26:201-216. E2,5a;A7 (2471)

Elandt-Johnson, R. C. [1967], Application of certain systematic designs in planning of factorial experiments (in Polish). Roczniki Nauk Rolniczych 94(A):665-678. E5ac;T2a (2472)

Elandt, R. C. and Andrew, G. M. [1964], Tables for application of the method of parabolic curves to a certain systematic arrangement. Sankhyā B 26:17-28. E2,5ac;A7 (2473)

Elliot, F. C., Darroch, J. G., and Wang, H. L. [1952], Uniformity trials with spring wheat. Agron. J. 44:524-528. E2,14a;A14 (2474)

Evans, T. [1960], Embedding incomplete latin squares. American Math. Monthly 67:958-961. E7,12c (2474a)

Evans, T. A. and Mann, H. B. [1951], On simple difference sets. Sankhyā 11:357-364. E2,7c (2474b)

Federer, W. T. [1950], The general theory of prime power lattice designs. IV. Analysis for p^4 treatments in blocks of p plots with four or more replicates. Cornell Univ. Agri. Expt. Sta. Memoir 299:1-29. E3ace (2475)

Federer, W. T. [1950], The general theory of prime-power lattice designs V. The analysis for a 6×6 incomplete lattice square illustrated with an example. Biometrics 6:34-58. Ellace (2476)

Federer, W. T. [1951], Evaluation of variance components from a group of experiments with multiple classifications. Iowa Agri. Expt. Sta. Res. Bull. 380:241-310. E2,6a;T9a;A13 (2477)

Federer, W. T. [1951], Application of Rao's general method of analysis to a simple or double lattice design. Biometrics Unit Mimeo. Ser. BU-16-M, Cornell Univ. E3ae (2478)

Federer, W.T. [1951], Application of Rao's general method of analysis to a cubic lattice design. Biometrics Unit Mimeo. Ser. BU-17-M, Cornell Univ. E3ae (2479)

Federer, W. T. [1951], A note on error (b) in the split plot design. Biometrics Unit Mimeo. Ser. BU-19-M, Cornell Univ. E5,12a (2480)

Federer, W. T. [1951], A note on degrees of freedom for the average effective error variance in two-dimensional one-restrictional lattices. Biometrics Unit Mimeo. Ser. BU-20-M, Cornell Univ. E2,3a (2481)

Federer, W. T. [1952], Analysis for a 5×5 lattice square in 2 replicates and in 4 replicates. Biometrics Unit Mimeo. Ser. BU-27-M, Cornell Univ. Ellac (2482)

Federer, W. T. [1953], Random arrangements for some three-dimensional lattice designs. Cornell Univ. Agri. Expt. Sta. Miscellaneous Bull. 19:1-178. E3c (2483)

Federer, W. T. [1954], Principles of experimentation. Biometrics Unit. Mimeo. Ser. BU-72-M, Cornell Univ. El,2,7ac;C (2484)

Federer, W. T. [1954], Covariance analysis in a two-way classification with unequal numbers in the subclasses. Biometrics Unit Mimeo. Ser. BU-51-M, Cornell Univ. E2a;T2a;A7 (2485)

Federer, W. T. [1954], A bivariate analysis for Snedecor's sugar beet example on covariance. Biometrics Unit Mimeo. Ser. BU-52-M, Cornell Univ. E2a;A7 (2486)

Federer, W. T. [1955], Query 116. Biometrics 11:239. E7ae;A7 (2487)

Federer, W. T. [1955], Covariance versus the method of fitting constants. Biometrics Unit Mimeo. Ser. BU-60-M, Cornell Univ. E7a (2488)

Federer, W. T. [1956], Augmented (or hoonuiaku) designs. Hawaiian Planters' Record 55:191-208 (also Biometrics Unit Mimeo. Ser. BU-74-M, Cornell Univ.) E4ace (2489)

Federer, W. T. [1956], Least squares estimates and sums of squares for a double change-over design. Biometrics Unit Mimeo. Ser. BU-70-M, Cornell Univ. E8a (2490)

Federer, W. T. [1956], Analysis for a 4×4 balanced lattice square with an additional replicate. Biometrics Unit Mimeo. Ser. BU-79-M, Cornell Univ. E11ac (2491)

Federer, W. T. [1957], Restrictions on the linear model. Biometrics Unit Mimeo. Ser. BU-83-M, Cornell Univ. E2a (2492)

Federer, W. T. [1957], Covariance analyses for unbalanced two-way classifications (summary). JASA 52:369. E2a;T2a;A7,10 (2493)

Federer, W. T. [1957], Variance and covariance analyses for unbalanced classifications. Biometrics 13:333-362. E2a;T2a;A7,10 (2494)

Federer, W. T. [1958], Balanced incomplete block designs for v treatments in b=v(v-1) incomplete blocks of size k=2. Biometrics Unit Mimeo Ser. BU-91-M, Cornell Univ. E2,7ace (2495)

Federer, W. T. [1958], Augmented designs (abstract). Biometrics 14:134. E4ac (2496)

Federer, W. T. [1959], Covariance analyses for unbalanced two-way classifications. Cornell Univ. Agri. Expt. Sta. Memoir 360:1-60. E2a;T2a;A7,10 (2497)

Federer, W. T. [1959], Rectangular lattice designs for $v=pk^n$ in incomplete blocks of size k (abstract). Biometrics 15:333-334. E3ac (2498)

Federer, W. T. [1959], A note on additivity. Biometrics Unit Mimeo. Ser. BU-101-M, Cornell Univ. E2a;A4 (2499)

Federer, W. T. [1960], Designs with one-way elimination of heterogeneity. Biometrics Unit Mimeo. Ser. BU-117-M, Cornell Univ. E3a (2500)

Federer, W. T. [1961], Augmented designs with one-way elimination of heterogeneity. Biometrics 17:447-473. E4ace (2501)

Federer, W. T. [1961], Augmented designs with two-, three- and higher-way elimination of heterogeneity (abstract). Biometrics 17:166. E9,12,13ac (2502)

Federer, W. T. [1961], Procedures and designs useful for screening material in selection work with particular reference to plants. Biometrics Unit Mimeo. Ser. BU-131-M, Cornell Univ. E4c;A2 (2503)

Federer, W. T. [1961], Design and analysis for a 2×4×4 nitrogen, potash, and phosphorous fertilizer trial on spinach. Biometrics Unit Mimeo. Ser. BU-132-M, Cornell Univ. E7c;T2a (2504)

Federer, W. T. [1962], Gradients within blocked experiments. Biometrics Unit Mimeo. Ser. BU-146-M, Cornell Univ. E7ace (2505)

Federer, W. T. [1963], Additional analyses for two-way layouts. Biometrics Unit Mimeo. Ser. BU-157-M, Cornell Univ. E7ae (2506)

Federer, W. T. [1963], Relationships between a three-way classification disproportionate numbers analysis of variance and several two-way classification and nested analyses. Biometrics 19:629-637 (also Biometrics Unit Mimeo. Ser. BU-152-M, Cornell Univ.) E2a;T2a;A10 (2507)

Federer, W. T. [1963], Procedures and designs useful for screening material in selection and allocation, with a bibliography. Biometrics 19:553-587 (also MRC Tech. Summary Report No. 368, U.S. Army Univ. Wisconsin). E4c;A2;B (2508)

Federer, W. T. [1964], Lecture notes on the "Design and Analysis of Experiments" for the Advanced Science Seminar in Mathematical Statistics at Colorado State Univ., 7/13/64 - 8/7/64. Biometrics Unit Mimeo., Cornell Univ., 14 Chapters. E1-12;T2-7;C (2509)

Federer, W. T. [1965], Incomplete data in split-plot, split-block, and other similar designs. Biometrics Unit Mimeo. Ser. BU-178-M, Cornell Univ. E10,12a (2510)

Federer, W. T. [1965], Research designs: Experimental and statistical controls. Proc. AHEA - Cornell Univ. Res. Conf. (also Biometrics Unit Mimeo. Ser. BU-198-M, Cornell Univ.). E7ac;A12 (2511)

Federer, W. T. [1966], Data collection and interpretation. Biometrics Unit Mimeo., Cornell Univ., 10 Chapters. E1,2,5,7,12,14ace; T2,3,4,8,9ac;C (2512)

Federer, W. T. [1967], Sequential design of experiments. Conf. Handbook, VI Inter. Biometric Conf., Sydney 5:80-116 (also Biometrics Unit Mimeo. Ser. BU-237-M, Cornell Univ.). E1,2,7a;S2 (2513)

Federer, W. T. and Atkinson, G. F. [1964], Tied-double-change-over designs. Biometrics 20:168-181 (also MRC Tech. Summary Report No. 391, U.S. Army Univ. Wisconsin). E8ace (2514)

Federer, W. T. and Brunk, M. E. [1960], Analyses for a balanced design in a marketing experiment. Biometrics Unit Mimeo. Ser. BU-127-M, Cornell Univ. E11ac;T7ac (2515)

Federer, W. T. and Farden, C. A. [1955], Analysis of variance set-up for Joint Project 69. Biometrics Unit Mimeo. Ser. BU-67-M, Cornell Univ. E13a;T2a (2516)

Federer, W. T. and Ferris, G. E. [1956], Least squares estimates of effects and sums of squares for a tied double change-over design. Biometrics Unit Mimeo. Ser. BU-73-M, Cornell Univ. E8ac (2517)

Federer, W. T. and Lowe, C. C. [1951], Expectation of mean squares from an experiment for perennial crops. Biometrics Mimeo. Ser. BU-25-M, Cornell Univ. E5,12,13a (2518)

Federer, W. T. and Moyer, J. C. [1965], Analyses of taste panel scores for grape juice samples. Biometrics Unit Mimeo. Ser. BU-203-M, Cornell Univ. E2ace;C (2519)

Federer, W. T. and Nissen, O. [1958], Means and variances of treatments from a randomized complete block design with missing plots. Biometrics Unit Mimeo. Ser. BU-99-M, Cornell Univ. E2,6a (2520)

Federer, W. T. and Plaisted, R. L. [1958], A rectangular lattice design for potato breeding experiments. Biometrics Unit Mimeo. Ser. BU-105-M, Cornell Univ. E3,4ace (2521)

Federer, W. T. and Plaisted, R. L. [1960], Estimates of specific and general combining ability from potato breeding experiments. Biometrics Unit Mimeo. Ser. BU-118-M, Cornell Univ. E3,4a;T9ac (2522)

Federer, W. T. and Plaisted, R. L. [1962], A method for estimating combining ability components of variance in incomplete block designs. American Potato J. 39:197-206. E3,4a;T9a (2523)

Federer, W. T. and Raktoe, B. L. [1966], Generalized lattice square designs. JASA 61:821-832 (also Biometrics Unit Mimeo. Ser. BU-168-M, Cornell Univ.) E11ace (2524)

Federer, W. T. and Raktoe, B. L. [1965], General theory of prime-power lattice designs. Lattice rectangles for $v=s^m$ treatments in s^r rows and s^c columns for $r+c=m$, $r \neq c$, and $v<1000$. JASA 60:891-904. E11ace (2525)

Federer, W. T. and Robson, D. S. [1952], General theory of prime-power lattice designs VI. Incomplete block design and analysis for p^5 varieties in blocks of p^2 plots. Cornell Univ. Agri. Expt. Sta. Memoir No. 309, pp. 1-37. E3ace (2526)

Federer, W. T., Robson, D. S., and Tukey, J. W. [1962], Tests for non-additivity. Biometrics Unit Mimeo. Ser. BU-147-M, Cornell Univ. E7ace;A4,11 (2527)

Federer, W. T. and Schlottfeldt, C. S. [1954], The use of covariance to control gradients in experiments. Biometrics 10:282-290 (errata 11:251). E7ae;A7 (2528)

Federer, W. T. and Tyler, E. A. [1954], Covariance analysis for an incomplete lattice square. Biometrics Unit Mimeo. Ser. BU-53-M, Cornell Univ. E11a;A7 (2529)

Federer, W. T. and Zelen, M. [1964], A catalogue of k-row by b-column latin rectangle designs for 2≤v≤10 with 2≤r≤10 replicates. Biometrics Unit Mimeo. Ser. BU-167-M, Cornell Univ. E7ac (2530)

Federer, W. T. and Zelen, M. [1964], Applications of the calculus for factorials arrangements. II.A. Designs with two-way elimination of heterogeneity. Biometrics Unit Mimeo. Ser. BU-197-M, Cornell Univ. E7ac (2531)

Federer, W. T. and Zelen, M. [1965], Design for two-way elimination of heterogeneity (abstract). Technometrics 7:270-271. E9c (2532)

Feng, X. N. and Dai, Z. D. [1964], Notes on finite geometries and the construction on PBIB designs. V. Some "Anzahl" Theorems in orthogonal geometry over finite fields of characteristic 2. Scientia Sinica 13:2005-2008. E3c (2533)

Ferguson, J. H. A. [1962], Nomogram voor het onderscheidingsvermogen van de F-toets gemodificeerd naar Keuls. (Nomographs for the power of the F-test, modified after Keuls.) Stat. Neerlandica 16:177-180. E2a;A3;S1 (2534)

Ferris, G. E. [1956], Three useful designs in taste-testing (abstract). Biometrics 12:234. E8a (2535)

Ferris, G. E. [1957], A modified Latin square design for taste-testing. Food Res. 22:251-258. E7ace (2536)

Ferris, G. E. [1957], The k-visit method of consumer testing (abstract). AMS 28:816. E5ac (2537)

Ferris, G. E. [1958], The k-visit method of consumer testing. Biometrics 14:39-49. E5ac;A2 (2538)

Finney, D. J. [1957], Stratification, balance and covariance. Biometrics 13:373-386. E1,2a;A7 (2539)

Finney, D. J. [1962], An unusual salvage operation. Biometrics 18:247-250. E2,6a (2540)

Finney, D. J. and Cope, F. W. [1956], The statistical analysis of a complex experiment involving unintentional constraints. Biometrics 12:345-368. E7ae (2541)

Finney, D. J. and Outhwaite, A. D. [1955], Serially balanced sequences. Nature 176:748. E8c (2542)

Fischbeck, G. [1955], Ein Beitrag zur anlage und Auswertung von Feldversuchen. Z. Pflanzenzüchtung 34:197-205. E2ae (2543)

Fleming, A. A. [1952], Field plot technique with hybrid corn under Alabama conditions (abstract). Proc. Assoc. Southern Agri. Workers 49:53. E3a;A14;S1 (2544)

Folks, J. L. [1959], Analysis of quadruple rectangular lattice designs. Biometrics 15:74-86 (acknowledgement 16:695). E3ace (2545)

Folks, J. L. and Graybill, F. A. [1957], Heterogeneity of error variances in a randomized block design. Biometrika 44:275-277. E2a;A11 (2546)

Folks, J. L. and Kempthorne, O. [1960], The efficiency of blocking in incomplete block designs. Biometrika 47:273-283. E3,14a;A5 (2547)

Folks, J. L. and West, D. L. [1959], On the exactness of the missing plot procedure in a randomized block design (abstract). AMS 30:1273-1274. E2,6a;A5 (2548)

Folks, J. L. and West, D. L. [1961], Note on the missing plot procedure in a randomized block design. JASA 56:933-941. E2,6a;A5 (2549)

Folsom, R. E. [1966], Balancing lot means as contrasted to covariance or constrained randomization for evaluating treatment differences in completely randomized designs. M. S. Thesis, Iowa State Univ. E2,4a;A7 (2550)

Fraga, C. G. and Costa, A. S. [1950], Análise de um experimento para combate de vira-cabeça do tomateiro. (The analyses of an experiment to control tomato spotted wilt.) Bragantia 10:305-316. E2e;A9 (2551)

Fraser, D. A. S. [1957], On the combining of interblock and intrablock estimates. AMS 28:814-816. E3a (2552)

Frazier, D. and Decker, R. [1954], Gasoline mileage in winter day-to-day use (summary). JASA 49:354-355. E7c (2553)

Freeman, G. H. [1957], Some experimental designs of use in changing from one set of treatments to another, Part I. JRSSB 19:154-162. E12ac (2554)

Freeman, G. H. [1957], Some experimental designs of use in changing from one set of treatments to another. Part II. Existence of the designs. JRSSB 19:163-165. E9,12c (2555)

Freeman, G. H. [1958], Some aspects of experimental designs of use in changing from one set of treatments to another. Ph. D. Thesis, London Univ. East Malling Res. Sta. E9,12ac (2556)

Freeman, G. H. [1957], Some further methods of constructing regular group divisible incomplete block designs. AMS 28:479-487. E3c (2557)

Freeman, G. H. [1958], Families of designs for two successive experiments. AMS 29:1063-1078. E5,12c (2558)

Freeman, G. H. [1959], The use of the same experimental material for more than one set of treatments. Appl. Stat. 8:13-20. E12ae (2559)

Freeman, G. H. [1961], Some further designs of type O:PP. AMS 32:1186-1190. E12c (2560)

Freeman, G. H. [1964], The addition of further treatments to Latin square designs. Biometrics 20:713-729 (abstract 20:383). E12ace (2561)

Freeman, G. H. [1964], The use of a systematic design for a spacing trial with a tropical tree crop. Biometrics 20:200-203. E5c (2562)

Freeman, G. H. [1966], Some non-orthogonal partitions of 4×4, 5×5 and 6×6 Latin squares. AMS 37:666-681. E7c (2563)

Freeman, G. H. [1967], The use of cyclic balanced incomplete block designs for directional seed orchards. Biometrics 23:761-778. E2,3c (2564)

Freeman, G. H. and Jeffers, J. N. R. [1962], Estimation of means and standard errors in the analysis of non-orthogonal experiments by electronic computer. JRSSB 24:435-446. E7,10ae;A10,15 (2565)

Fry, P. R. and Taylor, W. B. [1954], Analysis of virus local lesion experiments. Ann. Appl. Biology 41:664-674. E2,14a (2566)

Fulkerson, D. R. and Ryser, H. J. [1963], Width sequences for special classes of (0,1)-matrices. Canadian J. Math. 15:371-396. E2c (2567)

Fuller, W. A. and Cady, F. B. [1965], Estimation of asymptotic rotation and nitrogen effects. Agron. J. 57:299-302. E5,8a;A13 (2568)

Gaito, J. [1958], The single Latin square design in psychological research. Psychometrika 23:369-378. E7a (2569)

Gardner, C. O. [1951], Design and analysis of Bose's triangular singly linked block designs. Inst. Stat. Mimeo. Ser. No. 55, Univ. N. Carolina. E3ac (2570)

Gardner, M. [1959], How three modern mathematicians disproved a celebrated conjecture of Leonhard Euler. Sci. Amer. 201(Nov.):181-182, 184, 186, 188. E12c (2571)

Gassner, B. J. [1965], Equal-difference BIB designs. Proc. Amer. Math. Soc. 16:378-380. E2c (2572)

Gatty, R. [1960], Factorial designs for store sales experiments. Dept. Agri. Economics Tech. A. E. No. 1 Rutgers Univ. E7a;T2a (2573)

Gatty, R. [1965], Statistical models for experiments in merchandising. Proc. Business Economic Stat. Section, Amer. Stat. Assoc. pp. 227-235. E7a (2574)

Geidel, H. [1953], Vereinfachte Verrechnung eines "ausgewogenen Versuches in unvollständigen Blöcken", erläutert an einem Beispiel von A. Linder. Mitteilungsblatt Mathematische Statistik 5:64-69. E2ac (2575)

Geidel, H. [1953], Ein Beitrag zur Auswertung von Versuchsergebnissen mit der Varianzanalyse. Z. Pflanzenzüchtung 32:373-380. E2ae (2576)

Geidel, H. [1957], Zum Ausgleich von Bodenunterscheiden bei Blockanlagen. Z. Acker- Pflanzenbau 103:71-82. E2,7ae (2577)

Geidel, H. and Schuster, W. [1961], Zur Verrechnung von Feldversuchergebnissen nach dem lateinischen Quadrat und dem lateinschen Rechtek. Z. Acker- Pflanzenbau 113:425-432. E7,14a (2578)

Geisser, S. [1959], A method for testing treatment effects in the presence of learning. Biometrics 15:389-395. E7a;A7 (2579)

Geisser, S. [1961], The latin square as a repeated measurements design. Proc. Fourth Berkeley Symp. Math. Stat. Prob. 4:241-250. E7,8a (2580)

Ghosh, M.N. and Behari, V. [1965], Analysis of randomized block experiments when treatments have different errors. Calcutta Stat. Assoc. Bull. 14:93-105. E2a;A11 (2581)

Ghosh, S. P. [1967], On the construction of BIB designs using non-degenerated quadratics in finite Euclidean space (abstract). AMS 38:955. E2c (2582)

Gilbert, E. N. [1961], Design of mixed doubles tournament. American Math. Monthly 68:124-131. E2c;T11a (2583)

Gilbert, P. and Grossman, B. [1961], Comparison de deux méthodes d'analyse de données non orthogonales dans le cas d'expériences a deux facteurs. Biométrie-Praximétrie 2:19-36. E6ae (2584)

Girault, M. [1963], Rapport de synthèse (discussion). Colloques Inter. Centre Nat. Recherche Sci. No. 110, Le Plan d'Expériences pp. 275-288. E2,7,12ac;C (2585)

Giri, N. C. [1957], On a reinforced partially balanced incomplete block design. JISAS 9:41-51. E4ae (2586)

Giri, N. C. [1957], On row-balance in P.B.I.B. designs. JISAS 9:168-178. E7ac (2587)

Giri, N. C. [1958], Row balance P.B.I.B. design (abstract). Proc. Fourty-Fifth Indian Sci. Congress, Part III, p. 19. E7ac (2588)

Giri, N. C. [1959], On inter-block analysis of some p.b.i.b. designs having a missing block (abstract). JISAS 11:196. E3,6a (2589)

Giri, N. C. [1963], On the combined analysis of Youden squares and of latin square designs with some common treatments. Biometrics 19:171-174. E4,9a;A13 (2590)

Giri, N. C. [1965], On the F-test in the intrablock analysis of a class of two associate PBIB designs. JASA 60:285-293. E3a;A5 (2591)

Glass, S. and Kramer, C.Y. [1958], Analysis of variance of a Latin square design with missing observations (abstract). Virginia J. Sci. 9:452. E7,10a (2592)

Glenday, A. C. [1955], The mathematical separation of plant and weather effects in field growth studies. Australian J. Agri. Res. 6:813-822. E5ace (2593)

Glenn, W. A. [1960], A comparison of the effectiveness of tournaments. Biometrika 47:253-262 (also, Dept. Stat. Tech. Report No. 42, Virginia Polytechnic Inst.) E14;A2 (2594)

Glenn, W. A. and Kramer, C. Y. [1958], Analysis of variance of a randomised block design with missing observations. Appl. Stat. 7:173-185. E2,6a (2595)

Glenn, W. A. and Kramer, C. Y. [1958], Analysis of variance of a randomized block design with missing observations (abstract). Biometrics 14:570. E2,6a (2596)

Glenn, W. A. and Kramer, C. Y. [1958], Analysis of variance of a randomized block design with missing observations (abstract). Virginia J. Sci. 9:451-452. E2,6a (2597)

Goch, D. C. [1958], Examination of residuals in the analysis of variants. South African J. Sci. 54:67-69. E2a;A6 (2598)

Goebel, J. B. [1962], On the algebra of orthogonal Latin squares. Ph. D. Thesis, Oregon State Univ. E7,12,13c (2599)

Goff, M. M. [1960], Efficiency factors for balanced incomplete block designs under loss of experimental units. M. S. Thesis, Univ. N. Carolina. E2,6a (2600)

Golomb, S. W. and Posner, E. C. [1964], Rook domains, Latin squares, affine planes, and error-distributing codes. IEEE Trans. Information Theory IT-10:196-208. E7,12c (2601)

Gomes, F. P. [1951], A lei de Mitscherlich e a análise da variância em experiências de adubação. Anais Escola Superior Agri. "Luiz de Queiroz" 8:355-368. E1,2ae;A7 (2602)

Gomes, F. P. [1951], The interpolation of Mitscherlich's first approach law and the analysis of variance in experiments with fertilizers. Anais Escola Superior Agri. "Luiz de Queiroz" 8:195-204 (Spanish version 8:185-194). E7ae;A7 (2603)

Gomes, F. P. [1954], General method of analysis for lattice designs. Seminários Estatística (10°):33-54. E2,3a (2604)

Gomes, F. P. [1955], As geometrias finitas e sua aplicação nos delineamentos experimentais. (Finite geometries and their application in experimental designs.) An. I. Congr. Brasil, Estad, Agron. pp. 123-126. E2,3c (2605)

Gomes, F. P. [1955], A estimacão do efeito residual de fertilizantes por meio da Lei de Mitscherlich. Anais Escola Superior Agri. "Luiz de Queiroz" 12-13:69-75. E8a;A7 (2606)

Gomes, F. P. [1957], The analysis of factorial experiments in balanced incomplete blocks (abstract). Biometrics 13:239-240. E2a;T2a (2607)

Gomes, F. P. [1963], Diseño experimental en frutales. (Experimental design for fruit trees.) Revistade Agricultura Saõ Paulo 38:207-224. E2,7,13a;C;S1 (2608)

Gomes, F. P. and de Abreu, C. P. [1959], Novas variedades de cana-de-açúcar. Anais Escola Superior Agri. "Luiz de Queiroz" 16:199-210. E2ae;A13 (2609)

Gomes, F. P. and Guimaraes, R. F. [1958], Joint analysis of experiments in complete randomised blocks with some common treatments. Biometrics 14:521-526. E4ae;A13 (2610)

Gomes, F. P. and Malavolta, E. [1951], Pesquisas sôbre a análise estatistica de experiências de adubação com o auxílio da Lei de Mitscherlich. Anais de Escola Superior de Agri. "Luiz de Queiroz" 8:1-14. E2ae;A7 (2611)

Goodchild, N. A. [1966], Applications of nonorthogonal designs to situations where treatments or blocks are of unequal status or size. Biometrics 22:629-631. E4c (2612)

Gosslee, D. [1952], The use of finite fields in design of experiments (abstract). American Math. Monthly 59:440. E7c;T2c (2613)

Goswami, R. P. [1967], Efficiency of change over design in animal experimentation (abstract). JISAS 19(1):141-142. E8a (2614)

Gourlay, N. [1955], F-test bias for experimental designs in educational research. Psychometrika 20:227-248. E7a;T2a;A10 (2615)

Gourlay, N. [1955], F-test bias for experimental designs of the Latin square type. Psychometrika 20:273-287. E7a (2616)

Govinda Iyer, T. A. [1957], Quicker methods in the analysis of variance. Madras Agri. J. 44:326-336. E1,2a;A12 (2617)

Graybill, F. [1954], Variance heterogeneity in a randomized block design. Biometrics 10:516-520. E2ae;A7,11 (2618)

Graybill, F. [1955], Variance heterogeneity in a randomized block design (summary). JASA 50:577. E2a;A11 (2619)

Graybill, F. A. and Deal, R. B. [1959], Combining unbiased estimators. Biometrics 15:543-550. E2,3a (2620)

Graybill, F. A. and Pruitt, W. E. [1958], The staircase design: Theory. AMS 29: 523-533. E4ac (2621)

Graybill, F. A. and Seshadri, V. [1960], On the unbiasedness of Yates' method of estimation using interblock information. AMS 31:786-787 (abstract 31:815). E2a (2622)

Graybill, F. A. and Weeks, D. L. [1959], Combining inter-block and intra-block information in balanced incomplete blocks. AMS 30:799-805 (abstract 30:624). E2a (2623)

Greenberg, B. G. and Sarhan, A. E. [1958], Some applications of order statistics. Bull. ISI 36(3):172-183. E2ae;A12 (2624)

Greenberg, V. L. [1964], Robust inference in some experimental designs. Ph. D. Thesis, Univ. California. E3a (2625)

Greenberg, V. L. [1966], Robust estimation in incomplete blocks designs. AMS 37:1331-1337. E3a (2626)

Grizzle, J. E. [1965], The two-period change-over design and its use in clinical trials (abstract). Technometrics 7:272. E8c (2627)

Grizzle, J. E. [1965], The two-period change-over design and its use in clinical trials. Biometrics 21:467-480. E8ae (2628)

Gronow, D. G. C. [1961], Limitations of "square" experimental designs. Ph. D. Thesis, London Univ. E7ac (2629)

Grundy, P. M. [1950], The estimation of error in rectangular lattices. Biometrics 6:25-33. E3a (2630)

Grundy, P. M. [1951], A general technique for the analysis of experiments with incorrectly treated plots. JRSSB 13:272-283. E6,10,11ae (2631)

Grundy, P. M. and Healy, M. J. R. [1950], Restricted randomization and Quasi-Latin squares. JRSSB 12:286-291. E7ac;T2ac (2632)

Guérin, R. [1963], Aspects algèbriques du problème de Yamamoto. Comptes Rendus (premier semestre) 256:583-586. E7,12,13c (2633)

Guérin, R. [1963], Sur une généralisation de la méthode de Yamamoto pour la construction de carrés latin orthogonaux. Comptes Rendus (premier semestre) 256:2097-2100. E7,12,13c (2634)

Guérin, R. [1965], Vue d'ensemble sur les plans en blocs incomplets equilibres et partiellement equilibres. Rev. ISI 33:24-58. E2,3c;B (2635)

Guérin, R. [1966], Existence et properietés des carrés latins orthogonaux. Ph. D. Thesis, Univ. Paris. E7,12,13c (2636)

Guérin, R. [1966], Existence et properietés des carrés latins orthogonaux. Publications l'Inst. Stat. Univ. Paris 15:113-293. E7,12,13c;B (2637)

Guimarães, R. F., Pimentel Gomes, F., and Malavolta, E. [1959], Adubação em "torrão paulista" de Eucalyptus Saligna SM. Anais Escola Superior Agri. "Luiz de Queiroz". 16:211-218. E2ae;T2ae (2638)

Gurgel, J. T. A. and Mitidieri, J. [1955-1956], Estudos sôbre o quiabeiro II-Efeitos da autofecundação e do cruzamento. Anais Escola Superior Agri. "Luiz de Queiroz" 12-13:39-51. E2ae (2639)

Hader, R. J. [1956], Some experimental designs applicable to problems with wood (discussion). Forest Products J. 6:425-427. E2,3a (2640)

Haines, W. B., Crowther, E. M., and Thornton, G. J. [1954], Manuring hevea V: some long-term effects in the Dunlop (Malaya) experiments. Empire J. Expt. Agri. 22:203-210. E5a (2641)

Hall, M. [1952], A combinatorial problem on abelion groups. Proc. American Math. Soc. 3:584-587. E7,12c (2642)

Hall, M. [1956], A survey of difference sets. Proc. American Math. Soc. 7:975-986. E2c (2643)

Hall, M. [1958], A survey of combinatorial analysis, chapter 4, in "Some Aspects of Analysis and Probability" pp. 76-104. E2,3,12c (2644)

Hall, M. [1962], Automorphisms of Steiner triple systems. Bull. Symp. Pure Math., American Math. Soc. 6:47-66. E2c (2645)

Hall, M. [1964], Block designs, chapter 13, in "Applied Combinatorial Mathematics" pp. 369-405. E2,3,12,13c (2646)

Hall, M. [1965], Combinatorial analysis. Encyclopaedia Britannica pp. 120-123. E2,12,13c (2647)

Hall, M. [1966], Numerical analysis of finite geometries (with discussion). Proc. IBM Sci. Computing Symp. Combinatorial Problems. pp. 11-30. E7,12,13c (2648)

Hall, M. [1967], Combinatorial Theory, chapters 10, 11, 12, 13, 15, 16, and appendix I Blaisdell Publishing Co., Waltham, Mass., Toronto, and London. E2,3,7,12,13,14c (2649)

Hall, M. and Connor, W. S. [1954], An embedding theorem for balanced incomplete block designs. Canadian J. Math. 6:35-41. E2c (2650)

Hall, M. and Ryser, H. J. [1951], Cyclic incidence matrices. Canadian J. Math. 3:495-502. E2c (2651)

Hall, M. and Swift, J. D. [1955], Determination of Steiner triple systems of order 15. Math. Tables Other Aids Computation 9:146-152. E2c (2652)

Hall, M., Swift, J. D., and Walker, R. J. [1956], Uniqueness of the projective plane of order eight. Math. Tables Other Aids Computation 10:186-194. E7,12c (2653)

Hamaker, H. C. [1955], Experimental design in industry. Biometrics 11:257-286 (abstract 10:187-188). E2a;C (2654)

Hamaker, H. C. [1955], Naar efficiënte experimenten. (Towards efficient experiments.) Stat. Neerlandica 9:7-25. E5,7ae;C (2655)

Hamaker, H. C. [1959], A note on ANOVA in the transistor industry. IQC 16(1):12-14. E12ae (2656)

Hamming, G. [1952], Interactie-problemen. Landbouwkundig Tijdschrift (Wageningen) 64:162-173. E6,10a (2657)

Hanani, H. [1960], Tactical configurations, I incomplete balanced block designs. Math. Res. Center Tech. Summary Report No. 137, U.S. Army Univ. Wisconsin. E2c (2658)

Hanani, H. [1960], Construction of balanced incomplete block designs. Math. Res. Center Tech. Summary Report No. 165, U.S. Army Univ. Wisconsin. E2c (2659)

Hanani, H. [1961], The existence and construction of balanced incomplete block designs. AMS 32:361-386. E2ac (2660)

Hanani, H. [1963], On some tactical configurations. Canadian J. Math. 15:702-722. E2c (2661)

Hanani, H. [1964], On covering of balanced incomplete block designs. Canadian J. Math. 16:615-625. E2c (2662)

Hanani, H. [1965], A balanced incomplete block design. AMS 36:711. E2c (2663)

Hancock, J. and McMeekan, C. P. [1954], Studies of grazing behaviour in relation to grassland management. III. Rotational compared with continuous grazing. J. Agri. Sci. 45:96-103. E5a (2664)

Hänsel, H. [1959], Die Vermendung der Variationsbreite beider Schätzung der Standardabweichung und bei der Varianzanalyse ungeordneter Blockanlagen, sowie eine Weitere Vereinfachung des Hartley'schen Verfahrens durch direkte Bestimmung von Grenzdifferenzen. Bodenkultur 10:148-158. E2a;A12 (2665)

Hanson, H. L., Kline, L. and Lineweaver, H. [1951], Application of balanced incomplete block design to scoring of ten dried egg samples. Food Tech. 5:9-13. E2e (2666)

Harshbarger, B. [1951], Near balance rectangular lattices. Virginia J. Sci. (New Series) 2:13-27. E3ace (2667)

Harshbarger, B. [1951], Latinized rectangular lattice designs (abstract). Virginia J. Sci. 2:377. E11a (2668)

Harshbarger, B. [1953], Some lattice designs. Committe on Experimental Design, U.S. Dept. Agri., USDA Lecture Series, 9 pp. (Mimeo). E3ac (2669)

Harshbarger, B. [1953], Paired comparisons in a lattice design (abstract). Virginia J. Sci. 4:287. E11a;T11a (2670)

Harshbarger, B. [1955], The 2^3 factorial in a Latinized rectangular lattice design (summary). JASA 50:577 (also, Dept. Stat. Tech. Report No. 1, OOR Project No. 1166, Virginia Polytechnic Inst.). E11ac;T2a (2671)

Harshbarger, B. [1964], Triangular designs (abstract). Biometrics 20:385. E3a (2672)

Harshbarger, B. and Davis, L. L. [1952], Latinized rectangular lattices. Biometrics 8:73-84 (also, in Bi-Annual Report No. 3, Stat. Methods Sensory Differences Tests Food Quality, Va. Agri. Expt. Sta.). E11ace (2673)

Harter, H. L. [1960], On the analysis of split-plot experiments (abstract). AMS 31:525. E5a (2674)

Harter, H. L. [1961], On the analysis of split-plot experiments. Biometrics 17: 144-149. E5a (2675)

Harter, H. L. and Lum, M. D. [1962], An interpretation and extension of Tukey's one degree of freedom for non-additivity. Aeronautical Res. Laboratory, ARL 62-313, U.S. Air Force. E2a;T2a;A4 (2676)

Hartigan, J. A. [1967], Distribution of the residual sum of squares in fitting inequalities. Biometrika 54:69-84. E2ae (2677)

Hartley, H. O. [1950], The use of range in analysis of variance. Biometrika 37: 271-280. E2ae;A12 (2678)

Hartley, H. O. [1951], Double balancing of incomplete block designs (abstract). Biometrics 7:117-118. E7c (2679)

Hartley, H. O. [1958], Estimation from incomplete data in industrial research. Proc. Stat. Tech. Missile Evaluation Symp. pp. 179-198. E6a (2680)

Hartley, H. O. [1960], Analysis of variance. Ch. 20 in Mathematical Methods for Digital Computers, pp. 221-230. E2,5,7,12a;T2a (2681)

Hartley, H. O., Shrikhande, S. S., and Taylor, W. B. [1953], A note on incomplete block designs with row balance. AMS 24:123-126. E7c (2682)

Hatamura, M., Okuno, T., and Sasaki, T. [1954], The design and analysis of field experiments (in Japanese). Nogyo Gijutsu Kenkyusho Hokohu/Bull. Nat. Inst. Agri. Sci. No. 3(A):12-14. E1,2,3,7,11ae;C (2683)

Heady, E. O. and Dillon, J. L. [1961], Agricultural Production Functions, Iowa State Univ. Press, Ames, Iowa, Chapters 2-7,14. E2a;T2,4a (2684)

Healy, M. J. R. [1951], Latin rectangle designs for a 2^n factorial experiments on 32 plots. J. Agri. Sci. 41:315-316. E7ac;T2a (2685)

Healy, M. J. R. [1952], The analysis of lattice designs when a variety is missing. Empire J. Expt. Agri. 20:220-226. E2,3,6a (2686)

Healy, M. J. R. [1959], The analysis of field experiments on an electronic computer. Biom. Zeit. 1:210-214 E2,7a;T2a;A15 (2687)

Healy, M. J. R. [1959], Multivariate growth data (abstract). Biometrics 15:149. E5a;A7 (2688)

Healy, M. J. R. and Leech, F. B. [1950], Statistical analysis of results for successive tests on the same organism. Nature 166:319. E5a (2689)

Healy, M. and Westmacott, M. [1956], Missing values in experiments analysed on automatic computers, Appl. Stat. 5:203-206. E6a;A15 (2690)

Hedlin, R. A. and Ridley, A. O. [1964], Effect of crop sequence and manure and fertilizer treatments on crop yields and soil fertility. Agron. J. 56:425-427. E5,8e (2691)

Heinisch, O. [1958], Die Bedeutung der Biometrie für die Landwirtschaftswissenschaften im Allgemeinen und für die Pflanzenzüchtung im Besonderen. SB Deutsche Akad. Landwirtschaftwissen Berlin, 7, No. 14, 45 pp. E2a;A12 (2692)

Heite, H. J. and Linder, A. [1962], Über die Planung und Auswertung einer Rechts-links-Behandlung bei dermato-therapeutischen Untersuchungen. Dermatalogica 125:65-80. E2ae;C (2693)

Henderson, P. L. [1952], Methods of research in marketing. Application of the double change-over design to measure carry-over effects of treatments in controlled experiments. Dept. Agri. Economics Paper No. 3, Cornell Univ. E8e (2694)

Heuzé, G. [1964], Equivalences dans les schémas d'association, Comptes Rendus (premiere semestre) 258:5349-5351. E3c (2695)

Heuzé, G. [1964], Critère permettant de minimiser le nombre de classes associées d'un plan en blocs incomplets partiellement équilibrés, Comptes Rendus (premiere semestre) 258:2970-2971. E3ac (2696)

Heuzé, G. [1966], Contribution a l'étude des schémas d'asssociation. Publications Inst. Stat. Univ. Paris 15:1-59. E3ac (2697)

Hicks, C. R. [1955], Some applications of Hotelling's T. IQC 11(9):23-26. E1ae;A7 (2698)

Hicks, C. R. [1956], Fundamentals of analysis of variance, part I - The analysis of variance (ANOVA) model. IQC 13(2):17-20. E1ae;T2ae (2699)

Hicks, C. R. [1956], Fundamentals of analysis of variance, part II - The components of variance and the mixed model. IQC 13(3):5-8. Elae;T2ae (2700)

Hicks, C. R. [1956], Fundamentals of analysis of variance, part III - Nested designs in analysis of variance. IQC 13(4):13-16. Elae;T2ae (2701)

Hicks, C. R. [1959], Randomized incomplete blocks. ASQC Nat. Conv. Trans. 13:51-62. E2ae (2702)

Hicks, C. R. [1965], The analysis of covariance. IQC 22:282-286. Elae;A7 (2703)

Hill, B. M. [1967], Correlated errors in the random model. JASA 62:1387-1400. Ela (2704)

Hill, H. M. [1960], Experimental designs to adjust for time trends. Technometrics 2:67-82. E5ace;T4ace (2705)

Hill, H. M. and Wheeler, D. [1958], A designed experiment to evaluate seven yarn lubricants. ASQC Nat. Conv. Trans. 12:245-253. E3ae;T2ae (2706)

Hill, K. W. [1954], Wheat yields and soil fertility after a half century of farming. Soil Sci. Soc. America Proc. 18:182-184. E5a (2707)

Hinkelmann, K. [1964], Extended group divisible partially balanced incomplete block designs. AMS 35:681-695. E3ac (2708)

Hinkelmann, K. [1967], Circulant partial triallel crosses. Biom. Zeit. 9:22-33. E3ac;T9ac (2709)

Hirotsu, C. [1965], Research for a set of minimal sufficient statistics and Yates' combined estimator for BIBD and GDPBIBD. Reports Stat. Appl. Res. 12(3): 115-126. E2,3a (2710)

Hirotsu, C. [1966], Types of incomplete block designs for factorials. Reports Stat. Appl. Res. 13(2):12-28. E3a;T2a (2711)

Hirotsu, C. [1966], A note on the design with two sets of block constraints. Reports Stat. Appl. Res. 13(4):238-241. E7c (2712)

Hoblyn, T. N. and Pearce, S. C. [1954], Some considerations in the design of successive experiments on fruit plantations. Biometrics 10:176-177. E7,12c;C (2713)

Hoblyn, T. N., Pearce, S. C., and Freeman, G. H. [1954], Some considerations in the design of successive experiments in fruit plantations. Biometrics 10: 503-515. E7,12c;C (2714)

Hodnett, G. E. [1956], The analysis of a 3×6 experiment arranged in a quasi-latin square. Biometrics 12:245-258. E7ae;T2ae (2715)

Hoffman, A. J. [1960], On the uniqueness of the triangular association scheme. AMS 31:492-497. E3c (2716)

Hoffman, A. J. [1963], On the duals of symmetric partially-balanced incomplete block designs. AMS 34:528-531. E3c (2717)

Hoffman, A. J. and Ray-Chaudhuri, D. K. [1964], On the line graph of a symmetric balanced incomplete block design (abstract). AMS 35:1837. E2c (2718)

Hoffman, A. J. and Richardson, M. [1961], Block design games. Canadian J. Math. 13:110-128. E2,3c (2719)

Hogben, D. [1959], Test of difference between treatment and control with multiple replications of control and a missing plot. Biometrics 15:486-487. E6a (2720)

Hollander, M. [1965], Rank tests for randomized blocks when the alternatives have an a priori ordering. Ph. D. Thesis, Stanford Univ. (also, Dept. Stat. Tech. Report No. 9, Stanford Univ.). E2a;A12 (2721)

Hollander, M. [1967], Rank tests for randomized blocks when the alternatives have an a priori ordering. AMS 38:867-877 (abstract 36:1082-1083). E2a;A12 (2722)

Holmes, J. C. [1952], The establishment of experiments. Bull. Edinburgh College Agri. No. 32:35-49. E2,5,7,12a;A14 (2723)

Hoofnagle, W. S. [1965], Experimental designs in measuring the effectiveness of promotion. J. Marketing Res. 2:154-162. E7ae;C (2724)

Hopkins, J. W. [1954], Incomplete block rank analysis: Some taste test results. Biometrics 10:391-399. E2ae;A12 (2725)

Horváth, A. [1954], Az eltérés elemzés. (The analysis of difference.) Növénytermelés 3:103-114. E2,7a (2726)

Horváth, A. and Sváb, J. [1955], A fajtakísérletek módszerei. (Methods of variety tests.) Országos Növenyfajtakísérleti Intézet, Budapest, pp. 54-65. E2a (2727)

Horváth, A. and Sváb, J. [1955], Uj elgondolások a szántóföldi kísérletek értékelési módszére. (New concepts in methods of evaluating field trials.) Növénytermelés 4:43-66. E2,7a (2728)

Horváth, A. and Sváb, J. [1956], Megjegyzések Schnell Lászlóné "Hozzászólásához". (Comments on Mrs. László Schnell's remarks.) Növénytermelés 5:372. E2,7a (2729)

Horváth, A. and Sváb, J. [1957], Válasz Schnell Lászlóne "Hozzászólás" - ára. (Reply to Mrs. László Schnell's remarks.) Növenytermelés 6:181-182. E2,7a (2730)

Horváth, A. and Sváb, J. [1957], Fajtakísérletek módszerei. (Methods of variety tests.) Országos Növényfajtakísérleti Intézet, Budapest, pp. 67-80. E2a (2731)

Houston, T. R. [1966], Sequential counterbalancing in Latin squares. AMS 37: 741-743. E7c (2732)

Howes, D. R. [1960], Multi-dimensional staircase designs for reliability studies. Proc. Fifth Conf. Design Expt. Army Res. Dev. Testing. pp. 191-197. E4a;S2 (2733)

Hoyle, B. J. and Baker, G. A. [1961], Stability of variety response to extensive variations of environment and field plot design. Hilgardia 30:365-394E. E2a;A14;C (2734)

Hrubý, K. and Konvička, O. [1954], Polní pokusy, jejich zakládání a hodnocení. (Field experiments, their design and computation.) Olomouc: Kčs.35, 276 pp., 68 figs., 177 tables. E1,2a;C (2735)

Huang, C. [1955], Balanced incomplete block experiments. I. (in Chinese). Nung-lin-hsueh Pao/J. Agri. For. Taichung 4:241-260. E2a (2736)

Hudson, D. J. [1963], The design and analysis of experiments. Ph. D. Thesis, London Univ. Imperial College. E (2737)

Hughes, D. R. [1962], Combinatorial analysis, t-designs and permutation groups. Bull. Symp. Pure Math., American Math. Soc. 6:39-41. E2c (2738)

Hultquist, R. A. [1965], Minimal sufficient statistics for the two-way classification mixed model design. JASA 60:182-192. E2a (2739)

Hultquist, R. A. and Graybill, F.A. [1960], Minimal sufficient statistics for the two-way classification mixed model design (abstract). AMS 31:816. E2a;A10 (2740)

Husain, Q. M. [1958], Star design. Calcutta Stat. Assoc. Bull. 8:110-118 (addenda 8:169). E 4c (2741)

Hussain, Q. M. [1961], A note on the symmetrical balanced incomplete block (S.B.I.B designs with k=9, λ=2. Bull. ISI 38(4):11-16. E2c (2742)

Ikeda, S. [1967], A method of constructing PBIB designs of T_m type. Inst. Stat. Mimeo. Ser. No. 508, Univ. N. Carolina. E3c (2743)

Ikeda, S. [1967], On a method of construction of group divisible designs. Inst. Stat. Mimeo. Ser. No. 521, Univ. N. Carolina. E3c (2744)

Ikeda, S. and Ogawa, J. [1966], On the non-null distribution of the F-statistic for casing a partial null hypothesis in a random PBIB design with m associate classes of the Neyman model. Inst. Stat. Mimeo. Ser. No. 466, Univ. N. Carolina. E3,4a (2745)

Ikeda, S., Ogawa, J., and Ogasawara, M. [1965], On the asymptotic distribution of F-statistic under the null-hypothesis in a randomized PBIB design and M associate classes under the Neyman model. Inst. Stat. Mimeo. Ser. No. 454, Univ. N. Carolina. E3a (2746)

Iyer, T. A. G. [1957], Quicker methods in the analysis of variance. Madras Agri. J. 44:326-336. E1,2ae (2747)

Jacob, W. C. [1953], Split-plot half-plaid squares for irrigation experiments. Biometrics 9:157-175. E9,12ae;T2ae (2748)

Jacobsen, R. L. [1967], The geometric interpretation of missing observations. Biometrics Unit Mimeo Ser. BU-243-M, Cornell Univ. E6a (2749)

Jaech, J. L. [1966], An alternate approach to missing value estimation. Amer. Stat. 20(5):27-29. E2,6,7,10a (2750)

James, A. T. [1957], The relationship algebra of an experimental design. AMS 28:993-1002. E2,14a (2751)

James, E. and Bancroft, T. A. [1951], The use of half-plants in a balanced incomplete block in investigating the effect of calcium, phosphorus, and potassium, at two levels each, on the production of hard seed in crimson clover, <u>Trifolium incarnatum</u>. Agron. J. 43:96-98. E2ae (2752)

Jebe, E. H. [1962], A series of two-phase experiments. Proc. Seventh Conf. Design Expt. Army Res. Dev. Testing. pp. 739-760. E7,8ac (2753)

Jeffers, J. N. R. [1966], General analysis of non-orthogonal experiments. International Advisory Group Forest. Stat., Second Conf. Institutionen För Skoglig Matematisk Statistik, Stockholm, pp. 160-169. E2ae;A10 (2754)

Jenkins, G. M. and Chanmugam, J. [1960], Autocorrelation analysis and the design of experiments. Dept. Math. Tech. Report No. 37, Princeton Univ. E5a;C (2755)

John, A. J. [1966], The design and analysis of unbalanced incomplete block designs. Ph. D. Thesis, London Univ. E4ac (2756)

John, J. A. [1965], A note on the analysis of incomplete block experiments. Biometrika 52:633-636. E2,3a (2757)

John, J. A. [1966], Cyclic incomplete block designs. JRSSB 28:345-360. E2,3c (2758)

John, J. A. [1967], Reduced group divisible paired comparison designs. AMS 38:1887-1893. E3c;T11c (2759)

John, P. W. M. [1961], An application of a balanced incomplete block design. Technometrics 3:51-54. E2ae (2760)

John, P. W. M. [1962], Testing two treatments when there are three experimental units in each block. Appl. Stat. 11:164-169. E4ae (2761)

John, P. W. M. [1962], Augmented block designs (abstract). Biometrics 18:631. E4c (2762)

John, P. W. M. [1963], Analysis of diallel cross experiments in a split plot situation. Australian J. Biological Sci. 16:681-687. E12a;T9a (2763)

John, P. W. M. [1963], Using a two factor analysis of variance programme for experiments with several factors. Appl. Stat. 12:129-132. E12a;T2a;A15 (2764)

John, P. W. M. [1963], Extended complete block designs. Australian J. Stat. 5:147-152. E4ace (2765)

John, P. W. M. [1964], Balanced designs with unequal numbers of replicates. AMS 35:897-899. E2,4ac;A10 (2766)

John, P. W. M. [1966], An extension of the triangular association scheme to three associate classes. JRSSB 28:361-365. E3c (2767)

John, P. W. M. [1967], On obtaining balanced incomplete block designs from partially balanced association schemes. AMS 38:618-619. E2c (2768)

John, P. W. M. [1967], The folded cubic association scheme (preliminary report) (abstract). AMS 38:1311-1312. E3c (2769)

Johnson, D. M., Dulmage, A. L., and Mendelsohn, N. S. [1961], Orthomorphisms of groups and orthogonal latin squares I. Canadian J. Math. 13:356-372. E7,12,13c (2770)

Johnson, J. R. [1957], An application of the design of experiments to the surveillance of ammunition. Proc. First Conf. Design Expt. Army Res. Dev. Testing. pp. 145-155. E1,3e (2771)

Johnson, N. L. [1953], Some notes on the application of sequential methods in the analysis of variance. AMS 24:614-623. E1a;S2 (2772)

Jones, R. M. [1959], On a property of incomplete blocks. JRSSB 21:172-179. E2a (2773)

Jowett, G. H. [1952], The accuracy of systematic sampling from conveyor belts. Appl. Stat. 1:50-59. E1ae (2774)

Justesen, S. H. and Keuls, M. [1958], Note on the use of non-orthogonal designs. Bull. ISI 36(3):269-276. E4ac (2775)

Kälin, A. [1966], Versuchsanordnungen in unvollständingen Blöcken mit zusätzlichen Kontrollbehandlungen in jedem Block. Metrika 10:182-218. E4ace (2776)

Kanjo, A. I. [1965], Recovery of interblock information. Ph. D. Thesis, Virginia Polytechnic Inst. E2,3a (2777)

Kapadia, C. H. [1962], Minimal sufficient statistics for the partially balanced incomplete block (PBIB) design with two associate classes under an Eisenhart model II. Ann. Inst. Stat. Math. Tokyo 14:63-71. E3a (2778)

Kapadia, C. H. [1962], Minimal sufficient statistics for the partially balanced incomplete block design with two associate classes under an Eisenhart model II. (abstract). AMS 33:819. E3a (2779)

Kapadia, C. H. [1962], Variance components in two-way classification models with interaction (abstract). AMS 33:304. E2,3a (2780)

Kapadia, C. H. [1964], On the block structure of singular group divisible partially balanced incomplete block (SGDPBIB) designs (abstract). AMS 35:1401. E3c (2781)

Kapadia, C. H. [1966], On the block structure of singular group divisible designs. AMS 37:1398-1400. E3c (2782)

Kapadia, C. H. [1966], Combining intra and inter block analysis of group divisible designs. Technometrics 8:189-191. E3a (2783)

Kapadia, C. H. and Weeks, D. L. [1963], Variance components in two-way classification models with interaction. Biometrika 50:327-334. E2a (2784)

Kapadia, C. H. and Weeks, D. L. [1964], On the analysis of group divisible designs. JASA 59:1217-1224. E3a (2785)

Kapse, Y. S. [1953], Efficiency of different experimental designs with special referenc to intra-class correlations. JISAS 5:179-189. E2,5,14a (2786)

Katti, S. K. [1960], A direct method for constructing the intrablock subgroup. Biometrics 16:691-694. E3c (2787)

Keller, K. R. [1951], Relative efficiency of rectangular and triple rectangular lattice designs using hop uniformity trial data. Agron. J. 43:93-96. E3,14a (2788)

Kempthorne, O. [1953], A class of experimental designs using blocks of two plots. AMS 24:76-84. E2ac (2789)

Kempthorne, O. [1956], The efficiency factor of an incomplete block design. AMS 27:846-849. E2,3,14a (2790)

Kempthorne, O. and Barclay, W. D. [1953], The partition of error in randomized blocks. JASA 48:610-614. E2a;A5 (2791)

Kempthorne, O., Zyskind, G., Addelman, S., Throckmorton, T. N., and White, R. F. [1961], Analysis of variance procedures. Stat. Lab. Aeronautical Res. Lab. ARL-149, Iowa State Univ. and U.S. Air Force. Ellac;T7ac;A5 (2792)

Kenning, W. [1967], Bloques aleatorios. Instituto Nacional de Technologia Agropecuria Tucumán, Argentina, Publicación Miscelánea, 82 pp. E2ae;C (2793)

Keuls, M. [1952], The use of the "studentized range" in connection with an analysis of variance. Euphytica 1:112-122. E2e;A3 (2794)

Keuls, M. [1960], Tabellen en nomogrammen voor het onderscheidingsvermogen van de 5% en 10%-F-toets voor het gebruik bij de gewarde blokkenproef. (Tables and graphs of the power of 5% and 10%-F-tests for use in randomized block experiments.) Stat. Neerlandica 14:127-150. E2a (2795)

Keuls, M. [1960], La puissance du critère F dans l'analyse de la variance de plans en blocs au hasard. Nomogrammes pour le choix du nombre de répétitions. Bull. Inst. Agronomique Stations Recherhes Gembloux (Hors Série) 1:256-271 (also, in Biométrie-Praximétrie 1(3):65-80). E2a;Sl (2796)

Khatri, C. G. and Shah, S. M. [1962], An inequality for balanced incomplete block design. Ann. Inst. Stat. Math. 14:95-96. E2c (2797)

Kiefer, J. C. [1957], On the (nonrandomized) optimality of symmetrical designs (abstract). AMS 28:1058. E14 (2798)

Kiefer, J. C. [1957], On the non-optimality of symmetrical designs among randomized designs (abstract). AMS 28:1058. E14;T10 (2799)

Kiefer, J. [1958], On the nonrandomized optimality and randomized nonoptimality of symmetrical designs. AMS 29:675-699. E14 (2800)

Kiefer, J. [1959], Optimum experimental designs (with discussion). JRSSB 21:272-319. E14;T1,10;C (2801)

Kiefer, J. [1960], Optimum experimental designs (abstract). AMS 31:245. E14;T10 (2802)

Kiefer, J. [1961], Optimum experimental designs V, with applications to systematic and rotatable designs. Proc. Fourth Berkeley Symp. Math. Stat. Prob. 1:381-405. E5c;T1,4,10c (2803)

Kiefer, J. [1962], Two more criteria equivalent to D-optimality of designs. AMS 33:792-796. E14;T10 (2804)

Kiesselbach, T. A. and Lyness, W. E. [1952], Crop rotation experiments. Nebraska Agri. Expt. Sta. Bull. 416:1-33. E5,8ae (2805)

Kishen, K. [1958], Presidental address. Recent developments in experimental design. Proc. Fourty-Fifth Indian Sci. Congress, Part II, pp. 28-59. E2,3,4ac;T2,3,4ac; B;C (2806)

Kishen, K. and Rao, C. R. [1952], An examination of various inequality relations among parameters of the balanced incomplete block design. JISAS 4:137-144. E2c (2807)

Kishen, K. and Srivastava, J. N. [1960], On a class of resolvable partially balanced incomplete block designs (abstract). JISAS 12:150-151. E3ac (2808)

Kitagawa, T. and Mitome, M. [1955], A contribution to a notation system of the confounded factorial experiments. Bull. Math. Stat. 6(1&2):1-10. E2,3,9,10c;T2c (2809)

Knapp, R. H. [1957], A note on the t test. Amer. Stat. 11(1):17. E1a (2810)

Knowles, E.A. G. and Roseman, C. [1950], Graphical analysis of variations as a production department tool in Statistical Method in Industrial Production. Royal Statistical Society, pp. 20-29. E7a (2811)

Knowles, R. P. [1952], The use of lattice designs for testing forage crops. Sci. Agri. 32:614-617. E14;A14 (2812)

Koch, G. [1966], The analysis of incomplete blocks experiments with particular emphasis on the relationship between the recovery of interblock information and generalized least squares when block effects are assumed random. Inst. Stat. Mimeo. Ser. No. 464, Univ. N. Carolina. E2,3a (2813)

Koch, G. G. [1967], Some aspects of the statistical analysis of "split plot" experiments Part I: Completely randomized layouts. Inst. Stat. Mimeo. Ser. No. 527 Univ. N. Carolina. E5a (2814)

Koch, G. G. [1967], Some aspects of the statistical analysis of 'split plot' experiments (abstract). Biometrics 23:381-382. E5a (2815)

Kogan, L. S. [1953], Variance designs in psychological research. Psychological Bull. 50:1-40. E1,2,7ae;T2a;B (2816)

Kramer, C. Y. [1956], Factorial treatments in incomplete block designs. Ph. D. Thesis, Virginia Polytechnic Inst. E3a;T2a (2817)

Kramer, C. Y. and Bradley, R. A. [1956], Factorials in near-balanced incomplete block designs for k(k-1) treatments (preliminary report) (abstract). AMS 27:545-546. E3a;T2a (2818)

Kramer, C. Y. and Bradley, R. A. [1957], Intra-block analysis for factorials in two associate class group divisible designs. AMS 28:349-361. E3a;T2a (2819)

Kramer, C. Y. and Bradley, R. A. [1957], Examples of intra-block analysis for factorials in group divisible, partially balanced, incomplete block designs. Biometrics 13:197-224. E3ae;T2ae (2820)

Kramer, C. Y. and Bradley, R. A. [1957], Factorial treatments in group divisible incomplete block designs (abstract). AMS 28:816. E2ac;T2ac (2821)

Kramer, C. Y. and Glass, S. [1960], Analysis of variance of a Latin square design with missing observations. Appl. Stat. 9:43-50. E7,10ae (2822)

Krishnaiah, P. R. [1961], On simultaneous tests in nested designs (preliminary report) (abstract). AMS 32:628-629. E1a;A3 (2823)

Krishnaiah, P. R. [1963], Simultaneous tests and the efficiency of generalized balanced incomplete block designs. Ph. D. Thesis, Univ. Minnesota (also, Aerospace Res. Lab. ARL 63-174, U.S. Air Force.). E2,3,5,14a;A3,7 (2824)

Kruskal, W. [1952], A nonparametric analogue based upon ranks of one-way analysis of variance (abstract). AMS 23:140. E1a;A12 (2825)

Kshirsagar, A. M. [1957], A note on the total relative loss of information in any design. Calcutta Stat. Assoc. Bull. 7:78-81. E14 (2826)

Kshirsagar, A. M. [1957], On balancing in designs in which heterogeneity is eliminated in two directions. Calcutta Stat. Assoc. Bull. 7:161-166. E7c (2827)

Kshirsagar, A. M. [1958], A note on incomplete block designs. AMS 29:907-910. E2,3a (2828)

Kshirsagar, A. M. [1959], Distribution of the "blocks adjusted for treatments" sum of squares in incomplete block designs. AMS 30:246-249. E3a (2829)

Kshirsagar, A. M. [1962], Distribution of the adjusted row sum of squares in two-way designs. Calcutta Stat. Assoc. Bull. 11:155-159. E7a (2830)

Kshirsagar, A. M. [1966], Balanced factorial designs. JRSSB 28:559-567. E3,4ac;A8 (2831)

Kuebler, R. R. [1958], On the construction of a class of error-correcting binary signalling codes. Ph. D. Thesis, Univ. N. Carolina. E3,4c (2832)

Kuiper, N. H. [1952], Variantie-analyse. (Analysis of variance.) Statistica 6:149-194. E2,7a;T2a (2833)

Kuiper, N. H. and Corsten, L. C. A. [1953], Open plaatsen in variantieschema's. Mededelingen van de Landbouwhogeschool te Wageningen 53:25-30. E6ae (2834)

Kulkarni, G. A. [1960], A problem in reinforced incomplete block design. JISAS 12:143-145. E4a (2835)

Kulkarni, G. A. [1961], Incomplete block designs for slope-ratio assays (abstract). JISAS 13:239. E2a;T8ac (2836)

Kurkjian, B. M. [1959], Use of partially balanced block designs with three associate classes for confounded, asymmetrical factorial arrangements (abstract). AMS 30:615. E3c;T2c (2837)

Kurkjian, B. [1960], A general theory for asymmetrical confounded factorial experiments. Ph. D. Thesis, American Univ. E2,3a;T2a;A8 (2838)

Kurkjian, B. and Zelen, M. [1961], A general theory for analysis of asymmetrical, confounded factorial experiments (abstract). Biometrics 17:171. E3,4a;T2a (2839)

Kurkjian, B. and Zelen, M. [1963], Applications of the calculus of factorial arrangements. I. Block and direct product designs. Biometrika 50:63-73 (also MRC Tech. Summary Report No. 330, U.S. Army Univ. Wisconsin). E3,4ac;A8 (2840)

Kurkjian, B. and Zelen, M. [1963], Applications of the calculus of factorial arrangements I. Block and direct product designs (abstract). Proc. Eight Conf. Design Expt. Army Res. Dev. Testing. E4a (2841)

Kurtz, T. E. [1956], An extension of a multiple comparisons procedure. Ph. D. Thesis, Princeton Univ. E1,2a;A3 (2842)

Kurtz, T. E., Link, R. F., Tukey, J. W., and Wallace, D. L. [1965], Short-cut multiple comparisons for balanced single and double classifications: Part 1, Results (with discussion). Technometrics 7:95-169. E1,2ae;A3 (2843)

Kurtz, T. E., Link, R. F., Tukey, J. W., and Wallace, D. L. [1965], Short-cut multiple comparisons for balanced single and double classifications. Part 2, Derivations and approximations. Biometrika 52:485-498. E1,2a;A3 (2844)

Kurtzer, C. L. [1965], Balanced or near-balanced k row by b column designs for v treatments in r replicates with $2 \leq v$, $r \leq 10$. B.S. Thesis, Cornell Univ. E7ac (2845)

Kusumoto, K. [1964], On a design for two-way elimination of heterogeneity and its analysis. J. Sci. Hiroshima Univ. Ser. A-I 28:237-258. E7ac (2846)

Kusumoto, K. [1965], A necessary condition for the existence of regular and symmetrical PBIB designs of T_3 type. Ann. Inst. Stat. Math. 17:149-165.
E3ace (2847)

Kusumoto, K. [1965], Hypercubic designs. Wakayama Medical Reports 9:123-132.
E3c (2848)

Kuzmin, W. R. [1959], Experiments to expose marginal reliability designs. Nat. Symp. Reliability Quality Control. Proc. 5:55-64. E7ae;T2ae (2849)

Lafcn, M. [1957], Construction de blocs incomplets partiellement équilibrés à s+l classes associées. Comptes Rendus 244(premier semestre):1714-1717. E3c (2850)

Lafon, M. [1959], Coefficient d'efficacité d'un bloc incomplet partiellement équilibré. Comptes Rendus 248(premier semestre):3114-3115. E3c (2851)

Lafon, M. [1963], Recherche des plans d'expérience du type à blocs incomplets partiellement équilibrés à cinq répétitions. Comptes Rendus (deuxième semestre) 257:830-832. E3c (2852)

Lafon-Augé, M. [1957], Conditions d'existence d'un bloc incomplet partiellement équilibré. Comptes Rendus (deuxième semestre) 245:1774-1775. E3c (2853)

Lafon-Augé, M. [1960], Application de l'algèbra aux plans d'expériences: construction de blocs incomplets partiellement équilibrés et analyse de la variance. Publications Inst. Stat. Univ. Paris 9:193-287. E3ac (2854)

Lakshminarayan, [1958], The construction and combinatorial properties of Latin squares and balanced sequences. Ph. D. Thesis, Univ. Aberdeen. E7,13c (2855)

Langer, W. and Stern, K. [1955], Versuchstechnische Probleme bei der Analage von Klonplantagen. Z. Forstgenetik Forstpflanzenzüchtung 4:81-88. E2a;T9a;C (2856)

Lashof, T. W., Mandel, J., and Worthington, V. [1956], Use of the sensitivity criterion for the comparison of the Bekk and Sheffield smoothness testers. Tappi 39:532-543. E14 (2857)

Lathwell, D. J. and Evans, C. E. [1951], Nitrogen uptake from solution by soybeans at successive stages of growth. Agron. J. 43:264-270. E2e;T7e (2858)

Leech, F. B. [1950], Statistical analysis of results for successive tests on the same organism. Nature 165:323-324. E5,8a (2859)

Leech, F. B. and Healy, M. J. R. [1959], The analysis of experiments on growth rate. Biometrics 15:98-106 (abstract 15:151-152; correction 15:631). E5ae (2860)

Lees, K. A. and Tootill, J. P. R. [1955], Microbiological assay on large plates Part I. General considerations with particular reference to routine assay. Analyst 80:95-110. E9ae;T8a (2861)

Lees, K. A. and Tootill, J. P. R. [1955], Microbiological assay on large plates Part II. Precise assay. Analyst 80:110-123. E7,9ae;T8a (2862)

Lees, K. A. and Tootill, J. P. R. [1955], Microbiological assay on large plates Part III. High throughput, low precision assays. Analyst 80:531-535.
E2,7,11ae;T8a (2863)

Lemmer, H. H. and Stoker, D. J. [1967], Distribution-free analysis of variance for the two-way classification. South African Stat. J. 1:67-74.
E2a;A12 (2864)

Lemus, F. [1960], A mixed model factorial experiment in testing electrical connectors. IQC 17(6):12-16.
E1e;T2e (2865)

le Roux, D. P. [1965], Intra-block estimation of a missing observation in partially balanced incomplete block designs with two associate classes. Suid-Afrikaanse Tydskrif vir Landbouwetenskap 8:479-485.
E3,6a (2866)

le Roux, D. P. and Kramer, C. Y. [1963], Intra-block analysis of variance of a semi-regular group divisible partially balanced incomplete block design with missing observations. Suid-Afrikaanse Tydskrif vir Landbouwetenskap 6:689-700.
E3,6ae (2867)

Le Roy, H. L. and Landis, J. [1960], Das Schätzen fehlender Werte im a.b.c- bzw. a.b-Versuch. (The estimating of missing values in the a.b.c- or a.b-design.) Biom. Zeit. 2:98-107.
E2,6ae;T2ae (2868)

Lessman, K. J. and Atkins, R. E. [1963], Optimum plot size and relative efficiency of lattice designs for grain sorghum yield tests. Crop Sci. 3:477-481.
E3,14a;A14 (2869)

Levy, P. S. [1965], A new method of estimating treatment effects or treatment differences in balanced incomplete block designs (abstract). AMS 36:1325.
E2a (2870)

Li, L. [1962], Studies on the efficiency of lattice and randomized complete block designs in horticultural station trials of sweet potatoes (in Chinese). Chung-hua Nung-hsueh Hui Pao/J. Agri. Assoc. China, New Series 40:40-49.
E2,3a;A14 (2871)

Linder, A. [1952], Ein einfacher "ausgewogener Versuch in unvollständigen Blöcken" Mitteilungsblatt für Mathematische Statistik 4:129-138.
E2ac (2872)

Linder, A. [1964], Design and analysis of experiments. Inst. Stat. Mimeo. Ser. No. 398, Univ. N. Carolina.
E2,3a;T2,3a (2873)

Lindley, D. V. [1956], On a measure of the information provided by an experiment. AMS 27:986-1005.
E14 (2874)

Lindley, D. V. [1961], An experiment in the marking of an examination (with discussion). JRSSA 124:285-313.
E2,7ace (2875)

Linhart, H. [1966], Streuungszerlegung für Paar-Vergleiche. Metrika 10:16-38. E2,3a;T1ace (2876)

Linnerud, A. C., Gates, C. E., and Donker, J. D. [1962], Significance of carry-over effects with extra period Latin square change-over design (abstract). J. Dairy Sci. 45:675. E8a (2877)

Lintner, J. [1958], Random block and Quasi-Latin square designs used to determine fertilizer requirements of sugarcane. South African Sugar Technologists' Assoc. Proc. pp. 1-22. E2,7a;T2a (2878)

Liu, C. W. [1963], A method of constructing certain symmetrical partially balanced designs. Scientia Sinica 12:1935-1937. E3c (2879)

Liu, C. W. and Chang, L. C. [1964], Some PBIB(2) designs induced by association schemes. Scientia Sinica 13:840-841. E3c (2880)

Liu, W. R. and Chang, L. C. [1964], Group divisible incomplete block designs with parameters v≤10 and r≤10. Scientia Sinica 13:839-840. E3c (2881)

Lochow, J. v. and Schuster, W. [1961], Anlage und Auswertung von Feldversuchen. DLG-Verlags-GmbH., Frankfurt am Main, 130 pp. E2,5,7ae;T2ae (2882)

Lubin, A. [1960], On the repeated-measurements design in biological experiments. Proc. Fifth Conf. Design Expt. Army Res. Dev. Testing. pp. 123-131. E5,8a (2883)

Lucas, H. L. [1950], Statistics and research on pasture and grazing. Committee on Experimental Design Lecture Series I, No. 3, U.S. Dept. Agri. Agri. Res. Administration. E5,8a;C;S1 (2884)

Lucas, H. L. [1951], Bias in estimation of error in change-over trials with dairy cattle. J. Agri. Sci. 41:146-148. E8a (2885)

Lucas, H. L. [1956], Switchback trials for more than two treatments. J. Dairy Sci. 39:146-154. E8ace (2886)

Lucas, H. L. [1957], Extra-period Latin-square change-over design. J. Dairy Sci. 40:225-239. E8ace (2887)

Lucas, H. L. [1959], Experimental designs and analyses for feeding efficiency trials with dairy cattle. Chapter 15 in Nutritional and Economic Aspects of Feed Utilization by Dairy Cows, pp. 177-192. E8a;C (2888)

Lucas, H. L. 1960], Planification des experiences d'alimentation des vaches laitieres. Biométrie-Praximétrie 1(3-4):17-32. E7,8ac;C (2889)

Lucas, H. L. [1960], Design of dairy feeding experiments. Bull. Inst. Agronomique Stations Recherches Gembloux, (Hors série)1:212-224. E7,8ac;C (2890)

Lum, M. [1955], Rules for determining error terms in hierarchical and partially hierarchical models (abstract). AMS 26:151. Ela (2891)

Lynch, L. [1958], The analysis of paired ranked observations. Ph. D. Thesis, Virginia Polytechnic Inst. E2a;A12 (2892)

Lynch, P. B. [1960], Conduct of field experiments. Bull. 399, New Zealand Dept. Agri. E2,3,7,12ae;A14;C (2893)

Maag, U. R. [1965], Contributions to the design of experiments. Ph. D. Thesis, Univ. Toronto. E (2894)

Mack, A. R. and Cairns, R. R. [1959], Statistical procedures: A. Illustrated Sta. Div., Central Expt. Farm, Ottawa, Ontario, pp. 14+29 (mimeo.). E2,6ae;T2ae (2895)

Mack, A. R. and Cairns, R. R. [1959], Statistical procedures: B. Illustrated Sta. Div., Central Expt. Farm, Ottawa, Ontario, pp. 26 (mimeo.). E2,3ae (2896)

MacLean, J. A. R. and Wickens, R. [1951], Application of an incomplete block design to the assessment of quality in Cacao. Nature 168:434-435. E2,3a;T11a (2897)

MacLeod, L. B., Bishop, R. F., Jackson, L. P., MacEachern, C. R., and Goring, E. T. [1960], A long-term field experiment with commercial fertilizers and manure. I. Fertility levels and crop yields in a rotation of swedes, oats and hay. Canadian J. Soil Sci. 40:136-145. E5,8a (2898)

MacWilliams, F. J. and Mann, H. B. [1967], On the p-rank of the design matrix of a difference set. MRC Tech. Summary Report No. 803, U.S. Army and Univ. Wisconsin. E2c (2899)

Majindar, K. N. [1962], On the parameters and intersection of blocks of balanced incomplete block designs. AMS 33:1200-1205. E2c (2900)

Majindar, K. N. [1966], On integer matrices and incidence matrices of certain combinatorial configurations, I: square matrices II and III: rectangular matrices. Canadian J. Math. 18:1-17. E2c (2901)

Majumdar, K. N. [1953], On some theorems in combinatorics relating to incomplete block designs. AMS 24:377-389. E2c (2902)

Majumdar, K. N. [1954], On combinatorial arrangements. Proc. American Math. Soc. 5:662-664. E2c (2903)

Majumdar, K. N. [1961], An investigation of properties of incomplete block designs. Ph. D. Thesis, Purdue Univ. E2,3,14ac (2904)

Malavolta, E. *et al*. [1953], Competição entre adubos fosfatados em milho (*Zea mays* L Anais Escola Superior Agri. "Luiz de Queiroz" 10:109-120. E7,12ae (2905)

Mallows, C. L. [1959], The information in an experiment. JRSSB 21:67-72. E14 (2906)

Malone, M. J. [1967], Algorithms for analysis of experiments with incomplete structures. M. S. Thesis, Iowa State Univ. E3a;A15 (2907)

Manceau, J. N. [1957], Application of the covariance analysis to the comparative study of two anthelmintics. Bull. ISI 35(2):339-352. E2e;A7 (2908)

Mandel, J. [1954], Chain block designs with two-way elimination of heterogeneity. Biometrics 10:251-272. E7ace (2909)

Mandel, J. [1959], The analysis of Latin squares with a certain type of row-column interaction. Technometrics 1:379-387. E7ae (2910)

Mandel, J. and Stiehler, R. D. [1954], Sensitivity - A criterion for the comparison of methods of test. J. Res. NBS 53:155-159. E14 (2911)

Mandel, J. and Zelen M. [1954], New types of easily constructed partially balanced incomplete block designs (abstract). AMS 25:807. E3c (2912)

Mann, H. B. [1960], The algebra of a linear hypothesis. AMS 31:1-15. E2,14a (2913)

Mann, H. B. [1964], Balanced incomplete block designs and Abelian difference sets. Illinois J. Math. 8:252-261. E2c (2914)

Mann, H. B. and Menon, M. V. [1960], Intrablock and interblock estimates. Contributions to Probability and Statistics pp. 293-298. E2,3a (2915)

Mann, H. H. and Boyd, D. A. [1958], Some results of an experiment to compare ley and arable rotations at Woburn. J. Agri. Sci. 50:297-306. E5,8a (2916)

Martin, L. [1952], Statistical methods in radiochemistry (with discussion). Analyst 77:892-896. E2e;C (2917)

Martinez, A. [1965], Some considerations on combined analysis of experiments. M. S. Thesis, Iowa State Univ. E2,3,12a;A13 (2918)

Masuyama, M. [1957], On the optimality of balanced incomplete block designs. Reports Stat. Appl. Res. 5:4-8 (correction 5:71-72). E2,14c (2919)

Masuyama, M. [1957], On difference sets of constructing orthogonal arrays of index two and of strength two. Reports Stat. Appl. Res. 5:27-34. E2c (2920)

Masuyama, M. [1959], On cyclic difference sets which generate orthogonal arrays. Reports Stat. Appl. Res. 6:47-53. E2,3c;T2,3,7c (2921)

Masuyama, M. [1960], On the existence theorem of cyclic difference sets for the orthogonal arrays ($9\lambda, 3\lambda+1, 3, 2$). Reports Stat. Appl. Res. 7:53-55. E2,3c (2922)

Masuyama, M. [1961], Calculus of blocks and a class of partially balanced incomplete block designs. Reports Stat. Appl. Res. 8:56-69. E3c (2923)

Masuyama, M. [1961], On cyclic difference sets which generate orthogonal arrays, Part II: Reports Stat. Appl. Res. 8:70-76. E2,3c (2924)

Masuyama, M. [1962], Decomposition by bilateral cosets and its generalization. Sankhyā, A, 24:41-46. E3c (2925)

Masuyama, M. [1962], La décomposition périodique dans le calcul des blocs le raffinement des décompositions. Bull. ISI 39(3):155-160. E3c (2926)

Masuyama, M. [1962], On the classes of blocks. Reports Stat. Appl. Res. 9: 83-87. E2,3ac (2927)

Masuyama, M. [1962], On the uniqueness of the solution of a block equation $\mathfrak{X}\mathfrak{X}^*=\mathfrak{S}$ for $s(\mathfrak{X})=2,3$ and 4, elements being residues modulo v. Reports Stat. Appl. Res. 9:88-92. E2,3ac (2928)

Masuyama, M. [1963], Calculus of non-commutative blocks and its applications to experimental designs. Reports Stat. Appl. Res. 10:237-240. E3c;A8 (2929)

Masuyama, M. [1964], Construction of PBIB designs by fractional development. Reports Stat. Appl. Res. 11:47-54. E3c (2930)

Masuyama, M. [1963], A geometrical problem related to cyclic designs. Contributions to Statistics pp. 239-244. E3c (2931)

Masuyama, M. [1963], Le calcul des blocs et ses applications aux plans d'expérience. Colloques Internationaux du C.N.R.S., No. 110; Le Plan d'Expériences pp. 51-60. E2,3c (2932)

Masuyama, M. [1964], Linear graphs of PBIB designs. Reports Stat. Appl. Res. 11:147-151. E3c (2933)

Masuyama, M. [1965], Cyclic generation of triangular PBIB designs. Reports Stat. Appl. Res. 12:73-81 (errata 13:242). E3c (2934)

Masuyama, M. [1965], Construction of certain PBIB designs from a BIB design. Reports Stat. Appl. Res. 12:143-145. E3c (2935)

Masuyama, M. [1965], A class of triangular designs. Reports Stat. Appl. Res. 12:146-148. E3ac (2936)

Masuyama, M. [1966], Construction cyclique des BIPE par developpement fractionnaire Bull. ISI 41:951-956. E3c (2937)

Masuyama, M. [1966], Symbolic calculus of blocks applied to PBIB designs. Reports Stat. Appl. Res. 13(2):1-11. E3a;A8 (2938)

Masuyama, M. [1966], Relation diagrams of SBIB and other designs. Reports Stat. Appl. Res. 13(2):41-42. E3,4ac (2939)

Masuyama, M. [1966], A combinatorial proof of Bose-Connor inequality for PBIB designs of GD scheme. Reports Stat. Appl. Res. 13:236-237. E3c (2940)

Masuyama, M. [1967], Construction cyclique de blocs incomplets partiellement equilibres. Rev. ISI 35:107-124. E3c (2941)

Masuyama, M. and Okuno, T. [1957], On the optimality of Latin-, Youden- and Shrikhande square designs. Reports Stat. Appl. Res. 5:17-19. E7,14c (2942)

Mathai, A. M. [1964], Missing values in statistical analysis. Ph. D. Thesis, Univ. Toronto. E6a;A10 (2943)

Mathai, A. M. [1965], Approximate analysis for a two-way layout (abstract). AMS 36:365. E2a;A10 (2944)

Mathai, A. M. [1965], An approximate method of analysis for a two-way layout. Biometrics 21:376-385. E2a;A10 (2945)

Mathen, K. K. [1954], Note on the design of experiments and testing the efficiency of drugs having local healing power. Sankhyā 14:175-179. E2ae (2946)

Maurin, F. [1964], Sur une généralisation de la méthode des différences pour la construction de tableaux orthogonaux. Comptes Rendus 259:4490-4491. E3,7c (2947)

Maurin, F. [1965], Sur certains groupes de substitutions entraînant l'existence de tableaux orthogonaux. Comptes Rendus 260:52-55. E3,7c (2948)

Maurin, F. [1966], Automorphismes propres de groupes et groupes abélians finis sons-transitifs d'ordre t et d'indice unité. Application à la construction des tableaux orthogonaux. Comptes Rendus 262:1194-1197. E3,7c (2949)

Maurin, F. [1967], Automorphismes propres de groupes abélians finis et orthomorphismes de groupes finis. Application à la construction de tableaux orthogonaux. Comptes Rendus 264:702-704. E3,7c (2950)

Mazumdar, S. [1965], On the construction of cyclic collineations for obtaining a balanced set of prime-powered lattice designs. Biometrics Unit Mimeo. Ser. BU-208-M, Cornell Univ. E2,3,11,13c (2951)

Mazumdar, S. [1967], On the construction of cyclic collineations for obtaining a balanced set of L-restrictional prime-powered lattice designs. AMS 38:1293-1295. E2,3,11,13c (2952)

Mazurak, A. P., Valassis, V. T., and Harris, L. C. [1954], Water-stability of aggregates from potato plots as affected by different rotation systems under irrigation in western Nebraska. Proc. Soil Sci. Soc. America 18:243-247. E5,8a (2953)

Mazuy, K. K. and Connor, W. S. [1962], Student's t in an n-way classification with unequal variances (preliminary report) (abstract). AMS 33:820. E2,13a;All (2954)

Mazuy, K. K. and Connor, W. S. [1965], Student's t in a two-way classification with unequal variances. AMS 36:1248-1255. E2a;All (2955)

McIntyre, G. A. [1955], Design and analysis of two phase experiments. Biometrics 11:324-334. E7,12ae;S2 (2956)

McIntyre, G. A. [1956], Query 123. Biometrics 12:527-532. E7,12a;S2 (2957)

McNemar, Q. [1951], On the use of latin squares in psychology. Psychological Bull. 48:398-401. E7a (2958)

Mehra, K. L. [1962], Rank tests for incomplete block designs, paired-comparison case. Ph. D. Thesis, Univ. California. E2,3a;T11a (2959)

Meier, P. [1951], The variance of a weighted average using estimated weights (abstract). AMS 22:607. E3a;All (2960)

Meier, P. [1951], Weighted means and lattice designs. Ph. D. Thesis, Princeton Univ. E2,3a (2961)

Meier, P. [1952], The estimation of error in simple lattice designs (abstract). Biometrics 8:93. E3a (2962)

Meier, P. [1953], Variance of a weighted mean. Biometrics 9:59-73. E3a (2963)

Meier, P. [1954], Analysis of simple lattice designs with unequal sets of replications. JASA 49:786-813. E3a (2964)

Meier, P. and Free, S. M. [1961], Further consideration of methodology in studies of pain relief. Biometrics 17:576-583. E3a;C (2965)

Meister, F. [1952], Magische Quadrate. Verlag von Ernst Wurzel, Zurich, 71 pp. E12ac;B (2966)

Melton, B. and Finkner, M. D. [1967], Relative efficiency of experimental designs with systematic control plots for alfalfa yield tests. Crop Sci. 7:305-307. E5,14a (2967)

Mendelsohn, N. S. and Dulmage, A. L. [1958], Some generalizations of the problem of distinct representatives. Canadian J. Math. 10:230-241. E7c (2968)

Menon, M. V. K. [1959], Combined interblock and interblock estimates. Ph. D. Thesis, Ohio State Univ. E2,3a (2969)

Menon, P. K. [1961], Method of constructing two mutually orthogonal latin squares of order 3n+1. Sankhyā A 23:281-282. E7,12c (2970)

Menon, P. K. [1963], Combinatorial problems. Res. Training School Publication No. SM 63-2, Indian Stat. Inst. E2c (2971)

Merrill, L. B. [1954], A variation of deferred rotation grazing for use under southwest range conditions. J. Range Management 7:152-154. E5,8c (2972)

Mesner, D. M. [1955], Uniqueness of Latin square association schemes for partially balanced incomplete block designs (preliminary report) (abstract). AMS 26:151. E3c (2973)

Mesner, D. M. [1956], An investigation of certain combinatorial properties of partially balanced incomplete block designs and association schemes, with a detailed study of designs of Latin square and related types. Ph. D. Thesis, Michigan State Univ. E3c (2974)

Mesner, D. M. [1956], A new class of PBIB designs (abstract). AMS 27:1185. E3c (2975)

Mesner, D. M. [1957], The structure of incidence matrices of partially balanced incomplete block designs (abstract). Biometrics 13:244. E3c (2976)

Mesner, D. M. [1963], A note on parameters of PBIB association schemes. Inst. Stat. Mimeo. Ser. No. 375, Univ. N. Carolina. E3c (2977)

Mesner, D. M. [1964], Negative latin square designs. Inst. Stat. Mimeo. Ser. No. 410, Univ. N. Carolina. E3c (2978)

Mesner, D. M. [1965], A note on the parameters of PBIB association schemes. AMS 36:331-336. E3a (2979)

Mesner, D. M. [1966], The block structure of certain PBIB designs of partial geometric types. Inst. Stat. Mimeo. Ser. No. 457, Univ. N. Carolina. E3c (2980)

Mesner, D. M. [1967], A new family of partially balanced incomplete block designs with some latin square design properties. AMS 38:571-581. E3c (2981)

Mexas, A. G. [1967], Algorithms for computer analysis of variance of balanced complete structures. M. S. Thesis, Iowa State Univ. E2a;T2a;A15 (2982)

Mikhail, W. F. [1960], An inequality for balanced incomplete block designs. AMS 31:520-522 (abstract 31:535; acknowledgement of priority 31:1213). E2c (2983)

Mitchell, J. A., Bockman, C. D., and Lee, A. V. [1957], Determination of acetyl content of cellulose acetate by near infrared spectroscopy. Anal. Chem. 29: 499-502. E12e (2984)

Mitra, S. K. [1959], Some remarks on the missing plot analysis. Sankhyā 21:337-344. E2,6,7,10a (2985)

Mitra, S. K. [1960], On the F-test in the intrablock analysis of a balanced incomplete block design. Sankhyā 22:279-284. E2a (2986)

Mitscherlich, E. A. [1950], Über die Fehler bei Ertragsversuchen. Dtsch. Akad. Wiss. Heft 37, Akademie-Verlag, Berlin, pp.26. E5c;C (2987)

Mitscherlich, E. A. [1954], Über die Anlage von Feldversuchen und die Verarbeitung ihrer Ergebnisse. Z. Pflanzenzüchtung 33:17-22. E5a;C (2988)

Möbius, H.-J. [1956], Bemerkungen zu eininigen Fragen der Feldversuchsmethodik. Z. Landw. Vers.- Untersuch Wes. 2:282-294. E2a (2989)

Moonan, W. J. [1952], The generalization of the principles of some modern experimental designs for educational and psychological research. Ph. D. Thesis, Univ. Minnesota. E2,7ae;A7 (2990)

Moonan, W. J. [1954], Multivariate analysis of covariance for a Latin square (summary). JASA 49:349. E7a;A7 (2991)

Moraes, J. M., Almeida, J. R. et. al. [1951], Influência do cloro, sôbre a compasição do caldo da cana de açúcar Co 290, aplicado no solo, na forma de cloreto de sódio Anais Escola Superior Agri. "Luiz de Queiroz" 8:115-151. E7ae (2992)

Moriguti, S. [1954], Optimality of orthogonal designs. Reports Stat. Appl. Res. 3:75-97. E14 (2993)

Moser, W. O. J. [1967], The number of very reduced 4×n Latin rectangles. Canadian J. Math. 19:1011-1017. E7c (2994)

Mote, V. L. [1958], On a minimax property of a balanced incomplete block design. AMS 29:910-914. E2a (2995)

Movshin, J. [1959], Experimental design - Part I. Why design experiments? ASQC Nat. Conv. Trans. 13:509-522. E2a;C (2996)

Mudra, A. [1954], Zur Bestimmung der Einzelfehler in Sortenversuchen. Z. Pflanzenzüchtung 33:23-30. E2,7ae (2997)

Mudra, A. [1954], Ein Vergleich verschiedener Versuchsmethoden. Z. Pflanzenzüchtung 33:419-423. E14 (2998)

Mudra, A. [1955], Bemerkungen zu der Arbeit von N. Atanasiu: Zur Frage der Anlage und Auswertung der Feldversuche. Z. Pflanzenzüchtung 34:1-6. E2a (2999)

Muller, E. R. [1965], A method of constructing balanced incomplete block designs. Biometrika 52:285-288. E2c (3000)

Müller, K.-H. [1954], Eine graphische Methode der varianzanlytischen Versuchsauswertung. Z. Pflanzenzüchtung 33:427-436. C (3001)

Müller, K.-H. [1954], Kritische Bemerkungen zu der Arbeit von N. Atanasiu „Über die statistischen Auswertungsmethoden der Feldversuche." Z. Planzenernährung Düngung Bodenkunde 64:129-134. E5a (3002)

Müller, K.-H. [1955], Zur Bodenstreuung bei Feldversuchen. Z. Acker- und Pflanzenbau 99:119-125. E2ae;C (3003)

Müller, K.-H. [1955], Zum Bodenausgleich bei der Blockanlage des Feldversuches. Z. Landw. Versuchs Untersuch. Wes. 1:233-241. E2,7a (3004)

Müller, K.-H. [1956], Exakte Auswertungsverfahren des kontrollierten Anbauvergleiches. Z. Landw. Versuchs Untersuch. Wes. 2:153-161. E2e (3005)

Müller, K.-H. [1957], Die Anlage und Auswertung faktorieller Feldversuche. Z. Landw. Versuchs Untersuchungswesen 2:117-135. E2,12a;T2a (3006)

Müller, K.-H. [1957], Einfache Auswertungsmöglichkeiten von Versuchen mit fehlenden Angaben. Z. Landw. Versuchs Untersuchungswesen 3:391-400. E6a (3007)

Müller, K.-H. [1959], Vereinfachte Anlagen bei Komplexversuchen. Z. Landw. Versuchs Untersuchungswesen 5:5-13. E5,12a (3008)

Muller, M. E. and Watson, G. S. [1959], Randomization and linear least squares estimation. Dept. Math. Tech. Report No. 32, Princeton Univ. E2a;C (3009)

Mulligan, B. W. and Haught, A. F. [1956], Experimental design in flame photometry determinations (summary). JASA 51:520. E5ac (3010)

Murphy, E. F.; Covell, M. R., and Dinsmore, J. S. [1957], An examination of three methods for testing palatability as illustrated by strawberry flavor differences. Food Res. 22:423-439. E2e;C (3011)

Murthy, J. S. N. [1964], On the exploitation of a new series of Pairwise Balanced Designs of index unity for the construction of mutually orthogonal Latin Squares (abstract). JISAS 16:169-170. E7,12c (3012)

Murty, J. S. [1965], On the construction of mutually orthogonal latin squares of non-prime-power orders. JISAS 17:224-229. E7,12,13c (3013)

Murty, J. S. and Das, M. N. [1967], Balanced n-ary block design and their uses. JISA 5:73-82. E5ac;T6c (3014)

Murty, V. N. [1952], On a new approach to the analysis of lattice designs (abstract). Proc. Thirty-Ninth Indian Sci, Congress, Part III, p. 15. E3a (3015)

Murty, V. N. [1953], Analysis of a triple rectangular lattice design. Biometrics 9:422-424. E3ae (3016)

Murty, V. N. [1954], On a new approach to the analysis to lattice designs (abstract). Ikushugaku Zasshi/Japanese J. Breeding 3(3&4):15. E2,3a (3017)

Murty, V. N. [1956], Short-cut methods in the analysis of variance. Agri. J. Fiji 3:123-128. E2a;A12 (3018)

Murty, V. N. [1957], On a parametric relation between the parameters of a balanced incomplete block design (abstract). Proc. Fourty-Fourth Indian Sci. Congress, Part III, p. 19. E2c (3019)

Murty, V. N. [1957], Analysis of simple rectangular lattice designs. Summary Proc. First Conf. Tobacco Res. Workers, Bangalore, 31 January and 1 February, 1957, Indian Centennial Tobacco Comm., pp. 74-75. E3a (3020)

Murty, V. N. [1957], Separation of error sums of squares in a split-plot design. Summary Proc. First Conf. Tobacco Res. Workers, Bangalore, 31 January and 1 February, 1957, Indian Centennial Tobacco Comm., p. 75. E12a (3021)

Murty, V. N. [1961], A note on Fisher's inequality for balanced incomplete block designs (abstract). JISAS 13:237. E2a (3022)

Murty, V. N. [1961], An inequality for balanced incomplete block designs. AMS 32:908-909. E2c (3023)

Murty, V. N. [1964], On the block structure of PBIB designs with two associate classes. Sankhyā A, 26:381-382. E3c (3024)

Nair, C. R. [1962], On the methods of block section and block intersection applied to certain PBIB designs. Calcutta Stat. Assoc. Bull. 11:49-54. E3c (3025)

Nair, C. R. [1963], A note on the analysis of serially balanced sequences. JISA 1:93-96. E8a (3026)

Nair, C. R. [1964], The impossibility of certain PBIB designs. Calcutta Stat. Assoc. Bull. 13:87-88. E3c (3027)

Nair, C. R. [1964], A new class of designs. JASA 59:817-833 (corrigendum 60:1250). E3ace (3028)

Nair, C. R. [1966], On partially linked block designs. AMS 37:1401-1406. E4c (3029)

Nair, C. R. [1967], Sequences balanced for pairs of residual effects. JASA 62:205-225. E8ace (3030)

Nair, K. R. [1950], Partially balanced incomplete block designs involving only two replication. Calcutta Stat. Assoc. Bull. 3:83-86. E3c (3031)

Nair, K. R. [1951], Rectangular lattices and partially balanced incomplete block designs (abstract). Partially balanced incomplete block designs involving only two replicates (abstract). Proc. Thirty-Eighth Indian Sci. Congress, Part III, p. 13. E3ac (3032)

Nair, K. R. [1951], Rectangular lattices and partially balanced incomplete block designs. Biometrics 7:145-154. E3a (3033)

Nair, K. R. [1951], Some two-replicate partially balanced designs. Calcutta Stat. Assoc. Bull. 3:174-176. E3c (3034)

Nair, K. R. [1951], Some three-replicate partially balanced designs. Calcutta Stat. Assoc. Bull. 4:39-42. E3c (3035)

Nair, K. R. [1952], Relation between efficiency of incomplete block designs and the intra-class correlations associated with incomplete and complete blocks. JISAS 4:149-152. E2,3,14a (3036)

Nair, K. R. [1952], Analysis of partially balanced incomplete block designs illustrated on the simple square and rectangular lattices. Biometrics 8:122-155. E3ac (3037)

Nair, K. R. [1953], Design and analysis of singly linked blocks (abstract). Proc. Fortieth Indian Sci. Congress, Part III, p. 125. E3ac (3038)

Nair, K. R. [1953], A note on group divisible incomplete block designs. Calcutta Stat. Assoc. Bull. 5:30-35. E3c;T2c (3039)

Nair, K. R. [1953], Some unsolved problems in experimental designs. Calcutta Stat. Assoc. Bull. 4:156-160. E3c;T2c (3040)

Nair, K. R. [1953], A note on rectangular lattices. Biometrics 9:101-106 (abstract 8:93). E3c (3041)

Nair, K. R. [1953], Design and analysis of triangular singly linked blocks. Biometrics 9:141-156. E4ace (3042)

Nair, K. R. [1954], Design and analysis of triangular singly linked blocks (abstract). Ikushugaku Zasshi/Japanese J. Breeding 3(3&4):125. E4ac (3043)

Nair, K. R. [1954], The so-called almost-balanced incomplete block designs. Calcutta Stat. Assoc. Bull. 5:181-184. E3c (3044)

Nair, K. R. [1956], Simplified analysis of singly linked blocks (abstract). Proc. Forty-Third Indian Sci. Congress, Part III, pp. 17-18. E4a (3045)

Nair, K. R. [1956], Simplified analysis of singly linked blocks. Biometrics 12: 369-380. E4ae (3046)

Nair, K. R. [1958], A note on reinforced incomplete block designs. JISAS 10:150-156. E4a (3047)

Nair, K. R. and Kishen, K. [1960], Recent developments in experimental design with special reference to the work in India. Bull. ISI 37(3):161-177. E2,3,4,7,12ac; T2,3,4a;B (3048)

NaNagara, P. [1957], Lattice rectangle designs. Ph. D. Thesis, Cornell Univ. Ellace (3049)

Nandi, H. K. [1951], On the efficiency of experimental designs. Calcutta Stat. Assoc Bull. 3:167-171. E14 (3050)

Nandi, H. K. [1952], Analysis of covariance. Calcutta Stat. Assoc. Bull. 4:79-82. E2a;A7 (3051)

Nandi, H. K. [1959], Current news. Euler's conjecture on existence of orthogonal Latin squares. Calcutta Stat. Assoc. Bull. 9:76. E7,12c (3052)

Nandi, H. K. and Adhikary, B. [1966], On the definition of Bose-Shimamoto's cyclical association scheme. Calcutta Stat. Assoc. Bull. 15:165-168. E3c (3053)

Narayan, L. [1958], The construction and combinatorial properties of Latin squares and balanced sequences. Ph. D. Thesis, Univ. Aberdeen. E7,12,13c (3054)

Nelder, J. A. [1951], A note on the statistical independence of quadratic forms in the analysis of variance. Biometrika 38:482-483. E2a (3055)

Nelder, J. A. [1954], The interpretation of negative components of variance. Biometrika 41:544-548. E5,12a (3056)

Nelder, J. A. [1954], A note on missing-plot values. Biometrics 10:400-401. E6a (3057)

Nelder, J. A. [1962], New kinds of systematic designs for spacing experiments. Biometrics 18:283-307. E5c (3058)

Nelder, J. A. [1965], The analysis of randomized experiments with orthogonal block structure, I. Block structure and the null analysis of variance. II. Treatment structure and the general analysis of variance. Proc. Royal Soc., Ser. A, 283: 147-162, 163-178. E2,7,12,14a;T2a; A5,15 (3059)

Nissen, Ø. [1951], Detaljplaner og beregningseksempler for forsøk med mange-forsøksledd. Skrivemaskinstua, Oslo, July, Mimeo. E2,3,5,7,11ae (3060)

Nissen, Ø. [1951], Nyere metoder for forsøk med et stort antall forsøksledd. (Modern methods for experiments with a large number of treatments.) Forskning Forsøk Landbruket (Oslo) 2:192-202. E2,3,7,11a (3061)

Nissen, Ø. [1951], En plan for faktorielle forsøk med hovedvekten på bestemmelse av samspillene. (A design for factorial experiments when the determination of the interactions is of major interest). Forskning Forsøk Landbruket (Oslo) 2: 203-214. E12ac;T2ac (3062)

Nissen, Ø. [1951], The use of systematic 5×5 squares. Biometrics 7:167-170. E7ac (3063)

Noether, G. E. [1963], Efficiency of the Wilcoxon two-sample statistic for randomized blocks. JASA 58:894-898. E2a;A12 (3064)

Norton, D. A. [1952], Groups of orthogonal row-latin squares. Pacific J. Math. 2:335-341. E7,12c (3065)

Norton, D. A. and Stein, S. K. [1956], An integer associated with Latin squares. Proc. American Math. Soc. 7:331-334. E7c (3066)

Norton, H. W. [1955], A further note on missing data. Biometrics 11:110. E6a (3067)

O'Carroll, F. M. [1963], A method of generating randomized Latin squares. Biometrics 19:652-653. E7c (3068)

Ogasawara, M. [1965], A necessary condition for the existence of regular and symmetrical PBIB designs of T_m type. Inst. Stat. Mimeo. Ser. No. 418, Univ. N. Carolina. E3c (3069)

Ogasawara, T. and Takahashi, M. [1953], Orthogonality relation in the analysis of variance I. J. Sci. Hiroshima Univ. Ser. A-I 16:457-470. E2,14a (3070)

Ogawa, J. [1958], On the relationship algebra and the association algebra of the partially balanced incomplete block design (abstract). AMS 29:1279. E3c (3071)

Ogawa, J. [1959], A necessary condition for existence of regular and symmetrical experimental designs of triangular type, with partially balanced incomplete blocks. AMS 30:1063-1071 (abstract 30:615; Inst. Stat. Mimeo. Ser. No. 219, Univ. N. Carolina). E3c (3072)

Ogawa, J. [1959], The theory of the association algebra and the relationship algebra of a partially balanced incomplete block design. Inst. Stat. Mimeo. Ser. No. 224, Univ. N. Carolina. E3ac (3073)

Ogawa, J. [1959], A note on the analysis of a randomized block design. Inst. Stat. Mimeo. Ser. No. 218, Univ. N. Carolina. E2a (3074)

Ogawa, J. [1961], On a unified method of deriving necessary conditions for existence of symmetrical partially balanced incomplete block designs of certain types. Bull. ISI 38(4):43-57. E3c (3075)

Ogawa, J. [1961], The effect of randomization on the analysis of randomized block design. Ann. Inst. Stat. Math. 13:105-117. E2a;A5 (3076)

Ogawa, J. [1963], On the null-distribution of the F-statistic in a randomized balanced incomplete block design under the Neyman model. AMS 34:1558-1568. E2a;A5 (3077)

Ogawa, J. and Ishii, G. [1964], On the analysis of partially balanced incomplete block designs in the regular case. Inst. Stat. Mimeo. Ser. No. 412, Univ. N. Carolina. E3a (3078)

Ogawa, J. and Ishii, G. [1965], The relationship algebra and the analysis of variance of a partially balanced incomplete block design. AMS 36:1815-1828. E3ae (3079)

Ogawa, J., Ikeda, S., and Ogasawara, M. [1964], On the null-distribution of the F-statistic in a randomized partially balanced incomplete block design with two associate classes under the Neyman model. Inst. Stat. Mimeo. Ser. No. 414, Univ, N. Carolina. E3a (3080)

Ogawa, J. and Pasternack, B. S. [1958], On the problem of incomplete data (abstract). AMS 29:1279-1280. E6a;A10 (3081)

Ogawa, J. and Pasternack, B. S. [1961], Formulation of a model containing a chance mechanism according to which observations are missed: The randomized block design (abstract). AMS 32:922. E2,6a;A10 (3082)

Ogawa, J., Sadao, I., and Ogasawara, M. [1966], Sur la distribution-nulle de la statistique F pour tester une hypothese-nulle partielle dans un projet de bloc incomplet "randomize" et partiellement equilibre aux m classes associées sous le modele de Neyman (resume). Bull. ISI 41:960. E3a (3083)

Ogilvie, J. C. [1963], A simple method for the elimination of individual trends in the analysis of balanced sets of Latin squares. Biometrics 19:264-272. E7ae (3084)

Okuno, C. and Okuno, T. [1961], On the construction of a class of partially balanced incomplete block design by calculus of blocks. Reports Stat. Appl. Res. 8:113-139. E3c (3085)

Okuno, T. [1958], On the design and analysis of field experiments. II. Split-plot and split-block designs (in Japanese). Bull. Nat. Inst. Agri. Sci. 6(A):81-146. E12a (3086)

Okuno, T. [1961], The model and interpretations for analysis of covariance. Bull. ISI 38(4):555-565. E1ae;A7 (3087)

Okuno, T. [1962], Mathematical models and robust criteria in the study of the analysis of variance. Bull. Nat. Inst. Agri. Sci. 9(A):153-211. E2,12a;A7 (3088)

Ortlepp, H. [1955], Zur Anwendung einiger Versuchsmethoden im Feldversuchswesen. Z. Landw. Vers.- Untersuch Wes. 1:300-311. E2,5,7a;A14 (3089)

Ortlepp, H. [1957], Überlegungen und Vorschläge zur Anordnung der Teilstücke in Feldversuchsanlagen. Z. Landw. Vers. Untersuch Wes. 2:136-150. E2,7a;A14 (3090)

Ostle, B. [1952], Question 31: How F estimates unity, Amer. Stat. 6(3):32. E1a (3091)

Ostrowski, R. T. and Van Duren, K. D. [1961], On a theorem of Mann on Latin squares. Math. Computation 15:293-295. E7c (3092)

Outhwaite, A. D. and Rutherford, A. [1955], Covariance analysis as an alternative to stratification in the control of gradients. Biometrics 11:431-440. E7ae;A7 (3093)

Paige, L. J. and Tompkins, C. B. [1960], The size of the 10×10 latin square problem. Proc. Tenth Symp. Appl. Math. American Math. Soc. 10:71-83.
E7,12,13c (3094)

Paik, U. B. [1960], Field Experiment and Statistical Methods. New Agricultural Techniques. H. M. Publishing Co., Seoul, Korea. E2,3a;T2a;C (3095)

Paik, U. B. [1962], Analysis of lattice design for the farm experiment. Theses Collection Chungnam National Univ., Vol. II. E2,3a (3096)

Panse, V. G. and Abraham, T. P. [1958], Analysis of a long-term manurial-cum-rotational experiment (abstract). JISAS 10:160. E5,8a;A13 (3097)

Panse, V. G. and Abraham, T. P. [1960], Simple scientific experiments on farmers' land. Bull. ISI 37(3):179-189. E3ac;T3ac;C (3098)

Panse, V. G. and Sukhatme, P. V. [1954], Statistical Methods for Agricultural Workers. Indian Council Agri. Res., New Delhi, Rs 15: pp. xvi+361.
E1,2,3,5,7,12ace;
T2a (3099)

Parker, E. T. [1957], On collineations of symmetric designs. Proc. American Math. Soc. 8:350-351. E2c (3100)

Parker, E. T. [1958], Construction of some sets of pairwise orthogonal latin squares (preliminary report) (abstract). American Math. Soc. Notices 5:815.
E7,12c (3101)

Parker, E. T. [1959], Construction of some sets of mutually orthogonal latin squares. Proc. American Math. Soc. 10:946-949. E7,12,13c (3102)

Parker, E. T. [1959], Orthogonal latin squares (abstract 557-63). American Math. Soc. Notices 6:276. E7,12c (3103)

Parker, E. T. [1959], Nonextendible sets of mutually orthogonal Latin squares (preliminary report) (abstract 558-54). American Math. Soc. Notices 6:390-391.
E7,12,13c (3104)

Parker, E. T. [1959], Completion of disproof of Euler's conjecture (abstract 558-55). American Math. Soc. Notices 6:391. E7,12c (3105)

Parker, E. T. [1959], A computer search for latin squares orthogonal to latin squares of order ten (abstract). American Math. Soc. Notices 6:798.
E7,12c (3106)

Parker, E. T. [1959], Nonextendibility conditions on mutually orthogonal latin squares (abstract 564-77). American Math. Soc. Notices 6:801. E7,12,13c (3107)

Parker, E. T. [1959], Orthogonal latin squares. Proc. Nat. Academy Sci. (USA) 45:859-862. E7,12,13c (3108)

Parker, E. T. [1961], Computer searching for orthogonal latin squares of order 10. American Math. Soc. Notices 8:617. E7,12c (3109)

Parker, E. T. [1962], Nonextendibility conditions on mutually orthogonal latin squares. Proc. American Math. Soc. 13:219-221. E7,12,13c (3110)

Parker, E. T. [1962], On orthogonal latin squares. Proc. Symp. Pure Math., American Math. Soc. 6:43-46. E7,12,13c (3111)

Parker, E. T. [1963], Computer investigation of orthogonal latin squares of order ten. Proc. Symp. Appl. Math. American Math. Soc. 15:73-81. E7,12c (3112)

Parker, E. T. [1963], Remarks on balanced incomplete block designs. Proc. American Math. Soc. 14:729-730. E2c (3113)

Parker, E. T. [1967], A result in balanced incomplete block designs. J. Combinatorial Theory 3:282-285. E2c (3114)

Pascual, R. N. [1954], A comparison of the efficiency of randomized blocking and of complete randomization in pot tests. Philippine Agri. 38:435-443. E1,2,14a (3115)

Pasternack, B. S. [1960], Analysis of covariance for a 3×4 triple rectangular lattice design (3 associate P.B.I.B.). Biometrics 16:7-18. E3ae;A7 (3116)

Pasternack, B. S. and Ogawa, J. [1961], Stochastic bounds of the "F" statistic when data are incomplete one-way classification. Bull. ISI 38(4):189-200. E1,2ae (3117)

Patterson, H. D. [1950], The analysis of change-over trials. J. Agri. Sci. 40: 375-380. E8ae (3118)

Patterson, H. D. [1951], Change-over trials. JRSSB 13:256-271. E7,8ac;T2c (3119)

Patterson, H. D. [1952], The construction of balanced designs for experiments involving sequences of treatments. Biometrika 39:32-48. E8ac (3120)

Patterson, H. D. [1953], The analysis of the results of a rotation experiment on the use of straw and fertilizers. J. Agri. Sci. 43:77-88. E5,8a (3121)

Patterson, H. D. [1959], The analysis of a non-replicated experiment involving a single four-course rotation of crops. Biometrics 15:30-59. E5,8a (3122)

Patterson, H. D. [1964], Theory of cyclic rotation experiments (with discussion). JRSSB 26:1-45. E5,8ac;A13 (3123)

Patterson, H. D. [1965], The factorial combination of treatments in rotation experiments. J. Agri. Sci. 65:171-182. E5,8ac;T2a (3124)

Patterson, H. D. and Lucas, H. L. [1959], Extra-period change-over designs. Biometrics 15:116-132. E8ace (3125)

Patterson, H. D. and Lucas, H. L. [1962], Change-over designs. N. Carolina Agri. Expt. Sta. U.S. Dept. Agri. Tech. Bull. 147:1-52. E8ace (3126)

Patterson, H. D. and Williams, R. [1962], The residual effects of phosphorus fertilizers on yields of arable crops: preliminary results of six rotation experiments. Experimental Husbandry 8:85-103. E5,8a (3127)

Pavate, M. V. [1961], Combined analysis of balanced incomplete block designs with some common treatments. Biometrics 17:111-119. E4a;A13 (3128)

Peake, R. E. [1953], Planning an experiment in a cotton spinning mill. Appl. Stat. 2:184-192. E2e;C (3129)

Pearce, S. C. [1952], Some new designs of Latin square type. JRSSB 14:101-106. E7ace (3130)

Pearce, S. C. [1952], The design of calibration trials with three varieties. Annual Report, East Malling Res. Sta., 1951, pp. 105-107. E5c (3131)

Pearce, S. C. [1953], Field experimentation with fruit trees and other perennial plants Commonwealth Bureau Hort. Plantation Crops, Tech. Comm. 23, pp. x+131. E1,2,3,5,6,7,8,12ace T2a;A7,14;B;C;S1 (3132

Pearce, S. C. [1954], Experimentation with perennial plants. FAO/55/8/5513:Index/3 FAOUN, Rome, 51 pp. (Mimeo.) E5,6a;T2a;A9 (3133)

Pearce, S. C. [1955], The specific problems of experimental design and technique in perennial crops (abstract). Biometrics 11:538-539. E5a;C (3134)

Pearce, S. C. [1956], Some problems of experimental design and technique with perennial crops. Biometrics 12:330-337. E5a;A13 (3135)

Pearce, S. C. [1956], Problemas específicos de delineamento e de técnica experiment em culturas perenes. (Specific problems of design and experimental technique in perennial crops.) Agros, Rio Grande do Sul 9(2):12-21. E5,8a (3136)

Pearce, S. C. [1957], Experimenting with organisms as blocks. Biometrika 44: 141-149. E2ae (3137)

Pearce, S. C. [1959], Some recent applications of multivariate analysis to data from fruit trees. Annual Report, East Malling Res. Sta., 1958, pp. 73-76. E5a;A7 (3138)

Pearce, S. C. [1960], Supplemented balance. Biometrika 47:263-271 (corrigenda 48:475). E5,12ace (3139)

Pearce, S. C. [1963], The use and classification of non-orthogonal designs (with discussion). JRSSA 126:353-377. E2,3,4,5,7,14a (3140)

Pearce, S. C. [1964], Experimenting with blocks of natural size. Biometrics 20: 699-706. E4ac (3141)

Pearce, S. C. [1964], Non-orthogonal designs in biological research (abstract). Biometrics 20:393. E4ac (3142)

Pearce, S. C. [1967], The efficiency of non-orthogonal designs as used in biological research (abstract). Biometrics 23:596-597. E3,14a (3143)

Pearce, S. C. and Freeman, G. H. [1959], A review of experimental design at East Malling, 1953-1958. Annual Report East Malling Res. Sta., 1958, pp. 67-72. E5a (3144)

Pearce, S. C., Jolly, G. M., and Freeman, G. H. [1953], A review of experimental design at East Malling, 1948-1952. Annual Report East Malling Res. Sta., 1952, pp. 83-87. E5a (3145)

Pearce, S. C. and Taylor, J. [1950], The purpose and design of calibration trials. Annual Report East Malling Res. Sta., 1949, pp. 83-90. E5c (3146)

Phillips, J. P. N. [1964], The use of magic squares for balancing and assessing order effects in some analysis of variance designs. Appl. Stat. 13:67-73 (correction 15:151). E12c (3147)

Pires, A. J. C. and Baptista, J. G. [1955], Análise do delineamento em blocos casualizados quando existem dois e três talhões perdidos. (Analysis of randomized block layouts when two or three plots are missing.) Melhoramento 8:61-77. E6ae (3148)

Pompilj, G. [1963], Analisi delle medie (discussion). Colloques Internationaux du Centre National de la Recherche Scientifique, No. 110, Le Plan d'Expériences pp. 79-98. E2,3,12a;T2a (3149)

Potthoff, R. F. [1959], Multi-dimensional incomplete block designs. Ph. D. Thesis, Univ. N. Carolina. E3,7,12ac (3150)

Potthoff, R. F. [1962], Three-factor additive designs more general than the latin square. Technometrics 4:187-208. E7ace (3151)

Potthoff, R. F. [1962], Four-factor additive designs more general than the Greco-Latin square. Technometrics 4:361-366. E12ac (3152)

Potthoff, R. F. [1963], Three-dimensional incomplete block designs for interaction models. Biometrics 19:229-263. E7ac;T7ace (3153)

Potthoff, R. F. [1963], Some illustrations of 4 DIB design constructions. Calcutta Stat. Assoc. Bull. 12:19-30. E12c (3154)

Preece, D. A. [1966], Some row and column designs for two sets of treatments. Biometrics 22:1-25. E12ac (3155)

Preece, D. A. [1966], Classifying Youden rectangles. JRSSB 28:118-130. E7c (3156)

Preece, D. A. [1966], Some balanced incomplete block designs for two sets of treatments. Biometrika 53:497-506. E2,12ac (3157)

Preece, D. A. [1967], Incomplete block designs with v=2k. Sankhyā A 29:305-316. E2,3,14c (3158)

Preece, D. A. [1967], Cyclic generation of Robinson's balanced incomplete block designs. Biometrics 23:574-578. E9c (3159)

Preece, D. A. [1967], Nested balanced incomplete block designs. Biometrika 54:479-486. E9ac (3160)

Primrose, E. J. F. [1951], Quadrics in finite geometrics. Proc. Cambridge Philosophic Soc. 47:299-304. E2c (3161)

Primrose, E. J. F. [1952], Resolvable balanced incomplete block designs. Sankhyā 12:137-140. E2c (3162)

Proschan, F. [1954], Investigation of Latin squares. IQC 11(1):30-33. E7a (3163)

Proschan, F. and Babcock, A. B. [1955], How to design effective experiments. Chemical Eng. 62(August):191-198. E1,2a;T2a;A3;C (3164)

Purcell, W. R. [1951], Balancing and randomizing in experiments. IQC 7(4):7-14. E5e;C (3165)

Puri, M. L. and Sen, P. K. [1966], On some optimum nonparametric procedures in two-way layouts. Inst. Stat. Mimeo. Ser. No. 485, Univ. N. Carolina. E2a;A12 (3166)

Puri, M. L. and Sen, P. K. [1967], On robust estimation in incomplete block designs. AMS 38:1587-1591. E3a (3167)

Puri, M. L. and Sen, P. K. [1967], On some optimum nonparametric procedures in two-way layouts. JASA 62:1214-1229. E2a;A12 (3168)

Raghavarao, D. [1959], A note on the construction of Group Divisible designs from Hyper Graeco Latin Cubes of the first order. Calcutta Stat. Assoc. Bull. 9: 67-70. E3c (3169)

Raghavarao, D. [1960], On the block structure of certain PBIB designs with two associate classes having triangular and L_2 association schemes. AMS 31: 787-791. E3c (3170)

Raghavarao, D. [1960], A generalization of group divisible designs. AMS 31:756-771. E3c (3171)

Raghavarao, D. [1961], Some contributions to the design and analysis of experiments. Ph. D. Thesis, Bombay Univ. E (3172)

Raghavarao, D. [1962], Some results for GD m-associate designs. Calcutta Stat. Assoc. Bull. 11:150-154. E2,3c (3173)

Raghavarao, D. [1962], On the use of latent vectors in the analysis of group divisible and L_2 designs. JISAS 14:138-144. E3a (3174)

Raghavarao, D. [1962], Symmetrical unequal block arrangements with two unequal block sizes. AMS 33:620-633. E4ac (3175)

Raghavarao, D. [1962], On balanced unequal block designs. Biometrika 49:561-562. E4c (3176)

Raghavarao, D. [1963], A note on the block structure of BIB designs. Calcutta Stat. Assoc. Bull. 12:60-62. E2c (3177)

Raghavarao, D. [1966], Duals of partially balanced incomplete block designs and some nonexistence theorems. AMS 37:1048-1052. E3c (3178)

Raghavarao, D. [1966], An extended partial geometry. JISAS 18(1):99-107. E3c (3179)

Raghavarao, D. and Chandrasekhararao, K. [1964], Cubic designs. AMS 35:389-397. E3c (3180)

Raghavarao, D. and Tharthare, S. K. [1967], An inequality for doubly balanced incomplete block designs. Calcutta Stat. Assoc. Bull. 16:37-39. E2c (3181)

Raktoe, B. L. [1964], Application of cyclic collineations to the construction of balanced ℓ-restrictional prime powered lattice designs. Ph. D. Thesis, Cornell Univ. E2,11,13ac (3182)

Raktoe, B. L. [1964], Application of cyclic collineations to the construction of balanced ℓ-restrictional lattices. Biometrics Unit Mimeo. Ser. BU-172-M, Cornell Univ. E2,11,13c (3183)

Raktoe, B. L. [1965], Balanced ℓ-restrictional lattices (abstract). Technometrics 7:275. E2,11,13c;T2c (3184)

Raktoe, B. L. [1967], Application of cyclic collineations to the construction of balanced ℓ-restrictional prime powered lattice designs. AMS 38:1127-1141. E2,3,11,13c;T2c (3185)

Ramachandran, G. [1960], Analysis of covariance for a split-plot design-average variances for comparisons of adjusted means (abstract). JISAS 12:211. E5a;A7 (3186)

Ramachandran, G. [1964], Analysis of covariance for the split-plot design: Average variances. Biometrics 20:204-206. E5a;A7 (3187)

Ramakrishnan, C. S. [1956], On the dual of a PBIB design and a new class of designs with two replications. Sankhyā 17:133-142. E3ace (3188)

Ramakrishnan, C. S. [1957], The dual of a two associate PBIB design and derivation of a two parameter family of designs with two replications (abstract). Proc. Forty-Fourth Indian Sci. Congress, Part III, p. 19. E3ac (3189)

Ramanujacharyulu, C. [1965], Non-linear spaces in the construction of symmetric PBIB designs. Sankhyā A 27:409-414. E3c (3190)

Ramanujacharyulu, C. [1965], Classification of canonical form of quadrics in finite projective geometry. JISA 3:91-96. E2,3c (3191)

Ranzani, G. et al. [1953], Vinhaça e adubos minerais (I). Anais Escola Superior Agri. "Luiz de Queiroz" 10:97-108. E2ae (3192)

Rao, B. M. [1961], Cumulant component estimation in the balanced one-way population with finite sub-populations. Biometrics Unit Mimeo. Ser. BU-140-M, Cornell Univ E1a;A14 (3193)

Rao, B. M., Douglas, A. W., and Sheard, R. W. [1960], A numerical example of a response surface design. Biometrics Unit Mimeo. Ser. BU-120-M, Cornell Univ. E2ae;T4ae (3194)

Rao, C. R. [1956], On the recovery of inter block information in varietal trials. Sankhyā 17:105-114. E2,3,4a (3195)

Rao, C. R. [1956], A general class of quasifactorial and related designs. Sankhyā 17:165-174. E3,7ac (3196)

Rao, C. R. [1959], Expected values of mean squares in the analysis of incomplete block experiments and some comments based on them. Sankhya 21:327-336. E2,3a;A5 (3197)

Rao, C. R. [1960], Experimental designs with restricted randomisation. Bull. ISI 37(3):397-404. E4ac (3198)

Rao, C. R. [1961], Combinatorial arrangements analogous to orthogonal arrays. Sankhyā A 23:283-286. E2,7c (3199)

Rao, C. R. [1961], A study of BIB designs with replications 11 to 15. Sankhyā A 23: 117-127. E2c (3200)

Rao, C. R. [1961], A combinatorial assignment problem. Nature 191:100. E2c (3201)

Rao, K. S. [1961], BIB designs with one treatment missing. JISAS 13:159-168 (abstract 13:237-238). E2,6a (3202)

Rao, M. B. [1965], Solution to problem 37. JISA 3:60. E3c (3203)

Rao, M. B. [1965], A note on incomplete block designs with b=v. AMS 36:1877 E4a (3204)

Rao, M. B. [1966], Partially balanced block designs with two different number of replications. JISA 4:1-9. E4ac (3205)

Rao, M. B. [1966], Group divisible family of PBIB designs. JISA 4:14-28. E3c (3206)

Rao, M. B. [1966], A note on equi-replicate balanced designs with b=v. Calcutta Stat. Assoc. Bull. 15:43-44. E2c (3207)

Rao, P. V. [1960], On the construction of some partially balanced incomplete block designs with more than three associate classes. Calcutta Stat. Assoc. Bull. 9:87-92. E3c (3208)

Rao, P. V. [1960], The dual of a balanced incomplete block design. AMS 31:779-785. E3c (3209)

Rao, P. V. [1961], Analysis of a class of PBIB designs with more than two associate classes. AMS 32:800-808. E3a (3210)

Rao, P. V. [1963], The effect of truncation on the analysis of variance test for PBIB designs. Ph. D. Thesis, Univ. Georgia. E3a (3211)

Rao, P. V. [1963], The F-test in the intrablock analysis of a class of PBIB designs. Ann. Inst. Stat. Math. 15:25-36. E3a (3212)

Rao, P. V. [1963], The robustness of ANOVA for a class of 2-associate PBIB designs (abstract). AMS 34:684. E3,14a (3213)

Rao, P. V. [1965], The effect of truncation on the F-test for a class of PBIB designs (abstract). AMS 36:1605. E3a (3214)

Rao, V. R. [1958], A note on balanced designs. AMS 29:290-294. E2,14a (3215)

Rasch, D. and Kasdorff, K. [1964], Die statistische Auswertung von Gruppen-Perioden Versuchen (GPV). (On the statistical evaluation of group change-over designs.) Biom. Zeit. 6:113-122. E8ae (3216)

Raut, K. C. [1959], On generalized designs (abstract). JISAS 11:196. E4ac (3217)

Raut, K. C. [1960], Analysis of generalised non-orthogonal design with recovery of interblock information (abstract). JISAS 12:211. E4a (3218)

Raut, K. C. [1960], Generalised non-orthogonal design and its analysis with recovery of interblock information. JISAS 12:190-199. E3,4a;T4ac (3219)

Raut, K. C. and Das, M. N. [1959], Generalised partially balanced design. JISAS 11:145-162. E4ace (3220)

Rauterberg, E. [1954], Die Berechnung der Streuung von Versuchsergebnissen. Z. Pflanzenernährung Düngung Bodenkunde 64:244-258. E2,7ae (3221)

Rauterberg, E. [1956], Bemerkungen zu der Arbeit des Verfassers "Die Berechnung der Streuung von Versuchsergebnissen". Z. Pflanzenernährung Düngung Bodenkunde 72:220-224. E2,7a (3222)

Ray, W. D. [1956], Sequential analysis applied to certain experimental designs in the analysis of variance. Biometrika 43:388-403. E1,2a;S2 (3223)

Ray-Chaudhuri, D. K. [1959], On the application of the geometry of quadrics to the construction of partially balanced incomplete block designs and error correcting binary codes. Ph. D. Thesis, Univ. N. Carolina (also, Inst. Stat. Mimeo. Ser. No. 230, Univ. N. Carolina). E3c (3224)

Ray-Chaudhuri, D. K. [1962], Some results on quadrics in finite projective geometry based on Galois fields. Canadian J. Math. 14:129-138. E3c (3225)

Ray-Chaudhuri, D. K. [1962], Application of the geometry of quadrics for constructing PBIB designs. AMS 33:1175-1186 (also, Inst. Stat. Mimeo. Ser. No. 294, Univ. N. Carolina). E3c (3226)

Ray-Chaudhuri, D. K. [1963], On some connections between incomplete block designs and minimum covers (discussion). Colloques Internationaux du Centre National de la Recherche Scientifique, No. 110 Le Plan d'Expériences pp. 129-136. E2c (3227)

Ray-Chaudhuri, D. K. [1965], Some configurations in finite projective spaces and partially balanced incomplete block designs. Canadian J. Math. 17:114-123. E3c (3228)

Redman, C. E. and King, E. P. [1965], Group screening utilizing balanced and partially balanced incomplete block designs. Biometrics 21:865-874 (abstract 20:913). E2,3a;T7ac (3229)

Rees, D. H. [1966], The analysis of variance of designs with many non-orthogonal classifications. JRSSB 28:110-117. E7,12a;A10 (3230)

Rees, D. H. [1967], Some designs of use in serology. Biometrics 23:779-791. E5c (3231)

Reinach, S. G. [1965], A nonparametric analysis for a multi-way classification with one element per cell. Suid-Afrikaanse Tydskrif vir Landbouwetenskap 8:941-959. E2a,T2a;A12 (3232)

Renier, A. [1954], Classification et description des despositifs expérimentaux. Annales de l'Institut Experimental du Tabac de Bergerac 2:103-149. E1,2,7,12a;T2a (3233)

Ricciuti, C., Coleman, J. E. and Willits, C. O. [1955], Statistical comparison of three methods for determining organic peroxides. Anal. Chem. 27:405-407. E7ae (3234)

Rich, S. [1952], Using half-tree plots for increasing the efficiency of fungicide tests. Phytopathology 42:353-354. E2a (3235)

Rickmers, A. D. [1967], The missing data problems in factorial designs. ASQC Tech. Conf. Trans. 21:167-173. E6,10a (3236)

Riordan, J. [1952], A recurrence relation for three-line latin rectangles. American Math. Monthly 59:159-162. E7c (3237)

Rives, M. [1955], Méthodes d'analyse pour des essais incomplets. Annales Inst. Nat. Res. Agronomique 5(B):103-118. E7,10ae (3238)

Rives, M. [1959], Sur la comparaison des moyennes dans les essais variétaux. Annales Inst. Nat. Res. Agronomique 9(B):357-376. E2a;A3 (3239)

Robinson, J. [1966], Balanced incomplete block designs with double grouping of blocks into replications. Biometrics 22:368-373. E9ac (3240)

Robinson, J. [1967], Incomplete split plot designs. Biometrics 23:793-802. E12ac (3241)

Robson, D. S. [1958], Nonparametric estimation: Cumulant components in the balanced one-way classification. Biometrics Unit Mimeo. Ser. BU-93-M, Cornell Univ. E1a;A14 (3242)

Robson, D. S. [1961], Cumulant component analysis in balanced designs (abstract). Biometrics 17:175. E1a;A14 (3243)

Robson, D. S. [1961], Multiple comparisons with a control in balanced incomplete block designs. Technometrics 3:103-105 (also, Biometrics Unit Mimeo. Ser. BU-87-M, Cornell Univ.). E2a;A3 (3244)

Robson, D. S. and Atkinson, G. F. [1960], Individual degrees of freedom for testing homogeneity of regression coefficients in a one-way analysis of covariance. Biometrics 16:593-605. E1a;A4,7 (3245)

Rod, J, [1955], Diagramynejmen ších průkazných rozdílů mezi zkoušenými členy v polním pokuse, založeném v namátkových dílcích a řešeném analysou rozptylu. (Graphs of minimum significant differences between tested varieties in a field trial arranged in randomized blocks and carried out by the analysis of variance.) Sbornik Československé Akademie Zemědělských věd Praha 28:607-624. E2a (3246)

Rohatgi, V. K. [1966], A comparison of bounds for the number of common treatments between any two blocks of certain two-associate PBIB designs. Calcutta Stat. Assoc. Bull. 15:39-42. E3c (3247)

Rojas, B. and White, R. F. [1957], The modified latin square. JRSSB 19:305-317. E7ac (3248)

Romier, G. [1964], Automorphisms et Algèbre d'un schéma d'association. Caractérisat algébrique des correspondances partiellement équilibrées. Comptes Rendus 258: 5345-5348. E3c (3249)

Room, T. G. [1955], A new type of magic square. Math. Gazette 39:307. E7c (3250)

Roussel, N. [1954], Le problème de l'éxpérimentation des variétés de betterave et ses relations avec la biométrie. Publications Inst. Belge Amélioration Betterave 22:139-140. E2,3,7,11a (3251)

Roy, J. [1958], On the efficiency factor of block designs. Sankhyā 19:181-188. E2,14a (3252)

Roy, J. [1959], A class of two replicate incomplete block designs. Biometrics 15:259-269 (also, Inst. Stat. Mimeo. Ser. No. 201, Univ. N. Carolina). E3ace (3253)

Roy, J. and Laha, R. G. [1956], Classification and analysis of Linked Block designs. Sankhyā 17:115-132. E2,3,4ace (3254)

Roy, J. and Laha, R. G. [1956], Two associate partially balanced designs involving three replications. Sankhyā 17:175-184. E3c (3255)

Roy, J. and Laha, R. G. [1957], On partially balanced Linked Block designs. AMS 28:488-493. E3,4c (3256)

Roy, J. and Shah, K. R. [1961], Analysis of two-way designs (abstract). Proc. Forty-Eighth Indian Sci. Congress, Part III, pp. 24-25. E7ac (3257)

Roy, J. and Shah, K. R. [1961], Analysis of two-way designs. Sankhyā A 23:129-144. E7,14ae (3258)

Roy, J. and Shah, K. R. [1962], Recovery of interblock information. Sankhyā A 24: 269-280. E2,3a (3259)

Roy, P. M. [1952], A note on the resolvability of balanced incomplete block designs. Calcutta Stat. Assoc. Bull. 4:130-132. E2c (3260)

Roy, P. M. [1953], A note on the method of inversion of statistical designs. Sci. Culture 18:440-441. E2,3c (3261)

Roy, P. M. [1953], A note on the relation between bib and pbib designs. Sci. Culture 19:40-41. E2,3c (3262)

Roy, P. M. [1953], Hierarchial group divisible incomplete block designs with m-associate classes. Sci. Culture 19:210-211. E3c (3263)

Roy, P. M. [1953], A note on the unreduced balanced incomplete block designs. Sankhyā 13:11-16. E2c (3264)

Roy, P. M. [1954], Inversion of incomplete block designs. Bull. Calcutta Math. Soc. 46:47-58. E2,3c (3265)

Roy, P. M. [1954], On the relation between b.i.b. and p.b.i.b. designs. JISAS 6:30-47. E2,3c (3266)

Roy, P. M. [1954], Rectangular lattices and orthogonal group divisible designs. Calcutta Stat. Assoc. Bull. 5:87-98. E3c (3267)

Roy, P. M. [1954], On the method of inversion in the construction of partially balanced incomplete block designs from the corresponding b.i.b. designs. Sankhyā 14:39-52. E3c (3268)

Roy, P. M. [1955], Some combinatorial problems in the design of experiments. Ph. D. Thesis, Calcutta Univ. E2,3c (3269)

Roy, P. M. [1955], Analysis of p×(p-1), n-ple latinized rectangular lattices and their multiples. Calcutta Stat. Assoc. Bull. 6:113-131. E3ac (3270)

Roy, P. M. [1955], Difference theorems for the construction of Group Divisible Designs (abstract). On some simple properties of semi-regular Group Divisible Designs (abstract). Proc. Forty-Second Indian Sci. Congress, Part III, pp. 18-19. E3c (3271)

Roy, P. M. [1956], p×(p-1), (p-2)-ple rectangular lattices (abstract). Proc. Forty-Third Indian Sci. Congress, Part III, p. 18. E3c (3272)

Roy, P. M. [1957], On the distribution of varieties of p×(p-1), n-ple latinized rectangul lattices and weighted average variances. Calcutta Stat. Assoc. Bull. 7:101-114. E3a (3273)

Roy, P. M. [1962], On the properties and construction of HGD designs with m-associate classes. Calcutta Stat. Assoc. Bull. 11:10-38 (erratum 12:67). E3c (3274)

Roy, S. N. [1961], On the planning and interpretation of multifactor multiresponse experiments. Bull. ISI 38(4):59-72. E5,9ac;C (3275)

Roy, S. N. and Potthoff, R. F. [1958], Multi-dimensional incomplete block designs. Inst. Stat. Mimeo. Ser. No. 211, Univ. N. Carolina. E3,7,12c (3276)

Roy, S. N. and Roy, J. [1958], Analysis of variance with univariate or multivariate, fixed or mixed classical models. Inst. Stat. Mimeo. Ser. No. 208, Univ. N. Carolina. E2a;A7 (3277)

Roy, S. N.; Shrikhande, S. S. and Krishnaiah, P. R. [1960], On the efficiency of experimental designs (abstract). AMS 31:242. E14 (3278)

Roy, S. N. and Srivastava, J. N. [1962], Multivariate hierarchial designs (abstract). AMS 33:307. E5,9ac;A7 (3279)

Roy, S. N. and Srivastava, J. N. [1963], Hierarchial and p-block multiresponse designs and their analysis. Contributions to Statistics, pp. 419-428 (also, Inst. Stat. Mimeo. Ser. No. 341, Univ. N. Carolina). E5,9ac (3280)

Rundfeldt, H. [1953], Die Prüfung der wichtigsten Verfahren im Feldversuchswesen an Hand von Modellen. Z. Pflanzenzüchtung 32:300-354. E1,2,7ae;A14;C (3281)

Rundfeldt, H. [1955], Über die Auswertung von Blockversuchen. Der Züchter 25: 252-255. E2ae (3282)

Rundfeldt, H. [1960], On an improved method for the evaluation of single trials (abstract). Biometrics 16:313. E2,7a (3283)

Rundfeldt, H. [1963], Untersuchungen über den Einsatz elektronischer Rechenanlagen bei der Auswertung von Feldversuchen in der Pflanzenzüchtung. Z. Pflanzenzüchtur 49:253-386. E3a;A15 (3284)

Russell, T. S. [1956], Estimation of individual variations in an unreplicated two-way classification. Ph. D. Thesis, Virginia Polytechnic Inst. E2a;A11 (3285)

Russell, T. S. and Bradley, R. A. [1958], One-way variances in a two-way classification. Biometrika 45:111-129. E2a;A11 (3286)

Ryser, H. J. [1950], A note on a combinatorial problem. Proc. American Math. Soc. 1:422-424. E2c (3287)

Ryser, H. J. [1951], A combinatorial theorem with an application to latin rectangles. Proc. American Math. Soc. 2:550-552. E7c;T7a (3288)

Ryser, H. J. [1963], Combinatorial Mathematics. The Carus Math. Monographs, No. 14, Math. Assoc. Amer. and John Wiley and Sons, Inc. New York pp. xiv+ 154. E2,7,12,13c (3289)

Sade, A. [1951], Enumérations des carrés latins 1948. Omission dans les listes de Norton pour les carrés 7×7. J. Reine Angew. Math. pp. 189-190.
E7c (3290)

Sade, A. [1951], An omission in Norton's list of 7×7 squares. AMS 22:306-307.
E7c (3291)

Sade, A. [1958], Groupoïdes orthogonaux. Publicationes Mathematicae (Hungaria) 5:229-240.
E7,12,13c (3292)

Sade, A. [1958], Quasigroupes automorphes par le groupe linéaire et géométrié finie. J. Reine Angew. Math. 199:100-120.
E2c (3293)

Saha, G. M. [1967], Construction of partially balanced incomplete block designs through the confounded designs of factorial experiments (abstract). JISAS 19(1): 153.
E3c (3294)

Sahai, C. [1960], Generalized staircase designs (abstract). JISAS 12:151-152.
E4a (3295)

Sahasrabudhe, V. B. [1953], A statistical study of the factors affecting the efficiency of incomplete block designs. Indian Cotton Growing Rev. 7:129-142.
E3,14a (3296)

Sahasrabudhe, V. B. and Agarwal, K. N. [1958], A note on double rectangular lattice designs for cotton varietal trials, Indian Cotton Growing Rev. 12:95-97.
E3a (3297)

Sakai, K. [1953], Studies in competition in plants. I. Analysis of the competitional variance in mixed plant populations. Japanese J. Botany 14:161-168.
E2e;A14 (3298)

Sampford, M. R. [1957], Methods of construction and analysis of serially balanced sequences. JRSSB 19:286-304.
E8ac (3299)

Sampford, M. R. and Taylor, J. [1959], Censored observations in randomized block experiments. JRSSB 21:214-237.
E2,6ae;A6 (3300)

Sanders, P. G. and Duncan, D. B. [1953], Extensions of the multiple comparisons test to incomplete block designs. I (abstract). Virginia J. Sci. 4:290-291.
E2a;A3 (3301)

Saraf, W. S. [1961], On the structure and combinatorial properties of certain semi-regular group divisible designs. Sankhyā A 23:287-296.
E3c (3302)

Saunders, A. R. and Rayner, A. A. [1951], Statistical methods with special reference to field experiments. Union South Africa, Dept. Agri. Sci. Bull 200, 3rd Ed.
E (3303)

Saxena, P. N. [1950], A simplified method of enumerating Latin squares by MacMahon's differential operators. Part I. The 6×6 Latin squares. JISAS 2:161-188. E7c (3304)

Saxena, P. N. [1951], On the enumeration of Latin cubes of the first order. Ph. D. Thesis, Univ. Bombay. E12c (3305)

Saxena, P. N. [1951], A simplified method of enumerating Latin squares by MacMahon's differential operators Part II. The 7×7 Latin squares. JISAS 3:24-79. E7c (3306)

Saxena, P. N. [1960], On the Latin cubes of the second order and the fourth replication of the three-dimensional or cubic lattice designs. JISAS 12:100-140. E3,12ace (3307)

Scheid, F. [1960], A tournament problem. American Math. Monthly 67:39-41. E2c;T11c (3308)

Schmidt, W. [1962], Statistische Datenanalyse. Vereinfachte neuere Verfahren. Angewandte Botanik 36:63-85. E1ae;A11 (3309)

Schmidt, W. [1963], Statistische Datenanalyse. II. Vereinfachte Verfahren. Angewandte Botanik 37:187-269. E2ae;A6,12;C (3310)

Schnell, F. W. [1957], Zur Auswertung und Wirksamkeit teilweise balancierter Gitteranlagen. Moderne Methoden der Pflanzenzuchtung 44:119-132. E3,14a (3311)

Schnell, L. [1956], Hozzászólás Horváth Alajos és Sváb János dolgozatához. (Remarks on the work of Alajos Horváth and János Sváb.) Növénytermelés 5:367-372. E2,7a (3312)

Schröder, [1958], Missing plot estimates by means of covariance analysis in fertilizer field experiments on Majorana hortensis Moench (abstract). Biometrics 14:439. E6a (3313)

Schulze, J. [1958], Zur Bestimmung der Einzelfehler in Spaltanlagen. Z. Pflanzenzüchtung 40:37-58. E12ae (3314)

Schulze, J. [1960], Das Rechteckgitter. Wissenschaftliche Z. Humboldt Univ. 9:479-491. E3a (3315)

Schumann, D. E. W. [1956], The comparison of the sensitivities of experiments using different scales of measurement. Ph. D. Thesis, Virginia Polytechnic Inst. E14 (3316)

Schumann, D. E. W. and Bradley, R. A. [1957], The comparison of the sensitivities of similar experiments: Theory. AMS 28:902-920. E14 (3317)

Schumann, D. E. W. and Bradley, R. A. [1959], The comparison of the sensitivities of similar experiments: Model II of the analysis of variance (abstract). AMS 30:835. E14 (3318)

Schumann, D. E. W. and Bradley, R. A. [1959], The comparison of the sensitivities of similar experiments: Model II of the analysis of variance. Biometrics 15:405-416 (correction 15:631). E14 (3319)

Schutz, W. M. and Cockerham, C. C. [1966], The effect of field blocking on gain from selection. Biometrics 22:843-863. E3,14a (3320)

Schützenberger, M. P. [1951], An extension problem in the theory of incomplete block designs. JRSSB 13:120-125. E3c (3321)

Schützenberger, M. P. [1953], Remarques sur le problème du codage binaire. Publications Inst. Stat. Univ. Paris 2(1&2):125-128. E2c (3322)

Searle, S. R. [1965], Computing formulae for analyzing augmented random complete block designs. Biometrics Unit Mimeo. Ser. BU-207-M, Cornell Univ. E4a (3323)

Seber, G. A. F. [1964], Orthogonality in analysis of variance. AMS 35:705-710. E14;A10 (3324)

Seiden, E. [1954], On the problem of construction of orthogonal arrays. AMS 25: 151-156. E3c;T2c (3325)

Seiden, E. [1954], A remark on the geometrical method of construction of an orthogonal array (abstract). AMS 25:177-178. E3c (3326)

Seiden, E. [1955], Further remark on the maximum number of constraints of an orthogonal array. AMS 26:759-763. E3c (3327)

Seiden, E. [1955], On the maximum number of constraints of an orthogonal array. AMS 26:132-135 (correction 27:204). E2,3c (3328)

Seiden, E. [1960], On a geometrical method of construction of cyclic PBIB (preliminary report) (abstract). AMS 31:820. E3c (3329)

Seiden, E. [1961], On a geometrical method of construction of partially balanced designs with two associate classes. AMS 32:1177-1180. E3c (3330)

Seiden, E. [1963], On necessary conditions for the existence of some symmetrical and unsymmetrical triangular PBIB designs and BIB designs. AMS 34:348-351. E2,3c (3331)

Seiden, E. [1963], On the non-existence of balanced incomplete block designs BIBD, with parameters (46, 69, 9, 6, 1) and (51, 85, 10, 6, 1) (abstract). AMS 34:685 (correction 34:1629). E2c (3332)

Seiden, E. [1963], On the non-existence of some classes of P.B.I.B. based on triangular schemes (abstract). AMS 33:1210. E3c (3333)

Seiden, E. [1963], A supplement to Parker's "Remarks on balanced incomplete block designs". Proc. American Math. Soc. 14:731-732. E2c (3334)

Seiden, E. [1963], On a construction of a class of resolvable BIBD (abstract). AMS 34:685. E2c (3335)

Seiden, E. [1963], A method of construction of resolvable BIBD. Sankhyā A 25: 393-394. E2c (3336)

Seiden, E. [1963], On the non-existence of some classes of P.B.I.B. based on triangular schemes (summary). JASA 58:560. E3c (3337)

Seiden, E. [1965], On the non-existence of PBIBD with the parameters $v=28$, $n_1=12$, $n_2=18$, $p^2_{11}=4$, $b<v$ (abstract). AMS 36:361. E3c (3338)

Seiden, E. [1966], A note on construction of partially balanced incomplete block designs with parameters $v=28$, $n_1=12$, $n_2=15$ and $p^2_{11}=4$. AMS 37:1783-1789. E3c (3339)

Seiden, E. [1966], On a method of construction of partial geometries and partial Bolyai-Lobachevsky planes. American Math. Monthly 73:158-161. E3c (3340)

Sen, A. R. [1963], Use of pretreatment data in designing experiments on tea. Empire J. Expt. Agri. 31:41-49. E2,8,14ac (3341)

Seshadri, V. [1961], Estimation in the balanced incomplete block design. Ph. D. Thesis, Oklahama State Univ. E2a (3342)

Seshadri, V. [1963], Comparison of combined estimators in balanced incomplete blocks (abstract). AMS 34:1115. E2a (3343)

Seshadri, V. [1963], Combining unbiased estimators. Biometrics 19:163-170. E2a (3344)

Seshagiri Rao, K. [1961], B.I.B. designs with one treatment missing (abstract). JISAS 13:237. E2,6a (3345)

Seshagiri Rao, K. [1964], On the removal of a treatment from some non-symmetric BIBD (abstract). JISAS 16:169. E2,3c (3346)

Shah, B. V. [1958], On balancing in factorial experiments. AMS 29:766-779 (correction 30:1267). E3ac;T2ac (3347)

Shah, B. V. [1958], A note on orthogonality in experimental designs. Calcutta Stat. Assoc. Bull. 8:73-80. E14 (3348)

Shah, B. V. [1959], Some aspects of partially balanced incomplete block designs. Ph. D. Thesis, Bombay Univ. E3ac (3349)

Shah, B. V. [1959], On a generalization of the Kronecker product designs. AMS 30:48-54. E4c (3350)

Shah, B. V. [1959], A generalization of partially balanced incomplete block designs. AMS 30:1041-1050. E3ac (3351)

Shah, B. V. [1960], A matrix substitution method of constructing partially balanced designs. AMS 31:34-42. E3,4c:T2c (3352)

Shah, B. V. [1961], Asymmetrical factorials in incomplete blocks (abstract). Biometrics 17:176. E3ac;T2ac (3353)

Shah, K. R. [1960], On certain optimality criteria for incomplete block designs (abstract). Proc. Forty-Seventh Indian Sci. Congress, Part III, pp. 32-33. E14 (3354)

Shah, K. R. [1960], Optimality criteria for incomplete block designs. AMS 31: 791-794. E2,14a (3355)

Shah, K. R. [1960], A note on the variance in reinforced incomplete block designs. JISAS 12:146-149. E4a (3356)

Shah, K. R. [1962], An estimate of inter-group variance in one and two-way designs. Sankhyā A 24:281-286. E2,7a (3357)

Shah, K. R. [1964], Use of inter-block information to obtain uniformly better estimators AMS 35:1064-1078. E2,3a (3358)

Shah, K. R. [1964], On a local property of combined inter- and intra-block estimators. Sankhya A 26:87-90. E2,3a (3359)

Shah, K. R. [1964], Analysis of two-period experiments. JISA 2:97-108. E7,9,12a (3360)

Shah, K. R. [1967], On the recovery of inter-group information in one- and two-way designs. Ph. D. Thesis, Indian Stat. Inst. E2,3,7a (3361)

Shah, K. R. [1967], Uniformly better combined estimators in factorial arrangements wit confounding. JASA 62:638-642. E3a;T2a (3362)

Shah, S. M. [1963], An upper bound for the number of blocks in balanced incomplete block designs. JISA 1:91-92. E2c (3363)

Shah, S. M. [1963], Solution of problem 8. JISA 1:181. E2c (3364)

Shah, S. M. [1963], On the upper bound for the number of blocks in balanced incomplete block designs having a given number of treatments common with a given block. JISA 1:219-220. E2,3,11,13c (3365)

Shah, S. M. [1964], An upper bound for the number of disjoint blocks in certain PBIB designs. AMS 35:398-407. E3c (3366)

Shah, S. M. [1965], Bounds for the number of common treatments between any two blocks of certain PBIB designs. AMS 36:337-342. E3ac (3367)

Shah, S. M. [1966], On the block structure of certain partially balanced incomplete block designs. AMS 37:1016-1020 (correction 38:624). E3c (3368)

Shah, S. M. [1967], A note on the block structure of incomplete block designs. JISA 5:13-15. E2,3c (3369)

Sheehe, P. R. and Bross, I. D. J. [1961], Latin squares to balance immediate residual, and other order, effects. Biometrics 17:405-414. E8ac (3370)

Shih, C. S. [1966], Interval estimation for the exponential model and the analysis of rotation experiments. Ph. D. Thesis, Iowa State Univ. E5,8a (3371)

Shimamoto, T. [1952], Classification and analysis of partially balanced incomplete block designs with two kinds of associates. M. A. Thesis, Univ. N. Carolina. E3ac (3372)

Shrikhande, S. S. [1950], Some combinatorial problems in the design of experiments. Ph. D. Thesis, Univ. N. Carolina. E2,3,7ac (3373)

Shrikhande, S. S. [1950], The impossibility of certain symmetrical balanced incomplete block designs. AMS 21:106-111. E2c (3374)

Shrikhande, S. S. [1950], Construction of partially balanced designs with two accuracies (abstract). Biometrics 6:193. E3c (3375)

Shrikhande, S. S. [1950], Construction of partially balanced designs with two accuracies (abstract). AMS 21:313. E7c (3376)

Shrikhande, S. S. [1950], Designs for animal feeding experiments (abstract). Biometrics 6:193. E8c (3377)

Shrikhande, S. S. [1950], Designs for animal feeding experiments (abstract). AMS 21:314. E8c (3378)

Shrikhande, S. S. [1950], Designs for two-way elimination of heterogeneity (abstract). Biometrics 6:193. E7c (3379)

Shrikhande, S. S. [1951], Designs for two-way elimination of heterogeneity. AMS 22:235-247 (abstract 21:313-314). E7ac (3380)

Shrikhande, S. S. [1951], On the non-existence of certain difference sets for incomplete group designs. Sankhyā 11:183-184. E3c (3381)

Shrikhande, S. S. [1951], On the non-existence of affine resolvable balanced incomplete block designs. Sankhyā 11:185-186. E2c (3382)

Shrikhande, S. S. [1951], The impossibility of certain affine resolvable balanced incomplete block designs (abstract). AMS 22:609. E2c (3383)

Shrikhande, S. S. [1951], The nonexistence of difference sets for group designs (abstract). AMS 22:488. E3c (3384)

Shrikhande, S. S. [1952], On the dual of some balanced incomplete block designs. Biometrics 8:66-72. E3c (3385)

Shrikhande, S. S. [1952], An inequality for orthogonal arrays of strength 2 (abstract). AMS 23:141. E3c (3386)

Shrikhande, S. S. [1952], A series of group divisible designs for two-way elimination of heterogeneity (abstract). AMS 23:140. E7c (3387)

Shrikhande, S. S. [1953], The non-existence of certain affine resolvable balanced incomplete block designs. Canadian J. Math. 5:413-420. E2c (3388)

Shrikhande, S. S. [1953], Cyclic solutions of symmetrical group divisible designs (abstract). AMS 24:146. E3c (3389)

Shrikhande, S. S. [1953], Cyclic solutions of symmetrical group divisible designs. Calcutta Stat. Assoc. Bull. 5:36-39. E3c (3390)

Shrikhande, S. S. [1954], Affine resolvable balanced incomplete block designs and non-singular group divisible designs. Calcutta Stat. Assoc. Bull. 5:139-141. E2,3c (3391)

Shrikhande, S. S. [1959], The uniqueness of the L_2 association scheme. AMS 30:781-798 (abstract 29:938; Inst. Stat. Mimeo. Ser. No. 204, Univ. N. Carolina). E3c (3392)

Shrikhande, S. S. [1959], On a characterization of the triangular association scheme. AMS 30:39-47 (abstract 29:1283, Inst. Stat. Mimeo. Ser. No. 206, Univ. N. Carolina). E3c (3393)

Shrikhande, S. S. [1959], Group divisible designs and the construction of pairwise orthogonal sets of Latin squares. Preliminary report (abstract 557-17). Notices American Math. Soc. 6:181-182. E7,12c (3394)

Shrikhande, S. S. [1959], Relation between certain incomplete block design (abstract) AMS 30:622 (Inst. Stat. Mimeo. Ser. No. 207, Univ. N. Carolina). E3c (3395)

Shrikhande, S. S. [1960], Relations between certain incomplete block designs. Contributions to Probability and Statistics pp. 388-395. E2,3c (3396)

Shrikhande, S. S. [1961], A note on mutually orthogonal Latin Squares. Sankhyā A 23:115-116 (corrigendum 23:426). E7,12,13c (3397)

Shrikhande, S. S. [1961], Combinatorial problems. Res. Training School Publication No. SM61-5, Indian Stat. Inst. E2c (3398)

Shrikhande, S. S. [1962], On a two-parameter family of balanced incomplete block designs. Sankhyā A 24:33-40. E2c (3399)

Shrikhande, S. S. [1963-64], Some recent developments on mutually orthogonal latin squares. Math. Student 31(3&4):167-177. E7,12,13c (3399a)

Shrikhande, S. S. [1965], On a class of partially balanced incomplete block designs. AMS 36:1807-1814. E3c (3400)

Shrikhande, S. S. and Bhagwandas, [1965], Duals of incomplete block designs. JISA 3:30-37. E2,3c (3401)

Shrikhande, S. S. and Jain, N. C. [1962], The non-existence of some partially balanced incomplete block designs with Latin Square type association scheme. Sankhyā A 24:259-268. E3c (3402)

Shrikhande, S. S. and Raghavarao, D. [1963], A method of construction of incomplete block designs. Sankhyā A 25:399-402. E2,3c (3403)

Shrikhande, S. S. and Raghavarao, D. [1963], Affine α-resolvable incomplete block designs. Contributions to Statistics pp. 471-480. E2,3c (3404)

Shrikhande, S. S. and Raghavarao, D. [1964], A note on the non-existence of symmetric balanced incomplete block designs. Sankhyā A 26:91-92. E2c (3405)

Shrikhande, S. S., Raghavarao, D., and Tharthare, S. K. [1963], Non-existence of some unsymmetrical partially balanced incomplete block designs. Canadian J. Math. 15:686-701. E3c (3406)

Shrikhande, S. S. and Singh, N. K. [1962], On a method of constructing symmetrical balanced incomplete block designs. Sankhyā A 24:25-32. E2c (3407)

Shrikhande, S. S. and Singh, N. K. [1963], A note on balanced incomplete block designs. JISA 1:97-101. E2c (3408)

Sillitto, G. P. [1957], An extension property of a class of balanced incomplete block designs. Biometrika 44:278-279. E2c (3409)

Sillitto, G. P. [1964], Note on Takeuchi's table of difference sets generating balanced incomplete designs. Rev. ISIS 32:251. E2c (3410)

Shrikhande, S. S. [1960], Relations between certain incomplete block designs. Contributions to Probability and Statistics pp. 388-395. E2,3c (3396)

Shrikhande, S. S. [1961], A note on mutually orthogonal Latin Squares. Sankhyā A 23:115-116 (corrigendum 23:426). E7,12,13c (3397)

Shrikhande, S. S. [1961], Combinatorial problems. Res. Training School Publication No. SM61-5, Indian Stat. Inst. E2c (3398)

Shrikhande, S. S. [1962], On a two-parameter family of balanced incomplete block designs. Sankhyā A 24:33-40. E2c (3399)

Shrikhande, S. S. [1965], On a class of partially balanced incomplete block designs. AMS 36:1807-1814. E3c (3400)

Shrikhande, S. S. and Bhagwandas, [1965], Duals of incomplete block designs. JISA 3:30-37. E2,3c (3401)

Shrikhande, S. S. and Jain, N. C. [1962], The non-existence of some partially balanced incomplete block designs with Latin Square type association scheme. Sankhyā A 24:259-268. E3c (3402)

Shrikhande, S. S. and Raghavarao, D. [1963], A method of construction of incomplete block designs. Sankhyā A 25:399-402. E2,3c (3403)

Shrikhande, S. S. and Raghavarao, D. [1963], Affine α-resolvable incomplete block designs. Contributions to Statistics pp. 471-480. E2,3c (3404)

Shrikhande, S. S. and Raghavarao, D. [1964], A note on the non-existence of symmetric balanced incomplete block designs. Sankhyā A 26:91-92. E2c (3405)

Shrikhande, S. S., Raghavarao, D., and Tharthare, S. K. [1963], Non-existence of some unsymmetrical partially balanced incomplete block designs. Canadian J. Math. 15:686-701. E3c (3406)

Shrikhande, S. S. and Singh, N. K. [1962], On a method of constructing symmetrical balanced incomplete block designs. Sankhyā A 24:25-32. E2c (3407)

Shrikhande, S. S. and Singh, N. K. [1963], A note on balanced incomplete block designs. JISA 1:97-101. E2c (3408)

Sillitto, G. P. [1957], An extension property of a class of balanced incomplete block designs. Biometrika 44:278-279. E2c (3409)

Sillitto, G. P. [1964], Note on Takeuchi's table of difference sets generating balanced incomplete designs. Rev. ISI 32:251. E2c (3410)

Silverman, R. [1960], A metrization for power sets with applications to combinatorial analysis. Canadian J. Math. 12:158-176. E2,12,13c (3411)

Singer, J. [1960], A class of groups associated with Latin squares. American Math. Monthly 67:235-240. E12,13c (3412)

Singh, D. [1962], A method of working out the sum of totals over blocks containing a specified variety in the analysis of a n×n balanced lattice design. JISAS 14:151-156. E2a (3413)

Singh, M. [1950], Confounding in split-plot designs with restricted randomization of sub-plot treatments. Empire J. Expt. Agri. 18:190-202. E9,12ae;T2ace (3414)

Singh, N. [1963], Estimation of missing data in a latin square (abstract). JISAS 15:274. E7,10a (3415)

Singh, N. K. and Singh, K. N. [1964], The non-existence of some partially balanced incomplete block designs with three associate classes. Sankhyā A 26:239-250. E3c (3416)

Singh, N. K. and Shukla, G. C. [1963], The non-existence of some partially balanced incomplete block designs with three associate classes. JISA 1:71-77. E3c (3417)

Skolem, T. [1957], On certain distributions of integers in pairs with given differences. Mathematica Scandinavica 5:57-68. E2c (3418)

Skolem, T. [1958], Some remarks on the triple systems of Steiner. Mathematica Scandinavica 6:273-280. E2c (3419)

Smith, C. A. B. [1965], Personal probability and statistical analysis (with discussion). JRSSA 128:469-499. E1,2a;C (3420)

Smith, F. L. [1958], Effects of plot size, plot shape, and number of replications on the efficiency of bean yield trials. Hilgardia 28:43-63. E2,7a;A14 (3421)

Smith, H. F. [1950], Letter: Replicated experiments. Forestry 23:56-58. E14;T1ac (3422)

Smith, H. F. [1950], Error variance of treatment contrasts in an experiment with missing observations (with special reference to incomplete latin squares). JISAS 2:111-124. E7,10ae (3423)

Smith, H. F. [1952], Perennial experiments with tree crops. Inst. Stat. Mimeo. Ser. No. 62, Univ. N. Carolina. E8a (3424)

Smith, H. F. [1957], Missing plot estimates. Biometrics 13:115-118. E6a (3425)

Snedecor, G. W. [1950], Query 77. Biometrics 6:164-166. E5ae (3426)

Snedecor, G. W. [1952], Query 94. Biometrics 8:169-171. E7a;C (3427)

Snedecor, G. W. [1952], Query 96. Biometrics 8:383. E2,6ae;A10 (3428)

Snedecor, G. W. [1952], Query 97. Biometrics 8:384-385. E7a (3429)

Snedecor, G. W. [1953], Query 103. Biometrics 9:425-427. E2,6ae (3430)

Snedecor, G. W. [1954], Query 108. Biometrics 10:298-299. E2,6ae (3431)

Snedecor, G. W. [1954], Query 109. Biometrics 10:299-301. E2a (3432)

Southern Cooperative Group, [1951], Studies of sampling techniques and chemical analyses of vegetables. Southern Cooperative Ser. Bull. 10:1-143. E1,5,7e (3433)

Sparks, J. N. [1963], Expository notes on the problem of making multiple comparisons in a completely randomized design. J. Expt. Education 31:343-349. E1a;A3 (3434)

Sprent, P. [1955], A note on design and analysis of soil insecticide experiments. Biometrics 11:427-430. E3a (3435)

Sprott, D. A. [1954], A note on balanced incomplete block designs. Canadian J. Math. 6:341-346. E2c (3436)

Sprott, D. A. [1955], Balanced incomplete block designs. Ph. D. Thesis, Univ. Toronto. E2c (3437)

Sprott, D. A. [1955], Some series of partially balanced incomplete block designs. Canadian J. Math. 7:369-381. E3c (3438)

Sprott, D. A. [1955], Balanced incomplete block designs and tactical configurations. AMS 26:752-758. E2c (3439)

Sprott, D. A. [1956], Some series of balanced incomplete block designs. Sankhyā 17:185-192. E2c (3440)

Sprott, D. A. [1956], A note on combined interblock and intrablock estimation in incomplete block designs. AMS 27:633-641 (correction 28:269). E3a (3441)

Sprott, D. A. [1959], A series of symmetrical group divisible incomplete block designs. AMS 30:249-251. E3c (3442)

Sprott, D. A. [1962], Listing of BIB designs from r=16 to 20. Sankhyā A 24:203-204. E2c (3443)

Sprott, D. A. [1964], Generalizations arising from a family of difference sets. JISA 2:197-209. E3c (3444)

Srivastava, J. N. [1962], Contributions to the construction and analysis of designs. Ph. D. Thesis, Univ. N. Carolina (Inst. Stat. Mimeo. Ser. No. 301, Univ. N. Carolina). E2,3ac;T3ac (3445)

Srivastava, J. N. [1962], A generalized partially balanced association scheme (abstract). AMS 33:296. E3c (3446)

Srivastava, J. N. [1962], Incomplete multiresponse designs. Inst. Stat. Mimeo. Ser. No. 340, Univ. N. Carolina, 20 pp. E5c (3447)

Srivastava, J. N. [1963], An algorithm for the analysis of multidimensional partially balanced designs (abstract). AMS 34:1116. E3,13a;T7a (3448)

Srivastava, J. N. [1963], Some classes of designs in multiresponse experiments (summary). JASA 58:562. E5,9c;A7 (3449)

Srivastava, J. N. [1964], On a general class of designs for multiresponse experiments. Inst. Stat. Mimeo Ser. No. 402, Univ. N. Carolina. E5ac (3450)

Srivastava, J. N. [1965], Some necessary conditions for the existence of partially balanced arrays (abstract). AMS 36:1079. E3c (3451)

Srivastava, J. N. [1966], Incomplete multiresponse designs. Sankhyā A 28:377-388. E5,9ac (3452)

Srivastava, J. N. [1967], A new class of conditions for the existence of partially balanced arrays, including BIBD's and orthogonal arrays (preliminary report) (abstract). AMS 38:641. E2c (3453)

Srivastava, J. N. and Chopra, D. V. [1967], Partially balanced arrays with μ_2=2 and 3 (abstract). AMS 38:1937. E3c (3454)

Srivastava, J. N. and Maik, R. L. [1966], On a new property of PBIB designs useful in an application of MANOVA in psychometrics (abstract). AMS 37:546-547 (Aerospace Res. Lab. ARL 66-0105, U.S. Air Force). E3a (3455)

Srivastava, J. N. and Roy, S, N. [1960], On a generalization of balanced incomplete block designs (abstract). AMS 31:538. E2c (3456)

Stanley, J. C. [1955], Statistical analysis of scores from counterbalanced tests. J. Expt. Education 23:187-207. E7,12ace (3457)

Stanley, J. C. [1961], Analysis of doubly nested design. Educational Psychological Measurement 21:831-837. E12a (3458)

Stanton, R. G. [1957], A note on BIBDS. AMS 28:1054-1055 (acknowledgement of priority 30:254). E2,14c (3459)

Stanton, R. G. and Sprott, D. A. [1958], A family of difference sets. Canadian J. Math. 10:73-77. E2c (3460)

Stanton, R. G. and Sprott, D. A. [1964], Block intersections in balanced incomplete block designs. Canadian Math. Bull. 7:539-548. E2c (3461)

Stanton, R. G. and Mullin, R. C. [1966], Inductive methods for balanced incomplete block designs. AMS 37:1348-1354. E2,4c (3462)

Starr, S. [1960], Some results in the analysis of variance I (preliminary report) (abstract). AMS 31:529. E1a;T2a (3463)

Staude, H. [1959], Abkürzung des Range-Verfahrens von H. O. Hartley zur Auswertung von Blockversuchen. (Shortcut of H. O. Hartley's range-method for evaluating block experiments.) Biom. Zeit. 1:261-275. E2ae (3464)

Staude, H. [1961], Die Verbindung des Neuen Multiplen Rangetests von Duncan mit Hartleys Range-Verfahren zur Auswertung von Blockversuchen. (The combination of Duncan's new multiple range test with Hartley's range-method for evaluating block experiments.) Biom. Zeit. 3:47-53. E2a;A3 (3465)

Stearman, R. L. and Ward, T. G. [1957], Variability of specific activities and specific-activity ratios. Archives Biochemistry Biophysics 68:249-254. E1e;T2e (3466)

Steel, R. G. D. [1958], A class of augmented designs. MRC Tech. Summary Report No. 56:1-16, U.S. Army and Univ. Wisconsin. E4,7ac (3467)

Steel, R. G. D. and Federer, W. T. [1955], Yield-stand analyses. JISAS 7:27-45. E2ae;A3,7 (3468)

Stevens, W. L. [1951], Experiências de rotação. Bragantia 11:317-330. E5,8ac (3469)

Stevens, W. L. [1956], Rotation experiment in Brazil. J. Agri. Sci. 47:257-261. E5,8ac (3470)

Stiehler, R. D. [1956], Application of statistics to tire testing (summary). JASA 51:524. E7c (3471)

Still, H. A. [1961], Analysis of multiple covariance when the regression coefficients depend on the blocks. Ph. D. Thesis, Virginia Polytechnic Inst. E2,3a;A7 (3472)

Stone, M. [1961], Non-equivalent comparisons of experiments and their use for experiments involving location parameters. AMS 32:326-332. E14 (3473)

Storch, J. M. [1957], Open plaatsen in orthogonale proefschema's. (Missing plot analysis.) Stat. Neerlandica 11:131-139. E2,6a (3474)

Strand, E. [1959], Studies on the efficiency of experimental designs in small grain experiments. Acta Agri. Scandinavica 9:321-340. E2,3,5,14a;A14 (3475)

Strand, L. [1966], The use of analysis of covariance. An example with data from a manuring experiment. International Advisory Group of Forest Statisticians, Second Conf. Inst. Skoglig Matematisk Stat. pp. 308-318. E2ae;T2ae;A7 (3476)

Sutter, G. J. [1962], Some aspects of constrained randomization. M. S. Thesis, Iowa State Univ. E5ac;C (3477)

Sutter, G. J., Zyskind, G., and Kempthorne, O. [1963], Some aspects of constrained randomization. Dept. Stat. Tech. Report and Aeronautical Res. Lab. ARL 63-18, Iowa State Univ. and U.S. Air Force. E5ac;C (3478)

Sváb, J. [1957], A varianciaanalizis megközelítése négyzetreemelés nélkül különös tekintettel blokkelrendezésü szántóföldi kísérletek értékelésére. (A method requiring no squaring approximating to analysis of variance, with special reference to the evaluation of block-design field trials.) Növenytermeles 6:77-90. E2a (3479)

Sváb, J. [1958], Kísérletsorozat értékelése variancia-analízissel, teljes bontásban; a kölcsönhatás értelmezése. (The evaluation of a series of trials by means of variance analysis with complete partitioning; interpretation of the interaction.) Növenytermeles 7:121-142. E2a;A13 (3480)

Sváb, J. [1959], Módszer egysorozatos kísérletek elrendezésére és értékelésére. (Method for the layout and evaluation of unreplicated experiments.) Agrobotanika, Tapioszele 1:29-53 (Mimeo.). E5ac (3481)

Swaminathan, S. S. [1963], On designs obtainable by combining different designs (abstract). JISAS 15:259. E2,3,4c (3482)

Swaminathan, S. S. and Das, M. N. [1964], Combined analysis of incomplete block designs. JISAS 16:296-303. E2,3a;A13 (3483)

Swaminathan, S. S. and Das, M. N. [1965], Construction and analysis of several series of incomplete block designs, JISAS 17:43-57. E3,4,9c (3484)

Sweeny, H. C. [1955], Analyses of experiments with correlated observations and heterogeneous variances (summary). JASA 50:597. E5a;T2a;A11 (3485)

Sweeny, H. C. [1956], Some results on experimental designs when the usual assumptions are invalid. Ph. D. Thesis, Virginia Polytechnic Inst. E2,3,5a;T2a;A11 (3486)

Sweeny, H. C. [1959], The power of the analysis of variance test under constrained randomization (abstract). Biometrics 15:340. E5a (3487)

Taguchi, G. [1959], Linear graphs for orthogonal arrays and their applications to experimental designs with the aid of various techniques. Reports Stat. Appl. Res. 6:133-175. E2,3c;T2c (3488)

Taguchi, G. [1960], Tables of orthogonal arrays and linear graphs. Reports Stat. Appl. Res. 7:1-52. E2,3c;T2c (3489)

Takahashi, M. [1955], On the analysis of variance for the split-plot design. J. Sci. Hiroshima Univ. Ser. A-I 19:321-325. E5,12a (3490)

Takeuchi, K. [1961], On a special class of regression problems and its applications: some remarks about general models. Reports Stat. Appl. Res. 8:7-17. E2,ac;A7 (3491)

Takeuchi, K. [1961], On a special class of regression problems and its application: random balanced incomplete block designs. Reports Stat. Appl. Res. 8:18-28. E2,7ac;A7 (3492)

Takeuchi, K. [1961], On the optimality of certain type of PBIB designs. Reports Stat. Appl. Res. 8:140-145. E3,14c (3493)

Takeuchi, K. [1962], A table of difference sets generating balanced incomplete block designs. Rev. ISI 30:361-366. E2c (3494)

Takeuchi, K. [1963], A remark added to "On the optimality of certain type of PBIB designs." Reports Stat. Appl. Res. 10:225. E3,14c (3495)

Takeuchi, K. [1963], On the construction of a series of BIB designs. Reports Stat. Appl. Res. 10:226. E2c (3496)

Tan, W.Y. and Yuan, C.H. [1965], On the biometrical analysis of quantitative inheritance. Botanical Bull. Academia Sinica 6:189-196. E4ac;T9a (3497)

Tanaka, W. [1967], Quality design on orthogonal arrays. Reports Stat. Appl. Res. 14:1- E2,3c (3498)

Taylor, J. [1950], The comparison of pairs of treatments in split-plot experiments. Biometrika 37:443-444. E5,12a (3499)

Taylor, J. [1967], The value of orthogonal polynomials in the analysis of change-over trials with dairy cows. Biometrics 23:297-311 (abstract 20:396). E8a (3500)

Taylor, W. B. [1952], Some general classes of experimental designs involving two-way elimination of heterogeneity. Incomplete block design with row balance. M.S. Thesis, London Univ. College. E7ac (3501)

Taylor, W. B. [1957], Incomplete block designs with row balance and recovery of inter-block information. Biometrics 13:1-12. E7ae (3502)

Taylor, W. B. and Armstrong, P. J. [1953], The efficiency of some experimental designs used in dairy husbandry experiments. J. Agri. Sci. 43:407-412. E7,8a (3503)

Tharthare, S. K. [1963], Right angular designs. AMS 34:1057-1067. E3c (3504)

Tharthare, S. K. [1965], Generalized right angular designs. AMS 36:1535-1553. E3ac (3505)

Thayer, F. D., Bianco, E. G., and Wilcoxon, F. [1951], The application of statistics in the tanning laboratory. I. The use of a Youden square (with discussion). J. American Leather Chemists Assoc. 46:669-684. E7ae (3506)

Thompson, F. C. [1960], A long-term manurial experiment with hops. The effect of heavy manuring on soil fertility and on the yield, leaf mineral content and resin content of the variety Fuggle. J. Horticultural Sci. 35:185-201. E5a (3507)

Thompson, H. R. [1956], On a new class of partially balanced incomplete block designs. Calcutta Stat. Assoc. Bull. 6:193-195. E3,7c (3508)

Thompson, H. R. [1956], Extensions to missing plot techniques. Biometrics 12:241-244 (correction 13:543). E6a (3509)

Thompson, H. R. and Seal, K.E. [1964], Serial designs for routine quality control and experimentation. Technometrics 6:77-98. E4ace (3510)

Thompson, N. R., Blaser, R. E., Graf, G. C., and Kramer, C. Y. [1955], Application of a Latin square change-over design to dairy cattle grazing experiments. J. Dairy Sci. 38:991-996. E7ace (3511)

Thompson, W. A. [1954], On the ratio of variances in the mixed incomplete block model. Ph.D. Thesis, Univ. N. Carolina. E2,3a (3512)

Thompson, W. A. [1955], On the ratio of variances in the mixed incomplete block model. AMS 26:721-733. E3a (3513)

Thompson, W. A. [1955], The relative size of the inter- and intra-block error in an incomplete block design. Biometrics 11:406-426. E3ae (3514)

Thompson, W. A. [1956], A note on the balanced incomplete block designs. AMS 27:842-846. E2c (3515)

Thompson, W. A. [1958], A note on PBIB design matrices. AMS 29:919-922. E3a (3516)

Throckmorton, T. N. [1958], An analysis of a modified cross-over design. M.S. Thesis, Iowa State Univ. E8ac (3517)

Throckmorton, T. N. [1961], Structures of classification data. Ph.D. Thesis, Iowa State Univ. E11a;A5 (3518)

Tiao, G. C. [1965], Bayesian comparison of means of a mixed model with application to regression analysis. Dept. Stat. Tech. Report No. 49, Univ. Wisconsin. E2a;T2a;A7 (3519)

Tiao, G. C. and Draper, N. R. [1967], Bayesian analysis of linear models with two random components (with speacial reference to the balanced incomplete block design). Dept. Stat. Tech. Report No. 102, Univ. Wisconsin. E2a (3520)

Tiao, G. C. and Tan, W. Y. [1965], Bayesian analysis of random-effect models in the analysis of variance. II. Effect of autocorrelated errors. Dept. Stat. Tech. Report No. 54, Univ. Wisconsin. E2a;T2a (3521)

Timon, W. E. [1962], The Slipped-Block Design. Ph.D. Thesis, Oklohoma State Univ. E4a (3522)

Timon, W. E. [1962], The slipped-block design. (abstract) Biometrics 18:632. E4c (3523)

Titus, J. [1955], Testing significance of main plot treatments adjusted for split plot regression. M.S. Thesis, Univ. N. Carolina. E12a;A7 (3524)

Tocher, K. D. [1952], The design and analysis of block experiments (with discussion). JRSSB 14:45-100. E2,3,5,6,7, 10ace (3525)

Tocher, K. D. [1952], The design and statistical analysis of experiments. Ph.D. Thesis, London Univ. E (3526)

Torrens-Ibern, J. [1962], Plans d'expériences pour l'étude des régressions multiples. Bull. ISI 39(3):161-171. E7ac;T2a (3527)

Torrens-Ibern, J. [1963], Plans d'expériences pour l'étude de régression multiples. Collogues Inter. Centre Nat. Recherche Sci. No. 110, Le Plan d'Expériences pp 257-268. E12ac;A7 (3528)

Trehan, A. M. [1963], On the bounds of the number of common treatments between blocks of balanced incomplete block design. J. Indian Stat. Assoc., 1:102-103. E2c (3529)

Truitt, J. T. and Smith, H. F. [1956], Adjustment by covariance and consequent tests of significance in split-plot experiments. Biometrics 12:23-39.
E5a;A7 (3530)

Tsan, W. -L. [1957], The theory and practice of the multivarietal randomized arrangement pairing method (in Chinese). Nung-yeh-hsueh Pao/Act. Agri. Sinica 8:103-122. E4,5ac (3531)

Tsukibayashi, S. [1963], On analysis of covariance with product-moment and range methods. Reports Stat. Appl. Res. 10:177-206. E1,2,5,12a;A7 (3532)

Tukey, J. W. [1953], Some selected quick and easy methods of statistical analysis. Trans. New York Academy Sci., Series II, 16:88-97.
E1e;A3,12 (3533)

Tukey, J. W. [1955], Querry 113. Biometrics 11:111-113. E7a;A4 (3534)

Turyn, R. J. [1965], Character sums and difference sets. Pacific J. Math. 15:319-346. E2c;T3c (3535)

Urquhart, N. S. [1962], The repeated and slipped block design. M.S. Thesis, Colorado State Univ. E4ace (3536) van den Driessche, R. [1961], Blocs complets de deux unités. Biométrie-Praximétrie 2:131-132. E2a;A7 (3537)

van den Heiden, J. A. [1953], Dupliceerbaarheid en reproduceerbaarheid van waarnemingen. Stat. Neerlandica 7:15-22. E14 (3538)

van der Reyden, D. [1954], Design and analysis of soil insecticide field experiments. Biometrics 10:291-297. E3ae (3539)

van der Vaart, H. R. [1962], A critical appraisal of some aspects of the concept of optimal design in biology (abstract). Biometrics 18:626. E14 (3540)

van Elteren, P. and Noether, G. E. [1959], The asymptotic efficiency of the χ_r^2- test for a balanced incomplete block design. Biometrika 46:475-477. E2a;T11a (3541)

Varma, A. A. O. and Chilton, N. W. [1965], Studies in the design and analysis of dental experiments. 11. Sequental analysis (two dichotomous populations where the data are collected in groups.) J. Oral Therapeutics Pharmacology 2:44-51. E2ae;S2 (3542)

Vartak, M. N. [1955], On an application of Kronecker product of matrices to statistical designs. AMS 26:420-438. E4c (3543)

Vartak, M. N. [1958], On the Hasse-Minkowski invariant of the Kronecker product of matrices. Canadian J. Math. 10:66-72. E4c (3544)

Vartak, M. N. [1959], The non-existence of certain PBIB designs. AMS 30:1051-1062. E3,4c (3545)

Vartak, M. N. [1960], Relations among the blocks of the Kronecker product of designs. AMS 31:772-778. E4c;A8 (3546)

Vartak, M. N. [1961], On some applications of Kronecker product of matrices to statistical designs. Ph. D. Thesis, Univ. Bombay. E4,5c (3547)

Vartak, M. N. [1963], Connectedness of Kronecker product designs. JISA 1:215-218. E4,9c;T7c (3548)

Vartak, M. N. [1963], Disconnected balanced designs. JISA 1:104-107. E2,14c (3549)

vaz de Arruda, H. [1952], Análise de uma experiência sôbre variedades de sojá. (Analysis of an experiment with soybean varieties.) Bragantia 12:65-73. E2e;A7 (3550)

vaz de Arruda, H. [1954], Efficiência do delineamento em blocos ao acaso, em experiências comparativas de variedades e hibridos de milho. (Efficiency of randomized block designs in experiments to test corn varieties and hybrids.) Bragantia 13:217-222. E2,14a (3551)

vaz de Arruda, H. [1955], Delineamento em blocos incompletos balanceados com o número de variedades igual ao número de blocos. (Analysis of balanced, incomplete block designs when number of varieties equals number of blocks.) Bragantia 14:129-136 E2ace (3552)

vaz de Arruda, H. [1959], Aplicacão da transformacão raiz quadrada, análise da variância de dados experimentais. (Square root transformation in the analysis of certain field experimental data.) Bragantia 18:XV-XIX E2a;A9 (3553)

vaz de Arruda, H. [1960], Aplicacão da análise da covariância, num estudo sôbre tamanho de canteiro para experiências com cafeeiros. (The use of covariance analysis to determine plot size in coffee experiment). Bragantia 19:I-V E2e;A7,14 (3554)

Venekamp, J. T. N., Hamming, G. and Vervelde, G. J. [1952], A 4^3 factorial design with confounding in 8 × 8 quasi-Latin squares. Landbouwkundig Tijdschrift (Wageningen) 64:325 E12ac;T2a (3555)

Venkateswaran, A. N., Seshu, K.A., and Murugarajendran, C. [1963], An incomplete block design for large scale trials in groundnut breeding. Madras Agri. J. 50:80-81 E2a (3556)

Venturini, W. R. and Jorge, J. P. N. [1962], Efficiência do delineamento fatorial 3^3, em blocos de 9, em uma série de experimentos de adubacão do algodoeiro. (Efficiency of a 3^3 factorial design for cotton fertilizer experiments.) Bragantia 21:631-637 E14;T2a;A13 (3557)

Verdooren, L. R. [1965], "Change-over" -proven. (Change-over -trials). Stat. Neerlandica 19:323-333 E8ae (3558)

Vessereau, A. [1960], Controle d'un enceinte conditionnée. Revue Statistique Appliquée 8(2):53-75 E12ae;T2ae (3559)

Vittum, M. T., Peck, N. H., and Carruth, A. F. [1959], Response of sweet corn to irrigation, fertility level, and spacing N. Y. State Agri. Expt. Stat. Cornell Univ. Bull. No. 786 E12e;T2e (3560)

Vittum, M. T., Tapley, W. T., and Peck, N. H. [1958], Response of tomato varieties to irrigation and fertility level. N. Y. State Agri. Expt. Stat. Cornell Univ. Bull. No. 782 E12e;T2e (3561)

Vos, B. J. [1950], Statistics in biological assay: an example of the graded response. Ann. New York Academy Sci. 52:920-921 E7a;T8a (3562)

Walpole, R. E. [1958], Combined intra- and inter-block analysis for factorials in incomplete block designs. Ph. D. Thesis, Virginia Polytechnic Inst. E2a;T2a (3563)

Walpole, R. E. [1958], Combined intra- and inter-block analysis for factorials in incomplete block designs (abstract). Virginia J. Sci. 9:448. E3a;T2a (3564)

Walsh, J. E. [1959], Exact nonparametric tests for randomized blocks. AMS 30:1034-1040 (abstract 30:255). E2a;A12 (3565)

Wang, C. -C. [1964], Effects of number of buds and different size of seed pieces on the germination, growth and production of sugar cane. (in Chinese) Taiwan Tahsueh Nunghsuehyuan Yenchiu Paokao/Mem. College Agri. Nat. Taiwan Univ. 8(1):1-11. E2e;T2e (3566)

Wang, C. -C., Cheng, C. -P., and Hsu, J. -S. [1961], Experiments on the competition of Para-, Pangola-, and Napier-grasses. I. Competition between forage grasses and weeds. (in Chinese) Taiwan Tahsueh Nunghsuehyuan Yenchiu Paokao/Mem. College Agri. Nat. Taiwan Univ. 6(2):21-35. E2ae;A14 (3567)

Wang, C. -M. [1950], Notes on multi-varietal trials. I. Simple lattice. Memoirs College Agri. Nat. Taiwan Univ. 2(No. 3):1-31. E2,3ae (3568)

Wang, C. -M. [1953], Notes on the multi-varietal trials (II) Triple lattice. (in Chinese) Taiwan Tahsueh Nunghsuehyuan Yenchiu Paokao/Mem. College Agri. Nat. Taiwan Univ. 3(1):1-19. E3ae (3569)

Wang, C. -M. [1958], Comparison of eight statistical treatments of experimental data in fractions (or percentages) (in Chinese) Taiwan Tahsueh Nunghsuehyuan Yenchiu Paokao/Mem. College Agri. Nat. Taiwan Univ. 5(2):1-24. E2ae;A4,9 (3570)

Wang, H. -L. [1958], Studies on the analysis of pairing experiments. (in Chinese) Nung-yeh-hsueh Pao (J. Agri.)/Act. Agri. Sinica 9:88-97. E4,5ae (3571)

Wang, Y. [1964], A note on the maximal number of pairwise orthogonal Latin squares of a given order. Scientia Sinica 13:841-843. E7,12c (3572)

Ward, G. C. and Dick, I. D. [1952], Non-additivity in randomized block designs and balanced incomplete block designs. New Zealand J. Sci. Tech. B, 33:430-435. E2a;A4 (3573)

Watson, G. S. [1956], Missing and "mixed-up" frequencies in contingency tables. Biometrics, 12:47-50 (abstracts, 11:242 and 517). E6a (3574)

Watson, G. S. [1966], The statistics of orientation data. J. Geology 74:786-797 E1a (3575)

Wattier, J. B. [1954], Efficiencies of alternative designs in estimating the corn yield in Iowa. M.S. Thesis, Iowa State Univ. E14 (3576)

Weeks, D. L. and Graybill, F. A. [1959], Minimal sufficient statistics in incomplete block designs, Model II (abstract). AMS 30:624-625. E2,3a (3577)

Weeks, D. L. and Graybill, F. A. [1961], Minimal sufficient statistics for the balanced incomplete block design under an Eisenhart Model II. Sankhyā A 23:261-268. E2a (3578)

Weeks, D. L. and Graybill, F. A. [1962], A minimal sufficient statistic for a general class of designs. Sankhyā A 24:339-354. E2,3a (3579)

Weeks, E. [1952], Balanced incomplete design used as a means of determining the acceptability of seven types of clothing fabrics (abstract). Virginia J. Sci. 3:359-360. E2a (3580)

Wells, M. B. [1967], The number of Latin squares of order eight. J. Combinatorial Theory 3:98-99. E7c (3581)

West, O. [1954], Improvements in the technique and design of veld management experiments. South African J. Sci. 50:222-227. E5ac;C (3582)

White, R. F. [1963], Randomization analysis of the general experiment. Ph. D. Thesis, Iowa State Univ. E3,4a;A5 (3583)

Wiesen, J. M. [1951], The design and analysis of experiments involving sensory impressions. M.S. Thesis, Iowa State Univ. E2a;C (3584)

Wiggers, B. G. [1959], Efficiency in de statistiek. (Efficiency in statistics.) Stat. Neerlandica 13:261-280. E14;A12 (3585)

Wijnhoven, J. [1958], Statistische methoden en het ontwerpen van experimenten (Statistical methods and experimental design). Agricultura, Louvain 6:585-594. E2,3,7,13a;C (3586)

Wilcoxon, F. [1955], Significance tests by rank methods. ASQC Nat. Conv. Trans. 9:135-140. E1e;A12 (3587)

Wilcoxon, F. and Wilcox, R. A. [1964], Some rapid approximate statistical procedures. Lederle Laboratories, Pearl River, New York, pp. 1-60. E1,2ae;T2ae;A12 (3588)

Wilk, M. B. [1953], The randomization analysis of a generalized randomized block design. M.S. Thesis, Iowa State Univ. E2a (3589)

Wilk, M. B. [1955], Linear models and randomized experiments. Ph. D. Thesis, Iowa State Univ. E2,7,12a;A5 (3590)

Wilk, M. B. [1955], The randomization analysis of a generalized randomized block design. Biometrika 42:70-79. E2a;A5 (3591)

Wilk, M. B. and Kempthorne, O. [1956], Some aspects of the analysis of factorial experiments in a completely randomized design. AMS 27:950-985. E1a;T2a;A5 (3592)

Wilk, M. B. and Kempthorne, O. [1956], Analysis of Variance: Preliminary tests, pooling, and linear models. Derived linear models and their use in the analysis of randomized experiments. Stat. Lab. and Aernautical Res. Lab. WADC Tech. Report 55-244, vol.II, Iowa State Univ. and U.S. Air Force. E1,2,5,7,12a;A5 (3593)

Wilk, M. B. and Kempthorne, O. [1957], Non-additivities in a Latin square design. JASA 52:218-236 (summary 51:526). E7a;A4 (3594)

Wilk, R. E. [1963], The use of a cross-over design in a study of student teachers' classroom behaviors. J. Exptl. Education 31:337-341. E8a (3595)

Wilkinson, G. N. [1958], Estimation of missing values for the analysis of incomplete data. Biometrics, 14:257-286 E2,6ae (3596)

Wilkinson, G. N. [1958], The analysis of variance and derivation of standard errors for incomplete data. Biometrics, 14:360-384. E2,6,7,10ae (3597)

Wilkinson, G. N. [1960], Comparison of missing value procedures. Australian J. Stat. 2:53-65. E2,6,7,10ae (3598)

Wilkinson, J. W. [1966], Comparing designs for experiments that may be prematurely terminated (abstract). Technometrics 8:211. E5 (3599)

Willcocks, T. H. [1967], Some squared squares and rectangles. J. Combinatorial Theory 3:54-56. E7c (3600)

Williams, D. A. [1966], Errors of treatment comparisons when observations are missing from a randomized block experiment with additional replication of a control treatment. Biometrics 22:632-633. E2,6a (3601)

Williams, E. J. [1950], Experimental designs balanced for pairs of residual effects. Australian J. Sci. Res., Series A, 3:351-363. E8ac (3602)

Williams, E. J. [1950], Statistical analysis of results for successive tests on the same organism. Nature 166:319. E5a (3603)

Williams, R. M. [1952], Experimental designs for serially correlated observations. Biometrika 39:151-167. E5ac (3604)

Wilsie, C. P. [1954], Relative efficiency of lattice and randomized block designs for forage crop trials. Agron. J. 46:355-357. E2,14a (3605)

Winder, C. V. [1950], Some examples of the use of statistics in pharmacology. Ann. New York Academy Sci. 52:838-861. E7ae;T2,8a;C (3606)

Wishart, J. [1950], Field trials II: The analysis of covariance. Commonwealth Bur. Plant Breeding Genetics, Tech. Comm. 15. E2,6,7ae;A7 (3607)

Wishart, J. and Sanders, H. G. [1955], Principles and practice of field experimentation. Commonwealth Bur. Plant Breeding Genetics, Tech. Comm. 18. E2,5,7,12ae;T2ae; A14;C (3608)

Wishart, J., Thawani, V. D., Kishen, K., and Nair, K. R. [1954], Statistical methods. FAO/55/9/6340: In Dex/1, FAOUN, Rome, 111 pp (mimeo.). E2,3a;T2a;C (3609)

Wolock, F. W. [1965], Cyclic designs. Ph. D. Thesis, Virginia Polytechnic Inst. E3ac (3610)

Wolock, F. and David, H. A. [1965], Cyclic designs (abstract). Biometrics 21:259. E3c (3611)

Wood, S. R. [1963], Analysis of a rocket nozzle ablation experiment. ASQC, Ann. Conv. Trans. 17:329-332. E2e (3612)

Woodhouse, W. W. and Rigney, J. A. [1954], A unique application of a balanced lattice design. Agron. J. 46:181. E2,14a (3613)

Wooding, W. M. [1966], A balanced incomplete blocks design for consumer fragrance testing. ASQC, Ann. Tech. Conf. Trans. 20:815-822. E2ae (3614)

Wright, E. C. [1965], Field plans for a systematically designed polycross. Rec. Agri. Res. N. Ireland 14:31-41. E5c;T9ac (3615)

Yalavigi, C. C. [1963], A tournament problem. Math. Student 31:51-64. E2c;T11c (3616)

Yamamoto, K. [1950], An asymptotic series for the number of three-line latin rectangles. J. Math. Soc. Japan 1(4):226-241. E7c (3617)

Yamamoto, K. [1951], On the asymptotic number of Latin rectangles. Japanese J. Math. 21:113-119. E7c (3618)

Yamamoto, K. [1952], Note on the enumeration of 7 × 7 Latin squares. Bull. Math. Stat. 5(1&2):1-8. E7,12c (3619)

Yamamoto, K. [1954], Euler squares and incomplete Euler squares of even degrees. Memoirs Faculty Sci. Kyushu Univ. Series A, 8:161-180. E7,12c (3620)

Yamamoto, K. [1956], Structure polynomial of Latin rectangles and its application to a combinatorial problem. Memoirs Faculty Sci. Kyushu Univ. Series A, 10:1-13. E7,12c (3621)

Yamamoto, K. [1961], Generation principles of Latin squares. Bull. ISI 38(4):73-76. E7,12,13c (3622)

Yamamoto, K. [1965], A necessary condition for the existence of partially balanced incomplete block designs with an m-subset association scheme. Memoirs Faculty Sci. Kyushu Univ. Series A, 19:76-98. E3,4c (3623)

Yamamoto, K. [1965], On an orthogonal basis of the eigenspaces associated with partially balanced incomplete block designs of Latin square type association scheme. Memoirs Faculty Sci. Kyushu Univ. Series A, 19:99-104. E3,4c (3624)

Yamamoto, S. [1964], Some aspects for the composition of relationship algebras of experimental designs. J. Sci. Hiroshima Univ. Series A-I, 28:167-197. E2,3,4,7ac (3625)

Yamamoto, S. and Fujii, Y. [1963], Analysis of partially balanced incomplete block designs. J. Sci. Hiroshima Univ. Series A-I, 27:119-135. E3a (3626)

Yamamoto, S., Fujii, Y., and Hamada, N. [1965], Composition of some series of association algebras. J. Sci. Hiroshima Univ. Series A-I, 29:181-215. E2,3,4ac (3627)

Yamamoto, S., Fukuda, T., and Hamada, N. [1966], On finite geometries and cyclically generated incomplete block designs. J. Sci. Hiroshima Univ. Series A-I, 30:137-149. E2c (3628)

Yang, S.-J. [1958], A method of raising the degree of precision of trials involving a check variety and other varieties. (in Chinese). Nung-yeh-hsueh Pao (J. Agri.)/Act. Agri. Sinica 9:291-293. E4,5ac (3629)

Yates, F. [1952], Anãlise de uma experiẽncia de rotação. (The analysis of a rotation experiment.) Bragantia 12:213-235. E5,8ace (3630)

Yates, F. [1954], The analysis of experiments containing different crop rotations. Biometrics 10:324-346. E5,8a;A13 (3631)

Yates, F. [1967], A fresh look at the basic principles of the design and analysis of experiments. Proc. Fifth Berkeley Symp. Math. Stat. Prob. IV:777-790. E7a;T2,4a;C (3632)

Yates, F. and Patterson, H. D. [1958], A note on the six-course rotation experiments at Rothamsted and Woburn. J. Agri. Sci. 50:102-109. E5,8a (3633)

Yeh, S.-F. [1963], A study on the control of disturbing factors in an experiment by analysis of covariance. Taiwan Tahsueh Nunghsuehyuan Yenchiu Paokao/ Memoirs College Agri. Nat. Taiwan Univ. 7(1):1-11. E2,7ae;A7 (3634)

Youden, W. J. [1950], Comparative tests in a single laboratory. Statistical Method in Industrial Production, Royal Stat. Soc. pp 30-35. E2ae (3635)

Youden, W. J. [1950], A note on the four by four Latin Squares. Biometrics 6:289-290. E7a (3636)

Youden, W. J. [1951], Linked blocks: A new class of incomplete block designs (abstract). Biometrics 7:124. E4c (3637)

Youden, W.J. [1951], Single link designs (abstract). Virginia J. Sci. 2:380. E4a (3638)

Youden, W. J. [1952], The interpretation of chemical data. IQC 8(6):90-93 E2a;C (3639)

Youden, W. J. [1953], Experimental designs for the physical sciences (summary). JASA 48:625. E3a (3640)

Youden, W. J. [1954], Statistical design achieves important savings by reducing the number of runs using reference controls. Ind. Eng. Chemistry 46(8):99A-102A E2ac (3641)

Youden, W. J. [1954], Test prodecures may be improved by using statistical design while conducting tests. Ind. Eng. Chemistry 46(10):119A-122A E5ace (3642)

Youden, W. J. [1954], Long established experimental prodecures illustrate the principles of statistical design. Ind. Eng. Chemistry 46(12):105A-106A, E5ae (3643)

Youden, W. J. [1955], Statistical design does not necessarily reduce the experimental error of comparisons. Ind. Eng. Chemistry 47(2):97A-99A E5a (3644)

Youden, W. J. [1955], Gun problem illustrates the importance of proper scheduling of measurements. Ind. Eng. Chemistry 47(8):103A-105A E5c;C (3645)

Youden, W. J. [1955], Statistical design brings organization and direction into research and development. Ind. Eng. Chemistry 47(12):89A-90A E7c;C (3646)

Youden, W. J. [1956], Engineering vs. classical test patterns. Ind. Eng. Chemistry 48(8):59A-60A E5;C (3647)

Youden, W. J. [1956], Spark plug testing discloses operating characteristics of engines. Ind. Eng. Chemistry 48(10):61A-62A E2ac (3648)

Youden, W. J. [1956], Randomization and experimentation (abstract). AMS 27:1185-1186 E5ac;C (3649)

Youden, W. J. [1957], National physical standards and design of experiment. Bull. ISI 35(2):191-198 E3ac (3650)

Youden, W. J. [1958], Product specifications and test procedures. Ind. Eng. Chemistry 50(10):91A-92A E2ae (3651)

Youden, W. J. [1958], Circumstances alter cases. Ind. Eng. Chemistry 50(12):77A-78A E2a (3652)

Youden, W. J. and Connor, W. S. [1953], The chain block design. Biometrics 9:127-140 E4ace (3653)

Youden, W. J. and Connor, W. S. [1954], New experimental designs for paired observations. J. Res. NBS 53:191-196 E3ace;T11ace (3654)

Youden, W. J. and Hunter, J. S. [1955], Partially replicated Latin squares. Biometrics 11:399-405 E7ace (3655)

Zana, J. [1957], Die vom ungarischen Forschungsinstitut der Zuckerindustrie angewondten Feldversuchsmethoden und die mit ihnen gewonnenen Erfahrungen. Sbornik cukrovarnicko-řeparšké konference v Praze 14-19.X.1955 (Papers of the conference for sugar beet in Prague, 14-19 October, 1955) I and II, pp 283-295. E2,7,12a (3656)

Zana, J. [1957], A (k+1) típusú 5×5 - ös rácsnégyzet alkalmazása cukorrépa fajtakísérleteknél. (The use of the (k+1) 5x5 type of lattice square in sugar-beet variety trials.) Növénytermelés 6:119-130. E11a (3657)

Zelen, M. [1954], Analysis for some partially balanced incomplete block designs having a missing block. Biometrics 10:273-281 (abstract 9:263). E3,6ae (3658)

Zelen, M. [1954], Exact tests of significance for combining inter- and intra-block information in incomplete block designs (preliminary report) (abstract). AMS 25:810. E3a (3659)

Zelen, M. [1954], A note on partially balanced designs. AMS 25:599-602. E3c (3660)

Zelen, M. [1955], The use of incomplete block designs for factorial experiments (summary). JASA 50:599. E3a;T2a (3661)

Zelen, M. [1957], The analysis of covariance for incomplete block designs. Biometrics 13:309-332. E3ae;A7 (3662)

Zelen, M. [1957], Combining independent tests of significance for incomplete block designs. Ph. D. Thesis, American Univ. E2,3a (3663)

Zelen, M. [1957], The analysis of covariance for incomplete block designs (summary). JASA 52:383. E3a;A7 (3664)

Zelen, M. [1957], The analysis of incomplete block designs. JASA 52:204-217. E2,3a (3665)

Zelen, M. [1957], The use of incomplete block designs for asymmetrical factorial arrangements (abstract). AMS 28:526. E2,3ac;T2ac (3666)

Zelen, M. [1957], The use of incomplete block designs for asymmetrical factorial arrangements (abstract). Biometrics 13:422. E3a;T2a (3667)

Zelen, M. [1958], The use of group divisible designs for confounded asymmetrical factorial arrangements. AMS 29:22-40. E3a;T2a (3668)

Zelen, M. [1962], Introductory lectures on the statistical design of experiments. Math. Res. Center Orientation Lecture Series 1, U. S. Army Univ. Wisconsin. E2,7ae;T2ae;C (3669)

Zelen, M. and Federer, W. T. [1963], Applications of the calculus for factorial arrangements III: Two way elimination of heterogeneity. Math. Res. Center Tech. Summary Report No. 410:1-29, U. S. Army Univ. Wisconsin. E7,9ac;A8 (3670)

Zelen, M. and Federer, W. T. [1964], Applications of the calculus for factorial arrangements II: Two way elimination of heterogeneity. AMS 35:658-672 (also Math. Res. Center Tech. Summary Report No. 410, U. S. Army Univ. Wisconsin, E7,9ac;A8 (3671)

Zelen, M. and Joel, L. S. [1959], The weighted compounding of two independent significance tests. AMS 30:885-895. E3a;A11 (3672)

Zemach, R. B. [1965], On orthogonal arrays of strength four and their applications. Ph. D. Thesis, Michigan State Univ. E2,7,12,13c (3673)

Zimmermann, K. F. [1955], Feldversuchswesen: die Gittenquadratmethode in praktischer Anwendung. Der Züchter 25:132-138. E11ace (3674)

Zimmermann, K. F. [1957], Feldversuchswesen: Vereinfachtes Rechenschema für Blockversuche und Lateinische Quadrate. Der Züchter 27:89-92. E2,7ae (3675)

Zoellner, J. A. [1953], Experimental designs with blocks of two plots. M.S. Thesis, Iowa State Univ. E2,3ace (3676)

Zoellner, J. A. and Kempthorne, O. [1954], Incomplete block designs with blocks of two plots. Iowa State College Agri. Expt. Sta. Res. Bull. 418. E2ace (3677)

Zyskind, G. [1958], Error structures in experimental designs. Ph. D. Thesis, Iowa State Univ. E2,5,12a;A5 (3678)

Zyskind, G. [1960], Some randomization consequences in balanced incomplete blocks (abstract). AMS 31:245. E2a;A5 (3679)

Zyskind, G. [1963], Some consequences of randomization in a generalization of the balanced incomplete block design. AMS 34:1569-1581. E2a;A5 (3680)

Zyskind, G. and Kempthorne, O. [1959], The role of treatment error in comparative experiments (summary). JASA 54:508-509. E1,2a;T2a;C (3681)

Zyskind, G. and Kempthorne, O. [1960], Treatment errors in comparative experiments. Aeronautical Res. Lab. and Iowa State Univ., U. S. Air Force WADC Technical TN59-19. E1,2a;T2a;C (3682)

Zyskind, G., Kempthorne, O., White, R. F., Dayhoff, E. E., and Doerfler, T. E. [1964], Research on the analysis of variance and related topics. Aeronautical Res. Lab. and Iowa State Univ. U. S. Air Force. E2a;A5 (3683)

Co-authors list for category "E"

Name	Ref. No.
Abraham, T.P.	2284
	3097
	3098
Abreu, C.P.	2609
Addelman, S.	2792
Adhikary, B.	3053
Agarwal, K.N.	3297
Almeida, J.R.	2992
Andrew, G.M.	2473
Armstrong, P.J.	3503
Atkins, R.E.	2869
Atkinson, G.F.	2514
	3245
Babcock, A.B.	3164
Baeder, H.A.	2095
Baker, F.B.	2327
	2328
Baker, G.A.	2734
Bancroft, T.A.	2200
	2752
Banneick, A.	2069
Baptista, J.G.	2359
	3148
Barclay, W.D.	2791
Bearse, G.E.	2097
Behari, V.	2581
Bhagwandas	3401
Bhargava, P.N.	2407
	2408
	2409
Bhattacharya, K.N.	2191
Bianco, E.G.	3506
Bishop, R.F	2898
Blaser, R.E.	3511
Bockman, C.D.	2984
Bose, R.C.	2447
Boyd, D.A.	2916
Bradley, R.A.	2818
	2819
	2820
	2821

Name	Ref. No.
Bradley, R.A.	3286
	3317
	3318
	3319
Briggs, F.N.	2076
Bross, I.D.J.	3370
Brunk, M.E.	2515
Bush, K.A.	2161
Bush, S.	2008
Cady, F.B.	2568
Cairns, R.R.	2895
	2896
Cameron, J.M.	2162
	2163
Carmer, S.G.	2096
Carruth, A.F.	3560
Chakravarti, I.M.	2164
	2165
	2166
	2167
	2168
Chandrasekhararao, K.	3180
Chang, L.C.	2880
	2881
Chanmugam, J.	2755
Cheng, C.-P.	3567
Chilton, N.W.	3542
Chopra, D.V.	3454
Clatworthy, W.H.	2169
	2170
	2171
	2343
	2344
	2345
Cockerham, C.C.	3320
Coleman, J.E.	3234
Collier, R.O.	2074
Connor, W.S.	2172
	2173
	2650
	2954

VIII. BIBLIOGRAPHY FOR TREATMENT DESIGN

As has been stated previously, in order to make the treatment design bibliography complete, listings in Section VII that were also classified with a T code are repeated in the bibliography on treatment design. As may be observed from the listings, this area of research has been very active in the 1950 to 1967 period. One quite active area has been response surface designs discussed by G. E. P. Box and others and classified under T4. Another area in which there has been considerable activity has been the one on fractional replication as classified under T3 and T7. Confounding in factorial experiments (under T2), selection of X values and variables (under T1), weighing designs (under T6), dosage response designs (under T8), and genetic designs (under T9) have also received considerable attention in the 1950 to 1967 period.

The work in fractional replication appears to be using all available mathematical theory in several areas. In order to make substantially new progress in these areas, new mathematical theory will need to be developed. This is another example of where advances in one field of endeavor requires another field to make advances.

Listings under T8 and T9 are considered to be incomplete. However, sufficient listings are given to obtain a fairly complete coverage using those given under T8 or T9 and those listed in the references from the papers cited.

Abelson, R. M. and Bradley, R. A. [1954], A 2×2 factorial with paired comparisons. Biometrics 10:487-502 (abstract 10:303-304). T2,11ae;A12 (3684)

Abelson, R. M. and Bradley, R. A. [1954], A 2×2 factorial with paired comparisons (abstract). AMS 25:414. T2,11a;A12 (3685)

Abou-Yousef, M. I. [1963], Selection of regression variates for prediction. M. S. Thesis, Durham Univ. T1ac (3686)

Abraham, T. P. and Rao, V. Y. [1959], Economic analysis of fertilizer trial data on rice and wheat (abstract). JISAS 11:196-197. T4a (3687)

Abraham, T. P. and Rao, V. Y. [1966], An investigation on functional models for fertilizer response surfaces. JISAS 18(1):45-61. T2,4a;A7,13 (3688)

Abt, K. [1962], A method for the derivation of expected mean squares in the analysis of variance (abstract). AMS 33:814. T2a (3689)

Abt, K. [1963], Mean square expectations for orthogonal contrasts in mixed model analyses of variance (abstract). AMS 34:1118. T2a (3690)

Abt, K. [1963], Table of expectations of mean squares in the analysis of variance for crossed classifications. U.S. Naval Weapons Lab. Dahlgren, Va. NWL Report No. 1833. T2a (3691)

Abt, K. [1967], On the identification of the significant independent variables in linear models. Metrika 12:1-15, 81-96. T1a (3692)

Addelman, S. [1960], Fractional factorial plans. Ph. D. Thesis, Iowa State Univ. T7ac (3693)

Addelman, S. [1960], Main-effect designs for asymmetrical factorial experiments (abstract). AMS 31:230. T3,7c (3694)

Addelman, S. [1961], Irregular fractions of factorial experiments. Iowa State Univ. Aeronautical Res. Lab. ARL 9, U.S. Air Force. T7ac (3695)

Addelman, S. [1961], Irregular fractions of the 2^n factorial experiments. Technometrics 3:479-496. T7ac (3696)

Addelman, S. [1962], Orthogonal main-effect plans for asymmetrical factorial experiments. Technometrics 4:21-46 (erratum 4:440). T7ac (3697)

Addelman, S. [1962], Symmetrical and asymmetrical fractional factorial plans. Technometrics 4:47-58. T7ac (3698)

Addelman, S. [1962], The construction and analysis of non-orthogonal plans for 2^n factorial experiments. Proc. Seventh Conf. Design Expt. Army Res. Dev. Testing. pp. 27-38. T7ac (3699)

Addelman, S. [1962], Fractional replicate plans for symmetrical and asymmetrical factorial experiments (summary). JASA 57:485. T7c (3700)

Addelman, S. [1962], Augmenting factorial plans to accommodate additional two-level factors. Biometrics 18:308-322. T3,7ace (3701)

Addelman, S. [1963], Proportional frequency designs. Proc. Eighth Conf. Design Expt. Army Res. Dev. Testing. pp. 287-302. T7ac (3702)

Addelman, S. [1963], Techniques for constructing fractional replicate plans. JASA 58:45-71 (Research Triangle Inst., Tech. Report No. 2, Univ. N. Carolina). T3,4,7c (3703)

Addelman, S. [1964], Designs for the sequential application of factors. Technometrics 6:365-370 (Proc. Ninth Conf. Design Expt. Army Res. Dev. Testing. pp. 385-394). T3,7ac;S2 (3704)

Addelman, S. [1964], Some two-level factorial plans with split plot confounding. Technometrics 6:253-258. T3,7ac (3705)

Addelman, S. [1965], The construction of a 2^{17-9} resolution V plan in eight blocks of 32. Technometrics 7:439-443. T3c (3706)

Addelman, S. [1967], Equal and proportional frequency squares. JASA 62:226-240. E7ace;T7ace (3707)

Addelman, S. [1967], The selection of sequences of two-level factorial plans (abstract). Technometrics 9:185. T3,7c (3708)

Addelman, S. [1967], The selection of sequences of two-level fractional factorial plans. Aeronautical Res. Lab. ARL 67-0013, U.S. Air Force. T3c (3709)

Addelman, S. [1967], Research in sequential factorial designs. Aeronautical Res. Lab. ARL 67-0141, U.S. Air Force. T3ac (3710)

Addelman, S., Gaylor, D. W., and Bohrer, R. E. [1966], Sequences of combination chemotherapy experiments. Biometrics 22:730-746. T7ace (3711)

Addelman, S. and Kempthorne, O. [1961], Orthogonal main-effect plans. Iowa State Univ. Aeronautical Res. Lab. ARL 79, U.S. Air Force. T7ac (3712)

Addelman, S. and Kempthorne, O. [1961], Some main-effect plans and orthogonal array of strength two. AMS 32: 1167-1176. T3,7c (3713)

Adichie, J. N. [1967], Asymptotic efficiency of a class of non-parametric tests for regression parameters. AMS 38:884-893. T1a;A12 (3714)

Agrawal, H. [1963], A note on omission of variables in multiple regression analyses. JISA 1:226-229. T1a;A7 (3715)

Aird, P. L. [1956], Fertilizers in forestry and their use in hardwood population establishment. Pulp Paper Mag. Canada, Convention Issue. E2ae;T2a (3716)

Aitchison, J. and Silvey, S. D. [1957], The generalization of probit analysis to the case of multiple responses. Biometrika 44:131-140. T8a;A9 (3717)

Albasiny, E. L. [1965], The numerical solution of some non-linear equations, useful in the design of experiments. JRSSB 27:466-472. T5a (3718)

Alexander, G. and Watson, R. H. [1951], The assay of oestrogenic activity of Trifolium subterraneum L. by increase in uterine weight in the spayed guinea pig. I. Characteristics of the dose-response relationship. II. Assays. Australian J. Agri. Res. 2:457-479, 480-493. T8a (3719)

Alexander, M. J. [1964], Multivariate sensitivity experiments with non-interacting stimuli. Rocketdyne Res. Report RR 64-15, N. American Aviation. T4,7ac (3720)

Alexander, M. J. [1965], Design, testing and estimation in complex experimentation. III. The design and analysis of multivariate sensitivity experiments. Aeronautical Res. Lab. ARL 65-116, U. S. Air Force. T8a;A12;S2 (3721)

Alexander, M. J. [1965], Models for the analysis of multivariate experiments with interacting stimuli. Rocketdyne Res. Memorandum RM 1121-351, N. American Aviation. T4,7ac (3722)

Alexander, M. J., Goodman, N. R., Locks, M. O., and Webb, S. R. [1963], Some results on designs for regression experiments, design of experiments with auto-correlated errors present, and decision theory approch to complex experimentation. Aeronautical Res. Lab. ARL 63-107, U.S. Air Force. T1ac (3723)

Alexander, M. J. and Rothman, D. [1966], Designs and analyses for inverse response problems in sensitivity testing. Proc. Eleventh Conf. Design Expt. Army Res. Dev. Testing. pp. 337-365. T8ace (3724)

Allan, J. S, and Robertson, A. [1964], The effect of initial reverse selection upon total selection response. Genetical Res. 5:68-79. T9a (3725)

Allard, R. W. [1956], The analysis of genetic-environmental interactions by means of diallel crosses. Genetics 41:305-318. T9ae (3726)

Allen, R. S., Homeyer, P. G., and Jacobson, N. L. [1955], Use of activated glycerol dichlorohydrin for estimating vitamin A in dairy calf blood plasma. Iowa State College J. Sci. 29:721-734. E2ae;T2ae (3727)

Alling, D.W. [1967], Tests of relatedness. Biometrika 54:459-469. T2ae (3728)

Alvarez, R., Segalla, A. L., Wutke, A. C. P., and Freire, E. S. [1963], Adubação da cana-de-açúcar. VIII - Adubação mineral em solos massapê-salmourão (1957-58). (Fertilizer experiments with sugar cane in "Massapê-salmourão" soils.) Bragantia 22:657-675. T2e (3729)

Alway, G. G. [1962], The distribution of the number of circular triads in paired comparisons. Biometrika 49:265-269. T11c (3730)

Andersen, S. L. [1959], Statistics in the strategy of chemical experimentation. Chemical Eng. Progress 55(4):61-67. T2,4a;C (3731)

Anderson, R. L. [1952], Some applications of least-squares to research in agricultural economics and rural sociology (abstract). Econometrica 20:94-95. T2a;A7 (3732)

Anderson, R. L. [1956], A comparison of discrete and continuous models in agricultural production analysis. Methodological Procedures in the Economic Analysis of Fertilizer Use Data, pp. 39-61. T2,3ac;C (3733)

Anderson, R. L. [1957], Some statistical problems in the analysis of fertilizer respons data. Economic and Technical Analysis of Fertilizer Innovations and Resource Use ch. 17, pp. 187-206. T2,4a (3734)

Anderson, R. L. [1958], Some aspects of fractional experimentation (summary). JASA 53:569. T2,4a (3735)

Anderson, R. L. [1958], Complete factorials, fractional factorials and confounding. Experimental Designs in Industry, pp. 59-107. T2,3ace (3736)

Anderson, R. L. [1960], Some remarks on the design and analysis of factorial experiments. Contributions to Probability and Statistics, pp. 35-56. T2ac (3737)

Anderson, T. W. [1964], On Bayes procedures for a problem with choice of observations. AMS 35:1128-1135. T1a (3738)

Anderson, V. L. [1953], A model for the study of quantitative inheritance. Ph. D. Thesis, Iowa State Univ. T9a (3739)

Andrews, F. C. and Chernoff, H. [1955], A large-sample bioassay design with random doses and uncertain concentration. Biometrika 42:307-315. T8c (3740)

Andrus, C. F. [1963], Plant breeding systems. Euphytica 12:205-228. T9a;C (3741)

Anonymous [1958], Interlocked factorials: fewer runs. Chemical Eng. News 36(May 5):40-41. T7ac (3742)

Anscombe, F. J. [1959], Quick analysis methods for random balance screening experiments. Technometrics 1:195-209. T7ac (3743)

Anscombe, F. J. [1963], Bayesian inference concerning many parameters, with reference to super-saturated designs (with discussion). Bull. ISI 40:721-733. T7ac (3744)

Anscombe, F. J. [1967], Topics in the investigation of linear relations fitted by the method of least squares (with discussion). JRSSB 29:1-52. T2a;A7,14 (3745)

Antle, C. E. [1962], The N response problem. Ph. D. Thesis, Oklahoma State Univ. T1ac (3746)

Antle, C. E. [1962], The uniqueness of the spacing of observations in polynomial regression for minimax variance of the fitted values. AMS 33:810-811. T1c (3747)

Aoyama, H. [1953], On a test in paired comparisons. Ann. Inst. Stat. Math. 4:83-87. T11a (3748)

Armitage, P. [1966], Recent developments in medical statistics. Rev. ISI 34: 27-42. T8a;B;C (3749)

Armitage, P. and Allen, I. [1950], Methods of estimating the LD50 in quantal response data. J. Hygiene 48:298-322. T8a (3750)

Armitage, P. and Spicer, C. C. [1956], The detection of variation in host susceptibility in dilution counting experiments. J. Hygiene 54:401-414. T8a (3751)

Armstrong, R. J. and Thompson, A. E. [1967], A diallel analysis of tomato fruit cracking. Proc. American Soc. Hort. Sci. 91:505-513. T9a (3752)

Asano, C. [1960], Some studies on the optimum choice of dosage in biological assay. Reports Stat. Appl. Res. 7:65-78. T8ac (3753)

Ashford, J. R. and Smith, C. S. [1964], General models for quantal response to the joint action of a mixture of drugs. Biometrika 51:413-428. T8ae (3754)

Ashford, J. R. and Smith, C. S. [1965], An alternative system for the classification of mathematical models for quantal responses to mixtures of drugs in biological assay. Biometrics 21:181-188. T8a (3755)

Ashford, J. R. and Smith, C. S. [1965], An analysis of quantal response data in which the measurement of response is subject to error. Biometrics 21:811-825. T8a (3756)

Ashford, J. R. and Smith, C. S. [1966], Models for the non-interactive joint action of a mixture of stimuli in biological assay. Biometrika 53:49-59.
T8a (3757)

Ashford, J. R., Smith, C. S., and Brown, S. [1960], The quantal response analysis of a series of biological assays on the same subjects. Biometrika 47:23-32.
T8a (3758)

Ashton, G. C., Rennie, J. C., and Etter, E. [1958], Interpretation of interaction in the analysis of variance of a factorial experiment. Canadian J. Animal Sci. 38:181-186. T2a (3759)

Avrahami, R. [1967], On query No. 20. Technometrics 9:491-492.
T2a (3760)

Baasel, W. D. [1965], Exploring response surfaces to establish optimum conditions. Chemical Eng. 72(Oct.25):147-152. T4a (3761)

Bahn, E. [1959], Betrachtungen zum faktoriellen Versuch vom Typ 2^2. Z. Landw. Vers. Untersuch Wes. 5:232-254. E12a;T2a (3762)

Bailey, N. T. J. [1959], The use of linear algebra in deriving prime power factorial designs with confounding and fractional replication. Sankhyā 21:345-354. T2,3c;A8 (3763)

Bainbridge, J. R. [1951], Factorial experiments in pilot plant studies. Ind. Eng. Chemistry 43:1300-1306. E2e;T2ae (3764)

Bainbridge, J. R., Grand, A. M., and Radok, U. [1956], Tabular analysis of factorial experiments and the use of punch cards. JASA 51:149-158. T2ae;A15 (3765)

Baird, B. L. and Fitts, J. W. [1957], An agronomic procedure involving the use of a central composite design for determining fertilizer response surfaces. Ch. 13, Economic and Technical Analysis of Fertilizer Innovations and Resource Use. Iowa State College Press, Ames, Iowa, pp. 135-143. T4a (3766)

Baird, B. L. and Mason, D. D. [1959], Multi-variable equations describing fertility-corn yield response surfaces and their agronomic and economic interpretation. Agron. J. 51:152-156. T4a (3767)

Baker, A. G. [1957], Analysis and presentation of the results of factorial experiments. Appl. Stat. 6:45-55. T2ae (3768)

Baker, G. A. Jr., Christy, J., and Baker, G. A. [1964], Analysis of genetic changes in finite populations composed of mixtures of pure lines. J. Theoretical Biology 7:68-85. T9a (3769)

Baker, G. A. Jr., Christy, J., and Baker, G. A. [1964], Stochastic processes and genotypic frequencies under mixed selfing and random mating. J. Theoretical Biology 7:86-97. T9a (3770)

Baker, L. H. [1964], Diallel crosses of inbred lines of egg-type poultry repeated over locations and years II. Implications of results for design of a hybrid poultry breeding program (abstract). Biometrics 20:377-378. T9a (3771)

Balmer, E. et al. [1965], Contribuição ao estudo da influência dos fatôres físicos do solo, sôbre a incidência da murcha do algodoeiro, causada por Fusarium oxysporum f. vasinfectum (Atk.) Snyder and Hansen Anais Escola Superior Agri. "Luiz de Queiroz" 22:247-258. E2ae;T2ae (3772)

Bancroft, T. A. [1964], Analysis and inference for incompletely specified models involving the use of preliminary test(s) of significance. Biometrics 20:427-442. T2a;A11 (3773)

Bancroft, T. A. [1966], Inference for incompletely specified models in physical sciences (with discussion). Bull. ISI 41:497-515. T1a;A6,11;C (3774)

Banerjee, K. S. [1950], A note on the fractional replication of factorial arrangements. Sankhyā 10:87-94. T3,7ace (3775)

Banerjee, K. S. [1950], Some contributions to Hotelling's weighing designs. Sankhyā 10:371-382. T6c (3776)

Banerjee, K. S. [1950], How balanced incomplete block designs may be made to furnish orthogonal estimates in weighing designs. Biometrika 37:50-58. T6ace (3777)

Banerjee, K. S. [1950], Weighing designs. Calcutta Stat. Assoc. Bull. 3:64-76. T6ac (3778)

Banerjee, K. S. [1951], Weighing designs and partially balanced incomplete blocks. Calcutta Stat. Assoc. Bull. 4:36-38. T6c (3779)

Banerjee, K. S. [1951], Some observations on the practical aspects of weighing designs. Biometrika 38:248-251. E2ace;T6ace (3780)

Banerjee, K. S. [1961], A factorial approach to construction of true cost of living index and its application in studies of changes on national income. Sankhyā A 23:297-304. T7c (3781)

Banerjee, K. S. [1961], A unified statistical approach to the index number problem. Econometrica 29:591-601. T7c (3782)

Banerjee, K. S. [1963], Index numbers for factorial effects and their connection with a special kind of irregular fractional plans of factorial experiments. JASA 58: 497-508. T7c (3783)

Banerjee, K. S. [1963], Index numbers for factorial effects and their connection with a special kind of irregular fractional plans of factorial experiments (abstract). Biometrics 19:656. T7ac (3784)

Banerjee, K. S. [1963], Best linear unbiased index numbers and index numbers obtained through a factorial approach. Econometrica 31:712-718. T7c (3785)

Banerjee, K. S. [1965], On Hotelling's weighing designs under autocorrelation of errors. AMS 36:1829-1834. T6a (3786)

Banerjee, K. S. [1966], On nonrandomized fractional weighing designs. AMS 37:1836-1841. T6a (3787)

Banerjee, K. S. [1966], Singularity in Hotelling's weighing designs and a generalized inverse. AMS 37:1021-1032. T6a (3788)

Banerjee, K. S. and Federer, W. T. [1963], On estimates for fractions of a complete factorial experiment as orthogonal linear combinations of the observations (abstract). Biometrics 19:504. T3,7a (3789)

Banerjee, K. S. and Federer, W. T. [1963], On estimates for fractions of a complete factorial experiment as orthogonal linear combinations of the observations. AMS 34:1068-1078. T3,7a (3790)

Banerjee, K. S. and Federer, W. T. [1964], Estimates of effects for fractional replicates. AMS 35:711-715. T3,7a (3791)

Banerjee, K. S. and Federer, W. T. [1966], On estimation and construction in fractional replication. AMS 37:1033-1039. T3,7ac (3792)

Banerjee, K. S. and Federer, W. T. [1967], On a special subset giving an irregular fractional replicate of a 2^n factorial experiment. JRSSB 29:292-299. T7ac (3793)

Banga, O., Petiet, J., and von Bennekom, J. L. [1964], Genetical analysis of male-sterility in carrots. Euphytica 13:75-93. T9a (3794)

Bankier, J. D. [1960], An operational approach to the r-way crossed classification. AMS 31:16-22. T2a (3795)

Barnett, E. H. [1960], Introduction to evolutionary operation. Ind. Eng. Chemistry 52:500-503. T4ace (3796)

Barnett, M. K. and Mead, F. C. [1956], A 2^4 factorial experiment in four blocks of eight: A study in radioactive decontamination. Appl. Stat. 5:122-131. T2ae (3797)

Barraclough, C. G. [1955], Statistical analysis of multiple slope ratio assays. Biometrics 11:186-200. T8ae (3798)

Barron, C. L. [1962], Factorial experiments in reliability analysis. Proc. Nat. Symp. Reliability Quality Control 8:62-67. T2a (3799)

Baten, W. D. [1956], An analysis of variance applied to screw machines. IQC 12(10): 8-9. T2ae (3800)

Batson, H. C. [1950], Statistics in experimental immunology. Ann. New York Academy Sci. 52:862-876. T2e;C (3801)

Batson, H. C. [1956], Applications of factorial χ^2 analysis to experiments in chemistry. ASQC Nat. Conv. Trans. 10:9-23. T2ae (3802)

Batson, H. C. [1963], Bioassay: The quantal response assay. Proc. Eighth Conf. Design Expt. Army Res. Dev. Testing. pp. 551-577. T8a (3803)

Bauer, R. K. [1960], On the reduction of the number of characters in multivariate analysis (abstract). Biometrics 16:312. Tla;A7 (3804)

Bauer, T. W. [1957], Statistical analysis of some factors influencing the physical testing of pulp I. The influence of time and basis weight variations. Tappi 40:84-91. T2ae (3805)

Beale, E. M. L. [1958], On an iterative method for finding a local minimum of a function of more than one variable. Stat. Techniques Res. Group, Princeton Univ. Tech. Report No. 25, 44 pp. T4a (3806)

Beale, E. M. L., Kendall, M. G., and Mann, D. W. [1967], The discarding of variables in multivariate analysis. Biometrika 54:357-366. Tla;A7 (3807)

Beale, E. M. L. and Mallows, C. L. [1958], On the analysis of screening experiments. Stat. Techniques Res. Group, Princeton Univ., Tech. Report No. 20, pp. 1-29. Tla;A7 (3808)

Beazley, C. C. [1963], Factorial quality control. Chemical Eng. Progress Symp. Ser. 59(42):28-33. T2ac (3809)

Bechhofer, R. E. [1960], A multiplicative model for analyzing variances which are affected by several factors. JASA 55:245-264. T2a;A7 (3810)

Behnken, D. W. [1959], Simplex-sum designs - a class of second order rotatable designs derivable from those of first order. Ph. D. Thesis, Univ. N. Carolina. T4c (3811)

Behnken, D. W. [1959], Some new designs for exploring response surfaces (abstract). Biometrics 15:331. T4c (3812)

Behrens, W. U. [1960], Analysis of multi-factorial experiments (abstract). Biometrics 16:310. T2a (3813)

Bejar, J. [1957], Diseño de experimentos. Trabajos Estad. 8:91-108. El,2,7ae;T2ac (3814)

Bell, C. B. and Geller, H. [1967], Nonparametric tests for 2-factor experiments (preliminary report) (abstract). AMS 38:953. T2a;Al2 (3815)

Bellmann, K. and Ahrens, H. [1966], Modellpopulationen in der Selektionstheorie und einige Ergebnisse aus Simulationsstudien. Der Züchter 36:172-185. T9a (3816)

Bennett, B. M. [1954], Some further extensions of Fieller's theorem. Ann. Inst. Stat. Math. 5:103-106. T8a (3817)

Bennett, B. M. [1956], On confidence limits for the ratio of regression coefficients. Ann. Inst. Stat. Math. 8:41-43. T8a (3818)

Bennett, B. M. [1963], Slope ratio assays and confidence limits. Metrika 7:117-120. T8a (3819)

Bennett, B. M. [1963], On combining estimates of a ratio of means. JRSSB 25: 201-205. T8a (3820)

Bennett, B. M. [1966], Note on combining estimates in bioassay. Biom. Zeit. 8:5-9. T8a (3821)

Bennett, D. and Dudzinski, M. L. [1967], Bioassay responses of ewes to legume swards. I. Uterine weight response: variability, calibration, and prediction. Australian J. Agri. Res. 18:485-494. T8e (3822)

Berg, H. W., Filipello, F., Hinreiner, E., and Sawyer, F. M. [1955], Consumer wine-preference methodology studies at California fairs. Food Tech. 9:90-93. T11a;C (3823)

Berkson, J. [1950], Some observations with respect to the error of bio-assay. Biometrics 6:432-434. T8a (3824)

Berkson, J. [1953], A statistically precise and relatively simple method of estimating the bio-assay with quantal response, based on the logistic function. JASA 48: 565-599. T8a (3825)

Berkson, J. [1960], Problems recently discussed regarding estimating the logistic curve. Bull. ISI 37(3):207-211. T8a (3826)

Berry, J. W., Tucker, H., and Deutschman, A. J. [1963], Startch vinylation. Determination of optimum conditions by response surface designs. I & EC Process Design Dev. 2:318-322. T4ae (3827)

Beyer, W. H. [1962], Symmetrical complementation designs. Ph. D. Thesis, Virginia Polytechnic Inst. T5ac (3828)

Beyer, W. H. and Bargmann, R. E. [1961], Symmetrical complementation design (abstract). Biometrics 17:499-500. T7ac (3829)

Beyer, W. H. and Bargmann, R. E. [1962], Symmetrical complementation design (abstract). Biometrics 18:258. T7ac (3830)

Bhapkar, V. P. and Koch, G. [1965], On the hypothesis of "no interaction" in the three dimensional contingency tables. Inst. Stat. Mimeo. Ser. No. 440, Univ. N. Carolina. T2a (3831)

Bhapkar, V. P. and Koch, G. [1965], On the hypothesis of "no interaction" in the four dimensional contingency tables. Inst. Stat. Mimeo. No. 449, Univ. N. Carolina. T2a (3832)

Bhargava, P. N. [1959], Missing plots in asymmetrical factorial designs (abstract). JISAS 11:195. T7a (3833)

Bhargava, P. N. [1960], A note on the analysis of 3^n factor, fractional replication (abstract). JISAS 12:150. T3a (3834)

Bhattacharya, C. G. [1966], Fitting a class of growth curves. Sankhyā B 28:1-10. T1a (3835)

Bibile, S. W. [1953], The assay of cortical steroids by the mouse eosinophil test. J. Endocrinology 9:357-369. T8e (3836)

Bibile, S. W. and Vogt, M. [1953], Parallel biological assays of the cortisone equivalent of the dog's adrenal vein blood. J. Endocrinology 9:351-356. T8e (3837)

Bicking, C. A. [1952], Statistical methods in chemical development. IQC 8(4): 9-10, 12, 14, 16-17. T2ae;C (3838)

Bicking, C. A. [1958], Experiences and needs for design in ordnance experimentation. Experimental Designs in Industry, pp. 247-252. T2e;C (3839)

Bicking, C. A. [1962], Sequential experiment design for small samples from successiv production. Bull. ISI 39(3):393-397. T4,7ac;S2 (3840)

Bicking, C. A. and Gillespie, R. H. [1963], Exploring with experiment design. IQC 19(12):17-21. T4a;C (3841)

Biggers, J. D. [1952], The calculation of the dose-response line in quantal assays with special reference to oestrogen assays by the Allen-Doisy technique. J. Endocrinology 8:169-178. T8a;A9 (3842)

Biggers, J. D. [1953], The characteristics of the dose-response line in Allen-Doisy tests obtained by the intravaginal administration of oestrone in distilled water and in 50% aqueous glycerol. J. Endocrinology 9:145-154. T8e (3843)

Biggers, J. D. [1953], The effect of different protein solvents on the dose-response line for the intravaginal administration of oestrone in mice. J. Endocrinology 9:136-144. T8e (3844)

Binder, A. [1955], The choice of an error term in analysis of variance designs. Psychometrika 20:29-50. T2a (3845)

Binet, F. E., Leslie, R. T., Weiner, S., and Anderson, R. L. [1955], Analysis of confounded factorial experiments in single replications. N. Carolina Agri. Expt. Sta., Tech. Bull. No. 113, pp. 1-64. T2ace (3846)

Bingham, R. S. [1959], Design of experiments from a statistical viewpoint, part II. IQC 15(12):12-15. E7ae;T2ae (3847)

Bingham, R. S. [1963], EVOP for systematic process improvement. IQC 20(3):17-23. T4a (3848)

Birnbaum, A. [1956], Analysis of 2^n factorial experiments (summary). JASA 51:505. T2a (3849)

Birnbaum, A. [1958], On the analysis of factorial experiments without replication (abstract). AMS 29:1286. T2,3a (3850)

Birnbaum, A. [1959], On the analysis of factorial experiments without replication. Technometrics 1:343-357. T2,3a (3851)

Birnbaum, A. [1961], A multi-decision procedure related to the analysis of single degrees of freedom. Ann. Inst. Stat. Math. 12:227-236. T2a;A2 (3852)

Biser, E. [1962], General formulas and a positional index-algorithm for generating orthogonal contrasts in multi-variable statistical designs. Proc. Seventh Conf. Design Expt. Army Res. Dev. Testing. pp. 67-175. T2ae;A7 (3853)

Biser, E., Eisenberg, H., and Millman, G. [1959], The application of experimental design to a radar target acquisition system. Proc. Fourth Conf. Design Expt. Army Res. Dev. Testing. pp. 75-126. T2e;C (3854)

Black, A. N. [1950], Weighted probits and their use. Biometrika 37:158-167. T8a (3855)

Blackith, R. E. [1950], Bio-assay systems for the pyrethrins. III. Application of the twin cross-over design to crawling insect assays. Ann. Appl. Biology 37:508-515. E8ace;T8ae (3856)

Blanco, H. G., Venturini, W. R., Gargantini, H., and Cuiabano, N. [1965], Adubação mineral para o trigo no sul do estado de São Paulo. (Mineral fertilizers applied to wheat in the south of the State of São Paulo.) Bragantia 24:481-505. T2e (3857)

Bliss, C. I. [1950], The design of biological assays. Ann. New York Academy Sci. 52:877-888. T8ac;B (3858)

Bliss, C. I. [1952], The clinical assay of diuretic agents II. Estimation of the error in a clinical assay. Biometrics 8:237-248. E2ae;T8ae (3859)

Bliss, C. I. [1954], Insecticidal assays. Statistics and Mathematics in Biology, pp. 345-360. T8a (3860)

Bliss, C. I. [1956], The analysis of insect counts as negative binomial distributions (with discussion). Proc. Tenth Int'l. Congress Entomology 2:1015-1032. E2ae;T2ae;A9 (3861)

Bliss, C. I. [1957], Bioassays in U.S.P. XV and their precision. Bull. ISI 35(2): 333-338. T8a (3862)

Bliss, C. I. [1957], Some principles of bioassay. American Scientist 45:449-466. T8ac (3863)

Bliss, C. I. and Fleischer, D. A. [1964], The analysis in angles of a microbial experiment. Biometrics 20:883-891. T2ae;A9 (3864)

Bliss, C. I., Greenwood, M. L., and White, E. S. [1956], A rankit analysis of paired comparisons for measuring the effect of sprays on flavor. Biometrics 12:381-403. T11ae;A9,12 (3865)

Bliss, C. I. and Pabst, M. L. [1955], Assays for standardizing adrenal cortex extract in production. Bull. ISI 34(4):317-338. T8a (3866)

Blodgett, R. B. [1957], The advantages in industrial research of looking at the effects of many variables at the same time. Tappi 40(5):150A-156A. T2a (3867)

Bloemena, A. R. [1958], Roteerbare proefopzetten. (Rotatable designs.) Stat. Neerlandica 12:135-142. T4ac (3868)

Blomquist, M. O. [1966], Rank analysis of complete block factorial designs. Ph. D. Thesis, Univ. Minnesota. E2a;T2a;A12 (3869)

Blum, J. R. [1954], Multidimensional stochastic approximation methods. AMS 25: 737-744. T1ac (3870)

Bock, R. D. [1958], Remarks on the test of significance for the method of paired comparisons. Psychometrika 23:323-334. T11a (3871)

Boen, J. R. and Brown, B. W. [1965], Allocating of observations in parallel line assay (abstract). Biometrics 21:762. T8c (3872)

Boen, J. R. and Brown, B. W. [1967], Use of prior information to design a routine parallel line assay. Biometrics 23:257-267. T8ace (3873)

Boguslawski, E. v. [1958], Der Feldversuch mit polyfaktorieller Fragestellung. Z. Landw. Vers. Untersuch Wes. 4:316-342. E2,3,5ac;T2a (3874)

Bohidar, N. R. [1966], On the confounding of conditional effects and interactions in the symmetrical case (abstract). Biometrics 22:945. T2ac (3875)

Bohidar, N. R. [1967], Conditional effects and interactions in symmetrical factorial confounding with application to biology. Proc. Twelfth Conf. Design Expt. Army Res. Dev. Testing. pp. 207-220. T2ae (3876)

Bohidar, N. R. and Williams, D. [1965], The definitions and the estimation of the conditional effects and interaction in symmetrical factorial confounding (abstract). Biometrics 21:1021-1022. T2a (3877)

Bohren, B. B., Hill, W. G., and Robertson, A. [1966], Some observations on asymmetrical correlated responses to selection. Genetical Res. 7:44-57. T9a (3878)

Bojanowski, J. [1967], Recurrent selection for smut resistance in corn. Der Züchter 37:151-155. T9a (3879)

Booth, K. H. V. and Cox, D. R. [1962], Some systematic supersaturated designs. Technometrics 4:489-495. T7ac (3880)

Borden, R. J. [1952], Getting more information for less money through factorial design in experiments. Hawaiian Planters' Record 54:77-80. T2a (3881)

Bose, R. C. [1952], Mathematics of factorial designs. Proc. Int'l. Congress Math. 1:543-548. T2,3c (3882)

Bose, R. C. [1955], Paired comparison designs for testing concordance between judges. Inst. Stat. Mimeo. Ser. No. 134, Univ. N. Carolina. T11c (3883)

Bose, R. C. [1956], Paired comparison designs for testing concordance between judges. Biometrika 43:113-121. E2c;T11c (3884)

Bose, R. C. [1957], On a problem in Abelian groups and the construction of fractionally replicated designs (abstract). Biometrics 13:411-412. T3c (3885)

Bose, R. C. [1959], Some new cases of the packing problem in finite projective space with applications to fractionally replicated designs (abstract). AMS 30:615. T3c (3886)

Bose, R. C. [1961], Some ternary error correcting codes and fractionally replicated designs. Inst. Stat. Mimeo. Ser. No. 295, Univ. N. Carolina. T3,7c (3887)

Bose, R. C. [1961], On some connections between the design of experiments and information theory. Bull. ISI 38(4):257-271. T2,3c (3888)

Bose, R. C. [1963], Some ternary error correcting codes and fractionally replicated designs (discussion). Colloque International Centre National Recherche Scientifique, No. 110, Le Plans d'Expériences, pp. 21-32 (also Inst. Stat. Mimeo. Ser. No. 295, Univ. N. Carolina). T3,7c (3889)

Bose, R. C. [1965], Session chairman's prepared remarks. Proc. IBM Sci. Computing Symp. Stat., pp. 55-57. E12c;T4c (3890)

Bose, R. C. and Burton, R. C. [1957], On a problem in Abelian groups and the construction of fractionally replicated designs (abstract). AMS 28:533. T3c (3891)

Bose, R. C. and Carter, R. L. [1959], Complex representation in the construction of rotatable designs. AMS 30:771-780. T4c (3892)

Bose, R. C. and Carter, R. L. [1962], Response model coefficients and the individual degrees of freedom of a fractional design. Biometrics 18:160-171. T2,4ae (3893)

Bose, R. C. and Connor, W. S. [1960], Analysis of fractionally replicated 2^m3^n designs. Bull. ISI 37(3):141-160. T3ac (3894)

Bose, R. C. and Draper, N. R. [1958], Rotatable designs of second and third order in three or more dimensions. Inst. Stat. Mimeo. Ser. No. 197, Univ. N. Carolina. T4c (3895)

Bose, R. C. and Draper, N. R. [1958], Second order rotatable designs in three or more factors (abstract). AMS 29:621-622. T4c (3896)

Bose, R. C. and Draper, N. R. [1959], Second order rotatable designs in three dimensions. AMS 30:1097-1112. T4c (3897)

Bose, R. C. and Patel, M. S. [1961], Investigations on factorial designs. Inst. Stat. Mimeo. Ser. No. 285, Univ. N. Carolina. T2a (3898)

Bose, R. C. and Srivastava, J. N. [1963], On certain bounds useful in the theory of factorial experiments and error-correcting codes (abstract). AMS 34:679-680. T7c (3899)

Bose, R. C. and Srivastava, J. N. [1964], On a bound useful in the theory of factorial designs and error correcting codes. AMS 35:408-414 (also Inst. Stat. Mimeo. Ser. No. 359, Univ. N. Carolina). T2c (3900)

Bose, R. C. and Srivastava, J. N. [1964], Multidimensional partially balanced designs and their analysis, with applications to partially balanced factorial fractions. Sankhyā A 26:145-168 (also Inst. Stat. Mimeo. Ser. No. 376, Univ. N. Carolina). E3,7,13a;T7ac (3901)

Bose, R. C. and Srivastava, J. N. [1964], Analysis of irregular factorial fractions. Sankhyā A 26:117-144 (also Inst. Stat. Mimeo. Ser. No. 373, Univ. N. Carolina). T7ac (3902)

Bose, R. C. and Srivastava, J. N. [1964], Mathematical theory of factorial designs. Bull. ISI 40:780-794. T2,3,7ac (3903)

Bowlen, B. and Heady, E. O. [1955], Optimum combinations of competitive crops at particular locations (application of linear programming: 1). Iowa State Univ.Agri. Expt. Sta. Res. Bull. 426, pp. 373-400. T4a;A7 (3904)

Bowman, J. C. [1962], Recurrent selection. I. The detection of overdominance. Heredity 14:197-206. T9a (3905)

Bowman, J. C. [1962], Recurrent selection. II. An experimental study with mice and Drosophila. Genetical Res. 3:333-351. T9a (3906)

Bowman, J. C. and Falconer, D. S. [1960], Inbreeding depression and heterosis of litter size in mice. Genetical Res. 1:262-274. T9a (3907)

Box, G. E. P. [1952], Multi-factor designs of first order. Biometrika 39:49-57. T4ace (3908)

Box, G. E. P. [1952], Statistical design in the study of analytical methods (with discussion). Analyst 77:879-891. T2ae;C (3909)

Box, G. E. P. [1954], The exploration and exploitation of response surfaces: Some general considerations and examples. Biometrics 10:16-60. T4ace (3910)

Box, G. E. P. [1955], Plan statistique dans l'étude des méthodes de l'analyse chimique. Rev. Stat. Appl. 3(1):43-56. T2ae;C (3911)

Box, G. E. P. [1957], Integration of techniques in progress development. ASQC, Nat. Conv. Trans. 11:687-702 (also Stat. Tech. Res. Group. Tech. Report No. 2, Princeton Univ.). T3,4a;C (3912)

Box, G. E. P. [1957], Evolutionary operation: A method for increasing industrial productivity. Appl. Stat. 6:81-101. T4a (3913)

Box, G. E. P. [1958], Use of statistical methods in the elucidation of basic mechanisms. Bull. ISI 36(3):215-225 (also Stat. Tech. Res. Group Tech. Report No. 7, Princeton Univ.). T4a (3914)

Box, G. E. P. [1958], Some recent work on non-linear estimation and design (abstract). Biometrics 14:566. T7a (3915)

Box, G. E. P. [1958], Derivation of second order rotatable designs from those of first order. Stat. Tech. Res. Group, Tech. Report No. 17, Princeton Univ. T4c (3916)

Box, G. E. P. [1959], Replication of non-center points in the rotatable and near-rotatable central composite design. Biometrics 15:133-135. T4c (3917)

Box, G. E. P. [1960], Some general considerations in process optimization. J. Basic Eng., Ser. D, 82:113-119 (also Stat. Tech. Res. Group Tech. Report No. 13, Princeton Univ.). T4a (3918)

Box, G. E. P. [1960], Fitting empirical data. Ann. New York Academy Sci. 86: 792-816 (also Math. Res. Center Tech. Report No. 151, U. S. Army and Univ. Wisconsin). T4a;A7 (3919)

Box, G. E. P. [1961], The effects of errors in the factor levels and experimental design. Bull. ISI 38(3):339-355. T2,4a;A7 (3920)

Box, G. E. P. [1963], The effects of errors in the factor levels and experimental design. Technometrics 5:247-262. T1,2,3a (3921)

Box, G. E. P. [1964], A note on partial duplication of designs. Dept. Stat. Tech. Report No. 24, Univ. Wisconsin. T7ac (3922)

Box, G. E. P. [1964], An introduction to response surface methodology. Dept. Stat. Tech. Report No. 33, Univ. Wisconsin. T4a (3923)

Box, G. E. P. [1966], A simple system of evolutionary operation subject to empirical feedback. Technometrics 8:19-26 (also Dept. Stat. Tech. Report No. 40, Univ. Wisconsin). T4a (3924)

Box, G. E. P. [1966], A note on augmented designs. Technometrics 8:184-188. T7a (3925)

Box, G. E. P., Altpeter, R. J., and Kotnour, K. D. [1965], A discrete predictor controller applied to sinusoidal perturbation adaptive optimization. Dept. Stat. Tech. Report No. 58, Univ. Wisconsin. T4a (3926)

Box, G. E. P. and Behnken, D. W. [1958], Derivation of second order rotatable designs from those of first order. Stat. Tech. Res. Group Tech. Report No. 17, Princeton Univ. T4ac (3927)

Box, G. E. P. and Behnken, D. W. [1958], A class of three level second order designs for surface fitting. Stat. Tech. Res. Group Tech. Report No. 26, Princeton Univ. T4ac (3928)

Box, G. E. P. and Behnken, D. W. [1960], Simplex-sum designs: A class of second order rotatable designs derivable from those of first order. AMS 31:838-864 (also Inst. Stat. Mimeo. Ser. No. 232, Univ. N. Carolina). T4,5c (3929)

Box, G. E. P. and Behnken, D. W. [1960], Some new three level designs for the study of quantitative variables. Technometrics 2:455-475 (errata, 3:131, 576; also Math. Res. Center Tech. Summary Report No. 169, U.S. Army and Univ. Wisconsin). T4ace (3930)

Box, G. E. P. and Chanmugam, J. [1962], Adaptive optimization of continuous processes. I and EC Fundamentals 1(2):1-16. T4a (3931)

Box, G. E. P. and Coutie, G. A. [1956], Application of digital computers in the exploration of functional relationships. Proc. Inst. Elec. Eng., 103:100-107. T4a;A15 (3932)

Box, G. E. P. and Cox, D. R. [1964], An analysis of transformations (with discussion). JRSSB 26:211-252 (also Dept. Stat. Tech. Report No. 29, Univ. Wisconsin). T2ac;A9 (3933)

Box, G. E. P. and Draper, N. R. [1959], A basis for the selection of a response surface design. JASA 54:622-654 (also Stat. Tech. Group Tech. Report No. 23, Princeton Univ.). T4c (3934)

Box, G. E. P. and Draper, N. R. [1963], The choice of a second order rotatable design. Biometrika 50:335-352 (corrigendum, 52:305; also Dept. Stat. Tech. Report No. 10, Univ. Wisconsin). T4ac (3935)

Box, G. E. P. and Draper, N. R. [1965], The Bayesian estimation of common parameters from several responses. Biometrika 52:355-365 (also Dept. Stat. Tech. Report No. 31, Univ. Wisconsin). T4a (3936)

Box, G. E. P. and Draper, N. R. [1967], Isn't my process too variable for EVOP? Dept. Stat. Tech. Report No. 106, Univ. Wisconsin. T4a (3937)

Box, G. E. P. and Gardner, C. J. [1964], Designs subject to constraints (abstract). Biometrics 20:907-908. T7a (3938)

Box, G. E. P. and Gardner, C. J. [1966], Constrained designs - Part I. First order designs. Dept. Stat. Tech. Report No. 89, Univ. Wisconsin. T4,5ac (3939)

Box, G. E. P. and Guttman, I. [1966], Some aspects of randomization. JRSSB 28: 543-558 (also Dept. Stat. Tech. Report No. 64, Univ. Wisconsin. E2a;T4a;C (3940)

Box, G. E. P. and Hay, W. A. [1953], A statistical design for the efficient removal of trends occurring in a comparative experiment with an application in biological assay. Biometrics 9:304-319 (abstract 9:260-261). T8ace (3941)

Box, G. E. P. and Hill, W. J. [1965], Discrimination among methanistic models. Dept. Stat. Tech. Report No. 51, Univ. Wisconsin. T4a (3942)

Box, G. E. P. and Hunter, J. S. [1954], A confidence region for the solution of a set of simultaneous equations with an application to experimental design. Biometrika 41:190-199. T4a (3943)

Box, G. E. P. and Hunter, J. S. [1954], The study and exploitation of response regions (summary). JASA 49:328. T4ac (3944)

Box, G. E. P. and Hunter, J. S. [1957], Multi-factor experimental designs for exploring response surfaces. AMS 28:195-241 (also Inst. Stat. Mimeo. Ser. No. 92, Univ. N. Carolina). T4,10ac (3945)

Box, G. E. P. and Hunter, J. S. [1958], Experimental designs for the exploration and exploitation of response surfaces. Experimental Designs in Industry, pp. 138-190. T4ace (3946)

Box, G. E. P. and Hunter, J. S. [1959], Condensed calculations for evolutionary operation programs. Technometrics 1:77-95 (also Stat. Tech. Res. Group Tech. Report No. 27, Princeton Univ.). T4a (3947)

Box, G. E. P. and Hunter, J. S. [1961], The 2^{k-p} fractional factorial designs. Part I. Technometrics 3:311-351 (errata, 5:417; also Math. Res. Center Tech. Summary Report No. 218, U.S. Army and Univ. Wisconsin. T3ace (3948)

Box, G. E. P. and Hunter, J. S. [1961], The 2^{k-p} fractional factorial designs. Part II. Technometrics 3:449-458. T3ac (3949)

Box, G. E. P., Hunter, J. S., and Hader, R. J. [1954], The effect of inadequate models in surface fitting. Inst. Stat. Mimeo. Ser. No. 91, Univ. N. Carolina. T4ac (3950)

Box, G. E. P. and Hunter, W. G. [1962], A useful method for model-building. Technometrics 4:301-318. T2ace;C (3951)

Box, G. E. P. and Hunter, W. G. [1964], Non-sequential designs for the estimation of parameters in non-linear models. Dept. Stat. Tech. Report No. 28, Univ. Wisconsin. T1,4ace (3952)

Box, G. E. P. and Hunter, W. G. [1965], Sequential design of experiments for nonlinear models (with discussion). Proc. IBM Sci. Computing Symp. Stat., pp. 113-137 (also Dept. Stat. Tech. Report No. 21, Univ. Wisconsin). T4ac;A7 (3953)

Box, G. E. P. and Hunter, W. G. [1965], The experimental study of physical mechanisms. Technometrics 7:23-42 (also Dept. Stat. Tech. Report No. 29, Univ. Wisconsin). T4,7ac (3954)

Box, G. E. P. and Jenkins, G. M. [1962], Some statistical aspects of adaptive optimization and control. Dept. Stat. Tech. Report No. 8, Univ. Wisconsin. T4a (3955)

Box, G. E. P. and Lucas, H. L. [1959], Design of experiments in non-linear situations. Biometrika 46:77-90 (also Stat. Tech. Res. Group Tech. Report No. 15, Princeton Univ.). T4ac;A7 (3956)

Box, G. E. P. and Wilson, K. B. [1951], On the experimental attainment of optimum conditions (with discussion). JRSSB 13:1-45. T4ace (3957)

Box, G. E. P. and Youle, P. V. [1955], The exploration and exploitation of response surfaces: An example of the link between the fitted surface and the basic mechanism of the system. Biometrics 11:287-323. T4ace (3958)

Boyd, D. A., Hill, J. R., and Batey, T. [1968], The effect of yield of maincrop potatoes of different methods of fertilizer application. Experimental Husbandry 16:13-20. T2a (3959)

Braaten, M. O. [1965], The union of partial diallel mating designs and incomplete block environmental designs. Ph. D. Thesis, Univ. N. Carolina. T9ac (3960)

Braaten, M. O. and Cockerham, C. C. [1965], The union of partial diallel mating designs and incomplete block environmental designs (abstract). Biometrics 21:1022. T9c (3961)

Bradford, G. E., Chapman, A. B., and Grummer, R. H. [1958], Effects of inbreeding, selection, linecrossing and topcrossing in swine. III. Predicting combining ability and general conclusions. J. Animal Sci. 17:456-467. T9a (3962)

Bradley, R. A. [1952], Statistical methods for sensory difference tests of food quality. Va. Agri. Expt. Sta., Blacksburg, Va., Bi-Annual Report No. 5 (Mimeo.). E2,3a;T11a;C (3963)

Bradley, R. A. [1954], Incomplete block rank analysis: On the appropriateness of the model for a method of paired comparisons. Biometrics 10:375-390. E2ae;T11ae;A12 (3964)

Bradley, R. A. [1954], Rank analysis of incomplete block designs. II. Additional tables for the method of paired comparisons. Biometrika 41:502-537 (corrigendum, 51:288). E2a;T11a;A12 (3965)

Bradley, R. A. [1955], Rank analysis of incomplete block designs III. Some large-sample results on estimation and power for a method of paired comparisons. Biometrika 42:450-470. E2a;T11a (3966)

Bradley, R. A. [1955], Some large sample results on estimation and power for a method of paired comparisons. Biometrika 42:450-470. T11ae (3967)

Bradley, R. A. [1956], Mathematics and statistics fundamental to fitting of response surfaces. Proc. Rutgers Conf. American Soc. Quality Contributions, pp.33-42. T4a (3968)

Bradley, R. A. [1958], Recent research in statistical problems in subjective testing. Proc. Second Conf. Design Expt. Army Res. Dev. Testing. pp. 5-38. E3ae;T2ae;C (3969)

Bradley, R. A. [1958], Triangle, duo-trio, and difference-from-control tests in taste testing (abstract). Biometrics 14:566. T11a;C (3970)

Bradley, R. A. [1958], Triangle, duo-trio, and difference-from-control tests in taste testing (abstract). Virginia J. Sci. 9:448-449. T11a;C (3971)

Bradley, R. A. [1958], Determination of optimum operating conditions by experimental methods, Part I. Mathematics and statistics fundamental to the fitting of response surfaces. IQC 15(1):16-20. T2,3,4a (3972)

Bradley, R. A. [1961], Recherches récentes sur les problèmes statistiques dans les tests subjectifs. Biométrie-Praximétrie 2:57-68. T11ae (3973)

Bradley, R. A. [1963], Applications of the modified triangle test. Dept. Stat. Tech. Report No. 9, Florida State Univ. T11a (3974)

Bradley, R. A. [1963], Some relationships among sensory difference tests. Biometrics 19:385-397. E14;T11a;C (3975)

Bradley, R. A. [1964], Applications of the modified triangle test in sensory difference trials. J. Food Sci. 29:668-672. T11ae;C (3976)

Bradley, R. A. [1965], Another interpretation of a model for paired comparisons. Psychometrika 30:315-318 (also Dept. Stat. Report M 50, Florida State Univ.) T11a (3977)

Bradley, R. A. and Harmon, T. J. [1963], An analysis for modified triangle tests in sensory difference experimentation (abstract). AMS 34:679. T11a;C (3978)

Bradley, R. A. and Harmon, T. J. [1964], The modified triangle test. Biometrics 20:608-625 (also Dept. Stat. Tech. Report No. 8, Florida State Univ.). T11a;C (3979)

Bradley, R. A., Katti, S. K., and Coons, I. J. [1962], Optimal scaling for ordered categories (abstract). Biometrics 18:258. T2ae;A9,12 (3980)

Bradley, R. A., Katti, S. K., and Coons, I. J. [1962], Optimal scaling for ordered categories. Psychometrika 27:355-374 (also Dept. Stat. Tech. Report No. 2, Florida State Univ.). T2ae;A9,12 (3981)

Bradley, R. A. and Kramer, C. Y. [1958], Addenda to "Intra block analysis for factorials in two-associate class group divisible designs". AMS 29:933-935. E3a;T2a (3982)

Bradley, R. A. and Pendergrass, R. N. [1960], Ranking in triple comparisons. Bull. ISI 37(3):229-241. T11ae;A12 (3983)

Bradley, R. A. and Somerville, P. N. [1953], Statistical methods for sensory difference tests of food quality. Va. Agri. Expt. Sta. Blacksburg, Bi-Annual Report No. 7 (Mimeo). E2a;T11a;A2 (3984)

Bradley, R. A. and Somerville, P. N. [1954], Statistical methods for sensory difference tests of food quality. Va. Agri. Expt. Sta. Blacksburg, Bi-Annual Report No. 8 (Mimeo.). E2a;T11a;A3 (3985)

Bradley, R. A. and Terry, M. E. [1951], Statistical methods for sensory difference tests of food quality. Va. Agri. Expt. Sta. Blacksburg, Bi-Annual Report No. 2 (Mimeo.). E2,3a;T11a;C (3986)

Bradley, R. A. and Terry, M. E. [1951], Statistical methods for sensory difference tests of food quality. Va. Agri. Expt. Sta. Blacksburg, Bi-Annual Report No. 3 (Mimeo.). E2,3a;T11a;C (3987)

Bradley, R. A. and Terry, M. E. [1952], Rank analysis of incomplete block designs. I - The method of paired comparisons (abstract). Biometrics 8:95. E2a;T11a;A12 (3988)

Bradley, R. A. and Terry, M. E. [1952], Rank analysis of incomplete block designs. I - The method of paired comparisons (abstract). AMS 23:299-300. E2a;T11a;A12 (3989)

Bradley, R. A. and Terry, M. E. [1952], Rank analysis of incomplete block designs. I - The method of paired comparisons. Biometrika 39:324-345. E2ae;T11ae;A12 (3990)

Bradley, R. A. and Terry, M. E. [1952], Statistical methods for sensory difference tests of food quality. Va. Agri. Expt. Sta. Blacksburg, Bi-Annual Report No. 4 (Mimeo.). E2,3a;T11a;C (3991)

Bradley, R. A. and Terry, M. E. [1953], Statistical methods for sensory difference tests of food quality. Va. Agri. Expt. Sta. Blacksburg, Bi-Annual Report No. 6 (Mimeo.). E2,3a;T11a;C (3992)

Bradley, R. A., Walpole, R. E., and Kramer, C. Y. [1960], Intra- and inter-block analysis for factorials in incomplete block designs. Biometrics 16:566-581. E3a;T2a (3993)

Brandt, A. E. [1961], The analysis and interpretation of half-replicate experiments (abstract). Biometrics 17:500. T3a (3994)

Breese, E. L. [1956], The genetical consequences of assortative mating. Heredity 10:323-343. T9a (3995)

Breneman, W. R. [1951], A factorial analysis of pituitary-gonad-comb relationships in the chick. I. Action of pregnant-mare-serum and testosterone propionate on androgen secretion. Poultry Sci. 30:231-239. T2a (3996)

Brenna, L. S. [1958], Factorial treatments in lattice designs. Ph. D. Thesis, Virginia Polytechnic Inst. E3a;T2a (3997)

Brenna, L. S. and Kramer, C. Y. [1959], Factorial treatment combinations in lattices (abstract). Biometrics 15:633-634. E3a;T2a (3998)

Brenna, L. S. and Kramer, C. Y. [1961], Factorial treatments in rectangular lattice designs. JASA 56:368-378. E3ac;T2ac (3999)

Brigham, R. D. and Wilsie, C. P. [1955], Seed setting and vegetative vigor of ladino clover (Trifolium repens Leyss) clones and their diallel crosses. Agron. J. 47:125-127. T9e (4000)

Britten, E. J. [1960], A proposed classification of screening methods for plant breeding programs. Euphytica 9:293-303. T9a;A2 (4001)

Brock, N. and Schneider, B. [1965], A screening system for anti-cancer agents based on the therapeutic index. Biometrics 21:150-158. T8a;A2 (4002)

Brock, N. and Schneider, B. [1966], Quantitative Methoden bei der pharmako-therapeutischen Bewertung von Arzneimitteln - Auswertung, Planung, Screening. Biom. Zeit. 8:147-161. T8a;A2 (4003)

Broemeling, L. D. [1967], Confidence statements for unknown response surface ordinates (abstract). Technometrics 9:186. T4a (4004)

Bromfield, K. R. and Jones, M. W. [1962], Problems related to a bio-assay for spore-germination inhibitors associated with uredospores. Proc. Seventh Conf. Design Expt. Army Res. Dev. Testing. pp. 767-779. T8a (4005)

Brooks, S. H. [1955], Comparison of methods for estimating the optimal factor combination. Ph. D. Thesis, Johns Hopkins Univ. T2,4,10a (4006)

Brooks, S. H. [1959], A comparison of maximum-seeking methods. Operations Res. 7:430-457. T2,4,10a (4007)

Brooks, S. H. and Mickey, M. R. [1961], Optimum estimation of gradient direction in steepest ascent experiments. Biometrics 17:48-56. T4,10a (4008)

Bross, I. [1950], Estimates of the LD_{50}: A critique. Biometrics 6:413-423. T8a (4009)

Bross, I. D. J. [1962], An ametric approach to bioassay. Clinical Pharmacology and Therapeutics 3:369-373. T8a;C (4010)

Brown, B. W. [1966], Planning a quantal assay of potency. Biometrics 22:322-329 (abstract 21:762). T8ac (4011)

Brown, W. G. [1956], Practical applications of fertilizer production functions. Methodological Procedures in the Economic Analysis of Fertilizer Use Data, pp. 151-157. T2,4a (4012)

Brown, W. L. [1967], Results of non-selective inbreeding in maize. Der Züchter 37:155-159. T9a (4013)

Brownlee, K. A. [1950], A plant-scale planned experiment in penicillin production. Ann. New York Academy Sci. 52:820-826. T2,3ace (4014)

Brownlee, K. A. [1953], Experiments with many factors. Chemical Eng. Progr. 49:617-621. T2,3ae (4015)

Brownlee, K. A. [1957], The principles of experimental design. IQC 13(8):12-20. E1,2,6,7,12ae; T2ae;C (4016)

Brunk, H. D. [1959], Mathematical models for ranking from paired comparisons (abstract). AMS 30:1272-1273. T11a;A12 (4017)

Brunk, H. D. [1960], Mathematical models for ranking from paired comparisons. JASA 55:503-520. T11a (4018)

Brunk, M. E. and Federer, W. T. [1953], How marketing problems of the apple industry were attacked and the research results applied. Biometrics Unit Mimeo. Ser. BU-40-M, Cornell Univ. E7,8a;T3,7c (4019)

Brunk, M. E. and Federer, W. T. [1953], Experimental designs and probability sampling in marketing research. JASA 48:440-452. E7a;T7a;A13 (4020)

Budne, T. A. [1959], The application of random balance designs. Technometrics 1:139-155 (erratum 1:419). T7ae (4021)

Budne, T. A. [1959], The application of random balance designs (abstract). Biometrics 15:634. T7a (4022)

Budne, T. A. [1959], Random balance, part I - The missing statistical link in fact finding techniques. IQC 15(10):5-10. T7ac (4023)

Budne, T. A. [1959], Random balance, part II - Techniques of analysis. IQC 15(11):11-16. T7ae (4024)

Budne, T. A. [1959], Random balance, part III - Case histories. IQC 15(12):16-19. T7a (4025)

Buehler, R. J. [1960], Use of prior knowledge in finding the maximum response (abstract). AMS 31:232. T4c (4026)

Buehler, R. J., Shah, B. V., and Kempthorne, O. [1961], Steepest ascent PARTAN on ellipsoidal contours (abstract). AMS 32:623 (also Stat. Lab. Tech. Report No. 3, Iowa State Univ.). T4,7c (4027)

Buehler, R. J., Shah, B. V., and Kempthorne, O. [1961], Some properties of steepest ascent and related procedures for finding optimum conditions. Stat. Lab. Tech. Report No. 1, Iowa State Univ. T4ac (4028)

Buehler, R. J., Shah, B. V., and Kempthorne, O. [1961], Some further properties of the methods of parallel tangents and conjugate gradients. Stat. Lab. Tech. Report No. 3, Iowa State Univ. T4ac (4029)

Buehler, R. J., Shah, B. V., and Kempthorne, O. [1961], Iterated steepest ascent on ellipsoidal contours (abstract). AMS 32:622-623. T4,7c (4030)

Buehler, R. J., Shah, B. V., and Kempthorne, O. [1964], Methods of parallel tangents (with discussion). Chemical Eng. Progress Symp. Ser. 60:50, 1-7. T4,7ac (4031)

Bulfinch, A. [1960], Design of environmental experiments and reliability prediction. Proc. Fifth Conf. Design Expt. Army Res. Dev. Testing. pp. 171-189. T2ae;C (4032)

Burros, R. H. [1951], The application of the method of paired comparisons to the study of reaction potential. Psychological Rev. 58:60-66. T11a (4033)

Burt, C. [1953], The relative merits of ranks and paired comparisons. British J. Stat. Psychology 6:112-119. T11a (4034)

Burton, G. W. [1960], Principles of breeding vegetatively propagated plants. Proc. Tenth Congress Int'l. Soc. Sugar Cane Tech., pp. 661-670. T9a (4035)

Burton, R. C. [1956], On the construction of fractional factorial designs (preliminary report) (abstract). AMS 27:874. T3,7c (4036)

Burton, R. C. [1964], An application of convex sets to the construction of error correcting codes and factorial designs. Ph. D. Thesis, Univ. N. Carolina (also Inst. Stat. Mimeo. Ser. No. 393, Univ. N. Carolina). T3,7c (4037)

Burton, R. C. and Connor, W. S. [1957], On the identity relationship for fractional replicates of the 2^n series. AMS 28:762-767. T3c (4038)

Caliński, T. [1958], O zastosowaniu kontrastów ortogonalnych do badania interakcji. (On the application of orthogonal contrasts for investigating interaction.) Roczniki Nauk Rolniczych 79(A):635-675. T2a (4039)

Calvin, T. W. [1965], Individual comparisons in ANOVA. ASQC, Ann. Tech. Conf. Trans. pp. 559-569. T2ae (4040)

Cameron, J. M. [1965], Calibration designs (abstract). Technometrics 7:269. E4c;T6c (4041)

Carlborg, F. W. [1965], A procedure for selecting independent variables in multiple regression. Ph. D. Thesis, Univ. Chicago. T1c (4042)

Carlborg, F. W. [1965], A procedure for selecting independent variables in multiple regression (abstract). Technometrics 7:269. T1ac (4043)

Carmer, S. G., Seif, R. D., and Jacob, W. C. [1963], The use of computers for the analysis of data from variety trials. Agron. J. 55:585-587. T2a;A13,15 (4044)

Carmon, J. L., Stewart, H. A., Cockerham, C. C., and Comstock, R. E. [1956], Prediction equations for rotational crossbreeding. J. Animal Sci. 15:930-936. T9a (4045)

Carpenter, B. H. and Sweeny, H. C. [1965], Process improvement with "simplex" self-directing evolutionary operation. Chemical Eng. 72 (July 5):117-124, 126. T4a (4046)

Carpenter, R. G. [1963], Simplex evolutionary operation in relation to controlled clinical trials (abstract). Biometrics 19:503. T4a (4047)

Carr, J. M. and McCracken, E. A. [1960], Statistical program planning for process development. Chemical Eng. Prog. 56(11):56-61. T2,4ac (4048)

Carroll, M. B. and Dykstra, O. [1958], The application of fractional factorials in a food research laboratory. Experimental Designs in Industry, pp. 224-234. T3ae;C (4049)

Carter, F. L. [1960], Group testing in binomial and multinomial situations. Dept. Stat. Tech. Report No. 3, Virginia Polytechnic Inst. (Mimeo.). T7a (4050)

Carter, R. L. [1957], New designs for the exploration of response surfaces. Ph. D. Thesis, Univ. N. Carolina (also Inst. Stat. Mimeo. Ser. No. 172, Univ. N. Carolina). T4ac (4051)

Cassady, J. C. [1962], Individual degrees of freedom for non-additivity in two- and higher-way designs. M.S. Thesis, Cornell Univ. (also Biometrics Unit Mimeo. Ser. BU-136-M, Cornell Univ.). E2a;T2a;A4 (4052)

Caton, D. D. [1960], A method for selecting profitable levels of fertilization of range grass. Proc. Eighth Int'l. Grassland Congress, pp. 673-675. T4a;A7 (4053)

Catron, D. V., Bennison, R. W., Maddock, H. M., Ashton, G. C., and Homeyer, P. G. [1953], Effects of certain antibiotics and vitamin B_{12} on pantothenic acid requirements of growing-fattening swine. J. Animal Sci. 12:51-61. T2a;C (4054)

Chakrabarti, M. C. [1964], Query. JISA 2:55-59. T2ae (4055)

Chakravarti, I. M. [1956], Fractional replication in asymmetrical factorial designs and partially balanced arrays. Sankhyā 17:143-164. T3c (4056)

Chakravarti, I. M. [1960], Partially balanced arrays (abstract). AMS 31:531-532. T2,7c (4057)

Chakravarti, I. M. [1961], On some methods of construction of partially balanced arrays. AMS 32:1181-1185 (also Inst. Stat. Mimeo. Ser. No. 260, Univ. N. Carolina). T7c (4058)

Chakravarti, I. M. [1963], Orthogonal and partially balanced arrays and their application in design of experiments. Metrika 7:231-243. E7c;T3,7c (4059)

Chakravarty, M. C. [1964], Design of experiments. Res. Training School Pub. No. SM 64-3, Indian Stat. Inst. E;T (4060)

Chalbi, N. [1967], Biométrie et analyse quantitative de la compétition entre génotypes chez la Luzerne. Annales de l'Amélioration des Plantes 17(2): 119-158. T9a;A14 (4061)

Chambers, E. A. and Cox, D. R. [1967], Discrimination between alternative binary response models. Biometrika 54:573-578. T8a (4062)

Chand, U. [1953], Experiments on tree crops (abstract). Biometrics 9:429-430. E2a;T2a;C (4063)

Chand, U. and Abraham, T. P. [1957], Some considerations in the planning and analysis of fertilizer experiments in cultivators' fields. JISAS 9:101-134. E2a;T2a;A13;C (4064)

Chang, C. D., Kononenko, O. K., and Franklin, R. E. [1960], Maximum data through a statistical design. Ind. Eng. Chemistry 52:939-942. T4ae (4065)

Chanmugam. J. and Jenkins, G. M. [1963], Optimum experimenation in the process industries. Chemical Eng. Progress Symp. Ser. 59(42):108-117. T4a;C (4066)

Chapman, J. W. and Gates. C. E. [1963], Some aspects of hypothesis testing in response surface experiments (summary). JASA 58:548. T4a (4067)

Charbonnell, M. and van den Driessche, R. [1964], Analyse de la covariance d'expériences 3^3 en 27 unites sur ordinateur IBM 704. Biométrie-Praximétrie 5:20-25. T2a;A7,15 (4068)

Chase, G. R. [1966], An empirical Bayes approach in routine bioassay. Dept. Stat. Tech. Report No. 13, Stanford Univ. T8a (4069)

Chernoff, H. [1952], Optimal designs for estimating parameters (abstract). AMS 23:646. Tlc (4070)

Chernoff, H. [1953], Locally optimal designs for estimating parameters. AMS 24:586-602. Tlac (4071)

Chernoff, H. [1962], Optimal accelerated life designs for estimation. Technometrics 4:381-408. Tlace (4072)

Chernoff, H. [1963], Optimal design of experiments. Proc. Eighth Conf. Design Expt. Army Res. Dev. Testing. pp. 303-315. Tl,8ac (4073)

Chew, V. [1958], Basic experimental designs. Experimental Designs in Industry, pp. 3-57. El,2,6,7,12ae; T2a;B (4074)

Chew, V. [1962], Regression techniques in the analysis of variance. IQC 18 (12): 13-14. T2a (4075)

Chinloy, T., Innes, R. F., and Finney, D. J. [1953], An example of fractional replication in an experiment on sugar cane manuring. J. Agri. Sci. 43:1-11. T3ae (4076)

Chopra, D. V. [1966], On the non-existence of some partially balanced arrays with 2 symbols (preliminary report) (abstract). AMS 37:1858. T7c (4077)

Chopra, D. V. [1967], On the minimum number of assemblies required to contstruct some partially balanced arrays with 2 symbols (abstract). AMS 38:961. T4c (4078)

Chopra, D. V. [1967], Necesarry equations for the existence of some partially balanced arrays with 2 symbols (abstract). AMS 38:968. T4c (4079)

Chopra, D. V. [1967], On the trace comparison of some partially balanced arrays with 2 symbols (abstract). AMS 38:969. T4c (4080)

Chun, D. [1966], Interlaboratory tests -- short cuts. ASQC, Ann. Tech. Conf. Trans. pp. 147-149. T1la;A12 (4081)

Claringbold, P. J. [1955], A study of the individual median effective dose of oestrone in the ovariectomized mouse. J. Endocrinology 13:11-17. T8ae (4082)

Claringbold, P. J. [1955], Use of the simplex design in the study of joint action of related hormones. Biometrics 11:174-185. T5ace (4083)

Claringbold, P. J. [1956], A note on the 4^n series of factorial experiments. Biometrics 12:259-263. T2a (4084)

Claringbold, P. J. [1956], The within-animal bioassay with quantal responses. JRSSB 18:133-137. T8ace (4085)

Claringbold, P. J. [1958], Multivariate quantal analysis. JRSSB 20:398-405. T8a;A7 (4086)

Claringbold, P. J. [1959], Orthogonal contrasts in slope ratio investigations. Biometrics 15:307-322. T8ae;A7 (4087)

Claringbold, P. J., Biggers, J. D., and Emmens, C. W. [1953], The angular transformation in quantal analysis. Biometrics 9:467-482. T8a;A9 (4088)

Claringbold, P. J. and Emmens, C. W. [1961], Quantal responses. Quantitative Methods in Pharmacology, pp. 72-87. T8ae (4089)

Clark, V. A. [1963], Choice of levels for the several factor design problem. Ph. D. Thesis, Univ. California, Los Angeles. T1,4ac (4090)

Clark, V. [1964], Choice of levels in polynomial regression with one or two variables (abstract). Biometrics 20:657. T1ac (4091)

Clark, V. [1965], Choice of levels of polynomial regression with one or two variables. Technometrics 7:325-333. T1ac (4092)

Clarke, J. M., Smith, J. M., and Sondhi, K. C. [1961], Asymmetrical response to selection for rate of development in Drosophila subobscura. Genetical Res. 2: 70-81. T9a (4093)

Clarke, P. M. [1955], A note on the estimation of error in slope-ratio assays arranged in ramdomised blocks. Analyst 80:396-397. E2a;T8a (4094)

Clarke, P. M. [1956], A note on the combination of estimates of relative potency in multiple assays. Biometrics 12:40-46. T8ae (4095)

Clayton, G. A. and Robertson, A. [1957], An experimental check on quantitative genetical theory. II. The long term effects of selection. J. Genetics 55: 152-170. T9a (4096)

Clemens, W. R. [1963], Factorial and response surface techniques applied to semiconductor processes. ASQC, Ann. Conv. Trans. pp. 333-338. T2,4ae (4097)

Cochran, W. G. [1960], The design of experiments. Operations Research and Systems Engineering, pp. 508-553. E2,3,7a;T2a;C (4098)

Cochran, W. G. and Davis, M. [1963], Sequential experiments for estimating median lethal dose (discussion). Colleques Internationaux du Centre National de la Recherche Scientifique, No. 110, Le Plan d'Expériences, pp. 181-194. T8a;S2 (4099)

Cochran, W. G. and Davis, M. [1964], Stochastic approximation to the median effective dose in bioassay (with discussion). Stochastic Models in Medicine and Biology, pp. 281-300. T8a;S2 (4100)

Cochran, W. G. and Davis, M. [1965], The Robbins-Monro method for estimating the median lethal dose. JRSSB 27:28-44. T8ac;S2 (4101)

Cockerham, C. C. [1954], An extension of the concept of partitioning hereditary variance for analysis of covariances among relatives when epistasis is present. Genetics 39:859-882. T9a (4102)

Cockerham, C. C. [1956], Query 122. Biometrics 12:338-339. T2a;A10 (4103)

Cockerham, C. C. [1961], Implications of genetic variances in a hybrid breeding program. Crop. Sci. 1:47-52. T9ac (4104)

Cockerham, C. C. [1964], Diallelic competition (abstract). Biometrics 20:381. T9a (4105)

Cockerham, C. C. [1967], Prediction of double crosses from single crosses. Der Züchter 37:160-169. T9a (4106)

Cockerham, C. C. and Matzinger, D. F. [1966], Simultaneous selfing and partial diallel test crossing III. Optimum selection procedures. Australian J. Biological Sci. 19:795-805. T9ac (4107)

Cole, G. A. [1958], Air pollution with relation to agronomic crops: III. Vegetation survey methods in air pollution studies. Agron. J. 50:553-555. T8ac (4108)

Collier, R. O. [1963], A note on the multiple regression technique for deleting variables in the discriminant function. J. Expt. Education 31:351-353. T1a;A7 (4109)

Conagin, A. [1956], Aplicações dos experimentos fatoriais nas pesquisas de laboratório. (Applications of factorial experiments in laboratory investigations.) Revista Agri. São Paulo 31:235-248. T2,3a (4110)

Connor, W. S. [1954], New experimental designs for paired observations (summary). JASA 49:360. E2a;T11a (4111)

Connor, W. S. [1955], Some statistical problems encountered in industrial research. ASQC, Ann. Conv, Trans. 9:727-730. T2a (4112)

Connor, W. S. [1958], A fractional factorial experiment of the mixed 2^m3^n series. Trans. Fourteenth Annual Clinic Rochester Soc. Quality Control, pp. 59-70. T2,3ac (4113)

Connor, W. S. [1958], Use of the direct product of matrices in the analysis of factorial experiments (abstract). Biometrics 14:567. T2,7a;A8 (4114)

Connor, W. S. [1959], Some recent work on mixed fractional factorial designs (abstract). Biometrics 15:332-333. T7c (4115)

Connor, W. S. [1960], Factorial experiments in incomplete block designs. Ind. Eng. Chemistry 52(12):83A-84A. E4ace;T2ace (4116)

Connor, W. S. [1960], Fractional factorial experiment designs of mixed 2^m3^n series. Ind. Eng. Chemistry 52(6):93A-95A. T3ace (4117)

Connor, W. S. [1960], Fractional factorial experiment designs of mixed 2^m3^n series - Part II. Ind. Eng. Chemistry 52(8):79A-80A. T3ac (4118)

Connor, W. S. [1960], Construction of fractional factorial designs of the mixed 2^m3^n series. <u>Contributions to Probability and Statistics,</u> pp. 168-181. T2ac (4119)

Connor, W. S. [1961], Evaluation by overstress. Ind. Eng. Chemistry 53(6):73A-74A. T1a (4120)

Connor, W. S. [1961], Group screening designs. Ind. Eng. Chemistry 53(2):69A-70A. T7a (4121)

Connor, W. S. [1961], Group screening designs. Ind. Eng. Chemistry 53(4):75A-76A. T7ac (4122)

Connor, W. S. [1961], Group screening designs. Proc. Sixth Conf. Design Expt. Army Res. Dev. Testing. pp. 293-305. T7ac (4123)

Connor, W. S. [1965], On the use of fractional factorials to construct smallest designs for quadratic response surfaces. Paper presented at the meeting of the Inst. Math. Stat., Florida State Univ. T4,5ac (4124)

Connor, W. S., Clatworthy, W. H., and Zelen, M. [1957], Fractional factorial experimen designs for factors at two levels. Nat. Bur. Standards, Appl. Math. Ser. 48. T3c (4125)

Connor, W. S. and Young, S. [1961], Fractional factorial designs for experiments with factors at two and three levels. Nat. Bur. Standards Appl. Math. Ser. 58. T3ac (4126)

Connor, W. S. and Zelen, M. [1959], Fractional factorial experiment designs for factors at three levels. Nat. Bur. Standards Appl. Math. Ser. 54. T3c (4127)

Corkill, L. [1950], A comparison of methods of progeny testing for quantitative characters in ryegrass (Lolium sp.). New Zealand J. Sci. Tech. A 32(3):35-44. T9a (4128)

Corlett, E. N. and Gregory, G. [1960], The consistency of setting of a machine tool handwheel. Appl. Stat. 9:92-102. T3ae (4129)

Cornfield, J. [1954], Measurement and comparison of toxicities: The quantal response. Statistics and Mathematics in Biology, pp. 327-344. T8a (4130)

Cornfield, J. [1964], Comparative bioassays and the role of parallelism. J. Pharmacology Expt. Therapeutics 144:143-149. T8a;C (4131)

Cornfield, J. [1967], Letter to the Editor - The meaning of bioassay: A comment. Biometrics 23:160-162. T8a;C (4132)

Cornfield, J., Gordon, T., and Smith, W. W. [1961], Quantal response curves for experimentally uncontrolled variables. Bull. ISI 38(3):97-115. T8ae (4133)

Cornfield, J. and Tukey, J. W. [1956], Average values of mean squares in factorials. AMS 27:907-949. T2a (4134)

Coulter, B. S. [1964], A computer programme for analysis of variance in experiments with factorial treatments and one or more error terms. Appl. Stat. 13:110-117. T2a;A15 (4135)

Coutie, G. A. [1959], Use statistics for optimization. Chemical Eng. 66(November 16):190-191. T4a (4136)

Cox, C. P. [1961], A practical application of a theoretically inefficient design. Biometrics 17:646-649. E7ac;T2ac (4137)

Cox, C. P. [1967], Statistical analysis of log-dose response bioassay experiments with unequal dose ratios for the standard and unknown preparations. J. Pharmaceutical Sci. 56:359-364. T8a (4138)

Cox, C. P. and Leaverton, P. E. [1966], Statistical procedures for bioassays when the condition of similarity does not obtain. J. Pharmaceutical Sci. 55:716-723. T8a (4139)

Cox, C. P. and Ruhl, D. J. [1966], Simplified computation of confidence intervals for relative potencies using Fieller's theorem. J. Pharmaceutical Sci. 55:368-371. T8a (4140)

Cox, D. R. [1954], The design of an experiment in which certain treatment arrangements are inadmissible. Biometrika 41:287-295. T7ace (4141)

Cox, D. R. [1955], The design of an experiment in which some treatment arrangements are inadmissible (abstract). Biometrics 11:248-249. T7c (4142)

Cox, D. R., Herzberg, A., Daniel, C., and Finney, D. J. [1967], Query 20. Analysis of a factorial experiment (partially confounded 2^3). Technometrics 9:167-179. T2a (4143)

Cox, G. M. [1955], Some recent advances in experimental designs, with particular reference to estimating response surfaces (abstract). Biometrics 11:541. T4a;C (4144)

Cox, P. C. [1957], Some design techniques used for increasing cell size with special emphasis in the missile field. Proc. First Conf. Design Expt. Army Res. Dev. Testing. pp. 53-68. T7ace (4145)

Cox, P. C. [1957], The problem of grouped firing. Proc. First Conf. Design Expt. Army Res. Dev. Testing. pp. 213-216. T2ace (4146)

Cox, P. C. [1960], The application of fractional factorials in missile test programs. Proc. Fifth Conf. Design Expt. Army Res. Dev. Testing. pp. 285-289. T2e (4147)

Cox, P. C. [1963], Statistical design of experiments for continuous data (with comments). Proc. Eighth Conf. Design Expt. Army Res. Dev. Testing. pp. 53-84. T2a (4148)

Cragle, R. G., Myers, R. M., Waugh, R. K., Hunter, J. S., and Anderson, R. L. [1955], The effects of various levels of sodium citrate, glycerol, and equilibration time on survival of bovine spermatozoa after storage at -79°C. J. Dairy Sci. 38:508-514. T4ae (4149)

Cramer, E. M. [1964], Some comparisons of methods of fitting the dosage response curve for small samples. JASA 59:779-793. T8a (4150)

Crane, E., Smith, C., and Bulfinch, A. [1960], A statistical evaluation of the pyrotechnics electrostatic sensitivity tester. Proc. Fifth Conf. Design Expt. Army Res. Dev. Testing. pp. 239-262. T2e (4151)

Creasy, M. A. [1957], Analysis of variance as an alternative to factor analysis. JRSSB 19:318-325. T2a;A7 (4152)

Cress, C. E. [1966], A comparison of recurrent selection systems. Genetics 54: 1371-1379. T9a (4153)

Cress, C. E. [1967], Reciprocal recurrent selection and modifications in simulated populations. Crop Sci. 7:561-567. T9a (4154)

Crockett, J. B. and Chernoff, H. [1955], Gradient methods of maximization. Pacific J. Math. 5:33-50. T4c (4155)

Curnow. R. N. [1963], Sampling the diallel cross. Biometrics 19:287-306 (abstract 20:381). T9ace (4156)

Curnow, R. N. [1964], The effect of continued selection of phenotypic intermediates on gene frequency. Genetical Res. 5:341-353. T9a (4157)

Daday, H. [1965], General and specific combining ability for forage yield in lucerne (Medicago sativa L.). Australian J. Agri. Res. 16:293-299. T9a (4158)

Dalebroux, M. [1961], Exemple d'analyse d'un plan factoriel $3^2 \times (2)$ en parcelles découpées (split-plot). Biométrie-Praximétrie 2:121-129. E12ae;T2ae (4159)

Daniel, C. [1951], Design of experiments for most precise slope estimation. Ind. Eng. Chemistry 43:1298-1300. T1ac (4160)

Daniel, C. [1956], Fractional replication in industrial research. Proc. Third Berkeley Symp. Math. Stat. Prob. 5:87-98. T2ac (4161)

Daniel, C. [1956], Application of statistical methods in chemical engineering. Ind. Eng. Chemistry 48:1392-1402. T2,4ac;A7;C (4162)

Daniel, C. [1957], Fractional replication in industrial experimentation. ASQC, Ann. Conv. Trans. 11:229-233. T3ac (4163)

Daniel, C. [1958], Factorial experiment (summary). JASA 53:574. T3a (4164)

Daniel, C. [1958], Industrial experience with fractional replicates (abstract). AMS 29:1286. T3a;A6;C (4165)

Daniel, C. [1958], On varying one factor at a time. Biometrics 14:430-431. T8c (4166)

Daniel, C. [1959], Use of half-normal plots in interpreting factorial two-level experiments. Technometrics 1:311-341. T2,3ae (4167)

Daniel, C. [1960], Locating outliers in factorial experiments. Technometrics 2:149-156. T2a;A6 (4168)

Daniel, C. [1960], Locating outliers in factorial experiments (summary). JASA 55:359. T2a;A6 (4169)

Daniel, C. [1960], Parallel fractional replicates. Technometrics 2:263-268. T7ac (4170)

Daniel, C. [1961], Critique et analyse graphique des résultats de plans d'expériences factoriels 2^p. Rev. Stat. Appl. 9(3):11-26. T2,3a (4171)

Daniel, C. [1962], Sequences of fractional replicates in the 2^{p-q} series. JASA 57:403-429 (corrigenda 57:919). T3c (4172)

Daniel, C. [1963], Factor screening in process development. Ind. Eng. Chemistry 55(5):45-48. T7a (4173)

Daniel, C. [1964], Minimal augmentation of orthogonal main-effect plans to remove bias from two-factor interactions (abstract). AMS 35:1846. T7c (4174)

Daniel, C. [1966], Minimal two-level main-effect-clear plans (preliminary report) (abstract). AMS 37:313-314. T7c (4175)

Daniel, C. [1966], A second look at half-normal plots (abstract). Technometrics 8:204. T2a;A4 (4176)

Daniel, C. and Heerema, N. [1950], Design of experiments for most precise slope estimation or linear extrapolation. JASA 45:546-556. T1c (4177)

Daniel, C. and Riblett, E. W. [1954], A multifactor experiment. Ind. Eng. Chemistry 46:1465-1471. T3ace (4178)

Daniel, C. and Wilcoxon, F. [1966], Factorial 2^{p-q} plans robust against linear and quadratic trends. Technometrics 8:259-278. E5ace;T3ace (4179)

Darroch, J. G., Nordskog, A. W., and Van Horn, J. L. [1950], The effect of feeding concentrates to range ewes on lamb and wool productivity. J. Animal Sci. 9:431-44 T2e (4180)

Das, B. C. and Krishnamurthy, H. [1961], Survival rates of Indian carp (Catla catla, Labeo rohita, Cirrhina mrigala) from first to fourth week of life under experimental treatments isolating vitamin B_{12} from vitamin B complex. Sankhyā B 23:317-328. T2ae (4181)

Das, M. N. [1957], Bio-assays with non-orthogonal data. JISAS 9:67-81. E4ace;T8a;A10 (4182)

Das, M. N. [1960], Augmented fields and factorial designs in general (abstract). JISAS 12:152. E7c;T2c (4183)

Das, M. N. [1960], Fractional replicates as asymmetrical factorial designs (abstract). Proc. Forty-Seventh Indian Sci. Congress, part III, p. 32. T2,3c (4184)

Das, M. N. [1960], Fractional replicates as asymetrical factorial designs. JISAS 12:159-174. T3c (4185)

Das, M. N. [1960], On the analysis of asymmetrical factorial designs through that of symmetrical factorial designs (abstract). JISAS 12:209. T2a (4186)

Das, M. N. [1961], Construction of Rotatable Designs from factorial designs (abstract) Proc. Forty-Eighth Indian Sci. Congress, part III, p.24. T4c (4187)

Das, M. N. [1961], Construction of rotatable designs from factotial designs. JISAS 13:169-194. T4ac (4188)

Das, M. N. [1961], On problem connected with the asymmetrical factorial design (abstract). JISAS 13:240-241. T2c (4189)

Das, M. N. [1963], On construction of second order rotatable designs through balanced incomplete block designs with blocks of unequal sizes. Calcutta Stat. Assoc. Bull. 12:31-46. T4c (4190)

Das, M. N. [1964], A somewhat alternative approach for construction of symmetrical factorial designs and obtaining maximum number of factors. Calcutta Stat. Assoc. Bull. 13:1-17. T2c (4191)

Das, M. N. [1964], An alternative method for the construction of confounded symmetrical factorial designs (abstract). JISAS 16:168. T2c (4192)

Das, M. N. [1967], A de-coding method of linear codes (abstract). JISAS 19(1): 144-145. T2c (4193)

Das, M. N. and Bhargava, P. N. [1962], Designs based on constant frequency differences and their application. JISAS 14:157-164. T8c (4194)

Das, M. N. and Gill, B. S. [1963], Augmentation of second-order response surface designs through non-central points (abstract). JISAS 15:256-257. T4c (4195)

Das, M. N. and Gill, B. S. [1963], Response surface designs for agricultural experimentation (abstract). JISAS 15:276. T2,4a (4196)

Das, M. N. and Gill, B. S. [1964], Construction of second-order rotatable designs in blocks of equal size (abstract). JISAS 16:170-171. T4c (4197)

Das, M. N. and Kartha, C. P. [1966], An alternative method of construction of symmetrical confounded fractionally replicated designs (abstract). JISAS 18(2): 115-116. T3c (4198)

Das, M. N. and Kulkarni, G. A. [1960], A series of designs for parallel line bioassays with two or more preparations (abstract). JISAS 12:209-210. E3a;T8a (4199)

Das, M. N. and Kulkarni, G. A. [1966], Incomplete block designs for bio-assays. Biometrics 22:706-729. E3ace;T8a (4200)

Das, M. N. and Maini, J. S. [1965], Role of transformation for construction of design of experiments (abstract). JISAS 17:132-133. T4,7c (4201)

Das, M. N. and Narasimha, V. L. [1962], Construction of rotatable design through balanced incomplete block design with unequal sizes (abstract). Proc. Forty-Ninth Indian Sci. Congress, part III, p. 20. T4c (4202)

Das, M. N. and Narasimham, V. L. [1962], Construction of rotatable designs through balanced incomplete block designs. AMS 33:1421-1439. T4c (4203)

Das, M. N. and Rao, P. S. [1967], Construction and analysis of some new series of confounded asymmetrical factorial designs. Biometrics 23:813-822 (abstract 22: 210-211). T2c (4204)

Das, M. N. and Raut, K. C. [1966], Construction and analysis of $p\times 2^a$ confounded asymmetrical factorial designs where p is odd (abstract). JISAS 18(2):114. T2c (4205)

Das, M. N., Sukla, G. K., and Kartha, C. P. [1967], On a method of construction of symmetrical fractional factorials and related error correcting codes. Calcutta Stat. Assoc. Bull. 16:164-179. T3c (4206)

Davern, C. I., Peak, J.W., and Morley, F. H. W. [1957], The inheritance of flowering time in Trifolium subterraneum L. Australian J. Agri. Res. 8:121-134. T9a (4207)

David, H. A. [1958], Paired comparisons and tournaments (abstract). Biometrics 14:568. T11a;A2 (4208)

David, H. A. [1959], Tournaments and paired comparisons. Biometrika 46:139-149. T11a;A2 (4209)

David, H. A. [1960], The method of paired comparisons. Proc. Fifth Conf. Design Expt. Army Res. Dev. Testing. pp. 1-16. T11ae (4210)

David, H. A. [1963], The Method of Paired Comparisons. No. 12 Griffins Statistical Monographs and Courses. Chas. Griffin & Co. Ltd., London, and Hafner Publishing Co., New York. E2,3ac;T11ac;A2;B (4211)

David, H. A. [1963], The structure of cyclic paired-comparison designs. J. Australian Math. Soc. 3:117-127. E2,3c;T11c (4212)

David, H. A. [1965], Enumeration of cyclic paired-comparison designs. Amer. Math. Monthly 72:241-248. E3c;T11c (4213)

David , H. A. and Arens, B. E. [1959], Optimal spacing in regression analysis (abstract). Biometrics 15:333. T1ac (4214)

David, H. A. and Arens, B. E. [1959], Optimal spacing in regression analysis. AMS 30:1072-1081 (also Dept. Stat. Tech. Report No. 38, Virginia Polytechnic Inst.) T1ac (4215)

David, H. A. and Trivedi, M. C. [1962], Pair, triangle, and duo-trio tests. Dept. Stat. Tech. Report. No. 55, Virginia Polytechnic Inst. (Mimeo.). T11a (4216)

David, H. A. and Wolock, F. W. [1965], Cyclic designs. Proc. Tenth Conf. Design Expt. Army Res. Dev. Testing. pp. 283-297. E4ac;T11ac (4217)

Davidson, R. R. [1966], Multivariate paired comparisons. Ph. D. Thesis, Florida State Univ. T11a (4218)

Davidson, R. R. [1967], On a multivariate extension of the Bradley-Terry model for paired comparisons (abstract). AMS 38:1311. T11a (4219)

Davies, G. P. and Sears, G. W. [1955], Some makeshift methods of analysis applied to complex experimental results. Appl. Stat. 4:47-63. T2ae;A10 (4220)

Davies, O. L. [1958], The design of screening tests in the pharmaceutical industry. Bull. ISI 36(3):226-241. T8a;S2 (4221)

Davies, O. L. [1963], The design of screening tests. Technometrics 5:481-489. T8ace;A2 (4222)

Davies, O. L. and Hay, W. A. [1950], The construction and uses of fractional factorial designs in industrial research. Biometrics 6:233-249. T3ace (4223)

Davies, O. L., van Dun, F. J., and Hamaker, H. C. [1955], Design and analysis of industrial experiments. Stat. Neerlandica 9:189-207. T2,3,4e;C (4224)

Davies, P. [1967], The choice of variables in the design of experiments for polynomial regression. Ph. D. Thesis, Univ. Reading. T1ac (4225)

Davis, M. [1963], Comparison of sequential experiments for estimating the dosage-response curve (abstract). Biometrics 19:504. T8,10a;S2 (4226)

Davis, M. [1965], Sequential experiments in bioassay. Dept. Stat. Tech. Report No. 5, Harvard Univ. pp. 1-45. T8a;S2 (4227)

Davis, M. [1965], Sequential experiments in bioassay (abstract). Biometrics 21: 763. T8a;S2 (4228)

Davis, M. [1966], Comparison of sequential experiments for estimating the median lethal dose. Ph. D. Thesis, Harvard Univ. T8ac;S2 (4229)

Davis, R. S. [1962], Statistically designed pilot plant experiments. Chemical Eng. Progress 58(2):60-63. T4a (4230)

Dawson, E. H. and Dochterman, E. F. [1951], A comparison of sensory methods of measuring differences in food qualities. Food Tech. 5:79-81. T11a;C (4231)

Dean, B. V. and Marks, E. S. [1965], Optimal design of optimization experiments. Operations Res. 13:647-673. T4a;C;S1,2 (4232)

Dean, B. V. and Marks, E. S. [1965], Diseño optimo en experimentos de optimización Estadistica 23:714-744. T4a (4233)

Dean, L. A. [1954], Yield-of-phosphorus curves. Soil Sci. Soc. Amer. Proc. 18: 462-466. T4a (4234)

De Baun, R. M. [1956], Block effects in the determination of optimum conditions. Biometrics 12:20-22. T4c (4235)

De Baun, R. M. [1958], Experiences with response surface experimentation (summary). JASA 53:575. T4a (4236)

De Baun, R. M. [1958], An experimental design for three factors at three levels. Nature 181:209-210. T7c (4237)

De Baun, R. M. [1959], Response surface designs for three factors at three levels. Technometrics 1:1-8 (erratum 1:419). T4ace (4238)

De Baun, R. M. and Chew, V. [1960], Optimum allocation in regression experiments with two components of error. Biometrics 16:451-463. T1ace (4239)

De Baun, R. M. and Schneider, A. M. [1958], Experiences with response surface designs. Experimental Designs in Industry, pp. 235-246. T4ae (4240)

DeBusk, R. E. [1962], Experience in evolutionary operations at Tennessee Eastman Company. IQC 19(4):15-21. T4a (4241)

DeGroot, M. H. [1966], Optimal allocation of observations. Ann. Inst. Stat. Math. 18:13-28. E14;T1c (4242)

de la Garza, A. [1954], Spacing of information in polynomial regression. AMS 25: 123-130. T1ac (4243)

Del Priore, F. R. and Kommers, W. J. [1954], An example of a fractional replication in a bearing abrasive wear test (summary). JASA 49:354. T4a (4244)

DeLury, D. B. [1954], Designing experiments to isolate sources of variation. IQC 11(2):22-24. T2ac;C (4245)

DeLury, D. B. [1954], On the design of experiments. IQC 10(4):24-29. E1a;T2a;C (4246)

Dempster, A. P. [1960], Random allocation designs I: On general classes of estimation methods. AMS 31:885-905. T7ac (4247)

Dempster, A. P. [1961], Random allocation designs II: Approximate theory for simple random allocation. AMS 32:387-405. T7ac (4248)

Dempster, E. R., Lerner, I. M, and Lowry, D. C. [1952], Continuous selection for egg production in poultry. Genetics 37:693-708. T9a (4249)

Denenberg, V. H. [1954], Remark on "A qualification in the use of analysis of variance". Psychological Bull. 51:169-171. T2a (4250)

Denison, F. C., Federer, W. T., and Tanaka, R. K. [1955], Harvesting experimental plots of sugar cane. Slash-cut versus cut-and-pull. Hawaiian Sugar Planters' Assoc. Expt. Sta. Spec. Release 133. T2a;A14 (4251)

de Smet, F. [1966], Applications de la statistique en cardiologie. Biométrie-Praximétrie 7:89-115. T2ae;A7 (4252)

Dews, P. B. and Berkson, J. [1954], On the error of bio-assay with quantal response. Statistics and Mathematics in Biology, pp. 361-370. T8ae (4253)

Dey, A. [1967], Group divisible rotable designs (abstract). JISAS 19(1):137. T4c (4254)

Diamond, W. J. [1967], Three dimensional models of extreme vertices designs for four component mixtures. Technometrics 9:472-475. T5c (4255)

Dickerson, G. E. [1952], Inbred lines for heterosis tests? Heterosis, ch. 21, pp. 330-351. T9ac (4256)

Dickinson, A. G. and Jinks, J. L. [1956], A generalized analysis of diallel crosses. Genetics 41:65-78. T9a (4257)

Ditchburne, N. [1955], A method of analysis for a double classification arranged in a triangular table. Biometrics 11:453-464. T7ae (4258)

Djokoto, R. K. and Stephens, D. [1961], Thirty long-term fertilizer experiments under continuous cropping in Ghana. I. Crop yields and responses to fertilizers and manures. Empire J. Expt. Agri. 29:181-195. E5a;T2a (4259)

Djokoto, R. K. and Stephens, D. [1961], Thirty long-term fertilizer experiments under continuous cropping in Ghana. II. Soil studies in relation to the effects of fertilizers and manures on crop yields. Empire J. Expt. Agri. 29:245-258. E5a;T2a (4260)

Doerfler, T. E. [1962], The compounding of gradient error in the method of parallel tangents. M. S. Thesis, Iowa State Univ. T4ac (4261)

Doerfler, T. E. and Kempthorne, O. [1963], The compounding of gradient error in the method of parallel tangents. Iowa State Univ. Aerospace Res. Lab. ARL 63-144, pp. 1-68. T4ac;S2 (4262)

Dolby, J. L. [1963], On Kron's animated polyhedral models. Chemical Eng. Progress Symp. Ser. 59(42):44-48. T5ac (4263)

Draper, N. R. [1958], Rotatable designs of second and third order in three or more dimensions. Ph. D. Thesis, Univ. N. Carolina. T4ac (4264)

Draper, N. R. [1958], A third order rotatable design in four dimensions. Inst. Stat. Mimeo Ser. No, 198, Univ. N. Carolina. T4ac (4265)

Draper, N. R. [1960], Missing value formulae for certain three factor, second order response surface designs. Math. Res. Center, Tech. Summary Report No. 201, U.S. Army Univ. Wisconsin. T7a (4266)

Draper, N. R. [1960], Some rotatable designs of third order in three factors. Math. Res. Center Tech. Summary Report No. 179, U.S. Army Univ. Wisconsin. T4ac (4267)

Draper, N. R. [1960], Second order rotatable designs in four or more dimensions. AMS 31:23-33. T4c (4268)

Draper, N. R. [1960], Third order rotatable designs in three dimensions. AMS 31:865-874. T4c (4269)

Draper, N. R. [1960], A third order rotatable design in four dimensions. AMS 31:875-877. T4c (4270)

Draper, N. R. [1961], Missing values in response surface designs. Technometrics 3:389-398. T7a (4271)

Draper, N. R. [1961], Third order rotatable designs in three dimensions: Some specific designs. AMS 32:910-913. T4c (4272)

Draper, N. R. [1962], Third order rotatable designs in three factors: Analysis. Technometrics 4:219-234. T4ae (4273)

Draper, N. R. [1963], "Ridge analysis" of response surfaces. Technometrics 5:469-479. T4a (4274)

Draper, N. R. [1963], "Ridge analysis" of response surfaces. Dept. Stat. Tech. Report No. 12, Univ. Wisconsin. T4ac (4275)

Draper, N. R. and Hunter, W. G. [1966], Design of experiments for parameter estimation in multiresponse situations. Biometrika 53:525-533 (Dept. Stat. Tech. Report No. 63, Univ. Wisconsin). T4c (4276)

Draper, N. R. and Hunter, W. G. [1966], Transformations: Some examples revisited. Dept. Stat. Tech. Report No. 100, Univ. Wisconsin. T2ae;A9 (4277)

Draper, N. R. and Hunter, W. G. [1967], The use of prior distributions in the design of experiments for parameter estimation in non-linear situations. Biometrika 54:147-153 (Dept. Stat. Tech. Report No. 68, Univ. Wisconsin). T1ace (4278)

Draper, N. R. and Hunter, W. G. [1967], The use of prior distributions in the design of experiments for parameter estimation in non-linear situations: multiresponse case. Biometrika 54:662-665. T4ac (4279)

Draper, N. R. and Lawrence, W. E. [1965], Designs which minimize model inadequacies: cuboidal regions of interest. Biometrika 52:111-118 (also Dept. Stat. Tech. Report No. 36, Univ. Wisconsin). T4c (4280)

Draper, N. R. and Lawrence, W. [1965], Mixture designs for three factors. JRSSB 27:450-465. T5c (4281)

Draper, N. R. and Lawrence, W. E. [1965], Mixture designs for four factors. JRSSB 27:473-478 (also Dept. Stat. Tech. Report No. 48, Univ. Wisconsin). T5c (4282)

Draper, N. R. and Lawrence, W. E. [1966], The use of second-order 'spherical' and 'cuboidal' designs in the wrong regions. Biometrika 53:596-599 (also Dept. Stat. Tech. Report No. 57, Univ. Wisconsin). T4c (4283)

Draper, N. R. and Lawrence, W. E. [1967], Sequential designs for spherical weight functions. Technometrics 9:517-529 (also Dept. Stat. Tech. Report No. 97, Univ. Wisconsin). T4ac;S2 (4284)

Draper, N. R. amd McGregor, J. F. [1967], Some forty point four factor second order rotatable designs. Dept. Stat. Tech. Report No. 132, Univ. Wisconsin. T4ac (4285)

Draper, N. R. and Mitchell, T. J. [1966], Maximal 2^{k-p} designs of resolution V (preliminary report) (abstract). AMS 37:772. T7c (4286)

Draper, N. R. and Mitchell, T. J. [1966], Construction of the unique saturated resolution V design in 256 runs. Dept. Stat. Tech. Report No. 81, Univ. Wisconsin. T3c (4287)

Draper, N. R. and Mitchell, T. J. [1967], The construction of saturated 2_R^{k-p} designs. AMS 38:1110-1126 (also Dept. Stat. Tech. Report No. 80, Univ. Wisconsin). T7c (4288)

Draper, N. R. and Stoneman, D. M. [1964], Estimating missing values in unreplicated two-level factorial and fractional factorial designs. Biometrics 20: 443-458 (abstract 20:382; also Dept. Stat. Tech. Report No. 20, Univ. Wisconsin) T7a (4289)

Draper, N. R. and Stoneman, D. M. [1966], Factor changes and linear trends in eight-run two-level factorial designs. Dept. Stat. Tech. Report No. 82, Univ. Wisconsin. E5c;T3c (4290)

Draper, N. R. and Stoneman, D. M. [1966], Alias relationship for two-level Plackett and Burman designs. Dept. Stat. Tech. Report No. 96, Univ. Wisconsin. T3c (4291)

Draper, N. R. and Stoneman, D. M. [1966], Response surface designs for factors at two and three levels. Dept. Stat. Tech. Report No. 70, Univ. Wisconsin. T4c (4292)

Draper, N. R. and Stoneman, D. M. [1966], Response surface designs for factors at two and four levels. Dept. Stat. Tech. Report No. 73, Univ. Wisconsin. T4c (4293)

Dudley, J. W. and Powers, L. [1962], Population genetic studies on sodium and potassium in sugar beets (Beta vulgaris L.). J. Amer. Soc. Sugar Beet Tech. 11:97-127. T9a (4294)

Duffett, J. R. [1954], Some experience with the design of experiments. IQC 11(3): 36-38, 40. T7a (4295)

Duncan, D. B. and Rhodes, R. C. [1952], Multiple regression with a quantal response (abstract). Virginia J. Sci. 3:358. T8a (4296)

Dvoretsky, A. [1956], On stochastic approximation. Proc. Third Berkeley Symp. Math. Stat. Prob. 1:39-55. T1ac;S2 (4297)

Dyke, G. V. and Patterson, H. D. [1952], Analysis of factorial arrangements when the data are proportions. Biometrics 8:1-12. T2ae;A9 (4298)

Dykstra, O. [1956], Incomplete block rank analysis: 2^{p-q} fractional factorials using a method of paired comparisons (abstract). AMS 27:871. E2a;T3,11a;A12 (4299)

Dykstra, O. [1957], Applications of Scheffé's method of paired comparisons in food quality evaluation. ASQC Nat. Conv. Trans. 11:167-175. T11a;A12;C (4300)

Dykstra, O. [1958], Factorial experimentation in Scheffé's analysis of variance for paired comparisons. JASA 53:529-542. T2,11ae (4301)

Dykstra, O. [1959], Partial duplication of factorial experiments. Technometrics 1:63-75. T7ace (4302)

Dykstra, O. [1960], Partial duplication of response surface designs. Technometrics 2:185-195. T4ac (4303)

Dykstra, O. [1960], Rank analysis of incomplete block designs: A method of paired comparisons employing unequal repetitions on pairs. Biometrics 16:176-188. E2a;T11a;A12 (4304)

Dykstra, O. [1966], The orthogonalization of undesigned experiments. Technometrics 8:279-290 (erratum 8:731). T7ac (4305)

Eberhart, S. A. [1964], Theoretical relations among single, three-way, and double cross hybrids. Biometrics 20:522-539. T9a (4306)

Eberhart, S. A. and Gardner, C. O. [1966], A general model for genetic effects. Biometrics 22:864-881. T9a (4307)

Eberhart, S. A., Harrison, M. N., and Ogada, F. [1967], A comprehensive breeding system. Der Züchter 37:169-174. T9a (4308)

Eckhardt, R. C. [1952], Predicting yields of missing single crosses of corn. Agron. J. 44:215-216. T9a (4309)

Eder, M., Warner, R., and Keene, F. [1954], Statistically designed experiment of the factorial type applied to point-contact transistors. Nat. Symp. Reliability Quality Control, Proc. 1:1-11. T2ae (4310)

Edwards, A. L. and Horst, P. [1950], The calculation of sums of squares for interactions in the analysis of variance. Psychometrika 15:17-24. T2a (4311)

Ehrenfeld, S. [1955], Essentially complete classes of experiments (abstract). AMS 26:154. T7,8c (4312)

Ehrenfeld, S. [1956], Complete class theorems in design of experiments; part I: Complete class theorems in experimental design; part II: On the efficiency of experimental design. Ph. D. Thesis, Columbia Univ. T (4313)

Ehrenfeld, S. [1956], Complete class theorems in experimental design. Proc. Third Berkeley Symp. Math. Stat. Prob. 1:57-67. E14;T10 (4314)

Ehrenfeld, S. [1959], Randomization and factorial experiments (abstract). AMS 30:621. T3,7ac (4315)

Ehrenfeld, S. [1962], Some experimental design problems in attribute life testing. JASA 57:668-679 (summary 57:488). T8c;S1 (4316)

Ehrenfeld, S. [1966], On a minimal essentially complete class of experiments. AMS 37:435-440. T1c (4317)

Ehrenfeld, S. and Zacks, S. [1961], Randomization and factorial experiments. AMS 32:270-297. T3,7ac (4318)

Ehrenfeld, S. and Zacks, S. [1962], Comparison of randomized and non-randomized factorial experiments (abstract). AMS 33:831-832. T3,7ac (4319)

Ehrenfeld, S. and Zacks, S. [1963], Optimal strategies in factorial experiments. AMS 34:780-791. T2,3,7a (4320)

Ehrenfeld, S. and Zacks, S. [1967], Testing hypotheses in randomized factorial experiments. AMS 38:1494-1507 (abstract 38:635-636). T3,7a (4321)

Einbinder, S. K. and Olkin, I. [1963], Evaluation of performance reliability using regression models. Proc. Eighth Conf. Design Expt. Army. Res. Dev. Testing. pp. 473-501. T4ae;A7 (4322)

Eisenklam, P. [1953], Statistical methods in engineering experimentation. Parts I and II. Research 6:195-201, 231-237. T2,3ace (4323)

Eisler, H. [1964], A choice model for paired comparison data based on imperfectly nested sets. Psychometrika 29:363-370. T1la (4324)

Elandt, R. C. [1956], O pewnych testach interakcji w doswiadczeniach wieloletnich i wielokrotnych. Zagadnienie rejonizacji. (On certain interaction tests in multiple experiment. The problem of stratification.) Zastosowantia Matematyki 3(1):8-45. T2a;C (4325)

Elandt-Johnson, R. C. [1967], Some two- and three-dimensional systematic designs in factorial experiments (abstract). Technometrics 9:189. T2a (4326)

Elandt-Johnson, R. C. [1967], Optimal economy in planning experiments (abstract). Proc. Twelfth Conf. Design Expt. Army Res. Dev. Testing. pp. 119-120. T1a;A7 (4327)

Elandt-Johnson, R. C. [1967], Application of certain systematic designs in planning of factorial experiments (in Polish). Roczniki Nauk Rolniczych 94(A):665-678. E5ac;T2a (4328)

Elfving, G. [1952], Optimum allocation in linear regression theory. AMS 23:255-262. T1ac (4329)

Elfving, G. [1956], Selection of nonrepeatable observations for estimation. Proc. Third Berkeley Symp. Math. Stat. Prob. 1:69-75. T1c (4330)

Ellis, S. R. M., Jeffreys, G. V., and Wharton, J. T. [1964], Raschig synthesis of hydrazine. Investigation of chloramine formation reaction. I & EC Process Design Development 3:18-22. T3ae (4331)

Elston, R. C. and Bush, N. [1964], The hypotheses that can be tested when there are interactions in an analysis of variance model. Biometrics 20:681-698 (abstract 18:261). T2a (4332)

Emmens, C. W. [1950], The intravaginal assay of naturally occurring oestrogens. J. Endocrinology 6:302-307. T8ace (4333)

Falconer, D. S. [1953], Selection for large and small size in mice. J. Genetics 51:470-501. T9a (4334)

Farlie, D. J. and Keen, J. [1967], Quick ways to the top: A team game illustrating steepest ascent techniques. Appl. Stat. 16:75-80. T4c (4335)

Farrell, R. H., Kiefer, J., and Walbran, A. [1967], Optimum multivariate designs. Proc. Fifth Berkeley Symp. Math. Stat. Prob. 1:113-138. T1c (4336)

Federer, W. T. [1951], Evaluation of variance components from a group of experiments with multiple classifications. Iowa Agri. Expt. Sta. Res. Bull. 380:241-310. E2,6a;T9a;A13 (4337)

Federer, W. T. [1952], Additional treatments in factorial experiments (abstract). Biometrics 8:387-388. T7c (4338)

Federer, W. T. [1953], The value of summarizing results from individual experiments. Committee Expt. Design Agri. Res. Administration Mimeo. U.S.D.A., Biometrics Unit Mimeo. Ser. No. 41, Cornell Univ. T9a;A13;C (4339)

Federer, W. T. [1953], A 1/8 replicate of a 2^9 factorial. Biometrics Unit Mimeo. Ser. BU-42-M, Cornell Univ. T3,7ac (4340)

Federer, W. T. [1953], Some confounded arrangements. Biometrics Unit Mimeo. Ser. BU-43-M, Cornell Univ. T2,3,7ac (4341)

Federer, W. T. [1954], Covariance analysis in a two-way classification with unequal numbers in the subclasses. Biometrics Unit. Mimeo. Ser. BU-51-M, Cornell Univ. E2a;T2a;A7 (4342)

Federer, W. T. [1956], A method for evaluating genetic progress in a sugar cane breeding program. Hawaiian Planters' Record 55:177-190. T9a (4343)

Federer, W. T. [1957], Covariance analyses for unbalanced two-way classifications (summary). JASA 52:369. E2a;T2a;A7,10 (4344)

Federer, W. T. [1957], Variance and covariance analyses for unbalanced classifications. Biometrics 13:333-362. E2a;T2a;A7,10 (4345)

Federer, W. T. [1959], Covariance analyses for unbalanced two-way classifications. Cornell Univ. Agri. Expt. Sta. Memoir 360:1-60. E2a;T2a;A7,10 (4346)

Federer, W. T. [1959], Treatment design and the interpretation of experimental results (abstract). Biometrics 15:153 (Biometrics Unit Mimeo. Ser. BU-92-M, Cornell Univ.). T2,7a;C (4347)

Federer, W. T. [1960], Some illustrative examples of "treatment design" and the interpretation of experimental results. Proc. American Soc. Hort. Sci. 75: 811-818. T2,9a;A4,9 (4348)

Federer, W. T. [1961], Design and analysis for a 2x4x4 nitrogen, potash, and phosphorous fertilizer trial on spinach. Biometrics Unit Mimeo. Ser. BU-132-M, Cornell Univ. E7c;T2a (4349)

Federer, W. T. [1963], Discussion of "Statistical designs of experiments for continuous data". Proc. Eighth Conf. Design Expts. Army Res. Dev. Testing. pp. 67-70. T2a;S1;C (4350)

Federer, W. T. [1963], Relationships between a three-way classification disproportionate numbers analysis of variance and several two-way classification and nested analyses. Biometrics 19:629-637 (also Biometrics Unit Mimeo. Ser. BU-152-M, Cornell Univ.). E2a;T2a;A10 (4351)

Federer, W. T. [1964], Fractional alias in fractional replication (abstract). Biometrics 20:222. T7ac (4352)

Federer, W. T. [1964], Selection models (abstract). Biometrics 20:222 (Biometrics Unit Mimeo. Ser. BU-145-M, Cornell Univ.). T9a (4353)

Federer, W. T. [1964], Lecture notes on the Design and Analysis of Experiments for the Advanced Science Seminar in Mathematical Statistics at Colorado State University, 7/13/64-8/7/64. Biometrics Unit Mimeo Ser. BU-338-M, Cornell Univ., 14 chapters. E1-12;T2-7;C (4354)

Federer, W. T. [1966], Applications and construction of fractionally replicated designs Biometrics Unit Mimeo. Ser. BU-232-M, Cornell Univ. T7ac (4355)

Federer, W. T. [1966], Data collection and interpretation. Biometrics Unit Mimeo. Ser. BU-337-M, Cornell Univ., 10 chapters. E1,2,5,7,12,14ace T2,3,4,8,9ac;C (4356)

Federer, W. T. [1967], Analyses of diallel crossing and matched pair schemes when individual contributions are measurable. Biometrics Unit Mimeo. Ser. BU-246-M, Cornell Univ. T9a (4357)

Federer, W. T. [1967], Diallel cross designs and their relation to fractional replication. Der Züchter 37:174-178. T3,7,9ac (4358)

Federer, W. T. and Brunk, M. E. [1960], Analyses for a balanced design in a marketing experiment. Biometrics Unit Mimeo. Ser. BU-127-M, Cornell Univ. E11ac;T7ac (4359)

Federer, W. T. and Farden, C. A. [1955], Analysis of variance set-up for Joint Project 69. Biometrics Unit Mimeo. Ser. BU-67-M, Cornell Univ. E13a;T2a (4360)

Federer, W. T. and Henderson, C. R. [1956], A note on general and specific combining ability. Biometrics Unit Mimeo. Ser. BU-78-M, Cornell Univ. T9a (4361)

Federer, W. T. and Henderson, C. R. [1962], Some aspects of general and specific combining ability. Biometrics Unit ONR Tech. Report No. 8, Cornell Univ. T9a (4362)

Federer, W. T., Humbert, R. P. and Tanaka, R. K. [1955], Yield data from the six American Factors plantations and Olokele Sugar Company tests for the spring and fall plantings of the plant crop. Hawaiian Sugar Planters' Assoc. Expt. Sta. Spec. Release 116A, 45 pp. T2a;A13 (4363)

Federer, W. T. and Plaisted, R. L. [1960], Estimates of specific and general combining ability from potato breeding experiments. Biometrics Unit Mimeo. Ser. BU-118-M, Cornell Univ. E3,4a;T9ac (4364)

Federer, W. T. and Plaisted, R. L. [1962], A method for estimating combining ability components of variance in incomplete block designs. American Potato J. 39: 197-206. E3,4a;T9a (4365)

Federer, W. T., Powers, L. and Payne, M. G. [1963], Studies on statistical procedure applied to chemical genetic data from sugar beets. Colorado State Univ. Agri. Expt Sta. Tech. Bull. 77. T9a (4366)

Federer, W. T. and Raktoe, B. L. [1965], Some remarks on the generalized construction and analysis of fractional replicates (abstract). Biometrics 21:763-764. T3,7c (4367)

Federer, W. T. and Tanaka, R. K. [1955], A statistical study of the NxK interaction for yield characters in group test No. 11. Hawaiian Sugar Planters' Assoc. Expt. Sta. Spec. Release 116B, 11 pp. T2a;A13 (4368)

Federer, W. T. and Zelen, M. [1963], Applications of the factorial calculus to general unequal numbers analyses (abstract). AMS 34:1618. T2a;A8,10 (4369)

Federer, W. T. and Zelen, M. [1963], Applications of the factorial calculus to general unequal numbers analyses. Math. Res. Center Tech. Summary Report No. 393, U.S. Army Univ. Wisconsin. T2a;A8,10 (4370)

Federer, W. T. and Zelen, M. [1964], Applications of the calculus for factorial arrangements II: unequal numbers in the analysis of variance. Math. Res. Center Tech. Summary Report No. 411, U.S. Army Univ. Wisconsin. T2a; A8,10 (4371)

Federer, W. T. and Zelen, M. [1966], Analysis of multifactor classifications with unequal numbers of observations. Biometrics 22:525-552. T2a;A8,10 (4372)

Feldt, L. S. [1958], A comparison of the precision of three experimental designs employing a concomitant variable. Psychometrika 23:335-353. T2ac;A7 (4373)

Feller, W. [1967], On fitness and the cost of natural selection. Genetical Res. 9:1-15. T9a (4374)

Ferguson, J. H. A. [1956], Some applications of Binomial Probability Paper in genetic analyses. Euphytica 5:329-338. T9a (4375)

Fernelius, A. L. [1960], Design of an experiment to evaluate a bio-assay with non-parallel slopes. Proc. Fifth Conf. Design Expt. Army Res. Dev. Testing. pp. 328-331. T8ae (4376)

Ferrante, G. R. [1962], Laboratory evaluation of new creaseproofing agents using statistical techniques. American Dyestuff Reporter 51(2):41-43. T4a (4377)

Ferwerda, F. P. [1954], A tentative breeding method for robusta and other allogamous coffee species. Euphytica 3:12-19. T9c (4378)

Ferwerda, F. P. [1956], Recurrent selection as a breeding procedure for rye and other cross-fertilized plants. Euphytica 5:175-184. T9a (4379)

Fields, P. E. [1955], Factorial designs and the guidance of downstream migrant salmon and steelhead trout (abstract). Biometrics 11:508. T2a (4380)

Filho, V. C. and Freire, E. S. [1959], Adubação da mamoneira . II, Experiências de espaçamento × adubação. (Fertilizer experiments with castor beans. II. Plant density × fertility level.) Bragantia 18:77-99. T2e (4381)

Filho, J. B., Freire, E. S., and Abramides, E. [1966], Adubação da batata-doce com nitrogênio, fósforo e potássio em terrenos de "cerrado". (Effect of nitrogen, phosphorus and potassium on sweet potatoes cultivated on "cerrado" soils.) Bragantia 25:241-251. T2e (4382)

Finlay, K. W. and Wilkinson, G. N. [1963], The analysis of adaptation in a plant-breeding programme. Australian J. Agri. Res. 14:742-754. T9a (4383)

Finney, D. J. [1950], Two new uses of the Behrens-Fisher distribution. JRSSB 12:296-300. T8a;All (4384)

Finney, D. J. [1950], The fractional replication of factorial experiments — A correction. Ann. Eugenics 15:276. T3c (4385)

Finney, D. J. [1951], Subjective judgement in statistical analysis: An experimental study. JRSSB 13:284-297. T8a (4386)

Finney, D. J. [1952], The estimation of the ED50 for a logistic response curve. Sankhyā 12:121-136. T8a (4387)

Finney, D. J. [1952], On the precision of biological assay. Acta Pharmacologica Toxicologica 8:55-70. T8a (4388)

Finney, D. J. [1952], Graphical estimation of relative potency from quantal responses. J. Pharmacology Expt. Therapeutics 104:440-444. T8a (4389)

Finney, D. J. [1953], The choice of levels (abstract). Biometrics 9:428. T8c (4390)

Finney, D. J. [1955], Cross-over and single-subject designs for 4-point assays (abstract). Biometrics 11:548-549. T8c (4391)

Finney, D. J. [1958], Statistical problems in the selection of crop varieties (abstract). Heredity 12:399. T9a;A2 (4392)

Finney, D. J. [1958], Statistical problems of plant selection. Bull. ISI 36(3): 242-268. T9a;S1,2 (4393)

Finney, D. J. [1960], A simple example of the external economy of varietal selection. Bull. ISI 37(3):91-106. T9a;S1,2 (4394)

Finney, D. J. [1962], Genetic gains under three methods of selection. Genetical Res. 3:417-423. T9a (4395)

Finney, D. J. [1965], The meaning of bioassay. Biometrics 21:785-798 (abstract 21:264). T8a;C (4396)

Finney, D. J. [1966], The vital statistics of a drug. Biom. Zeit. 8:15-31. T8a;C (4397)

Finney, D. J. [1967], Letter to the Editor. Biometrics 23:587-588. T8a (4398)

Finney, D. J. and Wood, E. C. [1951], Intra-litter replication in biological assays. Nature 167:903-904. T8a (4399)

Fisher, A. G. [1966], Some techniques for analysis and interpretation of estimable functions in incomplete fixed-factor design. Ph. D. Thesis Rutgers Univ. T3,7a (4400)

Fisher, R. A. [1951], Query 91. Biometrics 7:433-434. T2a (4401)

Fleckenstein, M., Freund, R. A., and Jackson, J. E. [1958], A paired comparison test of typewriter carbon papers. Tappi 41:128-130. T11ae (4402)

Foata, D. C. [1962], Sur la construction des plans factoriels fractionnés et certains codes correcteurs à l'aide des caractères des groupes abéliens. Publications Inst. Stat. Univ. Paris 11:57-66. T3c (4403)

Foata, D. C. [1963], Sur la construction des plans factoriels fractionnés et certains codes correcteurs à l'aide des caractères des groupes abéliens. Colloques Internationaux du Centre National de la Recherche Scientifique No. 110 Le Plan d'Expériences, pp. 137-146. T3c (4404)

Fogelmanis, A. [1961], Application of composite designs in fitting quadratic response surfaces. M.S. Thesis, Iowa State Univ. T4a (4405)

Folks, J. L. [1958], Comparison of designs for exploration of response relationships. Ph. D. Thesis, Iowa State Univ. T4,10a (4406)

Folks, J. L. [1959], Comparison of designs for exploration of response relationships (summary). JASA 54:494. T4ac (4407)

Folks, L. [1963], The response surface concept as a basis for economic studies of fertilizer trials (summary). JASA 58:550. T4a (4408)

Ford, L. R. [1957], Solution of a ranking problem from binary comparisons. American Math. Monthly 64(8):28-33. T11c;A2 (4409)

Foreman, C. F., Curry, N. H., Homeyer, P. G., and Porter, A. R. [1958], A comparison of different types of stalls for dairy cattle. Iowa State College J. Sci. 33:43-53. T2ae (4410)

Foster, W. D. and Wolfe, E. K. [1959], Response surface techniques versus factorial analysis in a development application (abstract). Biometrics 15:334. T10 (4411)

Frankel, O. H. [1950], The development and maintenance of superior genetic stocks. Heredity 4:89-102. T9a (4412)

Franklin, N. L., Pinchbeck, P. H., and Popper, F. [1956], A statistical approach to catalyst development. Part I: The effect of process variables on the vapour phase oxidation of naphthalene. Trans. Inst. Chemical Eng. 34:280-293. T4a (4413)

Franklin, N. L., Pinchbeck, P. H., and Popper, F. [1958], A statistical approach to calalyst development. Part II: The integration of process and catalyst variables in the vapour phase oxidation of naphthalene. Trans. Inst. Chemical Eng. 36: 259-269. T4a (4414)

Freireich, E. J., Gehan, E. A., Rall, D. P. Schmidt, L. H., and Skipper, H. E. [1966], Quantitative comparison of toxicity data of anticancer agents in mouse, rat, hamster, dog, monkey, and man. Cancer Chemotherapy Reports 50:219-244. T8a;C (4415)

Frey, K. J. [1967], Mass selection for seed width in oat populations. Euphytica 16:341-349. T9a (4416)

Friend, J. K. [1958], Probits: An application to observational data. Appl. Stat. 7:25-41. T8ae (4417)

Fry, R. E. [1961], Finding new fractions of factorial experimental designs. Technometrics 3:359-370 (errata 5:134). T7ac (4418)

Fujii, Y. [1967], Geometrical association schemes and fractional factorial designs. J. Sci. Hiroshima Univ. Ser. A-I 31:195-209. T3c (4419)

Fujii, A. and Tsutagawa, N. [1965], Experimental analysis of process by the method of random combination of fractional factorial design. Reports Stat. Appl. Res. 12(4):165-177. T3,7a (4420)

Fuller, W. A. [1965], Stochastic fertilizer production functions for continuous corn. J. Farm Economics 47:105-119. T4a;All (4421)

Funk, E. M., Forward, J., and Lorah, M. [1954], Minimizing spoilage in shell eggs by thermostabilization. Poultry Sci. 33(3):532-538. T2,7e (4422)

Furukawa, N. [1960], The point estimation of the parameters in the mixed model. Kumamoto J. Sci. Ser. A 5:1-43. T2a (4423)

Fyfe, J. L. and Gilbert, N. [1963], Partial diallel crosses. Biometrics 19:278-286. T9ace (4424)

Gaddum, J. H. [1953], Simplified mathematics for bioassays. J. Pharmacy Pharmacology 5:345-358. T8a (4425)

Gaddum, J. H. [1953], Bioassays and mathematics. Pharmacological Rev. 5: 87-134. T8a;B;C (4426)

Gaido, J. J. and Terhune, H. D. [1961], Evaluation of variables in the pressure-kier-bleaching of cotton. American Dyestuff Reporter 50(21):23-26, 32. T4a (4427)

Gardiner, D. A. [1956], Some third order rotatable designs. Ph. D. Thesis, Univ. N. Carolina. T4c (4428)

Gardiner, D. A. and Cowser, K. E. [1961], Optimization of radionuclide removal from low-level process wastes by the use of response surface methods. Health Physics 5:70-78. T4a (4429)

Gardiner, D. A., Grandage, A. H. E., and Hader, R. J. [1956], Some third order rotatable designs. Inst. Stat. Mimeo. Ser. No. 149, Univ. N. Carolina. T4a (4430)

Gardiner, D. A., Grandage, A. H. E., and Hader, R. J. [1959], Third order rotatable designs for exploring response surfaces. AMS 30:1082-1096. T4c (4431)

Gardner, C. O. and Eberhart, S. A. [1966], Analysis and interpretation of the variety cross diallel and related populations. Biometrics 22:439-452. T9ae (4432)

Gargantini, H., Conagin, A., and Purchio, M. J. [1958], Ensaio de adubação N-P-K em cultura de trigo. (N-P-K experiment on wheat.) Bragantia 17:13-27. T2e (4433)

Gargantini, H., de Tella, R., and Conagin, A. [1958], Ensaio de adubação N-P-K em amendoim. (N-P-K fertilizer test with peanuts.) Bragantia 17:1-12. T2e (4434)

Gargantini, H., Nóbrega, S. A., Hungria, L. S., Wutke, A. C. P., Scivittaro, A., and Freire, E. S. [1965], Adubação mineral de batatinha. II. Vale do Paraíba. (Mineral fertilization of potatoes. II. Paraíba Valley.) Bragantia 24:29-40. T2e (4435)

Garside, M. J. [1965], The best sub-set in multiple regression analysis. Appl. Stat. 14:196-200. T1a (4436)

Gatty, R. [1960], Factorial designs for store sales experiments. Dept. Agri. Economics Tech. A.E. No. 1, Rutgers Univ. T2a (4437)

Gayle, J. B. [1964], Computer simulation study of Bruceton and Probit methods of sensitivity testing. Proc. Ninth Conf. Design Expt. Army Res. Dev. Testing. pp. 205-227. T8a (4438)

Gaylor, D. W. and Sweeny, H. C. [1965], Design for optimal prediction in simple linear regression. JASA 60:205-216. Tlc (4439)

George, K. C. and Das, M. N. [1965], A type of non-rotatable response surface design (abstract). JISAS 17:126. T4c (4440)

George, K. C. and Das, M. N. [1966], A type of central composite response surface designs. JISAS 18(2):21-29. T4c (4441)

Gerard, H. B. and Shapiro, H. N. [1958], Determining the degree of inconsistency in a set of paired comparisons. Psycometrika 23:33-46. Tllae (4442)

Ghosh, B. K. [1964], Sequential analysis of variance under random-effects and mixed models (abstract). AMS 35:928. T2a;S2 (4443)

Ghosh, B. K. [1967], Sequential analysis of variance under random and mixed models. JASA 62:1401-1417. T2a;S2 (4444)

Ghosh, M N. and Srivastava, S. K. [1963], A mixed model for fertilizer experiments in cultivators' fields. Calcutta Stat. Assoc. Bull. 12:113-123. T2a;A13 (4445)

Ghosh, M. N. and Swain, A. K. P.C. [1966], Analysis of covariance in Scheffé's mixed model. Calcutta Stat. Assoc. Bull. 15:137-146. T2a;A7 (4446)

Giesbrecht, F. and Kempthorne, O. [1965], Examination of a repeat mating design for estimating environmental and genetic trends. Biometrics 21:63-85. T9ace (4447)

Gilbert, E. N. [1961], Design of mixed doubles tournament. American Math. Monthly 68:124-131. E2c;Tlla (4448)

Gilbert, N. E. G. [1958], Diallel cross in plant breeding. Heredity 12:477-492. T7,9,llac (4449)

Gilbert, N. [1961], Polygene analysis. Genetical Res. 2:96-105. T9a (4450)

Gilbert, N. [1961], A tomato selection experiment. Genetical Res. 2:361-372. T9a (4451)

Gilbert, N. [1961], Polygene analysis II. Selection. Genetical Res. 2:456-460. T9a (4452)

Gilbert, N. [1967], Additive combining abilities fitted to plant breeding data. Biometrics 23:45-49. T9a (4453)

Gilbert, S. [1955], Statistical design of experiments in metallurgical research. IQC 12(5):13-18. T2,3a;C (4454)

Gill, B. S. and Das, M. N. [1965], Construction and analysis of truncated factorial designs (abstract). JISAS 17:128-129. T7c (4455)

Gill, B. S. and Jha, M. P. [1964], Construction and analysis of truncated and irregular fractional designs (abstract). JISAS 16:172-173. T7c (4456)

Gill, J. L. [1965], Effects of finite size on selection advance in simulated genetic populations. Australian J. Biological Sci. 18:599-617. T9a;A2 (4457)

Gill, J. L. [1965], A Monte Carlo evaluation of predicted selection response. Australian J. Biological Sci. 18:999-1007. T9a;A2 (4458)

Gill, J. L. [1965], Selection and linkage in simulated genetic populations. Australian J. Biological Sci. 18:1171-1187. T9a;A2 (4459)

Gillett, B. E. [1964], The determination of some optimal experimental designs. Ph. D. Thesis, Oklahoma State Univ. T1ac (4460)

Gillet, M. [1963], Un plan systématique de polycross pour les nombres de base pairs. Annales Amélioration Plantes 13:269-276. T9ac (4461)

Gilliland, D. C. [1967], On blocking a $(p^m)^n$ factorial. Dept. Stat. Prob. RM-199 DCG-4, Michigan State Univ. T2a (4462)

Gillis, M. B., Norris, L. C., and Heuser, G. F. [1951], The influence of particle size on the utilization of phosphates by the chick. Poultry Sci. 30:396-398. T2a (4463)

Ginsburg, M. and Heller, H. [1953], The antidiuretic assay of vasopressin by intravenous injection into unanaesthetized rats. J.Endocrinology 9:267-273. T8e (4464)

Glaser, R. H. [1958], An application of the Box technique to the evaluation of electrical components. Proc. National Symp. Reliability Quality Control 4:161-177. T4a (4465)

Glenn, W. A. [1959], Some aspects of paired-comparison experiments. Ph. D. Thesis, Virginia Polytechnic Inst. T11a (4466)

Glenn, W. A. and David, H. A. [1960], Ties in paired-comparison experiments using a modified Thurstone-Mosteller model. Biometrics 16:86-109. T11ae;A12 (4467)

Goldin, A. and Mantel, N. [1957], The employment of combinations of drugs in the chemotherapy of neoplasia: a review. Cancer Res. 17:635-654. T8ac (4468)

Goldman, A. S. [1962], Estimating the parameters in the model $y_{ijk} = a_i - b_j + e_{ijk}$ (abstract). AMS 33:311. T2a (4469)

Golub, A. and Grubbs, F. E. [1956], Analysis of sensitivity experiments when the levels of stimulus cannot be controlled. JASA 51:257-265 (corrigenda 62:1519). T8a (4470)

Gomes, F. P. [1951], A adubação da cana-de-açúcar, em pernambuco, determinada pel lei de Mitscherlich. Revista Agricultura São Paulo 26:357-364. T2a;A7 (4471)

Gomes, F. P. [1957], The analysis of factorial experiments in balanced incomplete blocks (abstract). Biometrics 13:239-240. E2a;T2a (4472)

Gomes, F. P. [1957-1958], A lei de Mitscherlich aplicada a experimentos de adubação com vinhaça. Anais Escola Superior Agri. "Luiz de Queiroz" 14-15:107-112. T2a;A7 (4473)

Gomes, F. P. [1961], On a formula for the estimation of the optimum dressing of a fertilizer. Biometrics 17:492-494. T4a (4474)

Gomes, F. P. and de Abreu, C. P. [1959], Sôbre ume fórmula para o cálculo da dose mais economica de adubo. Anais Escola Superior Agri. "Luiz de Queiroz" 16:191-198. T2,4a (4475)

Good, I. J. [1958], The interaction algorithm and practical Fourier analysis. JRSSB 20:361-372 (addendum 22:372-375). T2a (4476)

Good, I. J. [1967], Checks on Yates's algorithm. Biometrics 23:573. T2a (4477)

Goodwin, K., Dickerson, G. E., and Lamoureux, W. F. [1959], An experimental design for separating genetic and environmental changes in animal populations under selection (abstract). Biometrics 15:143-144. T9c (4478)

Goodwin, K., Dickerson, G. E., and Lamoreux, W. F. [1960], An experimental design for separating genetic and environmental changes in animal populations under selection. Biometrical Genetics, pp. 117-138. T9ac (4479)

Gopinath, D. M., Ramanarao, V. V., Subrahmanyam, M., and Narayana, C. L. [1966] A study of diallel crosses between varieties of Nicotiana tobacum L. for yield components. Euphytica 15:171-178. T9a (4480)

Gorman, J. W. and Hinman, J. E. [1962], Simplex lattice designs for multicomponent systems. Technometrics 4:463-487. T5ace (4481)

Gorman, J. W. and Toman, R. J. [1966], Selection of variables for fitting equations to data. Technometrics 8:27-51. T1,7ae (4482)

Gosslee, D. [1952], The use of finite fields in design of experiments (abstract). American Math. Monthly 59:440. E7c;T2c (4483)

Gorlay, N. [1955], F-test bias for experimental designs in educational research. Psychometrika 20:227-248. E7a;T2a;A10 (4484)

Gowe, R. S., Johnson, A. S., Downs, J. H., Gibson, R., Mountain, W. F., Strain, J. H., and Tinney, B. F. [1959], Environment and poultry breeding problems. 4. The value of a random-bred control strain in a selection study. Poultry Sci. 38:443-462. T9a (4485)

Gowe, R. S., Robertson, A., and Latter, B. D. H. [1959], Environment and poultry breeding problems. 5. The design of poultry control strains. Poultry Sci. 38:462-471. T9a (4486)

Graff, Y. [1960], Les levures sélectionnées en cidrerie. Variations de certains de leurs caractères selon les espèces de pommes utilisées. Biométrie-Praximétrie 1(2):1-34. T2ae;A10 (4487)

Grafius, J. E. [1963], Vector analysis applied to crop eugenics and genotype-environment interaction (with discussion). Statistical Genetics and Plant Breeding, pp. 197-217. T9a;A14 (4488)

Grafius, J. E. and Kiesling, R. L. [1958], Vector representation of biologic fields of force. Agron. J. 50:757-760. T9a (4489)

Grafius, J. E. and Kiesling, R. L. [1960], The prediction of relative yields of different oat varieties based on known environmental variables. Agron. J. 52:396-399. T9a;A7 (4490)

Grafius, J. E. and Wiebe, G. A. [1959], Expected genetic gain in yield in small grain. A geometrical interpretation. Agron. J. 51:560-562. T9a (4491)

Grant, R. L. and Van Dolah, R. W. [1962], Use of the up-and-down method with factorial designs. Proc. Seventh Conf. Design Expt. Army Res. Dev. Testing. pp. 39-65. T2ae;S2 (4492)

Gregson, R. A. M. [1960], Bias in the measurement of food preferences by triangular tests. Occupational Psychology 34:249-257. T11a (4493)

Greiner, T., Gold, H., Bliss, C. I., Gluck, J., Marsh, R., Mathes, S. B., Modell, W., Otto, H., Kwit, N. T., and Warshaw, L. [1951], Bioassay of diuretic agents in patients with congestive failure. J. Pharmacology Expt. Therapeutics 103:431-440. T8a (4494)

Greiner, T., Gold, H., and Bliss, C. I. [1952], The clinical assay of diuretic agents. I. Biologic considerations which determine the design. II. Estimation of the error in a clinical assay (with discussion). Biometrics 8:232-248. T8ae (4495)

Grey, E. F. [1962], Statistical methods as design tools. Proc. Nat. Symp. Reliability Quality Control 8:68-73. T4,5a (4496)

Gridgeman, N. T. [1950], Design and evaluation of biological assays. Nature 165:843-844. T8ac (4497)

Gridgeman, N. T. [1955], Taste comparisons: two samples or three? Food Tech. 9:148-150. T11a;A14;C (4498)

Gridgeman, N. T. [1959], Pair comparison, with and without ties. Biometrics 15: 382-388. T11a;A12 (4499)

Gridgeman, N. T. [1963], Significance and adjustment in paired comparisons. Biometrics 19:213-228. T11a (4500)

Gridgeman, N. T. [1964], Sensory comparisons: The 2-stage triangle test with sample variability. J. Food Sci. 29:112-117. T11ae;C (4501)

Griffing, B. [1956], Concept of general and specific combining ability in relation to diallel crossing systems. Australian J. Biological Sci. 9:463-493. T9ae (4502)

Griffing, B. [1956], A generalised treatment of the use of diallel crosses in quantitative inheritance. Heredity 10:31-50. T9a (4503)

Griffing, B. [1961], Accommodation of gene-chromosome configuration effects in quantitative inheritance and selection theory. Australian J. Biological Sci. 14:402-414. T9a;A2 (4504)

Griffing, B. [1962], Consequences of truncation selection based on combinations of individual performance and general combining ability. Australian J. Biological Sci. 15:333-351. T9a;A2 (4505)

Griffing, B. [1962], Prediction formulae for general combining ability selection methods utilizing one or two random mating populations. Australian J. Biological Sci. 15:650-665. T9a;A2 (4506)

Griffing, B. [1963], Comparisons of potentials for general combining ability selection methods utilizing one or two random-mating populations. Australian J. Biological Sci. 16:838-862. T9a;A2 (4507)

Griffing, B. [1965], Influence of sex on selection I. Contribution of sex-linked genes. Australian J. Biological Sci. 18:1157-1170. T9a (4508)

Griffing, B. [1966], Influence of sex on selection II. Contributions of autosomal genotypes having different values in the two sexes. Australian J. Biological Sci. 19:593-606. T9a (4509)

Griffing, B. [1966], Influence of sex on selection III. Joint contributions of sex-linked and autosomal genes. Australian J. Biological Sci. 19:775-793. T9a (4510)

Grizzle, J. E. [1959], An application of the logistic model in analyzing a factorial experiment when the data are proportions (preliminary report) (abstract). Biometrics 15:490-491. T2a (4511)

Grohskopf, H. [1953], Factorial design in the study of chemical reactions. Ind. Eng. Chemistry 45:1260-1266. T2a (4512)

Grohskopf, H. and Wilcoxon, F. [1953], A factorial design applied to a specific chemical process and development problem (summary). JASA 48:631. T2a (4513)

Gross, A. J. and Mantel, N. [1967], The effective use of both positive and negative controls in screening experiments. Biometrics 23:285-295. T8ac (4514)

Grové, C. C. [1965], The influence of temperature on the scouring of raw wool. American Dyestuff Reporter 54(1):13-16. T4a (4515)

Gruber, R. [1961], Une application de l'adjustement au moyen d'un système de polynomes orthogonaux en pisciculture. Biométrie-Praximétrie 2:107-120. T2ae (4516)

Grundy, P. M. and Healy, M. J. R. [1950], Restricted randomization and Quasi-Latin squares. JRSSB 12:286-291. E7ac;T2ac (4517)

Guest, P. G. [1958], The spacing of observations in polynomial regression. AMS 29:294-299. T1c (4518)

Guimarães, R. F., Gomes, F. P., and Malavolta, E. [1959], Adubação em "torrão paulista" de Eucalyptus Saligna SM. Anais Escola Superior Agri. "Luiz de Queiroz" 16:211-218. E2ae;T2ae (4519)

Gulliksen, H. [1956], A least squares solution for paired comparisons with incomplete data. Psychometrika 21:125-134. T11ae (4520)

Gulliksen, H. and Tucker, L. R. [1961], A general procedure for obtaining paired comparisons from multiple rank orders. Psychometrika 26:173-183. T11ac (4521)

Gulliksen, H. and Tukey, J. W. [1958], Reliability for the law of comparative judgment. Psychometrika 23:95-110. T11ae (4522)

Gun, A. [1965], The use of a preliminary test for interactions in the estimation of factorial means. Ph. D. Thesis, Univ. N. Carolina. T2a (4523)

Gun, A. M. [1967], Use of a preliminary test in the estimation of factorial means. Calcutta Stat. Assoc. Bull. 16:49-72. T2a;A11 (4524)

Gupta, S. S. and Shah, B. K. [1965], Exact moments and percentage points of the order statistics and the distribution of the range from the logistic distribution. AMS 36:907-920. T8a (4525)

Gupta, S. S. and Shah, B. K. [1965], Best linear unbiased estimators of the parameters of the logistic distribution using order statistics. Dept. Statistics Mimeo. Ser. No. 52, Purdue Univ. pp. 1-39. T8a (4526)

Gurland, J. [1961], Determination of minute insecticidal residues through biological assay (with discussion). Quantitative Methods in Pharmacology, pp. 342-356 (Math. Res. Center Tech. Summary Report No. 191, U.S. Army and Univ. Wisconsin). T8a (4527)

Gurland, J., Lee, I., and Dahm, P. A. [1960], Polychotomous quantal response in biological assay. Biometrics 16:382-398. T8ae (4528)

Gurley, W. R. [1965], Optimal symmetrical designs. Ph. D. Thesis, Oklahoma State Univ. T1,7,10c (4529)

Gurnow, R. N. [1965], A note on G. S. Watson's paper 'A study of the group screening method'. Technometrics 7:444-446. T7a (4530)

Haberman, S. [1958], A point of view in the analysis of simulation data. Proc. Third Conf. Design Expt. Army Res. Dev. Testing. pp. 191-219. T2ae (4531)

Hackler, W. C., Kriegel, W. W., and Hader, R. J. [1956], Effect of raw-material ratios on absorption of whiteware compositions. J. American Ceramic Soc. 39(1):20-25. T4a (4532)

Hader, R. J. [1958], Planning and analyzing experiments to determine the effects of several factors. Textile Quality Control Papers 5: T2,3,4a (4533)

Hader, R. J., Box, G. E. P., and Hunter, J. S. [1953], Experimental designs for multifactorial experiments. Inst. Stat. Mimeo. Ser. No. 71, Univ. N. Carolina. T4ac (4534)

Hader, R. J., Harward, M. E., Mason, D. D., and Moore, D. P. [1957], An investigation of some of the relationships between copper, iron, and molybdenum in the growth and nutrition of lettuce: I. Experimental design and statistical methods for characterizing the response surface. Soil Sci. Soc. America Proc. 21:59-64. T4ae (4535)

Hahn, G. J. and Shapiro, S. S. [1967], A catalog and computer program for use with symmetric and asymmetric fractional factorial experiments (abstract). Technometrics 9:188. T3,7a;A15 (4536)

Haldane, J. B. S. [1963], The design of experiments on mutation rates. J. Genetics 58:232-236. T9a;S1 (4537)

Haley, K. D. C. [1953], Estimation of the dosage mortality relationship when the dose is subject to error. Ph. D. Thesis, Stanford Univ. T8a (4538)

Hall, W. B. [1964], Incomplete 2^n factorial designs (abstract). JISAS 16:173. T7ac (4539)

Hallauer, A. R. and Wright, J. A. [1967], Genetic variances in the open-pollinated variety of maize, Iowa Ideal. Der Züchter 37:178-185. T9a (4540)

Hamaker, H. C. [1955], Proefopzetten in de industrie. (Experimental designs in industry.) Stat. Neerlandica 9:209-232. T2ae;C (4541)

Hamaker, H. C. [1958], Statistiek en experiment. (Statistics and experiments.) Stat. Neerlandica 12:119-130. T2ae;C (4542)

Hamaker, H. C. [1958], De recente ontwikkeling in proefopzetten met kwantitatieve factoren. (Recent developments in experimental designs for dealing with quantitative factors.) Stat. Neerlandica 12:201-212. T4,5,6ac (4543)

Hamaker, H. C. [1961], Examples of designed experiments. IQC 17(9):16-20. T2ae;C (4544)

Hansche, P. E. and Beres, V. [1966], An analysis of environmental variability in sweet cherry. Proc. American Soc. Hort. Sci. 88:167-172. T9c;S1 (4545)

Hansche, P. E., Beres, V., and Brooks, R. M. [1966], Heritability and genetic correlation in the sweet cherry. Proc. American Soc. Hort. Sci. 88:173-183. T9a (4546)

Hansen, E. D., Wiebe, H. H., and Thorne, W. [1958], Air pollution with relation to agronomic crops: VII. Fluoride uptake from soils. Agron. J. 50:565-568. T8ac (4547)

Hanson, W. D. and Brim, C. A. [1963], Optimum allocation of test material for two-stage testing with an application to evaluation of soybean lines. Crop Sci. 3:43-49. T9c;S2 (4548)

Harrington, E. C. [1955], Statistical experiment in plastics fabrication. Soc. Plastic Eng. J. 11(6):19-21. T2,4a (4549)

Harshbarger, B. [1953], Paired comparisons in a lattice design (abstract). Virginia J. Sci. 4:287. E11a;T11a (4550)

Harshbarger, B. [1955], The 2^3 factorial in a Latinized rectangular lattice design (summary). JASA 50:577 (also Dept. Stat. Tech. Report. No. 1, OOR Project No. 1166, Virginia Polytechnic Inst.). E11ac;T2a (4551)

Harshbarger, B. [1958], A statistical design for a surveillance test. Proc. Second Conf. Design Expt. Army Res. Dev. Testing. pp. 110-117. T2ae (4552)

Harter, H. L. and Lum, M. D. [1962], An interpretation and extension of Tukey's one degree of freedom for non-additivity. Aeronautical Res. Laboratory, ARL 62-313, U.S. Air Force. E2a;T2a;A4 (4553)

Hartigan, J. A. [1966], Probabilistic completion of a knockout tournament. AMS 37:495-503. T11ac;A2 (4554)

Hartley, H. O. [1959], Smallest composite designs for quadratic response surfaces. Biometrics 15:611-624. T4ac (4555)

Hartley, H. O. [1960], Analysis of variance. Ch. 20, Mathematical Methods for Digital Computers, pp. 221-230. E2,5,7,12a;T2a (4556)

Hartley, H. O. [1966], Maximum likelihood estimation for unbalanced factorial data. Proc. Eleventh Conf. Design Expt. Army Res. Dev. Testing. pp. 597-606. T2a;A10 (4557)

Harvey, W. R. [1960], Least-squares analysis of data with unequal subclass numbers. Agri. Res. Service, ARS-20-8, U.S. Dept. Agri. T2,9a;A10 (4558)

Hawkins, D. F. [1964], Observations on the application of the Robbins-Monro process to sequential toxicity assays. British J. Pharmacology Chemotherapy 22:392-402. T1,8a;S2 (4559)

Hawkins, D. F. [1964], Designs for sequential toxicity assays using the Robbins-Monro process (abstract). Biometrics 20:914-915. T8c;S2 (4560)

Hayashi, C. [1964], Multidimensional quantification of the data obtained by the method of paired comparison. Ann. Inst. Stat. Math. 16:231-245. T11a (4561)

Hayman, B.I. [1954], The theory and analysis of diallel crosses. Genetics 39:789-809. T9ae (4562)

Hayman, B. I. [1957], Interaction, heterosis and diallel crosses. Genetics 42: 336-355. T9a (4563)

Hayman, B. I. [1958], The theory and analysis of diallel crosses. II. Genetics 43:63-85. T9ae (4564)

Hayman, B. I. [1960], The theory and analysis of diallel crosses. III. Genetics 45:155-172. T9a (4565)

Hayman, B.I. [1963], Notes on diallel-cross theory (with discussion). Statistical Genetics and Plant Breeding, pp. 571-578. T9a (4566)

Hayman, B.I. and Mather, K. [1955], The description of genic interactions in continuous variation. Biometrics 11:69-82. T9a (4567)

Heady, E. O. [1956], Methodological problems in fertilizer use. Methodological Procedures in the Economic Analysis of Fertilizer Use Data, pp. 3-21. T4a (4568)

Heady, E. O. [1956], Fertilization in relation to conservation farming and allocation of resources within the farm. Methodological Procedures in the Economic Analysis of Fertilizer Use Data, pp. 188-200. T4a (4569)

Heady, E. O. and Dillon, J. L. [1961], Agricultural Production Functions, chapters 2-7, 14. Iowa State Univ. Press, Ames, Iowa. E2a;T2,4a (4570)

Heady, E. O. and Pesek, J. [1954], A fertiliser production surface with specification of economic optima for corn grown in calcareous Ida silt loam. J. Farm Economics 36:466-482. T4a (4571)

Heady, E. O. and Pesek, J. T. [1957], Some methodological considerations in the Iowa-TVA research project on economics of fertilizer use. Economic and Technical Analysis of Fertilizer Innovations and Resource Use, ch. 14, pp. 144-167. T2a (4572)

Heady, E. O., Pesek, J. T., and Brown, W. G. [1955], Crop response surfaces and economic optima in fertilizer use. Iowa State Univ. Agri. Expt. Sta. Res. Bull. 424:292-332. T4a (4573)

Healy, M. J. R. [1950], The planning of probit assays. Biometrics 6:424-431. T8c (4574)

Healy, M. J. R. [1951], Latin rectangle designs for a 2^n factorial experiments on 32 plots. J. Agri. Sci. 41:315-316. E7ac;T2a (4575)

Healy, M. J. R. [1956], Principles of biological assay. Modern Methods of Plant Analysis, Vol. 1. T8a;C (4576)

Healy, M. J. R. [1956], The analysis of a factorial experiment with additional treatments. J. Agri. Sci. 47:205-206. T2,7ae (4577)

Healy, M. J. R. [1959], The analysis of field experiments on an electronic computer. Biom. Zeit. 1:210-214. E2,7a;T2a;A15 (4578)

Healy, M. J. R. [1960], Analysis of factorial experiments on an electric computer (abstract). Biometrics 16:314-315. T2a;A15 (4579)

Healy, M. J. R. and Gower, J. C. [1961], Aliasing in partially confounded factorial experiments. Biometrika 48:218-220. T3a (4580)

Hecker, R. J. [1967], Evaluation of three sugar beet breeding methods. J. Amer. Soc. Sugar Beet Tech. 14:309-318. T9a (4581)

Hecker, R. J., Federer, W. T., Powers, L., and Payne, M. G. [1967], Relationships of four yield characters with five chemical characters in sugar beets. J. Amer. Soc. Sugar Beet Tech. 14:555-573. T9a;A7 (4582)

Henderson, C. R. [1952], Specific and general combining ability. Heterosis, ch. 22, pp. 352-370. T9a (4583)

Henderson, C. R. [1963], Selection index and expected genetic advance. Statistical Genetics and Plant Breeding, pp. 141-163. T9a;A7 (4584)

Hermanson, H. P. [1965], Maximization of potato yield under constraint. Agron. J. 57:210-213. T4a (4585)

Hermanson, H. P., Gates, C. E., Chapman, J. W., and Farnham, R. S. [1964], An agronomically useful three-factor response surface design based on dodecahedrc symmetry. Agron. J. 56:14-17. T4ace (4586)

Herzberg, A. M. [1963], Certain third order rotatable designs in four dimensions. M. A. Thesis, Univ. Saskatchewan. T4ac (4587)

Herzberg, A. M. [1964], Two third order rotatable designs in four dimensions. AMS 35:445-446. T4c (4588)

Herzberg, A. M. [1966], On rotatable and cylindrically rotatable designs. Ph. D. Thesis, Univ. Saskatchewan. T4,10c (4589)

Herzberg, A. M. [1966], Cylindrically rotatable designs. AMS 37:242-247. T4c (4590)

Herzberg, A. M. [1967], The behaviour of the variance function of the difference between two estimated responses. JRSSB 29:174-179. T4a (4591)

Herzberg, A. M. [1967], Cylindrically rotatable designs of types 1, 2 and 3. AMS 38:167-176. T4c (4592)

Herzberg, A. M. [1967], A method for the construction of second order rotatable designs in k dimensions. AMS 38:177-180. T4c (4593)

Hewlett, P. S. and Plackett, R. L. [1950], Statistical aspects of the independent joint action of poisions, particularly insecticides. II. Examination of data for agreement with the hypothesis. Ann. Appl. Biology 37:527-552. T8ae (4594)

Hewlett, P. S. and Plackett, R. L. [1952], Similar joint action of insecticides. Nature 169:198-199. T8ae (4595)

Hewlett, P. S. and Plackett, R. L. [1956], The relation between quantal and graded responses to drugs. Biometrics 12:72-78. T8a (4596)

Hewlett, P. S. and Plackett, R. L. [1959], A unified theory for quantal responses to mixtures of drugs: non-interactive action. Biometrics 15:591-610. T8a (4597)

Hewlett, P. S. and Plackett, R. L. [1960], Models for quantal responses to mixtures of drugs. Biometrics 16:488. T8a (4598)

Hewlett, P. S. and Plackett, R. L. [1961], Models for quantal responses to mixtures of two drugs (with discussion). Quantitative Methods of Pharmacology, pp. 328-339. T8a (4599)

Hewlett, P. S. and Plackett, R. L. [1964], A unified theory for quantal responses to mixtures of drugs: competitive action. Biometrics 20:566-575. T8a (4600)

Hext, G. R. [1963], The estimation of second-order tensors, with related tests and designs. Biometrika 50:353-373. T4ace (4601)

Hicks, C. R. [1956], Fundamentals of analysis of variance, part I - The analysis of variance (ANOVA) model. IQC 13(2):17-20. Elae;T2ae (4602)

Hicks, C. R. [1956], Fundamentals of analysis of variance, part II - The components of variance and the mixed model. IQC 13(3):5-8. Elae;T2ae (4603)

Hicks, C. R. [1956], Fundamentals of analysis of variance, part III - Nested designs in analysis of variance. IQC 13(4):13-16. Elae;T2ae (4604)

Hildreth, C. G. [1956], Discrete models with qualitative restrictions. Methodologica Procedures in the Economic Analysis of Fertilizer Use Data, pp. 62-75. T2a;A7 (4605)

Hildreth, C. [1957], Possible models for agronomic-economic research. Economic and Technical Analysis of Fertilizer Innovations and Resource Use, ch. 16, pp. 176-186. T2,4a (4606)

Hill, A. C., Transtrum, L. G., Pack, M. R., and Winters, W. S. [1958], Air pollution with relation to agronomic crops. VI. An investigation of the "hidden injury" theory of fluoride damage to plants. Agron. J. 50:562-563. T8ac (4607)

Hill, H. M. [1960], Experimental designs to adjust for time trends. Technometrics 2:67-82. E5ace;T4ace (4608)

Hill, H. M. and Wheeler, D. [1958], A designed experiment to evaluate seven yarn lubricants. ASQC, Nat. Conv. Trans. 12:245-253. E3ae;T2ae (4609)

Hill, R. R. [1966], Designs to estimate effects of clone substitution in alfalfa synthetics. Crop Sci. 6:471-473. T9ac (4610)

Hill, W. G. and Robertson, A. [1966], The effect of linkage on limits to artificial selection. Genetical Res. 8:269-294. T9a (4611)

Hill, W. J. and Hunter, W. G. [1966], A review of response surface methodology: A literature survey. Technometrics 8:571-590 (also Dept. Stat. Tech. Report No. 62, Univ. Wisconsin). T4a;B (4612)

Hill, W. J. and Hunter, W. G. [1966], Design of experiments for model discrimination in multiresponse situations. Dept. Stat. Tech. Report No. 65, Univ. Wisconsin. T4a (4613)

Hill, W. J. and Hunter, W. G. [1966], A joint design criterion for the dual problem of model discrimination and parameter estimation. Dept. Stat. Tech. Report No. 69, Univ. Wisconsin. T4ace (4614)

Hinchen, J. D. [1965], Experimentation without data. ASQC, Ann. Tech. Conf. Trans. pp. 533-538. T4ae (4615)

Hinkelmann, K. H. [1963], Design and analysis of multi-way genetic cross experiments. Ph. D. Thesis, Iowa State Univ. T7,9ac (4616)

Hinkelmann, K. [1963], A commonly occurring incomplete multiple classification model. Biometrics 19:105-117. T7a (4617)

Hinkelmann, K. [1965], Partial triallel crosses. Sankhyā A 27:173-196. T9ac (4618)

Hinkelmann, K. [1965], Unvollständige triallele. Biom. Zeit. 7:222-229. T9ac (4619)

Hinkelmann, K. [1966], Unvollständige diallele Kreuzungspläne. Biom. Zeit. 8: 242-265. T9a (4620)

Hinkelmann, K. [1967], Circulant partial triallel crosses. Biom. Zeit. 9:22-33. E3ac;T9ac (4621)

Hinkelmann, K. and Kempthorne, O. [1963], Two classes of group divisible partial diallel crosses. Biometrika 50:281-291. T9c (4622)

Hinkelmann, K. and Stern, K. [1960], Kreuzungspläne zur selektionszüchtung bei waldbaumen. Silvae Genetica 9:121-133. T9ae (4623)

Hirotsu, C. [1966], Types of incomplete block designs for factorials. Reports Stat. Appl. Res. 13(2):12-28. E3a;T2a (4624)

Hirotsu, C. [1967], A note on the analysis of variance for a complete n-way layout. Reports Stat. Appl. Res. 14:169-172. T2a (4625)

Hocking, R. R. and Leslie, R. N. [1967], Selection of the best subset in regression analysis. Technometrics 9:531-540 (abstract 9:188). Tlac (4626)

Hodges, J. L. [1960], A two stage sequential design for bio-assay (abstract). Biometrics 16:127. T8c;S2 (4627)

Hodges, J. L. and Lehmann, E. L. [1954], Matching in paired comparisons. AMS 25:787-791. Tllc (4628)

Hodnett, G. E. [1956], The use of response curves in the analysis and planning of series of experiments with fertilizers. Empire J. Exptl. Agri. 24:205-212. Tl,4a (4629)

Hodnett, G. E. [1956], The analysis of a 3x6 experiment arranged in a quasi-latin square. Biometrics 12:245-258. E7ae;T2ae (4630)

Hoel, P. G. [1958], Efficiency problems in polynomial estimation. AMS 29:1134-1145. Tlc (4631)

Hoel, P. G. [1961], Some properties of optimal spacing in polynomial estimation. Ann. Inst. Stat. Math. 13:1-8. Tlac (4632)

Hoel, P. G. [1965], Minimax designs in two dimensional regression. AMS 36: 1097-1106. Tlc (4633)

Hoel, P. G. [1965], Optimum designs for polynomial extrapolation. AMS 36: 1483-1493. T1c (4634)

Hoel, P. G. and Levine, A. [1964], Optimal spacing and weighting in polynomial prediction. AMS 35:1553-1560. T1c (4635)

Hoepner, P. H. and Lutz, J. A. [1966], An application of multiple covariance analysis to the estimation of fertilizer response functions. Agron. J. 58: 66-69. T4a;A7 (4636)

Hoerl, A. E. [1959], Optimum solution of many variables equations. Chemical Eng. Progress 55(11):69-78. T4ac (4637)

Hoerl, A. E. [1962], The application of ridge analysis to regression problems. Chemical Eng. Progress 58(3):54-59. T4ae;A7 (4638)

Hofmann, J. [1955], Life testing in controlled environmental conditions. Ph. D. Thesis, Iowa State Univ. T2,3a (4639)

Holloway, C. [1957], A systematic method of finding defining contrasts. JASA 52:46-52. T3a (4640)

Holms, A. G. [1966], Multiple-decision procedures for the ANOVA of two-level factorial replication-free experiments. Ph. D. Thesis, Western Reserve Univ. T2a;A2 (4641)

Homeyer, P. G., Preston, R. L., and Burroughs, W. [1959], The design and interpretation of a bioassay for detecting small quantities in the presence of interfering substances (abstract). Biometrics 15:152-153. T8ac (4642)

Hooke, R. and Jeeves T. A. [1961], "Direct search" solution of numerical and statistic problems. J. Assoc. Computing Machinery 8:212-229. T4ac;A15;S2 (4643)

Hopp, H. [1952], Some experimental designs useful for exploratory research in underdeveloped areas. Office Foreign Agri. Relations, U.S. Dept. Agri. T2ae (4644)

Horner, T. W. [1953], Non-allelic gene interaction and the interpretation of quantitative genetic data. Ph. D. Thesis, Univ. N. Carolina. T9a (4645)

Horner, T. W., Comstock, R. E., and Robinson, H. F. [1955], Non-allelic gene interactions and the interpretation of quantitative genetic data. N. Carolina Agri. Expt. Sta. Tech. Bull. 118. T9a (4646)

Horner, T. W. and Lana, E. P. [1957], A three-year study of general and specific combining ability in tomatoes. Proc. American Soc. Hort. Sci. 69:378-387. T9a (4647)

Horton, W. H. [1958], Experiences with fractional factorials. Experimental Designs in Industry, pp. 207-223. T3ace (4648)

Hromi, J. D. [1955], Application of the analysis of variance to problems in metallurgical research. ASQC, Nat. Conv. Trans. 9:731-741. T2a (4649)

Hromi, J. D. [1957], Some fractional factorial experiments. (Cited by V. Chew, Experimental Designs in Industry, p.266.) T3a (4650)

Huber, P. J. [1963], A remark on a paper of Trawinski and David entitled: "Selection of the best treatment in a paired-comparison experiment". AMS 34:92-94. T11a;A2 (4651)

Hughes, M. C. [1966], A systematic method for obtaining predicted values in mixed-level factorial experiments. IQC 22:348-354. T2ae (4652)

Hunter, J. S. [1954], Multi-factor experimental designs. Ph. D. Thesis, N. Carolina State College. T4ac (4653)

Hunter, J. S. [1954], Searching for optimum conditions. Trans. New York Academy Sci. Ser. II, 17:124-132. T2,4a (4654)

Hunter, J. S. [1954], Some third order composite designs (abstract). Biometrics 10:306. T4a (4655)

Hunter, J. S. [1956], Statistical methods for determining optimum conditions. ASQC, Nat. Conv. Trans. 10:415-428. T4ae (4656)

Hunter, J. S. [1957], Experimental methods of determining optimum conditions. Proc. First Conf. Design Expt. Army Res. Dev. Testing. pp. 17-27. T4ae (4657)

Hunter, J. S. [1958], A discussion on rotatable designs. ASQC, Nat. Conv. Trans. 12:531-543. T4ae (4658)

Hunter, J. S. [1958], Determination of optimum operating conditions by experimental methods, part II-1 - Models and methods. IQC 15(6):16-24. T4ae (4659)

Hunter, J. S. [1959], Determination of optimum operating conditions by experimental methods Part II-2, Models and methods. IQC 15(7):7-15. T4,5ace (4660)

Hunter, J. S. [1959], Determination of optimum operating conditions by experimental methods Part II-3, Models and methods. IQC 15(8):6-14. T4ace (4661)

Hunter, J. S. [1960], Some applications of statistics to experimentation. Chemical Eng. Progress Symp. Ser. 56(31):10-26. T4ace (4662)

Hunter, J. S. [1960], Optimize your chemical process with evolutionary operations. Chemical Eng. 67(September 19):193-202. T4a (4663)

Hunter, J. S. [1962], Supersaturated fractions (summary). JASA 57:492. T7c (4664)

Hunter, J. S. [1963], Sequential factorial designs (abstract). Biometrics 19:661. T8ac;S2 (4665)

Hunter, J. S. [1964], Sequential factorial estimation. Technometrics 6:41-55 (errata 7:93). T3,7ace;S2 (4666)

Hunter, J. S. [1966], The inverse Yates algorithm. Technometrics 8:177-183. T2a (4667)

Hunter, J. S. [1966], Letter to the Editor on "The inverse Yates algorithm". Technometrics 8:559. T2a (4668)

Hunter, J. S. [1966], Estimation and design for non-linear models. Proc. Eleventh Conf. Design Expt. Army Res. Dev. Testing. pp. 1-27. T4a;A7 (4669)

Hunter, J. S. [1966], Factorial, one run at a time, experiments. IQC 22:391-396. T2,4a (4670)

Hunter, W. G. and Hoff, M. L. [1966], Planning experiments to increase research efficiency. Dept. Stat. Tech. Report No. 78, Univ. Wisconsin. T2,4a;C (4671)

Hunter, W. G. and Hoff, M. L. [1967], Planning experiments to increase research efficiency. Ind. Eng. Chemistry 59(3):43-48. T2,3,4,7ae (4672)

Hunter, W. G. and Kittrell, J. R. [1966], Evolutionary operation: A review. Technometrics 8:389-397. T4a;B (4673)

Hunter, W. G. and Mezaki, R. [1964], Catalyst selection: An application of group screening. Dept. Stat. Tech. Report No. 27, Univ. Wisconsin. T7a (4674)

Hunter, W. G. and Mezaki, R. [1964], Catalyst selection by group screening. Ind. Eng. Chemistry 56(3):38-40. T7a (4675)

Hunter, W. G. and Reiner, A. M. [1965], Designs for discriminating between two rival models. Technometrics 7:307-323 (also Dept. Stat. Tech. Report No. 32, Univ. Wisconsin). T7ac;S2 (4676)

Hurst, D. C. [1962], Modifications of response surface techniques for biological use. Ph. D. Thesis, Univ. N. Carolina. T4,7ac (4677)

Hurst, D. C. [1958], The use of response surface analysis on biological material (abstract). Virginia J. Sci. 9:448. T4,8a (4678)

Hurst, D. C. and Mason, D. D. [1957], Some statistical aspects of the TVA-North Carolina cooperative project on determination of yield response surfaces for corn. Economic and Technical Analysis of Fertilizer Innovations and Resource Use, ch. 18, pp. 207-216. T2,4a (4679)

Huth, R. F. [1958], Factorial chi-square as a search technique. ASQC Nat. Conv. Trans. 12:167-176. T2a (4680)

Hutton, C. E., Robertson, W. K., and Hanson, W. D. [1956], Crop response to different soil fertility levels in a 5 by 5 by 5 by 2 factorial experiment: I. Corn. Soil Sci. Soc. America Proc. 20:531-537. T2a (4681)

Ibach, D. B. [1956], Evaluating response to fertilizer using standard yield curves. Methodological Procedures in the Economic Analysis of Fertilizer Use Data, pp. 142-147. T4a;A7 (4682)

Ibach, D. B. [1958], Use of standard exponential yield curves. USDA, Agri. Res. Service 43-69. T4a;A7 (4683)

Ibach, D. B. and Adams, W. E. [1965], An economic analysis of fertility experiments with coastal and common Bermudagrasses. (Cynodon dactylon (L.) Pers.) Agron. J. 57:15-20. T2a;A7 (4684)

Iizuka, K. [1963], Design and analysis of an experiment on the accuracy of shore hardness scale, realized by the conversion of Vickers hardness. Reports Stat. Appl. Res. 10:248-256. T2ae (4685)

Imhof, J. P. [1960], A mixed model for the complete three-way layout with two random-effects factors. AMS 31:906-928. T2a (4686)

Imhof, J. P. [1962], Testing the hypothesis of no fixed main-effects in Sheffé's mixed model. AMS 33:1085-1095. T2a (4687)

Inkson, R. H. E. [1961], The analysis of a $3^2 \times 2^2$ factorial experiment with confounding Appl. Stat. 10:98-107. T2ae (4688)

Inkson, R.H. E. [1964], The precision of estimates of the soil content of phosphate using the Mitscherlich response equation. Biometrics 20:873-882. T4a (4689)

Ipsen, J. [1955], Appropriate scores in bio-assays using death-times and survivor symptoms (abstract). Biometrics 11:240-241. T8a (4690)

Ishida, S. [1962], Application of the orthogonal design method to quantal response in factorial design. Reports Stat. Appl. Res. 9:24-33. T2,8a (4691)

Izzo, J. L. [1952], A clinical comparison of modified insulins I. The clinical problem. Biometrics 8:206-209, 215-217. T8ae;C (4692)

Izzo, J. L. and Crumb, S. L. [1950], A clinical comparison of modified insulins. J. Clinical Investigation 29:1514-1527. T8ae;C (4693)

Jackson, J. E. [1963], Evaluation of large experiments using incomplete paired comparisons (summary). JASA 58:554. T11ac (4694)

Jackson, J. E. and Fleckenstein, M. [1957], An evaluation of some statistical techniques used in the analysis of paired comparison data. Biometrics 13:51-64. T11a;A12 (4695)

Jacob, W. C. [1953], Split-plot half-plaid squares for irrigation experiments. Biometrics 9:157-175. E9,12ae;T2ae (4696)

Jain, R. C. and Das, M. N. [1967], Analysis of fractionally replicated experiments and linking linear and quadratic contrasts with the usual contrasts used in symmetrical factorial experiments (abstract). JISAS 19(1):152. T2,3a (4697)

James, A. N. [1963], Factorial experiments. The Statistician, Journal of the Institute of Statisticians 13:203-209. T2a (4697a)

James, J. W. [1962], Conflict between directional and centripetal selection. Heredity 17:487-499. T9a;A2 (4698)

James, J. W. [1965], Response curves in selection experiments. Heredity 20: 57-63. T9a;A2,7 (4699)

James, J. W. [1966], Selection from one or several populations. Australian J. Agri. Res. 17:583-589. T9ac (4700)

Jardine, R. [1958], Animal breeding and the estimation of genetic value. Heredity 12:499-511. T9a (4701)

Jebe, E. H. [1967], Some statistical applications in the testing of military vehicle rubber components. Proc. Twelfth Conf. Design Expt. Army Res. Dev. Testing. pp. 51-90. T2,7ae (4702)

Jebe, E. H. and Brown, W. A. [1961], Problems in the analysis and interpretation of information processing experiments. Proc. Sixth Conf. Design Expt. Army Res. Dev. Testing. pp. 91-109. T3a;C (4703)

Jenkins, G. M. and Chanmugam, J. [1962], The estimation of slope when the errors are autocorrelated. JRSSB 24:199-214. T4ace (4704)

Jensen, D. and Pesek, J. [1959], Generalization of yield equations in two or more variables: I. Theoretical considerations. II. Application to yield data. Agron. J. 51:255-263. T2a;A7 (4705)

Jensen, N. F. and Federer, W. T. [1965], Competing ability in wheat. Crop. Sci. 5:449-452. T9a;A14 (4706)

Jensen, S. [1959], Combining ability of unselected inbred lines of corn from incomplete diallel and top-cross tests. Ph. D. Thesis, Iowa State Univ. T9ac (4707)

Jinks, J. L. [1952], The analysis of quantitative data from a diallel cross between inbred varieties of Nicotiana rustica. Ph. D. Thesis, Univ. Birmingham. T9a (4708)

Jinks, J. L. [1954], The analysis of continuous variation in a diallel cross of Nicotiana rustica varieties. Genetics 39:767-788. T9a (4709)

Jinks, J. L. [1955], A survey of the genetical basis of heterosis in a variety of diallel crosses. Heredity 9:223-238. T9a (4710)

Jinks, J. L. [1956], The F_2 and backcross generations from a set of diallel crosses. Heredity 10:1-30. T9a (4711)

Jinks, J. L. and Hayman, B. I. [1953], The analysis of diallel crosses. Maize Genetics Coöperation News Letter 27:48-54. T9a (4712)

John, J. A. [1967], Reduced group divisible paired comparison designs. AMS 38: 1887-1893. E3c;T11c (4713)

John, P. W. M. [1960], Three-quarter replicates of 2^3 and 2^4 designs (abstract). AMS 31:816-817. T7c (4714)

John, P. W. M. [1961], A $3(2^{8-4})$ design of resolution V (abstract). AMS 32:1347. T7c (4715)

John, P. W. M. [1961], Three-quarter replicates of 2^4 and 2^5 designs. Biometrics 17:319-321. T7c (4716)

John, P. W. M. [1961], Three-quarter replicates of 2^n designs (abstract). AMS 32:636. T7c (4717)

John, P. W. M. [1962], Three quarter replicates of 2^n designs. Biometrics 18: 172-184. T7ace (4718)

John, P. W. M. [1963], Confounding in the $3(2^{4-2})$ designs (abstract). AMS 34: 689. T7c (4719)

John, P. W. M. [1963], Confounding $3(2^{5-2})$ designs of resolution V. AMS 34: 1125-1126. T7c (4720)

John, P. W. M. [1963], Using a two factor analysis of variance programme for experiments with several factors. Appl. Stat. 12:129-132. E12a;T2a;A15 (4721)

John, P. W. M. [1963], Analysis of diallel cross experiments in a split-plot situation. Australian J. Biological Sci. 16:681-687. E12a;T9a (4722)

John, P. W. M. [1964], Blocking of $3(2^{n-k})$ designs. Technometrics 6:371-376. T7ac (4723)

John, P. W. M. [1965], Semifolding s^{n-k} design (abstract). AMS 36:1076. T7c (4724)

John, P. W. M. [1966], Augmenting 2^{n-1} designs. Technometrics 8:469-480. T7ace (4725)

John, P. W. M. [1966], On identity relationships for 2^{n-r} designs having words of equal length. AMS 37:1842-1843. T3c (4726)

John, P. W. M. [1966], Nested fractions of 2^n designs (abstract) (with discussion). Bull. ISI 41:842. T3c (4727)

Johnson, A. F. [1959], Sensitivity and efficiency of experimental design. ASQC Nat. Conv. Trans. 13:447-450. T1c (4728)

Johnson, A. F. [1966], Properties of second order designs: Effect of transformation or truncation on prediction variance. Appl. Stat. 15:48-50. T4a (4729)

Johnson, A. F. [1967], On query No. 20. Technometrics 9:490. T2a (4730)

Johnson, C. H. [1963], Some properties of the methods of steepest ascent. Ph. D. Thesis, Oklahoma State Univ. T4,10a (4731)

Johnson, C. H. and Folks, J. L. [1964], A property of the method of steepest ascent. AMS 35:435-437. T4c (4732)

Johnson, G. L. [1956], Interdisciplinary considerations in designing experiments to study the profitability of fertilizer use. Methodological Procedures in the Economic Analysis of Fertilizer Use Data. pp. 22-36. T2a;C (4733)

Johnson, G. L. [1957], Planning agronomic-economic research in view of results to date. Economic and Technical Analysis of Fertilizer Innovations and Resource Use, ch. 19, pp. 217-225. T2,4a;C (4734)

Johnson, L. P. V. [1963], Applications of the diallel-cross techniques to plant breeding. Statistical Genetics and Plant Breeding, pp. 561-570. T9a (4735)

Johnson, P. A. [1966], Four and five level designs for discriminating between quadratic and exponential models. M.S. Thesis, Iowa State Univ. T4a (4736)

Johnson, R. C., Ball, W. E., Bruggrabe, W. F., Heiny, R. L., Russell, J. L., and Sweeny, R. F. [1960], Mathematics, computers, operations research, and statistics. Ind. Eng. Chemistry 52(4):359-367. T4ac (4737)

Jolly, G. M. [1950], The use of probits in combining precentage kills. Ann. Appl. Biology 37:597-606. T8ae;A9,13 (4738)

Juvancz, I., Fischer, J., and Csáki, P. [1964], Evaluation of response curves by the use of extreme values (abstract). Biometrics 20:386-387. T4a (4739)

Kadane, J. B. [1965], Some equivalence classes important in paired comparisons. Dept. Stat. Tech. Report No. 8, Stanford Univ. T11a (4740)

Kadane, J. B. [1966], Some equivalence classes in paired comparisons. AMS 37: 488-494. T11a (4741)

Kahlon, A. S. and Saxena, P. N. [1958], Economic analysis of bajra fertilizer rate experiments. A study of production functions. Indian J. Agron. 2:153-165. T4a (4742)

Kalabukhov, N. I. [1958], A denial of the laws of variation of organisms. (An English translation by W. D. C. Scott of the Russian periodical "J. Microbiology, Epidemiology and Immunobiology", Vol. 28, No. 3, 1957). Amer. Stat. 12(2): 20 and 26. T8a (4743)

Kalotay, A. J. [1966], Some connections between the design of experiments and coding theory. M. S. Thesis, Queen's Univ. T3ac (4744)

Karlin, S. and Studden, W. J. [1966], Optimal experimental designs. AMS 37: 783-815. T1ac (4745)

Karson, M. J. [1967], Bias and variance criteria for estimators and designs for fitting polynomial responses. Ph. D. Thesis, Univ. N. Carolina. T4ac (4746)

Karson, M. J., Manson, A. R., and Hader, R. J. [1966], Bias and variance criteria for estimators and designs for fitting polynomial responses (preliminary report) (abstract). AMS 37:1067. T1ac (4747)

Kartha, C. P. [1967], On construction of symmetrical fractional factorials and their tabulation (abstract). JISAS 19(1):149. T3c (4748)

Katsanis, D. J. and Fulton, C. L. [1961], Application of factorial experiment and Box technique to ballistic devices. Proc. Sixth Conf. Design Expt. Army Res. Dev. Testing. pp. 187-223. T2,4ae (4749)

Kehrberg, E. W. [1956], Some problems involved in fitting production functions to data recorded by soil-testing laboratories. Methodological Procedures in the Economic Analysis of Fertilizer Use Data, pp. 134-141. T4a;C (4750)

Kelleher, T. M. [1956], Analysis and interpretation of variation of inbred lines and F_1 crosses of corn. Ph. D. Thesis, Univ. N. Carolina. T9a (4751)

Kelleher, T. [1958], A synthesis of diallel cross methodology (abstract). Biometrics 14:570-571. T9a (4752)

Kelley, R. B. [1950], Inheritence of skin folding (wrinkling) of sheep. Australian J. Agri. Res. 1:471-495. T9e (4753)

Kempthorne, O. [1951], A review of the principles of experimental design and some applications. Comm. Exptl. Design, U.S. Dept. Agri., USDA Lecture Ser., Mimeo. T2a;C (4754)

Kempthorne, O. [1956], The theory of the diallel cross. Genetics 41:451-459. T9a (4755)

Kempthorne, O. [1961], The partial diallel cross (abstract). Biometrics 17:170. T9a (4756)

Kempthorne, O. [1965], Development of the design of experiments over the past ten years. Proc. Tenth Conf. Design Expt. Army Res. Dev. Testing. pp. 19-46. T1,4ac;B (4757)

Kempthorne, O. and Curnow, R. N. [1961], The partial diallel cross. Biometrics 17:229-250 (correction 18:128). T9ac (4758)

Kempthorne, O. and Tischer, R. G. [1953], An example of the use of fractional replication. Biometrics 9:295-303. T3ae (4759)

Kempthorne, O., Zyskind, G., Addelman, S., Throckmorton, T. N., and White, R. F. [1961], Analysis of variance procedures. Stat. Lab. and Aeronautical Res. Lab. ARL-149, Iowa State Univ. and U.S. Air Force. E11ac;T7ac;A5 (4760)

Kendall, M. G. [1955], Further contributions to the theory of paired comparisons. Biometrics 11:43-62. T11ac;A12 (4761)

Kenworthy, I. C. [1967], Some examples of simplex evolutionary operation in the paper industry. Appl. Stat. 16:211-224. T4a (4762)

Kenworthy, O. O. [1963], Factorial experiments with mixtures using ratios. IQC 19(12):24-26. T2,4a (4763)

Kesten, H. [1958], Accelerated stochastic approximation. AMS 29:41-59. T1ac (4764)

Kesting, L. W. [1964], An analysis of factorial experimental designs. Proc. Ninth Conf. Design Expt. Army Res. Dev. Testing. pp. 569-604. T3ae (4765)

Keuls, M. and Wiskunde, L. H. S. [1956], Het gebruik van de variantie-analyse bij het opsporen van de beste combinaties van ouders in de maisveredeling. (The use of variance analysis in detecting the best combination of parents in maize breeding.) Studkring. Plant Veredel., Wageningen Versl. 51:707-731 (Mimeo.). T9a (4766)

Kidwell, J. F. and Kempthorne, O. [1966], An experimental test of quantitative genetic theory. Der Züchter 36:163-167. T9a (4767)

Kiefer, J. C. [1957], On the non-optimality of symmetrical designs among randomized designs (abstract). AMS 28:1058. E14;T10 (4768)

Kiefer, J. [1957], Optimum sequential search and approximation methods under minimum regularity assumptions. J. Soc. Ind. Appl. Math. 5:105-136. T1c;S2 (4769)

Kiefer, J. [1959], Optimum experimental designs (with discussion). JRSSB 21: 272-319. E14;T1,10c (4770)

Kiefer, J. [1960], Optimum experimental designs (abstract). AMS 31:245. E14;T10 (4771)

Kiefer, J. [1961], Optimum designs in regression problems, II. AMS 32:298-325. T1c (4772)

Kiefer, J. [1961], Optimum experimental designs V, with applications to systematic and rotatable designs. Proc. Fourth Berkeley Symp. Math. Stat. Prob. 1:381-405. E5c;T1,4,10c (4773)

Kiefer, J. [1962], An extremum result. Canadian J. Math. 14:597-601. T1c (4774)

Kiefer, J. [1962], Two more criteria equivalent to D-optimality of designs. AMS 33:792-796. E14;T10 (4775)

Kiefer, J. and Wolfowitz, J. [1952], Stochastic estimation of the maximum of a regression function. AMS 23:462-466. T1ac (4776)

Kiefer, J. and Wolfowitz, J. [1959], Optimum designs in regression problems. AMS 30:271-294 (abstract 29:938). T1c (4777)

Kiefer, J. and Wolfowitz, J. [1959], The equivalence of two extremum problems (abstract). Notices American Math. Soc. 6:522. T1c (4778)

Kiefer, J. and Wolfowitz, J. [1960], The equivalence of two extremum problems. Canadian J. Math. 12:363-366. T1c (4779)

Kiefer, J. and Wolfowitz, J. [1964], Optimum extrapolation and interpolation designs, I. Ann. Inst. Stat. Math. 16:79-108. T1c (4780)

Kiefer, J. and Wolfowitz, J. [1964], Optimum extrapolation and interpolation designs, II. Ann. Inst. Stat. Math. 16:295-303. T1c (4781)

Kimball, A. W. [1950], Studies in the statistical design and analysis of microbiological assays of amino acids. Ph. D. Thesis, Univ. N. Carolina. T8ac (4782)

Kishen, K. [1950], Expression of unitary components of the highest order interactions in 3^5, 3^6, 4^4 and 5^3 designs in terms of sets for these interactions. JISAS 2:196-211. T2ac (4783)

Kishen, K. [1958], Presidential address. Recent developments in experimental design. Proc. Forty-Fifth Indian Sci. Congress, Part II, pp. 28-59. E2,3,4ac;T2,3,4ac; B;C (4784)

Kishen, K. [1959], A note on the construction of the $(2^{16}, 2^{11})$ and other associated confounded designs, keeping up to second order interactions unconfounded. JISAS 11:180-186. T2c (4785)

Kishen, K. [1959], On the maximum number of factors that can be accommodated in a 2^m design with fixed block size to 2^5 (abstract). JISAS 11:194-195. T2c (4786)

Kishen, K. [1960], On a class of asymmetrical factorial designs. Current Sci. 29:465-466. T2c (4787)

Kishen, K. and Srivastava, J. N. [1959], Mathematical theory of confounding in asymmetrical and symmetrical factorial designs. JISAS 11:73-110. T2c (4788)

Kishen, K. and Srivastava, J. N. [1959], Confounding in asymmetrical factorial designs in relation to finite geometrics. Current Sci. 28:98-100. T2c (4789)

Kishen, K. and Srivastava, J. N. [1960], On the general theory of construction and analysis of asymmetrical confounded factorial designs (abstract). JISAS 12:151. T2ac (4790)

Kishen, K. and Tyagi, B. N. [1961], On some methods of construction of asymmetrical factorial designs. Current Sci. 30:407-409. T2c (4791)

Kishen, K. and Tyagi, B. N. [1963], Partially balanced asymmetrical factorial designs. Contributions to Statistics, pp. 147-158. T2ac (4792)

Kishen, K. and Tyagi, B. N. [1963], Partially balanced asymmetrical factorial designs (abstract). JISAS 15:261-262. T2c (4793)

Kishen, K. and Tyagi, B. N. [1964], On the construction and analysis of some balanced asymmetrical factorial designs. Calcutta Stat. Assoc. Bull. 13:123-149. T2ac (4794)

Kishen, K. and Tyagi, B. N. [1966], On the construction and analysis of a class of balanced asymmetrical factorial designs (abstract). JISAS 18(2):116. T2c (4795)

Kitagawa, T. [1959], Successive process of statistical inferences applied to linear regression analysis and its specializations to response surface analysis. Bull. Math. Stat. 8:80-114. T4ac;A7;S2 (4796)

Kitagawa, T. [1960], A mathematical formulation of the evolutionary operation program. Memoirs Faculty Science, Kyushu Univ., Ser. A 15:21-71. T4ac (4797)

Kitagawa, T. [1961], Successive processes of statistical optimizing procedures. Proc. Fourth Berkeley Symp. Math. Stat. Prob. 1:407-434. T4ac (4798)

Kitagawa, T. [1962], A mathematical formulation of the evolutionary operation programs. Bull. ISI 39(2):293-309. T4a (4799)

Kitagawa, T. and Mitome, M. [1955], A contribution to a notation system of the confounded factorial experiments. Bull. Math. Stat. 6(1&2):1-10. E2,3,9,10c;T2c (4800)

Klein, I. and Marshall, D. I. [1964], Starch vinylation. Determination of optimum conditions by response surface designs. I & EC Process Design Development 3:287-288. T4a (4801)

Klingel, A. R. and McIntyre, R. G. [1962], An experimental strategy for investigating commercial processes. Appl. Stat. 11:79-92. T6ace (4802)

Knowlden, N. F. [1958], Compensating for systematic effects detected by an analysis of covariance in a microbiological assay. ASQC, Nat. Conv. Trans. 12:111-127. T8a;A7 (4803)

Knudsen, L. F. [1950], Statistics in microbiological assay. Ann. New York Academy Sci. 52:889-902. T8a (4804)

Kobayashi, T., Kawabe, Y., and Tamura, H. [1967], Statistical steepest ascent method for process optimizing control. Reports Stat. Appl. Res. 14:79-97. T4a (4805)

Kobwig, W. and Knappen, F. [1967], Die faktorielle varianzanalyse (2^2) mit ungleichen Zellenhäufigkeiten (abstract). Biometrics 23:868. T2a (4806)

Koch, G. G. and Sen, P. K. [1966], Some aspects of the statistical analysis of the "mixed model". Inst. Stat. Mimeo Ser. No. 500, Univ. N. Carolina. T2a;A12 (4807)

Koehler, T. L. [1959], Evolutionary operation: Its methods and application. Tappi 42:261-264. T4ae (4808)

Koehler, T. L. [1959], Evolutionary operation: A program for optimizing plant operation through the application of statistics to scale-up problems. Chemical Eng. Progress 55(10):76-79. T4a (4809)

Koehler, T. L. [1960], How statistics apply to chemical processes. Chemical Eng. 67(December 12):142-152. T4a;C (4810)

Kogan, L. S. [1953], Variance designs in psychological research. Psychological Bull. 50:1-40. El,2,7ae;T2a;B (4811)

Kojima, K. [1961], Effects of dominance and size of population on response to mass selection. Genetical Res. 2:177-188. T9a (4812)

Kôno, K. [1962], Optimum design for quadratic regression on the k-cube. Memoirs Faculty Sci., Kyushu Univ., Ser. A., 16:114-122 (correction 18:120). Tl,4c (4813)

Kramer, C. Y. [1956], Factorial treatments in incomplete block designs. Ph. D. Thesis, Virginia Polytechnic Inst. E3a;T2a (4814)

Kramer, C. Y. and Bradley, R. A. [1956], Factorials in near-balanced incomplete block designs for k(k-1) treatments (preliminary report) (abstract). AMS 27: 545-546. E3a;T2a (4815)

Kramer, C. Y. and Bradley, R. A. [1957], Intra-block analysis for factorials in two associate class group divisible designs. AMS 28:349-361. E3a;T2a (4816)

Kramer, C. Y. and Bradley, R. A. [1957], Examples of intra-block analysis for factorials in group divisible, partially balanced, incomplete block designs. Biometrics 13:197-224. E3ae;T2ae (4817)

Kramer, C. Y. and Bradley, R. A. [1957], Factorial treatments in group divisible incomplete block designs (abstract). AMS 28:816. E2ac;T2ac (4818)

Krane, S. A. [1963], Half-normal plots for multi-level factorial experiments. Proc. Eighth Conf. Design Expt. Army Res. Dev. Testing. pp. 261-285. T2ae (4819)

Krishnan, K. S. [1963], Use of factorial designs in large-scale field trials (abstract). JISAS 15:259-260. T2a (4820)

Krishnan, T. [1965], Truncation in quantal assay. Ann. Inst. Stat. Math. 17:211-231. T8ae (4821)

Kruskal, J. B. [1965], Analysis of factorial experiments by estimating monotone transformations of the data. JRSSB 27:251-263. T2ae;A9 (4822)

Kshirsagar, A. M. and Gupta, R. P. [1965], A note on the use of canonical analysis in factorial experiments (abstract). Biometrics 21:262. T2a (4823)

Kshirsagar, A. M. and Gupta, R. P. [1965], A note on the use of canonical analysis in factorial experiments. JISA 3:165-169. T2a;A7 (4824)

Kudô, A. and Furukawa, N. [1958], A model in probit analysis. Bull. Math. Stat. 8:1-7. T8ac (4825)

Kudô, A. and Katayama, T. [1965], On so-called secondary association in rice plants. II. Statistical analysis. Japanese J. Genetics 40(1):33-44. T9a (4826)

Kuehl, R. O. [1964], Estimators for genetic parameters of populations derived from parents of a diallel mating. Ph. D. Thesis, Univ. N. Carolina. T9a (4827)

Kuiper, N. H. [1952], Variantie-analyse. (Analysis of variance.) Statistica 6:149-194. E2,7a;T2a (4828)

Kuiper, N. H. [1953], On 4^3 factorial designs with confounding. Netherlands J. Agri. Sci. 1:11-14. T2ac (4829)

Kuipers, H. A. [1965], Variantieanalyse met behulp van een computer. (Analysis of variance with the aid of a computer.) Stat. Neerlandica 19:241-247. T2a;A7 (4830)

Kulkarni, G. A. [1960], Generalisation of Yates' method for analysing the factorial designs of any type (abstract). JISAS 12:210. T2a (4831)

Kulkarni, G. A. [1960], On adjustment of components of second order interaction confounded in a 3^3 design (abstract). JISAS 12:150. T2a (4832)

Kulkarni, G. A. [1961], Incomplete block designs for slope-ratio assays (abstract). 13:239. E2a;T8ac (4833)

Kullback, S. and Ku, H. H. [1967], Interaction in multi-dimensional contingency tables (preliminary report) (abstract). AMS 38:297. T2a (4834)

Kulshreshtha, A. C. and Das, M. N. [1967], Fitting of response surface in presence of a concomitant variable (abstract). JISAS 19(1):138-139. T2a;A7 (4835)

Kumar, S. [1965], A group-testing problem (abstract). AMS 36:727-728 (correction 36:1318). T7a (4836)

Kumar, S. [1965], A group-testing problem (abstract). AMS 36:1079. T7a (4837)

Kurkjian, B. M. [1959], Use of partially balanced block designs with three associate classes for confounded, asymmetrical factorial arrangements (abstract). AMS 30:615. E3c;T2c (4838)

Kurkjian, B. [1960], A general theory for asymmetrical confounded factorial experiments. Ph. D. Thesis, American Univ. E2,3a;T2a;A8 (4839)

Kurkjian, B. and Zelen, M. [1961], A general theory for analysis of asymmetrical, confounded factorial experiments (abstract). Biometrics 17:171. E3,4a;T2a (4840)

Kurkjian, B. and Zelen, M. [1962], Factorial designs and the direct product. Bull. ISI 39(2):509-519. T2a;A8 (4841)

Kurkjian, B. and Zelen, M. [1962], A calculus for factorial arrangements. AMS 33:600-619. T2a;A8 (4842)

Kurotori, I. S. [1966], Experiments with mixtures of components having lower bounds. IQC 22:592-596. T5ace (4843)

Kusaba, I. [1967], Research and application of EVOP in Japan. Reports Stat. Appl. Res. 14:61-64. T4a (4844)

Kuzmin, W. R. [1959], Experiments to expose marginal reliability designs. Nat. Symp. Reliability Quality Control, Proc. 5:55-64. E7ae;T2ae (4845)

Kyle, W. H. and Chapman, A. B. [1953], Experimental check of the effectiveness of selection for a quantitative character. Genetics 38:421-443. T9a (4846)

Lackamp, J. W. [1966], Some remarks on the polycross test method applied in grasses. Euphytica 15:291-296. T9a (4847)

Lambrakis, D. P. [1966], Designs for experiments with mixtures. Ph. D. Thesis, Univ. Aberdeen. T5,7ac (4848)

Lamond, D. R. and Southcott, W. H. [1962], Bioassay of oestrogen using sheep. Australian J. Biological Sci. 15:379-385. T8a (4849)

Lange, H. B. [1957], Investigating chemical plant variables. Chemical Eng. Progress 53(6):304-307. T2ae (4850)

Langer, W. and Stern, K. [1955], Versuchstechnische Probleme bei der Analage von Klonplantagen. Z. Forstgenetik Forstpflanzenzüchtung 4:81-88. E2a;T9a;C (4851)

Larson, H. J. and Bancroft, T. A. [1963], Sequential model building for prediction in regression analysis. I. AMS 34:462-479. T1a (4852)

Lascols, X. [1959], Sélection réciproque et maïs hybrides précoces franco-américains Ann. Inst. Nat. Recherche Agronomique 9(B):377-401. T9a (4853)

Last, K. W. [1962], Statistical design of complex experimental programs, final report. I. Optimum experimental designs obtained by minimizing a loss function. Rocketdyne Report R-3393-1 and Aeronautical Res. Lab. ARL 62-373, U.S. Air Force. T1,7,10ac (4854)

Lathwell, D. J. and Evans, C. E. [1951], Nitrogen uptake from solution by soybeans at successive stages of growth. Agron. J. 43:264-270. E2e;T7e (4855)

Latter, B. D. H. [1960], Natural selection for an intermediate optimum. Australian J. Biological Sci. 13:30-35. T9a (4856)

Latter, B. D. H. [1965], Quantitative genetic analysis in Phalaris tuberosa. I. The statistical theory of open-pollinated progenies. II. Assortative mating and maternal effects in the inheritance of date of ear emergence, seed weight and seedling growth rate. Genetical Res. 6:360-386. T9a (4857)

Latter, B. D. H. [1966], The interaction between effective population size and linkage intensity under artificial selection. Genetical Res. 7:313-323. T9a (4858)

Latter, B. D. H. and Robertson, A. [1960], Experimental design in the estimation of heritability by regression methods. Biometrics 16:348-353. T9a (4859)

Latter, B. D. H. and Robertson, A. [1962], The effects of inbreeding and artificial selection on reproductive fitness. Genetical Res. 3:110-138. T9a (4860)

Lawrence, W. E. [1964], Experimental designs which minimize model inadequacies. Ph. D. Thesis, Univ. Wisconsin. T1,4c (4861)

Leaverton, P. [1963], Statistical procedures applicable in the analysis of bioassays when the usual assumptions are not fulfilled. Ph. D. Thesis, Iowa State Univ. T8a (4862)

Leaverton, P. E. [1965], A test for monotonicity of the dose-response curve in bioassays (preliminary report) (abstract). AMS 36:728. T8a (4863)

Lee, I. [1963], Bioassay with quantal response observed at different times. Ph. D. Thesis, Iowa State Univ. T8a (4864)

Lees, K. A. and Tootill, J. P. R. [1955], Microbiological assay on large plates Part I. General considerations with particular reference to routine assay. Part II. Precise assay. Analyst 80:95-123. E7,9ae;T8a (4865)

Lees, K. A. and Tootill, J. P. R. [1955], Microbiological assay on large plates Part III. High throughput, low precision assays. Analyst 80:531-535. E2,7,11ae;T8a (4866)

Leffel, R. C. and Weiss, M. G. [1958], Analysis of diallel crosses among ten varieties of soybeans. Agron. J. 50:528-534. T9ae (4867)

Lemack, N. I., Livingston, G. E., Parkinson, L. R., Fellers, C. R., and Anderson, D. L. [1958], Exploratory rat and chick bioassays of scales from ocean perch and herring as animal feed. Food Res. 23:684-692. T8a (4868)

Lemus, F. [1960], A mixed model factorial experiment in testing electrical connectors. IQC 17(6):12-16. E1e;T2e (4869)

Leng, E. R. [1961], Predicted and actual responses during long-term selection for chemical composition in maize. Euphytica 10:368-378. T9a (4870)

Lenger, A. [1957], Analyse d'un essai factoriel sur la betterave sucrière par la méthode des surfaces de réponse. Publications Inst. Belge Amelioration Betterave 25:267-302. T2,4a (4871)

Leone, F. C. [1955], A designed experiment in engineering (summary). JASA 50: 586-587. T2a (4872)

Leone, F. C. [1959], Experimental design and analysis of variance. III. Additional designs and analyses. ASQC, Nat. Conv. Trans. 13:541-554. T2ae (4873)

Leppink, G. J. [1958], Het experimenteel bepalen van optimale condities. (Experimental determination of optimum conditions.) Stat. Neerlandica 12:143-148. T4a (4874)

Le Roy, H. L. [1960], La sélection artificielle suivant des caractères quantitatifs. Bull. Inst. Agronomique Stations Recherches Gembloux (Hors série) 1:225-241 (also, Biométrie-Praximétrie 1(3-4):33-49, 1960). T9a (4875)

Le Roy, H. L. and Landis, J. [1960], Das Schätzen fehlender Werte im a.b.c- bzw. a.b-Versuch. Biom. Zeit. 2:98-107. E2,6ae;T2ae (4876)

Levine, A. [1963], Spacing problems in minimum variance extrapolation (abstract). AMS 34:1627. Tlc (4877)

Levine, A. [1964], Optimal weighting and choice of observation points in prediction problems. Ph. D. Thesis, Univ. California. Tlac (4878)

Lewontin, R. C. and Hull, P. [1967], The interaction of selection and linkage III. Synergistic effect of blocks of genes. Der Züchter 37:93-98. T9a (4879)

Li, C. H. [1958], Worksheet gives optimum conditions. Chemical Eng. 65 (April 7): 151-156. T4ae (4880)

Li, C. H. [1962], A sequential method for screening experimental variables. JASA 57:455-477. T7,8ac (4881)

Lind, E. E., Golding, J., and Hickman, J. B. [1960], Fitting yield and cost response surfaces. Chemical Eng. Progress 56(11):62. T2,4ae (4882)

Lind, E. E. and Young, W. R. [1965], Con-Man: A 3-D device for the representation of response surfaces. ASQC, Ann. Tech. Conf. Trans. 19:545-551. T4a (4883)

Lind, E. E. and Young, W. R. [1967], Con-Man: A 3-D device for displaying response surfaces. IQC 23:436-439. T4a (4884)

Linder, A. [1964], Design and analysis of experiments. Inst. Stat. Mimeo. Ser. No. 398, Univ. N. Carolina. E2,3a;T2,3a (4885)

Linder, A. [1964], Statistics of bioassay. Inst. Stat. Mimeo. Ser. No. 404, Univ. N. Carolina. T8a (4886)

Lindley, D. V. [1953], Estimation of a functional relationship. Biometrika 40:47-49. Tla (4887)

Linhart, H. [1960], A criterion for selecting variables in a regression analysis. Psychometrika 25:45-58. Tlae (4888)

Linhart, H. [1966], Streuungszerlegung für Paar-Vergleiche. Metrika 10:16-38. E2,3a;Tllace (4889)

Lintner, J. [1958], Random block and Quasi-Latin square designs used to determine fertilizer requirements of sugarcane. South African Sugar Technologists' Assoc. Proc. pp. 1-22. E2,7a;T2a (4890)

Lochow, J. v. and Schuster, W. [1961], Anlage und Auswertung von Feldversuchen. DLG-Verlags-GmbH., Frankfurt am Main, 130 pp. E2,5,7ae;T2ae (4891)

Locks, M. O. [1963], The decision theory approach to complex experimentation (summary). JASA 58:556. T7c (4892)

Lonnquist, J. H. [1967], Mass selection for prolificacy in maize. Der Züchter 37:185-188. T9a (4893)

Lonnquist, J. H. and Rumbaugh, M. D. [1958], Relative importance of test sequence for general and specific combining ability in corn breeding. Agron. J. 50:541-544. T9ac (4894)

Louis, J. [1962], Coordinated Phase I studies for cooperative chemotherapy groups. Cancer Chermotherapy Reports 16:99-105. T8a (4895)

Lowe, C. W. [1964], Some techniques of evolutionary opertaion. Trans. Inst. Chem. Eng. 42(11):T334-T344. T4a (4896)

Lubin, A. and Summerfield, A. [1951], A square root method of selecting a minimum set of variable in multiple regression: II. A worked example. Psychometrika 16:425-437. T1a (4897)

Lucas, H. L. and Linnerud, A. C. [1967], Observations on the selection of predictors. Proc. Twelfth Conf. Design Expt. Army Res. Dev. Testing. pp. 395-402. T1a (4898)

Lum, M. D. [1957], Effects and the classical analysis of variance mixed model (abstract). AMS 28:1068. T2a;A11 (4899)

Lupton, F. G. H. [1961], Studies in the breeding of self-pollinating cereals 3. Further studies in cross prediction. Euphytica 10:209-224. T9a (4900)

Luykx, H. M. C. [1950], What is a mortality rate? Ann. New York Academy Sci. 52:935-942. T8a (4901)

Lyle, P. [1954], The construction of nomograms for use in statistics. Part II. The graphical analysis of the results of a factorial experiment. Appl. Stat. 3:184-195. T2ae (4902)

Mack, A. R. and Cairns, R. R. [1959], Statistical procedures: A. Illustrated Sta. Div., Central Expt. Farm, Ottawa, Ontario, pp. 14+29 (Mimeo.). E2,6ae;T2ae (4903)

MacLean, J. A. R. and Wickens, R. [1951], Application of an incomplete block design to the assessment of quality in Cacao. Nature 168:434-435. E2,3a;T11a (4904)

MacQueen, J. [1964], A mathematical approach to the problem of achieving selective biological effects. Biometrics 20:130-142. T5a (4905)

Madhava, K. B. [1956], Sequential approach in factorial designs. Rev. ISI 24: 64-72. T2,4ae;S2 (4906)

Mallows, C. L. [1959], Reduced analysis. Dept. Math. Tech. Report No. 28, Princeton Univ. T1,4a (4907)

Mallows, C. L. [1964], Choosing the variables in a linear regression: a graphical aid (abstract). Biometrics 20:911-912. T1a (4908)

Mallows, C. L. [1967], Choosing a subset regression (abstract). Technometrics 9:190. T1ac (4909)

Maloney, C. J. [1958], Methods of estimating lethal dose for man. Proc. Second Conf. Design Expt. Army Res. Dev. Testing. pp. 85-92. T8a (4910)

Mann, G. E. [1956], The technique of reciprocal recurrent selection. Agri. Rev. May, pp. 40-43. T9a (4911)

Mann, H. B. [1962], Main effects and interactions. Sankhyā A 24:185-202 (MRC Tech. Summary Report No. 182, U.S. Army and Univ. Wisconsin). T2a (4912)

Mann, H. B. [1963], Main effects and interactions in factorial designs. Colloques Internationaux du Centre National de la Recherche Scientifique No. 110, Le Plan d'Expériences, pp. 33-37. T2a (4913)

Manning, H. L. [1956], Yield improvement from a selection index technique with cotton. Heredity 10:303-322. T9a;A7 (4914)

Manson, A. R. [1966], Minimum bias designs for an exponential response. Ph. D. Thesis, Virginia Polytechnic Inst. T1ac (4915)

Manson, A. R. [1965], Minimum bias designs for the use of polynomial approximations to univariate exponential models (preliminary report) (abstract). AMS 36:1613. T1c (4916)

Mantel, N. [1958], An experimental design in combination chemotherapy (with discussion). Ann. New York Academy Sci. 76:909-931. T4ac (4917)

Margolin, B. H. [1966], Systematic methods for analyzing 2^n3^m factorial experiments. Proc. Eleventh Conf. Design. Expt. Army Res. Dev. Testing. pp. 31-59. T2a (4918)

Margolin, B. H. [1966], Orthogonal main-effect plans permitting estimation of all two-factor interactions for the 2^n3^m factorial series (abstract). AMS 37:1076-1077. T7c (4919)

Margolin, B. H. [1967], Systematic methods for analyzing 2^n3^m factorial experiments with applications. Technometrics 9:245-259. T2ae (4920)

Marks, B. L. [1962], Some optimal sequential schemes for estimating the mean of a cumulative normal quantal response curve. JRSSB 24:393-400. T8ac;S2 (4921)

Martin, F. G. and Cockerham, C. C. [1959], High speed selection studies. Biometrical Genetics, pp. 35-45. T9a;A2 (4922)

Martin, G. A. and Bell, A. E. [1960], An experimental check on the accuracy of prediction of response during selection. Biometrical Genetics, pp. 178-187. T9a (4923)

Martin, L. [1957], Expérimentation factorielle et analyse des surfaces de réponse. Pub. Inst. Belge Amélioration Betterave 25:209-266. T2,4a (4924)

Martin, L., Bourgain, R., and Reuse, J. [1957], Action anticoagulante d'un extrait lipoidique de cerveau de boeuf - etude biometrique. Pharmaceutisch Weekblad 92:962-972. T2ae (4925)

Mason, D. M. [1956], Functional models and experimental designs for characterizing response curves and surfaces. Methodological Procedures in the Economic Analysis of Fertilizer Use Data, pp. 76-98. T4a;A7 (4926)

Mastenbroek, C. [1960], A breeding programme for resistance to anthracnose in dry shell Haricot beans, based on a new gene. Euphytica 9:177-184. T9ace (4927)

Masuda, T. [1959], An example of a study of an analysis method by the application of the design of experiments. Reports Stat. Appl. Res. 6:37-43. T2ae (4928)

Masuyama, M. [1951], A new mathematical model of the superposition method of assay and its fundamental formulas. Reports Stat. Appl. Res. 1(1):24-25. T8a (4929)

Masuyama, M. [1956], An elementary method of construction of punched cards for p^n - and other designs. Reports Stat. Appl. Res. 4:78-84. T2a;A15 (4930)

Masuyama, M. [1958], Short note on C. R. Rao's method of constructing punched cards for factorial designs. Reports Stat. Appl. Res. 5:69-70. T2a;A15 (4931)

Masuyama, M. [1959], On cyclic difference sets which generate orthogonal arrays. Reports Stat. Appl. Res. 6:47-53. E2,3c;T2,3,7c (4932)

Masuyama, M. [1962], Adjoinable designs in factory experiments. Reports Stat. Appl. Res. 9:4-9. T4ace (4933)

Matérn, B. [1957], A routine for computing the degrees of freedom in analysis of variance. Biometrics 13:541-543. T2a (4934)

Mattson, V. [1956], An investigation of the role of recycled black liquor in sulphate pulping. Tappi 39:77-83. T2e (4935)

Matzinger, D. F. [1956], Components of variance of diallel crosses of maize in experiments repeated over locations and years. Ph. D. Thesis, Iowa State Univ. T9a;A13 (4936)

Matzinger, D. F. [1963], Experimental estimates of genetic parameters and their applications in self-fertilizing plants (with discussion). Statistical Genetics and Plant Breeding, pp. 253-279. T9a (4937)

Matzinger, D. F. and Cockerham, C. C. [1963], Simultaneous selfing and partial diallel test crossing. I. Estimation of genetic and environmental parameters. Crop Sci. 3:309-314. T9a (4938)

Matzinger, D. F. and Kempthorne, O. [1956], The modified diallel table with partial inbreeding and interactions with environment. Genetics 41:822-833. T9a (4939)

Matzinger, D. F., Mann, T. J., and Robinson, H. F. [1960], Genetic variability in flue-cured varieties of Nicotiana tabacum. I. Hicks Broadleaf × Coker 139. Agron. J. 52:8-11. T9a (4940)

Matzinger, D. F., Sprague, G. F., and Cockerham, C. C. [1959], Diallel crosses of maize in experiments repeated over locations and years. Agron. J. 51:346-350. T9a;A13 (4941)

Maurice, R. J. [1958], Selection of the population with the largest mean when comparisons can be made only in pairs. Biometrika 45:581-586. T11a;A2 (4942)

Mazumdar, S. [1958], Use of confounded designs and fractional replications in testing variety × fertilizer interactions. FAO/58/1/71; FAOUN, Rome, 9 pp. (Mimeo.). T2,3a (4943)

McAdams, H. T. [1954], Factorial experiments as determinants (abstract). Bull. American Math. Soc. 60:557-558. T2a (4944)

McArthur, D. S. [1955], Practical experimental designs in chemical research. ASQC Nat. Conv. Trans. 9:705-711. T2a (4945)

McArthur, D. S. and Heigl, J. J. [1957], Strategy in research. ASQC Nat. Conv. Trans. 11:1-18. T4a;C (4946)

McBride, G. and Robertson, A. [1963], Selection using assortative mating in Drosophila melanogaster. Genetical Res. 4:356-369. T9a (4947)

McCormick, E. J. and Bachus, J. A. [1952], Paired comparison ratings. I. The effect on ratings of reductions in the number of pairs. J. Appl. Psychology 36:123-127. T1la (4948)

McCune, D. C. [1967], Variables selection in multiple regression. ASQC Tech. Conf. Trans. 21:27-32. T1ac (4949)

McElrath, G. W. [1959], Experimental designs and analysis of variance. IV. Some factorial designs. ASQC Nat. Conv. Trans. 13:555-564. T2,3a (4950)

McGilchrist, C. [1967], Analysis of plant competition experiments for different ratios of species. Biometrika 54:471-477. T2a;A14 (4951)

McLaren, A. and Mitchie, D. [1954], Are inbred strains suitable for bio-assay? Nature 173:686-687. T8ac (4952)

McLean, R. A. and Anderson, V. L. [1966], Extreme vertices design of mixture experiments. Proc. Eleventh Conf. Design Expt. Army Res. Dev. Testing. pp. 274-284. T5ac (4953)

McLean, R. A. and Anderson, V. L. [1966], Extreme vertices design of mixture experiments (with discussion). Technometrics 8:447-456 (abstract 9:190; Dept. Stat. Mimeo. Ser. No. 44, Purdue Univ.). T5ac (4954)

McNemar, Q. [1958], Attenuation and interaction. Psychometrika 23:259-265. T2a (4955)

Mehra, K. L. [1962], Rank tests for incomplete block designs, paired-comparison case. Ph. D. Thesis, Univ. California. E2,3a;T1la (4956)

Mehra, K. L. [1962], Rank tests for paired-comparison experiments involving K treatments (abstract). AMS 33:827. T1la;A12 (4957)

Mehra, K. L. [1963], On multitreatment rank-order tests for paired-comparisons (abstract). AMS 34:683. T1la;A12 (4958)

Mehra, K. L. [1964], Rank tests for paired-comparison experiments involving several treatments. AMS 35:122-137. T1la;A12 (4959)

Mehta, J. S. and Das, M. N. [1964], On asymmetrical rotatable designs (abstract). JISAS 16:169. T4c (4960)

Mellits, E. D. [1966], Estimation and design for intersecting regressions. Ph. D. Thesis, Johns Hopkins Univ. T1ac (4961)

Mello, F. A. F., Brasil Sobº, M. O. C., and Haag, H. P. [1959], Contribuição ao estudo da aplicação do método da diagnose foliar ao algodoeiro (Grossypium hirsutum L., var I.A.C.817) II. Anais Escola Superior Agri. "Luiz de Queiroz" 16:123-133. T2ae (4962)

Menard, L. N., Crocomo, O. J., Gomes, F. P., and de Campos, H. [1961], Pulverização foliar em cafeeiro (Coffea arabica L.) II - Aplicação de adubos potássicos. Anais Escola Superior Agri. "Luiz de Queiroz" 18:277-285. T2ae (4963)

Metropolis, N. and Stein, P. R. [1967], On a class of (0,1) matrices with vanishing determinants. J. Combinatorial Theory 3:191-198. T7c (4964)

Meyer, D. L. [1961], Response surface designs with integer-valued factors. Ph. D. Thesis, Univ. Minnesota. T4ac (4965)

Meyer, D. L. [1963], Response surface methodology in education and psychology. J. Exptl. Education 31:329-336. T4ae (4966)

Meynell, G. G. [1963], Interpretation of distribution of individual response times in microbial infections. Nature 198:970-973. T8a (4967)

Mexas, A. G. [1967], Algorithms for computer analysis of variance of balanced complete structures. M. S. Thesis, Iowa State Univ. E2a;T2a;A15 (4968)

Michaels, S. E. and Pengilly, P. J. [1963], Maximum yield for specified cost. Appl. Stat. 12:189-193. T4a (4969)

Miller, L. C. [1950], Biological assays involving quantal responses. Ann. New York Academy Sci. 52:903-919. T8a;B (4970)

Miller, L. C. [1954], The quantal response in toxicity tests. Statistics and Mathematics in Biology, pp. 315-326. T8a (4971)

Miller, M. C. [1961], An investigation of the application of the method of steepest ascent in medical research. Ph. D. Thesis, Univ. Oklahoma. T4a (4972)

Miller, M. H. and Ashton, G. C. [1960], The influence of fertilizer placement and rate of nitrogen on fertilizer phosphorus utilization by oats as studied using a central composite design. Canadian J. Soil Sci. 40:157-167. T4a (4973)

Millington, A. J., Francis, C. M., and McKeown, N. R. [1964], Wether bioassay of annual pasture legumes. I. Oestrogenic activity in Medicago Tribuloides Desr. Var. Cyprus relative to four strains of Trifolium Subterraneum L. II. The oestrogenic activity of nine strains of Trifolium Subterraneum L. Australian J. Agri. Res. 15:520-536. T8a (4974)

Miranda, L. T., Viégas, G. P., and Freire, E. S. [1964], Adubação do milho. XXIV - Resultados de um ensaio permanente com estêrco, calcário e NPK mineral. (Fertilizer experiments with corn. XXIV - First 9-year results of a long term trial with manure, limestone, and NPK fertilizer.) Bragantia 23: 153-177. T2ae (4975)

Mitchell, J., Dion, H. G., Kristjanson, A. M., and Spinks, J. W. T. [1953], Crop and variety response to applied phosphate and uptake of phosphorous from soil and fertilizer. Agron. J. 45:6-11. T2e (4976)

Mitchell, J. W. [1956], Time-errors in the paired comparison taste preference test. Food Tech. 10:218-220. T11a;C (4977)

Mitchell, T. J. [1966], Construction of saturated 2^{k-p} designs of resolutions V and VI. Ph. D. Thesis, Univ. Wisconsin. T3c (4978)

Mitton, R. G. and Morgan, F. R. [1959], The design of factorial experiments: a survey of some schemes requiring not more than 256 treatment combinations. Biometrika 46:251-259. T3c (4979)

Miyasaka, S., Freire, E. S., Alves, S., and Rocha, T. R. [1966], Adubação mineral do feijoeiro. III - Efeitos de N, P, K, da calagem e de uma mistura de enxôfre e micronutrientes, em solo massapê-salmourão. (Mineral fertilizers for dry beans. III - Effects of N, P, K, liming and a mixture containing sulfur and micro-nutrients on "massapê-salmourão" soils.) Bragantia 25:179-188. T2ae (4980)

Miyasaka, S., Freire, E. S., Igue, T., and Campana, M. [1966], Adubação mineral do feijoeiro. II - Efeitos de N, P, K, da calagem e de uma mistura de enxôfre e micronutrientes, em terra-roxa-misturada. (Mineral fertilizers for dry beans. II - Effects of N, P, K, liming and a mixture containing sulfur and micronutrients on "terra-roxa-misturada" soils.) Bragantia 25:145-159. T2e (4981)

Miyasaka, S., Freire, E. S., Igue, T., Schmidt, N. C., and Leite, N. [1966], Adubação mineral do feijoeiro. V - Efeitos de N, P, K, S e de uma mistura de micronutrientes, em dois solos do Vale do Paraíba. (Mineral fertilizers for dry beans. V - Effects of N, P, K, S and a mixture of micronutrients on two soils of the Paraíba Valley.) Bragantia 25:307-316. T2e (4982)

Miyasaka, S., Freire, E. S., Mascarenhas, H. A. A., and Alcover, M. [1966], Adubação mineral do feijoeiro. VII - Efeitos de N, P, K, S, da calagem e de uma mistura de micronutrientes, no sul do planalto Paulista. (Mineral fertilizers for dry beans. VII - Effects of N, P, K, S, liming and a mixture of micronutrients in the southern section of the São Paulo Plateau.) Bragantia 25:385-392. T2e (4983)

Miyasaka, S., Freire, E. S., Mascarenhas, H. A. A., Pettinelli, A., and Igue, T. [1966], Adubação mineral do feijoeiro. VIII - Efeitos de N, P, K, S e de uma mistura de micronutrientes, em novas experiências conduzidas em Tatuí e Tietê. (Mineral fertilizers for dry beans. VIII - Effects of N, P, K, S and a mixture of micro-nutrients in the Tatuí and Tietê counties, State of São Paulo.) Bragantia 25:393-405. T2e (4984)

Miyasaka, S., Mascarenhas, H. A. A., Freire, E. S., Rocha, T. R., Alves, S., and Issa, E. [1966], Adubação mineral do feijoeiro. VI - Efeitos de N, P, K, S e de uma mistura de micronutrientes, em solo massapê-salmourão. (Mineral fertilizers for dry beans. VI - Effects of N, P, K, S and a mixture containing micronutrients on "massapê-salmourão" soils.) Bragantia 25:371-384. T2e (4985)

Miyasaka, S., Pettinelli, A., Freire, E. S., and Igue, T. [1966], Adubação mineral do feijoeiro. IV - Efeitos de N, P, K, da calagem e de uma mistura de enxôfre e micronutrientes, em Tietê e Tatuí. (Mineral fertilizers for dry beans. IV - Effects of N, P, K, liming and a mixture containing sulfur and micro-nutrients, in Tietê and Tatuí.) Bragantia 25:297-305. T2ae (4986)

Moder, J. J. [1956], A teaching aid for regression, correlation, analysis of variance, and other statistical techniques. IQC 13(4):16-21. T2e (4987)

Monahan, I. P. [1961], Incomplete variable designs in multivariate experiments. Ph. D. Thesis, Virginia Polytechnic Inst. T1ac (4988)

Moore, D. P., Harward, M. E., Mason, D. D., Hader, R. J., Lott, W. L., and Jackson, W. A. [1957], An investigation of some of the relationships between copper, iron, and molybdenum in the growth and nutrition of lettuce: II. Response surfaces of growth and accumulations of Cu and Fe. Soil Sci. Soc. America Proc. 21:65-74. T4ae (4989)

Moore, R. H. and Zeigler, R. K. [1967], The use of non-linear regression methods for analyzing sensitivity and quantal response data. Biometrics 23:563-566. T8a (4990)

Moran, P. A. P. [1954], The dilution assay of viruses. J. Hygiene 52:189-193. T8a (4991)

Morice, E. [1961], Méthode d'analyse des observations par "tout ou rien". Rev. Stat. Appl. 9(3):33-46. T8ae (4992)

Moriguti, S. [1961], Notes on orthogonal design of experiments. Bull. ISI 38(4): 17-30. T10 (4993)

Morley, F. H. W., Daday, H., and Peak, J. W. [1957], Quantitative inheritance in Lucerne, Medicago sativa L. I. Inheritance and selection for winter yield. Australian J. Agri. Res. 8:635-651. T9a (4994)

Morris, R. E. [1953], The relative merits of ranks and paired comparisons. British J. Stat. Psychology 6:112. T11a (4995)

Morrison, M. [1956], Fractional replication for mixed series. Biometrics 12:1-19 (correction 12:339). T3ace (4996)

Morrissey, J. H. [1955], New method for the assignment of psychometric scale values from incomplete paired comparisons. J. Optical Soc. America 45:373-378. T11ac (4997)

Morrow, E. B., Comstock, R. E., and Kelleher, T. [1958], Genetic variances in strawberries. Proc. American Soc. Hort. Sci. 72:170-185. T9a (4998)

Mosteller, F. [1951], Remarks on the method of paired comparisons. I. The least squares solution assuming equal standard deviations and equal correlations. Psychometrika 16:3-9. T11a (4999)

Mosteller, F. [1951], Remarks on the method of paired comparisons. II. The effect of an aberrant standard deviation when equal standard deviations and equal correlations are assumed. III. A test of significance for paired comparisons when equal standard deviations and equal correlations are assumed. Psychometrika 16:203-218. T11a (5000)

Muller, E.-R. [1962], On confounding in asymmetrical factorial design. Ph. D. Thesis, Univ. Aberdeen. T2a (5001)

Müller, K.-H. [1957], Die Anlage und Auswertung faktorieller Feldversuche. Z. Landw. Vers.- Untersuch Wes. 2:117-135. E2,12a;T2a (5002)

Murty, B. R., Arunachalam, V., and Anand, I. J. [1966], Diallel and partial diallel analysis of some yield factors in Linum usitatissimum L. (abstract). JISAS 18(2):107-108. T9a (5003)

Murty, J. S. [1965], Design and analysis of experiments with mixtures (abstract). JISAS 17:126. T5ac (5004)

Murty, J. S. and Das, M. N. [1967], Balanced n-ary block design and their uses. JISA 5:73-82. E5ac;T6c (5005)

Murty, V. N. [1957], Analysis of a 3^3 factorial design. Calcutta Stat. Assoc. Bull. 7:124-126. T2ae (5006)

Murty, V. N. [1959], Note on the analysis of confounded factorial designs (abstract). JISAS 11:195. T2a (5007)

Murty, V. N. and Nandi, H. K. [1957], A comment on "Analysis of 3^n design by V. N. Murty". Calcutta Stat. Assoc. Bull. 7:172. T2a (5008)

Myers, R. H. [1964], Methods for estimating the composition of a three component liquid mixture. Technometrics 6:343-356. T5a (5009)

Myers, R. H. and Alley, B. J. [1965], The use of regression analysis for correcting for matrix effects in the X-ray fluorescence analysis of pyrotechnic compositions. Proc. Tenth Conf. Design Expt. Army Res. Dev. Testing. pp. 61-72. T4a;A7 (5010)

Nair, K. R. [1953], A note on group divisible incomplete block designs. Calcutta Stat. Assoc. Bull. 5:30-35. E3c;T2c (5011)

Nair, K. R. [1953], Some unsolved problems in experimental designs. Calcutta Stat. Assoc. Bull. 4:156-160. E3c;T2c (5012)

Nair, K. R. and Kishen, K. [1960], Recent developments in experimental design with special reference to the work in India. Bull. ISI 37(3):161-177. E2,3,4,7,12ac; T2,3,4a;B (5013)

Nakagami, S. [1961], An example of consumer's preference test: on the application of the method of paired comparisons. Reports Stat. Appl. Res. 8:165-171. T11a;C (5014)

Nakamura, N. [1951], On testing combining ability in the S_O in maize and on synthethic varieties. (in Japanese). Ikushugaku Zasshi/Japanese J. Breeding 1:10-14. T9a (5015)

Nakamura, N. [1953], Estimating of combining ability. Japanese J. Breeding 3(2):23-28. T9ae (5016)

Narasimham, V. L. [1961], Construction of rotatable designs through balanced incomplete block designs (abstract). JISAS 13:240. T4c (5017)

Narayana, C. L. and Sardana, M. G. [1965], Design and analysis of some 4-factor qualitative-cum-quantitative experiments (abstract). JISAS 17:127-128. T2ac (5018)

Narayana, C. L. and Sardana, M. G. [1967], Design and analysis of some asymmetrical qualitative-cum-quantitative experiments. JISAS 19(1):46-68. T2ac (5019)

Narula, S. S. [1960], Augmented symmetrical and asymmetrical factorial designs. JISAS 12:57-87 (abstract 12:210). T7a (5020)

Nasoetion, A. H. [1965], An evaluation of two procedures to estimate genetic and environmental parameters in a simultaneous selfing and partial diallel test crossing design. Ph. D. Thesis, Univ. N. Carolina. T9a (5021)

Nasoetion, A. H., Cockerham, C. C., and Matzinger, D. F. [1967], Simultaneous selfing and partial diallel test crossing II. An evaluation of two methods of estimation of genetic and environmental variance. Biometrics 23:325-334. T9ac (5022)

Nass, C. A. G. [1951], Meting van immuniteit tegen Toxoplasmose met behulp van vrije curven. (Free-hand curves in estimating the potency of human sera against Toxoplasma.) Statistica 5:123-144. T8a (5023)

Naylor, A. F. [1964], Comparisons of regression constants fitted by maximum likelihood to four common transformations of binomial data. Ann. Human Genetics 27:241-246. T2ae;A9 (5024)

Nelder, J. A. [1963], Identification of contrasts in fractional replicates of 2^n experiments. Appl. Stat. 12:38-43. T3a (5025)

Nelder, J. A. [1964], The use of response surfaces in the interpretation of groups of experiments (abstract). Biometrics 20:390-391. T4a (5026)

Nelder, J. A. [1965], The analysis of randomized experiments with orthogonal block structure, I. Block structure and the null analysis of variance. II. Treatment structure and the general analysis of variance. Proc. Royal Soc. Ser. A, 283: 147-162, 163-178. E2,7,12,14a; T2a;A5,15 (5027)

Nelder, J. A. [1966], Inverse polynomials, a useful group of multi-factor response functions. Biometrics 22:128-141. T4a (5028)

Nelson, A. C., Tommerdahl, J. B., and Mazuy, K. K. [1964], Some design problems associated with multiple regression (abstract). Biometrics 20:912. T1c (5029)

Nelson, L. S. [1957], Use of edge-notched cards in the analysis of variance. IQC 13(10):5-7, 10. T2a;A15 (5030)

Neuwirth, S. I. and Naphtali, L. M. [1957], New statistical method rapidly determines optimum process condition. Chemical Eng. 64(June):238-242. T4ac (5031)

Newitt, D. M., Dombrowski, N., and Knelman, F. H. [1954], Liquid entrainment. 1. The mechanism of drop formation from gas or vapour bubbles. Trans. Inst. Chemical Eng. 32:244-261. T2ae (5032)

Nichols, A. B. [1953], U.S.P. reference standards for biological assay. J. American Pharmaceutical Assoc. 42:215-225. T8a (5033)

Nigam, A. K. [1967], On third order rotatable designs controlling number of levels (abstract). JISAS 19(1):137-138. T4c (5034)

Nigam, A. K. [1967], On third order rotatable designs with smaller number of levels. JISAS 19(2):36-41. T4c (5035)

Nigam, A. K. and Das, M. N. [1964], Second-order rotatable designs in three levels (abstract). JISAS 16:172. T4c (5036)

Nigam, A. K. and Das, M. N. [1966], On a method of construction of rotatable designs with smaller number of points controlling the number of levels. Calcutta Stat. Assoc. Bull. 15:147-157. T4c (5037)

Nissen, Ø. [1951], En plan for faktorielle forsøk med hovedvekten på bestemmelse av sampspillene. (A design for factorial experiments when the determination of the interactions is of major interest.) Forskning Forsøk Landbruket (Oslo) 2: 203-214. El2ac;T2ac (5038)

Nissen-Meyer, S. [1964], Evaluation of screening tests in medical diagnosis. Biometrics 20:730-755. T8a;A2 (5039)

Noble, W. M. and Wright, L. A. [1958], Air pollution with relation to agronomic crops: II. A bio-assay approach to the study of air pollution. Agron. J. 50: 551-553. T8ac (5040)

Nóbrega, S. A., Scivittaro, A., Gargantini, H., Wutke, A. C. P., Venturini, W. R., and Santos, C. F. de O. [1964], Adubação mineral da batatinha. I. Região da Alta Sorocabana. (Fertilizer experiments with potato. I. Alta Sorocabana region.) Bragantia 23:83-93. T2e (5041)

Noether, G. [1960], Remarks about a paired comparison model. Psychometrika 25:357-367. T11a (5042)

Noland, P. R. and Scott, K. W. [1960], Effect of varying protein and energy intakes on growth and carcass quality of swine. J. Animal Sci. 19:67-74. T2e (5043)

Nordskog, A. W. [1959], Note on optimum group size for progeny tests. Biometrics 15:513-517. T9a;S1 (5044)

Nordskog, A. W. [1967], Genetic advance from inter-line selection in poultry. Der Züchter 37:200-205. T9a (5045)

Norrington-Davies, J. [1967], Application of diallel analysis to experiments in plant competition. Euphytica 16:391-406. T9a;A14 (5046)

Norton, J. [1959], Influence of weighting choices on tests of main effects and interactions (abstract). Biometrics 15:339. T2a (5047)

Novack, J., Lynn, R. O., and Harrington, E. C. [1962], Process scale-up by sequential experimentation and mathematical optimization. Chemical Eng. Progress 58(2):55-59. T4c;S2 (5048)

Ogawa, J. [1960], Determination of optimum spacings for the estimation of the scale parameter of an exponential distribution based on sample quantiles. Ann. Inst. Stat. Math. 12:135-141. T1c (5049)

Ogilvie, J. C. [1965], Paired comparison models with tests for interaction. Biometrics 21:651-664. T11ae (5050)

Ohsawa, W. [1950], Structure of the response surface. A fundamental system in mass physiology. J. Inst. Polytechnics, Osaka City Univ. Ser. D, 1:1-12. T4c (5051)

Oktaba, W. [1956], Metoda kompletnego uwiklania interakcyj z podblokami w doświadczeniach czynnikowych o dwu, trzech, czterech i pięciu poziomach każdego z czynników. (Method of complete confounding of interactions with subblocks in factorial experiments with two, three, four and five levels of each factor.) Annales Universitatis Marial Curie - Sklodowska, Lubin 11(E): 123-186. T2ae (5052)

Okuno, T., Haga, T., and Yajima, K. [1965], A Monte Carlo study on the effects of inequality of error variances in 2^4 factorial experiment. Reports Stat. Appl. Res. 12:131-141. T2a;A11 (5053)

Oliveira, D. de A., Montojos, J. C., Igue, T., Miranda, H. da S., and de Feritas, M. L. [1966], Ensaios preliminares de adubação do arroz de sequeiro. III - Cultivar 'Dourado Precoce'. (Preliminary fertilizer trials with upland rice. III- Cultivar 'Dourado Precoce'.) Bragantia 25:1-8. T2e (5054)

Ollagnier, M. [1955], Utilisation des fiches perforées a 80 colonnes pour l' interprétation des résultats des expériences agronomiques factorielles (abstract). Biometrics 11:114. T2a;A15 (5055)

Ollagnier, M. and Gros, D. [1955], Utilisation des fiches perforées a 80 colonnes pour l'interprétation des résultats des expériences agronomiques factorielles. Rev. Stat. Appl. 3(2):55-64. T2a;A15 (5056)

Olree, J. G. [1967], Experimental design and computer analysis in viscose optimization studies. Textile Quality Control Papers 14:63-78. T4a (5057)

Olson, T. N. T. and Lee, I. [1966], Application of statistical methodology in quality control functions of the pharmaceutical industry. A survey. J. Pharmaceutical Sci. 55:1-14. T8a (5058)

O'neill, J. A. [1966], Investigations into the use and effect of some response metameters on 6-point parallel-line assay. Ph. D. Thesis, Univ. Aberdeen. T8a (5059)

Oosterhoff, J. [1964], On the selection of independent variables in a regression equation (abstract). AMS 35:1845. T1ac (5060)

Opatowski, I. [1950], On the interpretation of the dose-frequency curve in radiogenetics. Genetics 35:56-59. T8a (5061)

Oslet-Conter, J. [1961], Analyse biométrique du dosage microbiologique de la cyanocobalamine. Biométrie-Praximétrie 2:69-86. T8ae (5062)

O'sváth, J. [1966], Die Pfadkoeffizientenmethode in faktoriellen Versuchen. Biom. Zeit. 8:179-191. T2a;A7 (5063)

Ott, R. L. [1967], Optimal designs for inverse estimation in regression analysis. Ph. D. Thesis, Virginia Polytechnic Inst. Tlac (5064)

Ough, C. S. and Baker, G. A. [1964], Linear dependency of scale structure in differential odor intensity measurements. J. Food Sci. 29:499-505. Tlla;C (5065)

Owen, A. R. G. [1954], Experimental design in genetics (abstract). Biometrics 10:183-184. T9a (5066)

Page, A. R. and Hayman, B. I. [1960], Mixed sib and random mating when homozygotes are at a disadvantage. Heredity 14:187-196. T9a (5067)

Paik, U. B. [1960], <u>Field Experiment and Statistical Methods</u>, <u>New Agricultural Techniques</u>. H. M. Publishing Co. Seoul, Korea. E2,3a;T2a;C (5068)

Paik, U. B. [1962], Formulas for the designs and analysis of factorial experiment. Res. Bull. Korean Agri.Soc. 8. T2a (5069)

Paik, U. B. [1963], Augmented factorial designs. Theses Collection Chungnam Nat. Univ., Vol. III. T7ac (5070)

Paik, U. B. [1966], A method of analysis for unequal numbers of replicates in a factorial experiment. Biometrics Unit Mimeo Ser. BU-221-M, Cornell Univ. T2a;A10 (5071)

Paik, U. B. [1966], Augmented factorial designs. M. S. Thesis, Cornell Univ. T7ace (5072)

Paik, U. B. and Federer, W. T. [1966], On the construction of fractional replicates with special reference to saturated designs. Biometrics Unit Mimeo Ser. BU-224-M Cornell Univ. T7c (5073)

Panse, V. G. and Abraham, T. P. [1960], Simple scientific experiments of farmers' land. Bull. ISI 37(3):179-189. E3ac;T3ac;C (5074)

Panse, V. G. and Sukhatme, P. V. [1954], <u>Statistical Methods for Agricultural Workers</u>. Indian Council of Agricultural Research, New Delhi. Rs 15:xvi+361. E1,2,3,5,7,12ace; T2a (5075)

Papa, K. E. and Federer, W. T. [1960], Crossing plan for neurospora experiment to determine effect of selection through recombination free of heterotic effects. Biometrics Unit Mimeo Ser. BU-128-M, Cornell Univ. T9c (5076)

Pappas, M. A. [1962], EVOP requires communications to work. Hydrocarbon Processing Petroleum Refiner 41(5):159-160. T4ae (5077)

Park, G. T. [1961], Sensory testing by triple comparisons. Biometrics 17:251-260. T11a;A12 (5078)

Parks, W. L. [1956], Methodological problems in agronomic research involving fertilizer and moisture variables. <u>Methodological Procedures in the Economic Analysis of Fertilizer Use Data</u>, pp. 113-133. T4a;C (5079)

Pasteelnick, L. A. and Leder, W. B. [1957], Statistical analysis in a polymerization process. Chemical Eng. Progress 53(8):392-395. T2ae (5080)

Patel, M. S. [1961], Group-screening with more than two stages (abstract). AMS 32:1348. T7c;S2 (5081)

Patel, M. S. [1961], On the performance of the group screening method (abstract). AMS 32:630. T7c;S2 (5082)

Patel, M. S. [1961], Investigations on factorial designs. Ph. D. Thesis, Univ. N. Carolina. T3,7ac (5083)

Patel, M. S. [1961], Fractional factorial 2^m and 3^n designs with and without blocks, preserving the main effects and the two-factor interactions (abstract). AMS 32:630. T3,7c (5084)

Patel, M. S. [1962], On constructing the fractional replicates of the 2^m designs with blocks. AMS 33:1440-1449. T3c (5085)

Patel, M. S. [1962], Group-screening with more than two stages. Technometrics 4:209-217. T7ac;S2 (5086)

Patel, M. S. [1962], Partially duplicated fractional factorial designs (abstract). AMS 33:305-306. T7c (5087)

Patel, M. S. [1963], Partially duplicated fractional factorial designs. Technometrics 5:71-83. T7ac (5088)

Patel, M. S. [1963], A note on Watson's paper. Technometrics 5:397-398. T7a (5089)

Patel, M. S. [1963], On selecting the factors for experimentation (abstract). AMS 34:1114. T7c (5090)

Patel, M. S. [1964], On some aspects of a two stage group-screening method (abstract). AMS 35:939. T7ae;S2 (5091)

Patel, R. M. [1962], Selection among factorially classified variables. Ph. D. Thesis, N. Carolina State Univ. T9a (5092)

Patel, R. M., Cockerham, C. C., and Rawlings, J. O. [1961], Selection among factorially classified variables. Inst. Stat. Mimeo. Ser. No. 317, Univ. N. Carolina, 40 pp. T9a (5093)

Patterson, H. D. [1951], Change-over trials. JRSSB 13:256-271. E7,8ac;T2c (5094)

Patterson, H. D. [1965], The factorial combination of treatments in rotation experiments. J. Agri. Sci. 65:171-182. E5,8ac;T2a (5095)

Patwary, K. M. and Haley, K. D. C. [1967], Analysis of quantal response assays with dosage errors. Biometrics 23:747-760. T8a (5096)

Pavate, M. V., Narayana, C. L., and Subrahmanyam, M. [1965], On selection of independent variables in multiple regression and other linear models (abstract). JISAS 17:127. T1c;A7 (5097)

Pavate, M. V. and Subrahmanyam, M. [1961], Analysis of p × q factorial experiment with unequal number of replications by the use of multiple covariance (abstract). JISAS 13:238. T2a;A7,10 (5098)

Payne, M. G., Powers, L., and Maag, G. W. [1959], Population genetic studies on the total nitrogen in sugar beets (Beta vulgaris L.). J. Amer. Soc. Sugar Beet Tech. 10:631-646. T9a (5099)

Pearce, S. C. [1953], Field experimentation with fruit trees and other perennial plants. Commonwealth Bureau Horticulture Plantation Crops, Tech. Comm. 23, pp. x+131. E1,2,3,5,6,7,8,12ace;T2a;A7,14;B;C;S1 (5100)

Pearce, S. C. [1954], Experimentation with perennial plants. FAO/55/8/5513: InDEX/3,FAOUN, Rome, 51 pp. (Mimeo.). E5,6a;T2a;A9 (5101)

Pendergrass, R. N. [1958], The rank analysis of triple comparisons. Ph. D. Thesis, Virginia Polytechnic Inst. T11a (5102)

Pendergrass, R. N. [1958], Ranking in triple comparisons (summary). JASA 53:587. T11a (5103)

Pendergrass, R. N. and Bradley, R. A. [1960], Ranking in triple comparisons. Contributions to Probability and Statistics, pp. 331-351 (also V.P.I.,ONR, Tech. Report No.1). T11a;A12 (5104)

Perry, W. L. M. [1950], The design of toxicity tests. Medical Res. Council Special Report Ser. No. 270, H.M.S.O., London. T8ace (5105)

Pesek, J. and Heady, E. O. [1956], A two-nutrient response function with determination of economic optima for the rate and grade of fertilizer for alfalfa. Soil Sci. Soc. American Proc. 20:240-246. T4ae (5106)

Phifer, L. H. and Maginnis, J. B. [1960], Dry ashing of pulp and factors which influence it. Tappi 43:38-44. T4e (5107)

Pilgrim, F. J. and Wood, K. R. [1955], Comparative sensitivity of rating scale and paired comparison methods for measuring consumer preference. Food Tech. 9:385-387. T11a;C (5108)

Pillet, L. [1955], L'analyse de la variance et les plans factoriels dans leurs applications à l'étude de la conservation des agrumes. Rev. Stat. Appl. 3(3):39-61. T2ae (5109)

Pinchbeck, P. H. [1957], The kinetic implications of an empirically fitted yield surface for the vapour-phase oxidation of naphthalene to phthalic anhydride. Chemical Eng. Sci. 6:105-111. T4a (5110)

Pinkel, D. [1958], The use of body surface area as a criterion of drug dosage in cancer chemotherapy. Cancer Res. 18:853-856. T8a;C (5111)

Pinnell, E. L., Rinke, E. H., and Hayes, H. K. [1952], Gamete selection for specific combining ability. Heterosis, Ch. 24, pp. 378-388. T9ac (5112)

Piserchia, P. V. [1966], Some results of rotatable designs of order two. Inst. Stat. Mimeo. Ser. No. 501, Univ. N. Carolina, 43 pp. T4ac (5113)

Piserchia, P. V. [1967], Some results in rotatable designs of order two. M.S. Thesis, Univ. N. Carolina. T4ac (5114)

Plackett, R. L. [1960], Models in the analysis of variance (with discussion). JRSSB 22:195-217. T2a;A5 (5115)

Plackett, R. L. and Hewlett, P. S. [1952], Quantal responses to mixtures of poisons (with discussion). JRSSB 14:141-163. T8ae (5116)

Plackett, R. L. and Hewlett, P. S. [1963], A unified theory for quantal responses to mixtures of drugs: the fitting to data of certain models for two non-interactive drugs with complete positive correlation of tolerances. Biometrics 19:517-531. T8ae (5117)

Plackett, R. L. and Hewlett, P. S. [1967], A comparison of two approaches to the construction of models for quantal responses to mixtures of drugs. Biometrics 23:27-44. T8a (5118)

Plaisted, R. L., Sanford, L., Federer, W. T., Kehr, A. E., and Peterson, L. C. [1962], Specific and general combining ability for yield in potatoes. American Potato J. 39:185-197. T9a (5119)

Poch, F. A. [1955], Un diseño factorial aplicado al estudio de la productividad. Trabajos Estad. 6:155-162. T2,3a (5120)

Polaneczky, A. J. [1959], Random balance - or - once upon a time -. Amer. Stat. 13(3):22-23. T7a (5121)

Pollak, E. [1966], Some consequences of selection by culling when there is superiority of heterozygotes. Genetics 53:977-988. T9a;A2 (5122)

Pompilj, G. [1963], Analisi delle medie (discussion). Colloques Internationaux du Centre National de la Recherche Scientifique, No. 110, Le Plan d'Expériences, pp. 79-98. E2,3,12a;T2a (5123)

Potthoff, R. F. [1963], Three-dimensional incomplete block designs for interaction models. Biometrics 19:229-263. E7ac;T7ace (5124)

Powers, L. [1950], Determining scales and the use of transformations in studies on weight per locule of tomato fruit. Biometrics 6:145-163. T9ae;A9 (5125)

Powers, L. [1951], Gene analysis by the partitioning method when interactions of genes are involved. Botanical Gazette 113:1-23. T9a (5126)

Powers, L. [1955], Components of variance method and partitioning method of genetic analysis applied to weight per fruit of tomato hybrid and parental populations. U.S. Dept. Agri. Tech. Bull. 1131. T9a (5127)

Powers, L. [1957], Identification of genetically-superior individuals and the prediction of genetic gains in sugar beet breeding programs. J. Amer. Soc. Sugar Beet Tech. 9:408-432. T9ac (5128)

Powers, L. [1963], The partitioning method of genetic analysis and some aspects of its application to plant breeding (with discussion). Statistical Genetics and Plant Breeding, pp. 280-318. T9ace (5129)

Powers, L., Finkner, R. E., Doxtator, C. W., and Swink, J. F. [1957], Preliminary studies on reciprocal recurrent selection in sugar beets. J. Amer. Soc. Sugar Beet Tech. 9:596-610. T9ac (5130)

Powers, L., Locke, L. F., and Garrett, J. C. [1950], Partitioning method of genetic analysis applied to quantitative characters of tomato crosses. U.S. Dept. Agri. Tech. Bull. 998. T9a (5131)

Powers, L., Remmenga, E. E., and Urquhart, N. S. [1964], The partitioning method of genetic analysis applied to a study of weight per root and percentage sucrose in sugar beets. Colorado State Univ. Agri. Expt. Sta. Tech. Bull. 84. T9a (5132)

Powers, L., Robertson, D. W., and Clark, A. G. [1958], Estimation by the partitioning method of the numbers and proportions of genetic deviates in certain classes of frequency distributions. J. Amer. Soc. Sugar Beet Tech. 9:677-696. T9a (5133)

Powers, L., Robertson, D. W., and Remmenga, E. E. [1958], Estimation of the environmental variances and testing reliability of residual variances for weight per root in sugar beets. J. Amer. Soc. Sugar Beet Tech. 9:697-708. T9a (5134)

Powers, L., Robertson, D. W., Whitney, R. S., and Schmehl, W. R. [1958], Population genetic studies with sugar beets (Beta vulgaris L.) at different levels of soil fertility. J. Amer. Soc. Sugar Beet Tech. 9:637-676. T9a (5135)

Powers, L., Schmehl, W. R., Federer, W. T., and Payne, M. G. [1963], Chemical genetic and soils studies involving thirteen characters in sugar beets. J. Amer. Soc. Sugar Beet Tech. 12:393-448. T2,9a (5136)

Prairie, R. R. and Zimmer, W. J. [1964], 2^p factorial experiments with the factors applied sequentially. Proc. Ninth Conf. Design Expt. Army Res. Dev. Testing. pp. 395-420. T2,7ace (5137)

Prairie, R. R. and Zimmer, W. J. [1964], 2^p factorial experiments with the factors applied sequentially. JASA 59:1205-1216. T2,7ace (5138)

Prairie, R. R. and Zimmer, W. J. [1965], Factorial experiments with the factors applied sequentially (abstract). Technometrics 7:275. T2,7ac (5139)

Prairie, R. R. and Zimmer, W. J. [1967], Fractional replications of 2^p factorial experiments with the factors applied sequentially (abstract). Technometrics 9:192. T3,7c (5140)

Preece, D. A. [1966], On Addelman's 2^{17-9} resolution V plan. Technometrics 8:705-707. T7c (5141)

Prins, H. J. [1956], Analysis of factorial design for Poisson-variates. Report S 192, Afdeling Mathematische Statistiek, Mathematisch Centrum, Amsterdam, The Netherlands. T2a (5141a)

Proschan, F. and Babcock, A. B. [1955], How to design effective experiments. Chemical Eng. 62(August):191-198. E1,2a; T2a;A3; C (5142)

Quenouille, M. H. [1954], The experimental utilization of identical twins, clones, and other genetic subgroups. Statistics and Mathematics in Biology, pp. 149-158.
T9a (5143)

Quenouille, M. H. [1954], Checks on the calculation of the main effects and interactions in a 2^s factorial experiment. Ann. Human Genetics 19:151-152.
T2a (5144)

Quenouille, M. H. [1959], Experiments with mixtures. JRSSB 21:201-202.
T5a (5145)

Radkins, A. P. [1957], Some statistical considerations in organoleptic research: triangle, paired, duo-trio tests. Food Res. 22:259-265. T11a;C (5146)

Radkins, A. P. [1958], Sequential analysis in organoleptic research: triangle, paired, duo-trio tests. Food Res. 23:225-234. T11a;C (5147)

Raghavarao, D. [1959], Some optimum weighing designs. AMS 30:295-303. T6c (5148)

Raghavarao, D. [1960], Some aspects of weighing designs. AMS 31:878-884. T6c (5149)

Raghavarao, D. [1963], Construction of second order rotatable designs through incomplete block designs. JISA 1:221-225. T4c (5150)

Raghavarao, D. [1964], Singular weighing designs. AMS 35:673-680. T6ac (5151)

Raghavarao, D. [1965], A note on fractions of 3^{4n+1} plans. Technometrics 7: 69-71. T7c (5152)

Rahnefeld, G. W., Boylan, W. J., Comstock, R. E., and Madho, S. [1963], Mass selection for post-weaning growth in mice. Genetics 48:1567-1583. T9a (5153)

Raktoe, B. L. [1962], Some combined linear models for single replicated experiments in cultivators' fields. M. S. Thesis, Cornell Univ. T2a (5154)

Raktoe, B. L. [1965], Balanced ℓ-restrictional lattices (abstract). Technometrics 7:275. E2,11,13c;T2c (5155)

Raktoe, B. L. [1967], Application of cyclic collineations to the construction of balanced ℓ-restrictional prime powered lattice designs. AMS 38:1127-1141. E2,3,11,13c;T2c (5156)

Raktoe, B. L. and Federer, W. T. [1965], Statistical analysis of fertilizer response for rice yields in Nickerie, Surinam-I. Experimental results for 1957. Overgenomen Surinaamse Landbouw 13:1-4. T2a (5157)

Raktoe, B. L. and Federer, W. T. [1965], Statistical analysis of fertilizer response for rice yields in Nickerie, Surinam-II-1958. Overgenomen Surinaamse Landbouw 13:250-259. T2a (5158)

Raktoe, B. L. and Federer, W. T. [1966], Statistical analysis of fertilizer response for rice yields in Nickerie, Surinam-III-1959. Overgenomen Surinaamse Landbouw 14:104-112. T2a (5159)

Ramachander, P. R. and Das, M. N. [1963], On response surface designs (abstract). JISAS 15:274. T4c (5160)

Rao, A. P. and Narain, P. [1967], Sampling the diallel cross (abstract). JISAS 19(1):138. T9c (5161)

Rao, B. M., Douglas, A. W., and Sheard, R. W. [1960], A numerical example of a response surface design. Biometrics Unit Mimeo. Ser. BU-120-M, Cornell Univ E2ae;T4ae (5162)

Rao, C. R. [1950], The theory of fractional replication in factorial experiments. Sankhyā 10:81-86. T3c (5163)

Rao, C. R. [1951], A simplified approach to factorial experiments and the punched card technique in the construction and analysis of designs. Bull. ISI 33(2): 1-28. T2ac;A15 (5164)

Rao, C. R. [1962], Problems of selection with restrictions. JRSSB 24:401-405. T1a (5165)

Rao, M. B. [1966], Weighing designs when n is odd. AMS 37:1371-1381. T6c (5166)

Rao, P. S. and Das, M. N. [1965], Interrelationship among factorial designs with different numbers of levels (abstract). JISAS 17:130. T2c (5167)

Rao, P. V. and Kupper, L. L. [1967], Ties in paired-comparison experiments: A generalization of the Bradley-Terry model. JASA 62:194-204. T11a;A12 (5168)

Rao, S. V. S. P. and Das, M. N. [1967], Incomplete weighing designs through balanced ternary designs (abstract). JISAS 19(1):143-144. T6c (5169)

Rao, V. Y. [1959], Fertiliser response surfaces and economics of manuring (abstract). JISAS 11:197. T4a (5170)

Raut, K. C. [1960], Generalised non-orthogonal design and its analysis with recovery of interblock information. JISAS 12:190-199. E3,4a;T4ac (5171)

Rawlings, J. O. [1960], Theoretical analyses of double-cross and three-way hybrid populations. Ph. D. Thesis, Univ. N. Carolina. T9a (5172)

Rawlings, J. O. and Cockerham, C. C. [1959], Theoretical analyses of double-cross and three-way hybrid populations. Inst. Stat. Mimeo. Ser. No. 248, Univ. N. Carolina. T9a (5173)

Rawlings, J. O. and Cockerham, C. C. [1962], Analysis of double cross hybrid populations. Biometrics 18:229-244. T9a (5174)

Rawlings, J. O. and Cockerham, C. C. [1962], Triallel analysis. Crop. Sci. 2:228-231. T9a (5175)

Rayner, A. A. [1953], Quality × quantity interaction. Biometrics 9:387-411. T2a (5176)

Rayner, A. A. [1967], The square-summing check on the main effects and interactions in a 2^n factorial experiment as calculated by Yates's algorithm. Biometrics 23:571-573. T2a (5177)

Rayner, A. A. and Murdock, M. G. [1962], Estimation of orthogonal single-degree-of-freedom components of the second-order interaction in a 3×3×3 factorial design with two replications and partial confounding. Suid-Afrikaanse Tydskrif Landbouwetenskap 5:265-272. T2a (5178)

Rechtschaffner, R. L. [1967], Saturated fractions of 2^n and 3^n factorial designs. Technometrics 9:569-575. T7ac (5179)

Redman, C. E. and King, E. P. [1965], Group screening utilizing balanced and partially balanced incomplete block designs. Biometrics 21:865-874 (abstract 20:913). E2,3a;T7ac (5180)

Reeve, E. C. R. [1961], A note on non-random mating in progeny tests. Genetical Res. 2:195-203. T9a (5181)

Reid, E. [1951], Assay of diabetogenic pituitary preparations. J. Endocrinology 7:120-142. T8ae (5182)

Reiher, W. and Röstel, H.-J. [1962], Die Anwendung der Faktoranalyse in der Pflanzenzüchtung. Z. Pflanzenzüchtung 48:14-28. T9a;A7 (5183)

Reinach, S. G. [1965], A nonparametric analysis for a multi-way classification with one element per cell. Suid-Afrikaanse Tydskrif vir Landbouwetenskap 8:941-959. E2a;T2a;A12 (5184)

Remage, R. and Thompson, W. A. [1966], Maximum-likelihood paired comparison rankings. Biometrika 53:143-149. T11a (5185)

Rendel, J. M. and Robertson, A. [1950], Estimation of genetic gain in milk yield by selection in a closed herd of dairy cattle. J. Genetics 50:1-8. T9a (5186)

Renier, A. [1954], Classification et description des despositifs expérimentaux. Ann. l'Inst. Expt. du Tabac de Bergerac 2:103-149. E1,2,7,12a; T2a (5187)

Rerup, C. [1961], On the validity of a cross-over assay (abstract). Biometrics 17:181. T8a;C (5188)

Rerup, C. [1961], Criteria of validity in a cross-over test (with discussion). <u>Quantitative Methods in Pharmacology</u>, pp. 101-115. T8ace (5189)

Revusky, S. H. [1966], Some inferential statistics which are relatively compatible with an individual organism methodology. Proc. Eleventh Conf. Design Expt. Army Res. Dev. Testing. pp. 299-312. T8ac (5190)

Rhian, M. [1960], Design for estimation by covariance technique. Proc. Fifth Conf. Design Expt. Army Res. Dev. Testing. pp. 317-325. T7ac (5191)

Richards, B. L., Middleton, J. T., and Hewitt, W. B. [1958], Air pollution with relation to agronomic crops: V. Oxidant stipple of grape. Agron. J. 50:559-561. T8ac (5192)

Rigney, J. A. [1956], Query 119. Biometrics 12:82-84. T2a;A14 (5193)

Rios, S. [1958], Sur les lignes de régression. Bull. ISI 36(3):64-70. T1a (5194)

Robbins, H. and Monro, S. [1951], A stochastic approximation method. AMS 22:400-407. T1ac;S2 (5195)

Roberts, R. C. [1966], The limits to artificial selection for body weight in the mouse. I. The limits attained in earlier experiments. II. The genetic nature of the limits. Genetical Res. 8:347-375. T9a (5196)

Roberts, R. C. [1967], The limits to artificial selection for body weight in the mouse. III. Selection from crosses between previously selected lines. IV. Sources of new genetic variance - irradiation and outcrossing. Genetical Res. 9:73-98. T9a (5197)

Robertson, A. [1955], Prediction equations in quantitative genetics. Biometrics 11:95-98. T9a (5198)

Robertson, A. [1956], The effect of selection against extreme deviants based on deviation or on homozygosis. J. Genetics 54:236-248. T9a (5199)

Robertson, A. [1959], Designs for measuring heritability and genetic correlations (abstract). Biometrics 15:142-143. T9ac (5200)

Robertson, A. [1959], Experimental design in the evaluation of genetic parameters. Biometrics 15:219-226. T9ac (5201)

Robertson, A. [1960], Experimental design on the measurement of heritabilities and genetic correlations. <u>Biometrical Genetics</u>, pp. 101-106. T9a (5202)

Robertson, A. [1960], A theory of limits in artificial selection. Royal Soc. London, Proc. 153(B):234-249. T9a (5203)

Robertson, A. [1961], Inbreeding in artificial selection programmes. Genetical Res. 2:189-194. T9a (5204)

Robertson, A. [1964], The effect of non-random mating within inbred lines of the rate of inbreeding. Genetical Res. 5:164-167. T9a (5205)

Robertson, A. and Rendel, J. M. [1950], The use of progeny testing with artificial insemination in dairy cattle. J. Genetics 50:21-31. T9a;A14;S1 (5206)

Robertson, L. S., Johnson, G. L., and Davis, J. F. [1957], Problems involved in the integration of agronomic and economic methodologies in economic optima experiments. Economic and Technical Analysis of Fertilizer Innovations and Resource Use, Ch. 20, pp. 226-240. T2,4a;C (5207)

Robertson, W. K., Hutton, C. E., and Hanson, W. D. [1956], Crop response to different soil fertility levels in a 5 by 5 by 5 by 2 factorial experiment: II. Peanuts. Soil Sci. Soc. America Proc. 20:537-543. T2a (5208)

Robinson, H. F., Comstock, R. E., and Harvey, P. H. [1955], Genetic variances in open pollinated varieties of corn. Genetics 40:45-60. T9a (5209)

Robinson, H. F., Mann, T. J., and Comstock, R. E. [1954], An analysis of quantitative variability in Nicotiana tabacum. Heredity 8:365-376. T9a (5210)

Robinson, J. and Balaam, L. N. [1967], Variance heterogeneity and error correlation in factorial experiments. Australian J. Stat. 9:126-130. T2a;A11 (5211)

Robinson, P. [1965], The analysis of a diallel crossing experiment with certain crosses missing. Biometrics 21:216-219. T9a (5212)

Robinson, P. and Nielsen, K. F. [1960], Composite designs in agricultural research. Canadian J. Soil Sci. 40:168-176. T2,4ae (5213)

Robson, D. S. [1956], A factorial analysis of variance model and its application in genetics. Biometrics Unit Mimeo. Ser. BU-64-M, Cornell Univ. T9a (5214)

Robson, D. S. [1956], An alternative to the Hayman-Mather model for genetic interaction under selfing. Biometrics Unit Mimeo. Ser. BU-63-M, Cornell Univ. T9a (5215)

Robson, D. S. [1956], Random mating among full sibs, an experimental technique for estimating genetic variance components in a haploid population. Biometrics Unit Mimeo. Ser. BU-65-M, Cornell Univ. T9ac (5216)

Robson, D. S. [1956], Partitioning genetic variances and covariances under recurrent backcrossing in haploids. Biometrics Unit Mimeo. Ser. BU-69-M, Cornell Univ. T9a (5217)

Robson, D. S. [1957], Some biometrical formulae for the analysis of quantative inheritance systems involving two haploid or inbred diploid systems. Genetics 42:487-498. T9a (5218)

Robson, D. S. [1958], Some statistical methods for the simple assortive mating problem. Biometrics Unit Mimeo. Ser. BU-94-M, Cornell Univ. T9a (5219)

Robson, D. S. [1960], Genetic and statistical theory for quantitative inheritance studies of haploids. Biometrical Genetics, pp. 46-64. T9ac (5220)

Robson, D. S. [1964], A model for additive cross-over units in genetic linkage analysis. Biometrics Unit ONR Tech. Report No. 17, Cornell Univ. T9a (5221)

Robson, D. S., Powers, L., and Urquhart, N. S. [1967], The proportion of genetic deviates in the tails of a normal population. Der Züchter 37:205-216. T9a (5222)

Rocha, J. L. V., Filho, V. C., Freire, E. S., Scaranari, H., and Pettinelli, A. [1964], Adubação da mamoneira. IV - Experiências de espaçamento × adubação. (Fertilizer experiments with castor beans. IV - Plant density × fertility level.) Bragantia 23:257-269. T2e;A14 (5223)

Rojas, B. A. [1951], Analysis of a group of experiments on combining ability in corn. M. S. Thesis, Iowa State Univ. T9a;A13 (5224)

Rojas, B. A. and Sprague, G. F. [1952], A comparison of variance components in corn yield trials: III. General and specific combining ability and their interaction with locations and years. Agron. J. 44:462-466. T9a;A13 (5225)

Roseberry, T. D. [1960], Approximate methods in estimation of the dosage-response relation in bioassay. M. S. Thesis, Iowa State Univ. T8a (5226)

Rosenberg, D. and Ennerson, F. [1952], Production research in the manufacture of hearing aid tubes. IQC 8(6):94-97. T3ae (5227)

Ross, E. G. [1961], Response surface techniques as a statistical approach to research and development in ultrasonic welding. ASQC, Ann. Conv. Trans. 15:445-456. T2,4ae (5228)

Ross, S., Ginsburg, B. E., and Denenberg, V. H. [1957], The use of the split-litter technique in psychological research. Psychological Bull. 57:145-151. T9a;C (5229)

Rothman, D. and Zimmerman, J. M. [1965], The design of complex sensitivity experiments. Proc. Tenth Conf. Design Expt. Army Res. Dev. Testing. pp. 575-593. T4a (5230)

Rotti, A. [1961], Analyse de données non orthogonales dans le cas d'une expérience à deux facteurs. Biométrie-Praximétrie 2:37-53. T2ae (5231)

Rowell, J. G. [1954], The analysis of a factorial experiment (with confounding) on an electronic calculator. JRSSB 16:242-246. T2a;A15 (5232)

Roy, K. P. [1961], A method of screening defectives by collective testing. Calcutta Stat. Assoc. Bull. 10:139-146. T7ac (5233)

Roy, S. N. [1960], Analysis of multifactor multiresponse experiments with mixed factor and response types (summary). JASA 55:370. T2a (5234)

Roy, S. N. and Bhapkar, V. P. [1960], Some non-parametric analogs of "normal" ANOVA, MANOVA, and of studies in "normal" association. <u>Contributions to Probability and Statistics</u>, pp. 371-387 (Inst. Stat. Mimeo. Ser. No. 215, Univ. N. Carolina). T2a;A12 (5235)

Roy, S. N. and Cobb, W. [1959], Contributions to univariate and multivariate analysis of variance with "fixed effects," normal error and "random effects" not necessarily normal (abstract). AMS 30:616. T2a (5236)

Roy, S. N. and Cobb, W. [1960], Mixed model variance analysis with normal error and possibly non-normal other random effects: Part I: The univariate case. Part II: The multivariate case. AMS 31:939-968. T2a;A11 (5237)

Roy, S. N. and Srivastava, J. N. [1961], Inference on treatment effects and design of experiments in relation to such inferences. Inst. Stat. Mimeo. Ser. 274, Univ. N. Carolina. T2a;C (5238)

Rudorf, W. [1956], Neue Grundlagen und Methoden zur Züchtung auf Leistung. Z. Pflanzenzüchtung 55:441-460. T9a (5239)

Rümke, C. L. [1961], An efficient design for comparing the effects of two treatments (abstract). Biometrics 17:179. T8c (5240)

Rümke, C. L. [1965], Ervaringen met een computer in de Medische Statistiek. (Experiences with a computer in Medical Statistics.) Stat. Neerlandica 19:227-240. T2e (5241)

Ryser, H. J. [1951], A combinatorial theorem with an application to latin rectangles. Proc. American Math. Soc. 2:550-552. E7c;T7c (5242)

Sacks, J. and Ylvisaker, D. [1966], Designs for regression problems with correlated errors. AMS 37:66-89. Tlc (5243)

Sakai, K. [1951], Theoretical studies on individual selection in plant-breeding with special reference to gene recombination. Annual Report Nat. Inst. Genetics, Japan, 1949-1950, pp. 43-45. T9a;Sl (5244)

Samuel, O. C. [1962], Experimenting on the plant process without interfering with production. Food Processing 23(2):31-33, 49. T4a (5245)

Sand, F. M. [1963], Investigations in residual analysis and a modification of the least-squares method for multiple regression. Ph. D. Thesis, Princeton Univ. Tla;A4 (5246)

Sanders, P. G. [1958], Recommendations for the design of experiments for estimating quadratic regression. Proc. Second Conf. Design Expt. Army Res. Dev. Testing. pp. 167-171. Tlac (5247)

Sandler, G. H. and Satterthwaite, F. E. [1960], Random balance designs: A tool for reliability. Nat. Symp. Reliability Quality Control, Proc. 6:50-54. T7a (5248)

Sardana, M. G. [1961], Design and analysis of some confounded qualitative-cum-quantitative experiments. JISAS 13:87-136 (abstract 11:196). T2ac (5249)

Sardana, M. G. [1965], Construction and analysis of $2q \times 2^2$ asymmetrical factorial designs in two replications. JISAS 17:111-115. T2c (5250)

Sardana, M. G. and Das, M. N. [1965], On the construction and analysis of some confounded asymmetrical factorial designs. Biometrics 21:948-956. T2ac (5251)

Sardana, M. G. and Sreenath, P. R. [1965], Construction and analysis of the design of the type $K^2 \times 2^n$ in $K^2 \times 2^p$ plot blocks. Calcutta Stat. Assoc. Bull. 14:74-79. T2ac (5252)

Sarma, T. C. S. R. [1967], On construction of symmetrical fractional factorial designs. JISAS 19(1):83-91. T3,7ac (5253)

Sarndal, C. [1961], Estimation of location and scale parameters by optimally selected observations (abstract). AMS 32:638. Tla (5254)

Sastry, N. S. [1966], Inter-relationships among asymmetrical factorials (abstract). JISAS 18(2):115. T2c (5255)

Satterthwaite, F. E. [1957], Random balance experimental designs (summary). JASA 52:379-380. T7ac (5256)

Satterthwaite, F. E. [1959], Random balance experimentation. Technometrics 1:111-137. T7ac (5257)

Satterthwaite, F. E. and Shainin, D. [1959], Polyvariable experimentation. Application Stat. Experimentation Northeastern Sec. American Chemical Soc. pp. 1-5. T7a (5258)

Savchenko, V. K. [1966], Estimation of the general and specific combining abilities of polyploid forms in diallel crossing systems. Genetika 2:17-22. T9a (5259)

Sayar, H. Z. [1950], Multigenes and multigenic inheritance in the carpel number character of Gossypium herbaceum. Dept. Plant Breeding, Cornell Univ. (mimeo.) 188 pp. T9a (5260)

Schaepman, H. [1952], Application of the polycross test to grass breeding. Euphytica 1:105-111. T9ac (5261)

Schaffer, H. E. [1965], Estimation of fixed effects in a 2×2 factorial experiment with unequal subclass numbers. Biometrics Unit Mimeo. Ser. BU-196-M, Cornell Univ. T2a;A10 (5262)

Scheffé, H. [1952], An analysis of variance for paired comparisons (abstract). AMS 23:136. T11a (5263)

Scheffé, H. [1952], An analysis of variance for paired comparisons. JASA 47: 381-400. T11ae (5264)

Scheffé, H. [1954], Statistical methods for evaluation of several sets of constants and several sources of variability. Chemical Eng. Progress 50(4):200-205. T2ae (5265)

Scheffé, H. [1956], A "mixed model" for the analysis of variance. AMS 27:23-36. T2a;A7 (5266)

Scheffé, H. [1956], Alternative models for the analysis of variance. AMS 27:251-271. T2a;A5 (5267)

Scheffé, H. [1958], Experiments with mixtures (abstract). AMS 29:325-326. T5c (5268)

Scheffé, H. [1958], Experiments with mixtures. JRSSB 20:344-360 (corrigendum 21:238). T5ac (5269)

Scheffé, H. [1961], Reply to Mr. Quenouille's comments about my paper on mixtures. JRSSB 23:171-172. T5c (5270)

Scheffé, H. [1962], A new design for experiments with mixtures (abstract). AMS 33:312. T5c (5271)

Scheffé, H. [1963], The simplex-centroid design for experiments with mixtures (with discussion). JRSSB 25:235-263. T5ac (5272)

Scheid, F. [1960], A tournament problem. American Math. Monthly 67:39-41. E2c;T11c (5273)

Schneider, A. M. and Stockett, A. L. [1963], An experiment to select optimum conditions on the basis of arbitrary preference ratings. Chemical Eng. Progress Symp. Series A.I. Ch. E. 59(42):34-38. T11ae (5274)

Schneiderman, M. A. [1965], How can we find an optimal dose? J. Toxicology Appl. Pharmacology 7(suppl.2):44-53. T8a;C (5275)

Schnell, F. W. [1961], Zweidimensionale erbliche Differenzierung in Spaltungspopulationen bei Mais. Z. Pflanzenzüchtung 46:1-12. T9a (5276)

Schultz, E. F. [1954], Query 110. Biometrics 10:407-411. T2a;A11 (5277)

Schultz, E. F. and Goggans, J. F. [1961], A systematic procedure for determining potent independent variables in multiple regression and discriminant analysis. Auburn Univ. Agri. Expt. Sta. Bull. 336, pp. 1-75. T1ace;A7 (5278)

Schutz, W. M. and Bernard, R. L. [1967], Genotype × environment interactions in the regional testing of soybean strains. Crop Sci. 7:125-130. T9a;A13 (5279)

Schwarz, G. [1960], A class of factorial designs with unequal cell-frequencies. AMS 31:749-755. T2ac;A10 (5280)

Scott, J. F. [1966], A computer technique for analyzing factorial experiments with missing values. Int. Advisory Group Forest Stat. 2nd Conf. Institutionen Skoglig Matematisk Stat. Stockholm, pp. 302-307. T7a;A15 (5281)

Searle, S. R. [1966], Extensions of computer programs for analyses of variance. Biometrics Unit Mimeo. Ser. BU-209-M, Cornell Univ. T2a;A15 (5282)

Seiden, E. [1950], A theorem in finite projective geometry and an application to statistics. Proc. American Math. Soc. 1:282-286. T2c (5283)

Seiden, E. [1953], On the problem of construction of orthogonal arrays (abstract). AMS 24:496. T7c (5284)

Seiden, E. [1954], On the problem of construction of orthogonal arrays. AMS 25:151-156. E3c;T2c (5285)

Seiden, E. and Zemach, R. [1965], On orthogonal arrays of strength four (preliminary report) (abstract). AMS 36:729. T2,7c (5286)

Seiden, E. and Zemach, R. [1966], On orthogonal arrays. AMS 37:1355-1370. T2, 7c (5287)

Seldmayr, K. [1957], Rekurrente Selection auf reciproke Kombinationsfähigkeit. Der Züchter 27:65-69. T9a (5288)

Sen, P. K. [1967], On a class of non-parametric tests of interactions and factorial experiments. Inst. Stat. Mimeo. Ser. No. 525, Univ. N. Carolina. T2a;A12 (5289)

Sen, P. K. and David, H. A. [1966], Paired comparisons for paired characteristics. Inst. Stat. Mimeo. Ser. No. 491, Univ. N. Carolina. T11ac (5290)

Shah, B. V. [1958], On balancing in factorial experiments. AMS 29:766-779 (correction 30:1267). E3ac;T2ac (5291)

Shah, B. V. [1960], Balanced factorial experiments. AMS 31:502-514. T2,7ac (5292)

Shah, B. V. [1960], On a 5×2^2 factorial design. Biometrics 16:115-118 (correction 16:695). T2ac (5293)

Shah, B. V. [1960], A matrix substitution method of constructing partially balanced designs. AMS 31:34-42. E3,4c;T2c (5294)

Shah, B.V. [1961], Asymmetrical factorials in incomplete blocks (abstract). Biometrics 17:176. E3ac;T2ac (5295)

Shah, B. V., Buehler, R. J., and Kempthorne, O. [1961], The method of parallel tangents (PARTAN) for finding an optimum (abstract). AMS 32:632 (also, Stat. Lab. Tech. Report No. 2, Iowa State Univ.). T4,7c (5296)

Shah, B. V., Buehler, R. J., and Kempthorne, O. [1964], Some algorithms for minimizing a function of several variables. J. Soc. Industrial Appl. Math. 12:74-92. T4ac (5297)

Shah, B. V. and Kempthorne, O. [1962], Randomization in fractional factorials (abstract). AMS 33:823. T3a;A5 (5298)

Shah, B. V. and Kempthorne, O. [1962], Some properties of random allocation designs (abstract). AMS 33:822. T7c (5299)

Shah, K. R. [1965], Uniformly better combined estimators in factorial arrangements with confounding. Dept. Stat. RM 130 KS-1, Michigan State Univ. T2a (5300)

Shah, K. R. [1967], Uniformly better combined estimators in factorial arrangements with confounding. JASA 62:638-642. E3a;T2a (5301)

Shah, K. R. [1967], Partial diallel cross designs (abstract). AMS 38:966. T9c (5302)

Sharma, D., Tandon, J. P., and Batra, J. N. [1967], Effect of tester on combining ability estimates of maize germplasm complexes. Euphytica 16:370-376. T9a (5303)

Sharma, R. B. D. [1961], On designs suitable for varietal-cum-factorial trials (abstract). JISAS 13:239-240. T2c (5304)

Sheffield, F. D. [1957], Comment on a distribution-free factorial-design analysis. Psychological Bull. 54:426-428. T2a;A12 (5305)

Sheps, M. C. and Munson, P. L. [1956], The role of between-assay error in biological assays (abstract). Biometrics 12:534. T8a (5306)

Sheps, M. C. and Munson, P. L. [1957], The error of replicated potency estimates in a biological assay method of the parallel line type. Biometrics 13:131-148. T8ae (5307)

Shewell, C. T. [1956], Paper studies in catalytic cracking. ASQC Nat. Conv. Trans. 10:1-7. T4ae (5308)

Shirafuji, M. [1959], A two stage sequential design in response surface analysis. Bull. Math. Stat. 8:115-126. T4ac;S2 (5309)

Shuford, E. H., Jones, L. V., and Bock, R. D. [1960], A rational origin obtained by the method of contingent paired comparisons. Psychometrika 25:343-356. T11a (5310)

Shukla, G. K. and Das, M. N. [1963], On maximising the number of factors in 2^n factorial designs and the fractional replication derivable from them (abstract). JISAS 15:275. T2,7c (5311)

Sibuya, M. [1962], On a model in probit analysis. Ann. Inst. Stat. Math. 13: 251-257. T8ae (5312)

Siddig, M. A. [1967], Evaluation of several selection procedures for predicting yield performance of progenies in two populations of cotton (Gossypium barbadense L.). Euphytica 16:377-384. T9a (5313)

Singh, J. and Thompson, W. A. [1967], Maximum likelihood rankings from paired comparisons when ties are allowed (preliminary report) (abstract). AMS 38: 966-967. T11a;A12 (5314)

Singh, K.-D. [1967], Vollständige Varianzen und Kovarianzen in Pflanzenbeständen III. Monte-Carlo-Versuche uber den Einflufs der Konkurrenz Zwischen Genotypen auf die Voraussage des Ausleseerfolgs. Z. Pflanzenzüchtung 57:189-253. T9a;A14;B (5315)

Singh, M. [1950], Confounding in split-plot designs with restricted randomization of sub-plot treatments. Empire J. Expt. Agri. 18:190-202. E9,12ae;T2ace (5316)

Siotani, M. [1958], Note on the utilization of the generalized Student ratio in the analysis of variance or dispersion. Ann. Inst. Stat. Math. 9:157-171. T2a;A7,11 (5317)

Slater, P. [1960], The analysis of personal preferences. British J. Stat. Psychology 13:119-135. T11a (5318)

Slater, P. [1961], Inconsistencies in a schedule of paired comparisons. Biometrika 48:303-312. T11a (5319)

Smith, C. [1959], A comparison of testing schemes for pigs. Animal Production 1:113-121. T9a;S1 (5320)

Smith, C. [1960], Efficiency of animal testing schemes. Biometrics 16:408-415. T9a (5321)

Smith, H. and Rose, A. [1963], Subjective responses in process investigation. Ind. Eng. Chemistry 55(7):25-28. T4ac (5322)

Smith, H. F. [1950], Letter: Replicated experiments. Forestry 23:56-58. E14;T1ac (5323)

Smith, H. F. [1951], Simplified calculation of a linear regression. Nature 167:367. T1ac (5324)

Smith, T. J. and Brunot, C. A. [1966], A cause-and-effect simulator for interpolation and prediction with multiple regression equations. ASQC Ann. Tech. Conf. Trans. pp. 803-814. T4ae (5325)

Snedecor, G. W. [1953], Query 102. Biometrics 9:256-258. T2a (5326)

Sobel, M. [1960], A modified procedure for group testing (abstract). AMS 31:243. T7c (5327)

Sobel, M. [1960], Group testing to classify efficiently all units in a binomial sample. Information and Decision Processes, pp. 127-161. T7ac (5328)

Sobel, M. [1964], Optimal group testing. Dept. Stat. Tech. Report No. 72, Stanford Univ. T7a (5329)

Sobel, M. and Groll, P. A. [1959], Group testing to eliminate efficiently all defectives in a binomial sample. Bell System Tech. J. 38:1179-1252. T7ac (5330)

Sobel, M. and Groll, P. A. [1960], A problem in restrictive group-testing (abstract). AMS 31:243. T7c (5331)

Sobel, M. and Groll, P. A. [1966], Binomial group-testing with an unknown proportion of defectives. Technometrics 8:631-656. T7a (5332)

Sobotnik, A., Bielorai, H., and Yaron, D. [1965], Empirical estimates of response functions of some summer crops (sorghum and peanuts) to irrigation intensities. Nat. Univ. Inst. Agri. Rehovot, Israel, Bull. No. 82. T4a (5333)

Soller, M. and Genizi, A. [1967], Optimum experimental designs for realized heritability estimates. Biometrics 23:361-365. T9a (5334)

Somermeijer, W. H. [1957], Substituut-variabelen in correlatieberekingen. (Substitute variables in correlation analysis.) Stat. Neerlandica 11:153-160. T1a (5335)

Sommers, R. W. [1963], An iterative method for regression analysis. Chemical Eng. Progress Symp. Series 59(42):78-83. T7e;A7 (5336)

Spendley, W., Hext, G. R., and Himsworth, F. R. [1962], Sequential application of simplex designs in optimisation and evolutionary operation. Technometrics 4:441-461. T4ac (5337)

Spickett, S. G. and Thoday, J. M. [1966], Regular responses to selection 3. Interaction between located polygenes. Genetical Res. 7:96-121. T9a (5338)

Sprague, G. F. [1952], Early testing and recurrent selection. Heterosis, Ch. 26, pp. 400-417. T9ac (5339)

Sprague, G. F. [1960], Inbreeding compared with recurrent selection in corn improvement. Proc. Tenth Congress Int. Soc. Sugar Cane Tech. pp. 653-661. T9a (5340)

Sprague, G. F. and Brimball, B. [1950], Relative effectiveness of two systems of selection for oil content of the corn kernel. Agron. J. 42:83-88. T9c (5341)

Sprague, G. F. and Federer, W. T. [1951], A comparison of variance components in corn yield trials: II. Error, year × variety, location × variety, and variety components. Agron. J. 43:535-541. T9a;A13 (5342)

Sreenath, P. R. [1963], Some results in asymmetrical factorial designs (abstract). JISAS 15:260-261. T2,7ac (5343)

Sreenath, P. R. [1965], On certain methods of construction of confounded asymmetrical factorial designs with smaller number of replications. JISAS 17:166-181. T2ac (5344)

Sreenath, P. R. [1967], Construction and analysis of confounded designs $q \times 3^n$ through balanced incomplete block designs. JISAS 19(2):11-24 (abstract 19(1): 149-150). T2ac (5345)

Sreenath, P. R. and Das, M. N. [1963], Construction of some series of asymmetrical factorial designs with small number of replications (abstract). JISAS 15:273. T2c (5346)

Sreenath, P. R. and Sardana, M. G. [1967], Construction and analysis of designs $(2q+1)\times 2^n$ in $(2q+2)\times 2^p$ plot blocks. Calcutta Stat. Assoc. Bull. 16:29-36. T2ac (5347)

Srivastava, J. N. [1962], Analysis of fractionally replicated asymmetrical or symmetrical factorial designs, I (abstract). AMS 33:308. T7a (5348)

Srivastava, J. N. [1962], A new approach to factorial experiments (abstract). AMS 33:1500. T2a (5349)

Srivastava, J. N. [1962], Contributions to the construction and analysis of designs. Ph. D. Thesis, Univ. N. Carolina (Inst. Stat. Mimeo. Ser. No. 301, Univ. N. Carolina). E2,3ac;T3ac (5350)

Srivastava, J. N. [1963], An algorithm for the analysis of multidimensional partially balanced designs (abstract). AMS 34:1116. E3,13a;T7a (5351)

Srivastava, J. N. [1966], Theory of optimal nonsingular semi-regular fractional factorial plans (abstract). Bull. ISI 41:959. T7,10c (5352)

Srivastava, J. N. [1966], On a class of partially orthogonal 3^n and $2^m\times 3^n$ fractional factorial designs (preliminary report) (abstract). AMS 37:1865. T7c (5353)

Srivastava, J. N. [1967], Investigations on the basic theory of 2^m3^n fractional factorial designs of resolution V and related orthogonal arrays (abstract). AMS 38:637. T7c (5354)

Srivastava, J. N. and Anderson, D. A. [1965], Optimal balanced fractional plans for main effects in the presence of odd ordered interactions (abstract). AMS 36:1607. T7c (5355)

Srivastava, J. N. and Bose, R. C. [1963], Economic partially balanced 2^n factorial fractions (abstract). AMS 34:1622. T7c (5356)

Srivastava, J. N. and Bose, R. C. [1966], Some economic partially balanced 2^m factorial fractions. Ann. Inst. Stat. Math. 18:57-73. T7a (5357)

Stanley, J. C. [1961], Analysis of unreplicated three-way classifications, with applications to rater bias and trait independence. Psychometrika 26:205-219. T2a (5358)

Starks, R. H. [1963], Some new small fractions of 2^k3^m factorials (summary). JASA 58:562. T7c (5359)

Starks, T. H. [1958], Significance tests in experiments involving paired comparisons. Ph. D. Thesis, Virginia Polytechnic Inst. T11a (5360)

Starks, T. H. [1964], A note on small orthogonal main-effect plans for factorial experiments. Technometrics 6:220-222. T7ac (5361)

Starks, T. H. and David, H. A. [1958], Significance tests in paired comparisons (abstract). Virginia J. Sci. 9:452-453. T11a (5362)

Starks, T. H. and David, H. A. [1959], Significance test in paired comparisons (abstract). Biometrics 15:495. T11a (5363)

Starks, T. H. and David, H. A. [1961], Significance tests for paired-comparison experiments. Biometrika 48:95-108 (corrigenda 48:475; Dept. Stat. Tech. Report No. 41, Virginia Polytechnic Inst.). T11a (5364)

Starr, S. [1960], Some results in the analysis of variance I (preliminary report) (abstract). AMS 31:529. E1a;T2a (5365)

Statistical Engineering Laboratory [1957], Fractional factorial experiment designs for factors at two levels. NBS Appl. Math. Ser. 48, Nat. Bur. Standards pp. 1-85. T3c (5366)

Stavrou, J. [1965], An agriculturally oriented modification of a response surface design. M. S. Thesis, Iowa State Univ. T4ac (5367)

Stavrou, J. and Cady, F. B. [1967], Confounding the triple cube response surface design to reduce block size. Soil Sci. Soc. America Proc. 31:126-128. T4ac (5368)

Stearman, R. L. and Ward, T. G. [1957], Variability of specific activities and specific-activity ratios. Archives Biochemistry Biophysics 68:249-254. E1e;T2e (5369)

Steinhaus, H. [1958], Some remarks about tournaments. Calcutta Math. Soc. Golden Jubilee Commemoration Vol. pp. 323-327. T11ac (5370)

Sterrett, A. [1956], An efficient method for the detection of defective members of large populations. Ph. D. Thesis, Univ. Pittsburgh. T7a;S1 (5371)

Sterrett, A. [1957], On the detection of defective members of large populations. AMS 28:1033-1036. T7a;S1 (5372)

Stevens, W. L. [1958], Dilution series: A statistical test of technique. JRSSB 20:205-214 (corrigendum 21:238). T8a (5373)

Stone, M. [1959], Application of a measure of information to the design and comparison of regression experiments. AMS 30:55-70. T10 (5374)

Stone, R. L., Wu, S. M., and Tiemann, T. D. [1965], A statistical experimental design and analysis of the extraction of silica from quartz by digestion in sodium hydroxide solutions. Trans. American Inst. Mining, Metallurgical and Petroleum Eng. 232:115-124. T4ae (5375)

Stoneman, D. M. [1966], Response surface designs for specified factor levels. Ph. D. Thesis, Univ. Wisconsin. T4c (5376)

Stoner, A. K. and Thompson, A. E. [1966], A diallel analysis of solids in tomatoes. Euphytica 15:377-382. T9a (5377)

Stoner, A. K. and Thompson, A. E. [1966], The potential for selecting and breeding for solids content of tomatoes. Proc. American Soc. Hort. Sci. 89:505-511. T9a (5378)

Storey, C. [1964], A review of what can be achieved by optimisation techniques, assuming they exist. Trans. Inst. Chemical Eng. 42:T345-T351. T4a (5379)

Stormont, C. [1954], Research with cattle twins. Statistics and Mathematics in Biology, pp. 407-418. T9a;A14 (5380)

Stowe, R. A. and Mayer, R. P. [1966], Efficient screening of process variables. Ind. Eng. Chemistry 58(2):36-40. T7a (5381)

Strand, L. [1966], The use of analysis of covariance. An example with data from a manuring experiment. Int. Advisory Group Forest Stat., 2nd Conf. Inst. Skoglig Matematisk Stat. Stockholm, pp. 308-318. E2ae;T2ae;A7 (5382)

Streeter, J. G. [1966], A versatile computer program for two factor analysis of variance. Agron. J. 58:241. T2a;A15 (5383)

Studden, W. J. [1967], Optimal designs on Tchebycheff points. Dept. Stat. Mimeo. Ser. No. 118, Purdue Univ. T1c (5384)

Summerfield, A. and Lubin, A. [1951], A square root method of selecting a set of variables in multiple regression: I. The method. Psychometrika 16:271-284. T1a (5385)

Suneson, C. A., Ramage, R. T., and Hoyle, B. J. [1963], Compatibility of evolutionary and mutation breeding methods. Euphytica 12:90-92. T9a (5386)

Sutcliffe, J. P. [1957], A general method of analysis of frequency data for multiple classification designs. Psychological Bull. 54:134-137. T2a;A12 (5387)

Swanson, E. R. [1956], Selecting fertilizer programs by activity analysis. Methodological Procedures in the Economic Analysis of Fertilizer Use Data, pp. 171-187. T2,4a (5388)

Sweeny, H. C. [1955], Analyses of experiments with correlated observations and heterogenous variances (summary). JASA 50:597. E5a;T2a;All (5389)

Sweeny, H. C. [1956], Some results on experimental designs when the usual assumptions are invalid. Ph. D. Thesis, Virginia Polytechnic Inst. E2,3,5a;T2a;All (5390)

Taguchi, G. [1959], Linear graphs for orthogonal arrays and their applications to experimental designs with the aid of various techniques. Reports Stat. Appl. Res. 6:133-175. E2,3c;T2c (5391)

Taguchi, G. [1960], Tables of orthogonal arrays and linear graphs. Reports Stat. Appl. Res. 7:1-52. E2,3c;T2c (5392)

Taguchi, G. [1962], A randomly combined design. Reports Stat. Appl. Res. 9:93-126. T3,7ace (5393)

Takeuchi, K. [1960], On a special class of regression problems and its applications: random combined fractional factorial designs. Reports Stat. Appl. Res. 7:166-198. T3,7ac;A7 (5394)

Takeuchi, K. [1961], On a special class of regression problems and its applications: random combined fractional factorial designs (continued). Reports Stat. Appl. Res. 8:1-6. T3,7ac;A7 (5395)

Takeuchi, K. [1962], A remark on the model of Bradley and Terry in paired comparisons. Reports Stat. Appl. Res. 9:198. T11a (5396)

Tan, W.-Y. and Yuan, C.-H. [1965], On the biometrical analysis of quantitative inheritance. Botanical Bull. Academia Sinica 6:189-196. E4ac;T9a (5397)

Tanaka, W. [1967], Quality design on orthogonal arrays. ASQC Ann. Tech. Conf. Trans. 21:739-743. T7ace (5398)

Tanaka, W. [1967], Quality design on orthogonal arrays. Reports Stat. Appl. Res. 14:1-4. E2,3c (5399)

Taylor, O. C. [1958], Air pollution with relation to agronomic crops: IV. Plant growth suppressed by exposure to air-borne oxidants (smog). Agron. J. 50:556-558. T8ac (5400)

Terhune, H. D. [1963], A statistical analysis of the effect of four variable operating conditions on bleaching results. American Dyestuff Reporter 52(7):33-38. T4a (5401)

Terry, M. E., Bradley, R. A., and Davis, L. L. [1952], New designs and techniques for organoleptic testing. Food Tech. 6:250-254. T11a;C (5402)

Thaker, P. J. [1960], Some methods for study of response surfaces (abstract). JISAS 12:212. T2a (5403)

Thaker, P. J. [1961], Investigation of response surfaces (abstract). Proc. Forty-Eighth Indian Sci. Congress, part III, p. 25. T4c (5404)

Thaker, P. J. [1962], Some infinite series of second order rotatable designs. JISAS 14:110-120. T4c (5405)

Thaker, P. J. [1963], Construction of second-order rotatable designs through partially balanced incomplete block designs (abstract). JISAS 15:258.
T4c (5406)

Thaker, P. J. and Das, M. N. [1961], Sequential third order rotatable designs up to eleven factors. JISAS 13:218-231. T4c (5407)

Thawani, V. D. [1952], A simple method of construction of symmetrical confounded factorial designs. JISAS 4:124-136. T2c (5408)

Thoday, J. M. [1958], Homeostasis in a selection experiment. Heredity 12:401-415.
T9a (5409)

Thoday, J. M. and Boam, T. B [1961], Regular responses to selection I. Description of responses. Genetical Res. 2:161-176. T9a (5410)

Thoday, J. M., Gibson, J. B., and Spickett, S. G. [1964], Regular responses to selection 2. Recombination and accelerated response. Genetical Res. 5:1-19.
T9a (5411)

Thomas, H. L. [1964], Partitioned factorials. Ph. D. Thesis, Oklahoma State Univ.
T2,7a (5412)

Thomas, M. [1952], Back crossing. The theory and practice of the backcross method in the breeding of some non-cereal crops. Commonwealth Bur. Plant Breeding Genetics, Tech. Comm. 16. T9a;B (5413)

Thomas, M. D. [1958], Air pollution with relation to agronomic crops: I. General status of research on the effects of air pollution on plants. Agron. J. 50:545-550.
T8ac (5414)

Thompson, H. R. [1952], Factorial designs with small blocks. New Zealand J. Sci. Tech. B 33:319-344. T2ace (5415)

Thompson, H. R. and Dick, I. D. [1951], Factorial designs in small blocks derived from orthogonal Latin squares. JRSSB 13:126-130. T2ac (5416)

Thompson, K. F. [1964], Triple-cross hybrid kale. Euphytica 13:173-177.
T9a (5417)

Thompson, K. H. [1958], The design of an experiment for studying quantitative inheritance in neurospora. Biometrics Unit Mimeo. Ser. BU-90-M, Cornell Univ. T9ac (5418)

Thompson, K. H. [1960], Techniques and related genetic results for a quantitative inheritance study in Neurospora crassa. Biometrics Unit Mimeo. Ser. BU-114-M, Cornell Univ. T9ac (5419)

Thompson, K. H. [1960], Regression estimates of genetic effects utilizing several specific population means from different generations. Biometrics Unit. Mimeo. Ser. BU-123-M, Cornell Univ. T9a (5420)

Thompson, K. H. [1962], Estimation of the proportion of vectors in a natural population of insects. Biometrics 18:568-578. T7a;S1 (5421)

Thompson, K. H. and Federer, W. T. [1962], Estimation of the proportion of genetic deviates under the assumption of normality. Biometrics Unit Mimeo. Ser. BU-143-M, Cornell Univ. T9a (5422)

Thompson, W. A. and Remage, R. [1964], Rankings from paired comparisons. AMS 35:739-747. T11a;A12 (5423)

Thompson, W. A. and Singh, J. [1967], The use of limit theorems in paired comparison model building. Psychometrika 32:255-264. T11a (5423a)

Thorner, R. M. and Remein, Q. R. [1961], Some aspects of screening tests for the detection of disease suspects (abstract). Biometrics 17:177. T8c;A2 (5424)

Throckmorton, T. N. [1962], Structure diagrams and their uses in experimental design (abstract). Biometrics 18:266. T4a (5425)

Tiao, G. C. [1965], Bayesian comparison of means of a mixed model with application to regression analysis. Dept. Stat. Tech. Report No. 49, Univ. Wisconsin. E2a;T2a;A7 (5426)

Tiao, G. C. and Tan, W. Y. [1965], Bayesian analysis of random-effect models in the analysis of variance. II. Effect of autocorrelated errors. Dept. Stat. Tech. Report No. 54, Univ. Wisconsin. E2a;T2a (5427)

Tidwell, P. W. [1960], Chemical process improvement by response surface methods. Ind. Eng. Chemistry 52:510-512. T4ae (5428)

Tischer, R. G., Jerger, E. W., Kempthorne, O., Carlin, A. F., and Zoellner, A. J. [1953], Influence of variety on quality of dehydrated sweet corn. Food Tech. 7:223-226. T4a;C (5429)

Tischer, R. G. and Kempthorne, O. [1951], Influence of variations in technique and environment on the determination of consistency of canned sweet corn. Food Tech. 5:200-203. T3a (5430)

Tocher, K. D. [1952], On the concurrence of a set of regression lines. Biometrika 39:109-117. T2ae;A7 (5431)

Tocher, K. D. [1952], A note on the design problem. Biometrika 39:189. T10 (5432)

Torrens-Ibern, J. [1962], Plans d'expériences pour l'étude des régressions multiples. Bull. ISI 39(3):161-171. E7ac;T2a (5433)

Torrie, J. H. [1957], Evaluation of general and specific combining ability in perennial ryegrass (Lolium perenne L.). New Zealand J. Sci. Tech., Sec. A. 38:1025-1035. T9a (5434)

Toxopeus, H. J. [1959], Problems involved in breeding for resistance. Euphytica 8:223-231. T9a;C (5435)

Tramel, T. E. [1957], A suggested procedure for agronomic-economic fertilizer experiments. Economic and Technical Analysis of Fertilizer Innovations and Resource Use, ch. 15, pp. 168-175. T4a (5436)

Tramel, T. E. [1962], Response surfaces. Experimental results with cotton (abstract). Biometrics 18:625. T4,7a (5437)

Tramel, T. E. [1963], Response surfaces: Experimental results with cotton (summary). JASA 58:563. T4,7a (5438)

Trawinski, B. J. [1961], Selection of the best treatment in a paired-comparison experiment. Ph. D. Thesis, Virginia Polytechnic Inst. T11a;A2 (5439)

Trawinski, B. J. [1963], General form of the probability function associated with paired-comparison experiments. Classical and Contagious Discrete Distributions pp. 459-464. T11a (5440)

Trawinski, B. J. [1965], An exact probability distribution over sample spaces of paired comparisons. Biometrics 21:986-1000. T11a (5441)

Trawinski, B. J. and David, H. A. [1963], Selection of the best treatment in a paired-comparison experiment. AMS 34:75-91 (abstract 32:632). T11a;A2 (5442)

Truax, H. M. [1957], Graphical analysis of two level factorial experiments: A decision procedure (summary). JASA 52:382. T2a (5443)

Truett, J. [1962], A summary discussion of the Box technique for exploring response surfaces with emphasis on rotatable designs. Aerospace Res. Lab. ARL 62-331, U.S. Air Force. T4a (5444)

Tsutakawa, R. K. [1967], Asymptotic properties of the block up-and-down method in bio-assay. AMS 38:1822-1828. T8a;S2 (5445)

Tsutakawa, R. K. [1967], Random walk design in bio-assay. JASA 62:842-856. T8a;S2 (5446)

Tukey, J. W. [1959], Little pieces of mixed factorials (abstract). Biometrics 15: 641-642. T7a (5447)

Turner, C. W. [1956], Biological assay of beef steer carcasses for estrogenic activity following the feeding of diethylstilbestrol at a level of 10mg. per day in the ration. J. Animal Sci. 15:13-24. T8e (5448)

Turner, N. [1955], Tests for type of action of hydrocarbon insecticides applied jointly. Connecticut Agri. Expt. Sta. Bull. 594. T8a (5449)

Turyn, R. J. [1965], Character sums and difference sets. Pacific J. Math. 15: 319-346. E2c;T3c (5450)

Tyagi, B. N. [1964], On construction of second and third order rotatable designs through pairwise balanced and doubly balanced designs. Calcutta Stat. Assoc. Bull. 13:150-162. T4c (5451)

Tyagi, B. N. [1964], A note on the construction of a class of second order rotatable designs. JISA 2:52-54. T4c (5452)

Umland, A. W. and Smith, W. N. [1959], The use of LaGrange multipliers with response surfaces. Technometrics 1:289-292. T7a (5453)

Underwood, W. M. [1962], Experimental methods for designing extrusion screws. Chemical Eng. Progress 58(1):59-65. T4ae (5454)

Ungar, P. [1960], The cutoff point for group testing. Comm. Pure Appl. Math. 13:49-54. T7a (5455)

Ura, S. [1957], On Scheffé's analysis of variance for paired comparisons. Reports Stat. Appl. Res. 4:132-146. T11a (5456)

Ura, S. [1960], Pair, triangle and duo-trio test. Reports Stat. Appl. Res. 7:107-119. T11ac;C (5457)

Uranisi, H. [1962], Symmetric designs for exploring response surfaces. Bull. Math. Stat. 10(3&4):17-30. T4ac (5458)

Ury, H. K. [1963], On a generalized weighing problem (preliminary report) (abstract). AMS 34:695. T6ac (5459)

Vajda, S. [1951], Relations between variously defined effects and interactions in analysis of variance. AMS 22:283-288. T2a (5460)

van.den Driessche, R. [1961], Analyse systématisée des expériences factorielles. Biométrie-Praximétrie 2:245-259. T2a (5461)

van den Heiden, J. A. [1952], On a correction term in the method of paired comparisons. Biometrika 39:211-212. T11a (5462)

van der Kley, F. K. [1955], The efficiency of some selection methods in populations resulting from crosses between self-fertilizing plants. Euphytica 4:58-66. T9ac (5463)

van der Reyden, D. [1957], The use of orthogonal polynomial contrasts in the confounding of factorial experiments. Ph. D. Thesis, Univ. N. Carolina. T2ac (5464)

van der Vaart, H. R. [1957], Bias in certain current procedures of response surface estimation (abstract). AMS 28:1070. T4a (5465)

van der Vaart, H. R. [1958], Some results on the probability distribution of the latent roots of a symmetric matrix of continuously distributed elements, and some applications to the theory of response surface estimation. Inst. Stat. Mimeo. Ser. No. 189, Univ. N. Carolina. T4a (5466)

van der Vaart, H. R. [1958], On certain types of bias in current methods of response surface estimation. Inst. Stat. Mimeo. Ser. No. 205, Univ. N. Carolina. T4a (5467)

van der Vaart, H. R. [1960], On certain types of bias in current methods of response surface estimation. Bull. ISI 37(3):191-203. T4a (5468)

van Eck, L. F. [1962], Evolutionary operations: A path to more effective use of process data. IQC 19(2):8-10. T4a (5469)

van Eeden, C. and Rumke, C. L. [1961], A k sample trend-test. Stat. Neerlandica 15:25-29. T8a (5470)

van Elteren, P. and Noether, G. E. [1959], The asymptotic efficiency of the χ_r^2- test for a balanced incomplete block design. Biometrika 46:475-477. E2a;T11a (5471)

van Elteren, P. and van Peype, W. F. [1956], Enige rangcorrelatieschema's. Stat. Neerlandica 10:177-195. T11ae (5472)

van Strik, R. [1961], A method of estimating relative potency and its precision in the case of semi-quantitative responses (with discussion). Quantitative Methods in Pharmacology pp. 88-100. T8ace (5473)

Vartak, M. N. [1963], Connectedness of Kronecker product designs. JISA 1:215-218. E4,9c;T7c (5474)

Vaurio, V. M. and Daniel, C. [1954], Evaluation of several sets of constants and several sources of variability. Chemical Eng. Progress 50(2):81-86. T2ae (5475)

vaz de Arruda, H. [1953], Análise de uma experiência fatorial 4^2, em blocos de 4 canteiros, apresentando parcial fusão de interação. (Analysis of a 4^2 factorial experiment with blocks of 4 plots with partial confounding.) Agros, Rio Grande do Sul 6:72-86. T2a (5476)

Venekamp, J. T. N., Hamming, G. and Vervelde, G. J. [1952], A 4^3 factorial design with confounding in 8×8 quasi-Latin squares. Lanbouwkundig Tijdscrift, Wageninge 64:325. El2ac;T2a (5477)

Venturini, W. R. and Jorge, J. P. N. [1962], Eficiência do delineamento fatorial 3^3, em blocos de 9, em uma série de experimentos de adubação do algodoeiro. (Efficiency of a 3^3 factorial design for cotton fertilizer experiments.) Bragantia 21:631-637. E14;T2a;A13 (5478)

Vessereau, A. [1960], Controle d'un enceinte conditionnée. Rev. Stat. Appl. 8(2):53-75. El2ae;T2ae (5479)

Vessereau, A. [1961], Note sur les plans factoriels 2^p. Rev. Stat. Appl. 9(3):7-9. T2,3a (5480)

Vittum, M. T., Peck, N. H., and Carruth, A. F. [1959], Response of sweet corn to irrigation, fertility level, and spacing. Cornell Univ. N.Y. State Agri. Expt. Sta. Bull. No. 786. El2e;T2e (5481)

Vittum, M. T., Tapley, W. T., and Peck, N. H. [1958], Response of tomato varieties to irrigation and fertility level. Cornell Univ. N.Y. State Agri. Expt. Sta. Bull. No. 782. El2e;T2e (5482)

Vos, B. J. [1950], Statistics in biological assay: an example of the graded response. Ann. New York Academy Sci. 52:920-921. E7a;T8a (5483)

Voss, R. and Pesek, J. [1962], Generalization of yield equations in two variables: III. Application of yield data from 30 initial fertility levels. Agron. J. 54:267-271. T2a;A7 (5484)

Wadley, F. M. [1961], Note on precision of graded vs. all-or-none response in bioassay. Proc. Sixth Conf. Design Expt. Army Res. Dev. Testing. pp. 279-283. T8a (5485)

Wadley, F. M. [1963], The independent action theory of mortality as tested at Fort Detrick. Proc. Eighth Conf. Design Expt. Army Res. Dev. Testing. pp. 85-88. T8a (5486)

Wadsworth, H. M. and Hardie, N. G. [1964], A screening procedure for the empirical exploration of response surfaces (abstract). Technometrics 6:124-125. T1,7a (5487)

Walker, A. C. [1957], Balanced incomplete diallel crosses (abstract). Biometrics 13:552. T9a (5488)

Walker, D. R., Hendershott, C. H., and Snedecor, G. W. [1958], A statistical evaluation of a growth substance bioassay method using extracts of dormant peach buds. Plant Physiology 33:162-166. T8a (5489)

Walker, W. M. and Carmer, S. G. [1967], Determination of input levels for a selected probability of response on a curvilinear regression function. Agron. J. 59:161-162. T1,4a (5490)

Walker, W. M., Pesek, J., and Heady, E. O. [1963], Effect of nitrogen, phosphorus and potassium fertilizer on the economics of producing bluegrass forage. Agron. J. 55:193-196. T4a (5491)

Walpole, R. E. [1958], Combined intra- and inter-block analysis for factorials in incomplete block designs. Ph. D. Thesis, Virginia Polytechnic Inst. E2a;T2a (5492)

Walpole, R. E. [1958], Combined intra- and inter-block analysis for factorials in incomplete block designs (abstract). Virginia J. Sci. 9:448. E3a;T2a (5493)

Wang, C.-C. [1964], Effects of number of buds and different size of seed pieces on the germination, growth and production of sugar cane (in Chinese). Taiwan Tahsueh Nunghsuehyuan Yenchiu Paokao/Mem. College Agri. Nat. Taiwan Univ. 8(1):1-11. E2e;T2e (5494)

Wang, Y. Y. [1966], Repeated selection with multivariate normal distributions. Biometrics Unit Mimeo. Ser. BU-231-M, Cornell Univ. T9a (5495)

Wang, Y. Y. [1967], Difference of proportions of deviates in selection. Biometrics Unit Mimeo. Ser. BU-234-M, Cornell Univ. T9a (5496)

Warner, B. T. [1964], Method of graphical analysis of a 2+2 and 3+3 biological assays with graded responses. J. Pharmacy Pharmacology 16:220-233. T8a (5497)

Watson, G. S. [1961], A study of the group screening method. Technometrics 3: 371-388. T7a (5498)

Wearden, S. [1965], Analysis for an augmented hierarchical genetic design (abstract). Biometrics 21:1031. T9ac (5499)

Webb, S. R. [1962], Incomplete factorial designs: orthogonality, non-orthogonality, and construction of designs using linear programming. Ph. D. Thesis, Univ. Chicago. T3,7,10ac (5500)

Webb, S. R. [1962], Some new incomplete factorial designs (abstract). AMS 33:296. T7c (5501)

Webb, S. [1965], Design, testing and estimation in complex experimentation. I. Expansible and contractible factorial designs and the application of linear programming to combinatorial problems. Aerospace Res. Lab. ARL 65-116, Rocketdyne and U.S. Air Force. T2,3,5,7ac;A15 (5502)

Webb, S. [1966], Construction and comparison of non-orthogonal incomplete factorial designs. Proc. Eleventh Conf. Design Expt. Army Res. Dev. Testing. pp. 61-75. T7ac;A15 (5503)

Webb, S. [1967], Non-orthogonal designs of even resolution (abstract). Technometric 9:194. T7a (5504)

Webb, S. R. [1967], Small incomplete factorial experiment designs for two and three level factors. Aerospace Res. Lab. ARL 67-0022, U.S. Air Force. T7a (5505)

Webb, S. R. [1967], Results on non-orthogonal incomplete factorial designs. Aerospa Res. Lab. ARL 67-0021, U.S. Air Force. T7ac (5506)

Webb, S. R. and Galley, S. W. [1965], Design, testing and estimation in complex experimentation. IV. A computer routine for evaluating incomplete factorial designs. Aeronautical Res. Lab. ARL 65-116, Rocketdyne and U.S. Air Force. T3,7,10ac;A15 (5507)

Webster, J. T. [1959], Query 139: Differential regression. Biometrics 15:326-329. T2a;A7 (5508)

Webster, J. T. [1960], A decision procedure for the inclusion of an independent variable in a linear estimator. Ph. D. Thesis, Univ. N. Carolina. T1a (5509)

Weeks, D. L. and Williams, D. R. [1964], A note on the determination of connectedness in an N-way cross classification. Technometrics 6:319-324. T2,7a (5510)

Weil, C. S. [1952], Tables for convenient calculation of median-effective dose (LD_{50} or ED_{50}) and instructions in their use. Biometrics 8:249-263. T8a (5511)

Weiler, H. [1961], A problem of optimum allocation arising in chemical analyses by multiple isotope dilution. Technometrics 3:509-518. T1ace (5512)

Weiling, F. [1966], Hat J. G. Mendel bei seinen Versuchen „zu genau" gearbeitet?- Der χ^2-Test und seine Bedeutung für die Beurteilung genetischer Spaltungsverhältnisse. Der Züchter 36:359-365. T9a (5513)

Weinstein, J. [1967], Single degree of freedom orthogonal components of a factor at 2^k levels in terms of linear combinations of the 2^k contrasts of K factors at 2 levels. Proc. Twelfth Conf. Design Expt. Army Res. Dev. Testing. pp. 195-205. T2a (5514)

Welch, L. F., Adams, W. E., and Carmon, J. L. [1963], Yield response surfaces, isoquants, and economic fertilizer optima for Coastal bermudagrass. Agron. J. 55:63-67. T2,4ae (5515)

Wellensiek, S. J. [1952], The theoretical basis of the polycross test. Euphytica 1:15-19. T9ac (5516)

Westlake, W. J. [1965], Composite designs based on irregular fractions of factorials. Biometrics 21:324-336. T4c (5517)

Wetherill, G. B. [1963], Sequential estimation of quantal response curves (with discussion). JRSSB 25:1-48. T8ac;S2 (5518)

Wetherill, G. B., Chen, H., and Vasudeva, R. B. [1966], Sequential estimation of quantal response curves: A new method of estimation. Biometrika 53:439-454. T8ac;S2 (5519)

Wetz, J. M. [1964], Criteria for judging adequacy of estimation by an approximating response function. Ph. D. Thesis, Univ. Wisconsin. T4,10a (5520)

Whidden, P. [1956], Design of experiment in metals processing. ASQC Nat. Conv. Trans. 10:677-683. T4a (5521)

White, D. [1964], Construction of confounding plans for mixed factorial designs. Ph. D. Thesis, Oklahoma State Univ. T2,3c (5522)

White, D. and Hultquist, R. A. [1965], Construction of confounding plans for mixed factorial designs. AMS 36:1256-1271. T2c (5523)

White, T. G. [1966], Diallel analyses of quantitatively inherited characters in Gossypium hirsutum L. Crop Sci. 6:253-255. T9a (5524)

Whitehouse, R. N. H., Thompson, J. B., and Do Valle Ribeiro, M. A. M. [1958], Studies on the breeding of self-pollinating cereals. 2. The use of a diallel cross analysis in yield prediction. Euphytica 7:147-169 (erratum 7:292). T9a (5525)

Whitwell, J. C. [1959], Practical application of evolutionary operation. ASQC Nat. Conv. Trans. 13:603-616. T4a (5526)

Whitwell, J. C. [1959], Evolutionary operation in chemical processes. Tappi 42:467-473. T4a (5527)

Whitwell, J. C. and Morbey, G. K. [1961], Reduced designs of resolution five. Technometrics 3:459-477. T7ac (5528)

Weidenhofer, H. [1966], The estimation of equipotent concentrations in replicated single-subject bioassays. M. S. Thesis, Iowa State Univ. T8a (5529)

Wilburn, N. T. [1963], Application of 2^{8-4} fractional factorials in screening of variables affecting the performance of dry process zinc battery electrodes. Proc. Eighth Conf. Design Expt. Army Res. Dev. Testing. pp. 17-49. T3ae (5530)

Wilcoxon, F. and Wilcox, R. A. [1964], Some rapid approximate statistical procedure Lederle Laboratories, Pearl River, New York, pp. 1-60. E1,2ae;T2ae;A12 (5531)

Wilde, D. J. [1965], A review of optimization theory. Ind. Eng. Chemistry 57(8): 18-31. T4ac (5532)

Wilk, M. B. and Gnanadesikan, R. [1961], Graphical analysis of multi-response experimental data using ordered distances. Proc. Nat. Academy Sci. U.S.A. 47:1209-1212. T2,4a (5533)

Wilk, M. B. and Gnanadesikan, R. [1961], A plotting procedure in MANOVA (abstract) AMS 32:632-633. T2a (5534)

Wilk, M. B., Gnanadesikan, R., and Ebner, A. R. [1961], On the estimation of error variance and number of significant effects in two-level factorial experiments (abstract). AMS 32:633. T2a (5535)

Wilk, M. B. and Kempthorne, O. [1955], Fixed, mixed, and random models. JASA 50:1144-1167 (summary 50:599). T2a;A5 (5536)

Wilk, M. B. and Kempthorne, O. [1956], Some aspects of the analysis of factorial experiments in a completely randomized design. AMS 27:950-985. E1a;T2a;A5 (5537)

Wilkie, D. [1961], Confounding in 3^3 factorial experiments in nine blocks. Appl. Stat. 10:83-92. T2ae (5538)

Wilkie, D. [1962], A method of analysis of mixed level factorial experiments. Appl. Stat. 11:184-195. T2ae (5539)

Wilkinson, G. N. [1962], A principle of similarity of response curves in plant nutrition (abstract). Biometrics 18:267-268. T4a (5540)

Wilkinson, J. W. [1956], Analysis of paired comparison designs with incomplete repetitions. Ph. D. Thesis, Univ. N. Carolina. T11a (5541)

Wilkinson, J. W. [1957], An analysis of paired comparison designs with incomplete repetitions. Biometrika 44:97-113. T11ae (5542)

Williams, D. R. [1963], A new approach to factorial experimentation. Ph. D. Thesis, Oklahoma State Univ. T2a (5543)

Williams, E. J. [1952], Some exact tests in multivariate analysis. Biometrika 39: 17-31. T2a;A7 (5544)

Williams, E. J. [1952], Use of scores for the analysis of association in contingency tables. Biometrika 39:274-289. T2ae (5545)

Williams, E. J. [1952], The interpretation of interactions in factorial experiments. Biometrika 39:65-81. T2ae (5546)

Williams, E. J. [1958], Optimum allocation for estimation of polynomial regression (abstract). Biometrics 14:573-574. T1c (5547)

Williams, J. S. [1961], An evaluation of the worth of some selection indices. Ph. D. Thesis, Univ. N. Carolina. T9a (5548)

Williams, J. S., Cockerham, C. C., and Roy, S. N. [1961], An evaluation of the worth of some selection indices. Inst. Stat. Mimeo. Ser. No. 281, Univ. N. Carolina. T9a (5549)

Williams, K. R. [1963], Comparing screening designs. Ind. Eng. Chemistry 55(6): 29-32. T7ac (5550)

Wilson, C. L. [1960], A designed experiment. Ind. Eng. Chemistry 52:504-506. T4ae (5551)

Wilson, K. V. [1956], A distribution-free test of analysis of variance hypotheses. Psychological Bull. 53:96-101. T2a;A12 (5552)

Wilson, S. P., Kyle, W. H., and Bell, A. E. [1965], The effects of mating systems and selection on pupa weight in Tribolium. Genetical Res. 6:341-351. T9a (5553)

Winder, C. V. [1950], Some examples of the use of statistics in pharmacology. Ann. New York Academy Sci. 52:838-861. E7ae;T2,8a;C (5554)

Winne, D. [1962], Die Bestimmung von Geraden, die sich in einem Punkt auf der Ordinatenachse schneiden, bei unsymmetrischer Anordnung der Versuchsergebnisse (unsymmetrical multiple slope ratio assay). Biom. Zeit. 4:217-238. T8a;A7 (5555)

Winters, L. M. [1952], Rotational crossbreeding and heterosis. Heterosis ch. 23, pp. 371-377. T9c (5556)

Wishart, J. and Sanders, H. G. [1955], Principles and practice of field experimentati Commonwealth Bur. Plant Breeding Genetics Tech. Comm. 18. E2,5,7,12ae;T2ae; A14;C (5557)

Wishart, J., Thawani, V. D., Kishen, K., and Nair, K. R. [1954], Statistical method FAO/55/9/6340:In Dex/1, FAOUN, Rome, 111 pp. (Mimeo.). E2,3a;T2a;C (5558)

Wöhrmann, K. [1967], Fremdbefruchtungsrate und genotypische „fitness". Der Züchter 37:142-145. T9a (5559)

Wong, S.-H. and Glenn, W. A. [1960], The analysis of paired-comparison experiments involving ties. Dept. Stat. Tech. Report No. 48, Virginia Polytechnic Inst. T11a (5560)

Wood, E. C. [1953], The efficient planning of microbiological assays (with discussio Analyst 78:451-460. T8a;C (5561)

Wood, S. R. [1964], Industrial application of blocked central composite rotatable design. ASQC Ann. Conv. Trans. 18:160-162. T4ae (5562)

Wood, S. R. and Hartvigsen, D. E. [1963], Statistical design and analysis of qualification test program for a small rocket engine. ASQC Ann. Conv. Trans. 17:345-348 (also, IQC 20(12):14-18). T2e (5563)

Woodson, G. S. [1957], Sequential bioassay? Proc. First Conf. Design Expt. Army Res. Dev. Testing. pp. 241-251. T8ae;S2 (5564)

Woodworth, R. C. [1956], Organizing fertilizer input-output data in farm planning. Methodological Procedures in the Economic Analysis of Fertilizer Use Data pp. 158-170. T2,4a;C (5565)

Wright, E. C. [1965], Field plans for a systematically designed polycross. Record Agri. Res. N. Ireland 14:31-41. E5c;T9ac (5566)

Wright, G. M. [1956], Missing values in factorial experiments. Nature 178:1481. T2,7a (5567)

Wright, G. M. [1958], The estimation of missing values in factorial experiments. New Zealand J. Sci. 1:1-8. T2,7ae (5568)

Wu, S. M. [1964], Tool-life testing by response surface methodology - Parts 1 and 2. J. Eng. Ind., Ser. B 86:105-116. T4ae (5569)

Wu, S. M. and Meyer, R. N. [1964], Cutting-tool temperature-predicting equation by response-surface methodology. J. Eng. Ind. Ser. B 86:150-156.
T4ae (5570)

Wu, S. M. and Meyer, R. N. [1964], A first-order five-variable cutting-tool temperature equation and chip equivalent. J. Eng. Ind. Ser. B 86:395-400.
T3ae (5571)

Yalavigi, C. C. [1963], A tournament problem. Math. Student 31:51-64. E2c;T11c (5572)

Yamakawa, N. [1965], Random combined fractional factorial designs 1; Theory and application of the random combined fractional factorial designs (RACOFFD). Bull. Math. Stat. 11(3&4):1-38. T7ace (5573)

Yamakawa, N. [1966], Random combined fractional factorial designs II: Sampling theory from the finite population. Bull. Math. Stat. 12(1&2):1-67. T7ac;A14 (5574)

Yang, C. H. [1966], Some designs of maximal (+1,-1)-determinant of order n≡2 (mod 4). Math. Computation 20:147-148. T6c (5575)

Yates, F. [1955], The use of transformations and maximum likelihood in the analysis of quantal experiments involving two treatments. Biometrika 42:382-403. T8ae;A9 (5576)

Yates, F. [1960], Utilisation de calculateurs électroniques dans l'analyse d'expériences avec répétitions et de groupes d'expériences de même plan expérimental. Biométrie Praximétrie 1(3):3-15. T2a;A13,15 (5577)

Yates, F. [1960], The use of electronic computers in the analysis of replicated experiments, and groups of experiments of the same design. Bull. Inst. Agronomique Sta. Recherches Gembloux (Hors série) 1:201-211. T2a;A13,15 (5578)

Yates, F. [1967], A fresh look at the basic principles of the design and analysis of experiments. Proc. Fifth Berkeley Symp. Math. Stat. Prob. 4:777-790. E7a;T2,4a;C (5579)

Yates, F. and Anderson, A. J. B. [1966], A general computer programme for the analysis of factorial experiments. Biometrics 22:503-524. T2a;A15 (5580)

Yates, F., Lipton, S., Sinha, P., and Das Gupta, K. P. [1959], An exploratory analysis of a large set of 3×3×3 fertilizer trials in India. Empire J. Exptl. Agri. 27:263-275. T2a;A13 (5581)

Yates, R. A. [1964], Yield depression due to phosphate fertilizer in sugar-cane. Australian J. Agri. Res. 15:537-547. T2e (5582)

Yoshida, Y. [1962], Theoretical studies on the methodological procedures of radiation breeding: I. New methods in autogamous plants following seed irridation. Euphytica 11:95-111. T9ac (5583)

Yoshida, Y. [1964], Theoretical studies on the methodological procedures of radiation breeding: II. Expectation of the number of desirable mutants with the new methods. Euphytica 13:65-74. T9a (5584)

Youden, W. J. [1956], Statistical design as guide for research programs. Ind. Eng. Chemistry 48(5):44A and 46A. T4a (5585)

Youden, W. J. [1956], Black box test experimenters. Ind. Eng. Chemistry 48(12): 57A-58A. T2,4c (5586)

Youden, W. J. [1957], Factorial experiments help to improve your food. Ind. Eng. Chemistry 49(2):85A-86A. T2ae (5587)

Youden, W. J. [1957], Statistical design. Ind. Eng. Chemistry 49(4):91A. T4c (5588)

Youden, W. J. [1957], Recent developments in factorial experimentation. Ind. Eng. Chemistry 49(6):73A-74A. T3ace (5589)

Youden, W. J. [1957], Interpretation of experimental results. Ind. Eng. Chemistry 49(12):73A-74A. T3ace (5590)

Youden, W. J. [1959], Evolutionary operation. Ind. Eng. Chemistry 51(6):79A-80A. T4a (5591)

Youden, W. J. [1959], Designs for multifactor experimentation. Ind. Eng. Chemistry 51(10):79A-80A. T6,7ae (5592)

Youden, W. J. [1959], Problems in statistical design. Ind. Eng. Chemistry 51(12): 85A-86A. T7ac;A2 (5593)

Youden, W. J. [1961], Partial confounding in fractional replication. Technometrics 3:353-358. T7ac (5594)

Youden, W. J. [1963], Ranking laboratories by round-robin tests. Materials Res. Standards 3:9-13. T11a;A2 (5595)

Youden, W. J. and Connor, W. S. [1954], New experimental designs for paired observations. J. Res. NBS 53:191-196. E3ace;T11ace (5596)

Youden, W. J., Connor, W. S., and Severo, N. C. [1959], Measurements made by matching with known standards. Technometrics 1:101-109. T1c (5597)

Youden, W. J., Kempthorne, O., Tukey, J. W., Box, G. E. P., Hunter, J. S., Satterthwaite, F. E., and Budne, T. A, [1959], Discussion of the papers of Messrs. Satterthwaite and Budne. Technometrics 1:157-193. T7a (5598)

Young, S. S. Y. [1961], A further examination of the relative efficiency of three methods of selection for genetic gains under less-restricted conditions. Genetical Res. 2:106-121. T9,10a (5599)

Young, S. S. Y. and Turner, H. N. [1965], Selection schemes for improving both reproduction rate and clean wool weight in the Australian Merino under field conditions. Australian J. Agri. Res. 16:863-880. T9a (5600)

Zacks, S. [1962], Minimax strategies, information and randomized factorial experiments. Ph. D. Thesis, Columbia Univ. T3a (5601)

Zacks, S. [1963], On a complete class of linear unbiased estimators for randomized factorial experiments. AMS 34:769-779. T2,7a (5602)

Zacks, S. [1966], Randomized fractional weighing designs. AMS 37:1382-1395 (abstract 37:1433). T7c (5603)

Zacks, S. and Ehrenfeld, S. [1961], Distribution function for randomized factorial experiments (abstract). AMS 32:926. T2a (5604)

Zagatto, A. G. and Gomes, F. P. [1960], O problema técnico-econômico da adubação. Anais Escola Superior Agri. "Luiz de Queiroz" 17:149-163. T4a (5605)

Zelen, M. [1955], The use of incomplete block designs for factorial experiments (summary). JASA 50:599. E3a;T2a (5606)

Zelen, M. [1957], The use of incomplete block designs for asymmetrical factorial arrangements (abstract). AMS 28:526. E2,3ac;T2ac (5607)

Zelen, M. [1957], The use of incomplete block designs for asymmetrical factorial arrangements (abstract). AMS 28:526 E,2,3ac;T2ac (5608)

Zelen, M.· [1957], The use of incomplete block designs for asymmetrical factorial arrangements (abstract). Biometrics 13:422. E2,3ac;T2ac (5609)

Zelen, M. [1958], The use of group divisible designs for confounded asymmetrical factorial arrangements. AMS 29:22-40. E3a;T2a (5610)

Zelen, M. [1958], Factorial analysis of life-tests (abstract). AMS 29:1290-1291. T2a (5611)

Zelen, M. [1959], Factorial experiments in life testing (summary). JASA 54:508. T2a (5612)

Zelen, M. [1959], Factorial experiments in life testing. Technometrics 1:269-288 (erratum 2:121). T3ae (5613)

Zelen, M. [1960], Analysis of two-factor classifications with respect to life tests. <u>Contributions</u> <u>to</u> <u>Probability</u> <u>and</u> <u>Statistics</u> pp. 508-517. T2a (5614)

Zelen, M. [1962], Introductory lectures on the statistical design of experiments. Math. Res. Center Orientation Lecture Ser. 1, U.S. Army Univ. Wisconsin. E2,7ae;T2ae;C (5615)

Zelen, M. and Connor, W. S. [1959], Multi-factor experiments. IQC 15(9):14-17. T2,3ace (5616)

Zelen, M. and Federer, W. T. [1965], Application of the calculus for factorial arrangements. III. Analysis of factorials with unequal numbers of observations. Sankhyā A 27:383-400. T2a;A8,10 (5617)

Zelen, M. and Kurkjian, B. [1960], A calculus for factorial arrangements (abstract). AMS 31:822. T2a;A8 (5618)

Zyskind, G. [1956], Query 121. Biometrics 12:224-225. T7ae;A10 (5619)

Zyskind, G. [1965], Missing values in factorial experiments. Technometrics 7:649-650. T2,7ae (5620)

Zyskind, G. and Kempthorne, O. [1959], The role of treatment error in comparative experiments (summary). JASA 54:508-509. E1,2a;T2a;C (5621)

Zyskind, G. and Kempthorne, O. [1960], Treatment errors in comparative experiments. Aeronautical Res. Lab. and Iowa State Univ. WADC Tech., TN 59-19, U.S. Air Force pp. 1-78. E1,2a;T2a;C (5622)

Co-authors list for category "T"

Name	Ref. No.
Abraham, T.P.	4064
	5074
Abramides, E.	4382
Abreu, C.P.	4475
Adams, W.E.	4684
	5515
Addelman, S.	4760
Ahrens, H.	3816
Alcover, M.	4983
Allen, I.	3750
Alley, B.J.	5010
Altpeter, R.J.	3926
Alves, S.	4980
	4985
Anand, I.J.	5003
Anderson, A.J.B.	5580
Anderson, D.A.	5355
Anderson, D.L.	4868
Anderson, R.L.	3846
	4149
Anderson, V.L.	4953
	4954
Arens, B.E.	4214
	4215
Arunachalam, V.	5003
Ashton, G.C.	4054
	4973
Babcock, A.B.	5142
Bachus, J.A.	4948
Baker, G.A.	3769
	3770
	5065
Balaam, L.N.	5211
Ball, W.E.	4737
Bancroft, T.A.	4852
Bargmann, R.E.	3829
	3830
Batey, T.	3959
Batra, J.N.	5303
Behnken, D.W.	3927
	3928
	3929
	3930

Name	Ref. No.
Bell, A.E.	4923
	5553
Bennison, R.W.	4054
Beres, V.	4545
	4546
Berkson, J.	4253
Bernard, R.L.	5279
Bhapkar, V.P.	5235
Bhargava, P.N.	4194
Bielorai, H.	5333
Biggers, J.D.	4088
Bliss, C.I.	4494
	4495
Boam, T.B.	5410
Bock, R.D.	5310
Bohrer, R.E.	3711
Bose, R.C.	5356
	5357
Bourgain, R.	4925
Box, G.E.P.	4534
	5598
Boylan, W.J.	5153
Bradley, R.A.	3684
	3685
	4815
	4816
	4817
	4818
	5104
	5402
Brasil Sobr°, M.O.C.	4962
Brim, C.A.	4548
Brimball, B.	5341
Brooks, R.M.	4546
Brown, B.W.	3872
	3873
Brown, S.	3758
Brown, W.A.	4703
Brown, W.G.	4573
Bruggrabe, W.F.	4737
Brunk, M.E.	4359
Brunot, C.A.	5325
Budne, T.A.	5598
Buehler, R.J.	5297
	5298

IX. BIBLIOGRAPHY ON TOPICS RELATED TO EXPERIMENT AND TREATMENT DESIGN

References in this part of the bibliography refer to topics related to the conduct and analyses of experiment and treatment design. The citations cover such topics as the philosophy of experimentation, the determination of sample sizes for both fixed and sequential procedures, and bibliographies related to experiment and treatment designs. As may be seen from the classification described in section IV, the statistical topics covered represent a rather wide selection.

Some of the major developments in the analysis of experiment and treatment designs have been in the areas of the concept of error rates (A1), multiple decision procedures (A2), multiple comparisons procedures (A3), nonadditivity (A4), randomization tests (A5), residual analyses (A6), multivariate analyses procedures (A7), calculus for analyses of experiments (A8), distribution-free procedures (A12) and data analyses procedures (C). The coverage in this part is nearly complete even though less emphasis was placed here than on parts one to three.

Abelson, R. P. [1953], A note on the Neyman-Johnson technique. Psychometrika 18:213-218. A7 (5623)

Abraham, T. P. and Kulkarni, G. A. [1963], Investigations on the optimum pre-experimental period in field experimentation on perennial crops. JISAS 15:175-184 (abstract 15:255-256). C (5624)

Abraham, T. P., Misra, R. K., and Ghose, R. L. M. [1956], Observations on progeny row trials with rice crop. Indian J. Genetics Plant Breeding 16:47-51. A7;S1 (5625)

Abraham, T. P. and Vachhani, M. V. [1964], Investigations on field experimental techniques with rice crop. I. Size and shape of plots and blocks in field experiments with transplanted rice crop. Indian J. Agri. Sci. 34:152-165. A14;S1 (5626)

Abramson, L. R. [1966], Asymptotic sequential design of experiments with two random variables. JRSSB 28:73-87. S2 (5627)

Abt, K. [1960], Analyse de covariance et analyse par différences. Metrika 3:26-45, 95-116, 177-211. A7 (5628)

Abt, K. [1961], Analysis of variance of differences versus analysis of covariance (abstract). Biometrics 17:164. A7 (5629)

Abt, K. [1967], A method of fitting constants for non-orthogonal layouts with interactions and empty cells (abstract). AMS 38:634. A10 (5630)

Adichie, J. N. [1967], Estimates of regression parameters based on rank tests. AMS 38:894-904. A7,12 (5631)

Afifi, A. A. and Elashoff, R. M. [1966], Missing observations in multivariate statistics. I. Review of the literature. JASA 61:595-604. A7 (5632)

Afifi, A. A. and Elashoff, R. M. [1967], Missing observations in multivariate statistics. II. Point estimation in simple linear regression. JASA 62:10-29 A7 (5633)

Agarwal, K. C. [1961], Review of uniformity trials (abstract). JISAS 13:241. A14 (5634)

Aggarwal, O. P. [1966], Bayes and minimax procedures for estimating the arithmetic mean of a population with two-stage sampling. AMS 37:1186-1195 S2 (5635)

Agrawal, K. C. [1963], Uniformity trial studies on mango (abstract). JISAS 15:256 A14 (5636)

Ahmad, M. [1967], Truncated logarithmic distribution. Bull. Inst. Stat. Res. Training, Univ. Dacca. 2(1):40-45 A2 (5637)

Ahmad, M. and Kudô, A. [1967], Modified and partially truncated Poisson distribution. Bull. Inst. Stat. Res. Training, Univ. Dacca. 1(2):82-90. A2 (5638)

Ahring, R. M., Morrison, R. D., and Wilhite, M. L. [1959], Uniformity trials on germination of switchgrass seed. Agron. J. 51:734-737. A14 (5639)

Alam, K. [1967], On selecting the largest category (abstract). AMS 38:634-635. A2 (5640)

Alam, K. and Rizvi, M. H. [1965], Selection from multivariate normal populations (abstract). AMS 36:1321. A2,7 (5641)

Alam, K. and Rizvi, M. H. [1966], Selection from multivariate normal populations. Ann. Inst. Stat. Math. Tokyo. 18:307-318. A2,7 (5642)

Albert, A. E. [1961], The sequential design of experiments for infinitely many states of nature. AMS 32:774-799 (abstract 31:811). S2 (5643)

Alexander, H. W. [1962], Vectorial aspects of analysis of variance (abstract). AMS 33:1485. A14 (5644)

Alling, D. W. [1963], Early decision in the Wilcoxon two-sample test. JASA 58:713-720. A12 (5645)

Alling, D. W. [1966], Closed sequential tests for binomial probabilities. Biometrika 53:73-84. S2 (5646)

Allmaras, R. R. and Gardner, C. O. [1956], Soil sampling for moisture determination in irrigation experiments. Agron. J. 48:15-17. S1 (5647)

Altman, I. B. [1954], Relationship between sample size and AOQL for attribute single-sampling plans. IQC 10(4):29-30. S1 (5648)

Alvey, N. G. [1955], Error in glasshouse experiments. Plant Soil 6:347-359. A14;C;S1 (5649)

Amble, V. N. [1953], Animal experiments (abstract). Biometrics 9:430. C (5650)

Amerine, M. A. and Roessler, E. B. [1952], Techniques and problems in the organoleptic examination of wines. Proc. Third Annual Meeting, American Soc. Enologists, pp 97-115. C (5651)

Anderson, E. [1956], Natural history, statistics, and applied mathematics. American J. Botany 43:882-889. C (5652)

Anderson, R. L. [1953], Proper planning reduces research errors. J. Farm Economics 35:572-581. C (5653)

Anderson, R. L. [1962], Designs for estimating variance components. Proc. Seventh Conf. Design Expt. Army Res. Dev. Testing. pp. 781-823. C;S1 (5654)

Anderson, T. W. [1951], Estimating linear restrictions on regression coefficients for multivariate normal distribution. AMS 22:327-351. A7 (5655)

Anderson, T. W. [1957], Maximum likelihood estimates for a multivariate normal distribution when some observations are missing. JASA 52:200-203. A7 (5656)

Anderson, T. W. [1960], A modification of the sequential probability ratio test to reduce the sample size. AMS 31:165-197. S2 (5657)

Anderson, T. W. [1962], The choice of the degree of a polynomial regression as a multiple decision problem. AMS 33:255-265 (abstract 33:297). A7 (5658)

Anderson, T. W. [1963], A test for equality of means when covariance matrices are unequal. AMS 34:671-672. A3 (5659)

Anderson, T. W. [1964], Sequential analysis with delayed observations. JASA 59:1006-1015. S2 (5660)

Anderson, T. W. [1965], Some properties of confidence regions and tests of parameters in multivariate distributions (with discussion). Proc. IBM Scientific Computing Symp. Stat. pp. 15-28. A7 (5661)

Anderson, T. W. [1965], Samuel Stanley Wilks, 1906-1964. AMS 36:1-27. B (5662)

Andrew, W. D. [1960], The effect of varying row spacing and seeding density within rows of the perennial grass component of a mixed sward. Australian J. Agri. Res. 11:686-692. A14 (5663)

Angelotti, R., Foter, M. J., Busch, K. A. and Lewis, K. H. [1958], A comparative evaluation of methods for determining the bacterial contamination of surfaces. Food Res. 23:175-185. C (5664)

Anonymous [1960], The Pudden clinal plot: thinning experiments without surrounds. Empire Forestry Rev. 39:168-171. A14 (5665)

Anscombe, F. J. [1950], Soil sampling for potato root eelworm cysts. A report presented to the Conference of Advisory Entomologists. Ann. Appl. Biology 37:286-295. S1 (5666)

Anscombe, F. J. [1952], Large-sample theory of sequential estimation. Proc. Cambridge Philosophical Soc. 48:600-607. S2 (5667)

Anscombe, F. J. [1953], Sequential estimation (with discussion). JRSSB 15:1-29. S2 (5668)

Anscombe, F. J. [1954], Fixed-sample-size analysis of sequential observations. Biometrics 10:89-100. (correction 13:543) S1 (5669)

Anscombe, F. J. [1959], Rejection of outliers (abstract). Biometrics 15:632. A6 (5670)

Anscombe, F. J. [1960], Rejection of outliers. Technometrics 2:123-147. A6 (5671)

Anscombe, F. J. [1960], Notes on sequential sampling plans. JRSSA 123:297-306. S2 (5672)

Anscombe, F. J. [1961], Examination of residuals. Proc. Fourth Berkeley Symp. Math. Stat. Prob. 1:1-36. A6 (5673)

Anscombe, F. J. [1961], Bayesian statistics. Amer. Stat. 15(1):21-24. C (5674)

Anscombe, F. J. [1963], Sequential medical trials. JASA 58:365-383. S2 (5675)

Anscombe, F. J. and Barron, B. A. [1966], Treatment of outliers in samples of size three. J. Res. NBS. Mathematics and Mathematical Physics Vol. 70B:141-147. A6 (5676)

Anscombe, F. J. and Page, E. S. [1954], Sequential tests for binomial and exponential populations. Biometrika 41:252-253. S2 (5677)

Anscombe, F. J. and Tukey, J. W. [1955], The criticism of transformations (abstract). JASA 50:566. A9 (5678)

Anscombe, F. J. and Tukey, J. W. [1963], The examination and analysis of residuals. Technometrics 5:141-160. A6 (5679)

Arbonnier, P. [1966], L'utilization du calcul automatique á la station de biométrie du C.N.R.F. Int. Advisory Group Forest Stat. Second Conf. Institutionen Skoglig Matematisk Stat. Stockholm pp. 1-19 A15 (5680)

Archbald, D. [1950], The effect of quadrat size and quadrat method on apparent plant dispersion (abstract). Ecol. Soc. American Bull. 31:52. A14 (5681)

Archibald, E. E. A. [1950], Plant populations. II. The estimation of the number of individuals per unit area of species in heterogeneous plant populations. Ann. Botany 14(53):7-21. A14 (5682)

Ariens, E. J. and Simonis, A. M. [1961], Analysis of the action of drugs and drug combinations (with discussion). Quantitative Methods in Pharmacology, pp. 286-311. C (5683)

Armiger, W. H., Dean, L. A., Mason, D. D., and Koch, E. J. [1958], Effect of size and type of pot on relative precision, yields, and nutrient uptake in greenhouse fertilizer experiments. Agron. J. 50:244-247. A14 (5684)

Armiger, W. H., and Fried, M. [1958], Effect of pot size and shape on yield and phosphorus uptake of millet. Agron. J. 50:462-465. A14 (5685)

Armitage, P. [1954], Sequential tests in prophylactic and therapeutic trials. Quarterly J. Medicine, New Series 23:255-274. S2 (5686)

Armitage, P. [1957], Restricted sequential procedures. Biometrika 44:9-26. S2 (5687)

Armitage, P. [1958], Numerical studies in the sequential estimation of a binomial parameter. Biometrika 45:1-9. S2 (5688)

Armitage, P. [1958], Sequential procedures for medical trials (abstract). Biometrics 14:132. S2 (5689)

Armitage, P. [1958], Sequential methods in clinical trials. American J. Public Health Nation's Health. 48:1395-1402. S2 (5690)

Armitage, P. [1961], Sequential analysis in medicine. Stat. Neerlandica 15:73-82. S2 (5691)

Armitage, P. [1961], Theoretical aspects of sequential designs. Stat. Neerlandica 15:83-89. S2 (5692)

Armitage, P. [1963], Sequential medical trials: Some comments on F. J. Anscombe's paper. JASA 58:384-387. S2 (5693)

Armitage, P. [1967], Some developments in the theory and practice of sequential medical trials. Proc. Fifth Berkeley Symp. Math. Stat. Prob. 4:791-804. S2 (5694)

Armstrong, J. S. [1967], Derivation of theory by means of factor analysis or Tom Swift and his electric factor analysis machine. Amer. Stat. 21(5):17-21. A7 (5695)

Arnáiz, G. [1958], Problemas de regression y correlacion. Trabajos Estadistica 9:43-56. A7 (5696)

Arnold, K. J. [1951], Tables to facilitate sequential t-tests. Nat. Bur. Standards, Applied Math. Series 7. S2 (5697)

Asano, C. [1962], On successive inferences and their multivariate applications in biometry (abstract). AMS 33:814. A11 (5698)

Asano, C. and Satô, S. [1962], A bivariate analogue of pooling of data. Bull. Math. Stat. 10(3&4):39-59. A11 (5699)

Ashton, E. H., Healy, M. J. R., Oxnard, C. E., and Spence, T. F. [1965], The combination of locomotor features of the primate shoulder girdle by canonical analysis. J. Zoology 147:406-429. A7 (5700)

Askovitz, S. I. [1957], A short-cut graphic method for fitting the best straight line to a series of points according to the criterion of least squares. JASA 52:13-17. A7 (5701)

Aspinall, D. and Milthorpe, F. L. [1959], An analysis of competition between barley and white persicaria. I. The effects on growth. Ann. Appl. Biology 47:156-172. A14 (5702)

Atanasiu, N. [1954], Über den Ausgleich bei Feldversuchen. Z. Acker - und Pflanzenbau, 98:255-264. C (5703)

Atiqullah, M. [1962], The estimation of residual variance in quadratically balanced least-squares problems and the robustness of the F-test. Biometrika 49:83-91. A11 (5704)

Atiqullah, M. [1964], The robustness of the covariance analysis of a one-way classification. Biometrika 51:365-372. A7 (5705)

Atkinson, G. F. and Robson, D. S. [1962], Sequential estimation of vaccine effectiveness. Biometrics 18:212-223. S2 (5706)

Auble, D. [1953], Extended tables for the Mann-Whitney statistic. Bull. Inst. Educational Res. Indiana Univ. 1(2):1-33. A12 (5707)

Bagai, O. P. [1962], Statistics proposed for various tests of hypotheses and their distributions in particular cases. Sankhyā A, 24:409-418.
A7 (5708)

Bagai, O. P. [1963], Addendum: Statistics proposed for various tests of hypotheses and their distributions in particular cases. Sankhyā A, 24:409-418.
A7 (5709)

Bailey, G. L., Broster, W. H., and Burt, A. W. A. [1958], Experiments on the nutrition of the dairy heifer. II. Experimental methods in short-term experiments. J. Agri. Sci. 50:1-7. A13;S1 (5710)

Bailey, N. T. J. [1952], The scope of medical statistics. Appl. Stat. 1:149-162.
C (5711)

Baker, F. B. and Collier, R. O. [1961], Analysis of experimental designs by means of randomization, a Univac 1103 program. Behavioral Sci. 6:369.
A4,15 (5712)

Baker, F. B. and Collier, R. O. [1966], Monte Carlo F-II: A computer program for analysis of variance F-tests by means of permutation. Educational Psychological Measurement. 26:169-173. A15 (5713)

Baker, G. A. [1950], Properties of some tests in sequential analysis. Biometrika 37:334-346. S2 (5714)

Baker, G. A. [1952], Field trial problems (abstract). Biometrics 8:271-272.
C (5715)

Baker, G. A. [1952], Field trial problems (abstract). AMS 23:480.
C (5716)

Baker, G. A. [1954], Organoleptic ratings and analytical data for wines analyzed into orthogonal factors. Food Research 19:575-580. A7;C (5717)

Baker, G. A. [1954], The effect of selection on linear functions of normally distributed correlated variables on the distributions of other linear functions. Ann. Inst. Stat. Math. Tokyo. 5:91-95. A2 (5718)

Baker, G. A. and Amerine, M. A. [1953], Organoleptic ratings of wines estimated from analytical data. Food Res. 18:381-389. C (5719)

Baker, G. A., Amerine, M. A., and Roessler, E. B. [1954], Errors of the second kind in organoleptic difference testing. Food Res. 19:206-210.
C (5720)

Baker, G. A. and Baker, R. E. [1953], Strawberry uniformity yield trials. Biometrics 9:412-421 (abstract 7:300). A14 (5721)

Baker, G. A., Huberty, M. R., and Veihmeyer, F. J. [1952], A uniformity trial on unirrigated barley of ten years' duration. Agron. J. 44:267-270. A13 (5722)

Baker, G. A., Mrak, V., and Amerine, M. A. [1958], Errors of the second kind in an acid threshold test. Food Res. 23:150-154. C (5723)

Baker, R. A. [1962], Subjective panel testing. IQC 19(3):22-28. C (5724)

Baker, R. E. and Baker, G. A. [1950], Experimental design for studying resistance of strawberry varieties to verticillium wilt. Phytopathology 40:477-482. A9,14 (5725)

Baker, R. J. and McKenzie, R. I. H. [1967], Use of control plots in yield trials. Crop Sci. 7:335-337. A14 (5726)

Baker, R. W. R. and Nissim, J. A. [1963], Expressions for combining standard errors of two groups and for sequential standard error. Nature 198:1020 A11 (5727)

Balaam, L. N. [1963], Multiple comparisons - A sampling experiment. Australian J. Stat. 5:62-84. A3 (5728)

Balaam, L. N. and Federer, W. T. [1965], Query 11: Error rate bases. Technometrics 7:260-262. A3 (5729)

Balaam, L. N. and Hunter, R. D. [1962], The analysis of a series of wheat varietal trials. Australian J. Stat. 4:61-70. A13 (5730)

Balba, M. A. and Bray, R. H. [1956], The application of the Mitscherlich equation for the calculation of plant composition due to fertilizer increments. Soil Sci. Soc. Amer. Proc. 20:515-518. A7 (5731)

Bancroft, T. A. [1954], Preliminary tests and pool rules (summary). JASA 49:348. A11 (5732)

Banerjee, S. [1961], On confidence interval for two-means problem based on separate estimates of variances and tabulated values of t-table. Sankhyā 10:359-378. A11 (5733)

Banerjee, S. [1962], The problem of testing linear hypothesis about population means when the population variances are not equal and M-test. Sankhyā A 24:363-376. A7,11 (5734)

Bangdiwala, I. S. and Monroe, R. J. [1958], Ordering population means by sequential procedures (abstract). Biometrics 14:563. A2;S2 (5735)

Bangdiwala, I. S. and Monroe, R. J. [1958], Ordering population regression coefficients by sequential procedures (abstract). Biometrics 14:564. A2;S2 (5736)

Bangdiwala, I. S. and Monroe, R. J. [1958], Ordering population variances by sequential procedures (abstract). Biometrics 14:565. A2;S2 (5737)

Banks, S. [1966], After the experiment, so what? (summary). JASA 61:541. C (5738)

Banneick, A. and Ehrenpfordt, V. [1962], Methodik und Probleme des exakten Leistungsvergleichs von diploidem und tetraploidem Roggen. Albrecht-Thaer. Arch. 6:31-47. C (5739)

Bannerot, H. [1958], Possibilités d'utilisation de l'analyse séquentielle pour le controlle des variétés - populations. Ann. Inst. Nat. Recherche Agronomique 8(B):59-74. S2 (5740)

Baptista, J. G., Semedo, J. L., and Mascarenhas, G. H. N. [1955], Acerca da amostragem em ensaios de milho para a determina ção da humidade do grão. Melhoramemto 8:5-21. S1 (5741)

Bargmann, R. E. [1958], Generalized distributions and the problem of confidence statements in multivariate analysis (abstract). Virginia J. Sci. 9:453. A7 (5742)

Bargmann, R. E. [1961], Multivariate statistical analysis in psychology and education. Bull. ISI 38(4):79-86. A7 (5743)

Barker, J. S. F. [1967], Modern problems of population genetics in animal husbandry. Der Züchter 37:309-323. B;C (5744)

Barker, R. [1964], Use of linear programming in making farm management decisions. Cornell Univ. Agri. Expt. Sta. Bull. 993. A7 (5745)

Barnard, G. A. [1950], On the Fisher-Behrens test. Biometrika 37:203-207. All (5746)

Barnard, G. A. [1952], The frequency justification of certain sequential tests. Biometrika 39:144-150. S2 (5747)

Barnett, V. D. [1962], Large sample tables of percentage points for Hartley's correction to Bartlett's criterion for testing the homogeneity of a set of variances. Biometrika 49:487-494. All (5748)

Barr, D. R. [1966], On the solution of Bechhofer's general goal in the indifference zone formulation of the ranking and selection problem (abstract). AMS 37:1417. A2 (5749)

Barr, D. R. and Rizvi, M. H. [1966], An introduction to ranking and selection procedures. JASA 61:640-646 (summary 61:541). A2 (5750)

Barten, A. P. [1962], Note on unbiased estimation of the squared multiple correlation coefficient. Stat. Neerlandica 16:151-163. A7 (5751)

Bartholomew, D. J. [1959], A test of homogeneity for ordered alternatives. Biometrika 46:36-48. A2,12 (5752)

Bartholomew, D. J. [1959], A test of homogeneity for ordered alternatives. II Biometrika 46:328-335. A2,12 (5753)

Bartholomew, D. J. [1961], Ordered tests in the analysis of variance. Biometrika 48:325-332. A2 (5754)

Bartholomew, D. J. [1961], A test of homogeneity of means under restricted alternatives (with discussion). JRSSB 23:239-281. A2 (5755)

Bartholomew, D. J. [1967], Hypothesis testing when the sample size is treated as a random variable (with discussion). JRSSB 29:53-82. A11;S2;C (5756)

Bartlett, M. S. [1950], Tests of significance in factor analysis. British J. Psych. (Stat. Sect.) 3:77-85. A7 (5757)

Bartlett, M. S. [1951], An inverse matrix adjustment arising in discriminant analysis. AMS 22:107-111. A7 (5758)

Bartlett, M. S. [1951], The goodness of fit of a single hypothetical discriminant in the case of several groups. Ann. Eugenics 16:199-214. A7 (5759)

Bartlett, M. S. [1956], Comment on Sir Ronald Fisher's paper: "On a test of significance in Pearson's Biometrika Tables (No. 11)." JRSSB 18:295-296 A11;C (5760)

Bartlett, M. S. [1965], R. A. Fisher and the last fifty years of statistical methodology. JASA 60:395-409 (summary 60:654). C (5761)

Bartlett, N. S. and Govindarajulu, Z. [1965], Some distribution-free statistics and their application to the selection problem (abstract). AMS 36:1597-1598. A2,12 (5762)

Bartlett, N. S. and Govindaralajulu, Z. [1967], Selecting a subset containing the best hypergeometric population (abstract). AMS 38:953. A2 (5763)

Barton, D. E. and David, F. N. [1958], A test for birth order effect. Ann. Human Genetics 22:250-257. A12 (5764)

Barton, D. E. and David, F. N. [1959], Combinatorial extreme value distributions. Mathematika 6:63-76. A6 (5765)

Barton, D. E. and David, F. N. [1960], Models of functional relationship illustrated on astronomical data. Bull. ISI 37(3):9-33. A7 (5766)

Barton, D. E., David, F. N., and Fix, E. [1960], The polykays of the natural numbers. Biometrika 47:53-59. A14 (5767)

Barylko-Pikielna, N. and Metelski, K. [1964], Determination of contribution coefficients in sensory scoring of over-all quality. J. Food Sci. 29:109-111. C (5768)

Basu, A. P. [1965], On some tests of hypotheses relating to the exponential distribution when some outliers are present. JASA 60:548-559. A6 (5769)

Basu, D. [1964], Recovery of ancillary information. Sankhyā A 26:3-16. A7;S1 (5770)

Baten, W. D. [1950], Reaction of age groups to organoleptic tests. Food Tech. 4:277-279. C (5771)

Bätz, G. [1957], Über die Anlage und Auswertung von Sortenprüfungen. Albrecht-Thaer-Arch. 2:239-309. C (5772)

Bätz, G. [1957], Zu einigen Fragen der Anlage und Auswertung mehrjähriger Sortenversuche. Z. Landw. Vers. Untersuch Wes. 3:407-423. A13;C (5773)

Bauer, T. W. [1957], The practical calculation of Hotelling's T. IQC 14(1):7-10. A7 (5774)

Bayton, J. A. [1955], Rating scales and psycological factors in taste preference research. ASQC Nat. Conv. Trans. 9:275-291. C (5775)

Beach, G. [1952], Plot technique with carnations. Proc. Amer. Soc. Hort. Sci. 60:479-486. A14 (5776)

Beale, E. M. L. [1960], Confidence regions in non-linear estimation (with discussion). JRSSB 22:41-88. A7 (5777)

Bechhofer, R. E. [1951], The effect of preliminary tests of significance on the size and power of certain tests of univariate linear hypotheses with special reference to the analysis of variance (Preliminary report) (abstract). AMS 22:143. A11 (5778)

Bechhofer, R. E. [1952], The probability of a correct ranking (Preliminary report) (abstract). AMS 23:139-140. A2 (5779)

Bechhofer, R. E. [1955], Multiple decision procedures for ranking means. ASQC Nat. Conv. Trans. 9:513-519. A2 (5780)

Bechhofer, R. E. [1960], Multiple-decision ranking problems arising from factorial experiments on variances of normal populations (Preliminary report) (abstract). AMS 31:231. A2 (5781)

Bechhofer, R. E. [1965], A Monte Carlo study of the performance characteristics of two competing sequential ranking procedures (abstract). Technometrics 7:268-269. A2;S2 (5782)

Bechhofer, R. E. [1967], A two-stage subsampling procedure for ranking means of finite populations with an application to bulk sampling problems. Technometrics 9:355-364. A2;S2 (5783)

Bechhofer, R. E., Dunnett, C. W., and Sobel, M. [1953], A two-sample multiple decision procedure for ranking means of normal populations with unknown variances (Preliminary report) (abstract). AMS 24:136. A2 (5784)

Bechhofer, R. E., Dunnett, C. W., and Sobel, M. [1954], A single-sample multiple decision procedure for ranking means of normal populations with known variances. A two-sample multiple decision procedure for ranking means of normal populations with a common unknown variance. A sequential multiple decision procedure for ranking means of normal populations with known variances (summary of paper). JASA 49:358-359. A2;S2 (5785)

Bechhofer, R. E., Dunnett, C. W., and Sobel, M. [1954], A two-sample multiple decision procedure for ranking means of normal populations with a common unknown variance. Biometrika 41:170-176. A2;S2 (5786)

Bechhofer, R. E., Elmaghraby, S., and Morse, N. [1959], A single-sample multiple decision procedure for selecting the multinomial event which has the highest probability. AMS 30:102-119. A2 (5787)

Bechhofer, R. E. and Sobel, M. [1953], A sequential multiple decision procedure for ranking means of normal populations with known variances (Preliminary report) (abstract). AMS 24:136-137. A2 (5788)

Bechhofer, R. E. and Sobel, M. [1954], On a sequential ranking procedure (Preliminary report) (abstract). Bull. Amer. Math. Soc. 60:34-35. A2;S2 (5789)

Bechhofer, R. E. and Sobel, M. [1954], A single-sample multiple decision procedure for ranking variances of normal populations. AMS 25:273-289. A2 (5790)

Bechhofer, R. E. and Sobel, M. [1956], A sequential multiple decision for selecting the multinomial event with the largest probability (Preliminary report) (Abstract). AMS 27:861. A2 (5791)

Bechhofer, R. E. and Sobel, M. [1958], Non-parametric multiple-decision procedures for selecting that one of k populations which has the highest probability of yielding the largest observation (abstract). AMS 29:325. A2,12 (5792)

Becker, G. M., DeGroot, M. H. and Marshak, J. [1963], Probabilities of choices among very similar objects: an experiment to decide between two models. Behavioral Sci. 8:306-311. C (5793)

Becker, W. A. [1961], Comparing entries in random sample tests. Poultry Sci. 40:1507-1514. A2,13 (5794)

Beckhart, G. H. [1961], Short cut statistical methods as applied to certain reliability problems. ASQC Nat. Conv. Trans. 15:425-435. A2;S1 (5795)

Beckman, F. S. and Quarles, D. A., [1956], Multiple regression and correlation analysis on the IBM type 701 and type 704 electronic data processing machines. Amer. Stat. 10(1):6-8 and 16. A7,15 (5796)

Bedell, B. J. [1950], Determination of the optimum number of items to retain in a test measuring a single ability. Psychometrika 15:419-430. S1 (5797)

Beecher, H. K. [1952], Clinical studies of analgestic drugs I. Experimental pharmacology and measurement of the subjective response (abstract). Biometrics 8:218-220, 227-231. C (5798)

Beecher, H. K. [1952], Experimental pharmacology and measurement of the subjective response. Science 116:157-162. C (5799)

Behari, V. [1964], Percentage points of Bartlett's criterion for testing the homogeneity of variances. JISA 2:119-125. A11 (5800)

Behrens, W. U. [1956], Die Gültigkeit des t-Testes. Z. Pflanzenzüchtung 36:214-227. A3 (5801)

Behrens, W. U. [1956], Zur abgekürzten Berechnung der Standardabweichung. Z. Acker- Pflanzenbau 101:459-464. A12 (5802)

Behrens, W. U. [1959], Beitrag zur Diskriminanzanalyse. Biometrische Zeitschrift 1:3-14. A7 (5803)

Behrens, W. U. [1963], Der Vergleich von Mittelwerten aus Normalverteilungen, deren Streuungen voneinander unabhängig sind. Biometrische Zeitschrift 5:219-230. A11 (5804)

Bejar, J. [1956], Regresion en mediana y la programacion lineal. Trabajos Estad. 7:141-158. A7 (5805)

Bejar, J. [1957], Calculo practico de la regresion en mediana. Trabajos Estad. 8:157-173. A7 (5806)

Bell, C. B. and Doksum, K. A. [1965], Some new distribution-free statistics. AMS 36:203-214. A12 (5807)

Bellman, R. [1956], A problem in the sequential design of experiments. Sankhyā 16:221-229. S2 (5808)

Bellucci, R. [1958], Experimental design to study the effect of balloon size on wind response. Proc. Third Conf. Design Expt. Army Res. Dev. Testing. pp. 151-165. C (5809)

Belson, I. [1964], Minimum sample size for accepting a criterion error or failure rate with a given degree of confidence. J. Electronics Div. ASQC 2(3):16-23. S1 (5810)

Belz, M. H. [1951], Statistical applications in industry, with reference to some recent work in Australia, 1951. Bull. ISI 33(5):29-34. C (5811)

Belz, M. H. [1957], Common errors in obtaining and evaluating experimental data. Physics Medicine Biol. 2:3-16. C (5812)

Benard, A. and van Elteren, Ph. [1953], A generalization of the method of m rankings. Indagationes Mathematicae 15:358-369. A12 (5813)

Beni, G. [1960], Interazione stazione × varietà e prove pluriannuali nella sperimentazione collegiale sul frumento per l'Italia settentrionale. (The interaction station × variety and tests extending over several years in the collective experiments on wheat for northern Italy.) Genetica Agraria, Rome 11:377-393. A13 (5814)

Bennett, B. M. [1951], Note on a solution of the generalized Behrens-Fisher problem. Ann. Inst. Stat. Math. 2:87-90. A11 (5815)

Bennett, B. M. [1952], Estimation of means on the basis of preliminary tests of significance. Ann. Inst. Stat. Math. 4:31-43. A11 (5816)

Bennett, B. M. [1956], On the variance stabilizing properties of certain logaritmic transformations. Trabajos Estad. 7:295-297. A9 (5817)

Bennett, B. M. [1956], On the use of preliminary tests in certain statistical procedures. Ann. Inst. Stat. Math. 8:45-52. A7 (5818)

Bennett, B. M. [1957], On the variance stabilizing properties of certain transformations II. Trabajos Estad. 8:69-74. A9 (5819)

Bennett, B. M. [1957], Tests for linearity of regression involving correlated observations. Ann. Inst. Stat. Math. 8:193-195. A7 (5820)

Bennett, B. M. [1962], On multivariate sign tests. JRSSB 24:159-161. A7,12 (5821)

Bennett, B. M. [1964], A bivariate signed rank test. JRSSB 26:457-461. A7,12 (5822)

Bennett, B. M. [1965], Note on a χ^2-Approximation for the multivariate sign test. JRSSB 27:82-85. A7,12 (5823)

Bennet, B. M. [1965], On multivariate signed rank tests. Ann. Inst. Stat. Math. 17:55-61. A7,12 (5824)

Bennet, B. M. [1966], Multiple regression analysis of binary and multinomial variates. Sankhyā 28:301-304. A7 (5825)

Bennett, B. M. [1967], Use of Haldane-Smith test in examining randomness of residuals. Metron 26:371-373. A6 (5826)

Bennett, B. M. [1967], On the bivariate generalization of the Wilcoxon two sample test and its relation to rank correlation theory. Rev. ISI 35:30-33. A7,12 (5827)

Bennett, B. M. and Hsu, P. [1961], Sampling studies on the Behrens-Fisher problem. Metrika 4:89-104. A11 (5828)

Bennett, C. A. [1951], Application of tests for randomness. Ind. Eng. Chemistry. 43:2063-2067. A6;C (5829)

Bennett, C. A. [1954], Effect of measurement error on chemical process control. IQC 10(4):17-20. A14;C (5830)

Bennett, G., Spahr, B. M. and Dodds, M. L. [1956], The value of training sensory test panel. Food Tech. 10:205-208. C (5831)

Berg, C. [1960], Optimization in process development. Chem. Eng. Prog., 56(8):42-47. C (5832)

Berg, H. W., Filipello, F., Hinreiner, E. and Webb, A. D. [1955], Evaluation of thresholds and minimum difference concentrations for various constituents of wines. I. Water solutions of pure substances. Food Tech. 9:23-26. C (5833)

Berg, H. W., Filipello, F., Hinreiner, E. and Webb, A. D. [1955], Evaluation of thresholds and minimum difference concentration for various constituents of wines. II. Sweetness: the effect of ethyl alcohol, organic acids and tannin. Food Tech. 9:138-140. C (5834)

Berkson, J. [1950], Are there two regressions? JASA 45:164-180. A7 (5835)

Berry, G. [1963], A transformation for the tenderometer-yield relationship in shelled peas. Biometrics 19:491-494. A9 (5836)

Berry, J. M., Witter, L. D. and Folinazzo, J. F. [1956], Growth characteristics of spoilage organisms in orange juice and concentrate. Food Tech. 10:553-556. A7 (5837)

Bessler, S. A. [1960], Theory and applications of the sequential design of experiments, k-actions and infinitely many experiments. Ph. D. Thesis, Stanford Univ. S2 (5838)

Bessler, S. A. [1960], Elements of the sequential design of experiments (abstract). AMS 31:812-813. S2 (5839)

Bhalli, M. A., Day, A. D., Tucker, H., Thompson, R. K. and Massey, G. D. [1964], End-border effects in irrigated barley yield trials. Agron. J. 56:346-348. A14 (5840)

Bhapkar, V. P. [1954], A note on t test for paired samples. Calcutta Assoc. Bull. 5:142-147. A11 (5841)

Bhat, B. R. [1959], On the distribution of various sums of squares in an analysis of variance table for different classifications with correlated and non-homogeneous errors. JRSSB 21:114-119. A11 (5842)

Bhate, D. H. [1955], Sequential analysis, with special reference to distributions of sample size. Ph. D. Thesis, London Univ. College. S2 (5843)

Bhate, D. H. [1959], Approximation to the distribution of sample size for sequential tests. Biometrika 46:130-138. S2 (5844)

Bhate, D. H. [1960], Approximation to the distribution of the sample size for sequential tests. II. Tests of composite hypotheses. Biometrika 47:190-193. S2 (5845)

Bhattacharya, C. G. [1966], Fitting a class of growth curves. Sankhyā B 28:1-10. A7 (5846)

Bhattacharya, P. K. [1961], Use of concomitant measurements in design and analysis of experiments. Inst. Stat. Mimeo. Series No. 309. Univ. N. Carolina (10 pp). A7;C (5847)

Bhattacharya, P. K. [1961], Selecting the "best" t out of k populations (abstract). AMS 32:915-916. A2 (5848)

Bhattacharyya, A. [1952], On the uses of the t-distribution in multivariate analysis. Sankhyā 12:89-104. A7 (5849)

Bhattacharyya, A. K. [1960], A note on the efficiency of sequential sampling plans based on gauging. Calcutta Stat. Assoc. Bull. 9:117-121. S2 (5850)

Bhattacharyya, P. K. [1956], Comparison of the means of k normal populations. Calcutta Stat. Assoc. Bull. 7:1-16. A2 (5851)

Bhattacharyya, P. K. [1958], On the large sample behaviour of a multiple-decision procedure. Calcutta Stat. Assoc. Bull. 8:43-47. A2 (5852)

Bicking, C. A. [1954], Some uses of statistics in the planning of experiments. IQC 10(4):20-24. C (5853)

Bicking, C. A. [1955], Some examples of designed experiments in industry. Tappi 38:174-178. C (5854)

Bicking, C. A. [1957], Selection of most efficient experimental designs. Tooling and Production 23, April, 1:119. C (5855)

Bicking, C. A. [1961], Designing valid experiments. ASQC, Nat. Conv. Trans. 15:439-444. C (5856)

Bicking, C. A. [1961], Applications of industrial statistics in research and development. IQC 17(8):14-16. C (5857)

Bicking, C. A. [1961], Primer on experimentation. IQC 17(11):20-25. C (5858)

Bicking, C. [1962], La validité des plans d'expérience. Rene Stat. Appliquée. 10(3):75-83. C (5859)

Bicking, C. A. [1964], Bibliography on sampling of raw materials and products in bulk. Tappi 47(5):147A-170A. B (5860)

Billewicz, W. Z. [1955], A simple non-parametric graphical test for the significance of a difference. Appl. Stat. 4:97-102. A12 (5861)

Billewicz, W. Z. [1956], Matched pairs in sequential trials for significance of a difference between proportions. Biometrics 12:283-300. S2 (5862)

Billewicz, W. Z. [1958], Some practical problems in a sequential medical trial. Bull. ISI 36(3):165-171. S2 (5863)

Billings, M. G. [1967], A moderately distribution free approach to reliability estimation based on the first order statistic. Proc. Twelfth Conf. Design Expt. Army. Res. Dev. Testing, pp. 337-349. A12 (5864)

Binet, F. E. and Watson, G. S. [1956], Algebraic theory of the computing routine for tests of significance on the dimensionality of normal multivariate systems. JRSSB 18:70-78. A7 (5865)

Bingham, R. S. [1958], Guide to the use of statistics in the chemical industry. ASQC, Nat. Conv. Trans. 12:159-166. C (5866)

Bingham, R. S. [1959], Design of experiments from a statistical viewpoint, part I. IQC 15(11):29-34. C (5867)

Bingham, R. S. [1962], What's behind experiment design? IQC 18(12):11-13. C (5868)

Birnbaum, A. [1952], Note on the problem of combining independent tests of significance (abstract). AMS 23:477. A11 (5869)

Birnbaum, A. [1962], Another view on the foundations of statistics. Amer. Stat. 16(1):17-21. C (5870)

Birnbaum, Z. W. and Tang, V. K.-T. [1964], Two simple distribution-free tests of goodness of fit. Rev. ISI 32:2-13. A12 (5871)

Biser, E. and Meyerson, M. [1958], The application of design of experiments and modeling techniques to complex weapons systems. Proc. Second Conf. Design Expt. Army Res. Dev. Testing. pp. 131-151. C (5872)

Black, C. A. and Kempthorne, O. [1954], Willcox's agrobiology: I. Theory of the nitrogen constant 318. Agron. J. 46:303-307. A7 (5873)

Black, C. A. and Kempthorne, O. [1954], Willcox's agrobiology: II. Application of the nitrogen constant 318. Agron. J. 46:307-310. A7 (5874)

Black, C. A. Kempthorne, O. and White, W. C. [1955], Willcox's agrobiology: IV. Review of Willcox's reply. Agron. J. 47:497-498. A7 (5875)

Black, J. N. [1957], Seed size as a factor in the growth of subterranean clover (Trifolium subterraneum L.) under spaced and sward conditions. Australian J. Agri. Res. 8:335-351. A14 (5876)

Black, J. N. [1958], Competition between plants of different initial seed sizes in swards of subterranean clover (Trifolium subterraneum L.) with particular reference to leaf area and the light microclimate. Australian J. Agri. Res. 9:299-318. A14 (5877)

Black, J. N. [1960], An assessment of the role of planting density in competition between red clover (Trifolium pratense L.) and lucerne (Medicago sativa L.) In the early vegetative stage. Oikos 11:26-42. A14 (5878)

Black, J. N. [1961], Border and orientation effects and their elimination in experimental swards of subterranean clover (Trifolium subterraneum L.) Australian J. Agri. Res. 12:203-211. A14 (5879)

Black, J. N. [1961], Competition between two varieties of subterranean clover (Trifolium subterraneum L.) as related to the proportions of seed sown. Australian J. Agri. Res. 12:810-820. A14 (5880)

Blackwell, D. and Hodges, J. L. [1957], Design for the control of selection bias. AMS 28:449-460. C (5881)

Blaim, K. [1955], Metodyka pobierania średnich próbek dla potrzeb dóswiadczalnictwa rolniczego (The technique of collecting average samples as required for agricultural experimentation). Roczniki Nauk Rolniczych 70(A):463-478. C (5882)

Blakemore, J. W. and Hoerl, A. E. [1963], Fitting nonlinear reaction rate equations to data. Chem. Eng. Progress Symp. Ser. 59(42):14-27. A7 (5883)

Bland, R. P. and Duncan, D. B. [1961], Bayes rules for the problem of choosing the largest mean (preliminary report) (abstract). AMS 32:622. A2 (5884)

Blaser, R. E. and Brady, N. C. [1950], Nutrient competition in plant associations. Agron. J. 42:128-135. A14 (5885)

Blaser, R. E., Taylor, T., Griffeth, W., and Skrdla, W. [1956], Seedling competition in establishing forage plants. Agron. J. 48:1-6. A14 (5886)

Bleasdale, J. K. A. [1960], Studies on plant competition. Biology of Weeds, pp. 133-142. A14 (5887)

Bleasdale, J. K. A. and Nelder, J. A. [1960], Plant population and crop yield. Nature 188:342. A14 (5888)

Bleicher, E. and Duncan, D. B. [1953], Extension of the multiple comparisons test to incomplete block designs. II. (abstract). Virginia J. Sci. 4:289. A3 (5889)

Bliss, C. I. [1955], The role of a biometrician in the pharmaceutical industry. IQC 11(7):10-15. C (5890)

Bliss, C. I. [1958], Periodic regression in biology and climatology. Conn. (State) Agri. Expt. Sta. Bull. 615, 55 pp., illus. A7 (5891)

Bliss, C. I. [1959], Some statistical aspects of preference and related tests. Proc. Fourth Conf. Design. Expt. Army Res. Dev. Testing, pp. 249-271. C (5892)

Bliss, C. I. [1959], Periodic regression in biology and climatology (abstract). Biometrics 15:159-161. A7 (5893)

Bliss, C. I. [1960], Some statistical aspects of preference and related tests. Appl. Stat. 9:8-19. C (5894)

Bliss, C. I. [1963], An analysis of some insect trap records. Classical and Contagious Discrete Distributions, pp. 385-394. A2,6 (5895)

Bliss, C. I. [1964], R. A. Fisher's contributions to medicine and bioassay. Biometrics 20:273-285. C (5896)

Bliss, C. I., Cochran, W. G. and Tukey, J. W. [1956], A rejection criterion based upon the range. Biometrika 43:418-422. A6 (5897)

Bliss, C. I. and Reinker, K. A. [1964], A lognormal approach to diameter distributions in even-aged stands. Forest Sci. 10(3):350-360. A9 (5898)

Block, J. [1960], On the number of significant findings to be expected by chance. Psychometrika 25:369-380. A3 (5899)

Blom, G. [1954], Transformations of the binomial, negative binomial, Poisson and χ^2 distributions. Biometrika 41:302-316. (corrigenda 43:235) A9 (5900)

Blumberg, M. S. [1957], Evaluating health screening procedures. Operations Res. 5:351-360. A2 (5901)

Blumen, I. [1957], On the bivariate sign test (abstract). AMS 28:1066. A7,12 (5902)

Blumen, I. [1958], A new bivariate sign test. JASA 53:448-456. A7,12 (5903)

Bock, R. D. [1956], The selection of judges for preference testing. Psychometrika 21:349-366. C (5904)

Bock, R. D. [1965], A computer program for univariate and multivariate analysis of variance (with discussion). Proc. IBM Scientific Computing Symp. Stat. pp. 69-111. A7,15 (5905)

Bodmer, W. F. [1959], A significantly extreme deviate in data with non-significant heterogeneity chi square. Biometrics 15:538-542. A6 (5906)

Bofinger, V. J. [1965], The k-sample slippage problem. Australian J. Stat. 7:20-31. A2 (5907)

Boggs, M. M. and Ward, A. C. [1950], Scoring techniques for sulfited foods. Food Tech. 4:282-284. C (5908)

Boguslawski, E. von. [1950], Zur Auswertung von Sortenversuchen und ähnlichen Versuchsfragen. Z. Acker-u. Pflanzenbau 92:397-415. C (5909)

Bohrer, R. E. [1966], On Bayes sequential design of experiments. Ph. D. Thesis, Univ. N. Carolina. S2 (5910)

Bohrer, R. [1966], On Bayes sequential design with two random variables. Biometrika 53:469-475. S2 (5911)

Bol, M., Gerritsen, A. G., and van Soest, J. [1959], Verhoging van de arbeidsproduktiviteit in de bosbouw door rationalisatieonderzoek. Stat. Neerlandica 13:197-202. C (5912)

Bombara, E. L. [1958], Applications of selecting sample sizes for F-tests. Proc. Second Conf. Design Expt. Army Res. Dev. Testing. pp. 153-165. S1 (5913)

Bombara, E. L. [1964], Probability that stress is less than strength at prescribed confidence levels, for normally distributed data. Proc. Ninth Conf. Design Expt. Army Res. Dev. Testing. pp. 111-189. S1 (5914)

Boot, J. C. G. [1965], Price determination based on quality. An application of minimax. Stat. Neerlandica 19:41-53. S1 (5915)

Borges, R. [1964], Varianzanalyse mit Kontrollbeobachtungen und Scheffé's S-Methode als kombiniertes Konfidenz- und Testverfahren. Biometrische Zeitschrift 6:1-9. A3 (5916)

Bormann, F. H. [1953], The statistical efficiency of sample plot size and shape in forest ecology. Ecology 34:474-487. A14 (5917)

Borojević, S. and Borojević, K. [1961], Uticaj ruba na genetski potencijal za prinos kod pšenice . (Genetic capacity for yield in wheat in relation to the border effect.) Savremen. Poljopr., Novi Sad 9:391-401. A14 (5918)

Bose, P. K. [1950], A modern approach to the analysis of human behaviour. Calcutta Stat. Assoc. Bull. 3:46-52. A7 (5919)

Bose, P. K. [1957], Presidential address: Statistical methods in psychometric research. Proc. Forty-fourth Indian Sci. Congress. Part II. pp. 20-50. B;C (5920)

Bose, R. C. [1950], Least squares aspects of analysis of variance. Inst. Stat. Mimeo Series No. 9, Univ. N. Carolina. pp. 1-86. C (5921)

Bose, R. C. and Roy, S. N. [1953], Simultaneous confidence interval estimation (abstract). AMS 24:144. A3 (5922)

Bose, R. C. and St. Pierre, J. [1954], On a decision procedure to select the population with the largest mean (preliminary report) (abstract). AMS 25:813-814. A2 (5923)

Bothun, R. E. [1961], Relative performance of four flax varieties using eight different types of experimental plots (abstract). Agron. Abstracts p. 48. A14 (5924)

Botts, R. R. [1957], "Extreme value" methods simplified. Agricultural Economics Res. 9:88-95. A6 (5925)

Bourdeau, P. F. [1953], A test of random versus systematic ecological sampling. Ecology 34:499-512. A14;S1 (5926)

Box, G. E. P. [1950], Problems in the analysis of growth and wear curves. Biometrics 6:362-389. A7 (5927)

Box, G. E. P. [1953], Non-normality and tests on variances. Biometrika 40:318-335. A11 (5928)

Box, G. E. P. [1957], Iterative experimentation (abstract). Biometrics 13:240-241. C (5929)

Box, G. E. P. [1957], Iterative experimentation (abstract). AMS 28:816-817. C (5930)

Box, G. E. P. [1959], Transformations of the independent variable (summary). ASQC Nat. Conv. Trans. 13:19. A7 (5931)

Box, G. E. P. [1966], Use and abuse of regression. Technometrics 8:625-629. A7 (5932)

Box, G. E. P. et al. [1961], Common pitfalls in the design and analysis of experiments. Proc. Sixth Conf. Design Expt. Army Res. Dev. Testing. pp. 243-245. C (5933)

Box, G. E. P., Jenkins, G. M. and Wichern, D. W. [1967], Least squares analysis with a dynamic model. Dept. Stat. Tech. Report No. 105, Univ. Wisconsin. A7 (5934)

Box, G. E. P. and Tiao, G. C. [1966], A Bayesian approach to some outlier problems. Dept. Stat. Tech. Report No. 87, Univ. Wisconsin. A6 (5935)

Box, G. E. P. and Tidwell, P. W. [1962], Transformation of the independent variables. Technometrics 4:531-550. A7,9 (5936)

Box, G. E. P. and Watson, G. S. [1962], Robustness to non-normality of regression tests. Biometrika 49:93-106. (corrigenda 52:669) A7,11 (5937)

Boyd, D. A. [1963], The relationship between crop response and the determination of soil phosphorus by chemical methods. II. Proc. Conf. on Soil Phosphorus. Tech. Bull. 13:94-102. A13 (5938)

Bozivich, H., Bancroft, T. A. and Hartley, H. O. [1956], Power of analysis of variance test procedures for certain incompletely specified models. I. AMS 27:1017-1043. A11 (5939)

Bradley, R. A. [1952], Corrections for non-normality in the use of the two-sample _t_- and _F_-tests at high significance levels. AMS 23:103-113. A11 (5940)

Bradley, R. A. [1953], Some statistical methods in taste testing and quality evaluation. Biometrics 9:22-38. B;C (5941)

Bradley, R. A. [1955], Statistical designs for taste test panels. ASQC, Nat. Conv. Trans. 9:621-626. C (5942)

Bradley, R. A. [1964], Sequential rank tests (abstract). Biometrics 20:379-380. A12;S2 (5943)

Bradley, R. A. [1967], Topics in rank-order statistics. Proc. Fifth Berkeley Symp. Math. Stat. Prob. 1:593-607. A2;S2 (5944)

Bradley, R. A. and Kramer, A. [1957], Addenda to 'A quick, rank test for significance of differences in multiple comparisons'. Food Tech. 11:412. A3,12 (5945)

Bradley, R. A., Martin, D. C., and Wilcoxon, F. [1965], Sequential rank tests I. Monte Carlo studies of the two-sample procedure. Technometrics 7:463-483. A12;S2 (5946)

Bradley, R. A., Merchant, S. D., and Wilcoxon, F. [1966], Sequential rank tests II. Modified two-sample procedures. Technometrics 8:615-623. A12;S2 (5947)

Bradt, R. N. [1954], On the design and comparison of certain dichotomous experiments. Ph. D. Thesis, Stanford Univ. S2 (5948)

Bradt, R. N. and Karlin, S. [1956], On the design and comparison of certain dichotomous experiments. AMS 27:390-409. S2 (5949)

Bradu, D. [1964], Main effect analysis of the general non-orthogonal layout with any number of factors. AMS 36:88-97 (abstract 35:1403). A10 (5950)

Brand, N. [1964], Taste response and poliomyelitis. Ann. Human Genetics 27:233-237. C (5951)

Brandt, A. E. [1964], Disproportionate sub-class frequencies (abstract). Biometrics 20:380. A10 (5952)

Brandt, A. E. and Fletcher, G. H. [1955], Design of a clinical investigation of very high voltage sources in the radiotherapy of cancer (abstract). Biometrics 11:552-553. C (5953)

Brandt, F. O. [1950], Fünfzigjahrige Erfahrungen mit der Kleinwanzlebener Doppelstandardmethode für Leistungsprufung von Beta-Rüben. Z. Pflanzenzüchtung 29:90-104. C (5954)

Brandt, J. W. A. [1951], Rate of early growth in domestic fowl. Poultry Sci. 30:343-361. A7 (5955)

Braverman, S. [1967], Sample size comparisons based on functions of order statistics. Amer. Stat. 21(5):22-24. S1 (5956)

Brearley, A. and Cox, D. R. [1953], An outline of statistical methods for use in the textile industry, 3rd ed. Wool Ind. Res. Assoc. C (5957)

Brearley, A. and Cox, D. R. [1951], On designing an industrial experiment. Tect. Recorder 69:87. C (5958)

Brearley, A. and Cox, D. R. [1952], Experimentation in the textile industry. Wool Review 23:39. C (5959)

Breaux, H. J. [1967], Computational considerations in multiple linear regression. Proc. Twelfth Conf. Design Expt. Army Res. Dev. Testing. pp. 37-47. A7 (5960)

Breny, H. [1955], L'etat actuel du probleme de Behrens-Fisher. Trabajos Estad. 6:111-131. A11 (5961)

Bresenham, J. E. [1966], Reliability growth models. ASQC, Ann. Tech. Conf. Trans. pp. 179-187. A7 (5962)

Brier, G. W. and Thom, H. [1960], Experimental design in weather modification experiments (summary). JASA 55:355. C (5963)

Briggle, L. W., Petersen, H. D., and Hayes, R. M. [1967], Performance of a winter wheat hybrid, F_2, F_3, and parent varieties at five population levels. Crop. Sci. 7:485-490. A14 (5964)

Brillinger, D. R. [1964], The asymptotic behaviour of Tukey's general method of setting approximate confidence limits (the Jackknife) when applied to maximum likelihood estimates. Rev. ISI 32:202-206. C (5965)

Brim, C. A. and Mason, D. D. [1959], Estimates of plot size for soybean yield trials. Agron. J. 51:331-334. A14 (5966)

Brock, N. and Schneider, B. [1966], Problems of evaluation in screening of anticancer agents. <u>Methods in Drug Evaluation</u>. pp. 411-417. A2;C (5967)

Brooks, S. H. [1955], The estimation of an optimum subsampling number. JASA 50:398-415. S1,2 (5968)

Bross, I. [1952], Sequential medical plans. Biometrics 8:188-205. S2 (5969)

Bross, I. D. J. [1955], Therapy for intellectual obesity or common sense in reducing figures. Amer. J. Obstetrics Gynecology, 69:372-377. C (5970)

Bross, I. D. J. [1958], Sequential clinical trials. J. Chronic Diseases 8:349-365. S2 (5971)

Bross, I. D. J. [1958], How to use ridit analysis. Biometrics 14:18-38. A9 (5972)

Bross, I. D. J. [1961], Query 166: On a graphical sequential test. Biometrics 17:649-651. C;S2 (5973)

Bross, I. D. J. [1961], Statistical dogma: a challenge. Amer. Stat. 15(3):14-15. C (5974)

Bross, I. D. J. [1961], Outliers in patterned experiments: A strategic appraisal. Technometrics 3:91-102. A6 (5975)

Bross, I. D. J. [1964], Taking a covariable into account. JASA 59:725-736. A7,12 (5976)

Brown, A. R. and Morris, H. D. [1960], Plot technique for grain sorghum yield trials (abstract). Agron. Abstracts p. 45. A14 (5977)

Brown, A. R. and Morris, H. D. [1967], Estimation of optimum plot size and shape for grain sorghum yield trials. Agron. J. 59:576-577. A14 (5978)

Brown, C. M., and Weibel, R. O. [1957], Border effects in winter wheat and spring oat tests. Agron. J. 49:382-384. A14 (5979)

Brown, G. W. and Mood, A. M. [1951], On median tests for linear hypotheses. Proc. Second Berkeley Symp. Math. Stat. Prob. pp. 159-166. A12 (5980)

Brownlee, K. A. [1951], Correlation methods applied to production process data. Ind. Eng. Chemistry 43:2068-2071. A7 (5981)

Brownlee, K. A., Hodges, J. L., and Rosenblatt, M. [1953], The up-and-down method with small samples. JASA 48:262-277. S2 (5982)

Brumbaugh, M. A. [1954], Principles of sampling in the chemical field. IQC 10(4):6-14. C;S1 (5983)

Brunk, M. E. [1953], Controlled experiments in retail merchandizing. J. Farm Economics 35:916-923. C (5984)

Brykczyński, J. [1957], O wielkości pólka w doświodczeniach odmianowych (on the size of plots in variety experiments). Hodowla Róslin Aklimatyz. Nasiennictwo 1:757-761. A14 (5985)

Buch, M. L., Dryden, E. C., Hills, C. H., and Oyler, J. R. [1956], Organoleptic evaluation of applesauce fortified with essence and citric acid. Food Tech. 10:560-562. C (5986)

Buck, S. F. [1960], A method of estimation of missing values in multivariate data suitable for use with an electronic computer. JRSSB 22:302-306. A7 (5987)

Buckland, W. R. and Fox, R. [1963], Bibliography of basic texts and monographs on statistical methods. Oliver and Boyd Ltd., Edinburgh and London. B (5988)

Bühlmann, H. and Huber, P. J. [1963], Pairwise comparison and ranking in tournaments. AMS 34:501-510 (abstract 33:825). A2 (5989)

Bulfinch, A. [1957], Sensitivity testing. Proc. First Conf. Design Expt. Army Res. Dev. Testing. pp. 167-173. C (5990)

Bulfinch, A. [1958], Design of experiment. Proc. Second Conf. Design Expt. Army Res. Dev. Testing. pp. 233-241. C (5991)

Bulfinch, A. [1959], Characteristics of various methods of collecting data in tests of increased severity. Proc. Fourth Conf. Design Expt. Army Res. Dev. Testing. pp. 137-156. C (5992)

Bulfinch, A. [1963], Evaluation of various laboratory methods for determining reliability. Proc. Eighth Conf. Design Expt. Army Res. Dev. Testing. pp. 503-528. C (5993)

Bullen, E. R. [1956], The interpretation of field trial results. Indian J. Agron. 1:133-140. C (5994)

Burr, E. J. [1960], Earthquakes and Uranus: Misuse of a statistical test of significance. Nature 186:336-337. C (5995)

Burros, R. H. [1958], Experimental designs for organization research using limited resources. Proc. Second Conf. Design Expt. Army Res. Dev. Testing. pp. 221-224. C (5996)

Burrows, G. L. [1958], Interpreting straight line relationships. IQC 15(4):15-18. A7 (5997)

Busch, T. [1965], The mess in measurement. ASQC, Ann. Tech. Conf. Trans. pp. 456-462. C (5998)

Butler, J. M. and Hook, L. H. [1966], Multiple factor analysis in terms of weighted regression. Educational Psychological Measurement 26:545-564. A7 (5999)

Butterbaugh, G. I. [1958], Applications of statistical methods in business administration. ASQC, Nat. Conv. Trans. 12:33-52. B (6000)

Butters, B. [1964], Some practical considerations in the conduct of field trials with Robusta coffee. J. Hort. Sci. 39:24-33. A9,14;S1 (6001)

Büttner, H. [1964], Analysenfehler und Kontrollmöglichkeiten im klinischer Laboratorium. Methods Information Medicine 3:105-109. C (6002)

Byer, A. J. and Abrams, D. [1953], A comparison of the triangular and two-sample taste-test methods. Food Tech. 7:185-187. C (6003)

Calinski, T. [1966], On the distribution of the F-type statistics in the analysis of a group of experiments. JRSSB 28:526-542. A11,13 (6004)

Callebaut, R. [1966], Statistiche studie van blancoproeven op grote proefvelden ter beoordeling van de homogeniteit van de bodem als groeimilieu. Biométrie - Praximétrie, 7:3-42. A14 (6005)

Calzada Benza, J. [1957], El error experimental y la precisión en los experimentos. (The experimental error and the precision of experiments.) Boletín de la Estación Experimental Agrícola de la Molina No. 67, 33 pages (Mimeo.) A14 (6006)

Cameron, J. M. [1957], Some examples of the use of high speed computers in statistics. Proc. First Conf. Design Expt. Army Res. Dev. Testing. pp. 129-135. A15 (6007)

Cameron, J. M. [1963], An algorithm for obtaining an orthogonal set of individual degrees of freedom for error. J. Res. NBS 67B:19-22. A11,14 (6008)

Caplan, F. [1955], Statistical design in electronics production-line experimentation. IQC 12(5):12-13. C (6009)

Carlborg, F. W. [1966], Multiple regression and historical explanation (summary). JASA 61:543. A7 (6010)

Carlin, A. F., Kempthorne, O., and Gordon, J. [1956], Some aspects of numerical scoring in subjective evaluation of foods. Food Res. 21:273-281. C (6011)

Carlson, F. D., Sobel, E. and Watson, G. S. [1966], Linear relationships between variables affected by errors. Biometrics 22:252-267. A7 (6012)

Carmer, S. G. [1964], Exponential regression and a computer program for the estimation of parameters. Agron. J. 56:515-518. A7,15 (6013)

Carmer, S. G. [1965], A computer program for estimation of parameters in linear and nonlinear regression models. Agron. J. 57:517-518. A7,15 (6014)

Carmer, S. G. and Jackobs, J. A. [1965], An exponential model for predicting optimum plant density and maximum corn yield. Agron. J. 57:241-244. A7,14 (6015)

Carmer, S. G. and Motto, H. L. [1966], A computer program for laboratory instrument calibration. Agron. J. 58:462-463. A7,15 (6016)

Carmer, S. G. and Seif, R. D. [1963], Calculation of orthogonal coefficients when treatments are unequally replicated and/or unequally spaced. Agron. J. 55:387-389. A7 (6017)

Carroll, J. B. [1953], An analytical solution for approximating simple structure in factor analysis. Psychometrika 18:23-38. A7 (6018)

Carter, H. B., Turner, H. N. and Hardy, M. H. [1958], The influence of various factors on some methods of estimating fibre and follicle population density in the skin of Merino sheep. I. Methods of delineating area of natural skin. Australian J. Agri. Res. 9:237-251. S1 (6019)

Castellano, V. [1965], Corrado Gini: a Memoir. Metron 24:3-84. B (6020)

Castellano, V. [1966], Sciences, method and Statistics. Metron 25:7-54. C (6021)

Castronovo, A. and Muller, A. [1959], El uso de la covariancia en ensayos de campo con Nabo "de Vertus". (The use of covariance in field trials of the Vertus turnip.) Revista de Investigaciones Agrícolas, Buenos Aires 13(1):57-62. A14 (6022)

Cattell, R. B. [1952], The three basic factor-analytic research designs - their interrelations and derivatives. Psychological Bull. 49:499-520. A7 (6023)

Cattell. R. B. [1965], Factor analysis: an introduction to essentials. I. The purpose and underlying models. Biometrics 21:190-215. A7;B;C (6024)

Cattell, R. B. [1965], Factor analysis: An introduction to essentials. II. The role of factor analysis in research. Biometrics 21:405-435. A7;B;C (6025)

Cavalli, L. L. [1951], Un metodo rapido di calcolo della correlazione media fra più caratteri. Metron 16(1-2):151-167. A7 (6026)

Cellier, K. M. [1960], Considerations in experiments for the control of codling moth (Cydia pomonella L.) in apples. Australian J. Agri. Res. 11:186-196. A14;S1 (6027)

Cellier, K. M. and Edwards, G. R. [1951], A potato uniformity trial. J. Dept. Agri. S. Australia 55:89-90. A14 (6028)

Chacko, V. J. [1963], Testing homogeneity against ordered alternatives. AMS 34:945-956. A2,12 (6029)

Chacko, V. J. [1966], Modified chi-square test for ordered alternatives. Sankhyā B 28:185-190. A2 (6030)

Chakravarti, S. [1965], The problem of testing the hypothesis of equality of means of several univariate normal populations when variances are different. Calcutta Stat. Assoc. Bull. 14:126-150. A11 (6031)

Chakravarti, S. [1966], A note on multivariate analysis of variance test when dispersion matrices are different and unknown. Calcutta Stat. Assoc. Bull. 15:75-92. A7 (6032)

Chakravarti, S. [1967], Effect of the inequality of variances in analysis of variance F-test. Calcutta Stat. Assoc. Bull. 16:103-120. A11 (6033)

Chalbi, N. [1967], La compétition entre génotypes et ses effets sur les caractères quantitatifs de la Luzerne. Annales de l'Amélioration des Plantes 17:67-82. A14 (6034)

Chambers, C. [1967], Extension of tables of percentage points of the largest variance ratio, s^2_{max}/s^2_0. Biometrika 54:225-227. A11 (6035)

Chambers, M. L. and Jarratt, P. [1964], Use of double sampling for selecting best population. Biometrika 51:49-64. A2;S2 (6036)

Chamblee, D. S. and Lovvorn, R. L. [1953], The effect of rate and method of seeding on the yield and botanical composition of alfalfa-orchardgrass and alfalfa-tall fescue. Agron. J. 45:192-196. A14 (6037)

Chang, L. -C. [1963], The minimum number of replicates required in field trials of rice in Taiwan (in Chinese). Chung-hua Nung-hsueh Hui Pao/J. Agri. Assoc. China 43:9-26. S1 (6038)

Chapanis, A. [1953], Notes on an approximation method for fitting parabolic equations to experimental data. Psychometrika 18:327-336. A7 (6039)

Chapas, L. C. [1961], Plot size and reduction of variability in oil-palm experiments. Empire J. Expt. Agri. 29:212-224. A14 (6040)

Chapman, D. G. [1950], Some two sample tests. AMS 21:601-606. S2 (6041)

Chassan, J. B. [1960], On a test for order. Biometrics 16:119-121. A2 (6042)

Chassan, J. B. [1962], An extension of a test for order. Biometrics 18:245-247. A2 (6043)

Chatterjee, S. K. [1959], On an extension of Stein's two-sample procedure to the multi-normal problem. Calcutta Stat. Assoc. Bull. 8:121-148. A7;S2 (6044)

Chatterjee, S. K. [1959], Some further results on the multinormal extension of Stein's two-sample procedure. Calcutta Stat. Assoc. Bull. 9:20-28. A7;S2 (6045)

Chatterjee, S. K. [1960], Sequential tests for the bivariate regression parameters with known power and related estimation procedures. Calcutta Stat. Assoc. Bull. 10:19-34. A7;S2 (6046)

Chatterjee, S. K. [1962], Simultaneous confidence intervals of predetermined length based on sequential samples. Calcutta Stat. Assoc. Bull. 11:144-149. A3 (6047)

Chatterjee, S. K. [1965], Some non-parametric tests for the bivariate two sample association problem. Calcutta Stat. Assoc. Bull. 14:14-35. A7,12 (6048)

Chatterjee, S. K. [1966], A multi-sample non-parametric scale test based on U-statistics. Calcutta Stat. Assoc. Bull. 15:109-119. A12 (6049)

Chatterjee, S. K. and Sen, P. K. [1964], Non-parametric tests for the bivariate two-sample location problem. Calcutta Stat. Assoc. Bull. 13:18-58. A12 (6050)

Chen, T. [1960], Multiple comparisons of population means. M. S. Thesis, Iowa State Univ. A3 (6051)

Chernoff, H. [1959], Sequential design of experiments. AMS 30:755-770. S2 (6052)

Chernoff, H. [1961], Sequential experimentation. Bull. ISI 38(4):3-9. S2 (6053)

Chernoff, H. [1967], Sequential models for clinical trials. Proc. Fifth Berkeley Symp. Math. Stat. Prob. 4:805-812. C;S2 (6054)

Chew, V. [1958], Bibliography. <u>Experimental</u> <u>Designs</u> <u>in</u> <u>Industry</u>. pp. 253-268. B (6055)

Chew, V. and Fiskell, J. G. A. [1957], Sieve analysis of some fertilizer mixtures. J. Assoc. Official Agri. Chemists, 40:936-948. A14 (6056)

Chiang, C. L. [1954], Competition and other interactions between species. <u>Statistics</u> <u>and</u> <u>Mathematics</u> <u>in</u> <u>Biology</u>, pp. 197-215. A14 (6057)

Chilton, N. W. and Fertig, J. W. [1953], The estimation of sample size in experiments. I. Using comparisons of averages. J. Dental Res. 32:530-540. S1 (6058)

Chilton, N. W. and Fertig, J. W. [1953], The estimation of sample size in experiments. II. Using comparisons of proportions. J. Dental Res. 32:606-612. S1 (6059)

Chipman, E. W. and MacKay, D. C. [1960], The interactions of plant populations and nutritional levels on the production of sweet corn. Proc. Amer. Soc. Hort. Sci. 76:442-447. A14 (6060)

Chipman, J. S. and Rao, M. M. [1964], The treatment of linear restrictions in regression analysis. Econometrica 32:198-209. A7 (6061)

Clarke, G. M. [1964], Tasting tests on soft fruits from field pathology trials (abstract). Biometrics 20:223. C (6062)

Clem, M. A., Mosier, C. C., and Sprague, G. F. [1956], Simplified punched card procedures for predicting double cross performance. Agron. J. 48:319-320. A15 (6063)

Clement, W. L. [1958], Problems in Army field experimentation. Proc. Second Conf. Design Expt. Army Res. Dev. Testing. pp. 225-230. C (6064)

Cochran, W. G. [1950], The present status of biometry. Bull. ISI 32(2):132-150. C (6065)

Cochran, W. G. [1951], Testing a linear relation among variances. Biometrics 7:17-32. A11 (6066)

Cochran, W. G. [1952], Determination of the size of experiments and samples. Committee on Experimental Design, U.S.D.A., USDA Lecture Series, 11 pp. (Mimeo.) S1 (6067)

Cochran, W. G. [1952], El estado actual de la biometria. Trabajos Estad. 3:171-195. C (6068)

Cochran, W. G. [1954], The combination of estimates from different experiments. Biometrics 10:101-129. A13 (6069)

Cochran, W. G. [1957], The philosophy underlying the design of experiments. Proc. First Conf. Design. Expt. Army Res. Dev. Testing. pp. 1-8. C (6070)

Cochran, W. G. [1957], Analysis of covariance: Its nature and uses. Biometrics 13:261-281. A7 (6071)

Cochran, W. G. [1962], The potential contribution of electronic machines to the field of statistics. Ann. Computation Lab. Harvard Univ. 31:230-238. A15 (6072)

Cochran, W. G. [1963], The Behrens-Fisher test when the range of the unknown variance ratio is restricted. Sankhyā A 25:353-362. A11 (6073)

Cochran, W. G. [1964], Comparison of two methods of handling covariates in discriminatory analysis. Ann. Inst. Stat. Math. 16:43-53. A7 (6074)

Cochran, W. G. [1967], Planning and analysis of non-experimental studies. Proc. Twelfth Conf. Design Expt. Army Res. Dev. Testing. pp. 319-336. C (6075)

Cochran, W. G. and Carroll, S. P. [1953], A sampling investigation of the efficiency of weighting inversely as the estimated variance. Biometrics 9:447-459. A13 (6076)

Chopra, A. S. [1965], Application of orthogonal polynomials in fitting an asymptotic regression curve. JISAS 17:76-82. A7 (6077)

Chow, Y. S. and Robbins, H. [1965], On the asymptotic theory of fixed-width sequential confidence intervals for the mean. AMS 36:457-462. A3;S2 (6078)

Chowdhury, S. B. [1954], The most powerful unbiased critical regions and the shortest unbiased confidence intervals associated with the distribution of classical D^2-statistic. Sankhyā 14:71-80. A7 (6079)

Christidis, B. G. [1955], Regional variety tests with cotton-results of 1954 (in Greek with English summary). Cotton Res. Inst. Sci. Bull. 3:1-60. A13 (6080)

Christidis, B. G. [1957], Regional cotton tests. Results of 1956. Cotton Res. Inst. Sci. Bull. 4:1-71. A13 (6081)

Christidis, B. G. [1959], Regional cotton tests. Results of 1957. Cotton Res. Inst. Sci. Bull. 5:1-66. A13 (6082)

Christidis, B. G. [1960], Regional cotton tests. Results of 1958 (in Greek with English summary). Cotton Res. Inst. Sci. Bull. 6:5-68. A13 (6083)

Christidis, B. G. [1961], Regional cotton tests. Results of 1959 (in Greek with English summary). Cotton Res. Inst. Sci. Bull. 7:5-63. A13 (6084)

Christidis, B. G. [1962], Regional cotton tests. Results of 1960 (in Greek with English summary). Cotton Res. Inst. Sci. Bull. 8:1-74. A13 (6085)

Christidis, B. G. [1963], Regional cotton tests. Results of 1961 (in Greek with English summary). Cotton Res. Inst. Sci. Bull. 9:1-80. A13 (6086)

Chu, J. T., Leone, F. C., and Topp, C. W. [1957], Some uses of quasi-ranges II (abstract). Biometrics 13:412-413. A3 (6087)

Chun, D. [1965], On an extreme rank sum test with early decision. JASA 60:859-863. A12 (6088)

Chung, J. H. and Fraser, D. A. S. [1958], Randomization tests for a multivariate two-sample problem. JASA 53:729-735. A5,7,12 (6089)

Chung, K. L. [1954], On the Robbins-Munro procedure (abstract). Bull. Amer. Math. Soc. 60:171. S2 (6090)

Churchman, C. W. [1954], The philosophy of experimentation. <u>Statistics and Mathematics in Biology</u> pp. 159-172. C (6091)

Cochran, W. G. and Hopkins, C. E. [1961], Some classification problems with multivariate qualitative data. Biometrics 17:10-32. A7 (6092)

Coda, N. [1955], An engineer evaluates statistical methods. IQC 12(6):13-18. C (6093)

Coey, W. E., Smyth, W. G., and Smyth, J. V. [1957], The value of discard rows in field experiments with barley. Res. Expt. Record Ministry Agri. N. Ireland 5:7-11. A14 (6094)

Cohen, A. [1961], Application of simultaneous confidence intervals to two regression problems (abstract). AMS 32:1348. A3 (6095)

Cohen, A. C. [1959], Simplified procedures for estimating parameters of a normal distribution from restricted samples. Proc. Fourth Conf. Design Expt. Army Res. Dev. Testing. pp. 177-213. A2 (6096)

Cohen, L. [1958], On mixed single sample experiments. AMS 29:947-971. S1 (6097)

Cole, L. C. [1962], A closed sequential test design for toleration experiments. Ecology 43:749-753. S2 (6098)

Collen, M. F., Rubin, L., Neyman, J., Dantzig, G. B., Baer, R. M., and Seigelaub, A. B. [1964], Automated multiphasic screening and diagnosis. Amer. J. Public Health 54:741-750. A2 (6099)

Collis-George, N. and Davey, B. G. [1960], The doubtful utility of present-day field experimentation and other determinations involving soil-plant interactions. Soils Fertilizers 23:307-310. C (6100)

Colton, T. [1963], A model for selecting one of two medical treatments. JASA 58:388-400. A2;S1,2 (6101)

Colton, T. [1965], A two-stage model for selecting one of two treatments. Biometrics 21:169-180 (abstract 19:657). A2;S2 (6102)

Colville, W. L., Dreier, A., McGill, D. P., Grabouski, P., and Ehlers, P. [1964], Influence of plant population, hybrid, and "productivity level" on irrigated corn production. Agron. J. 56:332-335. A14 (6103)

Conagin, A. [1950], Disposiçao sistemática dos canteiros. Sua influência sôbre a estimativa do êrro experimental. (Systematic arrangement of trial plots. Effect on estimate of experimental error.) Bragantia 10(7):203-207. A14 (6104)

Conagin, A. [1951], Delineamentos experimentais. Revista Agri. São Paulo 26:87-108. C (6105)

Conagin, A. [1956], New tests for comparison of means (abstract). Biometrics 12:230. A3 (6106)

Conagin, A. [1959], Testes modernos para a comparação de médias. (Recent tests for comparison of means.) Bragantia 18:1-14. A3 (6107)

Conagin, A. [1959], Determinação do número de repetições no planejamento de experimentos. (Determination of number of replications in the planning of experiments.) Bragantia 18:13-27. S1 (6108)

Conagin, A. and Fraga, C. G. [1955], Design and analysis of coffee experiments (abstract). Biometrics 11:539-540. C (6109)

Connor, W. S. [1960], Interpreting reliability by fitting theoretical distributions to failure data. Ind. Eng. Chemistry 52(2):75A-76A. C (6110)

Connor, W. S. [1960], Interpreting reliability by fitting theoretical distributions to failure data. Ind. Eng. Chemistry 52(4):71A-72A. C (6111)

Connor, W. S. [1961], Equivalent tolerances. Ind. Eng. Chemistry 53(10):61A-62A. C (6112)

Connor, W. S. [1961], Locating important sources of variation. Ind. Eng. Chemistry 53(12):73A-74A. A14 (6113)

Conners, H. [1951], Field plot techniques for sweet potatoes obtained from uniformity trial data. M. S. Thesis, Iowa State Univ. A14 (6114)

Cook, M. B. [1951], Bivariate k-statistics and cumulants of their joint sampling distribution. Biometrika 38:179-195. A7,14 (6115)

Cooke, D. [1962], The timing of harvests in horticultural experiments. Biometrics 18:478-498. C (6116)

Cooper, C. F. [1957], The variable plot method for estimating shrub density. J. Range Management 10:111-115. A14 (6117)

Cooper, C. F. [1963], An evaluation of variable plot sampling in shrub and herbaceous vegetation. Ecology 44:565-569. A14 (6118)

Coote, G. G. [1960], The principles of conducting variety trials involving subjective judgements. J. Australian Inst. Agri. Sci. 26:241-245. C (6119)

Corlett, T. [1963], Ballade of multiple regression. Appl. Stat. 12:145. A7 (6120)

Cornfield, J. [1966], A Bayesian test of some classical hypotheses--with applications to sequential clinical trials. JASA 61:577-594. S2 (6121)

Cornfield, J. [1966], Sequential trials, sequential analysis and the likelihood principle. Amer. Stat. 20(2):18-23. S2 (6122)

Cornfield, J. [1967], Discriminant functions. Rev. ISI. 35:142-153. A7 (6123)

Cornfield, J. and Greenhouse, S. W. [1967], On certain aspects of sequential clinical trials. Proc. Fifth Berkeley Symp. Math. Stat. Prob. 4:813-829. C;S2 (6124)

Cornish, E. A. [1957], An application of the Kronecker product of matrices in multiple regression. Biometrics 13:19-27. A7,14 (6125)

Cornish, E. A. [1960], Fiducial limits for parameters in compound hypotheses. Australian J. Stat. 2:32-40. A3 (6126)

Corsten, L. C. A. [1964], Identificeren op grond van waarnemingen. Stat. Neerlandica 18:1-14. A7 (6127)

Court, L. M. [1957], Three applications of statistics to electronics. Proc. First Conf. Design Expt. Army Res. Dev. Testing. pp. 69-83. C (6128)

Cowden, D. J. [1958], A procedure for computing regression coefficients. JASA 53:144-150. A7 (6129)

Cox, C. P. [1956], A geometrical derivation of the analyses of covariance and variance. JRSSA 19:333-335. A7 (6130)

Cox, C. P. [1958], A concise derivation of general orthogonal polynomials. JRSSB 20:406-407. A7 (6131)

Cox, C. P. [1963], Statistical principles for the line of best fit. Laboratory Practice, August. A7 (6132)

Cox, C. P. and Roseberry, T. D. [1966], A large sample sequential test, using concomitant information, for discrimination between two composite hypotheses. JASA 61:357-367. S2 (6133)

Cox, D. F. and Kempthorne, O. [1963], Randomization tests for comparing survival curves. Biometrics 19:307-317 (abstract 18:629). A5 (6134)

Cox, D. R. [1952], Estimation by double sampling. Biometrika 39:217-227. S2 (6135)

Cox, D. R. [1952], Sequential tests for composite hypotheses. Proc. Cambridge Philosophical Soc. 48:290-299. S2 (6136)

Cox, D. R. [1955], Some statistical methods connected with series of events (with discussion). JRSSB 17:129-164. A5,7 (6137)

Cox, D. R. [1958], The regression analysis of binary sequences (with discussion). JRSSB 20:215-242 (corrigendum 21:238). A7 (6138)

Cox, D. R. [1960], Regression analysis when there is prior information about supplementary variables. JRSSB 22:172-176. A7 (6139)

Cox, D. R. [1960], A note on tests of homogeneity applied after sequential sampling. JRSSB 22:368-371. A5;S2 (6140)

Cox, D. R. [1961], Design of experiments: The control of error. JRSSA 124:44-48. C (6141)

Cox, D. R. [1963], Large sample sequential tests for composite hypotheses. Sankhyā A 25:5-12. S2 (6142)

Cox, D. R. [1964], Some applications of exponential ordered scores. JRSSB 26:103-110. A7 (6143)

Cox, D. R. [1965], A remark on multiple comparison methods. Technometrics 7: 223-224. A3 (6144)

Cox, E. L. [1960], May some data be discarded? (abstract) Biometrics 16:132. A6 (6145)

Cox, G. M. [1950], The organization and functions of the Institute of Statistics of the University of North Carolina (with discussion). JRSSB 12:1-18. C (6146)

Cox, G. M. [1950], The function of designs in experiments. Ann. New York Acad. Sci. 52:800-807. C (6147)

Cox, G. M. [1950], A survey of types of experimental designs (abstract). Biometrics 6:301-302, 317-318. C (6148)

Cox, G. M. [1957], Statistical frontiers. JASA 52:1-12. C (6149)

Cox, G. M. [1960], El papel de estadistica en la investigacion cientifica. Revista Del Colegio Medico 11:104-110. C (6150)

Cox, P. C. [1957], Estimating an average or standard trajectory. Proc. First Conf. Design Expt. Army Res. Dev. Testing. pp. 229-231. C (6151)

Cox, W. E. [1965], Experimental designs for marketing analysis (summary). JASA 60:657. C (6152)

Cramer, E. M. [1965], A general multivariate and univariate analysis of variance computer program (abstract). Biometrics 21:1023-1024. A7,15 (6153)

Crasson, G. [1962], Disposition des champs d'essais pour la détermination de la résistance au virus de l'enroulement et au virus Y. European Potato J. 5:290-315. C (6154)

Creager, J. A. [1958], General resolution of correlation matrices into components and its utilization in multiple and partial regression. Psychometrika 23:1-8. A7 (6155)

Creasy, M. A. [1956], Confidence limits for the gradient in the linear functional relationship. JRSSB 18:65-69. A7 (6156)

Crews, J. W., Jones, G. L., and Mason, D. D. [1963], Field plot technique studies with flue-cured tobacco. I. Optimum plot size and shape. Agron. J. 55:197-199. A14 (6157)

Crouse, C. F. [1961], A non-null ranking model for a sequence of m alternatives. Biometrika 48:441-444. A12 (6158)

Crump, S. L. [1952], A clinical comparison of modified insulins. II. The biometrical aspect of the problem. Biometrics 8:210-217. C (6159)

Crump, L. [1954], Some aspects of experimental design. IQC 10(4):14-16. C (6160)

Cunia, T. [1964], Least squares estimates and parabolic regression with restricted location for the stationary point. JASA 59:564-571. A7 (6161)

Curtis, F. H. [1962], Linear programming the management of a forest property. J. Forestry 60:611-616. A7 (6162)

Cushen, W. E. [1957], Operational gaming and the design of experiments. Proc. First Conf. Design Expt. Army Res. Dev. Testing. pp. 47-51. C (6163)

Cyril, B. [1950], The factorial analysis of qualitative data. British J. Psychology (Statistical Section). 3:166-185. A7 (6164)

Dagnelie, P. [1960], Quelques problèmes statistiques posés par l'utilisation de l'analyse factorielle en phytosociologie. (Some statistical problems set by the use of factor analysis in phytosociology.) Bull. Inst. Agronomique Stations Recherches Gembloux, (Hors Série) 1:430-437. A7 (6165)

Dagnelle, P. [1962], L'application de l'analyse multi-variable à l'étude des conmunautés végétales. Bull. ISI 39(2):265-275. A7 (6166)

Dagnelie, P. [1964], La détermination du nombre de répétitions en vue de l'estimation d'une moyenne. Biométrie - Praximétrie 5:117-135. S1 (6167)

Dagnelie, P. [1965], A propos des transformations de variables. Biométrie - Praximétrie 6:59-78. A9 (6168)

Dagnelie, P. [1965], A propos de quelques méthodes de comparaisons multiples de moyennes. Biométrie - Praximétrie 6:115-124. A3 (6169)

Dagnelie, P. [1966], La régression multiple. Biométrie - Praximétrie 7:193-238. A7 (6170)

Dagnelie, P. [1966], La corrélation et la régression simples. Biométrie - Praximétrie 7:133-166. A7 (6171)

Dagnelie, P. [1966], Introduction a l'analyse statistique a plusieurs variables. Biométrie - Praximétrie 7:43-66. A7 (6172)

Dalenius, T. and Matérn, B. [1964], Is there a need for a unified theory of random experiments? Metrika 8:235-247. A5;C (6173)

Daniel, H. A., Cox, M. B., Tucker, B. B., and Viets, F. G. [1957], Design of plots conforming to the land for evaluating moisture conservation practices. Soil Sci. Soc. Amer. Proc. 21:347-350. A14 (6174)

Daniels, H. E. [1950], Rank correlation and population models (with discussion). JRSSB 12:171-191. A12 (6175)

Daniels, H. E. [1962], Processes generating permutation expansions. Biometrika 49:139-149. A9 (6176)

Danilova, L. [1965], Methods of field experimentation with vegetables (in Russian). Vestn. sel'skohozjajstv. Nauk. (Rep. Agri. Sci.) 10:141-145. C (6177)

Darmois, G. [1953], Analyse générale des liaisons stochastiques; etude particulière de l'analyse factorielle linéaire. Rev. ISI 21:2-8. A7 (6178)

Da Rocha, G. L. [1955], Grazing experiments in the State of São Paulo (abstract). Biometrics 11:542-544. C (6179)

Das, M. N. [1953], Analysis of covariance in two-way classification with disproportionate cell frequencies. JISAS 5:161-178. A7,10 (6180)

David, F. N. [1955], The transformation of discrete variables. Ann. Human Genetics 19:174-182. A9 (6181)

David, F. N. [1956], A note on Wilcoxon's and allied tests. Biometrika 43:485-488. A12 (6182)

David, F. N. and Fix, E. [1961], Rank correlation and regression in a nonnormal surface. Proc. Fourth Berkeley Symp. Math. Stat. Prob. 1:177-197. A7 (6183)

David, F. N. and Johnson, N. L. [1951], The effect of non-normality on the power function of the F-test in the analysis of variance. Biometrika 38:43-57. A11 (6184)

David, F. N. and Johnson, N. L. [1951], A method of investigating the effect of nonnormality and heterogeneity of variance on tests of the general linear hypothesis. AMS 22:382-392. A11 (6185)

David, F. N. and Johnson, N. L. [1956], Some tests of significance with ordered variables (with discussion). JRSSB 18:1-31. A6;B (6186)

David, H. A. [1951], Further applications of range to the analysis of variance. Biometrika 38:393-409. A12 (6187)

David, H. A. [1956], The ranking of variances in normal populations. JASA 51:621-626. A3,11 (6188)

David, H. A. [1956], On the application to statistics of an elementary theorem in probability. Biometrika 43:85-91. A3 (6189)

David, H. A. [1956], Revised upper percentage points of the extreme studentized deviate from the sample mean. Biometrika 43:449-451. A3 (6190)

David, H. A. [1962], Multiple decisions and multiple comparisons. Contributions to Order Statistics, chapter 9, pp. 144-162. A3;B (6191)

David, H. A. [1967], Tests for outliers. Proc. Twelfth Conf. Design Expt. Army Res. Dev. Testing. pp. 151-161. A6 (6192)

David, H. A. and Paulson, A. S. [1965], The performance of several tests for outliers. Biometrika 52:429-436. A6 (6193)

David, H. T., Davidson, D. T., and O'Flaherty, C. A. [1961], Detecting outliers in soil-additive strength tests. Materials Res. Standards 1:947-950. A6 (6194)

David, S. T., Kendall, M. G., and Stuart, A. [1951], Some questions of distribution in the theory of rank correlation. Biometrika 38:131-140. A12 (6195)

Davidson, B. R. [1965], Significance of the differences between variety yields under experimental and farm conditions. Nature 207:1009. C (6196)

Davies, O. L. [1950], Biometric methods in the chemical industry. Biometrics 6:228-230. C (6197)

Davieson, I. D. [1961], A new strategy inexperimental design. B.H.P. Tech. Bull. (Australia) 5:6-11. C (6198)

Davis, D. E. and Zippin, C. [1954], Planning wildlife experiments involving percentages. J. Wildlife Management 18:170-178. S1 (6199)

Davis, J. G. and Hanson, H. L. [1954], Sensory test methods. I. The triangle intensity (T-I) and related test systems for sensory analysis. Food Tech. 8:335-339. C (6200)

Dawson, E. H. and Harris, B. L. [1951], Sensory methods for measuring differences in food quality. Agri. Information Bull. 34. U.S.D.A. pp. I-IV, 1-34. B;C (6201)

Dawson, J. H. [1964], Competition between irrigated field beans and annual weeds. Weeds 12:206-208. A14 (6202)

Day, B. B. [1954], The design of experiments in naval engineering. Amer. Stat. 8(2):7-12 and 23. C (6203)

Day, B. B., Del Priore, F. R., and Sax, E. [1955], The technique of regression analysis. IQC 12(2):10-19. A7 (6204)

Dayhoff, E. [1964], On the equivalence of polykays of the second degree and Σ's. AMS 35:1663-1672. A14 (6205)

DeBaun, R. M. [1964], Letters to the editor: Analysis of residuals. Technometrics 6:127. A6 (6206)

deCandole, C. A. and Richardson, B. A. [1961], A trial comparing certain side effects of two nerve gas antidotes, using human subjects. Proc. Sixth Conf. Design Expt. Army Res. Dev. Testing. pp. 329-338. C (6207)

Dedolph, R. R. [1960], A suggested method for handling data obtained with an exponential (variable dosage) sprayer. Proc. Amer. Soc. Hort. Sci. 75:789-798. A7 (6208)

Deely, J. [1966], Sequential selection of the best of k population (abstract). AMS 37:1858-1859. A2;S2 (6209)

Deemer, W. L. and Olkin, I. [1951], The Jacobians of certain matrix transformations useful in multivariate analysis. Biometrika 38:345-367. A7 (6210)

de Finetti, B. [1961], The Bayesian approach to the rejection of outliers. Proc. Fourth Berkeley Symp. Math. Prob. 1:199-210. A6 (6211)

de Boer, J. [1953], Sequential test with three possible decisions for testing an unknown probability. Appl. Sci. Res. Section B, 3:249-259. S2 (6212)

de Groot, M. J. W. [1961], Van ziektegeval tot ziektengetal. Stat. Neerlandica 15:419-429. C (6213)

de Jonge, H. [1963], Beslissen zonder (ver)gissen bij de keuze van een toets. Stat. Neerlandica 17:449-477. A5,7,12 (6214)

de Jonge, H. [1964], Communicatie van medische statistiek: de communicatie tussen medicus en statistiek. (Communication in medical statistics: communication between the physician and statistician.) Stat. Neerlandica 18:417-424. C (6215)

de Jongh, D. K. [1958], Design for decision in het clinische experiment. (Design for decision in the clinical trial). Stat. Neerlandica 12:213-229. C (6216)

de Leve, G. [1957], Enige statistische aspecten van de factoranalyse. (Some statistical aspects of factor analysis.) Stat. Neerlandica 11:201-209. A7 (6217)

Del Priore, F. R. and Day, B. B. [1958], An example in statistical planning of laboratory experiments, or the engineer and statistician can meet. ASQC Nat. Conv. Trans. 12:217-238. C (6218)

DeLury, D. B. [1951], On the planning of experiments for the estimation of fish populations. J. Fishery Res. Board Canada 8(4):281-307. C (6219)

DeLury, D. B. [1952], Designing experiments to isolate sources of variation. ASQC Ann. Conv. Trans. 6:147. C (6220)

DeLury, D. B. [1953], On the design of experiments. ASQC Ann. Conv. Trans. 7:439. C (6221)

DeLury, D. B. [1954], On the design of experiments. IQC 10(4):24-29. C (6222)

DeLury, D. B. [1956], Elements of the analysis of covariance (abstract). Biometrics 12:533. A7 (6223)

Dembiczak, C. M., Eaton, H. D., Beall, G., and Lucas, H. L. [1957], Design and conduct of calf nutrition studies. I. One-vs. two- and three-day growth measurements. J. Dairy Sci. 40:1133-1151. A14 (6224)

Deming, L. S. [1960], Selected bibliography of statistical literature, 1930 to 1957: I. Correlation and regression theory. II. Time series. J. Res. NBS 64B:55-82. B (6225)

Deming, L. S. [1960], Selected bibliography of statistical literature, 1930 to 1957: III. Limit theorems. J. Res. NBS 64B:175-192. B (6226)

Deming, L. S. [1962], Selected bibliography of statistical literature, 1930 to 1957: V. Frequency functions, moments, and graduation. J. Res. NBS 66B:15-28. B (6227)

Deming, L. S. [1962], Selected bibliography of statistical literature, 1930 to 1957: VI. Theory of estimation and testing of hypotheses, sampling distributions, and theory of sample surveys. J. Res. NBS 66B:109-151. B (6228)

Deming, L. S. [1963], Selected bibliography of statistical literature: Supplement, 1958 to 1960. J. Res. NBS 67B:91-133. B (6229)

Deming, L. S. and Gupta, D. [1961], Selected bibliography of statistical literature, 1930 to 1957: IV. Markov chains and stochastic processes. J. Res. NBS 65B:61-93. B (6230)

Deming, W. E. [1961], Uncertainties in statistical data, and their relation to the design and management of statistical surveys and experiments. Bull. ISI 38(4): 365-383. C (6231)

Deming, W. E. [1966], Code of professional conduct. Sankhyā B 28:11-18. C (6232)

Dempster, A. P. [1963], Multivariate theory for general stepwise methods. AMS 34:873-883. A7 (6233)

Dempster, A. P. [1963], Stepwise multivariate analysis of variance based on principal variables. Biometrics 19:478-490. A7 (6234)

Dempster, A. P. [1964], Tests for the equality of two covariance matrices in relation to a best linear discriminator analysis. AMS 35:190-199. A7,11 (6235)

Dempster, A. P. [1967], Estimation in multivariate analysis. <u>Multivariate Analysis</u> pp. 315-334. A7 (6236)

De Munter, P. [1960], Introduction à la nation de traitment et au problème de la comparaison simultanée de plus de deux traitements. Bull. Inst. Agronomique Stations Recherches Gembloux (Hors Série)1:417-429. A3;C (6237)

de Oliveira, A. J. [1957], Analysis of a group of experiments on oats. I. Complete data. Agronomia Lusitana 19:329-356. A13 (6238)

Derman, C. [1957], Non-parametric up-and-down experimentation. AMS 28:795-798. A12;S2 (6239)

Deshpande, M. V. [1963], Three sample non-parametric test (abstract). JISAS 15:262. A12 (6240)

Deshpande, J. V. [1963], Exact small sample significance points for the non-parametric V-test for problem of several samples. JISA 1:167-171. A12 (6241)

Deshpande, J. V. [1965], A non-parametric test based on U-statistics for the problem of several samples. JISA 3:20-29. A12 (6242)

de V. Weir, J. B. [1966], Table of 0.1 percentage points of Behren's d. Biometrika 53:267-268. A11 (6243)

de Vries, M. [1953], Kwantificering in taal en crypto-analyse. (Quantification in language and crypto-analysis.) Statistica 7:233-256. C (6244)

de Vries, M. [1957], Experiment en waarneming in de economische statistiek. (Experiment and observation in economic research.) Stat. Neerlandica 11:77-87. C (6245)

Dewey, D. R. and Lu, K. H. [1959], A correlation and path-coefficient analysis of components of crested wheatgrass seed production. Agron. J. 51:515-518. A7 (6246)

de Wit, C. T. [1960], On competition. Versl. Landbouwk Onderzoek 66, 8, 1-82. A14 (6247)

de Wit, C. T., Ennik, G. C., van den Bergh, J. P., and Sonneveld, A. [1960], Competition and non-persistence as factors affecting the composition of mixed crops and swards. Proc. Eighth Intl. Grassland Congress, pp. 736-741. A14 (6248)

de Wolff, P. [1957], Een eenvoudige grafische methode voor de aanpassing van een rechtlijnige trend volgens de methode van de kleinste kwadraten aan een historische reeks. (A short cut graphic method for fitting the best straight line to a series of points according to the criteria of least squares.) Stat. Neerlandica 11:169-170. A7 (6249)

d'Herbemont, G. [1962], Sur un aspect géométrique des analyses de variance. Publ. Inst. Statistique Université Paris 11:195-219. A14 (6250)

Dick, I. D. [1952], The equivalence of factor analysis and confluence analysis in problems of multicollinearity. New Zealand J. Sci. Tech. B 33:345-347. A7 (6251)

Dickinson, A. W. [1963], Simultaneous regression equations. Chemical Eng. Progress Symp. Series 59(42):84-89. A7 (6252)

Didio, V. A. [1958], Some statistical aspects of fatique test planning. Proc. Second Conf. Design Expt. Army. Res. Dev. Testing. pp. 93-100. C (6253)

Divers, C. K. [1961], Statistical aids to visual inspection. ASQC Nat. Conv. Trans. 15:525-531. A2 (6254)

Dixon, W. J. [1950], Analysis of extreme values. AMS 21:488-506 (abstract 21:145). A6 (6255)

Dixon, W. J. [1953], Processing data for outliers. Biometrics 9:74-89. A6 (6256)

Dixon, W. J. [1960], Statistics in medical research. Proc. Fifth Conf. Design Expt. Army Res. Dev. Testing. pp. 265-275. C (6257)

Dixon, W. J. [1962], Rejection of observations. Contributions to Order Statistics. Chapter 10H, pp. 299-342. A6 (6258)

Dixon, W. J. [1964], Query 4: Rejection of outlying values. Technometrics 6:228. A6 (6259)

Doig, A. [1957], A bibliography on the theory of queues. Biometrika 44:490-514. B (6260)

Doksum, K. [1966], Distribution-free statistics based on normal deviates in analysis of variance. Rev. ISI 34:376-388. A12 (6261)

Dolby, J. L. [1963], A quick method for choosing a transformation. Technometrics 5:317-325. A9 (6262)

Dominick, B. A. [1953], Merchandising McIntosh apples in retail stores. Cornell Univ. Agri. Expt. Sta. Bull. 895:22. C (6263)

Donald, C. M. [1951], Competition among pasture plants. I. Intra-specific competition among annual pasture plants. Australian J. Agri. Res. 2:355-376. A14 (6264)

Donald, C. M. [1954], Competition among pasture plants. II. The influence of density on flowering and seed production in annual pasture plants. Australian J. Agri. Res. 5:585-597. A14 (6265)

Donald, C. M. [1958], The interaction of competition for light and for nutrients. Australian J. Agri. Res. 9:421-435. A14 (6266)

Donald, H. P. and Purser, A. F. [1956], Competition in utero between twin lambs. J. Agri. Sci. 48:245-249. A14 (6267)

Doney, D. L., Plaisted, R. L., and Peterson, L. C. [1965], Genotypic competition in progeny performance evaluation of potatoes. Crop Sci. 5:433-435. A14 (6268)

Doney, J. M. and Weiler, H. [1959], The total number of fibres on sheep. I. Estimation by clean fleece and fibre weight. Australian J. Agri. Res. 10:287-298. A14 (6269)

Doornbos, R. [1956], Significance of the smallest of a set of estimated normal variances. Stat. Neerlandica 10:117-126. A2,11 (6270)

Doornbos, R. [1959], Statistische methoden voor het aanwijzen van uitbijters. (Statistical methods for the rejection of outlying observations.) Stat. Neerlandica 13:453-462. A6,12 (6271)

Doornbos, R. and Prins, H. J. [1956], Slippage tests for a set of gamma-varieties. Indagationes Mathematicae 18:329-337. A2,6 (6272)

Doornbos, R. and Prins, H. J. [1958], On slippage tests. Indagationes Mathematicae 20:38-55. A2,6 (6273)

Doornbos, R. and Prins, H. J. [1958], On slippage tests III. Two distribution-free slippage tests and two tables. Indagationes Mathematicae 20:438-447. A2 (6274)

Dorff, M. and Gurland, J. [1960], Small sample behavior of slope estimators in a linear functional relation. Math. Res. Center, Tech. Summary Report No. 214, U.S. Army and Univ. Wisconsin. A7 (6275)

Dorff, M. and Gurland, J. [1961], Estimation of the parameters of a linear functional relation. JRSSB 23:160-170 (Math. Res. Center Tech. Summary Report No. 200, U.S. Army and Univ. Wisconsin). A7 (6276)

Dorst, J. C. [1958], Use of variation or replications in breeding work. Euphytica 7:111-122. C (6277)

Drapala, W. J. and Johnson, C. M. [1961], Border and competition effects in millet and sudangrass plots characterized by different levels of nitrogen fertilization. Agron. J. 53:17-19. A14 (6278)

Drion, E. F. [1951], Estimation of the parameters of a straight line and of the variances of the variables, if they are both subject to error. Koninklijke Nederlandse Akademie van Wetenschappen, Proc. Series A, 54:256-260. A7 (6279)

Duarte, A. J. S. [1959], Ensaios com microtalhões. (Trials with microplots.) Agronomia Lusitana, 21:75-90. A14 (6280)

Duckworth, W. E. and Wyatt, J. K. [1958], Rapid statistical techniques for operations research workers. Operations Res. Quarterly 9:218-233. C;S1 (6281)

Dudzinski, M. L. [1960], Estimation of regression slope from tail regions with special reference to the volume line. Biometrics 16:399-407. A7 (6282)

Dugué, D. [1960], Valeurs extrêmes en statistique (with discussion). J. Société Stat. Paris 101:194-203. A6 (6283)

Dumas, M. M. [1960], Forme canonique du critère χ^2 et plan d'expérience. J. Société Stat. Paris 101:112-115. C (6284)

Duncan, D. B. [1951], A significance test for differences among ranked treatments in an analysis of variance (abstract). AMS 22:142. A3 (6285)

Duncan, D. B. [1951], A significance test for differences between ranked treatments in the analysis of variance. Virginia J. Sci. 2:171-189. A3 (6286)

Duncan, D. B. [1951], A significance test for differences among ranked treatments in an analysis of variance (abstract). Virginia J. Sci. 2:378. A3 (6287)

Duncan, D. B. [1952], On the properties of the multiple comparisons test. Virginia J. Sci. (new series) 3:49-67. A3 (6288)

Duncan, D. B. [1953], Testing the homogeneity of treatment means in an analysis of variance of engineering data (summary). JASA 48:641. A3 (6289)

Duncan, D. B. [1953], Multiple range tests and the multiple comparisons test (Preliminary report) (abstract). Biometrics 9:262. A3 (6290)

Duncan, D. B. [1953], Multiple range tests and the multiple comparisons test (abstract). Virginia J. Sci. 4:288. A3 (6291)

Duncan, D. B. [1953], Multiple range tests and the multiple comparisons test (Preliminary report) (abstract). AMS 24:498. A3 (6292)

Duncan, D. B. [1954], Statistical methods applied to highway-research experimentation (with discussion). Proc. Thirty-third Annual Meeting Highway Res. Board. pp. 112-120. C (6293)

Duncan, D. B. [1955], Multiple range and multiple F tests. Biometrics 11:1-42. A3 (6294)

Duncan, D. B. [1957], Multiple range tests for correlated and heteroscedastic means. Biometrics 13:164-176. (also Institute of Statistics Mimeo. Series No. 161, Univ. N. Carolina) A3,11 (6295)

Duncan, D. B. [1958], Useful Bayes solutions for multiple comparisons problems. I. (preliminary report) (abstract). AMS 29:622-623. A2,3 (6296)

Duncan, D. B. [1958], Useful Bayes solutions for multiple comparisons problems. I. (abstract) Biometrics 14:568-569. A2,3 (6297)

Duncan, D. B. [1959], A simple minimum-average-risk procedure for the multiple comparisons problem (abstract). AMS 30:621. A2,3 (6298)

Duncan, D. B. [1959], A simple minimum average risk procedure for the multiple comparisons problem. Biometrics 15:636. A2,3 (6299)

Duncan, D. B. [1961], Bayes rules for a common multiple comparisons problem and related student-t problems. AMS 32:1013-1033. A3 (6300)

Duncan, D. B. [1964], On the simultaneous estimation of a missile trajectory and error components including the error power spectra of several tracking systems. Proc. Ninth Conf. Design Expt. Army Res. Dev. Testing. pp. 661-681.
A7 (6301)

Duncan, D. B. [1965], A Bayesian approach to multiple comparisons. Technometrics 7:171-222.
A3 (6302)

Duncan, D. B. and Bonner, R. G. [1955], Simultaneous confidence intervals derived from multiple range and multiple F tests (summary). JASA 50:574-575.
A3 (6303)

Duncan, D. B. and Jones, R. H. [1966], Multiple regression with stationary errors. JASA 61:917-928.
A7 (6304)

Duncan, D. B. and Schneider, R. E. [1952], Multiple regression with a polychotomous criterion (abstract). Virginia J. Sci. 3:358.
A7 (6305)

Dunn, O. J. [1959], Confidence intervals for the means of dependent, normally distributed variables. JASA 54:613-621.
A3 (6306)

Dunn, O. J. [1961], Multiple comparisons among means (summary). JASA 56:396.
A3 (6307)

Dunn, O. J. [1964], Multiple comparisons using rank sums. Technometrics 6:241-252.
A3,12 (6308)

Dunn, O. J. and Massey, F. J. [1962], Estimation of multiple contrasts using a multivariate t-distribution (abstract). AMS 33:1489.
A3 (6309)

Dunn, O. J. and Massey, F. J. [1965], Estimation of multiple contrasts using t-distributions. JASA 60:573-583.
A3 (6310)

Dunnett, C. W. [1955], A multiple comparison procedure for comparing several treatments with a control. JASA 50:1096-1121.
A3 (6311)

Dunnett, C. W. [1955], Multiple comparisons with a standard. ASQC Nat. Conv. Trans. 9:485-491.
A3 (6312)

Dunnett, C. W. [1956], Multiple decision procedure (abstract). Biometrics 12:534.
A2 (6313)

Dunnett, C. W. [1960], The statistical theory of drug screening and other problems connected with the screening, testing and selection of drugs. Ph. D. Thesis, Univ. Aberdeen.
A2 (6314)

Dunnett, C. W. [1960], On selecting the largest of k normal population means (with discussion). JRSSB 22:1-40.
A2 (6315)

Dunnett, C. W. [1961], Query 162. Multiple comparisons between treatments and a control. Biometrics 17:324-326. A3 (6316)

Dunnett, C. W. [1964], New tables for multiple comparisons with a control. Biometrics 20:482-491. A3 (6317)

Dunnett, C. W. and Sobel, M. [1954], A bivariate generalization of Student's t-distribution, with tables for certain special cases. Biometrika 41:153-169. A7 (6318)

Dunnett, C. W. and Sobel, M. [1955], Approximations to the probability integral and certain percentage points of a multivariate analogue of Student's t-distribution. Biometrika 42:258-260. A7 (6319)

Durbin, J. [1953], A note on regression when there is extraneous information about one of the coefficients. JASA 48:799-808. A7 (6320)

Durbin, J. [1954], Errors in variables. Rev. ISI 22:23-32. A7 (6321)

Dutka, S. [1967], Experimental designs in communication research (summary). JASA 62:728. C (6322)

Dutoit, E. and Webster, R. [1966], A simplified technique for estimating degrees of freedom for a two population t test when the standard deviations are unknown and not necessarily equal. Proc. Eleventh Conf. Design Expt. Army Res. Dev. Testing. pp. 415-447. A11 (6323)

Dutton, A. M. [1954], Application of some multivariate analysis techniques to data from radiation experiments. Statistics and Mathematics in Biology pp. 81-91. A7 (6324)

Dwass, M. [1955], A note on simultaneous confidence intervals. AMS 26:146-147. A3 (6325)

Dwass, M. [1959], Multiple confidence procedures. Ann. Inst. Stat. Math. 10:277-282. A3 (6326)

Dwass, M. [1960], Some k-sample rank-order tests. Contributions to Probability and Statistics pp. 198-202. A12 (6327)

Dwyer, D. D. [1958], Competition between forbs and grasses. J. Range Management. 11:115-118. A14 (6328)

Eberhart, S. A., Penny, L. H. and Sprague, G. F. [1964], Intra-plot competition among maize single crosses. Crop Sci. 4:467:471. A14 (6329)

Ebert, D. and Schnee, M. [1957], Zum problem der Erntetecknik im Feldversuchswesen Einzel Pflanzen- und Reihenernte. Z. Landw. Vers.- Untersuch Wes. 3:245-267. C (6330)

Eckles, A. J. [1958], Some differences in experimental data. Proc. Second Conf. Design Expt. Army Res. Dev. Testing. pp. 125-129. C (6331)

Eckles, A. J. [1960], The conduct of military field research on a shoestring. Proc. Fifth Conf. Design Expt. Army Res. Dev. Testing. pp. 403-416. C (6332)

Edgett, G. L. [1956], Multiple regression with missing observations among the independent variables. JASA 51:122-131. A7 (6333)

Edwards, A. W. F. and Cavalli-Sforza, L. L. [1965], A method for cluster analysis. Biometrics 21:362-375. A7 (6334)

Efron, B. [1967], The two sample problem with censored data. Proc. Fifth Berkeley Symp. Math. Stat. Prob. 4:831-853. A2,12 (6335)

Ehrenberg, A. S. C. [1950], The unbiased estimation of heterogeneous error variances. Biometrika 37:347-357. A11 (6336)

Ehrenberg, A. S. C. [1952], On sampling from a population of rankers. Biometrika 39:82-87. A12 (6337)

Ehrenberg, A. S. C. [1963], Bivariate regression is useless. Appl. Stat. 12:161-179. A7 (6338)

Ehrenberg, P. [1950], Zur Beurteilung der Zuver lässigkeit unserer Versuchsergebnisse. Landw. Jb. Bayern 27(5&6):69-84. C (6339)

Ehrenfeld, S. [1957], Sensitivity testing. Proc. First Conf. Design Expt. Army Res. Dev. Testing. pp. 201-204. C (6340)

Ehrenpfordt, V. [1961], Teilstückgrösse bei statischen Versuchen, deren Parzellen einzeln gepflügt werden müssen. Z. Landw. Vers .- Untersuch Wes. 7:100-108. A14 (6341)

Ehrenpfordt, V. [1961], Beeinflussung der Parzellenerträge durch Randwirkung bei Kartoffelsortenprüfungen und Konsequenzen für die praktische Versuchsdurchführung. Z. Landw. Vers. Untersuch Wes. 7:179-189. A14 (6342)

Eicker, F. [1967], Limit theorems for regressions with unequal and dependent errors. Proc. Fifth Berkeley Symp. Math. Stat. Prob. 1:59-82. A7 (6343)

Eid, M. T., Black, C. A., Kempthorne, O., and Zoellner, J. A. [1954], Significance of soil organic phosphorus to plant growth. Iowa Agri. Expt. Sta. Res. Bull. 406: 745-776. A7 (6344)

Eindhoven, J., Peryam, D., Heiligman, F., and Hamman, J. W. [1964], Effects of sample sequence on food preferences. J. Food Sci. 29:520-524. C (6345)

Eisen, J. N. [1964], A note on orthogonal polynomials applied to treatment levels with unequal replications. J. Food Sci. 29:105-108. A7,10 (6346)

Eisenhart, C. [1960], Some canons of sound experimentation. Bull. ISI 37(3):339-350. C (6347)

Eisenhart, C. [1957], The principle of randomization in the design of experiments. Proc. First Conf. Design Expt. Army Res. Dev. Testing. pp. 15-16. C (6348)

Eisenhart, C. [1963], Realistic evaluation of the precision and accuracy of instrument calibration systems. J. Res. NBS 67C:161-187. B;C (6349)

Eisenhart, C. [1964], Realistic evaluation of the precision and accuracy of instrument calibration systems. Proc. Ninth Conf. Design Expt. Army Res. Dev. Testing. pp. 469-536. B;C (6350)

Eisenhart, C. [1964], The meaning of "least" in Least Squares. J. Washington Academy Sci. 54:24-33. A7;C (6351)

Eisenhart, C. [1964], Expertise or experiments - A letter to the Editor of Science. Science 146:997-998. C (6352)

Eisler, S. L. [1957], Long term exposure tests of various ordnance materials. Proc. First Conf. Design Expt. Army Res. Dev. Testing. pp. 233-234. C (6353)

Eisler, S. L. [1958], Experimental design for determining specification limits for manganese-aluminum bronze. Proc. Second Conf. Design Expt. Army Res. Dev. Testing. pp. 205-206. C (6354)

Elandt, R. C. [1953], Eksperymentalny blad metody. (Experimental error of the method.) Wiadomości Chemiczne 3(70):97-122. C (6355)

Elandt, R. [1955], Metody syntezy statystycznej doświadczeń jednopowtórzeniowych. (Methods of statistical syntheses of single replication experiments.) Roczniki Nauk Rolniczych 70(A):443-462. A11,13 (6356)

Elandt, R. [1958], O stosowanin analizy wariancji. I. Uwagi metodyczne. (On the use of analysis of variance. I. Methodological observations.) Roczniki Nauk Rolniczych 78(A):697-717. A9,10 (6357)

Elandt, R. [1959], O stosowaniu analizy wariancji. II. Uwagi metodyczne. (Some remarks on the use of analysis of variance. Part II.) Roczniki Nauk Rolniczych 80:171-186. A3 (6358)

Elandt, R. C. [1960], Zagadnienie zmniejszania ilości powtórzeń w doświadczeniach rejonizacyjnych. (Problem of reduction of replication number in multiple experiments.) Roczniki Nauk Rolniczych 83(A-1):203-214. S1 (6359)

Elandt, R. C. [1963], Optimal and sufficient allocation of multiple varietal experiments. Biometrics 19:615-628. S1 (6360)

Elandt-Johnson, R. C. [1966], Multi-dimensional orthogonal polynomials for certain models in multivariate analysis. Sankhyā B 28:191-198. A7 (6361)

El-Sayyad, G. M. [1967], Fixed sample and sequential sampling schemes. Ph. D. Thesis, Univ. College Wales. S1,2 (6362)

Elston, R. C. [1961], On additivity in the analysis of variance. Biometrics 17:209-219 (abstract 17:166; correction 17:509). A4 (6363)

Elston, R. C. [1963], A weight-free index for the purpose of ranking or selection with respect to several traits at a time. Biometrics 19:85-97. A2,7 (6364)

Elston, R. C. [1964], On estimating time-response curves. Biometrics 20:643-647. A7 (6365)

Elston, R. C. and Grizzle, J. E. [1962], Estimation of time-response curves and their confidence bands. Biometrics 18:148-159. A7 (6366)

Emmens, C. W. [1960], The role of statistics in physiological research. Biometrics 16:161-175. C (6367)

Emmens, C. W. [1961], The planning and analysis of some field trials with cattle (abstract). Biometrics 17:332. C (6368)

Emerson, P. L. [1965], Orthogonal polynomials for unequally weighted means. Biometrics 21:226-230. A7 (6369)

Enters, J. H. [1958], Statistics in small plants with special reference to quality control and related activities. Bull. ISI 36(3):477-498. A12;C (6370)

Ergun, S. [1956], Application of the principle of least squares to families of straight lines. Ind. Eng. Chemistry. 48:2063-2068. A7 (6371)

Ericson, W. A. [1967], On the economic choice of experiment sizes for decision regarding certain linear combinations. JRSSB 29:503-512. S1 (6372)

Erlander, S. and Gustavsson, J. [1965], Simultaneous confidence regions in normal regression analysis with an application to road accidents. Rev. ISI 33:364-377. A3,7 (6373)

Escribano, I. F. [1960], El porcentaje de "fallos" y las cosechas. (The percentage of "blanks" and harvest yields.) Anales Inst. Nacional Investigaciones Agrónomicas, Madrid 9:385-420. A14 (6374)

Evans, T. C., Barber, J. C., and Squillace, A. E. [1961], Some statistical aspects of progeny testing. Proc. Sixth Southern Conf. Forest Tree Improvement, Univ. Fla., Gainesville, June 7 and 8, pp. 73-79. A14 (6375)

Fabius, J. [1959], Stochastische approximatie. Stat. Neerlandica 13:445-452. A7 (6376)

Fairweather, W. R. [1967], Some extensions of Somerville's procedure for ranking means of normal populations (abstract). AMS 38:961. A2 (6377)

Federer, W. T. [1951], Testing proportionality of covariance matrices. AMS 22:102-106. A7,11 (6378)

Federer, W. T. [1953], Plot technique in greenhouse experiments. Proc. American Soc. Horticultural Sci. 62:31-34. A14 (6379)

Federer, W. T. [1954], Survey of experimental design. Statistics and Mathematics in Biology, pp. 143-148. B;C (6380)

Federer, W. T. [1955], Sample size. Biometrics Unit Mimeo. Ser. BU-59-M, Cornell Univ. S1 (6381)

Federer, W. T. [1960], Experimental error rates (abstract). Biometrics 16:132-133. A1 (6382)

Federer, W. T. [1961], Experimental error rates. Proc. American Soc. Horticultural Sci. 78:605-615. A1 (6383)

Federer, W. T. [1964], Literature review of experimental design through 1949. Math. Res. Center Tech. Summary Report No. 405, U.S. Army Univ. Wisconsin. B (6384)

Federer, W. T. [1966], Randomization and sample size in experimentation. Biometrics Unit Mimeo. Ser. BU-236-M, Cornell Univ. C;S1 (6385)

Feldman, D. [1962], Contribution to the "two-armed bandit" problem. AMS 33:847-856 (abstract 32:1348-1349). S2 (6386)

Feldt, L. A. and Mahmoud, M. W. [1958], Power function charts for specification of sample size in analysis of variance. Psychometrika 23:201-210. S1 (6387)

Ferenez, P. and Lloyd, B. H. [1953], Statistics can help in solving process problems. Chem. Eng. 60(May):207-210. C (6388)

Ferenez, P. and Lloyd, B. H. [1953], Next time, use statistics. Chem. Eng. 60 (April):219-222,226. C (6389)

Ferguson, J. H. A. [1963], Random variabilty in horticultural experiments. Euphytica 11:213-220. A14 (6390)

Ferguson, T. S. [1961], Rules for rejection of outliers. Rev. ISI 29(3):29-43. A6 (6391)

Ferguson, T. S. [1961], On the rejection of outliers. Proc. Fourth Berkeley Symp. Math. Stat. Prob. 1:253-287. A6 (6392)

Ferris, G. E. [1959], Sensory testing. ASQC, Nat. Conv. Trans. 13:21-32. C (6393)

Fertig, J. W., Chilton, N. W., and Varma, A. A. O. [1964], Studies in the design and analysis of dental experiments. 9. Sequential analysis (sign test). J. Oral Therapeutics Pharmacology 1:45-56. A12;S2 (6394)

Fieller, E. C. and Smith, C. A. B. [1951], Note on the analysis of variance and intraclass correlation. Annals Eugenics 16:97-104. A10 (6395)

Filho, V. C. [1954], Resultados de experiências de espaçamento da mamoneira anã, variedade I. A. 38. (Results of spacing trials for a dwarf variety of castor bean, I. A. 38.) Bragantia 13:297-305. A14 (6396)

Filipello, F. [1956], A critical comparison of the two-sample and triangular binomial designs. Food Res. 21:235-241. C (6397)

Filipello, F. [1957], Organoleptic wine-quality evaluation. I. Standards of quality and scoring vs. rating scales. Food Tech. 11:47-51. C (6398)

Filipello, F. [1957], Organoleptic wine-quality evaluation. II. Performance of judges. Food Tech. 11:51-53. C (6399)

Finch, D. J. [1950], The effect of non-normality on the z-test, when used to compare the variances in two populations. Biometrika 37:186-189. A11 (6400)

Finney, D. J. [1954], Functional relationships in experimentation (their role in the design and analysis of experiments) (abstract). Biometrics 10:177. C (6401)

Finney, D. J. [1956], The statistician and the planning of field experiments (with discussion). JRSSA 119:1-27. C (6402)

Finney, D. J. [1956], Incomplete determination of a measure of quality in a series of experiments. J. Agri. Sci. 48:124-128. C (6403)

Finney, D. J. [1956], Multivariate analysis and agricultural experiments. Biometrics 12:67-71. A7 (6404)

Finney, D. J. [1958], The efficiencies of alternative estimators for an asymptotic regression equation. Biometrika 45:370-388. A7 (6405)

Finney, D. J. [1963], Plant breeding, variety trials, and statistical methods. Empire Cotton Growing Rev., London 40:161-169. C (6406)

Finney, D. J. [1963], Statistical experience and the design of investigations (with discussion). Colloques Internationaux du Centre National de la Recherche Scientifique, No. 110 Le Plan d'Expériences pp. 61-78. C (6407)

Finney, D. J. [1964], Sir Ronald Fisher's contributions to biometric statistics. Biometrics 20:322-329. C (6408)

Finney, D. J. [1964], The replication of variety trials. Biometrics 20:1-15. C;S1 (6409)

Finney, D. J. and Nissen, Ø. [1956], Incomplete determination of a measure of quality in a series of experiments. J. Agri. Sci. 48:124-128. A13 (6410)

Finney, D. J. and Palca, H. [1950], The elimination of bias due to edge-effects in forest sampling. Forestry 23:31-47. A14 (6411)

Finney, D. J. and Sampford, M. R. [1961], The comparison of linear regressions for sets of correlated variates. Australian J. Stat. 3:1-19. A7 (6412)

Fisch, K. R. [1957], An application of analysis of variance to the evaluation of the effect of test variables and reproducibility of a newly developed laboratory apparatus. Proc. First Conf. Design Expt. Army Res. Dev. Testing. pp. 205-211. C (6413)

Fischer, J. and Csáki, P. [1962], Analysis of reaction curves by extreme values. Acta Medica Academiae Scientiarum Hungaricae 18:363-370. A6 (6414)

Fisher, G. R. [1957], Maximum likelihood estimators with heteroscedastic errors. Rev. ISI 25:52-55. A7,11 (6415)

Fisher, G. R. [1962], Iterative solutions and heteroscedasticity in regression analysis. Rev. ISI 30:153-159. A7,11 (6416)

Fisher, R. A. [1952], Sequential experimentation. Biometrics 8:183-187. S2 (6417)

Fisher, R. A. [1953], The expansion of statistics. JRSSA 116:1-6. C (6418)

Fisher, R. A. [1954], The analysis of variance with various binomial transformations. Biometrics 10:130-151. A9 (6419)

Fisher, R. A. [1955], Statistical methods and scientific induction. JRSSB 17:69-78. C (6420)

Fisher, R. A. [1955], The contribution of biometry to plant breeding (abstract). Biometrics 11:535. C (6421)

Fisher, R. A. [1956], On a test of significance in Pearson's Biometrika Tables (No.11) JRSSB 18:56-60. All;C (6422)

Fisher, R. A. [1957], Comment on the notes by Neyman, Bartlett, and Welch in this journal. JRSSB 19:179. All;C (6423)

Fisher, R. A. [1960], Scientific thought and the refinement of human reasoning. J. Operations Res. Soc. Japan 3:1-10. C (6424)

Fisher, R. A. [1961], Sampling the reference set. Sankhyā A 23:3-8. A11 (6425)

Fisher, R. A. [1961], The weighted mean of two normal samples with unkown variance ratio. Sankhyā A 23:103-114. A11 (6426)

Fisher, R. A. [1963], The place of design of experiments in the logic of scientific inference (with discussion). Colloques Internationaux du Centre National de la Recherche Scientifique, No. 110, Le Plan d'Expériences, pp. 13-19. C (6427)

Fisher, R. A. [1965], The place of the design of experiments in the logic of scientific inference. Sankhyā A 27:33-38 (also in Contributions to Statistics pp. 101-106). C (6428)

Fisher, R. A. and Healy, M. J. R. [1956], New tables of Behrens' test of significance. JRSSB 18:212-216. A11 (6429)

Fisk, P. R. [1967], Models of the second kind in regression analysis. JRSSB 29: 266-281. A7 (6430)

Fiske, D. W. and Jones, L. V. [1954], Sequential analysis in psychological research. Psychological Bull. 51:264-275. S2 (6431)

Fisser, H. G. [1961], Variable plot, square foot plot, and visual estimate for shrub crown cover measurements, J. Range Management 14:202-207. A14 (6432)

Fix, E. and Hodges, J. L. [1955], Significance probabilities of the Wilcoxon test. AMS 26:301-312. A12 (6433)

Flagle, C. D. [1967], A decision theoretical comparison of three procedures of screening for a single disease. Proc. Fifth Berkeley Symp. Math. Stat. Prob. 4:887-901. A2;B;C (6434)

Flapper, P. [1967], Transformatie van niet-normale verdelingen. (Transformation of non-normal distributions.) Stat. Neerlandica 21:151-155. A9 (6435)

Fleming, A. A., Rogers, T. H., and Bancroft, T. A. [1957], Field plot technique with hybrid corn under Alabama conditions. Agron. J. 49:1-4. A14 (6436)

Fletcher, C. M. [1964], The problem of observer variation in medical diagnosis with special reference to chest diseases. Methods Information Medicine 3:98-103. C (6437)

Flynn, A. J. [1959], A five variable multiple correlation study on UNIVAC I. ASQC Nat. Conv. Trans. 13:657-658. A7,15 (6438)

Folks, J. L. and Antle, C. E. [1965], Optimum allocation of sampling units to strata when there are R responses of interest. JASA 60:225-233. S1 (6439)

Forgy, E. W. [1965], Cluster analysis of multivariate data: efficiency vs. interpretability of classifications (abstract). Biometrics 21:768-769. A7 (6440)

Forshaw, R. P., Maddock, H. M., and Catron, D. V. [1950], A proposed growth curve for nursing Duroc pigs raised in dry lot (abstract). J. Animal Sci. 9:655-656. A7 (6441)

Forsythe, G. E. [1957], Generation and use of orthogonal polynomials for data-fitting with a digital computer. J. Soc. Industrial Appl. Math. 5:74-88. A7 (6442)

Fortmann, H. R. [1951], Observations on "selection" of data. Agron. J. 43:560-561. A14 (6443)

Fortuin, G. J. and van Beek, A. [1960], Een smaak-onderzoek. (A flavor trial.) Stat. Neerlandica 14:175-185. C (6444)

Foster, W. D. [1963], Analysis of a function in collaborative experimentation. Proc. Eight Conf. Design Expt. Army Res. Dev. Testing. pp. 89-97. A13 (6445)

Foster, W. D. [1967], Trial variability interpreted as differences in translation or rotation in function analysis of variance. Proc. Twelfth Conf. Design Expt. Army Res. Dev. Testing. pp. 243-249. A7 (6446)

Fraga, C. G. and Meirelles de Miranda, R. [1957], Analysis of a non-orthogonal experiment (abstract). Biometrics 13:238. A10 (6447)

Fraser, A. S. [1951], Competition between skin follicles in sheep. Nature 167: 202-203. A14 (6448)

Fraser, A. S. and Short, B. F. [1952], Competition between skin follicles in sheep. Australian J. Agri. Res. 3:445-452. A14 (6449)

Fraser, D. A. S. [1967], Data transformations and the linear model. AMS 38:1456-1465. A9 (6450)

Free, S. M. [1964], Statistical methods as applied to dermatology. Methods Information Medicine 3:38-40. C (6451)

Freeman, G. H. [1956], The selection and use of a panel for taste sensitivity tests with fruits. Annual Report, E. Malling Res. Sta., 1955, pp. 86-88. C (6452)

Freeman, G. H. [1963], The combined effect of environmental and plant variation. Biometrics 19:273-277. A14;C (6453)

Freeman, H., Kuzmack, A., and Maurice, R. [1967], Multivariate t and the ranking problem. Biometrika 54:305-308. A2 (6454)

Freeman, M. F. and Tukey, J. W. [1950], Transformations related to the angular and the square root. AMS 21:607-611 (abstract 21:305). A9 (6455)

Freis, E. D. and Williams, J. H. [1962], Practical considerations in cooperative therapeutic trials. Methods Information Medicine 1:9-15. C (6456)

French, M. H. [1956], Minimum size of experimental plots for the assessment of pasture yield. East African Agri. Forestry Res. Organization, Annual Report, 1956, pp. 82-83. A14 (6457)

French, M. H. [1960], Errores asociados con al uso de pequeñas parcelas de prueba en la evaluacion de rendimiento de pastos. (Errors resulting from the use of small test plots in the evaluation of pasture yield.) Agronomia Tropical, Venezuela 10:71-76. A14 (6458)

French, M. H. and Santiago Rodríguez, C. [1960], Variaciones en los rendimientos de diferentes pastos en los tropicos. (Variation in the yield of different grasses in the tropics.) Agronomia Tropical, Venezuela 10:77-86. A14 (6459)

Freund, J. E. and Miller, I. [1956], Some statistical aspects of a problem in scientific philosophy. Amer. Stat. 10(2):7-11. C (6460)

Freund, R. J. [1963], A warning of roundoff errors in regression. Amer. Stat. 17(5): 13-15. A7 (6461)

Freund, R. J., Vail, R. W., and Clunies-Ross, C. W. [1961], Residual analysis. JASA 56:98-104. A6 (6462)

Frey, K. J. [1953], The effect of cost functions on number of replications and samples per plot in field plot experiments of oats. Agron. J. 45:265-267. S1 (6463)

Frey, K. J. [1965], The utility of hill plots in oat research. Euphytica 14:196-208. A14 (6464)

Frey, K. J., and Baten, W. D. [1953], Optimum plot size for oat yield tests. Agron. J. 45:502-504. A14 (6465)

Frey, K. J. and Rodgers, P. [1961], Yield components in oats. V. Optimum number of replicates and samples per plot for spikelet counts. Agron. J. 53:28-29. S1 (6466)

Friedman, H. P. and Rubin, J. [1967], On some invariant criteria for grouping data. JASA 62:1159-1178. A7 (6467)

Fries, J. [1966], Introduction to a general discussion on: Statistical problems in the construction of yield tables. Int. Advisory Group Forest Stat. Second Conf. Institutionen Skoglig Matematisk Stat. Stockholm, pp. 80-84. C (6468)

Fries, J. and Matérn, B. [1966], On the use of multivariate methods for the construction of tree taper curves. Int. Advisory Group Forest Stat. Second Conf. Institutionen Skoglig Matematisk Stat. Stockholm, pp. 85-117. A7 (6469)

Fry, P. R. and Taylor, W. B. [1954], Analysis of virus local lesion experiments. Ann. Appl. Biology 41:664-674. C (6470)

Fujita, H. and Utida, S. [1953], The effect of population density on the growth of an animal population. Ecology 34:488-498. A14 (6471)

Fülgraff, G. [1965], Sequentielle statistische Prüfverfahren in der Pharmakologie. Arzneimittel-Forschung 15:382-387. S2 (6472)

Fulkerson, R. S. [1959], The effects of seeding rate and row width in relation to seed production in orchard grass, Dactylis glomerata L. Canadian J. Plant Sci. 39:355-363. A14 (6473)

Gabriel, K. R. [1962], The model of ante-dependence for data of biological growth. Bull. ISI 39(2):253-264. A7 (6474)

Gabriel, K. R. [1963], Statistical design of an artificial rainfall stimulation experiment in Israel (with discussion). Colloques Internationaux Centre National de la Recherche Scientifique, No. 110 Le Plan d'Expériences, pp. 147-163. C (6475)

Gabriel, K. R. [1963], Analysis of variance of proportions with unequal frequencies. JASA 58:1133-1157. A10 (6476)

Gabriel, K. R. [1964], A procedure for testing the homogeneity of all sets of means in analysis of variance. Biometrics 20:459-477. A3 (6477)

Gabriel, K. R. [1966], Simultaneous test procedures for multiple comparisons on categorical data. JASA 61:1081-1096. A3 (6478)

Gadzinski, C. [1961], Data fun - data frustration - data failure. IQC 17(8):8-9. C (6479)

Gaito, J. [1959], Multiple comparisons in analysis of variance. Psychological Bull. 56:392-393. A3 (6480)

Gall, H. [1957], Untersuchungen zur Auswertung langjähriger Sortenversuche bei Kartoffeln. Z. Landw. Vers.- Untersuch Wes. 3:319-390. A13 (6481)

Games, P. A. and Lucas, P. A. [1966], Power of the analysis of variance of independent groups on non-normal and normally transformed data. Educational Psychological Measurement 26:311-327. A9 (6482)

Gardner, A. L. [1960], A technique for the investigation of inter-cultivar competition in grass species. Proc. Eight Int. Grassland Congress, pp. 322-324. A14 (6483)

Garner, W. R. [1950], A note on the use of the method of successive differences in empirical curve fitting. Psychological Bull. 47:260-262. A7 (6484)

Garner, W. R. and McGill, W. J. [1956], The relation between information and variance analyses. Psychometrika 21:219-228. A10 (6485)

Garruti, R. S. and Conagin, A. [1961], Escala de valores para a avaliação da qualidade da bebida do café. (A scoring scale for quality evaluation of the coffee beverage.) Bragantia 20:557-562. C (6486)

Garruti, R. S., Pigatti, A., and Orlando, P. A. [1961], Qualidades organolépticas de purês de batatinhas procedentes de culturas tratadas com inseticidas. (Application of insecticides and their influence on the potato puree flavor.) Bragantia 20:857-865. C (6487)

Gart, J. J. [1963], A median test with sequential application. Biometrika 50:55-62. A12;S2 (6488)

Gärtner, M. [1955], Grafické vyhodnocovanie pol'nohospodárskych a biologických pokusov. (Graphical evaluation of agricultural and biological experiments.) Slovenská Akadémia Vied, 74 pp. C (6489)

Gastwirth, J. L. [1965], Asymptotically most powerful rank tests for the two-sample problem with censored data. AMS 36:1243-1247. A6,12 (6490)

Gastwirth, J. L. [1965], Percentile modifications of two sample rank tests. JASA 60:1127-1141. A12 (6491)

Gatty, R. [1966], Multivariate analysis for marketing research: an evaluation. Appl. Stat. 15:157-172. A7 (6492)

Gatty, R. and Wolf, D. [1961], Merchandising prepackaged cut flowers in supermarkets. New Jersey Agri. Expt. Sta. Bull. 799:1-20. C (6493)

Gayen, A. K. [1950], The distribution of the variance ratio in random samples of any size drawn from non-normal universes. Biometrika 37:236-255. A11 (6494)

Gayen, A. K. [1950], Significance of difference between the means of two non-normal samples. Biometrika 37:399-408. A11 (6495)

Gayen, A. K. [1952], The inverse hyperbolic sine transformation on Student's t for non-normal samples. Sankhyā 12:105-108. A9 (6496)

Gaylor, D. W. [1953], A statistical evaluation of techniques and choice of estimators in grazing experiments with feeder cattle in Iowa. M.S. Thesis. Iowa State Univ. A14;C (6497)

Gaylor, D. W. and Merrill, J. A. [1968], Augmenting existing data in multiple regression. Technometrics 10:73-81. A7 (6498)

Geary, R. C. [1966], A note on residual heterovariance and estimation efficiency in regression. Amer. Stat. 20(4):30-31. A7,11 (6499)

Geffroy, J. [1958], Contribution à la théorie des valeurs extrêmes. Publications Inst. Stat. Univ. Paris 7(3&4):37-121, 8(1):3-65. A6 (6500)

Gehan, E. A. [1962], An application of multivariate regression analysis to the problem of predicting survival in patients with acute leukemia. Bull. ISI 39(3):173-179. A7 (6501)

Gehan, E. A. [1965], A generalized Wilcoxon test for comparing arbitrarily singly-censored samples. Biometrika 52:203-223. A12 (6502)

Geidel, H. [1954], Mathematical fundamentals of the analysis of variance and the design of experiments (abstract). Biometrics 10:308. A11 (6503)

Geidel, H. [1955], Eine Möglichkeit der graphischen Ausführung des F- and t-Testes. Z. Pflanzenzüchtung 34:325-328. C (6504)

Geidel, H. [1956], Zur näherungsweisen Berechnung des Schätswertes des Standardfehlers von Beobachtungsreihen. Z. Acker- und Pflanzenbau 101:453-458. A12 (6505)

Geidel, H. [1957], Zur Methodik des Feldversuches in Moderne Methoden der Pflanzenzüchtung. Arb. dtsch. Landw. - Gesell. 44:72-81. C (6506)

Geidel, H. [1958], Zur Genauigkeit von Versuchsergebnissen. Z. Pflanzenzüchtung 39:438-446. A13;C (6507)

Geidel, H. [1958], Zur Variabilität der Einzelpflanze. Z. Acker- Pflanzenbau 106:49-57. A14 (6508)

Geidel, H. and Schuster, W. [1961], Untersuchungen über die Variabilität der Kartoffelpflanze. Z. Acker- Pflanzenbau 114:87-100. A14 (6509)

Geidel, H. and Wermke, H. [1958], Zur Auswertung von Feldversuchen mit Hilfe von Lochkartenmaschinen. Z. Pflanzenzüchtung 39:225-238. A15 (6510)

Geise, C. E. [1953], Influence of objective quality factors on subjective evaluation of canned sweet corn. Food Tech. 7:15-20. C (6511)

Geisler, G. and Staab, J. [1958], Versuchsanstellung im Weinbau. Vitis 1:257-281. A14;S1 (6512)

Geisser, S. [1963], Multivariate analysis of variance for a special covariance case. JASA 58:660-669 (corrigenda 59:1296). A7 (6513)

Georgescu-Roegen, N. [1965], Measure, quality and optimum scale. Sankhyā A 27:39-64 (corrigendum 28:99). C (6514)

Ghent, A. W. [1963], Studies of regeneration in forest stands devastated by the spruce budworm. III. Problems of sampling precision and seedling distribution. Forest Sci. 9:295-310. A14;S1 (6515)

Ghosh, B. [1950], A multi-stage stochastic model for some natural fields. Calcutta Stat. Assoc. Bull. 3:21-31. A14 (6516)

Ghosh, B. [1950], Multi-stage stochastic models for natural fields (abstract). Proc. Thirty-seventh Indian Sci. Congress, part III, pp. 8-9. A14 (6517)

Ghosh, B. [1954], A variance in areal sampling. Calcutta Stat. Assoc. Bull. 5: 73-81. A14 (6518)

Ghosh, B. [1956], Optimum structure of rectangular sample units. Calcutta Stat. Assoc. Bull. 6:176-180. A14 (6519)

Ghosh, B. [1956], A model for perimeter errors. Calcutta Stat. Assoc. Bull. 6:189-192. A14 (6520)

Ghosh, B. K. [1963], On sequential tests of ratio of variances based on range. Biometrika 50:419-430. A11;S2 (6521)

Ghosh, B. K. [1964], Simultaneous tests by sequential methods in hierarchial classification. Biometrika 51:439-450. S2 (6522)

Ghosh, M. N. [1950], Rank weighted mean and its use in descriptive statistics. Calcutta Stat. Assoc. Bull. 3:32-35. A6 (6523)

Ghosh, M. N. [1954], Simultaneous test of linear hypotheses by analysis of variance methods (summary). JASA 49:348-349. A3 (6524)

Ghosh, M. N. [1955], Simultaneous tests of linear hypotheses. Biometrika 42:441-449. A3 (6525)

Ghosh, M. N. [1961], Linear estimation of parameters in the case of unequal variances. Calcutta Stat. Assoc. Bull. 10:117-130. A7,11 (6526)

Ghosh, M. N. [1963], Hotelling's generalized T^2 in the multivariate analysis of variance. JRSSB 25:358-367. A7 (6527)

Ghosh, M. N. and Sharma, D. [1963], Power of Tukey's test for non-additivity. JRSSB 25:213-219. A4 (6528)

Ghosh, M. N. and Sharma, D. [1963], Power of Tukey's test for non-additivity (abstract). JISAS 15:274. A4 (6529)

Ghurye, S. G. and Robbins, H. [1954], Two-stage procedures for estimating the difference between means. Biometrika 41:146-152. S2 (6530)

Gibbons, J. D. [1964], A proposed two-sample rank test: the Psi test and its properties. JRSSB 26:305-312. A12 (6531)

Gibson, W. M. and Jowett, G. H. [1957], 'Three-group' regression analysis. Part I. Simple regression analysis. Appl. Stat. 6:114-122. A7 (6532)

Gibson, W. M. and Jowett, G. H. [1957], 'Three-group' regression analysis. Part II. Multiple regression analysis. Appl. Stat. 6:189-197. A7 (6533)

Giesbrecht, F. G. [1967], Study of residuals in planned experiments. Ph. D. Thesis, Iowa State Univ. A6,9 (6534)

Giesbrecht, J. [1961], Effect of incomplete hills and compensating treatments in comparative corn yield trials. Canadian J. Plant Sci. 41:91-96. A14 (6535)

Gilchrist, W. G. [1961], Some sequential tests using range. JRSSB 23:335-342. S2 (6536)

Gini, C. [1958], Logic in Statistics. Metron 19(1-2):1-77. C (6537)

Gini, C. [1960], Statistical methods with special reference to agriculture. Metron 20:319-420. C (6538)

Gini, G. [1964], Statistical methods with special reference to agriculture. Chapter XVII, Comparability of data and methods of elimination. Metron 23:325-374. C (6539)

Girardot, N. F., Peryam, D. R., and Shapiro, R. [1952], Selection of sensory testing panels. Food Tech. 6:140-143. C (6540)

Gjeddebaek, N. F. [1961], Contribution to the study of grouped observations. VI. Skandinavisk Aktuarietidskrift 45:55-73. C (6541)

Glass, D. N. [1965], Multiple regression analysis -- A method of process control, evaluation and comparison. ASQC Ann. Tech. Conf. Trans. pp. 570-576. A7 (6542)

Glasser, M. [1964], Linear regression analysis with missing observations among the independent variables. JASA 59:834-844. A7 (6543)

Glendinning, D. R. and Vernon, A. J. [1965], Inter-varietal competition in cocoa trials. J. Horticultural Sci. 40:317-319. A14 (6544)

Gleser, L. J. [1965], On the asymptotic theory of fixed-size sequential confidence bounds for linear regression parameters. AMS 36:463-467. S2 (6545)

Gleser, L. J. [1966], The comparison of multivariate tests of hypothesis by means of Bahadur efficiency. Sankhyā A 28:157-174. A7 (6546)

Gleser, L. J. and Zacks, S. [1966], On a fixed-width confidence interval for the common mean of two distributions having unequal variances: I. The case of one variance known. Dept. Stat. Tech. Report No. 58, Johns Hopkins Univ. A11 (6547)

Gnanadesikan, M. [1967], Some selection and ranking procedures for multivariate normal populations. Ph. D. Thesis, Purdue Univ. A2 (6548)

Gnanadesikan, M. R. [1966], Some selection and ranking procedures for multivariate normal populations (abstract). AMS 37:1418. A2 (6549)

Gnanadesikan, R. [1957], On the equality of the variances of several univariate normal populations and some multivariate extensions (abstract). AMS 28:1061-1062. A3,7 (6550)

Gnanadesikan, R. [1959], Equality of more than two variances and of more than two dispersion matrices against certain alternatives. AMS 30:177-184. A11 (6551)

Gnanadesikan, R. [1963], Multivariate statistical methods for analysis of experimental data. IQC 19(9):22-26, 31-32. A7 (6552)

Gnanadesikan, R. [1965], S. N. Roy's interests in and contributions to the analysis and design of certain quantitative multiresponse experiments (summary). JASA 60:660. A3,7 (6553)

Gnanadesikan, R. and Wilk, M. B. [1963], Statistical comparisons in data analysis (summary). JASA 58:551. A14 (6554)

Goldfeld, S. M. and Quandt, R. E. [1965], Some tests for homoscedasticity. JASA 60:539-547 (corrigenda 62:1518-1519). A7,11,12 (6555)

Goldman, A. [1963], Sample size for a specified width confidence interval on the ratio of variances from two independent normal populations. Biometrics 19:465-477. A3;S1 (6556)

Goldman, A. J. and Zelen, M. [1964], Weak generalized inverses and minimum varianc linear unbiased estimation. J. Res. NBS 68B(4):151-172. A7 (6557)

Golub, G. H. [1963], Comparison of the variance of minimum variance and weighted least squares regression coefficients. AMS 34:984-991. A7 (6558)

Gomes, F. P. [1953], The use of Mitscherlich's regression law in the analysis of experiments with fertilizers. Biometrics 9:498-516. A7 (6559)

Gomes, F. P. [1954], A comparação entre médias de tratamentos na análise da variânci (The comparison of treatment means in the analysis of variance.) Anais Escola Superior Agri. "Luiz de Queiroz" 11:1-12. A3 (6560)

Gomes, F. P. [1955], Methods of describing crop response fertilizers in perennial crops (abstract). Biometrics 11:540-541. A13 (6561)

Gomes, F. P. and Nogueira, I. R. [1951], Tabelas de polinomios para a interpolação da equação de Mitscherlich. Anais Escola Superior Agri. "Luiz de Queiroz" 8:57-67. A7 (6562)

Gonin, H. T. [1961], The use of orthogonal polynomials of the positive and negative binomial frequency functions in curve fitting by Aitken's method. Biometrika 48: 115-123 (corrigendum 48:476). A7 (6563)

Goodall, D. W. [1961], Objective methods for the classification of vegetation: IV. Pattern and minimal area (appendix by F. E. Binet). Australian J. Botany 9:162-196. A14;C (6564)

Goodman, L. A. [1954], Kolmogorov-Smirnov tests for psychological research. Psychological Bull. 51:160-168. A12 (6565)

Goodman, L. A. [1964], Simultaneous confidence intervals for contrasts among multinomial populations. AMS 35:716-725. A3 (6566)

Goodman, L. A. [1965], On simultaneous confidence intervals for multinomial proportions. Technometrics 7:247-254. A3 (6567)

Gordon, J. [1954], Uniformity trial on Gold Coast cacao (a note). Empire J. Expt. Agri. 22:332. A14 (6568)

Goslings, J. [1961], De medisch-ethische zijde van de sequente analyse. (The medical-ethical aspects of sequential analysis in controlled clinical trials.) Stat. Neerlandica 15:55-72. C;S2 (6569)

Gooslee, D. G. [1956], Level of significance and power of the unweighted means test. Ph. D. Thesis, Univ. N. Carolina. A10 (6570)

Gooslee, D. G. and Lucas, H. L. [1965], Analysis of variance of disproportionate data when interaction is present. Biometrics 21:115-133. A10 (6571)

Gotoh, K. [1956], Genetic analysis of varietal differentation in cereals III. Competitive ability of local strains of the barley variety "Hosogara No. 2". Japanese J. Genetics 31:1-8. A14 (6572)

Goudriaan, J. [1965], Naar aseptische en realistische methoden in de toegepaste statistiek. Stat. Neerlandica 19:3-13. C (6573)

Gourlay, N. [1953], Covariance analysis and its applications in psychological research. British J. Stat. Psychology Sec. 6:25-34. A7 (6574)

Govinda Iyer, T. A. [1958], Replications in field experiments. Madras Agri. J. 45:223-230. S1 (6575)

Govindarajulu, Z. [1964], A supplement to Mendenhall's bibliography on life testing and related topics. JASA 59:1231-1291 (corrigendum 60:1249-1250). B (6576)

Govindarajulu, Z. [1965], Normal approximations to the classical discrete distributions. Sankhyā A 27:143-172. B (6577)

Govindarajulu, Z., Haynam, G. E., and Leone, F. C. [1966], Extension of Mood's test to c-samples: null distribution and exact power against exponential and rectangular alternatives. JISA 4:170-188. A12 (6578)

Gower, J. C., Martin, A. H., and Simpson, H. R. [1965], An outline of a programming language for the analysis of surveys, experiments and multivariate data. Tagungsberichte Nr. 86, Deutsche Akademie, pp. 159-180. A15 (6579)

Grafius, J. E. [1956], Components of yield in oats: A geometrical interpretation. Agron. J. 48:419-423. A14 (6580)

Graham, J. E. [1960], Pooling procedures for analysis of survey data. M. S. Thesis, Iowa State Univ. A11 (6581)

Grandage, A. [1958], Query 130: Orthogonal coefficients for unequal intervals. Biometrics 14:287-289. A7 (6582)

Grandage, A. [1960], Analysis of variance and covariance. J. Farm Economics 42:1434-1438. A7,15 (6583)

Grandage, A., Casody, R. B., and Lucas, H. L. [1954], The analysis of errors in assaying fecal androgens in the dairy cow. J. Dairy Sci. 37:72-80. C (6584)

Grant, D. A. [1956], Analysis-of-variance tests in the analysis and comparison of curves. Psychological Bull. 53:141-154. A7 (6585)

Gray, K. B. [1966], A concept of sufficiency for the sequential design of experiments (Preliminary report) (abstract). AMS 37:1074. S2 (6586)

Gray, P. P. [1954], Modern taste testing from the viewpoint of the master brewer. Wallerstein Laboratories Communications 17:241-258. C (6587)

Graybill, F. A. [1958], Determining sample size for a specified width confidence interval. AMS 29:282-287 (abstract 29:326-327). S1 (6588)

Graybill, F. A. and Connell, T. L. [1964], Sample size required for estimating the variance within d units of the true value. AMS 35:438-440. S1 (6589)

Graybill, F. A. and Kneebone, W. R. [1959], Determining minimum populations for initial evaluation of breeding material. Agron. J. 51:4-6. S1 (6590)

Graybill, F. A. and Morrison, R. D. [1960], Sample size for a specified width confidence interval on the variance of a normal distribution (abstract). AMS 31:814-815. S1 (6591)

Graybill, F. A. and Morrison, R. D. [1960], Sample size for a specified width confidence interval on the variance of a normal distribution. Biometrics 16: 636-641. S1 (6592)

Greaves, M. J. [1959], Multiple correlation for processing quality control data on a digital computer. ASQC Nat. Conv. Trans. 13:627-633. A7,15 (6593)

Green, B. F. and Tukey, J. W. [1960], Complex analyses of variance: general problems. Psychometrika 25:127-152. C (6594)

Green, J. M. [1956], Border effects in cotton variety tests. Agron. J. 48:116-118. A14 (6595)

Green, P. E. [1962], Decision making in chemical marketing. Ind. Eng. Chemistry 54(9):30-34. S1 (6596)

Greenberg, B. G. [1951], Why randomize? Biometrics 7:309-322. C (6597)

Greenberg, B. G. [1953], The use of analysis of covariance and balancing in analytical surveys. American J. Public Health, Part I. 43:692-699. A7 (6598)

Greenberg, B. G. [1959], Conduct of cooperative field and clinical trials. Amer. Stat. 13(3):13-17 and 28. C (6599)

Greenberg, B. G. and Sarhan, A. E. [1958], Applications of order statistics to health data. American J. Public Health 48:1388-1394. A2,6 (6600)

Greenberg, B. G. and Sarhan, A. E. [1958], Applications of order statistics in medical experiments. Proc. Second Conf. Design Expt. Army Res. Dev. Testing. pp. 39-49. A2 (6601)

Greenberg, B. G. and Sarhan, A. E. [1959], Matrix inversion, its interest and application in analysis of data. JASA 54:755-766. A10 (6602)

Greenhouse, S. W. and Geisser, S. [1958], Application of analysis of variance techniques to profile analysis (summary). JASA 53:578. A7 (6603)

Greenwood, J. A. and Sandomire, M. M. [1950], Sample size required for estimating the standard deviation as a per cent of its true value. JASA 45:257-260. S1 (6604)

Greenwood, M. L., Potgieter, M., and Bliss, C. I. [1951], The effect of certain pre-freezing treatments on the quality of eight varieties of cultivated highbush blueberries. Food Res. 16:154-160. C (6605)

Gregson, R. A. M. [1963], Validation problems in interpreting preference responses to mixed food qualities. Appl. Stat. 12:1-13. C (6606)

Greiner, T. and Gold, H. [1952], The clinical assay of diuretic agents I. Biologic considerations which determine the design. Biometrics 8:232-237, 245-248. C (6607)

Gridgemen, N. T. [1955], The Bradley-Terry probability model and preference tasting. Biometrics 11:335-343. A12 (6608)

Gridgeman, N. T. [1956], A tasting experiment. Appl. Stat. 5:106-112. C (6609)

Gridgeman, N. T. [1956], Group size in taste sorting trials. Food Res. 21:534-539. C (6610)

Gridgeman, N. T. [1958], Psychophysical bias in taste testing by pair comparison, with special reference to position and temperature. Food Res. 23:217-220. C (6611)

Gridgeman, N. T. [1959], The lady tasting tea (abstract). Biometrics 15:157. C (6612)

Gridgeman, N. T. [1959], The lady tasting tea, and allied topics. JASA 54:776-783. C (6613)

Gridgeman, N. T. [1959], Sensory item sorting. Biometrics 15:298-306. C (6614)

Gridgeman, N. T. [1960], Statistics and taste testing. Appl. Stat. 9:103-112. C (6615)

Gridgeman, N. T. [1963], Matching trials by double criteria. Biometrics 19:398-405. C (6616)

Griffiths, J. C. [1954], Analysis of variance models in sedimentary petrology (summary). JASA 49:344-345. C (6617)

Grimm, H. [1960], Transformation von Zufallsvariablen. (Transformation of random variables.) Biom. Zeit. 2:164-182. A9;B (6618)

Grimm, H. and Malý, V. [1962], Anwendung der Sequenzanalyse auf die negative Binomialverteilung. (Application of sequential analysis to the negative binomial distribution.) Biom. Zeit. 4:182-192. S2 (6619)

Griswold, J. W. [1958], Determining durability of textile fabrics by means of controlle field testing. Proc. Third Conf. Design Expt. Army Res. Dev. Testing. pp. 311-314 C (6620)

Grizzle, J. E. [1960], Choosing a sample size in clinical trials. J. Indian Medical Profession 7(2):3149-3151. S1 (6621)

Gronow, D. G. C. [1951], Test for the significance of the difference between means in two normal populations having unequal variances. Biometrika 38:252-256. A11 (6622)

Gronow, D. G. C. [1953], Non-normality in two-sample t-tests. Biometrika 40: 222-225. A11 (6623)

Gross, H. D., Goode, L., and Petersen, R. G. [1965], Use of a uniformity trial in grazing experiments. Agron. J. 57:272-273. A14 (6624)

Grubbs, F. E. [1950], Sample criteria for testing outlying observations. AMS 21: 27-58. A6 (6625)

Grundy, P. M., Rees, D. H., and Healy, M. J. R. [1954], Decision between two alternatives ---how many experiments? Biometrics 10:317-323. S1,2 (6626)

Grundy, P. M., Healy, M. J. R., and Rees, D. H. [1956], Economic choice of the amount of experimentation (with discussion), JRSSB 18:32-55.S1 (6627)

Guest, P. G. [1950], Orthogonal polynomials in the least squares fitting of observations. Philosophical Magazine 41:124-137. A7 (6628)

Guest, P. G. [1950], Estimation of the error at a point on a least-squares curve. Australian J. Sci. Res., Ser. A, 3:173-182. A7 (6629)

Guest, P. G. [1951], The fitting of polynomials by the method of weighted grouping. AMS 22:537-548. A7 (6630)

Guest, P. G. [1952], Curve-fitting by the method of grouping. Autralian J. Sci. Res., Ser. A, 5:238-257. A7 (6631)

Guest, P. G. [1953], Note on the fitting of polynomials to equally-spaced observations. J. Math. Physics 32:68-71. A7 (6632)

Guest, P. G. [1953], On the standard errors in the fitting of polynomials to unequally spaced observations. Australian J. Physics 6:131-154. A7 (6633)

Guest, P. G. [1953], Efficiencies in the method of grouping . Australian J. Physics 6:361-370. A7 (6634)

Guest, P. G. [1953], The Doolittle method and the fitting of polynomials to weighted data. Biometrika 40:229-231. A7 (6635)

Guest, P. G. [1954], Grouping methods in the fitting of polynomials to equally spaced observations. Biometrika 41:62-76. A7 (6636)

Guest, P. G. [1956], Grouping methods in the fitting of polynomials to unequally spaced observations. Biometrika 43:149-160. A7 (6637)

Guion, R. M. [1954], Regression analysis: prediction from classified variables. Psychological Bull. 51:505-510. A7 (6638)

Gulliksen, H. and Wilks, S. S. [1950], Regression tests for several samples. Psychometrika 15:91-114. A7 (6639)

Gumbel, E. J. [1953], Probability tables for the analysis of extreme-value data. Nat. Bureau Standards, Appl. Math. Ser. 22. A6 (6640)

Gumbel, E. J. [1958], Statistical theory of extreme values. Bull. ISI. 36(3):12-14. A6 (6641)

Gumbel, E. J. [1960], Distributions des valeurs extrèmes en plusiers dimensions. Publications Inst. Stat. Univ. Paris 9:171-173. A6 (6642)

Gumbel, E. J. [1962], Une distribution à deux variables et généralisations de ses valeurs extremes. Rev. Stat. Appl. 10(1):97-98. A6 (6643)

Gumbel, E. J. [1962], Statistical theory of extreme values (main results). Contribution to Order Statistics, Chapter 6, pp. 56-93. A6;B (6644)

Gumbel, E. J. and Lieblein, J. [1954], Some applications of extreme-value methods. Amer. Stat. 8(5):14-17. A6 (6645)

Gupta, S. D. [1964], Non-parametric classification rules. Sankhyā A 26:25-30. A12 (6646)

Gupta, S. S. [1960], A single sample decision procedure for selecting a subset containing the best of several normal populations and some extensions (abstract). AMS 31:235. A2 (6647)

Gupta, S. S. [1963], Selection and ranking procedures and order statistics for the binomial distribution. Classical and Contagious Discrete Distributions, pp. 219-230. A2 (6648)

Gupta, S. S. [1963], On a selection and ranking procedure for gamma populations. Ann. Inst. Stat. Math. Tokyo, 14:199-216. A2 (6649)

Gupta, S. S. [1963], Bibliography on the multivariate normal integrals and related topics. AMS 34:829-838. B (6650)

Gupta, S. S. [1965], On some multiple decision (selection and ranking) rules. Technometrics 7:225-245. A2 (6651)

Gupta, S. S. [1966], On some selection and ranking procedures for multivariate normal populations using distance functions. Multivariate Analysis, pp. 457-475. A2,7 (6652)

Gupta, S. S., Huyett, M. J., and Sobel, M. [1957], Selection and ranking problems with binomial populations. ASQC Nat. Conv. Trans. 11:635-643. A2 (6653)

Gupta, S. S. and Nagel, K. [1967], On selection and ranking procedures and order statistics from the multinomial distribution. Sankhyā B 29:1-34. A2 (6654)

Gupta, S. S. and Sobel, M. [1957], On selecting a subset which contains all populations better than a standard (abstract). Biometrics 13:414-415. A2 (6655)

Gupta, S. S. and Sobel, M. [1960], On the distribution of the ratio of the smallest of several chi-squares to an independent chi-square (abstract). AMS 31:235-236. A2 (6656)

Gupta, S. S. and Sobel, M. [1960], On a single sample procedure for selecting from several normal populations a subset containing the population with the smallest variance (abstract). AMS 31:236. A2 (6657)

Gupta, S. S. and Sobel, M. [1960], Selecting a subset containing the best of several binomial populations. Contributions to Probability and Statistics, pp. 224-248. A2 (6658)

Gupta, S. S. and Sobel, M. [1960], On the distribution of the ratio of the largest of several chi-squares to an independent chi-square with application to ranking problems (abstract). AMS 31:815. A2 (6659)

Gupta, S. S. and Sobel, M. [1962], On the smallest of several correlated F statistics. Biometrika 49:509-523. A2 (6660)

Gupta, S. S. and Sobel, M. [1962], On selecting a subset containing the population with the smallest variance. Biometrika 49:495-507. A2 (6661)

Gurland, J. [1955], Query 117. Biometrics 11:395-398. All (6662)

Gurland, J. [1962], Note on a paper by Ray and Pitman. JRSSB 24:537-538. All (6663)

Gurland, J. and McCullough, R. S. [1962], Testing equality of means after a preliminary test of equality of variances. Biometrika 49:403-417. (Math. Res. Center Tech. Summary Report No. 267, U. S. Army and Univ. Wisconsin.) All (6664)

Gurland, J. and Rosenberg, L. [1957], Testing homogeneity of means in the presence of heterogeneity of variance (abstract). AMS 28:1056. All (6665)

Gutman, N. J. [1958], Determining whether a product meets taste specifications. Proc. Second Conf. Design Expt. Army Res. Dev. Testing. pp. 203-204. C (6666)

Guttman, I. [1962], A sequential selection procedure for the best population (preliminary report) (abstract). AMS 33:303. A2;S2 (6667)

Guttman, I. [1963], A sequential procedure for the best population. Sankhyā A 25:25-28. A2;S2 (6668)

Guttman, I. and Meeter, D. A. [1965], On Beale's measures of non-linearity. Technometrics 7:623-637. A7 (6669)

Guttman, I. and Smith, D. [1966], Investigation of rejection rules for outliers in small samples from the normal distribution. I: Discussion of the problem. Dept. Stat. Tech. Report No. 90, Univ. Wisconsin. A6 (6670)

Guttman, I. and Smith, D. [1966], Investigation of rejection rules for outliers in small samples from the normal distribution. II: Estimation of the mean (known variance). Dept. Stat. Tech. Report No. 91, Univ. Wisconsin. A6 (6671)

Guttman, I. and Smith, D. [1966], Investigation of rejection rules for outliers in small samples from the normal distribution. IV: Estimation in the case where both parameters are unknown. Dept. Stat. Tech. Report No. 93, Univ. Wisconsin. A6 (6672)

Guttman, I. and Smith, D. E. [1966], Investigation of rejection rules for outliers in small samples from the normal distribution (preliminary report) (abstract). AMS 37:1074-1075. A6 (6673)

Guttman, I. and Tiao, G. C. [1963], A Bayesian approach to some best population problems (abstract). Biometrics 19:506. A2 (6674)

Guttman, L. [1953], Image theory for the structure of quantitative variates. Psychometrika 18:277-296. A7 (6675)

Guttman, L. [1955], A generalized simplex for factor analysis. Psychometrika 20: 173-192. A7 (6676)

Gysel, A. [1951], Beiträge zur Tecknik des landwirtschaftlichen Versuchswesens. I. Geschichtliche Entwicklung, Wesen und Ziele der Versuchatechnik dargestellt am Feldversuch. Schweiz landw. Mh. 29:98-102. C (6677)

Hader, R. J. and Grandage, A. H. E. [1958], Simple and multiple regression analyses. Experimental Designs in Industry, pp. 109-137. A7 (6678)

Hader, R. J. and Youden, W. J. [1952], Experimental statistics. Analytical Chemistry 24:120-124. B;C (6679)

Hafley, W. L. and Lewis, J. S. [1963], Analyzing messy data. Ind. Eng. Chemistry 55(4):37-39. C (6680)

Haga, T. [1960], A two-sample rank test on location. Ann. Inst. Stat. Math. 11: 211-219. A12 (6681)

Haggstrom, G. W. [1964], Optimal stopping and experimental design. Ph. D. Thesis, Univ. Illinois. S2 (6682)

Haggstrom, G. W. [1966], Optimal stopping and experimental design. AMS 37:7-29. S2 (6683)

Haight, F. A. [1961], Index to the distributions of mathematical statistics. J. Res. NBS 65B:23-60. B (6684)

Haight, F. A. [1964], Annotated bibliography of scientific research in road traffic and safety. Operations Res. 12:976-1039. B (6685)

Haitovsky, Y. [1966], A note on regression on principal components. Amer. Stat. 20(4):28-29. A7 (6686)

Hajnal, J. [1961], A two-sample sequential t-test. Biometrika 48:65-75. S2 (6687)

Hakim, R., Soebijanto, and Jackson, R. I. [1964], Size of plots in relation to yield trials for rice and maize. Philippine Agri. 47:454-459. A14 (6688)

Haldane, J. B. S. [1956], Biometry. Sankhyā 17:207-214. C (6689)

Haldane, J. B. S. [1958], The scope of biological statistics. Sankhyā 20:195-206. C (6690)

Hale, R. W. [1952], Experimental errors in laying experiments. J. Agri. Sci. 42: 347-352. A14;C (6691)

Hall, C. A. [1966], Deleting observations from a least squares solution. Proc. Eleventh Conf. Design Expt. Army Res. Dev. Testing. pp. 450-467. A6 (6692)

Hall, W. J. [1958], An optimum property of some Bechhofer-type non-sequential multiple-decision rules (abstract). AMS 29:621. A2 (6693)

Hall, W. J. [1959], The most-economical character of some Bechhofer and Sobel decision rules. AMS 30:964-969. A2 (6694)

Hall, W. J. [1962], Some sequential analogs of Stein's two-stage test. Biometrika 49:367-378. S2 (6695)

Hall, W. J., Wijsman, R., and Ghosh, J. K. [1965], The relationship between sufficiency and invariance with applications in sequential analysis. AMS 36:575-614. S2 (6696)

Halperin, M. [1951], Normal regression theory in the presence of intra-class correlation. AMS 22:573-580. A7 (6697)

Halperin, M. [1963], Confidence interval estimation in non-linear regression. JRSSB 25:330-333. A7 (6698)

Halperin, M. [1964], Note on interval estimation in non-linear regression when responses are correlated. JRSSB 26:267-269. A7 (6699)

Halperin, M. and Greenhouse, S. W. [1958], Note on multiple comparisons for adjusted means in the analysis of covariance. Biometrika 45:256-259. A3,7 (6700)

Halperin, M., Greenhouse, S. W., Cornfield, J., and Zalokar, J. [1955], Tables of percentage points for the studentized maximum absolute deviate in normal samples. JASA 50:185-195. A6 (6701)

Halperin, M., Rastogi, S. C., Ho, I., and Yang, Y. Y. [1967], Shorter confidence bands in linear regression. JASA 62:1050-1067. A7 (6702)

Halvorson, H. W. [1960], Regression and systems of equations analysis on the IBM-650. J. Farm Economics 42:1450-1458. A7,15 (6703)

Hamaker, H. C. [1957], Designed experiments in industry. ASQC Nat. Conv. Trans. 11:355-356. C (6704)

Hamaker, H. C. [1958], De kunst van experimenteren. (How to make experiments.) Stat. Neerlandica 12:170-172. C (6705)

Hamaker, H. C. [1958], Quelques exemples concernant l'éstablissement de plans d'éxperiences. Rev. Stat. Appl. 8(1):9-27. C (6706)

Hamaker, H. C. [1961], Toegepaste statistiek. (Applied statistics.) Stat. Neerlandica 15:1-18. C (6707)

Hamaker, H. C. [1962], On multiple regression analysis. Stat. Neerlandica 16:31-56. A7 (6708)

Hameed, S. A. [1953], Choice of plot size in the objective estimation of corn yield. M. S. Thesis, Iowa State Univ. A14 (6709)

Hamman, J. W. and Eindhoven, J. [1963], Effectiveness of certain experimental plans utilized in sensory evaluations. Proc. Eighth Conf. Design Expt. Army Res. Dev. Testing. pp. 373-386. C (6710)

Hamy, M. A. [1952], Sur une nouvelle méthode d'éxpérimentation. Academie d' Agriculture de France. Comptes Rendus des Séances. 38:299-300. A14 (6711)

Hanajiri, A. [1966], An application of multiple regression analysis to cupola melting operations. ASQC Ann. Tech. Conf. Trans. pp. 156-166. A7 (6712)

Hanamoto, B. and Jebe, E. H. [1963], Size effects in the measurement of soil strength parameters. Proc. Eighth Conf. Design Expt. Army Res. Dev. Testing. pp. 349-372. C (6713)

Hancock, J. [1950], Studies in monozygotic cattle twins. III. Uniformity trials: milk and butterfat production. New Zealand J. Sci. Tech. A31(5):23-39. A14 (6714)

Hancock, J. [1950], Studies in monozygotic cattle twins. IV. Uniformity trials: grazing behaviour. New Zealand J. Sci. Tech. A32(4):22-59. A14 (6715)

Hancock, J. [1951], Studies in monozygotic cattle twins. V. Uniformity trials: growth. New Zealand J. Sci. Tech. A33(4):17-29. A14;C (6716)

Hancock, J. [1952], Studies in monozygotic cattle twins. VI. Uniformity trials: physiological and biochemical characteristics. New Zealand J. Sci. Tech. A34:131-152. A14 (6717)

Hanna, G. C. and Baker, G. A. [1951], Analysis of asparagus field trials on the basis of partial records. Proc. American Soc. Hort. Sci. 57:273-276. A14 (6718)

Hänninen, P. [1951], Koeympyröiden käytöstä kenttäkokeissa. (Using round plots for field experiments.) Valt. Maatalousk. Tiedon. No. 233. A14 (6719)

Hanson, W. D. [1959], Minimum family sizes for the planning of genetic experiments. Agron. J. 51:711-715. S1 (6720)

Hanson, W. D., Brim, C. A., and Hinson, K. [1961], Design and analysis of competition studies with an application to field plot competition in the soybean. Crop Sci. 1:255-258. A14 (6721)

Harper, R. [1956], Factor analysis as a technique for examining complex data on foodstuffs. Appl. Stat. 5:32-48. C (6722)

Harrington, E. C. [1961], Industrial applications of nonparametric statistics. ASQC Nat. Conv. Trans. 15:415-418. A12 (6723)

Harrington, E. C. [1963], A procedure for sequential experimentation. Chemical Eng. Progress Symp. Series, 59(42):1-7. S2 (6724)

Harris, C. W. [1955], Separation of data as a principle in factor analysis. Psychometrika 20:23-28. A7 (6725)

Harris, C. W. [1956], Relationships between two systems of factor analysis. Psychometrika 21:185-190. A7 (6726)

Harris, E. K. [1960], Analysis of experiments measuring threshold taste. Biometrics 16:245-260. C (6727)

Harris, T. E. [1950], Question 25: Regression using minimum absolute deviations. Amer. Stat. 4(1):14-15. A7 (6728)

Harrison, S. [1953], The use of statistics in food processing. IQC 10(2):26-30. C (6729)

Harrison, S. and Elder, L. W. [1950], Some applications of statistics to laboratory taste testing. Food Tech. 4:434-439. C (6730)

Harshbarger, B. et al. [1964], What type of statisticians are needed in research and development laboratories. Proc. Ninth Conf. Design Expt. Army. Res. Dev. Testing. pp. 547-568. C (6731)

Harte, C. [1965], Anwendung der Covarianzanalyse beim Vergleich von Regressionskoeffizienten. (Application of the analysis of covariance in comparing regression coefficients.) Biom. Zeit. 7:151-164. A7 (6732)

Harte, R. A. [1950], Statistics in nutrition research. Ann. New York Academy Sci. 52:827-837. C (6733)

Harter, H. L. [1955], Error rates and sample size in multiple comparisons (summary). JASA 50:577. A1,3;S1 (6734)

Harter, H. L. [1955], Error rates and sample sizes for multiple range tests (abstract). AMS 26:774. A1,3 (6735)

Harter, H. L. [1957], Error rates and sample sizes for range tests in multiple comparisons. Biometrics 13:511-536. A1,3;S1 (6736)

Harter, H. L. [1957], Critical values for Duncan's new multiple range (summary). JASA 52:372. A3 (6737)

Harter, H. L. [1960], Critical values for Duncan's new multiple range test. Biometrics 16:671-685. A3 (6738)

Harter, H. L. [1960], Tables of range and Studentized range. AMS 31:1122-1147. A3 (6739)

Harter, H. L. [1961], Corrected error rates for Duncan's new multiple range test. Biometrics 17:321-324. A3 (6740)

Harter, H. L. [1961], Use of tables of percentage points of range and Studentized range. Technometrics 3:407-411. A3 (6741)

Harter, H. L. and Lum, M. D. [1958], A note on Tukey's one degree of freedom for non-additivity (abstract). Biometrics 14:136-137. A4 (6742)

Harter, H. L. and Moore, A. H. [1965], Maximum-likelihood estimation of the parameters of Gamma and Weibull populations from complete and from censored samples. Technometrics 7:639-643. (also Aerospace Res. Lab. ARL 65-275, U. S. Air Force) A6 (6743)

Hartley, H. O. [1950], The maximum F-ratio as a short-cut test for heterogeneity of variance. Biometrika 37:308-312. A11 (6744)

Hartley, H. O. [1951], The fitting of polynomials to equidistant data with missing values. Biometrika 38:410-413. A7 (6745)

Hartley, H. O. [1954], Some significance test procedures for multiple comparisons (abstract). AMS 25:176. A3 (6746)

Hartley, H. O. [1955], Some recent developments in analysis of variance. Communications Pure Appl. Math. 8:47-72. A3 (6747)

Hartley, H. O. [1955], Query 118. Biometrics 11:504-505. A3 (6748)

Hartley, H. O. [1958], Changes in the outlook of statistics brought about by modern computers. Proc. Third Conf. Design Expt. Army Res. Dev. Testing. pp. 345-363. A15 (6749)

Hartley, H. O. [1959], The efficiency of internal regression for the fitting of the exponential regression. Biometrika 46:293-295. A7 (6750)

Hartley, H. O. [1961], The modified Gauss-Newton method for the fitting of non-linear regression functions by least squares. Technometrics 3:269-280. A7 (6751)

Hartley, H. O. [1963], In Dr. Bayes' consulting room. Amer. Stat. 17(1):22-24. C (6752)

Hartley, H. O. [1963], The design and analysis of experiments with the help of digital computers. <u>Applications of Digital Computers</u>. pp. 179-194. A15 (6753)

Hartley, H. O. [1964], Some small sample theory for nonlinear regression estimation. Proc. Ninth Conf. Design Expt. Army Res. Dev. Testing. pp. 655-660. A7 (6754)

Hartley, H. O. [1964], Exact confidence regions for the parameters in non-linear regression laws. Biometrika 51:347-353. A7 (6755)

Hartley, H. O. and Booker, A. [1965], Nonlinear least squares estimation. AMS 36:638-650. A7 (6756)

Hartley, H. O., Homeyer, P. G., and Kozicky, E. L. [1955], The use of log transformations in analyzing fall roadside pheasant counts. J. Wildlife Management 19:495-496. A9 (6757)

Hartley, H. O. and Pearson, E. S. [1950], Table of the probability integral of the t-distribution (with tables). Biometrika 37:168-172. A11 (6758)

Hartwig, E. E., Johnson, H. W., and Carr, R. B. [1951], Border effects in soybean test plots. Agron. J. 43:443-445. A14 (6759)

Harvey, J. C., Reed, J. W., and Thamer, M. A. [1962], Development of a multiphasic screening examination for medical care patients - I, II, and III. J. Chronic Diseases 15:827-856. C (6760)

Harvey, P. N. [1952], The practice of arable crop experimentation. Norfolk Agri. Sta. England, pp. 79. C (6761)

Harvey, W. R. [1960], Analysis of data with unequal subclass numbers when a set of orthogonal comparisons among the subclass means is desired (abstract). Biometrics 16:133. A10 (6762)

Harwood, P. D. [1963], Therapeutic dosage in small and large mammals (letter). Sci. 139:684-685. C (6763)

Hasty, W. L. [1962], Fitting the modified exponential function by the methods of multiple regression. Proc. Seventh Conf. Design Expt. Army Res. Dev. Testing. pp. 761-766. A7 (6764)

Hatheway, W. H. [1961], Convenient plot size. Agron. J. 53:279-280. A14 (6765)

Hatheway, W. H. and Williams, E. J. [1958], Efficient estimation of the relationship between plot size and the variability of crop yields. Biometrics 14:207-222 (also, Inst. Stat. Mimeo. No. 174, Univ. N. Carolina). A14 (6766)

Hayakawa, T. and Kabe, D. G. [1965], On testing the hypothesis that submatrices of the multivariate regression matrices of k populations are equal. Ann. Inst. Stat. Math. 17:67-73. A7 (6767)

Hayashi, C. [1950], On the quantification of qualitative data from the mathematico-statistical point of view (An approach for applying this method to the parole prediction). Ann. Inst. Stat. Math. 2:35-47. A9 (6768)

Hayashi, C. [1952], On the prediction of phenomena from qualitative data and the quantification of qualitative data from the mathematico-statistical point of view. Ann. Inst. Stat. Math. 3:69-98. A9 (6769)

Hayashi, C. [1954], Multidimensional quantification - with the applications to analysis of social phenomena. Ann. Inst. Stat. Math, 5:121-143.
A9 (6770)

Hays, V. W., Frape, D. L., Homeyer, P. G., Speer, V. C., and Catron, D. V. [1958], Factors contributing to within treatment variability in baby pig nutrition experiments (abstract). J. Animal Sci. 17:1160-1161. A14 (6771)

Heady, H. F., Gibbens, R. P., and Powell, R. W. [1959], A comparison of the charting, line intercept, and line point methods of sampling shrub types of vegetation. J. Range Management 12:180-188. A14 (6772)

Healy, M. J. [1954], Decision between two alternatives; how many experiments? (abstract). Biometrics 10:168-169. S1,2 (6773)

Healy, M. J. R. [1961], Experiments for comparing growth curves (abstract). Biometrics 17:333. A7 (6774)

Healy, M. J. R. [1963], Bibliography of the works of Sir Ronald Fisher. JRSSA 126:170-178. B (6775)

Healy, M. J. R. [1963], A subject index to the Kendall-Doig bibliography of statistical literature. JRSSA 126:270-275. B (6776)

Healy, M. J. R. [1965], Descriptive uses of discriminant functions. Mathematics and Computer Science in Biology and Medicine, Medical Res. Council, pp. 93-102. A7 (6777)

Healy, M. J. R. and Taylor, L. R. [1962], Tables for power-law transformations. Biometrika 49:557-559. A9 (6778)

Hebert, L. P. and Newton, B. S. [1960], Bunch planting of sugarcane seedlings at the U.S. Sugarcane Field Station, Houma, Louisiana. Proc. 10th Congress Int. Soc. Sugar Cane Tech., pp. 715-720. A14 (6779)

Hedlund, G. J. [1955], Use of taste panels in product development. ASQC Nat. Conv. Trans. 9:587-603. C (6780)

Heermann, E. F. [1963], Some least-squares transformations of regression estimators of orthogonal factors. Proc. Eighth Conf. Design Expt. Army Res. Dev. Testing. pp. 137-143. A7 (6781)

Heite, H. J. [1962], Über die therapeutische Versuchsplanung bei der äufzeren Behandlung von Hautkrankheiten Methods Information Medicine 1:52-55. C (6782)

Helgason, S. B. and Chebib, F. S. [1963], A mathematical interpretation of interplant competition effects. Statistical Genetics and Plant Breeding. pp. 535-545. A14 (6783)

Hemelrijk, J. [1950], A family of parameterfree tests for symmetry with respect to a given point. I and II. Koninklijke Nederlandse Akademie Wetenschappen Proc. Ser. A, 53:945-955 and 1186-1198. A12 (6784)

Hemelrijk, J. [1950], Rangcorrelatie en de schattingsproef van Varangot. (Rank correlation methods applied to an experiment on estimation of van Varangot.) Statistica 4:216-225. A12 (6785)

Hemelrijk, J. [1951], Parametrische en parametervrije methoden en hun toepassingen. (Parametric and non-parametric methods and their applications.) Statistica 5:171-184. A12 (6786)

Hemelrijk, J. [1952], Note on Wilcoxon's two-sample test when ties are present. AMS 23:133-135. A12 (6787)

Hemelrijk, J. [1952], A theorem on the sign test when ties are present. Koninklijke Nederlandse Akademie Wetenschappen, Proc. Ser. A, 55:322-326. A12 (6788)

Hemelrijk, J. [1958], Statistische proefopzetten: bewijs en detectie. (Statistics: proof and detection.) Stat. Neerlandica 12:111-118. C (6789)

Hemelrijk, J. [1958], Distribution-free tests against trend and maximum likelihood estimates of ordered parameters. Bull. ISI 36(3):15-25. A12 (6790)

Hemelrijk, J. [1959], In Memoriam Prof. Dr. D. van Dantzig. Stat. Neerlandica 13:416-432. B (6791)

Hemelrijk, J. and van Herk, C. G. G. [1958], Een geval van statistische detectie. (A case study of statistical detection.) Stat. Neerlandica 12:191-200. C (6792)

Hemmerle, W. J. [1964], Algebraic specification of statistical models for analysis of variance computations. J. Assoc. Computing Machinery 11:234-239. A15 (6793)

Henderson, P. L. [1964], Application of covariance analysis to marketing problems (with discussion). Proc. Business Economic Stat. Sec. American Stat. Assoc. pp. 422-428. A7 (6794)

Henderson, P. L. [1965], Application of covariance analysis to marketing problems (summary). JASA 60:662. A7 (6795)

Henry, J. -M. [1960], Développement des méthodes biométriques et statistiques dans la recherche agronomique au Congo Belge at au Ruanda-Urundi. Biométrie-Praximétrie 1(3):81-156. (also published Bull. Inst. Agronomique Sta. Recherches Gembloux (Hors Série) 1:272-293.) A7,14;C (6796)

Henson, W. R. [1954], A sampling system for poplar insects. Canadian J. Zoology 32:421-433. A14 (6797)

Herbst, L. J. [1965], The statistical Fourier Analysis of variances. JRSSB 27:159-165. A6,7 (6798)

Herbst, L. N. [1963], A test for variance heterogeneity in the residuals of Gaussian moving average. JRSSB 25:451-454. A11 (6799)

Hersey, M. D. and Eisenhart, C. [1965], A development of the theory of errors with reference to economy of time. J. Res. NBS 69B:139-146. C (6800)

Hey, E. N. and Hey, M. H. [1960], The statistical estimation of a rectangular hyperbola. Biometrics 16:606-617. A7 (6801)

Heymer, A. [1962], Die Planung therapeutischer Untersuchungen bei den akuten Pneumonien. Methods Information Medicine 1:86-90. C (6802)

Heyna, B. [1953], Omzetting van een in gehele graden Fahrenheit gegeven frequentieverdeling in een, welke een klassebreedte van een graad Celsius heeft. (Transformation of a frequency distribution.) Statistica 7:97-104. A9 (6803)

Hicks, C. R. [1954], Analysis of variance when the numbers of observations in the subgroups are unequal. IQC 10(6):52-53, 56, 60, and 62. A10 (6804)

Hicks, C. R. [1960], The experiment, the design, the analysis. IQC 17(6):17-20. C (6805)

Hill, A. B. [1951], The clinical trial. British Medical Bull. 7:278-282. C (6806)

Hill, B. M. [1960], A relationship between Hodges' bivariate sign test and a non-parametric test of Daniels. AMS 31:1190-1192 (correction 32:619). A7,12 (6807)

Hill, F. I. [1958], The design of controlled field experiments. Proc. Third Conf. Design Expt. Army Res. Dev. Testing. pp. 179-189. C (6808)

Hills, M. [1967], Discrimination and allocation with discrete data. Appl. Stat. 16:237-250. A7 (6809)

Hinchen, J. D. [1966], From opinion to fact through sequential experimentation. IQC 23:114-117. C (6810)

Hinreiner, E. [1956], Organoleptic evaluation by industry panels - the cutting bee. Food Tech. 10:203-205. C (6811)

Hinreiner, E., Filipello, F., Berg, H. W., and Webb, A. D. [1955], Evaluation of thresholds and minimum difference concentrations for various constituents of wines. IV. Detectable differences in wine. Food Tech. 9:489-490. C (6812)

Hinreiner, E., Filipello, F., Webb, A. D., and Berg, H. W. [1955], Evaluation of thresholds and minimum difference concentrations for various constituents of wines. III. Ethyl alcohol, glycerol and acidity in aqueous solution. Food Tech. 9:351-353. C (6813)

Hinson, K. and Hanson, W. D. [1962], Competition studies in soybeans. Crop Sci. 2:117-123. A14 (6814)

Hiorns, R. W. [1965], The fitting of growth and allied curves of the asymptotic regression type by Stevens's method. Tracts Computers No. 28, Cambridge Univ. Press. xv+52. A7 (6815)

Hiraga, Y., Morimura, H., and Watanabe, H. [1954], Tables for three-sample test. Ann. Inst. Stat. Math. 5:97-102. A12 (6816)

Hocking, R. R. [1965], Quadratic regression with inequality restraints on the parameters. JASA 60:914-919. A7 (6817)

Hodges, J. L. [1954], The up-and-down method with small samples (summary). JASA 49:354. S2 (6818)

Hodges, J. L. [1955], Galton's rank-order test. Biometrika 42:261-262. A12 (6819)

Hodges, J. L. [1955], A bivariate sign test. AMS 26:523-527. A7,12 (6820)

Hodges, J. L. [1967], Efficiency in normal samples and tolerance of extreme values for some estimates of location. Proc. Fifth Berkeley Symp. Math. Stat. Prob. 1:163-186. A6,12 (6821)

Hodges, J. L. and Lehmann, E. L. [1960], Comparison of the normal scores and Wilcoxon tests. Proc. Fourth Berkeley Symp. Math. Stat. Prob. 1:307-317. A12 (6822)

Hodges, J. L. and Lehmann, E. L. [1962], Rank methods for combination of independent experiments in analysis of variance. AMS 33:482-497. A12 (6823)

Hodnett, G. E. [1953], A uniformity trial on groundnuts. J. Agri. Sci. 43:323-328. A14 (6824)

Hoeffding, W. [1951], "Optimum" nonparametric tests. Proc. Second Berkeley Symp. Math. Stat. Prob. pp. 83-92. A12 (6825)

Hoel, P. G. [1951], Confidence regions for linear regression. Proc. Second Berkeley Symp. Math. Stat. Prob. pp. 75-81. A7 (6826)

Hoel, P. G. [1964], Methods for comparing growth type curves. Biometrics 20: 859-872. A7,12 (6827)

Hoffmann, R. G. [1963], Statistics in the practice of medicine. J. American Medical Assoc. 185:864-873. C (6828)

Hogben, D., Pinkham, R. S., and Wilk, M. B. [1962], Some properties of Tukey's test for non-additivity (abstract). AMS 33:1492-1493. A4 (6829)

Hogg, R. V. [1965], On models and hypotheses with restricted alternatives. JASA 60:1153-1162. A2 (6830)

Holiday, R. [1960], Plant population and crop yield. Nature 186:22-24. A14 (6831)

Holiday, R. [1960], Plant population and crop yield: Part I and II. Field Crop Abstracts 13:159-167, 247-254. A14;B (6832)

Hollander, M. [1966], An asymptotically distribution-free multiple comparison procedure-treatments vs. control. AMS 37:735-738 (abstract 36:1083). A3,12 (6833)

Holle, M. and Peirce, L. C. [1960], Plot technique for field evaluation of earliness, pod number, and total yield in the lima bean. Proc. American Soc. Hort. Sci. 76:403-408. A14 (6834)

Homeyer, P. G. [1954], Some problems of technique and design in animal feeding experiments. Statistics and Mathematics in Biology, pp. 399-406. C (6835)

Homeyer, P. G., Maddock, H. M., and Catron, D. V. [1950], More efficient design of swine feeding experiments (abstract). J. Animal Sci. 9:660. C (6836)

Homeyer, P. G. and Pauls, J. F. [1954], Statistical analysis and techniques in chick nutrition experiments (abstract). Poultry Sci. 33:1060. C (6837)

Hooke, R. [1956], On the use of randomization in the investigation of an unknown function (abstract). AMS 27:545. C (6838)

Hooke, R. [1958], Use of randomization in the investigation of unknown functions. JASA 53:176-186. C (6839)

Hooper, J. W. and Theil, H. [1958], The extension of Wald's method of fitting straight lines to multiple regression. Rev. ISI 26-47. A7 (6840)

Hopkins, J. W. [1953], Laboratory flavor scoring: Two experiments in incomplete blocks. Biometrics 9:1-21. C (6841)

Hopkins, J. W. [1954], Some statistical aspects of flavor and aroma testing. Statistics and Mathematics in Biology, pp. 383-392. C (6842)

Hopkins, J. W. [1954], Some observations on sensitivity and repeatability of triad taste difference tests. Biometrics 10:521-530. C (6843)

Hopkins, J. W. and Gridgeman, N. T. [1955], Comparative sensitivity of pair and triad flavor intensity difference tests. Biometrics 11:63-68. C (6844)

Horsnell, G. [1953], The effect of unequal group variances on the F-test for the homogeneity of group means. Biometrika 40:128-136. A11 (6845)

Hotelling, H. [1951], A generalized T test and measure of multivariate dispersion. Proc. Second Berkeley Symp. Math. Stat. Prob. pp 23-41. A7 (6846)

Hotelling, H. [1954], Multivariate analysis. Statistics and Mathematics in Biology, pp. (67-80. A7 (6847)

Hotelling, H. [1958], The statistical method and the philosophy of science. Amer. Stat. 12 (5):9-14. C (6848)

Hotelling, H. [1961], The behavior of some standard statistical tests under non-standard conditions. Proc. Fourth Berkeley Symp. Math. Stat. Prob. 1:319-359. A11;B (6849)

Hotelling, H. [1963], Different meanings of experimental design (with discussion). Colloques Internationaux Centre Nat. Recherche Sci. No. 110 Le Plan d'Expériences. pp. 39-49. C (6850)

Howes, D. R. [1965], Applications of dimensional analysis to multiple regression analysis. Proc. Tenth Conf. Design Expt. Army Res. Dev. Testing. pp. 47-58. A7 (6851)

Hoyle, B. J. and Baker, G. A. [1960], Factor analysis of twenty-eight independent field yield trials in nine strains of Hannchen barley (abstract). Biometrics 16: 127-128. A13 (6852)

Hoyle, B. and Baker, G. A. [1961], Game theory applied to field trials (abstract). Biometrics 17:167-168. C (6853)

Hoyle, B. J. and Baker, G. A. [1962], A binary system for evaluating large numbers of interactions. Biometrics 18:630-631. C (6854)

Hozumi, K., Asahira, T., and Kira, T. [1956], Intraspecific competition among higher plants. VI. Effect of some growth factors on the progress of competition. J. Inst. Polytechnics, Osaka City Univ. D, 7:15-34. A14 (6855)

Hozumi, K., Koyama, H., and Kira, T. [1955], Intraspecific competition among higher plants. IV. A preliminary account of the interaction between adjacent individuals. J. Inst. Polytechnics, Osaka City Univ. D, 6:121-130. A14(6856)

Hruschka, H. W. and Koch, E. J. [1964], A reason for randomisation within controlled environmental chambers. Proc. American Soc. Hort. Sci. 85:677-684. A14 (6857)

Hübner, R. [1957], Ein graphischer Ertragsvergleich unter Berücksichtigung der Grenzdifferenzen (GD), dargestellt an den Ergebnissen einer Futterroggen-Prüfung. Z. Acker- Pflanzenbau, 102:299-310. C (6858)

Huddleston, H. F. [1956], Use of order statistics in estimating standard deviations. Agri. Economics Res. 8:95-99. A12 (6859)

Hudimoto, H. [1956], Note on fitting a straight line when both variables are subject to error and some applications. Ann. Inst. Stat. Math. 7:159-167. A7 (6860)

Hudimoto, H. [1956], On the distribution-free classification of an individual into one of two groups. Ann. Inst. Stat. Math. 8:105-112. A7,12 (6861)

Hudimoto, H. [1957], A note on the probability of the correct classification when the distributions are not specified. Ann. Inst. Stat. Math. 9:31-36. A7,12 (6862)

Hudimoto, H. [1959], On a two-sample non-parametric test in the case that ties are present. Ann. Inst. Stat. Math. 11:113-120. A12 (6863)

Hudimoto, H. [1964], On a distribution-free two-way classification. Ann. Inst. Stat. Math. 16:247-253. A12 (6864)

Hummell, F. C. [1952], An experiment on the sampling of early thinning. Forestry 25(1):19-31. A14 (6865)

Huntoon, R. D. [1958], An example of design of experiments at the National Bureau of Standards. Proc. Second Conf. Design Expt. Army Res. Dev. Testing. pp. 1-4. C (6866)

Huntsberger, D. V. [1954], An extension of preliminary tests of significance permitting control of disturbances in statistical inferences. Ph. D. Thesis, Iowa State Univ. A11 (6867)

Huntsberger, D. V. [1954], An extension of preliminary tests for pooling data (summary). JASA 49:348. A11 (6868)

Huntsberger, D. V. [1955], A generalization of a preliminary testing procedure for pooling data. AMS 26:734-743. A11 (6869)

Hurst, R. L. [1960], The design, analysis and interpretation of greenhouse and laboratory experiments (abstract). Biometrics 16:698. C (6870)

Hurst, R. L. [1963], Disproportionate procedures (abstract). Biometrics 19:661.
A10 (6871)

Hutton, L. R. and Darroch, J. G. [1967], Examining some assumptions of the analysis of variance in experimental data (abstract). Biometrics 23:604.
A4,9,11 (6872)

Hyder, D. N. and Sneva, F. A. [1960], Bitterlich's plotless method for sampling basal ground cover of bunchgrasses. J. Range Management 13:6-9.
A14 (6873)

Ihm, P. [1955], Eine exakte Methode als Ersatz für die Varianzanalyse in bestimmten Fällen. Der Züchter 25:365-368. A3 (6874)

Ihm, P. [1957], Varianzanalyse und Konfidenzbehauptungen. Der Züchter 27:172-177. A11 (6875)

Ikusima, I., Shinozaki, K. and Kira, T. [1955], Intraspecific competition among higher plants. III. Growth of duckweed, with a theoretical consideration on the C-D effect. J. Inst. Polytechnics, Osaka City Univ., D, 6:107-119. A14 (6876)

Immich, H. [1964], Fehler bei der Erhebung und Dokumentation klinischer Befunde. Methods Information Medicine 3:95-98 C (6877)

Isaac, G. J. [1964], The impact of administrative limitations on the design of experiments. Proc. Ninth Conf. Design Expt. Army Res. Dev. Testing. pp. 229-233. C (6878)

Isaacson, S. L. [1954], Problems in classifying populations. Statistics and Mathematics in Biology pp. 107-117. A7 (6879)

Ishler, N. H., Laue, E. A. and Janisch, A. J. [1954], Reliability of taste testing and consumer testing methods. II. Code bias in consumer testing. Food Tech. 8:389-391. C (6880)

Isida, M. D. [1952], A remark on the linear regression estimate. Ann. Inst. Stat. Math. 4:7-9. A7 (6881)

Ito, K. [1961], On multivariate analysis of variance tests. Bull. ISI 38(4):87-98. A7 (6882)

Ito, K. and Schull, W. J. [1964], On the robustness of the T_0^2 test in multivariate analysis of variance when variance-covariance matrices are not equal. Biometrika 51:71-82. A7,11 (6883)

Ives, K. H. and Gibbons, J. D. [1967], A correlation measure for nominal data. Amer. Stat. 21(5):16-17. A12 (6884)

Jackson, J. E. [1959], Some multivariate statistical techniques used in color matching data. Optical Soc. America 49:585-592. A7 (6885)

Jackson, J. E. [1959], Multivariate sequential procedures for testing means (preliminary report) (abstract). Biometrics 15:493. A7;S2 (6886)

Jackson, J. E. [1960], Bibliography on sequential analysis. JASA 55:561-580. B (6887)

Jackson, J. E. [1961], Multivariate analysis illustrated by Nike-Hercules: I. Separation of product and measurement variability. II. Acceptance sampling. Proc. Sixth Conf. Design Expt. Army Res. Dev. Testing. pp. 307-327. A7;S2 (6888)

Jackson, J. E. and Lawton, W. H. [1967], Query 22. Regression residual analysis. Technometrics 9:339-340. A6 (6889)

Jacob, W. C. [1953], Some general considerations of plot technique in horticulture. Proc. American Soc. Hort. Sci. 62:35-45. A14;B (6890)

Jacob, W. C. [1955], Sampling potatoes for phosphorus analyses. Proc. American Soc. Hort. Sci. 65:307-312. A14 (6891)

Jacob, W. C. [1960], Interpretation of experimental results (abstract). Biometrics 16:133-134. C (6892)

Jacobs, R. M. [1962], Potential applications of reliability techniques in commercial product lines. IQC 19(2):11-14. C (6893)

Jacobson, J. E. [1963], The Wilcoxon two-sample statistic: Tables and bibliography. JASA 58:1086-1103. A12;B (6894)

Jaech, J. L. [1964], A note on the equivalence of two methods of fitting a straight line though cumulative data. JASA 59:863-866. A7 (6895)

Jaech, J. L. [1965], Understanding multiple regression. ASQC Ann. Tech. Conf. Trans. pp. 539-544. A7 (6896)

Jain, M. B. [1966], Size, shape and arrangement of plots for trials in range lands (abstract). JISAS 18(2):108-109. A14 (6897)

Jaiswal, M. C. [1967], Fitting a straight line to data from a truncated population by the method of moments. Metron 26:374-380. A7 (6898)

James, G. S. [1951], The comparison of several groups of observations when the ratios of the population variances are unknown. Biometrika 38:324-329. A11 (6899)

James, G. S. [1954], Tests of linear hypotheses in univariate and multivariate analysis when the ratios of the population variances are unknown. Biometrika 41:19-43. A7,11 (6900)

James, G. S. [1956], On the accuracy of weighted means and ratios. Biometrika 43: 304-321. A11 (6901)

James, G. S. [1959], The Behrens-Fisher distribution and weighted means. JRSSB 21:73-90. A11 (6902)

Janick, J. [1961], Sample size for the determination of strawberry fruit size. Proc. American Soc. Hort. Sci. 78:292-294. S1 (6903)

Jardine, R. [1958], Ranking methods and the measurement of attitudes. JASA 53: 720-728. A12 (6904)

Jardine, R., Moss, H. J., and Mullaly, J. V. [1963], Wheat quality: A factor analysis of some test data. Australian J. Agri. Res. 14:603-621. A7 (6905)

Jebe, E. H. [1959], Query 142: On a quail roadside count technique. Biometrics 15:628-631. S1 (6906)

Jebe, E. H. [1961], Multivariate analysis for Project Michigan experiments. Proc. Sixth Conf. Design Expt. Army Res. Dev. Testing. pp. 111-117. A7 (6907)

Jeffers, J. N. R. [1966], The design and analysis of forest experiments. Biométrie-Praximétrie 7:117-126. C (6908)

Jeffers, J. N. R. [1966], The statistician's use of electronic computers. Biometrie-Praximétrie 7:239-250. A15 (6909)

Jeffers, J. N. R. [1966], Principal component analysis in taxonomic research. Int. Advisory Group Forest Stat., Second Conf. Institutionen Skoglig Matematisk Stat. pp. 138-159. A7 (6910)

Jeffers, J. N. R. [1967], Two case studies in the application of principal component analysis. Appl. Stat. 16:225-236. A7 (6911)

Jeffrey, A. D. [1963], Values and choice. Univ. Rhode Island, Agri. Expt. Sta. Bull. 370. C (6912)

Jensen, N. F. [1965], Multiline superiority in cereals. Crop Sci. 5:566-568. A14 (6913)

Jensen, N. F. and Federer, W. T. [1964], Adjacent row competition in wheat. Crop Sci. 4:641-645. A14 (6914)

Jessen, R. J. [1955], Determining the fruit count on a tree by randomized branch sampling. Biometrics 11:99-109. A14;S1 (6915)

John, S. [1960], On some classification problems-I. Sankhyā 22:301-308. A7 (6916)

John, S. [1960], On some classification statistics. Sankhyā 22:309-316. A7 (6917)

John, S. [1963], A tolerance region for multivariate normal distributions. Sankhyā A 25:363-368. A7 (6918)

John, S. [1964], Further results on classification by W. Sankhyā A 26:39-46. A7 (6919)

John, S. [1964], Methods for the evaluation of probabilities of polygonal and angular regions when the distribution is bivariate t. Sankhyā A 26:47-54. A2 (6920)

Johnson, C. D. [1963], Comparison of approaches to obtaining a transformation matrix effecting a fit to a factor solution obtained in a different sample. Proc. Eighth Conf. Design Expt. Army Res. Dev. Testing. pp. 119-135. A7 (6921)

Johnson, F. A. and Hixon, H. J. [1952], The most efficient size and shape of plot to use for cruising in old-growth Douglas-fir timber. J. Forestry 50:17-20. A14 (6922)

Johnson, J. P. [1967], Pooling regressions and a statistical outlier methodology for lines. Ph. D. Thesis, Iowa State Univ. A6,7 (6923)

Johnson, L. P. V. and Keeping, E. S. [1952], Composite mean squares and their degrees of freedom. Appl. Stat. 1:202-205. A11 (6924)

Johnson, N. L. [1957], Sequentially determined confidence intervals. Biometrika 44:279-281. S2 (6925)

Johnson, N. L. [1961], Choosing a sequential test. ASQC Nat. Conv. Trans. 15:405-409. S2 (6926)

Johnson, N. L. [1961], Sequential analysis: A survey. JRSSA 124:372-411. S2;B (6927)

Johnson, N. L. [1961], Some notes on the investigation of heterogeneity in interactions (abstract). AMS 32:627. A11 (6928)

Johnson, N. L. [1961], On the choice of a sequential procedure. <u>Quantitative Methods in Pharmacology</u>, pp. 27-40. S2 (6929)

Johnson, N. L. [1962], Estimation of sample size. Technometrics 4:59-67. S1 (6930)

Johnson, N. L. [1967], Sample censoring. Proc. Twelfth Conf. Design Expt. Army Res. Dev. Testing. pp. 403-424. A6 (6931)

Johnson, N. L. and Maurice, R. J. [1963], A minimax-regret procedure for choosing between two populations using sequential sampling. JRSSB 25:297-304. A2;S2 (6932)

Johnson, W. C. [1959], A mathematical procedure for evaluating relationships between climate and wheat yields. Agron. J. 51:635-639. A7 (6933)

Johnsson, H. [1963], Arrangement and design of field experiments in progeny testing. Proc. World Consultation Forest Genetics Tree Improvement, Stockholm, Sweden, 23-30 August, No. 2a/1:iii+8. A14 (6934)

Joint Committee American Soc. Agronomy, American Dairy Sci. Assoc., American Soc. Animal Production, and American Soc. Range Management. [1952], Pasture and range research techniques. Agron. J. 44:39-50. B;C (6935)

Jonckheere, A. R. [1954], A distribution-free k-sample test against ordered alternatives. Biometrika 41:133-145. A2,12 (6936)

Jones, G. L., Matzinger, D. F., and Collins, W. K. [1960], A comparison of flue-cured tobacco varieties repeated over locations and years with implications on optimum plot allocation. Agron. J. 52:195-199. S1 (6937)

Jones, H. L. [1965], The jackknife method (with discussion). Proc. IBM Scientific Computing Symp. Stat. pp. 185-201. C (6938)

Jones, R. H. [1965], An experiment in non-linear prediction. J. Appl. Meteorology 4:701-705. A7 (6939)

Jones, W. W., Embleton, T. W., and Cree, C. B. [1957], Number of replications and plot sizes required for reliable evaluation of nutritional studies and yield relationships with citrus and avocado. Proc. American Soc. Hort. Sci. 69: 208-216. A14 (6940)

Jowett, G. H. [1962], Operational research techniques in mining. Australian Inst. Mining Metallurgy Proc. No. 202, pp. 103-118. C (6941)

Joyce, C. R. B. and Welldon, R. M. C. [1965], The objective efficacy of prayer: a double-blind clinical trial. J. Chronic Diseases 18:367-377. C (6942)

Joyce, R. J. V. and Roberts, P. [1959], The determination of the size of plot suitable for cotton spraying experiments in the Sudan Gezira. Ann. Appl. Biology 47: 287-305. A14 (6943)

Juska, F. V., Tyson, J., and Harrison, C. M. [1955], The competitive relationship of Merion bluegrass as influenced by various mixtures, cutting heights, and levels of nitrogen. Agron. J. 47:513-518. A14 (6944)

Juvancz, I. [1959], On recording the results of acute experiments. Acta Medica Academiae Scientiarum Hungaricae 13:167-177. A6 (6945)

Kabe, D. G. [1963], Estimation of a set of fixed variates for observed values of dependent variates with normal multivariate regression models subjected to linear restrictions. Ann. Inst. Stat. Math. 15:51-59. A7 (6946)

Kabe, D. G. [1966], On the exact distribution of the Fisher-Behren's-Welch statistic. Metrika 10:13-15. A11 (6947)

Kabe, D. G. [1966], Some results for the normal multivariate regression models. Australian J. Stat. 8:22-27. A7 (6948)

Kábrt, B. [1964], Ovplyvnenie pokusných výsledkov na malých parcelách ozimnej pšenice okrajovým efektom. (The experimental results from small plots of winter wheat influenced by the border effect.) Pol'nohospodárstvo 10:249-256. A14 (6949)

Kakwani, N. C. and Gupta, D. B. [1967], Note on the bias of the Prais and Aitchison's and Fisher's iterative estimators in regression analysis with heteroscedastic errors. Rev. ISI 35:291-295. A7,11 (6950)

Kalamkar, R. J. et al. [1956], The role of statistician in agricultural and animal husbandry research. JISAS 8:56-65. C (6951)

Kaltofen, H. [1952], Einige Bemerkungen zum Feldversuchswesen, insbesondere zur Prüfung von Verrechnungsmethoden mit Hilfe von Modellversuchen. Z. Acker-Pflanzenbau 94:345-352. C (6952)

Kaltofen, H. [1957], Über Feldversuche einfachster Struktur. Z. Acker- Planzenbau 103:315-332. C (6953)

Kaltofen, H. [1958], Über die Fehlerschätzung bei Feldversuchen einfachster Struktur. Z. Acker- Pflanzenbau 105:145-168. A14;C (6954)

Kamat, A. R. [1956], A two-sample distribution-free test. Biometrika 43:377-387. A12 (6955)

Kanda, M. [1951], On the area, shape and number of replications required for plots in field experiments. I. On the case provided by experimental trials of rice varieties for yielding capacity (in Japanese). Nihon Sakumotsugaku Kai Kiji (Proc. Crop. Sci. Soc. Japan) 20:161-162. A14;S1 (6956)

Kanda, M. and Kakizaki, Y. [1957], Studies on the spacing density of rice plants. Part 1. Density effects on yield and intraspecific competition. Sci. Reports Res. Inst., Tôhoku Univ. 8(D):107-126. A14 (6957)

Kanda, M. and Kakizaki, Y. [1959], Studies on spacing density of rice plant. 2. Interrelationships between spacing density and mode of hill arrangement. Sci. Reports Res. Inst., Tôhoku Univ. 10(D):35-59. A14 (6958)

Kapur, M. N. [1957], Some investigations on the problem of ranking of varieties (abstract). Stat. News Letter 7:16. A2 (6959)

Karas, J. and Savage, I. R. [1967], Publications of Frank Wilcoxon (1892-1965). Biometrics 23:1-10. B (6960)

Karn, M. N. [1964], Some further statistical information derived from Dr. Brand's data. Ann. Human Genetics 27:238-239. C (6961)

Karon, B. P. [1964], A note on the treatment of age as a variable in regression equations. Amer. Stat. 18(3):27-28. A7 (6962)

Karst, O. J. [1958], Linear curve fitting using least deviations. JASA 53:118-132. A7 (6963)

Kastenbaum, M. A. [1959], A confidence interval on the abscissa of the point of intersection of two fitted linear regressions. Biometrics 15:323-324. A7 (6964)

Katti, S. K. [1965], Multiple covariate analysis. Biometrics 21:957-974. A7 (6965)

Keeping, E. S. [1951], A significance test for exponential regression. AMS 22: 180-198. A7 (6966)

Keller, E. R. [1951], Beiträge zur Technik des landwirtschaftlichen Versuchswesens. II. Über die Anlage von Feldversuchen. Schweiz. Landw. Mh. 29:209-220. C (6967)

Keller, K. R. and Li, J. C. R. [1951], Further information on the relationship between the number of vines per hill and yield in hops (Humulus lupulus L.). Agron. J. 43:243-245. A14 (6968)

Kemp, C. D. [1960], The need for a dynamic approach to grassland experimentation. Proc. Eighth Int. Grassland Congress, pp. 728-731. C (6969)

Kemp, C. D. [1962], Some problems arising in animal grazing experiments (abstract). Biometrics 18:267. C (6970)

Kemp, C. D. [1964], Some aspects of the interpretation of pasture experiments (abstract). Biometrics 20:387-388. C (6971)

Kemp, K. W. [1958], Formulae for calculating the operating characteristic and the average sample number of some sequential tests. JRSSB 20:379-386. S2 (6972)

Kempthorne, O. [1955], The randomization theory of experimental inference. JASA 50:946-967 (abstract 48:634). A5 (6973)

Kempthorne, O. [1957], The contributions of statistics to agronomy. Advances in Agronomy 9:177-204. C (6974)

Kempthorne, O. [1957], Query 126. Arrangements of pots in greenhouse experiments. Biometrics 13:235-237. C (6975)

Kempthorne, O. [1959], Experiments versus surveys (summary). JASA 54:498. C (6976)

Kempthorne, O. [1963], Can multivariate data be analyzed by univariate methods? (summary) JASA 58:555. A7 (6977)

Kempthorne, O. [1966], Multivariate responses in comparative experiments. Multivariate Analysis, pp. 521-540. A7 (6978)

Kempthorne, O. [1966], Some aspects of experimental inference. JASA 61:11-34 (corrigendum 62:1520). A11;C (6979)

Kempthorne, O., Zyskind, G., Bosson, R. P., Martin, F. B., Doerfler, T. E., and Carney, E. J. [1966], Research on the analysis of variance and data interpretation. Stat. Lab. Iowa State Univ. and Aerospace Research Lab., ARL 66-0240 U.S. Air Force. A5,7,12,14 (6980)

Kendall, M. G. [1951], Regression, structure and functional relationship. Part I. Biometrika 38:11-25. A7 (6981)

Kendall, M. G. [1952], Regression, structure and functional relationship. Part II. Biometrika 39:96-108. A7 (6982)

Kendall, M. G. [1959], Hiawatha designs an experiment. Amer. Stat. 13(5):23-24. C (6983)

Kendall, M. G. [1961], Statistics, past, present and future. Stat. Neerlandica 15:333-349. C (6984)

Kendall, M. G. [1962], Ranks and measures. Biometrika 49:133-137. A9 (6985)

Kendall, M. G. [1963], Ronald Aylmer Fisher, 1890-1962. Biometrika 50:1-15. B (6986)

Kendall, M. G. and Doig, A. G. [1964], Bibliography of Statistical Literature, 1950-1958. Oliver and Boyd Ltd. Edinburgh and London, pp. xii+297. B (6987)

Kendall, M. G. and Doig, A. G. [1965], Bibliography of Statistical Literature, 1940-1949. Oliver and Boyd Ltd., Edinburgh and London, pp. 190. B (6988)

Kendall, M. G. and Lawley, D. N. [1956], The principles of factor analysis. JRSSA 119:83-84. A7 (6989)

Kendall, M. G. and Smith, B.B. [1950], Factor analysis (with discussion). JRSSB 12:60-94. A7 (6990)

Kendall, M. G. and Sundrum, R. M. [1953], Distribution-free methods and order properties. Rev. ISI 21:124-134. A12 (6991)

Kennard, R. W. [1963], Statistics and technology. IQC 19(12):21-23. C (6992)

Kerrich, J. E. [1955], Confidence intervals associated with a straight line fitted by least squares. Stat. Neerlandica 9:125-129. A7 (6993)

Kerrich, J. E. [1966], Fitting the line y=ax when errors of observations are present in both variables. Amer. Stat. 20(1):24. A7 (6994)

Keuls, M. [1954], Testing differences between means in an analysis of variance (abstract). Biometrics 10:167-168. A3 (6995)

Keuls, M. [1962], Weten in aansluiting op een variantie-analyse. Stat. Neerlandica 16:373-387. A3 (6996)

Keuls, M. [1963], Uitbijternegerende schatters bij duplobepalingen. (Outlier-ignoring estimators for measurement in duplo.) Stat. Neerlandica 17:299-317. A6 (6997)

Keuls, M. and Verdooren, L. R. [1964], Overzicht van de publikaties door prof. dr. M. J. van Uven. Stat. Neerlandica 18:19-25. B (6998)

Khanna, K. L. and Bandyopadhyay, K. S. [1950], Studies in sampling technique. Part III. Estimation of mite-incidence in sugarcane. Indian Academy Sci. Proc. Section B 31:111-119. A14 (6999)

Khanna, K. L., Nigam, L. N., and Bandyopadhyay, K. S. [1950], Studies in sampling technique. Part II. Estimation of Pyrilla incidence in sugarcane. Indian Academy Sci. Proc. Section B 31:34-44. A14 (7000)

Khatri, C. G. [1961], A note on the interval estimation related to the regression matrix. Ann. Inst. Stat. Math. 13:145-146. A7 (7001)

Khatri, C. G. [1962], Simultaneous confidence bounds on the departures from a particular kind of multicollinearity. Ann. Inst. Stat. Math. 13:239-242. A7 (7002)

Khatri, C. G. [1966], A note on a MANOVA model applied to problems in growth curve. Ann. Inst. Stat. Math. 18:75-86. A7 (7003)

Kidwell, J. F. and Chase, H. B. [1967], Fitting the allometric equation - A comparison of ten methods by computer simulation. Growth 31:165-179. A7 (7004)

Kiefer, J. and Sacks, J. [1962], Asymptotically optimum sequential procedures. II. Designs (abstract). AMS 33:1494. S2 (7005)

Kiefer, J. and Sacks, J. [1963], Asymptotically optimum sequential inference and design. AMS 34:705-750. A2,9;S2 (7006)

Kimball, A. W. [1950], Sequential sampling plans for use in psychological test work. Psychometrika 15:1-15. S2 (7007)

Kimball, A. W. [1951], On dependent tests of significance in the analysis of variance. AMS 22:600-604. A3 (7008)

Kimball, A. W. [1957], Errors of the third kind in statistical consulting. JASA 52: 133-142. C (7009)

Kimball, A. W. [1959], Errors of the third kind in statistical consulting. Proc. Fourth Conf. Design Expt. Army Res. Dev. Testing. pp. 1-10. C (7010)

Kimball, B. F. [1953], A multiple group least squares' problem and the significance of the associated orthogonal polynomials. JASA 48:320-335. A7 (7011)

Kimura, M. and Crow, J. F. [1962], The measurement of effective population number. Math. Res. Center Tech. Summary Report No, 335, U.S. Army and Univ. Wisconsin. S1 (7012)

King, E. P. [1964], Optimal replication in sequential drug screening. Biometrika 51:1-10. A2;S2 (7013)

King, H. E. [1957], A note on the design and analysis of replicated progeny row tests. West African Cotton Res. Conf., Regional Res. Sta., Ministry Agri., Samaru, Northern Nigeria, 18 to 23 Nov., 1957, pp. 105-106. A14 (7014)

King, J. R. [1959], Summary of extreme-value theory and its relation to reliability analysis. ASQC Nat. Conv. Trans. 13:163-177. A6 (7015)

King, J. R. [1966], Data analysis with probability papers. ASQC Ann. Tech. Conf. Trans. 20:224-234. A7 (7016)

King, S. C. and Bray, D. F. [1959], Competition between strains of chickens in separate versus intermingled flocks. Poultry Sci. 38:86-94. A14 (7017)

Kinsinger, F. E., Eckert, R. E., and Currie, P. O. [1960], A comparison of the line-interception, variable-plot and loop methods as used to measure shrub crown cover. J. Range Management 13:17-21. A14 (7018)

Kira, T., Ogawa, H., and Hozumi, K. [1954], Intraspecific competition among higher plants. II. Further discussions of Mitscherlich's law. J. Inst. Polytechnics Osaka City Univ., D,5:1-7. A14 (7019)

Kira, T., Ogawa, H., Hozumi, K., Koyama, H., and Yoda, K. [1956], Intraspecific competition among higher plants. V. Supplementary notes on the C-D effect. J. Inst. Polytechnics Osaka City Univ., D,7:11-14. A14 (7020)

Kira, T., Ogawa, H., and Sakazaki, N. [1953], Intraspecific competition among higher plants. I. Competition-yield-density interrelationship in regularly dispersed populations. J. Inst. Polytechnics Osaka Univ., D,4:1-16. A4 (7021)

Kirkpatrick, M. E., Lamb, J. C., Dawson, E. H., and Eisen, J. N. [1957], Selection of a taste panel for evaluating the quality of processed milk. Food Tech, 11:3-8. C (7022)

Kirkton, H. C. [1966], Non-additivity in the analysis of variance. M. S. Thesis, Univ. Aberdeen. A4 (7023)

Kishen, K. [1953], Some unsolved problems in experimental designs (abstract). Biometrics 9:428. C (7024)

Kitagawa, T. [1961], The logical aspects of successive processes of statistical inferences and controls. Bull. ISI 38(4):151-164. C (7025)

Kitagawa, T. [1967], Information science and its connection with statistics. Proc. Fifth Berkeley Symp. Math. Stat. Prob. 1:491-530. B;C (7026)

Kitagawa, T., Kitahara, T., Nomachi, Y. and Watanabe, N. [1953], On the determination of sample size from the two sample theoretical formulation. Bull. Math. Stat. 5(3&4):35-46. S1 (7027)

Kloek, T. and Bannink, R. [1962], Principal-component analysis applied to business test data. Stat. Neerlandica 16:57-69. A7 (7028)

Kloek, T., and Jochems, D. B. [1962], Alternative specifications of bivariate relations in the analysis of business test data. Stat. Neerlandica 16:71-87. A7 (7029)

Klotz, J. H. [1964], On the normal scores two-sample rank test. JASA 59:652-664. A12 (7030)

Knake, E. L. and Slife, F. W. [1962], Competition of <u>Setaria faberii</u> with corn and soybeans. Weeds 10:26-29. A14 (7031)

Knight, R. [1960], The growth of cocksfoot (<u>Dactylis glomerata</u> L.) under spaced plant and sward conditions. Australian J. Agri. Res. 11:457-472. A14 (7032)

Knight, W. [1963], The use of the range in place of the standard deviation in Stein's test. AMS 34:346-347. S2 (7033)

Knight, W. [1965], A method of sequential estimation applicable to the hypergeometric, binomial, Poisson, and exponential distributions. AMS 36:1494-1503. S2 (7034)

Kniss, J. R. and Wenger, W. S. [1965], Factors affecting sensitivity experiments. Proc. Tenth Conf. Design Expt. Army Res. Dev. Testing. pp. 595-611. A12;S2 (7035)

Knowles, E. A. G. [1954], Experimental designs in industry (with particular reference to production investigations) (abstract). Biometrics 10:186. C (7036)

Koch, E. J. [1953], Plot technique in small fruits. Proc. American Soc. Hort. Sci. 62:14-20. A14 (7037)

Koch, E. J. [1960], Presentation of experimental results (abstract). Biometrics 16:134. C (7038)

Koch, E. J. and Hunter, J. H. [1957], The use of experimental design and covariance techniques for increasing precision of pecan experiments. Proc. American Soc. Hort. Sci. 69:170-175. A7;C (7039)

Koch, E. J. and Rigney, J. A. [1951], A method of estimating optimum plot size from experimental data. Agron. J. 43:17-21. A14 (7040)

Kohnstamm, G. A. [1957], Statistiek en Marktanalyse. (Statistics and market research.) Stat. Neerlandica 11:55-62. C (7041)

Kokan, A. R. and Khan, S. [1967], Optimum allocation in multivariate surveys: An analytical solution. JRSSB 29:115-125. S1 (7042)

Koller, S. [1963], Einfuehrung in die Methoden der aetiologischen Forschung-Statistik und Dokumentation. Methods Information Medicine 2:1-13. C (7043)

Koller, S. [1964], Systematik der statistischen Schlufzfehler. Methods Information Medicine 3:113-117. C (7044)

Konijn, H. S. [1955], On some tests for treatment effects in paired replicates (abstract). Bull. American Math. Soc. 61:321. A12 (7045)

Konijn, H. S. [1957], Some non-parametric tests for treatment effects in paired replications. JISAS 9:145-167. A12 (7046)

Konijn, H. S. [1959], Basing decisions on an analysis of variance. Australian J. Stat. 1:57-68. A3 (7047)

Konijn, H. S. [1959], Models used in statistical studies of relations between uncontrolled variables. Australian J. Stat. 1:82-93. A7 (7048)

Konijn, H. S. [1960], Multiple comparison with controls. Australian J. Stat. 2: 16-18. A3 (7049)

Konijn, H. S. [1960], Some remarks on regression analysis with uncontrolled regressors. Australian J. Stat. 2:47-52. A7 (7050)

Konijn, H. S. [1961], Non-parametric, robust and short-cut methods in regression and structural analysis. Australian J. Stat. 3:77-86. A7,12 (7051)

Konüs, A. A. [1966], The single regression line. Metron 25:108-115. A7 (7052)

Konstantinov, P. N. and Plotnikov, N. I. [1959], The question of method in field experiments (in Russian). Vestnik sel'skohozjajstv. Nauk. (Rep. Agri. Sci.) 6:35-42. C (7053)

Konstantinov, P. N. and Plotnikov, N. I. [1960], Cu privire la metodica experientelor de cimp. (On method in field experiments.) An. Romîno - Soviet: Ser. Agri. Zooteh. 14(1):3-11. C (7054)

Kossack, C. F. [1959], The AASHO road test as an example of large scale tests. Proc. Fourth Conf. Design Expt. Army Res. Dev. Testing. pp. 11-29. C (7055)

Koyama, H. and Kira, T. [1956], Intraspecific competition among higher plants. VIII. Frequency distribution of individual plant weight as affected by the interaction between plants. J. Inst. Polytechnics Osaka City Univ. D,7:73-94. A14 (7056)

Kramer, A. [1952], New tri-metric test predicts canned corn quality. Food Eng. 24(4):86-88, 139, 141. C (7057)

Kramer, A. [1956], A quick, rank test for significance of differences in multiple comparisons. Food Tech. 10:391-392. A3,12 (7058)

Kramer, A. [1957], Inspection frequencies and sample numbers for raw materials procured for food processing. Food Tech. 11:176-179. S1 (7059)

Kramer, A. and Ditman, L. P. [1956], A simplified variables taste panel method for detecting flavor changes in vegetables treated with pesticides. Food Tech. 10: 155-159. C (7060)

Kramer, A., Kornetsky, A., Elehwany, N., Steinmetz, G., and Morin, E. L. [1956], A procedure for sampling and grading raw green asparagus. Food Tech. 10:212-214. A14 (7061)

Kramer, A. and Ogle, W. L. [1953], Sampling procedure for grading tomatoes. Food Tech. 7:353-355. A14 (7062)

Kramer, C. Y. [1955], A method of choosing judges for a sensory experiment. Food Res. 20:492-496. C (7063)

Kramer, C. Y. [1955], On the analysis of variance of a two-way classification with unequal sub-class numbers. Biometrics 11:441-452. A10 (7064)

Kramer, C. Y. [1956], Extension of multiple range tests to group means with unequal numbers of replications. Biometrics 12:307-310. A3 (7065)

Kramer, C. Y. [1956], Additional tables for a method of choosing judges for a sensory experiment. Food Res. 21:598-600. C (7066)

Kramer, C. Y. [1957], Simplified computations for multiple regression. IQC 13 (8): 1-4. A7 (7067)

Kramer, C. Y. [1957], Extension of multiple range tests to group correlated adjusted means. Biometrics 13:13-18. A3 (7068)

Kramer, C. Y. [1966], Simple regression in multiple regression notation. Amer. Stat. 20(3):25. A7 (7069)

Kramer, C. Y. and Duncan, D. B. [1953], On the analysis of variance of a two-way classification with unequal subclass numbers (abstract). Biometrics 9:264. A10 (7070)

Kramer, C. Y. and Duncan, D. B. [1953], On the analysis of variance of a multiway classification with unequal sub-class numbers (abstract). Virginia J. Sci. 4:286. A10 (7071)

Kramer, M. and Greenhouse, S. W. [1956], Determination of sample size and selection of cases. Psychopharmacology: Problems in Evaluation. S1 (7072)

Kraus, B. S. and Choi, S. C. [1958], A factorial analysis of the prenatal growth of the human skeleton. Growth 22:231-242. A7 (7073)

Krause, G. F. [1964], A stochastic representation of growth curves (abstract). Biometrics 20:911. A7 (7074)

Krishna Iyer, P. V. [1950], Further contributions to the theory of probability distributions of points on a line - I. JISAS 2:141-160. A6 (7075)

Krishna Iyer, P. V. [1951], A non-parametric method of testing k samples. Nature 167:33. A12 (7076)

Krishnaiah, P. R. [1962], Remarks on multiple comparison (summary). JASA 57:493. A3 (7077)

Krishnaiah, P. R. [1962], Multiple comparison tests in multi-response experiments (abstract). AMS 33:1495-1496. A3 (7078)

Krishnaiah, P. R. [1964], Multiple comparison tests in multivariate case. Aerospace Res. Lab. ARL 64-124, U.S. Air Force. A3 (7079)

Krishnaiah, P. R. [1964], Multiple comparison tests in multivariate case (abstract). Technometrics 6:120. A3,7 (7080)

Krishnaiah, P. R. [1965], Multiple comparison tests in multi-response experiments. Sankhyā A 27:65-72. A3,7 (7081)

Krishnaiah, P. R. [1965], Simultaneous tests for the equality of variances against certain alternatives. Australian J. Stat. 7:105-109. A11 (7082)

Krishnaiah, P. R. [1965], On the simultaneous ANOVA and MANOVA tests. Ann. Inst. Stat. Math. 17:35-53. A3,7 (7083)

Krishnaiah, P. R. [1965], On a multivariate generalization of the simultaneous analysis of variance test. Ann. Inst. Stat. Math. 17:167-173. A3,7 (7084)

Krishnaiah, P. R. and Armitage, J. V. [1966], Tables for multivariate t distribution. Sankhyā B 28:31-56. A3,7 (7085)

Krumbein, W. C. and Tukey, J. W. [1956], Multivariate analysis of mineralogic, lithologic, and chemical composition of rock bodies. J. Sedimentary Petrology 26:322-337. A7 (7086)

Kruskal, W. H. [1957], Historical notes on the Wilcoxon unpaired two-sample test. JASA 52:356-360. A12 (7087)

Kruskal, W. H. [1960], Some remarks on wild observations. Technometrics 2:1-3. A6 (7088)

Kruskal, W. [1961], The coordinate-free approach to Gauss-Markov estimation, and its application to missing and extra observations. Proc. Fourth Berkeley Symp. Math. Stat. Prob. 1:435-451. A6,7,14 (7089)

Kruskal, W. H., Ferguson, T. S., Tukey, J. W., and Gumbel, E. J. [1960], Discussion of the papers of Messrs. Anscombe and Daniel. Technometrics 2:157-166. A6 (7090)

Kruskal, W. H. and Wallis, W. A. [1952], Use of ranks in one-criterion variance analysis. JASA 47:583-621 (errata 48:907-911). A12;B (7091)

Kudô, A. [1956], On the testing of outlying observations. Sankhyā 17:67-76. A6 (7092)

Kudô, A. [1960], The symmetric multiple decision problems. Memoirs Faculty Sci. Kyushu Univ., Ser. A. 14:179-206. A2,3 (7093)

Kudô, A. [1961], Some problems of symmetric multiple decisions in multivariate analysis. Bull. ISI 38(4):165-171. A6,7 (7094)

Kudô, A. [1963], A multivariate analogue of the one-sided test. Biometrika 50: 403-418. A2,7 (7095)

Kudô, A. and Fujisawa, H. [1966], Some multivariate tests with restricted alternative hypotheses. Multivariate Analysis, pp. 73-85. A7 (7096)

Kuiper, N. H. [1954], Een opmerking over het aanpassen van functies aan een groot aantal waarnemingsuitkomsten. (Note on the fitting of a function to a large number of observations.) Stat. Neerlandica 8:1-6. A7 (7097)

Kulash, W. M. and Monroe, R. J. [1954], Laboratory tests for control of wireworms. J. Economic Entomology 47:341-345. C (7098)

Kulash, W. M. and Monroe, R. J. [1955], Field tests for control of wireworms attacking corn. J. Economic Entomology 48:11-19. C (7099)

Kullback, S. and Rosenblatt, H. M. [1957], On the analysis of multiple regression in k categories. Biometrika 44:67-83. A7 (7100)

Küppers-Sonnenberg, G. A. [1958], Vergleichbarkeit von Topinamburerträgen in der Leistungsprüfung. Saatgutwirtschaft 10:135-136. C (7101)

Kurlander, A. B., Hill, E. H., and Enterline, P. E. [1955], An evaluation of some commonly used screening tests for heart disease and hypertension. J. Chronic Diseases 2:427-439. A2;C (7102)

Kurlander, A. B., Iskrant, A. P., and Kent, M. E. [1954], Screening test for diabetes - A study of specificity and sensitivity. Diabetes 3:213-219. A2 (7103)

Kurtz, T. E. [1956], An extension of a method of making multiple comparisons (preliminary report) (abstract). AMS 27:547. A3 (7104)

Kurtz, T., Melsted, S. W., and Bray, R. H. [1952], The importance of nitrogen and water in reducing competition between intercrops and corn. Agron. J. 44:13-17. A14 (7105)

Kurup, R. S. [1965], Certain distribution-free tests of regression. JISAS 17:104-110. A7,12 (7106) la Bastide, J. G. A. [1967], A computer program for the layouts of seed orchards. Euphytica 16:321-323. A15 (7107)

Lachenbruch, P. A. and Mickey, M. R. [1967], Estimation of error rates in discriminant analysis (abstract). Proc. Twelfth Conf. Design. Expt. Army Res. Dev. Testing. pp. 49. A7 (7108)

Lana, E. P., Homeyer, P. G., and Haber, E. S. [1953], Field plot technique in vegetable crops. Proc. American Soc. Hort. Sci. 62:21-30. A14 (7109)

Lancaster, H. O. [1950], Statistical control in haematology. J. Hygiene 48:402-417. C (7110)

Lancaster, H. O. [1964], Bibliography of vital statistics in Australia and New Zealand. Australian J. Statistics 6:33-99. B (7111)

Landenna, G. [1963], Considerazioni sull'impostazione bayesiana dell'analisi sequenziale. Colloques Internationaux Centre Nat. Recherche Sci. No. 110, Le Plan d'Expériences pp. 269-274. S2 (7112)

Landi, R. [1957], Prove spaziali e prove pluriannuali. (Tests in space and tests over a number of years.) Genetica Agraria, Roma 7:39-55. C (7113)

Lang, A. L., Pendleton, J. W., and Dungan, G. H. [1956], Influence of population and nitrogen levels on yield and protein and oil contents of nine corn hybrids. Agron. J. 48:284-289. A14 (7114)

Lang, R. W. and Holmes, J. C. [1965], The effect of plant population and distribution on the yield and quality of swedes. J. Agri. Sci. 65:91-99. A14 (7115)

Lange, C. J. and Palmer, F. H. [1958], Experimental design for field studies in leadership. Proc. Third Conf. Design Expt. Army Res. Dev. Testing. pp. 173-178. C (7116)

Lange, J. and Schumacher, K. [1962], Die Versuchsplanung bei der therapeutischen Forschung im Bereich der Leukämien, Methods Information Medicine 1:95-99. C (7117)

Langlie, H. J. [1963], A reliability test method for "one-shot" items. Proc. Eighth Conf. Design Expt. Army Res. Dev. Testing. pp. 145-165. C (7118)

Langner, W. [1953], Die Klonanordnung im Samenplantagen. Z. Forstgentik Forstpflanzenzüchtung 2:119-121. C (7119)

Lark, P. D. [1954], Application of statistical analysis to analytical data. Analytical Chemistry 26:1712-1715. A6;C (7120)

Larson, E. W. and Foster, W. D. [1959], An appraisal of sequential analysis under conditions restricted by the requirement for advanced scheduling and programming. Proc. Fourth Conf. Design Expt. Army Res. Dev. Testing. pp. 173-176. C (7121)

Larson, H. J. and Bancroft, T. A. [1963], Biases in prediction by regression for certain incompletely specified models. Biometrika 50:391-402. A7 (7122)

Lasagna, L. [1955], The controlled clinical trial: theory and practice. J. Chronic Diseases 1:353-367. C (7123)

Lasagna, L. [1962], Controlled trials: nuisance or necessity? Methods Information Medicine 1:79-82. C (7124)

Laubscher, N. F. [1960], On the stabilization of the Poisson variance. Trabajos Estad. 11:199-207. A9 (7125)

Laue, E. A., Ishler, N. H., and Bullman, G. A. [1954], Reliability of taste testing and consumer testing methods. I. Fatigue in taste testing. Food Tech. 8:387-388. C (7126)

Laue, R. V. [1962], A multivariate approach to screening test data (summary). JASA 57:493. A6,7 (7127)

Laurent, A, G. [1964], Bombing problems, radial error and outliers (circular model) (abstract). Technometrics 6:121. A6 (7128)

Lawrence, F. N. [1957], Statistical approach to the evaluation of electric initiators. Proc. First Conf. Design Expt. Army Res. Dev. Testing. pp. 175-199. C (7129)

Lawrence, W. J. C. [1955], Techniques for experiments with pot plants. Plant Soil 6:332-346. C (7130)

Laycock, D. H. [1955], The effect of plot shape in reducing the errors of tea experiments. Tropical Agri. 32:107-114. A14 (7131)

Leckie, D. S. [1955], Applications of regression analysis to steel plant problems. ASQC Nat. Conv. Trans. 9:419-432. A7 (7132)

Leech, F. B. and Healy, M. J. [1959], The analysis of experiments on growth rate. Biometrics 15:98-106. A7 (7133)

Lehmann, E. L. [1951], Consistency and unbiasedness of certain nonparametric tests. AMS 22:165-179. A12 (7134)

Lehmann, E. L. [1957], A theory of some multiple decision problems. II. AMS 28: 547-572. A2,3 (7135)

Lehmann, E. L. [1963], Robust estimation in analysis of variance. AMS 34:957-966. A10,12 (7136)

Lehmann, E. L. [1963], A class of selection procedures based on ranks. Mathematische Annalen 150:268-275. A2,12 (7137)

Lehmann, E. L. [1964], Asymptotically nonparametric inference in some linear models with one observation per cell. AMS 35:726-734. A12 (7138)

Lein, A. [1950], "Schauversuche" mit exakter Anlage und Auswertung. Neue Mitt. Landw. 5:221-222. C (7139)

Lein, A. [1951], Bemerkungen zu neueren Arbeiten über Fragen des Feldversuches. Z. Pflanzenzuchtung 30:89-111. C (7140)

Lein, A. [1954], Application of Fisher's methods to the design and performance of agricultural experiments (abstract). Biometrics 10:309-310. C (7141)

Lellouch, J. [1965], Quelques aspects du problème des comparisons multiples (abstract). Biometrics 21:264. A3 (7142)

Leonard, W. H. [1954], Experimental designs in agronomy. Statistics and Mathematics in Biology pp. 133-141. C (7143)

Leone, F. C. [1961], Statistical programs for high speed computers. Technometrics 3:301-304. A15 (7144)

Leone, F. C., Nottingham, R. B., and Zucker, J. [1957], Significance tests and the dollar $ign. IQC 13(12):5-21. C (7145)

Leone, R. C. [1958], Significance test III - Variances. ASQC Nat. Conv. Trans. 12:503-515. A11 (7146)

Le Roy, H. L. [1956], Zwei bekannte statistische Prüfverfahren im Lichte neuester Erkenntnisse. Schweizerische Landwirtschaftliche Monatschefte 34:285-297. A3 (7147)

Le Roy, H. L. [1959], Grundlage und Anwendungsmöglichkeiten de Methode des Pfadkoeffizienten. Biom. Zeit. 1:30-43. A7 (7148)

Leslie, R. T. and Brown, B. M. [1966], Use of range in testing heterogeneity of variance. Biometrika 53:221-225. A11 (7149)

Lessells, W. J. and Webber, J. [1959-1962], The effect of nitrogen on spring cereals. Exptl. Husbandry 12:62-73. A13 (7150)

Lessells, W. J. and Webber, J. [1957-1963], The effect of nitrogen on winter wheat. Exptl. Husbandry 12:74-88. A13 (7151)

Lessman, K. J. and Atkins, R. E. [1963], Comparisons of planting arrangements and estimates of optimum hill plot size for grain sorghum yield tests. Crop Sci. 3: 489-492. A14 (7152)

Leti, G. [1961], Le distribuzioni degli estremi e del campo di variazione dei campioni di una popolazione discreta e finita. Metron 21:201-255. A6 (7153)

Levene, H. [1960], Robust tests for equality of variances. Contributions to Probability and Statistics pp. 278-292. A11 (7154)

Levert, C. [1951], Betrekkingen tussen de regressiecoëfficiënten van Galton en Frisch bij lineaire transformatie. (Relations between the regression coefficients of Galton and Frisch in case of linear transformation.) Statistica 5:33-45. A7 (7155)

Li, C. C. [1964], Two additional views of linear regression coefficients. Amer. Stat. 18(4):27-28. A7 (7156)

Li, J. C. R. and Keller, K. R. [1951], An application of serial correlation in field experiments. Agron. J. 43:201-203. A14 (7157)

Li, L. [1957], Studies on the technique of sweet-potato field trials (in Chinese). Chung-hua Nung-hsueh Hui Pao/J. Agri. Assoc. China 19:1-10. A14 (7158)

Li, L. [1964], Studies on the technique of experimentation with sweet potato varieties (in Chinese). Chung-hua Nung-hsueh Hui Pao/J. Agri. Assoc. China 45:11-17. A13 (7159)

Liang, G. H. L. and Walter, T. L. [1966], Genotype × environmental interactions from yield tests and their application to sorghum breeding programs. Canadian J. Genetics Cytology 8:306-311. A13 (7160)

Lider, L. A. [1955], A group of long-term, perennial and non-replicated root-stock trials (abstract). Biometrics 11:514-515. A13 (7161)

Lieberman, G. J. and Miller, R. G. [1963], Simultaneous tolerance intervals in regression. Biometrika 50:155-168. A3,7 (7162)

Lieberman, G. J., Miller, R. G., and Hamilton, M. A. [1967], Unlimited simultaneous discrimination intervals in regression. Biometrika 54:133-145. A7 (7163)

Lieblein, J. [1952], Properties of certain statistics involving the closest pair in a sample of three observations. J. Res. NBS 48:255-268. A6 (7164)

Lieblein, J. [1962], The closest two out of three observations. Contributions to Order Statistics, chapter 7B, pp. 129-135. A6 (7165)

Lieblein, J. [1962], Extreme-value distribution. Contributions to Order Statistics chapter 12C, pp. 397-406. A6 (7166)

Liebmann, A. J. and Panettiere, B. R. [1957], Quality control and consumer testing for distilled alcoholic beverages. Wallerstein Laboratories Comm. 20:27-39. C (7167)

Lienert, G. A. [1956], Die statistische Analyse medizinisch-klinischer Laboratoriumsuntersuchungen. Arztliche Forschung 10:398-404. C (7168)

Lienert, G. A. [1959], Prinzip und Methode der multiplen Faktorenanalyse, demonstriert an einem Beispiel. Biom. Zeit. 1:88-141. A7 (7169)

Lienert, G. A. [1962], Über die Anwendung von Variablen-Transformationen in der Psychologie. Biom. Zeit. 4:145-181. A9 (7170)

Linder, A. [1955], On a particular kind of grazing experiment (abstract). Biometrics 11:555. C (7171)

Linder, A., Chakravarti, I. M., and Vuagnat, P. [1963], Fitting asymptotic regression curves with different asymptotes. Contributions to Statistics pp. 221-228. A7 (7172)

Lindley, D. V. and Barnett, B. N. [1965], Sequential sampling: two decision problems with linear losses for binomial and normal random variables. Biometrika 52:507-532 S2 (7173)

Linhart, H. [1959], Techniques for discriminant analysis with discrete variables. Metrika 2:138-149. A7 (7174)

Linhart, H. [1960], Approximate tests for m rankings. Biometrika 47:476-480. A12 (7175)

Linnik, Y. V. [1963], On the Behrens-Fisher problem. Bull. ISI 40:833-841. A11 (7176)

Linnik, Y. V. [1963], Remarks on the Behrens-Fisher problem. Sankhyā A 25:377-380. A11 (7177)

Linnik, Y. V. [1966], Latest investigation on Behrens-Fisher problem. Sankhyā A 28:15-24. A11 (7178)

Lisman, J. H. C. [1954], Statistische analyse van muziek. Stat. Neerlandica 8:83-91. C (7179)

Litchfield, J. T. [1960], Sequential analysis, screening and serendipity. J. Medicinal Pharmaceutical Chemistry 2:469-492. A2;S2 (7180)

Litchfield, J. T. [1961], Forecasting drug effects in man from studies in laboratory animals. J. American Medical Assoc. 177:34-38. C (7181)

Loatman, P. J. [1958], Some problems encountered in the evaluation of erosion in cannon bores. Proc. Third Conf. Design Expt. Army. Res. Dev. Testing. pp. 289-309. C (7182)

Locascio, S. J., Martin, F. G., and Lundy, H. W. [1966], Plot size studies with watermelons. Proc. American Soc. Hort. Sci. 89:597-600. A14 (7183)

Lockhart, E. E. [1951], Binomial systems and organoleptic analysis. Food Tech. 5:428-431. C (7184)

Lockart, L. W. [1954], Sampling of fleeces for yield, staple length, and crimps per inch measurement. Australian J. Agri. Res. 5:555-567. S1 (7185)

Lockwood, V. M. [1958], Evaluation of the screening procedures. Part I - Engineering evaluation. ASQC Nat. Conv. Trans. 12:421-430. A2 (7186)

Lombardi, G. [1951], Sequential methods in subjective tests (abstract). Virginia J. Sci. 2:377. C (7187)

Loosli, J. K. [1954], Determination of the nutritional requirements of dairy animals and swine. <u>Statistics and Mathematics in Biology</u> pp. 393-398. C (7188)

Loraine, P. K. [1952], On a useful set of orthogonal comparisons. JRSSB 14:234-237. A7 (7189)

Love, H. T., Cowgill, W. H., and Hopp, H. [1952], Test of a proposed method of sampling cinchona bark for fluorometric analysis for quinine. Proc. American Soc. Hort. Sci. 60:279-282. S1 (7190)

Lower, R. L. and Thompson, A. E. [1966], Sampling variation of acidity and solids in tomatoes. Proc. American Soc. Hort. Sci. 89:512-522. S1 (7191)

Lubin, A. [1962], Statistics. Annual Rev. Psychology 13:345-370. B (7192)

Lucas, H. L. [1952], Magnitudes of experimental errors in grazing trials (abstract). J. Animal Sci. 11:784. C (7193)

Lucas, H. L. [1952], Methods of computing results of grazing trials (abstract). J. Animal Sci. 11:784. C (7194)

Lucas, H. L. [1960], Critical features of good dairy feeding experiments. J. Dairy Sci. 43:193-212. C (7195)

Lucas, H. L. [1960], Theory and mathematics in grassland problems. Proc. Eighth Int. Grassland Congress pp. 732-736. C (7196)

Lüdecke, H. and Müller, A.v. [1964], Anlage, Untersuchung und Auswertung von Zuckerrübenfeldversuchen. Arb. dtsch. Landw. -Gesell. C (7197)

Ludwig, O. [1959], Ungleichungen für extremwerte und andere Ranggrössen in Anwendun auf biometrische Probleme. Biom. Zeit. 1:203-209. A6 (7198)

Ludwig, O. [1960], An application of the theory of extreme values in experimental medicine (abstract). Biometrics 16:312. A6 (7199)

Ludwig, W. [1959], Über die Anfänge der Statistik und Biometrik. Biom. Zeit. 1:71-80. C (7200)

Lugg, J. W. H. and Whyte, J. M. [1955], Taste thresholds for phenylthiocarbamide of some population groups. I. The thresholds of some civilized ethnic groups living in Malaya. Ann. Human Genetics 19:290-311. C (7201)

Lum, M. D. [1963], On characterizing nonadditivity (abstract). AMS 34:357. A4 (7202)

Lunden, A. P. [1961], A present-day problem in breeding and testing new varieties. Proc. First Triennial Conf. European Assoc. Potato Res., Braunschweig-Völkenrode, Germany, 12-17 Sept., 1960, pp. 257-259. C (7203)

Lush, J. L. [1955], Statistics in investigations in animal production. JISAS 7:7-22. C (7204)

Lynd, J. Q., Graybill, F., and Totusek, R. [1956], Factors affecting results of grazing trials with yearling steers. Agron. J. 48:352-355. C (7205)

Lynd, J. Q., Graybill, F., and Totusek, R. [1957], Grazing trial evaluations using paired pastures with yearling steers. Agron. J. 49:488-492. C (7206)

Lyne, A. G. and Verhagen, A. M. W. [1957], Growth of the marsupial Trichosurus vulpecula and a comparison with some higher mammals. Growth 21:167-195. A7 (7207)

MacGillivray, J. H. and Zahara, M. [1960], The most economical length of picking row for hand harvest. Proc. American Soc. Hort. Sci. 76:710-716. A14 (7208)

Mackey, A. O. and Jones, P. [1954], Selection of members of a food tasting panel: Discernment of primary tastes in water solution compared with judging ability for foods. Food Tech. 8:527-530. C (7209)

MacKinnon, W. J. [1964], Table for both the sign test and distribution-free confidence intervals of the median for sample sizes to 1,000. JASA 59:935-956. A12 (7210)

Macoy, D. S. [1959], A five variable multiple correlation study on UNIVAC I. ASQC Nat. Conv. Trans. 13:653-656. A7,15 (7211)

MacQueen, J. [1967], Some methods for classification and analysis of multivariate observations. Proc. Fifth Berkeley Symp. Math. Stat. Prob. 1:281-297. A7 (7212)

Madansky, A. [1959], The fitting of straight lines when both variables are subject to error. JASA 54:173-205. A7 (7213)

Madansky, A. [1964], Instrumental variables in factor analysis. Psychometrika 29:105-113. A7 (7214)

Magwire, C. A. [1953], Sequential decisions involving the choice of experiments. Ph. D. Thesis, Stanford Univ. S2 (7215)

Mahalanobis, P. C. [1950], Why statistics? Sankhyā 10:195-228. C (7216)

Mahalanobis, P. C. and Sengupta, J. M. [1951], On the size of sample cuts in crop-cutting experiments in the Indian Statistical Institute: 1939-1950. Bull. ISI 33(2):359-404. A14 (7217)

Mahamunulu, D. M. [1964], Note on ranking with three populations (abstract). AMS 35:1851. A2 (7218)

Mahamunulu, D. M. [1965], A class of ranking and selection procedures (preliminary report) (abstract). AMS 36:728. A2 (7219)

Mahamunulu, D. M. [1967], On a generalized goal in fixed-sample ranking and selection problems. Ph. D. Thesis, Univ. Minnesota. A2 (7220)

Mahamunulu, D. M. [1967], Some fixed-sample ranking and selection problems. AMS 38:1079-1091. A2 (7221)

Mahoney, C. H., Stier, H. L., and Crosby, E. A. [1957], Evaluating flavor differences in canned foods. I. Genesis of the simplified procedure for making flavor difference tests, II. Fundamentals of the simplified procedure. Food Tech. 11:29-43. C (7222)

Mainland, D. [1950], Statistics in clinical research: some general principles. Ann. New York Academy Sci. 52:922-930. C (7223)

Mainland, D. [1960], The use and misuse of statistics in medical publications. Clinical Pharmacology Therapeutics 1:411-422. C (7224)

Mainland, D. [1960], The clinical trial - some difficulties and suggestions. J. Chronic Diseases 11:484-496. C (7225)

Mainland, D. [1961], Experiences in the development of multiclinic trials. J. New Drugs 1:197-205. C (7226)

Maisel, H. [1966], Best k of 2k-1 comparisons. JASA 61:329-344. A2 (7227)

Mallios, W. S. [1962], On linear regression systems. Proc. Seventh Conf. Design Expt. Army Res. Dev. Testing. pp. 629-653. A7 (7228)

Mallios, W. S. [1967], A structural regression approach to covariance analysis when the covariable is uncontrolled. JASA 62:1037-1049. A7 (7229)

Mallows, C. L. [1953], Sequential discrimination. Sankhyā 12:321-338. A7;S2 (7230)

Mallows, C. L. [1957], Non-null ranking models. I. Biometrika 44:114-130. A12 (7231)

Malm, O. J., Levenson, S. M., and Horowitz, R. E. [1960], Design of experiments using germfree animals. Proc. Fifth Conf. Design Expt. Army Res. Dev. Testing. pp. 133-149. C (7232)

Maloney, C. J. [1957], Punched card computing of analyses of variance. Proc. First Conf. Design Expt. Army Res. Dev. Testing. pp. 97-127. A15 (7233)

Maloney, C. J. [1962], Statistics in army research development and testing. Amer. Stat. 16(3):13-17. C (7234)

Malý, V. [1960], Sequenzprobleme mit mehreren Entscheidungen und Sequenzschätzung. I. Allgemeine Theorie und Anwendung auf die Binomialverteilung. Biom. Zeit. 2: 45-64. S2 (7235)

Malý, V. [1963], Sequenzprobleme mit mehreren entscheidungen und sequenzschätzung. VI. Sequenzprüfverfahren einer einfachen Hypothese gegen zusammengesetzte konstante oder einfache veränderliche Alternative. Biom. Zeit. 5:24-31. S2 (7236)

Malý, V. [1966], Zur Frage des Sequenzvergleichs von zwei relativen Häufigkeiten beim Paarversuch. Biom. Zeit. 8:162-178. S2 (7237)

Mandel, J. [1954], Statistical principles of testing (summary). JASA 49:340-341. C (7238)

Mandel, J. [1955], Statistical principles of testing. IQC 12 (5):18-21. C (7239)

Mandel, J. [1957], Fitting a straight line to certain types of cumulative data. JASA 52:552-566. A7 (7240)

Mandel, J. [1958], A note on confidence intervals in regression problems. AMS 29:903-907. A3 (7241)

Mandel, J. [1959], The measuring process. Technometrics 1:251-267. A13 (7242)

Mandel, J. [1961], Non-additivity in two-way analysis of variance. JASA 56:878-888. A4 (7243)

Mandel, J. [1964], Estimation of weighting factors in linear regression and analysis of variance. Technometrics 6:1-25. A7 (7244)

Mandel, J. and Linnig, F. J. [1956], Statistical methods in chemistry. Analytical Chemistry 28:770-777. B;C (7245)

Mandelson, J. [1957], The relation between the engineer and the statistician. IQC 13 (11):31-34. C (7246)

Mann, H. H. and Barnes, T. W. [1950], The competition between barley and certain weeds under controlled conditions. IV. Competition with *Stellaria media*. Ann. Appl. Biology 37:139-148. A14 (7247)

Mantel, N. and Haenszel, W. [1959], Statistical aspects of the analysis of data from retrospective studies of disease. J. Nat. Cancer Inst. 22:719-748. C (7248)

Manwani, A. H. [1960], Economic amount of experimentation for deciding between two alternatives (abstract). JISAS 12:153. S1 (7249)

Manwani, A. H. [1960], Further investigations on economic amount of experimentation (abstract). JISAS 12:214-215. A13 (7250)

Marani, A. [1963], Estimation of optimum plot size using Smith's procedure. Agron. J. 55:503. A14 (7251)

Marani, A. [1964], Some variety × environmental interactions in oriental tobacco and their implications on variety testing methods. Israel J. Agri. Res. 14:117-120. A13 (7252)

Mardia, K. V. [1967], A non-parametric test for the bivariate two-sample location problem. JRSSB 29:320-342. A7,12 (7253)

Maritz, J. S. [1953], Estimation of the correlation coefficient in the case of a bivariate normal population when one of the variables is dichotomized. Psychometrika 18:97-110. A7 (7254)

Maritz, J. S. [1962], Confidence regions for regression parameters. Australian J. Stat. 4:4-10. A7 (7255)

Marks, M. R. [1951], Two kinds of experiment distinguished in terms of statistical operations. Psychological Rev. 58:179-184. C (7256)

Marmorston, J., Weiner, J. M., and Hopkins, C. E. [1965], Multivariate analysis: A tool in medical research (abstract). Biometrics 21:771-772. A7 (7257)

Marquardt, D. W. [1963], An algorithm for least-squares estimation of nonlinear parameters. J. Soc. Ind. Appl. Math. 11:431-441. A7 (7258)

Marriott, F. H. C. [1952], Tests of significance in canonical analysis. Biometrika 39:58-64. A7 (7259)

Marshall, C. E. [1958], Greenhouse and laboratory experiments (summary). JASA 53:584. C (7260)

Martin, L. [1952], Statistical methods in radiochemistry. Analyst 77:892-896. C (7261)

Martin, L. [1953], Applications à la Biologie des méthodes biométriques et statistiques Ann. Soc. Royale Sci. Medicales Naturelles Bruxelles 6:133-149. A7;C (7262)

Martin, L. [1955], Justesse et precision des resultats experimentaux obtenus sur la betterave sucriere. Etude des facteurs pouvant influencer ces caracteres (with discussion). Sucrerie Belge 74:477-489. C (7263)

Martin, L. [1958], Analyse et ajustement des courbes de fréquences unimodales par la méthode des probits. Bull. ISI 36(3):43-59. A7 (7264)

Martin, L. [1960], Adjustment d'un faisceau de régressions curvilignes au moyen d'un systeme de polynomes orthogonaux. Biométrie-Praximétrie 1(1):35-52. A7 (7265)

Martin, L. [1960], Homométrie, allométrie et cograduation en biométrie générale. Biom. Zeit. 2:73-97. C (7266)

Martin, L. [1960], Développement des méthodes biométriques et statistiques dans la recherche agronomique en Belgique. Bull. Inst. Agronomique Stations Recherches Gembloux (Hors Série)1:294-303. C (7267)

Martin, L. [1962], Transformations of variables in clinical-therapeutical research. Methods Information Medicine 1:38-50. A6;B (7268)

Martin, L. and Lenger, A. [1954], Analyse biométrique des courbes d'exportation d'éléments minéraux du sol par les végétaux. Bull. Inst. Agronomique Stations Recherches Gembloux 22:272-286. A7 (7269)

Martini, P. [1962], Grundsätzliches zur therapeutisch-klinischen Versuchsplanung. Methods Information Medicine 1:1-5. C (7270)

Marutiram, B. [1967], Analysis of variance in two-way classification with disproportionate cell frequencies (abstract). JISAS 19(1):153-154. A10 (7271)

Mason, D. D. [1958], Field experiments (summary). JASA 53:585. C (7272)

Mason, D. D. and Koch, E. J. [1953], Some problems in the design and statistical analysis of taste tests. Biometrics 9:39-46. C (7273)

Massey, F. J. [1951], A note on a two sample test. AMS 22:304-306. A12 (7274)

Massey, F. J. [1953], On the analysis of data matched in pairs (abstract). AMS 23:475. A12 (7275)

Masuyama, M. [1953], Rapid methods of estimating the sum of specified areas in a field of given size. Reports Stat. Appl. Res. 2:113-119. A14 (7276)

Masuyama, M. [1953], A note to "Rapid methods of estimating the sum of specified areas in a field of given size". Reports Stat. Appl. Res. 3:54. A14 (7277)

Masuyama, M. [1954], On the problem of screening defective articles during production. Sankhyā 14:67-70. A2 (7278)

Masuyama, M. [1954], On the error in crop cutting experiment due to the bias on the border of grid. Sankhyā 14:181-186. A14 (7279)

Masuyama, M. and Hatamura, M. [1956], Recent advances in biometry in Japan. Biometrics 12:449-461. B;C (7280)

Masuyama, M. and Sengupta, J. M. [1955], On a bias in a crop-cutting experiment (application of integral geometry to areal sampling problems--Part V). Sankhyā 15:373-376. A14 (7281)

Matches, A. G. [1966], Sample size for mower-strip sampling of pastures. Agron. J. 58:213-215. A14;S1 (7282)

Mathai, A. M. [1965], Some useful results in analysis of data (abstract). AMS 36:365. A10 (7283)

Matusita, K. [1954], On the estimation by the minimum distance method. Ann. Inst. Stat. Math. 5:59-65. A7 (7284)

Matusita, K. [1954], A remark to "On the estimation by the minimum distance method." Ann. Inst. Stat. Math. 6:124. A7 (7285)

Matusita, K. [1967], Classification based on distance in multivariate Gaussian cases. Proc. Fifth Berkeley Symp. Math. Stat. Prob. 1:299-304. A7 (7286)

Matusita, K. et al. [1955], Some problems of sampling in the forest survey. Ann. Inst. Stat. Math. 7:1-23. A14 (7287)

Mauldin, W. P. and Ross, J. A. [1967], Demographic experiments: A review of designs (summary). JASA 62:737. C (7288)

Maurice, R. J. [1957], A minimax procedure for choosing between two populations using sequential sampling. JRSSB 19:255-261. A2;S1,2 (7289)

Maurice, R. [1959], A different loss function for the choice between two populations. JRSSB 21:203-213. A2 (7290)

Maxwell, A. E. [1959], Statistical methods in factor analysis. Psychological Bull. 56:228-235. A7 (7291)

Maxwell, A. E. [1961], Recent trends in factor analysis. JRSSA 124:49-59. A7 (7292)

May, J. and Lubin, A. [1951], Eliminating superfluous variables in an analysis of variance with disproportionate frequencies (abstract). Biometrics 7:118. A10 (7293)

May, J. M. [1952], Extended and corrected tables of the upper percentage points of the 'studentized' range. Biometrika 39:192-193. (Correction 40:236) A3 (7294)

McCall, C. H. [1960], Linear contrasts, part I. IQC 17(1):19-21. A3 (7295)

McCall, C. H. [1960], Linear contrasts, part II. IQC 17(2):12-16. A3 (7296)

McCall, C. H. [1960], Linear contrasts, part III. IQC 17(3):5-8. A3 (7297)

McCornack, R. L. [1965], Extended tables of the Wilcoxon matched pair signed rank statistic. JASA 60:864-871. A12 (7298)

McCullough, R. S. [1961], Testing equality of means under variance heterogeneity. Ph. D. Thesis, Iowa State Univ. A11 (7299)

McCullough, R. S., Gurland, J., and Rosenberg, L. [1960], Small sample behaviour of certain tests of the hypothesis of equal means under variance heterogeneity. Biometrika 47:345-353 (corrigendum 48:230). A11 (7300)

McDonald, B. J. and Thompson, W. A. [1965], Non-parametric multiple comparison techniques (abstract). Technometrics 7:274. A3,12 (7301)

McDonald, B. J. and Thompson, W. A. [1967], Rank sum multiple comparisons in one- and two-way classifications. Biometrika 54:487-497. A3,12 (7302)

McFerran, J. [1956], Plot technique studies with spinach. Ph. D. Thesis, Cornell Univ. A14 (7303)

McGilchrist, C. A. [1965], Analysis of competition experiments. Biometrics 21: 975-985. A14 (7304)

McGilchrist, C. A. [1967], Plant competition experiments. Ph. D. Thesis, Univ. New South Wales. A14 (7305)

McHugh, R. B. [1953], The comparison of two correlated sample variances. American J. Psychology 66:314-315. A11 (7306)

McHugh, R. B. [1961], Confidence interval inference and sample size determination. Amer. Stat. 15(2):14-17. S1 (7307)

McHugh, R. B. and Ellis, D. S. [1955], The "post-mortem" testing of experimental comparisons. Psychological Bull. 52:425-428. A3 (7308)

McKenzie, R. M. [1955], Sampling variations in the concentrations of elements in soils. Australian J. Agri. Res. 6:699-706. S1 (7309)

McKibben, E. G. and Berry, M. O. [1952], The value of replications in research. Agri. Eng. 33:792-798. S1 (7310)

McLaughlin, D. H. and Tukey, J. W. [1961], The variance of means of symmetrically trimmed samples from normal populations, and its estimation from such trimmed samples (trimming/Winsorization I). Stat. Tech. Res. Group Tech. Report No.42, Princeton Univ. A6 (7311)

McNemar, Q. [1957], On Wilson's distribution-free test of analysis of variance hypotheses. Psychological Bull. 54:361-362. A12 (7312)

Mead, R. [1967], A mathematical model for the estimation of inter-plant competition. Biometrics 23:189-205. A14 (7313)

Meeter, D. A. [1966], On a theorem used in nonlinear least squares. J. SIAM Appl. Math. 14:1176-1179. A7 (7314)

Meijs, L. T. J. [1955], Het oordeel van een bedrijfsleider over de toepassing van de statistiek in het bedrijf. (A manager's view on the application of statistics in the concern.) Stat. Neerlandica 9:79-83. C (7315)

Memoria, J. M. P. [1952], Comparação de groupos "versus" comparações emparelhadas (Group comparisons "versus" paired comparisons.) Ceres, Minas Gerais 9:105-116. A3 (7316)

Mendenhall, W. [1958], A bibliography on life testing and related topics. Biometrika 45:521-543. B (7317)

Merrill, J. A. and Gaylor, D. W. [1967], Augmenting existing data in multiple regression (abstract). Technometrics 9:191. A7 (7318)

Merritt, E. S., Aitken, J. R., and Stewart, I. J. [1957], Experimental errors in egg production experiments. Canadian J. Animal Sci. 37:143-151. C (7319)

Metakides, T. A. [1953], Calculation and testing of discriminant functions. Trabajos Estad. 4:339-368. A7 (7320)

Méwissen, D. J. and Betz, E. H. [1961], Sequential tests and protective effectiveness cystamine and chlorpromazine (with discussion). <u>Quantitative Methods in Pharmacology</u> pp. 56-69. S2 (7321)

Meyer, J. H., Lofgreen, G. P., and Garrett, W. N. [1960], A proposed method for removing sources of error in beef cattle feeding experiments. J. Animal Sci. 19:1123-1131. C (7322)

Michaels, S. E. [1964], The usefulness of experimental designs (with discussion). Appl. Stat. 13:221-235. C (7323)

Michaels, S. E. [1967], Multi-response multiple regression (abstract). Technometrics 9:191. A7 (7324)

Michelsen, P. F. [1957], Operational experiments. Proc. First Conf. Design Expt. Army Res. Dev. Testing. pp. 35-46. C (7325)

Michelson, L. F., Lachman, W. H., and Allen, D. D. [1958], The use of the "weighted-rankit" method in variety trials. Proc. American Soc. Hort. Sci. 71:334-338. A9 (7326)

Middleton, G. K., Hebert, T. T., and Murphy, C. F. [1964], Effect of seeding rate and row width on yield and on components of yield in winter barley. Agron. J. 56:307-308. A14 (7327)

Mielke, P. W. and McHugh, R. B. [1965], Non-orthogonality in the two-way classification for the mixed effects finite population model. Biometrics 21: 308-323. A10 (7328)

Miller, J. D. and Koch, E. J. [1962], A plot technique study with birdsfoot trefoil. Agron. J. 54:95-97. A14 (7329)

Miller, J. D. and Koch, E. J. [1966], Further studies on plot techniques with birdsfoot trefoil. Agron. J. 58:458-459. A14 (7330)

Miller, J. G. and Mountier, N. S. [1955], The border effect in wheat trials with different spacings between plots. New Zealand J. Sci. Tech. 37 (A):287-299. A14 (7331)

Miller, R. G. [1964], A trustworthy jackknife. AMS 35:1594-1605. C (7332)

Mills, W. T. and Mason, D. D. [1958], Statistics help evaluate functional components of farm machinery. Agri. Eng. 39 (1):31-33. A14;C (7333)

Milthorpe, F. L. [1961], Mechanisms in Biological Competition. Symposia Soc. Expt. Biology pp vi+365. A14 (7334)

Milton, R. C. [1964], An extended table of critical values for the Mann-Whitney (Wilcoxon) two-sample statistic. JASA 59:925-934. A12 (7335)

Minina, I. P. [1960], Spacing as a factor determining crop yields and interspecific competition in grass mixtures. Proc. Eight Int. Grassland Congress pp. 307-309. A14 (7336)

Mishra, M. [1966], Analysis of transformation in the "analysis of variance" (abstract). JISAS 18 (2):110. A9 (7337)

Mishra, M. [1967], Analysis of transformation in the "analysis of variance". JISA 5:57-62. A9 (7338)

Mitchell, J. W. [1956], Duration of sensitivity in trio taste testing. Food Tech. 10:201-203. C (7339)

Mitchell, J. W. [1957], Problems in taste difference testing. I. Test environment. II, Subject variability due to time of the day and day of the week. Food Tech. 11:476-479. C (7340)

Mitchell, J. W. [1957], The design of experiment in stability testing. Proc. First Conf. Design Expt. Army Res. Dev. Testing. pp. 157-165. C (7341)

Mitchell, J. W. [1958], Observation on the use of models in the design of experiment. Proc. Second Conf. Design Expt. Army Res. Dev. Testing. pp. 209-211. C (7342)

Mitchell, J. W. [1958], Design of an experiment in the reliability analysis of a complex component. Proc. Third Conf. Design Expt. Army Res. Dev. Testing. pp. 323-327. C (7343)

Mitscherlich, E. A. [1955], Grundlegendes zur Ausschaltung der Ungleichartigkeit des Bodens bei Feldversuchen. Z. Pflanzenzüchtung 34:209-212. A14 (7344)

Mittmann, O. M. J. [1960], Eine methode zur Auffindung ursächlicher Zusammenhänge mittels partieller Korrelationskoeffizienten. Metron 20:3-44. A7 (7345)

Moder, J. J. [1962], A sequential search procedure for locating a response jump. Technometrics 4:610-614. S2 (7346)

Monroe, R. J. and Mason, D. D. [1955], Problems of experimental inference with special reference to multiple location experiments and experiments with perennial crops. Proc. American Soc. Hort. Sci. 66:410-414. C (7347)

Monzón, D. [1956], Analisis e interpretacion de un ensayo de uniformidad con maíz. (Analysis and interpretation of a uniformity trial with maize.) Agronomía Tropical, Venezuela 6:15-22. A14 (7348)

Monzón Paiva, D. E. and Viso Rodriguez, A. [1958], Determinacion del tamaño optimo de unidad experimental mediante la ley de la Varianza de H. Fairfield Smith. (Determination of optimum plot size of an experimental unit by means of H. Fairfield Smith's variance law.). Agronomía Tropical, Venezuela 8:43-49. A14 (7349)

Mood, A. M. [1951], Erratum: "On the determination of sample sizes in designing experiments, JASA 43:391-402, 1948." JASA 46:515. S1 (7350)

Moonan, W. J. [1957], Linear transformation to a set of stochastically dependent normal variables. JASA 52:247-252. A9 (7351)

Moore, J. F. [1952], A study of field plot technique with sprouting broccoli. Proc. American Soc. Hor. Sci. 59:471-474. A14 (7352)

Moore, J. F. and Darroch, J. G. [1956], Field plot technique with Blue Lake pole beans, bush beans, carrots, sweet corn, spring and fall cauliflower. Washington Agri. Expt. Sta. Tech. Bull. 21. A14;S1 (7353)

Moore, P. G. [1957], Transformations to normality using fractional powers of the variable. JASA 52:237-246. A9 (7354)

Moore, P. G. [1962], Some applications of statistical methods to the paper industry. Bull. ISI 39(3):381-392. C (7355)

Moore, P. G. [1962], Regression as an analytical tool. Appl. Stat. 11:106-119. A7 (7356)

Moore, P. G. [1963], A statistical approach to the allocation of technical effort in some industrial situations (with discussion). JRSSA 126:493-536. C (7357)

Moore, P. G. and Tukey, J.W. [1954], Query 112. Biometrics 10:562-568. A4,9 (7358)

Moran, P. A. P. [1950], Recent developments in ranking theory (with discussion on pages 182-191). JRSSB 12:153-162. A12 (7359)

Moran, P. A. P. [1950], A test for the serial independence of residuals. Biometrika 37:178-181. A6 (7360)

Moran, P. A. P. [1951], Partial and multiple rank correlation. Biometrika 38:26-32. A7 (7361)

Morillon, Y. [1960], Théorie de la randomisation. Rev. Stat. Appl. 8(1):29-44. C (7362)

Moroney, M. J. [1959], Efficiency by statistics. Stat. Neerlandica 13:281-294. C (7363)

Morris, M. J. [1967], An abstract bibliography of statistical methods in grassland research. U.S. Dept. Agri. Miscellaneous Pub. No. 1030, pp. 4+222. B (7364)

Morse, P. M. and Bickle, A. [1967], The combination of estimates from similar experiments, allowing for inter-experiment variation. JASA 62:241-250. A13 (7365)

Moser, H. A., Dutton, H. J., Evans, C. D., and Cowan, J. C. [1950], Conducting a taste panel for the evaluation of edible oils. Food Tech. 4:105-109. C (7366)

Moses, L. E. [1952], Non-parametric statistics for psychological research. Psychological Bull. 49:122-143. A12 (7367)

Moses, L. E. [1953], General review of non-parametric methods with special emphasis on randomization tests (summary). JASA 48:631-632. A5,12 (7368)

Moses, L. E. [1963], Rank tests of dispersion. AMS 34:973-983. A12 (7369)

Moses, L. E. [1965], Query 10: Confidence limits from rank tests. Technometrics 7:257-260. A12 (7370)

Moshman, J. [1952], Testing a straggler mean in a two-way classification using the range. AMS 23:126-132. A6 (7371)

Moshman, J. [1958], A method for selecting the size of the initial sample in Stein's two sample procedure. AMS 29:1271-1275. S1,2 (7372)

Mosteller, F. [1952], Clinical studies of analgesic drugs II. Some statistical problems in measuring the subjective response to drugs. Biometrics 8:220-231. C (7373)

Mosteller, F. [1952], The world series competition. JASA 47:355-380. A2 (7374)

Mosteller, F. [1961], Optimal length of play for a binomial game. Math. Teacher 54:411-412. S1 (7375)

Mosteller, F. and Youtz, C. [1961], Tables of the Freeman-Tukey transformations for the binomial and Poisson distributions. Biometrika 48:433-440. A9 (7376)

Mott, G. O. [1955], The grazing trial for measuring the output of pasture (abstract). Biometrics 11:544. C (7377)

Mott, G. O. and Lucas, H. L. [1953], The design, conduct, and interpretation of grazing trials on cultivated and improved pastures. Proc. Sixth Int. Grassland Congress. C (7378)

Mountier, N. S. [1964], Plot size and guard rows in potato experiments. New Zealand J. Agri. Res. 7:180-197. A14 (7379)

Mowchan, W. L. [1966], Monte Carlo investigation of the probability distributions of Dixon's criteria for testing outlying observations. Proc. Eleventh Conf. Design Expt. Army Res. Dev. Testing. pp. 367-414. A6 (7380)

Mukerji, A. K. [1952], Size and shape of plots in wheat trials. Proc. Bihar Academy Agri. Sci. 1(1):19-24. A14 (7381)

Mukerji, B. *et al*. [1957], Symposium on "The role of statistics in biological assays". JISAS 9:61-66. C (7382)

Müller, A. [1959], The statistical evaluation of a long term series of sugar beet trials with the object of reducing trial area. Inst. Inter. Recherches Betteravières 22nd Winter Congress, Brussels, pp. 427-430. A13;S1 (7383)

Müller, A. and Castronovo, A. [1959], Tipo de parcela adecuado para ensayos de campo con nabo "de Vertus". (Plot type suitable for field trials of the Vertus turnip.) Rev. Investigaciones Agrícolas, Buenos Aires 13(1):47-55. A14 (7384)

Müller, K.-H. [1956], On the accounting for soil variability in field trial analysis of variance (abstract). Biometrics 12:228. C (7385)

Müller, K.-H. [1961], Zur Ausschaltung fraglicher Beobachtungswerte bei Feldversuchen. Z. Landw. Vers. Untersuch Wes. A6 (7386)

Mullis, C. W. [1960], The design and redesign of an experiment. Proc. Fifth Conf. Design Expt. Army Res. Dev. Testing. pp. 291-302. C (7387)

Mumaw, C. R. and Weber, C. R. [1957], Competition and natural selection in soybean varietal composites. Agron. J. 49:154-160. A14 (7388)

Munk, J. F. and Wilk, M. B. [1966], Detecting outliers in a two-way table (abstract). AMS 37:766. A6 (7389)

Munro, S. [1954], Some aspects of sequential experimentation (summary). JASA 49:360. S2 (7390)

Murphy, B. P. [1967], Some two-sample tests when the variances are unequal: a simulation study. Biometrika 54:679-683. A11 (7391)

Murphy, R. B. [1951], On tests for outlying observations. Ph. D. Thesis, Princeton Univ. A6 (7392)

Murphy, W. J. [1956], Math and statistics and us. Ind. Eng. Chemistry 48(3):69A. C (7393)

Murray, D. B. [1950], A uniformity trial with swamp rice. Tropical Agri. 27: 105-107. A14 (7394)

Murray, W. K. [1958], Evaluation of interlaboratory tests with limited controls and data. Proc. Second Conf. Design Expt. Army Res. Dev. Testing. pp. 231-232. C (7395)

Murthy, M. N. [1963], A note on determination of sample size. Sankhyā A 25:381-382. S1 (7396)

Myers, L. F. and Lipsett, J. [1958], Competition between skeleton weed (Chondrilla juncea L.) and cereals in relation to nitrogen supply. Australian J. Agri. Res. 9:1-12. A14 (7397)

Myers, M. H., Axtell, L. A., and Zelen, M. [1966], The use of prognostic factors in predicting survival for breast cancer patients. J. Chronic Diseases 19:923-933. C (7398)

Myers, M. H., Schneiderman, M. A., and Armitage, P. [1966], Boundaries for closed (wedge) sequential t test plans. Biometrika 53:431-437. S2 (7399)

Myers, R. H. and Womeldorph, D. [1967], A method for adjusting for particle size and matrix effects in the X-Ray fluorescence analysis procedure. Proc. Twelfth Conf. Design Expt. Army Res. Dev. Testing. pp. 251-264. A7 (7400)

Nacke, O. and Wagner, G. [1964], Bibliographie zum Thema "Die Rolle des Fehlers in der Medizin; Fehlerforschung als Aufgabe der medizinischen Dokumentation". Methods Information Medicine 3:132-150. B;C (7401)

Nagar, A. L. and Kakwani, N. C. [1965], Note on the use of prior information in statistical estimation of economic relations. Sankhyā A 27:105-112. A7 (7402)

Nagel, K. [1966], On selecting a subset containing the best of several discrete populations (abstract). AMS 37:543. A2 (7403)

Nair, K. R. [1952], Tables of percentage points of the 'Studentized' extreme deviate from the sample mean. Biometrika 39:189-191. A3 (7404)

Nair, K. R. [1953], Some unsolved problems in experimental designs (abstract). Biometrics 9:428. C (7405)

Nair, K. R. [1954], The fitting of growth curves. <u>Statistics and Mathematics in Biology</u> pp. 119-132. A7 (7406)

Nair, K. R. [1954], Presidential address: Place of statistics in scientific research. Calcutta Stat. Assoc. Bull. 5:101-109. C (7407)

Nandi, H. K. [1950], Queries and answers. Calcutta Stat. Assoc. Bull. 3:37-40. A13 (7408)

Nandi, H. K. [1951], On analysis of variance test. Calcutta Stat. Assoc. Bull. 3:103-114. A11 (7409)

Nandi, H. K. [1953], Analysis of serial experiments. Calcutta Stat. Assoc. Bull. 5:43-46. A13 (7410)

Nandi, H. K. [1959], Query: Estimation of mean by an auxiliary variable. Calcutta Stat. Assoc. Bull. 9:71-73. A7 (7411)

Nandi, H. K. [1961], Some aspects of simultaneous confidence interval estimation. Calcutta Stat. Assoc. Bull. 10:131-138. A3 (7412)

Nandi, H. K. [1965], Samarendra Nath Roy: 1906-1964. Calcutta Stat. Assoc. Bull. 14:1-8. B (7413)

Nandi, H. K. [1965], Manindra Naṭh Ghosh: 1918-1965. Calcutta Stat. Assoc. Bull. 14:89-92. B (7414)

Natrella, M. G. [1960], The relation between confidence intervals and tests of significance. Amer. Stat. 14(1):20-22, 38. C;S1 (7415)

Naylor, T. H., Burdick, D. S., and Sasser, W. E. [1967], Computer simulation experiments with economic systems: The problem of experimental design. JASA 62:1315-1337. C (7416)

Neave, H. R. [1966], A development of Tukey's quick test of location. JASA 61: 949-964. A12 (7417)

Nelder, J. A. [1961], Models and experiments for growth analysis (abstract). Biometrics 17:332. A7 (7418)

Nelder, J. A. [1962], An alternative form of a generalized logistic equation. Biometrics 18:614-616. A7 (7419)

Nelder, J. A. [1963], Models for the relation between crop yield and the spatial arrangement of the plants (abstract). Biometrics 19:198. C (7420)

Nelder, J. A. [1966], Approximations to the critical values for Duncan's multiple range test. Biometrics 22:179-182. A3 (7421)

Nelder, J. A. and Moss, N. [1956], The spacing of lettuce in heated glasshouses. J. Hort. Sci. 31:177-187. A14 (7422)

Nelemans, F. A., Zelvelder, W. G., and Leppink, G. J. [1961], Klinische evaluatie van pijnstillende middelen. (Clinical evaluation of the measurement of pain.) Stat. Neerlandica 15:431-438. C (7423)

Nelson, D. C. and Nylund, R. E. [1962], Competition between peas grown for processing and weeds, Weeds 10:224-229. A14 (7424)

Nelson, K. E., Baker, G. A., Winkler, A. J., Amerine, M. A., Richardson, H. B., and Jones, F. R. [1963], Chemical and sensory variability in table grapes. Hilgardia 34:1-42. C (7425)

Nemenyi, P. [1961], Some distribution-free multiple comparison procedures in the asymptotic case (abstract). AMS 32:921-922. A3,12 (7426)

Nemenyi, P. [1962], Distribution-free multiple comparisons (abstract). Biometrics 18:263-264. A3,12 (7427)

Nemenyi, P. [1965], Distribution-free slippage tests -- A correction (abstract). AMS 36:359. A3,12 (7428)

Newton, R. G. and Spurrell, D. J. [1967], A development of multiple regression for the analysis of routine data. Appl. Stat. 16:51-64. A7 (7429)

Newton, R. G. and Spurrell, D. J. [1967], Examples of the use of elements for clarifying regression analyses. Appl. Stat. 16:165-172. A7 (7430)

Neyman, J. [1956], Note on an article by Sir Ronald Fisher. JRSSB 18:288-294. C (7431)

Neyman, J. [1967], Experimentation with weather control (with discussion). JRSSA 130:285-326. C (7432)

Neyman, J. and Scott, E. L. [1951], On certain methods of estimating the linear structural relation. AMS 22:352-361. A7 (7433)

Neyman, J. and Scott, E. L. [1960], Unbiased estimation based on transformed variables, with particular reference to cloud seeding experiments. Proc. Fifth Conf. Design Expt. Army Res. Dev. Testing. pp. 353-372. A9 (7434)

Neyman, J. and Scott, E. L. [1960], Correction for bias introduced by a transformation of variables. AMS 31:643-655. A9 (7435)

Neyman, J. and Scott, E. L. [1961], Design of cloud seeding experiments. Bull. ISI 38(4):31-41. C (7436)

Nicholson, W. L. [1965], Background problems in the analysis of count data. ASQC Ann. Tech, Conf. Trans. pp. 552-558. A9 (7437)

Nieuwhof, M. [1959], Problems in the design of variety trials with spring cabbage. Euphytica 8:151-156. C (7438)

Njos, A. and Nissen, O. [1956], Standard errors of field experiments at the farm crop institute agricultural college of Norway. Agron. J. 48:416-418. A13 (7439)

Nogueira, I. R. [1950], Sôbre uma propriedade da equação utilizada para a interpolação da Lei de Mitscherlich. Anais Escola Superior Agri. "Luiz de Queiroz" 7:105-108. A7 (7440)

Nogueira, I. R. [1950], A técnica da resolução das equações relativas à interpolação da Lei de Mitscherlich pelo metodos dos quadrados minimos. Anais Escola Superior Agri. "Luiz de Queiroz" 7:109-113. A7 (7441)

Nohe, E. [1965], Möglichkeiten zur Vereinfachung bei Sortenversuchen. Saatgutwirtschaft 17:43-44. C (7442)

Nomachi, Y. [1967], A closed sequential procedure selecting the best population in a familiy of populations with one parameter exponential distributions. Bull. Math. Stat. 12(3&4):21-34. A2;S2 (7443)

Nonnecke, I. L [1959], The precision of field experiments with vegetable crops as influenced by plot and block size and shape. I. Sweet corn. Canadian J. Plant Sci. 39:443-457. A14 (7444)

Nordskog, A. W., David, H. T., and Eisenberg, H. B. [1961], Optimum sample size in animal disease control. Biometrics 17:617-625. S1 (7445)

Norris, F. G. [1955], Departures from randomness. ASQC Nat. Conv. Trans. 9:183-195. A14 (7446)

Nüesch, P. E. [1966], On the problem of testing location in multivariate populations for restricted alternatives. AMS 37:113-119. A2,7 (7447)

Oatman, H. T. [1959], General linear correlation with an IBM 604 electronic-calculating punch and plotting of the data with an electronic associates data plotter. ASQC Nat. Conv. Trans. 13:635-647. A7,15 (7448)

Oberhoffer, G. [1962], Die Versuchsplanung bei den bösartigen Tumoren. Methods Information Medicine 1:91-95. C (7449)

Oberhoffer, G. [1964], Vermeidung und Ausschaltung von Fehlern bei der Planung und Auswertung therapeutisher Prüfungen. Methods Information Medicine 3:5-10. C (7450)

O'Carroll, F. M. [1962], A computer programme for non-orthogonal block experiments. Irish J. Agri. Res. 1:196-199. A15 (7451)

Odell, R. T. [1950], A study of sampling methods used in determining the productivity of Illinois soils. Agron. J. 42:328-335. A14 (7452)

Oehmisch, W. [1959], Die Anwendung der regressionsrechnung auf die Untersuchung der Abhängigkeit der Weitsprungleistungen vol Alter und Körpermassen. Biom. Zeit. 1:15-29. A7 (7453)

Oehmisch, W. [1963], Eine Anwendung des Zeichen- und Wilsontestes auf Körperhöhenmessungen in der DDR. Biom. Zeit. 5:108-118. A12 (7454)

Ohata, H. [1962], Statistical analysis of the tests for ecological adaptability of bred lines of wheat and barley (in Japanese). Nogyo Gijutsu Kenkyusho Hokoku/ Bull. Nat. Inst. Agri. Sci. 9(A):69-151. A3,11,13 (7455)

Ohlrogge, A. J. [1954], Plot border effects in a liming experiment. Agron. J. 46:241-242. A14 (7456)

Olbrich, E. [1966], Algorithmische Notation für Varianzanalysen. Biom. Zeit. 8:32-41. A14 (7457)

Oliveira, A. J. [1957], Analysis of a group of variety experiments on oats. M. S. Thesis, Iowa State Univ. A13 (7458)

Olkin, I. [1953], Note on 'The Jacobians of certain matrix transformations useful in multivariate analysis.' Biometrika 40:43-46. A7 (7459)

Olkin, I. and Rubin, H. [1964], Multivariate Beta distributions and independence properties of the Wishart distribution. AMS 35(1):261-269. A7 (7460)

Ollagnier, M. [1951], Forme, dimension des parcelles et nombre de répétitions dans les essais culturaux sur arachide et sur palmier à huile. Oléagineux Rev. Gén. Corps Gras Dérivés 6:707-710. A14;S1 (7461)

Olmstead, P. S. [1960], Statistical evaluation. ASQC Nat. Conv. Trans. 14: 339-347. C (7462)

Olson, H. H. [1952], Uniformity and nutritional studies with monozygotic bulls (abstract). J. Dairy Sci. 35:489. A14 (7463)

Olson, H. H. and Petersen, W. E. [1951], Uniformity of semen production and behavior in monozygous triplet bulls (abstract). J. Dairy Sci. 34:489-490. A14 (7464)

Ostle, B. [1967], Industry use of statistical test design. IQC 24:24-34. B;C (7465)

Ostle, B. and Tischer, R. G. [1954], Statistical methods in food research. Advances Food Res. 5:161-259. C (7466)

Ough, C. S., Singleton, V. L., Amerine, M. A., and Baker, G. A. [1964], A compariso of normal and stressed-time conditions on scoring of quality and quantity attributes J. Food Sci. 29:506-519. C (7467)

Overall, J. E. and Dalal, S. N. [1965], Design of experiments to maximize power relative to cost. Psychological Bull. 64:339-350. S1 (7468)

Owens, A. H. [1962], Predicting anticancer drug effects in man from laboratory animal studies. J. Chronic Diseases 15:223-238. C (7469)

Pabst, W. R. [1957], Statistics in the physical sciences in the United States. Bull. ISI 35(2):211-222. C (7470)

Pachares, J. [1959], Table of the upper 10% points of the studentized range. Biometrika 46:461-466. A3 (7471)

Page, D. J. [1966], Some practical aspects of factory and field trials with fibreboard cases. Appl. Stat. 15:94-109. C (7472)

Palley, M. N. and Horwitz, L. G. [1961], Properties of some random and systematic point sampling estimators. Forest Sci. 7:52-65. A14 (7473)

Palley, M. N. and O'Regan, W. G. [1961], A computer technique for the study of forest sampling methods. I. Point sampling compared with line sampling. Forest Sci. 7:282-294. A14 (7474)

Pan, S.-C. [1964], Studies on the technique of field experimentation with summer soya beans in Taiwan (in Chinese). Chung-hau Nung-hsueh Hui Pao/J. Agri. Assoc. China 45:18-30. A14;S1 (7475)

Pangborn, R. M. and Chrisp, R. B. [1964], Taste interrelationships. VI. Sucrose, sodium chloride, and citric acid in canned tomato juice. J. Food Sci. 29:490-498. C (7476)

Pangborn, R. M. and Trabue, I. M. [1964], Taste interrelationships. V. Sucrose, sodium chloride, and citric acid in lima bean purée. J. Food Sci. 29:233-240. C (7477)

Panse, V. G. [1953], Long term experiments (abstract). Biometrics 9:429. C (7478)

Panse, V. G. [1954], Principles of the survey method of experimentation (abstract). Biometrics 10:174-176. C (7479)

Panse, V. G. [1963], Plot size again. JISAS 15:151-159. A14 (7480)

Panse, V. G. and Sukhatme, P. V. [1953], Experiments in cultivators' fields. JISAS 5:144-160. C (7481)

Park, T. [1954], Competition: An experimental and statistical study. <u>Statistics and Mathematics in Biology</u> pp. 175-195. A14 (7482)

Park, T. [1954], Experimental studies of interspecies competition. II. Temperature, humidity, and competition in two species of <u>Tribolium</u>. Physiological Zoology 27:177-238. A14 (7483)

Park, T. [1955], Ecological experimentation with animal populations. Scientific Monthly 81:271-275. C (7484)

Park, T. [1957], Experimental studies of interspecies competition. III. Relation of initial species proportion to competitive outcome in populations of Tribolium. Psychological Zoology 30(1):22-40. A14 (7485)

Park, T. and Lloyd, M. [1955], Natural selection and the outcome of competition. American Naturalist 89:235-240. A14 (7486)

Parkhurst, R. A. [1958], A wide band telemetering system. Proc. Second Conf. Design Expt. Army Res. Dev. Testing. pp. 173-189. C (7487)

Parks, A. B. [1953], Ranking versus scoring in palatability tests using small, trained panels (abstract). Biometrics 9:264. A12 (7488)

Parks, G. M. [1965], Extreme value statistics in time study. J. Ind. Eng. 16: 351-355. A6 (7489)

Parrini, P. and Toderi, G. [1964], Rilievi sperimentali sull' ampiezza parcellare in prove su colture diverse. (Results of experiments on plot size in trials with various crops.) Genetica Agraria, Roma 18:585-602. A14 (7490)

Patel, H. M. et al. [1955], Symposium on experiments in cultivators' fields. JISAS 7:98-110. C (7491)

Patel, J. K. [1967], Selecting a subset containing the best one of several IFRA populations (preliminary report) (abstract). AMS 38:957. A2 (7492)

Patel, K. M. [1966], A multivariate sign test (abstract). AMS 37:766. A7,12 (7493)

Patil, G. P. [1963], A selected bibliography of statistical literature on classical and contagious discrete distributions. Classical and Contagious Discrete Distributions pp. 469-552. B (7494)

Patil, G. P. [1965], On multivariate generalized power series distribution and its application to the multinomial and negative multinomial. Classical and Contagious Distributions pp. 183-194. A7 (7495)

Patil, G. P. and Joshi, S. W. [1966], Bibliography of classical and contagious discrete distributions. Pennsylvania State Univ., and Aerospace Res. Lab. ARL 66-0185, U.S. Air Force 270 pp. B (7496)

Patil, J. A. and Chavan, V. M. [1958], Influence of different spacings on yield and other agronomic characters of onion (Allium cepa L.). Indian J. Agron. 3:41-47. A14 (7497)

Patil, V. H. [1964], The Behrens-Fisher problem and its Bayesian solution. JISA 2:21-31. A11 (7498)

Patil, V. H. [1964], Difficulties involved in computing Behrens-Fisher densities, cumulative probabilities and percentage points from first principles. JISA 2:109-118. A11 (7499)

Patil, V. H. [1965], Approximation to the Behrens-Fisher distributions. Biometrika 52:267-271. A11 (7500)

Patterson, H. D. [1956], A simple method for fitting an asymptotic regression curve. Biometrics 12:323-329. A7 (7501)

Patterson, H. D. [1958], The use of autoregression in fitting an exponential curve. Biometrika 45:389-400. A7 (7502)

Patterson, H. D. [1960], A further note on a simple method for fitting an exponential curve. Biometrika 47:177-180. A7 (7503)

Patterson, H. D. and Lipton, S. [1959], An investigation of Hartley's method for fitting an exponential curve. Biometrika 46:281-292. A7 (7504)

Patterson, H. D. and Ross, G. J. S. [1963], The effect of block size on the errors of modern cereal experiments. J. Agri. Sci. 60:275-278. A13,14 (7505)

Patterson, R. E. [1950], A method of adjustment for calculating comparable yields in variety tests. Agron. J. 42:509-511. A13 (7506)

Paull, A. E. [1950], On a preliminary test for pooling mean squares in the analysis of variance. AMS 21:539-556 (abstract 22:134). A11 (7507)

Pauls, J. [1954], The effect of selection of the experimental units on experimental error. M. S. Thesis, Iowa State Univ. A14 (7508)

Paulson, E. [1951], Some slippage problems for the normal distribution (preliminary report) (abstract). AMS 22:484. A2 (7509)

Paulson, E. [1962], A sequential procedure for comparing several experimental categories with a standard or control. AMS 33:438-443. A3;S2 (7510)

Paulson, E. [1962], A sequential procedure for selecting the population with the largest mean from k normal populations with a common known variance (preliminary report) (abstract). AMS 33:306. A2;S2 (7511)

Paulson, E. [1963], A sequential decision procedure for choosing one of k hypotheses concerning the unknown mean of a normal distribution. AMS 34:549-554. A2 (7512)

Paulson, E. [1964], Some new developments in the sequential treatment of multiple decision problems (abstract). Technometrics 6:124. S2 (7513)

Paulson, E. [1964], A sequential procedure for selecting the population with the largest mean from k normal populations. AMS 35:174-180. A2;S2 (7514)

Paulson, E. [1967], Sequential procedures for selecting the best one of several binomial populations. AMS 38:117-123 (abstract 36:1322-1323). A2;S2 (7515)

Pearce, S. C. [1950], The interpretation of uniformity trials. Annual Report, East Malling Res. Sta., 1949, pp. 91-92. A14 (7516)

Pearce, S. C. [1952], Studies in the measurement of apple trees I. The use of trunk girths to estimate tree size. Annual Report, East Malling Res. Sta., 1951, pp. 101-104. A14 (7517)

Pearce, S. C. [1955], Some considerations in deciding plot size in field trials with trees and bushes. JISAS 7:23-26. A14 (7518)

Pearce, S. C. [1965], Statistical techniques in fruit tree research. Biométrie-Praximétrie 6:79-92. C (7519)

Pearce, S. C. and Brown, A. H. F. [1960], Improving fruit tree experiments by a preliminary study of the trees. J. Hort. Sci. 35:56-65. A14;C (7520)

Pearce, S. C. and Holland, D. A. [1957], Randomized branch sampling for estimating fruit number. Biometrics 13:127-130. A14;S1 (7521)

Pearce, S. C. and Holland, D. A. [1960], Some applications of multivariate methods in Botany. Appl. Stat. 9:1-7. A7 (7522)

Pearson, A. M., Baten, W. D. and Simon, M. [1958], The influence of salt upon panel scores of irradiated and unirradiated beef roasts. Food Res. 23:384-387. C (7523)

Pearson, E. S. [1956], Some aspects of the geometry of statistics. The use of visual presentation in understanding the theory and application of mathematical statistics (discussion). JRSSA 119:125-149. C (7524)

Pearson, E. S. [1957], John Wishart 1898-1956. Biometrika 44:1-8. B (7525)

Pearson, E. S. [1963], A statistician's place in assessing the likely operational performance of Army weapons and equipment. Proc. Eight Conf. Design Expt. Army Res. Dev. Testing. pp. 1-15. C (7526)

Pearson, E. S. [1966], Alternative tests for heterogeneity of variance; some Monte Carlo results. Biometrika 53:229-234. A11 (7527)

Peggs, A. D. [1951], Analysis of variance with unequal numbers. British J. Psychology Stat. Sec. 4:77-84. A10 (7528)

Pendleton, J. W. and Dungan, G. H. [1958], Effect of row direction on spring oat yields. Agron. J. 50:341-343. A14 (7529)

Pendleton, J. W. and Seif, R. D. [1962], Role of height in corn competition. Crop Sci. 2:154-156. A14 (7530)

Peritz, E. [1965], On inferring order relations in analysis of variance. Biometrics 21:337-344. A3 (7531)

Peritz, E. [1967], On some nonparametric tests of dispersion. Stat. Neerlandica 21:81-89. A12 (7532)

Peryam, D. R. [1957], Factors affecting the accuracy and reliability of sensory tests. ASQC Nat. Conv. Trans. 11:675-685. C (7533)

Peryam, D. R. and Girardot, N. F. [1952], Advanced taste test method. Food Eng. 24(7):58-61, 194. C (7534)

Peryam, D. R. and Pilgrim, F. J. [1957], Hedonic scale method of measuring food preferences. Food Tech. 11:9-14. C (7535)

Peryam, D. R. and Shapiro, R. [1955], Perception, preference, judgement-clues to food quality. IQC 11(7):15-19, 22-23. C (7536)

Pesek, J. T. [1956], Agronomic problems in securing fertilizer response data desirable for economic analysis. <u>Methodological Procedures in the Economic Analysis of Fertilizer Use Data</u>, pp. 101-112. C (7537)

Pestana, C. G. A. [1960-1], Determinação da mais promissora forma para os talhões de um ensaio de campo com batateiras (<u>Solanum tuberosum</u>). (Determination of the most promising shape for the plots in a field trial with potatoes.) Portugaliae Acta Biologica, Lisboa 6(A):133-140. A14 (7538)

Petersen, R. G. and Chamblee, D. S. [1955], Optimum size of sample for hand separation of forage crop mixtures into their component species in small plot experiments. Agron. J. 47:20-23. S1 (7539)

Petersen, R. C. and Lucas, H. L. [1960], Experimental errors in grazing trials. Proc. Eighth Int. Grassland Congress pp. 747-750. C (7540)

Peura, B. J. [1965], Analysis of the N-way classification design model: Non-orthogonal layout with one or more concomitant variables (preliminary report) (abstract). AMS 36:1080. A7,10 (7541)

Pfanzagl, J. [1959], Ein kombiniertes Test & Klassifikations-Problem. Metrika 2: 11-45. A2 (7542)

Pfanzagl, J. [1960], Tests und Konfidenzintervalle für exponentielle Verteilungen und deren Anwendung auf einige diskrete Verteilungen. Metrika 3:1-25. A3,5 (7543)

Phatarfod, R. M. [1965], Sequential analysis of dependent observations I. Biometrika 52:157-165. S2 (7544)

Pierce, W. L. [1963], How to design war games to answer research questions. Proc. Eighth Conf. Design Expt. Army Res. Dev. Testing. pp. 453-471. C (7545)

Pigden, W. J. and Greenshields, J. E. R. [1960], Interaction of design, sward and management on yield and utilization of herbage in Canadian grazing trials. Proc. Eighth Int. Grassland Congress pp. 594-597. C (7546)

Pillai, K. C. S. [1952], On the distribution of 'Studentized' range. Biometrika 39: 194-195. A3 (7547)

Pillai, K. C. S. [1959], Upper percentage points of the extreme studentized deviate from the sample mean. Biometrika 46:473-474. A3 (7548)

Pillai, K. C. S. and Tienzo, B. P. [1959], On the distribution of the extreme studentize deviate from the sample mean. Biometrika 46:467-472. A3 (7549)

Plackett, R. L. [1950], Some theorems in least squares. Biometrika 37:149-157. A10 (7550)

Plaisted, R. L. [1956], Onion field plot technique. Proc. American Soc. Hort. Sci. 67:390-397. A14 (7551)

Plaisted, R. L. [1956-57], Design and analysis of onion storage trials (abstract). Iowa State College J. Sci, 31:499-500. A14 (7552)

Plaisted, R. L. [1957], Some components of the variance of specific gravity of potatoes. Proc. American Soc. Hort. Sci. 70:391-396. A13;S1 (7553)

Plaisted, R. L., Horner, T. W., and Lana, E. P. [1957], Design and analysis of onion storage trials. Proc. American Soc. Hort. Sci. 69:412-420. A9;S1;C (7554)

Pointer, J. P. and Koch, E. J. [1961], Estimates of optimum plot size from uniformity data in Maryland tobacco. Tobacco Sci. 5:112-117. A14 (7555)

Pompilj, G. [1958], On entirely pseudo-linear regression. Bull. ISI 36(3):60-63. A7 (7556)

Popovich, M., Devcic, J., and Castronovo, A. [1955], Efectos de la competencia en los ensayos comparativos con zapallitos de tronco. (Effects of competition in comparative tests of bush gourds.) Rev. Investigaciones Agrícolos, Buenos Aires 9:373-378. A14 (7557)

Potter, G. F. [1953], Field plot technique for tree crops. Proc. American Soc. Hort. Sci. 62:4-13. A14 (7558)

Potter, G. F. [1960], Statistical problems encountered by the horticulturist (abstract). Biometrics 16:136. C (7559)

Potthoff, R. F. [1962], A test of whether two parallel regression lines are the same when the variances may be unequal. Inst. Stat. Mimeo. Ser. No. 334, Univ. N. Carolina. A11 (7560)

Potthoff, R. F. [1963], Illustrations of some Scheffé-type tests for some Behrens-Fisher type regression problems. Inst. Stat. Mimeo. Ser. No. 377, Univ. N. Carolina. A3,11 (7561)

Potthoff, R. F. [1965], Some Scheffé-type tests for some Behrens-Fisher type regression problems. JASA 60:1163-1190 (Inst. Stat. Mimeo. Ser. No. 374, Univ. N. Carolina). A3,7,11 (7562)

Potthoff, R. F. and Roy, S. N. [1964], A generalized multivariate analysis of variance model useful especially for growth curve problems. Biometrika 51:313-326. A7 (7563)

Prais, S. J. and Aitchison, J. [1954], The grouping of observations in regression analysis. Rev. ISI 22:1-22. A7 (7564)

Pratt, J. W. [1959], Remarks on zeros and ties in the Wilcoxon signed rank procedures. JASA 54:655-667. A12 (7565)

Pratt, J. W. [1966], The outer needle of some Bayes sequential continuation regions. Biometrika 53:455-467. S2 (7566)

Press, S. J. [1967], Structured multivariate Behrens-Fisher problems. Sankhyā A 29:41-48. A11 (7567)

Proschan, F. [1952], Use of random numbers. IQC 9(1):32-34. C (7568)

Proschan, F. [1957], Testing suspected observations. IQC 13(7):14-19. A6 (7569)

Proudfoot, F. G., Gowe, R. S., and Cheney, B. F. [1957], Studies in the design of comparative poultry tests. 1. A comparison of three strains of white leghorns housed in replicated fifty-bird pens and intermingled in a large pen. Canadian J. Animal Sci. 37:168-178. C (7570)

Puckridge, D. W. and Donald, C. M. [1967], Competition among wheat plants sown at a wide range of densities. Australian J. Agri. Res. 18:193-211. A14 (7571)

Purcell, W. R. [1963], Group randomizing. IQC 20(4):20, 29-32. C (7572)

Putter, J. [1967], Orthonormal bases of error spaces and their use for investigating the normality and variances of residuals. JASA 62:1022-1036.
A6,11 (7573)

Putter, J., Loebenstein, G., and Slomnicki, I. [1958], Determining sample size in field experiments with tomatoes. Ktavim, Rehovot 8:21-23.
A14;S1 (7574)

Putter, J., Yaron, D., and Bielorai, H. [1966], Quadratic equations as an interpretative tool in biological research. Agron. J. 58:103-104.
A7 (7575)

Puri, M. L. [1965], Some distribution-free k-sample rank tests of homogeneity against ordered alternatives. Comm. Pure Appl. Math. 18:51-63. A2,12 (7576)

Puri, M. L. [1965], On some tests of homogeneity of variances. Ann. Inst. Stat. Math. 17:323-330. A11,12 (7577)

Puri, M. L. and Puri, P. S. [1966], Some distribution free multiple decision procedures for certain problems in analysis of variance (abstract). AMS 37:554.
A2,12 (7578)

Puri, M. L. and Sen, P. K. [1966], On a class of multivariate multisample rank-order tests. Sankhyā A 28:353-376. A7,12 (7579)

Puri, P. S. and Puri, M. L. [1966], Selection procedures based on ranks: scale parameter case (abstract). AMS 37:1068. A2,12 (7580)

Quade, D. [1965], Rank analysis of covariance (abstract). Biometrics 21:1028-1029. A7,12 (7581)

Quade, D. [1966], On analysis of variance for the k-sample problem. AMS 37:1747-1758. A7,12 (7582)

Quade, D. [1967], Rank analysis of covariance. JASA 62:1187-1200. A7,12 (7583)

Quenouille, M. H. [1950], Computational devices in the application of least squares. JRSSB 12:256-272. A10 (7584)

Quenouille, M. H. [1950], Multivariate experimentation. Biometrics 6:303-316, 318-319. A7 (7585)

Quesenberry, C. P. [1962], Some tests for outliers. Ph. D. Thesis, Virginia Polytechnic Inst. A6 (7586)

Quesenberry, C. P. and David, H. A. [1961], Some tests for outliers. Proc. Sixth Conf. Design Expt. Army Res. Dev. Testing. pp. 247-277. A6 (7587)

Quesenberry, C. P. and David, H. A. [1961], Some tests for outliers. Biometrika 48:379-390. A6 (7588)

Quesenberry, C. P. and David, H. A. [1961], Some tests for outliers (abstract). AMS 32:637. A6 (7589)

Quintelier, G. [1950], Het nomografisch bepalen van de parameters van een steekproefschema. (Nomographic determination of the parameters of a sampling plan.) Statistica 4:204-215. S1 (7590)

Raatz, U. [1966], Eine Modifikation White-Tests bei grossen Stichproben. Biom. Zeit. 8:42-54. A12 (7591)

Radcliffe, J. [1966], Factorizations of the residual likelihood criterion in discriminant analysis. Proc. Cambridge Philosophical Soc. 62:743-752. A7 (7592)

Radford, P. J. [1967], Growth analysis formulae - their use and abuse. Crop Sci. 7:171-175. C (7593)

Radhakrishnan, T. V. [1954], A note on the technique of "ranking" in plant breeding. Madras Agri. J. 41:57-59. A12;C (7594)

Ramabhadran, V. K. [1950], Method of fitting growth curves. JISAS 2:125-140. A7 (7595)

Ramachandran, G. [1954], Some contributions to sequential analysis. Ph. D. Thesis, Univ. Madras. S2 (7596)

Ramachandran, K. V. [1954], On simultaneous analysis of variance test (abstract). AMS 25:617. A3 (7597)

Ramachandran, K. V. [1956], Contributions to simultaneous confidence interval estimation. Biometrics 12:51-56. A3 (7598)

Ramachandran, K. V. [1956], On the Studentized largest and smallest χ^2 (abstract). AMS 27:860. A3 (7599)

Ramachandran, K. V. [1956], On the Tukey test for the equality of means and the Hartley test for the equality of variances. AMS 27:825-831. A3,11 (7600)

Ramachandran, K. V. [1958], On a test for the equality of several means (abstract). AMS 29:620-621. A3 (7601)

Ramachandran, K. V. [1958], A test of variances. JASA 53:741-747. A11 (7602)

Ramanujacharyulu, C. [1964], Analysis of preferential experiments. Psychometrika 29:257-261. A2 (7603)

Rampton, H. H. and Petersen, R. G. [1962], Relative efficiency of plot sizes and numbers of replications as indicated by yields of orchardgrass seed in a uniformity test. Agron. J. 54:247-249. A14 (7604)

Ranga Rao, D. S., Kundalkar, O. G., and Satbhai, P. N. [1960], Statistical analysis of the data of simple manurial trials on paddy in cultivators' fields of Thana District, Bombay State (1954-55). JISAS 12:206-207. A13 (7605)

Rao, B. M. [1961], A property of the coefficient of variation and its use in transformation of data. Biometrics Unit Mimeo. Ser. BU-137-M, Cornell Univ. A9 (7606)

Rao, B. M. [1962], Some properties of the coefficient of variation and F statistics with respect to transformations of the form X^k. M. S. Thesis, Cornell Univ. A9 (7607)

Rao, B. M., Federer, W. T, and Whitlock, J. H. [1963], A study of five transformations on a natural enzootic of Trichostrongylidosis. IV. Weight data. Biometrics Unit Mimeo. Ser. BU-155-M, Cornell Univ. A9 (7608)

Rao, C. R. [1950], Sequential tests of null hypotheses. Sankhyā 10:361-370. S2 (7609)

Rao, C. R. [1950], Statistical inference applied to classificatory problems. Sankhyā 10:229-256. A7 (7610)

Rao, C. R. [1950], A note on the distribution of $D^2_{p+q} - D^2_p$ and some computational aspects of D^2 statistic and discriminant function. Sankhyā 10:257-268. A7 (7611)

Rao, C. R. [1951], Statistical inference applied to classificatory problems. Part II. The problem of selecting individuals for various duties in a specified ratio. Sankhyā 11:107-116. A2;S2 (7612)

Rao, C. R. [1953], Discriminant functions for genetic differentiation and selection. Sankhyā 12:229-246. A2,7 (7613)

Rao, C. R. [1955], Estimation and tests of significance in factor analysis. Psychometrika 20:93-111. A7 (7614)

Rao, C. R. [1955], Analysis of dispersion for multiply classified data with unequal numbers in cells. Sankhyā 15:253-280. A7,10 (7615)

Rao, C. R. [1958], Some statistical methods for the comparison of growth curves. Biometrics 14:1-17. A7 (7616)

Rao, C. R. [1959], Some problems involving linear hypotheses in multivariate analysis. Biometrika 46:49-58. A7 (7617)

Rao, C. R. [1960], Presidential address. Multivariate analysis: An indispensable statistical aid in applied research. Proc. Forty-Seventh Indian Sci. Congress, part II, pp. 16-40. A7;B (7618)

Rao, C. R. [1960], Multivariate analysis: An indispensable statistical aid in applied research. Sankhyā 22:317-338. A7;B (7619)

Rao, C. R. [1961], Generation of random permutations of given number of elements using random sampling numbers. Sankhyā A 23:305-308. A15 (7620)

Rao, C. R. [1961], Some observations on multivariate statistical methods in anthropological research. Bull. ISI 38(4):99-109. A7 (7621)

Rao, C. R. [1962], A note on a generalized inverse of a matrix with application to problems in mathematical statistics. JRSSB 24:152-158. C (7622)

Rao, C. R. [1962], Use of discriminant and allied functions in multivariate analysis. Sankhyā A 24:149-154. A7 (7623)

Rao, C. R. [1963], The use and interpretation of principal component analysis in applied reserach. Sankhyā A 26:329-358. A7 (7624)

Rao, C. R. [1964], Sir Ronald Aylmer Fisher - The architect of multivariate analysis. Biometrics 20:286-300. A7;B (7625)

Rao, C. R. [1965], Problems of selection involving programming techniques. Proc. IBM Scientific Computer Symp. Stat. pp. 29-51. A2,7 (7626)

Rao, C. R. [1965], The theory of least squares when the parameters are stochastic and its application to the analysis of growth curves. Biometrika 52:447-458. A7 (7627)

Rao, C. R., Bose, P. K., Kishen, K., and Iyer, S. S. [1953], Discussion on statistical methods in biological sciences. Calcutta Stat. Assoc. Bull. 4: 143-148. C (7628)

Rao, C. R. _et al_. [1963], Scientific contributions of Professor P. C. Mahalanobis. _Contributions to Statistics_ pp. 495-516. B (7629)

Rao, M. B. [1965], Application of Greenberg and Sarhan's method of inversion of partitioned matrices in the analysis of non-orthogonal data. JASA 60:1200-1202. A10 (7630)

Rao, M. M. [1960], Some results on transformations in the analysis of variance (abstract). AMS 31:819-820. A9 (7631)

Rao, M. M. [1963], Discriminant analysis. Ann. Inst. Stat. Math. 15:11-24. A7 (7632)

Rasch, D. [1960], Probleme der Varianzanalyse bei ungleicher Klassenbesetzung. Biom. Zeit. 2:194-203. A10 (7633)

Rasch, D. [1962], Die Faktoranalyse und ihre Anwendung in der Tierzucht. Biom. Zeit. 4:15-39. A7 (7634)

Rasch, D. and Stammberger, A. [1967], Verschiedene Schätzverfahren für die Parameter de Funktion $\eta = \alpha + \beta e^{\gamma x}$. Biom. Zeit. 9:34-49. A7 (7635)

Rasch, G. [1960], On general laws and the meaning of measurement in psychology. Proc. Fourth Berkeley Symp. Math. Stat. Prob. pp. 321-333. C (7636)

Rasmusson, D. C. and Lambert, J. W. [1961], Comparison of rod-row with field plots in barley varietal testing. Crop. Sci. 1:259-260. A14 (7637)

Ratkowsky, D. A. [1964], Applications of nonparametric statistics. Parts 1 and 2. British Chem. Eng. 9:305-310, 527-531. A12 (7638)

Rau, A. A. [1953], Some difficulties in experimentation with coffee - a plantation crop (abstract). Biometrics 9:428-429. C (7639)

Ray, W. D. [1957], Sequential confidence intervals for the mean of a normal population with unknown variance. JRSSB 19:133-143. S2 (7640)

Ray, W. D. and Pitman, E. N. T. [1961], An exact distribution of the Fisher-Behrens-Welch statistic for testing the difference between the means of two normal populations with unknown variances. JRSSB 23:377-384. A11 (7641)

Read, D. R. [1954], The design of chemical experiments. Biometrics 10:1-15. C (7642)

Reid, D. D. [1950], Statistics in clinical research. Ann. New York Academy Sci. 52:931-934. C (7643)

Reiersøl, O. [1950], Identifiability of a linear relation between variables which are subject to error. Econometrica 18:375-389. A7 (7644)

Reiersøl, O. [1950], On the identifiability of parameters in Thurstone's multiple factor analysis. Psychometrika 15:121-149. A7 (7645)

Reiersøl, O. [1961], Linear and non-linear multiple comparisons in logit analysis. Biometrika 48:359-365 (corrigenda 49:284). A3 (7646)

Reimer, C. [1957], Some applications of rank order statistics in sensory panel testing. Food Res. 22:629-634. A12;C (7647)

Reiner, A. M. and Hunter, W. G. [1964], Designs to distinguish between two rival models (abstract). Biometrics 20:914. S2 (7648)

Reinhardt, H. E. [1963], The heuristics of data transformation. Math. Res. Center Tech. Summary Report No. 355. U.S. Army and Univ. Wisconsin. A9 (7649)

Reisch, J. S. and Webster, J. T. [1965], Independent sums of squares for more than one factor in an analysis of covariance (abstract). Biometrics 21:255-256. A7 (7650)

Remmenga, E. [1955], The nature in magnitude of experimental errors in grazing trials. Ph. D. Thesis, Purdue Univ. C (7651)

Rendel, J. M. [1959], Optimum group size in half-sib family selection. Biometrics 15:376-381. S1 (7652)

Reuter, G. E. H. [1961], Competition processes. Proc. Fourth Berkeley Symp. Math. Stat. Prob. 2:421-430. A14 (7653)

Rhyne, A. L. [1964], Some multiple comparison sign tests. Ph. D. Thesis, Univ. N. Carolina. A3,12 (7654)

Rhyne, A. L. [1964], A multiple comparison sign test for comparing all treatments versus a control (abstract). Biometrics 20:914. A3,12 (7655)

Rhyne, A. L. and Steel, R. G. D. [1965], Tables for a treatments versus control multiple comparisons sign test. Technometrics 7:293-306. A3,12 (7656)

Rhyne, A. L. and Steel, R. G. D. [1967], A multiple comparisons sign test: all pairs of treatments. Biometrics 23:539-549. A3,12 (7657)

Rice, E. L. and Penfound, W. T. [1955], A evaluation of the variable-radius and paired-tree methods in the blackjack-post oak forest. Ecology 36:315-320. A14 (7658)

Richards, F. J. [1959], A flexible growth function for empirical use. J. Exptl. Botany 10:290-300. A7 (7659)

Richter, D. L. [1958], A problem in two-stage experimentation (abstract). AMS 29:1283. S2 (7660)

Richter, D. [1960], Two-stage experiments for estimating a common mean. AMS 31:1164-1173. S2 (7661)

Riewe, M. E. [1961], Use of the relationship of stocking rate to gain of cattle in an experimental design for grazing trials. Agron. J. 53:309-313. S1 (7662)

Riffenburgh, R. H. [1959], A note on the graphic method for obtaining a least squares fit. Amer. Stat. 13(1):21. A7 (7663)

Rigney, J. A. [1956], Sampling soils for composition studies. Proc. American Soc. Hort. Sci. 68:569-575. A14;S1 (7664)

Rigney, J. A. and Nelson, W. L. [1951], Some factors affecting the accuracy of sampling cotton for fiber determinations. Agron. J. 43:531-535.
A14;S1 (7665)

Rijkoort, P. J. [1952], A generalization of Wilcoxon's test. Koninklijke Nederlandse Akademie Wetenschapen, Ser. A, 55:394-404 (errata 56:407).
A12 (7666)

Rios, S. [1957], Sobre las lineas de regression modal. Trabajos Estad. 8:147-156.
A7 (7667)

Rives, M. [1960], Sur l'analyse de la variance. I. Emploi de transformations. Annales Inst. Nat. Recherche Agronomique 10(B):309-331. A9,10 (7668)

Rizvi, M. H. [1963], On selecting a subset of normal populations containing the population whose mean has the largest absolute value (abstract). AMS 34: 1627-1628. A2 (7669)

Rizvi, M. H. [1963], Ranking normal populations by the absolute values of their means: fixed sample size case (abstract). AMS 34:1628. A2 (7670)

Rizvi, M. H. and Sobel, M. [1967], Nonparametric procedures for selecting a subset containing the population with the largest α-quantile. AMS 38:1788-1803.
A2,12 (7671)

Rizvi, M. H., Sobel, M., and Woodworth, G. [1966], Nonparametric ranking procedures for comparison with a control, I and II (preliminary report) (abstract). AMS 37:774. A3,12 (7672)

Robbins, H. [1952], Some aspects of the sequential design of experiments. Bull. Amer. Math. Soc. 58:527-535. S2 (7673)

Robbins, H. [1960], A statistical screening problem. Contributions to Probability and Statistics pp. 352-357. A2;S2 (7674)

Roberts, C. D. [1963], An asymptotically optimal sequential design for comparing several experimental categories with a standard or control. Ph. D. Thesis, Univ. N. Carolina (Inst. Stat. Mimeo. Ser. No. 344, Univ. N. Carolina).
A3;S2 (7675)

Roberts, C. D. [1963], An asymptotically optimal sequential design for comparing several experimental categories with a control. AMS 34:1486-1493.
A3;S2 (7676)

Roberts, C. D. [1964], An asymptotically optimal fixed sample size procedure for comparing several experimental categories with a control. AMS 35:1571-1575.
A2,3 (7677)

Roberts, H. R., McCall, C. H., and Thomas, R. E. [1958], Some statistical considerations for small sample evaluation in triangular taste tests. Food Res. 23:388-395. C (7678)

Robertson, A. [1950], Some observations on experiments with identical twins in dairy cattle. J. Genetics 50:32-35. A14 (7679)

Robertson, A. [1957], Optimum group size in progeny testing and family selection. Biometrics 13:442-450. S1 (7680)

Robertson, A. [1960], On optimum family size in selection programmes. Biometrics 16:296-298. S1 (7681)

Robertson, A. [1962], Weighting in the estimation of variance components in the unbalanced single classification. Biometrics 18:413-417. A11 (7682)

Robertson, J. D. and Armitage, P. [1959], Comparison of two hypotensive agents. Anaesthesia 14:53-64. C (7683)

Robocker, W. C. and Miller, B. J. [1955], Effects of clipping, burning and competition on establishment and survival of some native grasses in Wisconsin. J. Range Management 8:117-120. A14 (7684)

Robson, D. S. [1959], A simple method for constructing orthogonal polynomials when the independent variable is unequally spaced. Biometrics 15:187-191. A7 (7685)

Robson, D. S. [1964], A rank-sum test of whether two multivariate samples were drawn from the same population. Biometrics Unit ONR Tech. Report No. 14, Cornell Univ. A12 (7686)

Robson, D. S. and Vithayasai, C. [1964], Choosing between Tukey's hsd and Scheffé's procedure for testing a set of non-orthogonal linear contrasts. Biometrics Unit Mimeo. Ser. BU-169-M, Cornell Univ. A3 (7687)

Rod, J. and Vágnerová, V. [1962], Praktikum ze šlechtění rostlin. Polní pokusnictví. (The practice of plant breeding. Field testing.) Státní pedagogické nakladatelství, n.p., Praha (Mimeo.). C (7688)

Roessler, E. B., Baker, G. A., and Amerine, M. A. [1956], One-tailed and two-tailed tests in organoleptic comparisons. Food Res. 21:117-121. C (7689)

Rojas, B. [1958], The analysis of groups of similar experiments (abstract). Biometrics 14:293. A13 (7690)

Rojas, B. A. [1958], The analyses of groups of similar experiments. Ph. D. Thesis, Iowa State Univ. A13 (7691)

Rojas, B. A. [1958], The analysis of tillage experiments. Stat. Lab. Final Report, Iowa State Univ. A11,13;S1 (7692)

Romani, J. [1956], Tests no parametricos en forma secuencial. Trabajos Estad. 7:43-96. A12;S2 (7693)

Romero, S., Carney, E. J., and Rojas, B. [1966], The power of test on experimental designs (in Spanish). Agrociencia 1:31-50. S1 (7694)

Roosje, G. S. [1961], Experimentation on trees and other perennial crops. Stat. News Letter 11(2):2-6. A14 (7695)

Roosje, G. S. [1963], Overwegingen bij het aanleggen van proefvelden en het opzetten van veldproeven in de fruitteelt. (Considerations on the planning of field trials and the establishment of field trials in fruit growing.) Mededeelingen Directeur van de Tuinbouw.'s-Gravenhage 26:136-150. A14 (7696)

Rosado, F. B. [1964], Sobre los metodos estadisticos en fitosociologia. Trabajos Estad. 15:185-202. B (7697)

Rosenbaum, S. [1951], The variance of least-square estimates under linear restraints. JRSSB 13:250-255. A7 (7698)

Rosenbaum, S. [1953], Tables for a nonparametric test of dispersion. AMS 24:663-668. A12 (7699)

Rosenbaum, S. [1965], On some two-sample non-parametric tests. JASA 60:1118-1126. (corrigenda 61:1249). A12 (7700)

Rosenberg, L. [1957], Testing the difference between means in the presence of variance heterogeneity. M.S. Thesis, Iowa State Univ. A11 (7701)

Rosengard, A. [1962], Étude des lois-limites jointes et marginales de la moyenne et des valeurs extrêmes d'un échantillon. Publications Inst. Stat. Univ. Paris 11:3-56. A6 (7702)

Rosenthal, I. and Ferguson, T. S. [1956], An asymptotically distribution-free multiple comparison method with application to the problem of n rankings of m objects (abstract). AMS 27:868. A2,3 (7703)

Rosenthal, I. and Ferguson, T. S. [1965], An asymptotically distribution-free multiple comparison method with application to the problem of n rankings of m objects. British J. Stat. Psychology 18:243-254. A3,12 (7704)

Ross, W. M. [1958], A comparison of grain sorghum varieties in plots with and without border rows. Agron. J. 50:344-345. A14 (7705)

Ross, W. M. and Miller, J. D. [1955], A comparison of hill and conventional yield tests using oats and spring barley. Agron J. 47:253-255.
A14 (7706)

Rotti, A. [1962], Auswahl einer Regressionsfläche zur Anpassung von beleuchtungsstärken in einem Klima-Schrank. Biom. Zeit. 4:263-273. A7 (7707)

Rotti, A. [1963], Etude statistique de la repartition spatiale de l'eclairement lumineux au laboratoire. Biométrie-Praximétrie 4:133-164. A14 (7708)

Rotti, A. [1964], Aanpassing van een regressierechte door de oorsprong. Biométrie-Praximétrie 5:80-92. A7 (7709)

Roy, J. [1956], On some quick decision methods in multivariate and univariate analysis. Sankhyā 17:77-88. A7,9 (7710)

Roy, J. and Mitra, S. [1954], A method of selection for improvement. Calcutta Stat. Assoc. Bull. 5:82-86. A2 (7711)

Roy, J. and Murthy, V. K. [1960], Percentage points of Wilks' L_{mvc} and L_{vc} criteria. Psychometrika 25:243-250. A7,11,12 (7712)

Roy, S. N. [1954], Simultaneous confidence bounds on canonical regressions (abstract). AMS 25:413. A3 (7713)

Roy, S. N. [1954], Some further results in simultaneous confidence interval estimation. AMS 25:752-761 (abstract 25:173). A3 (7714)

Roy, S. N. [1956], A note on "Some further results in simultaneous confidence interval estimation". AMS 27:856-858. A3 (7715)

Roy, S. N. [1963], Some remarks on normal multivariate analysis of variance. Colloques Internationaux Centre Nat. Recherche Scientifique, No. 110, Le Plan d'Expériences, pp. 99-113. A7 (7716)

Roy, S. N. and Bargmann, R. E. [1958], Tests of multiple independence and the associated confidence bounds. AMS 29:491-503. A3,7 (7717)

Roy, S. N. and Bose, R. C. [1953], On a set of simultaneous confidence interval statements in multivariate analysis of variance (abstract). AMS 24:144.
A3,7 (7718)

Roy, S. N. and Bose, R. C. [1953], On a relevant set of simultaneous confidence interval statements in discriminant analysis (abstract). AMS 24:145.
A3,7 (7719)

Roy, S. N. and Bose, R. C. [1953], Simultaneous confidence interval estimation. AMS 24:513-536 (Inst. Stat. Mimeo. Ser. No. 67, Univ. N. Carolina).
A3 (7720)

Roy, S. N. and Gnanadesikan, R. [1957], Further contributions to multivariate confidence bounds. Biometrika 44:399-410. A3,7 (7721)

Roy, S. N. and Gnanadesikan, R. [1959], Some contributions to ANOVA in one or more dimensions: I. AMS 30:304-317. A7 (7722)

Roy, S. N. and Gnanadesikan, R. [1959], Some constributions to ANOVA in one or more dimensions: II. AMS 30:318-340. A7 (7723)

Roy, S. N. and Krishnaiah, P. R. [1960], On dependent tests in analysis of variance (abstract). AMS 31:537. A3 (7724)

Roy, S. N. and Mitra, S. K. [1956], An introduction to some non-parametric generalizations of analysis of variance and multivariate analysis. Biometrika 43:361-376. A7,12 (7725)

Roy, S. N. and Roy, J. [1959], On testability in normal ANOVA and MANOVA with all "fixed effects" (abstract). AMS 30:616. A7 (7726)

Royall, R. M. [1966], A class of non-parametric estimates of a smooth regression function. Dept. Stat. Tech. Report No. 14, Stanford Univ. A7,12 (7727)

Rozeboom, W. W. [1960], The fallacy of the null-hypothesis significance test. Psychological Bull. 57:416-428. C (7728)

Ruben, H. [1960], On the distribution of the weighted difference of two independent Student variables. JRSSB 22:188-194. A11 (7729)

Ruben, H. [1961], Studentisation of two-stage sample means from normal populations with unknown common variance. Sankhyā A 23:231-250. S2 (7730)

Ruben, H. [1962], Studentisation of two stage sample means from normal populations with unknown variances. I. General theory and application to the confidence estimation and testing of the difference in population means. Sankhyā A 24: 157-180. A11;S2 (7731)

Ruben, H. [1962], Studentisation of two-stage sample means from normal populations with unknown variances. II. Confidence estimation for the mean of a stratified population. Sankhyā A 24:251-254. A11;S2 (7732)

Ruben, H. [1962], Studentisation of two-stage sample means from normal populations with unknown variances. III. Joint confidence estimation of a set of means. Sankhyā A 24:255-258. A11;S2 (7733)

Ruben, H. [1966], On the simultaneous stabilization of variances and covariances. Ann. Inst. Stat. Math. 18:203-210. A9,11 (7734)

Rubin, E. [1961], Questions and answers, Bibliographies for statisticians (and others). Amer. Stat. 15(5):19. B (7735)

Rubin, E. [1963], Questions and answers. Recent developments in statistical bibliography. Amer. Stat. 17(1):25. B (7736)

Rubin, E. [1964], Questions and answers. Developments in statistical bibliography. Amer. Stat. 18(1):24-25. B (7737)

Rubin, E. [1965], Questions and answers. Developments in statistical bibliography, 1963-64. Amer. Stat. 19(5):39-40. B (7738)

Ruhl, D. B. [1967], Some pooling procedures for two correlated means based upon a preliminary test of significance (abstract). Biometrics 23:607. A11 (7739)

Rümke, C. L. [1953], Kwantificering in medisch onderzoek. (Quantification in medical research.) Stat. Neerlandica 7:223-232. C (7740)

Rümke, C. L. [1958], De taak van de medische statistiek. (The task of medical statistics.) Stat. Neerlandica 12:1-16. C (7741)

Rümke, C. L. [1958], Toepassing van statistiek bij experimenten in een farmacologisch laboratorium. (The application of statistical methods on experiments in a pharmacological laboratory.) Stat. Neerlandica 12:149-158. C (7742)

Rümke, C. L. [1961], An efficient design for comparing the effects of two treatments. Quantitative Methods in Pharmacology pp. 41-48. S2 (7743)

Rümke, C. L. [1961], Een algemene inleiding tot de sequente analyse. (A general introduction to sequential analysis.) Stat. Neerlandica 15:47-54. S2 (7744)

Rümke, C. L. [1962], Enkele opmerkingen over de opzet van onderzoekingen naar de werkzaamheid van cytostatica in de kliniek. (Some remarks on the design of clinical trials of cancer chemotherapeutic agents.) Stat. Neerlandica 16:261-268. C (7745)

Rümke, C. L. [1964], Communicatie en medische statistiek. (Communication and medical statistics.) Stat. Neerlandica 18:425-431. C (7746)

Rümke, C. L. and de Jonge, H. [1965], De toetsen van Dunnett en van Steel voor het vergelijken van de uitkomsten van k proefgroepen met die van één controlegroep. Stat. Neerlandica 19:15-24. A3;12 (7747)

Rundfeldt, H. [1956], Review of methods usually applied in field plot technique (abstract). Biometrics 12:228. C (7748)

Rundfeldt, H. [1957], Zur Berechnung eines optimalen Verhältnisses zwischen der Anzahl der Prüfjahre, der Prüforte und der Vergleichsteilstücke bei Feldversuchen. Z. Pflanzenzüchtung 37:192-201. A13 (7749)

Rundfeldt, H. [1957], Über die Vorteile einer erweiterten Auswertung von Feldversuchen. Moderne Methoden der Pflanzenzüchtung 44:97-118. A13;S1 (7750)

Rush, H. R. [1959], Applications of sequential type designs and analyses to field tests. Proc. Fourth Conf. Design Expt. Army Res. Dev. Testing. pp. 223-239. C (7751)

Rushton, S. [1950], On a sequential t-test. Biometrika 37:326-333. S2 (7752)

Rushton, S. [1951], On least square fitting by orthonormal polynomials using the Choleski method. JRSSB 13:92-99. A7 (7753)

Rushton, S. [1952], On a two-sided sequential t-test. Biometrika 39:302-308. S2 (7754)

Rushton, S. [1952], On sequential tests of the equality of variances of two normal populations with known means, Sankhyā 12:63-78. A11;S2 (7755)

Rushton, S. [1954], Sequential procedures in the analysis of variance. Proc. Int. Congress Math. pp. 300-301. S2 (7756)

Rűther, H. [1958], Die Feldversuchstechnik, 2nd edition (1st edition, 1952). Deutscher Bauernverlag. C (7757)

Rutstein, D. D. [1951], Screening tests in mass surveys and their use in heart disease case finding. Circulation 4:659-665. A2 (7758)

Ryan, T. A. [1959], Multiple comparisons in psychological research. Psychological Bull. 56:26-47. A3 (7759)

Ryan, T. A. [1959], Comments on orthogonal components. Psychological Bull. 56:394-396. A3 (7760)

Ryan, T. A. [1960], Significance tests for multiple comparison of proportions, variances, and other statistics. Psychological Bull. 57:318-328. A3 (7761)

Ryan, T. A. [1962], The experiment as the unit for computing ratio of error. Psychological Bull. 59:301-305. A3 (7762)

Ryser, G. K. and Crandall, B. H. [1952], Adaptation of the IBM card system to analysis of sugar-beet varietal data. Proc. American Soc. Sugar Beet Tech. pp. 421-425. A15 (7763)

Sabherwal, V. C. [1965], The consist routine. A computer programme for solving simultaneous equations. Sankhyā B 27:145-158. A10,15 (7764)

Sachs, V. [1962], Die Sequenzanalyse als statistische Prüfmethode im Rahmen Medizinischer experimenteller, insbesondere klinischer Untersuchungen. Arztliche Forschung 16:331-345. S2 (7765)

Saegusa, T. [1962], Factorial experiment on the breakage of electronic tube grid winding wire. Reports Stat. Appl. Res. 9:61-68. C (7766)

Sahasrabudhe, V. B. [1955], Block efficiency and number of replications for progeny row trials of cotton. Indian Cotton Growing Rev. 9:106-111. A14;S1 (7767)

Sakai, K. [1951], On variance due to competition between plants of different types in plant populations. Annual Report Nat. Inst. Genetics, Japan, 1949-1950 pp, 41-43. A14 (7768)

Sakai, K. [1955], Competition in plants and its relation to selection (with discussion). Cold Spring Harbor Symp. Quantitative Biology 20:137-157. A14 (7769)

Sakai, K. [1956], Studies on competition in plants. VI. Competition between autotetraploids and their diploid prototypes in _Nicotiana tabacum_ L. Cytologia 21: 153-156. A14 (7770)

Sakai, K. [1957], Studies on competition in plants. VII. Effect on competition of a varying number of competing and non-competing individuals. J. Genetics 55:227-234. A14 (7771)

Sakai, K. and Mukaide, H. [1967], Estimation of genetic, environmental, and competitional variances in standing forests. Silvae Genetica 16:149-152. A14 (7772)

Sakai, K. and Suzuki, Y. [1954], Studies on competition in plants. III. Competition and spacing in one dimension. Japanese J. Genetics 29:197-201. A14 (7773)

Sakai, K. and Suzuki, Y. [1955], Studies on competition in plants. V. Competition between allopolyploids and their diploid parents. J. Genetics 53:585-590. A14 (7774)

Sakai, K. and Suzuki, Y. [1955], Studies on competition in plants. II. Competition between diploid and autotetraploid plants of barley. J. Genetics 53:11-20. A14 (7775)

Sakai, K. and Utiyamada, H. [1957], Studies on competition in plants. VIII. Chromosome number, hybridity and competitive ability in _Oryza sativa_ L. J. Genetics 55:235-240. A14 (7776)

Sakai, M. and Marumine, S. [1954], Studies on a method of correcting for missing plants in sweet-potato experimental plots (in Japanese). Nihon Sokumotsugaku Kai Kiji (Proc. Crop Sci. Soc. Japan) 24:195-196. A14 (7777)

Salati, E. [1955], Experimentação agrícola. (Agricultural experimentation.) An. I. Congr. Brasil. Estud. Agron. pp. 71-80. C (7778)

Salmon, S. C. [1951], Analysis of variance and long-time variety tests of wheat. Agron. J. 43:562-570. A13 (7779)

Salmon, S. C. [1953], Random versus systematic arrangement of field plots. Agron. J. 45:459-462. C (7780)

Salmon, S. C. [1955], Random versus systematic arrangements in non-Latin square field experiments. Agron. J. 47:289-294. C (7781)

Salmon, S. C. and Hanson, A. A. [1964], The Principles and Practice of Agricultural Research. Leonard Hill, London. A14;C (7782)

Sampford, M. R. [1955], The use of litter-mates in response-time experiments (abstract). Biometrics 11:249. C (7783)

Sampford, M. R. [1960], Some statistical problems in the estimation of herbage yields and consumption. Proc. Eighth Int. Grassland Congress pp. 742-746. C (7784)

Samuel, E. [1965], Sequential compound estimators. AMS 36:879-889. S2 (7785)

Sandelius, M. [1950], On non-sequential estimation when the sample size is a random variable. Ann. Royal Agri. College Sweden 17:400-406. S1 (7786)

Sandison, A. [1959], Influence of site and season on agricultural variety trials. Nature 184:834. A13 (7787)

Sandison, A. and Bartlett, B. O. [1958], Comparison of varieties for yield. J. Nat. Inst. Agri. Botany 8:351-357. A13 (7788)

Sands, D. E. [1954], A determination of the optimum plot size for predicting cotton yield. M. S. Thesis, Univ. N. Carolina. A14 (7789)

Sankaran, M. [1966], Note on statistical screening procedure of Robbins. JISA 4:94-98. A2 (7790)

Sathe, Y., Zelen, M., and Zweifel, J. [1965], The interpretation of clinical trials. J. Chronic Diseases 18:385-395. C (7791)

Sato, R. [1951], "r" distributions and "r" tests. Ann. Inst. Stat. Math. 2:91-124 (erratum 3:127-128). A7 (7792)

Sato, R. [1951], The r tests relating to the regression. Ann. Inst. Stat. Math. 3:45-56. A7 (7793)

Sato, S. [1962], A multivariate analogue of pooling of data. Bull. Math. Stat. 10(3&4):61-76. A11 (7794)

Sauger, L. and Tourte, R. [1951], Contribution à la technique des essais culturaux au Sénégal. Forme et dimensions des parcelles. Nombre de répétitions. Agron. Trop. 6:29-37. A14;S1 (7795)

Săulescu, N., Popa, T., and Rădoi, A. [1960], Contribuţii la metodica experienţelor cu soiuri la porumb. (Contributions to experimental method with maize varieties.) Lucrăr. sti. Inst. Agron. N. Bălcescu Bucureşti pp. 155-164. A14 (7796)

Savage, I. R. [1953], Bibliography of nonparametric statistics and related topics. JASA 48:844-906. A12;B (7797)

Saw, J. G. [1966], A conservative test for the concurrence of several regression lines and related problems. Biometrika 53:272-275. A7 (7798)

Saw, J. G. [1966], A non-parametric comparison of two samples one of which is censored. Biometrika 53:599-602. A12 (7799)

Sawrey, W. L. [1958], A distinction between exact and approximate nonparametric methods. Psychometrika 23:171-177. A12 (7800)

Sax, E. [1958], Evaluation of tube screening procedures. Part II - Statistical analyses of data. ASQC, Nat. Conv. Trans. 12:431-442. S2 (7801)

Saxena, K. M. L. [1966], A multiple decision procedure for ranking the means of normal populations: dependent sample case (abstract). AMS 37:1860. A2 (7802)

Saxena, P. N. [1963], On some general properties of the 'minimum recommended rate' in fertilizer application. JISAS 15:160-174. C (7803)

Scadding, J. G. [1962], The planning of trials of new drugs in pulmonary tuberculosis. Methods Information Medicine 1:130-132. C (7804)

Scheffé, H. [1952], On judging all contrasts in the analysis of variance (preliminary report) (abstract). AMS 23:477. A3 (7805)

Scheffé, H. [1953], A method for judging all contrasts in the analysis of variance. Biometrika 40:87-104. A3 (7806)

Scheffé, H. [1958], Fitting straight lines when one variable is controlled. JASA 53:106-117. A7 (7807)

Scheffé, H. [1961], Simultaneous interval estimates of linear functions of parameters. Bull. ISI 38(4):245-253. A3 (7808)

Schlösser, L. A., Haufe, W., and Geidel, H. [1961], Zum rechnerischen Ausgleich von Fehlstellen, insbesondere bei Zuckerrübenversuchen. Z. Pflanzenzüchtung 44:348-379. A14 (7809)

Schmid, C. F. [1956], What price pictoral charts? Estadística 14:12-25. C (7810)

Schneider, A. M. and Stockett, A. L. [1963], An experiment to select optimum operating conditions on the basis of arbitrary preference ratings. Chemical Eng. Progress Symp. Series 59(42):34-38. C (7811)

Schneider, B. [1960], Covariance adjustments by orthogonal polynomials (abstract). Biometrics 16:310-311. A7 (7812)

Schneider, B. [1963], Die Bestimmung der Parameter im Ertragsgesetz von E. A. Mitscherlich. I. Die Methode der kleinsten Quadrate. (On the determination of parameters in E. A. Mitscherlich's regression law. I. The method of least squares.). Biom. Zeit. 5:78-95. A7 (7813)

Schneider, B. [1965], Die statistischen Schlussweisen und ihre Bedeutung für die Biometrie. (The problem of statistical inference and its significance for biometrics.) Biom. Zeit. 7:102-115. C (7814)

Schneider, B. [1967], Einführung in die multivariate analyse. (Introduction to multivariate analysis.) Biom. Zeit. 9:269-284. A7 (7815)

Schneiderman, M. A. [1962], The clinical excursion into 5-fluorouracil. J. Chronic Diseases 15:283-295. C (7816)

Schneiderman, M. A. [1964], The proper size of a clinical trial: "Grandma's strudel" method. J. New Drugs 4:3-11. C;S1 (7817)

Schneiderman, M. A. [1967], Mouse to man: statistical problems in bringing a drug to clinical trial. Proc. Fifth Berkeley Symp. Math. Stat. Prob. 4:855-866. C (7818)

Schneiderman, M. A. and Armitage, P. [1962], A family of closed sequential procedures. Biometrika 49:41-56. S2 (7819)

Schneiderman, M. A. and Armitage, P. [1962], Closed sequential t-tests. Biometrika 49:359-366. S2 (7820)

Schneiderman, M. A., Myers, M. H., Sathe, Y. S., and Koffsky, P. [1964], Toxicity, the therapeutic index, and the ranking of drugs. Science 144:1212-1214. C (7821)

Schnell, W. [1956], On the sphere of permissible generalization of field trial results (abstract). Biometrics 12:229. C (7822)

Scholl, J. M. and Staniforth, D. W. [1957], Establishment of birdsfoot trefoil as influenced by competition from weeds and companion crops. Agron. J. 49:432-435. A14 (7823)

Schreiner, W. [1954], Rand- und Nachbarwirkungen bei Kartoffelversuchen. Z. Pflanzenzüchtung 33:169-178. A14 (7824)

Schultz, A. M. [1956], The use of regression in range research. J. Range Management 9:41-46. A7 (7825)

Schultz, A. M., Gibbens, R. P., and Debano, L. [1961], Artificial populations for teaching and testing range techniques. J. Range Management 14:236-242. A14 (7826)

Schultz, E. F. [1955], Optimum allocation of experimental material with an illustrative example in estimating fruit quality. Proc. American Soc. Hort. Sci. 66:421-433. A14;S1 (7827)

Schultz, E. F. and Schneider, G. W. [1955], Sample size necessary to estimate size and quality of fruit, growth of trees, and pre cent fruit set of apples and peaches. Proc. American Soc. Hort. Sci. 66:36-43. S1 (7828)

Schumann, T. E. W. [1950], Die probleem van die aanpassing van polinome aan 'n reeks ewewydige gegewens. (The adjustment of polynomials to a series of equidistant data.) Statistica 4:158-167. A7 (7829)

Schutz, W. M. and Brim, C. A. [1967], Inter-genotypic competition in soybeans. I. Evaluation of effects and proposed field plot design. Crop Sci. 7:371-376. A14 (7830)

Schwartz, N. and Foster, D. [1957], Methods for rating quality and intensity of the psychological properties of foods. Food Tech. 11:15-20. C (7831)

Scott, A. J. [1966], A location of effort in the design of selection procedures. Ph. D. Thesis, Univ. Chicago. A2 (7832)

Seal, H. L. [1967], Studies in the histroy of probability and statistics. XV. The historical development of the Gauss linear model. Biometrika 54:1-24. A7;B;C (7833)

Seal, K. C. [1954], An optimum decision procedure for ranking means of normal populations (abstract). AMS 25:618. A2,3 (7834)

Seal, K. C. [1954], On a property of a class of decision procedures for ranking means of normal populations (preliminary report) (abstract). AMS 25:413. A2 (7835)

Seal, K. C. [1958], Selection of a subset superior, inferior, or equivalent to a control population. Calcutta Stat. Assoc. Bull. 8:20-30. A2 (7836)

Searle, S. R. [1962], Number of paired observations needed for estimating repeatability and heritabilty from regression analyses. J. Animal Sci. 21:426-427. S1 (7837)

Searls, D. T. [1963], On the probability of winning with different tournament procedures. JASA 58:1064-1081. A2;S2 (7838)

Searls, D. T. [1963], On the probability of winning with different tournament procedures (abstract). Biometrics 19:663. A2;S2 (7839)

Seder, L. A. [1951], A new science of trouble shooting. Ind. Eng. Chemistry 43: 2053-2059. C (7840)

Seelbinder, B. M. [1953], On Stein's two-stage sampling scheme. AMS 24(4): 640-649. S2 (7841)

Seif, R. D. [1957], Optimum field plot size and shape for lima bean yields. Ph. D. Thesis, Cornell Univ. A14 (7842)

Sekar, C. C. and Chakraborty, P. N. [1952], On the concept and use of orthogonal semi-polynomials. Sankhyā 12:141-150. A7 (7843)

Sen, A. R. [1963], Some techniques of experimentation with clonal tea based on a uniformity trial. Empire J. Expt. Agri. 31:296-310. A14 (7844)

Sen, A. R. and Biswas, A. K. [1966], Some techniques of experimentation with tea in north-east India. Exptl. Agri. 2:89-100. A14 (7845)

Sen, P. K. [1962], On studentized non-parametric multi-sample location tests. Ann. Inst. Stat. Math. 14:119-131. A12 (7846)

Sen, P. K. [1962], Query: Wilcoxon's test when variables are subject to error. Calcutta Stat. Assoc. Bull. 11:107-111. A12 (7847)

Sen, P. K. [1963], On weighted rank-sum tests for dispersion. Ann. Inst. Stat. Math. 15:117-135. A12 (7848)

Sen, P. K. [1965], Some further applications of non-parametric methods in dilution (-direct) assays. Biometrics 21:799-810. A12;C (7849)

Sen, P. K. [1966], On nonparametric simultaneous confidence regions and tests for the one criterion analysis of variance problem. Ann. Inst. Stat. Math. 18:319-336. A3,12 (7850)

Sen, P. K. [1966], On some nonparametric tests for symmetry in twoway tables. JISA 4:125-142. A12 (7851)

Sen, P. K. [1967], On pooled estimation and testing of heterogeneity of shift parameters by distributionfree methods. Calcutta Stat. Assoc. Bull. 16:139-152. A12 (7852)

Sen, P. K. [1967], A class of permutation tests for stochastic independence, I. Sankhyā A 29:157-174. A12 (7853)

Sen, P. K. [1967], Nonparametric tests for multivariate interchangeability. Part I: Problems of location and scale in bivariate distributions. Sankhyā A 29:351-372. A7,12 (7854)

Sen, P. K. [1967], On a class of two-sample bivariate nonparametric tests. Proc. Fifth Berkeley Symp. Math. Stat. Prob. 1:639-656. A7,12 (7855)

Sen, P. K. [1967], On a class of nonparametric tests for MANOVA in two way layouts. Proc. Twelfth Conf. Design Expt. Army Res. Dev. Testing. pp. 121-150. A7,12 (7856)

Sen, P. K. and Govindarajulu, Z. [1966], On a class of c-sample weighted rank-sum tests for location and scale. Ann. Inst. Stat. Math. 18:87-105. A12 (7857)

Sengupta, J. M. [1964], On perimeter bias in sample cuts of small size. Sankhyā B 26:53-68. A14 (7858)

Sengupta, S. and Rao, J. S. [1966], Statistical analysis of cross-bedding azimuths from the Kamthi formation around Bheemaraṃ, Pranhita-Godavari Valley. Sankhyā B 28:165-174. A9 (7859)

Seth, G. R. [1950], On the distribution of the two closest among a set of three observations. AMS 21:299-301. A6 (7860)

Seth, G. R. et al. [1955], Symposium on the present status of agricultural experiments. JISAS 7:87-97. C (7861)

Seth, G. R., Sukhatme, B. V., and Ram, B. M. [1958], A survey of agronomic research programmes in India. Indian J. Agri. Sci. 38(3):409-468. C (7862)

Shah, B. K. [1961], A simple method of fitting the regression curve $y = \alpha + \delta x + \beta\rho^x$ Biometrics 17:651-653. A7 (7863)

Shah, B. K. and Khatri, C. G. [1963], Further investigation in fitting the regression curve of the type $y = \alpha + \delta x + \beta\rho^x$. JISA 1:202-214. A7 (7864)

Shah, B. K. and Khatri, C. G. [1965], A method of fitting the regression curve $E(y) = \alpha + \delta x + \beta\rho^x$. Technometrics 7:59-65. A7 (7865)

Shah, B. K. and Patel, I. R. [1960], The least square estimates of the constants for the Makeham second modification of Gompertz's Law. J. Maharaja Sayojirao Univ. Baroda 9(2):1-10. A7 (7866)

Shah, S. M. and Ramachandran, K. V. [1960], Curve fitting using least absolute deviations. Calcutta Stat. Assoc. Bull. 10:48-53. A7 (7867)

Sharpe, R. H. and Blackmon, G. H. [1950], A study of plot size and experimental design with pecan yield data. Proc. American Soc. Hort. Sci. 56:236-241. A14 (7868)

Shepard, R. N. [1962], The analysis of proximities: multidimensional scaling with an unknown distance function. Psychometrika 27:125-140, 219-246. A14 (7869)

Sherman, E. [1965], A note on multiple comparisons using rank sums. Technometrics 7:255-256. A3,12 (7870)

Shimada, Y. [1958], Statistical studies on the design of yield survey and field experiment in natural grassland. Part 1. Estimation of yield, especially with reference to size and shape, and replication of field experimental plot in natural Miscanthus grassland. Part 2. Estimation of number of bracken, Pteridium aquilinum (L.) Kuhn., in the Miscanthus grassland. Sci. Reports Res. Inst. Tôhoku Univ. 9(D): 117-136. A14;S1 (7871)

Shimada, Y. [1959], Statistical studies on the design of yield survey and field experiment in natural grassland. Part 3. Estimation of yield, especially with reference to size and shape, and replication of field experimental plot in natural Zoysia grassland. Sci. Reports Res. Inst., Tôhoku Univ. 10(D):87-107. A14;S1 (7872)

Shinozaki, K. and Kira, T. [1956], Intraspecific competition among higher plants. VII. Logistic theory of the C-D effect. J. Inst. Polytech., Osaka City Univ., D, 7: 35-72. A14 (7873)

Shirafuji, M. [1956], Note on the determination of the replication numbers for the slippage problem in r-way layout. Bull. Math. Stat. 7:46-51. S1 (7874)

Shiue, C.-J, and John, H. H. [1962], A proposed sampling design for extensive forest inventory: Double systematic sampling for regression with multiple random starts. J. Forestry 60:607-610. A14 (7875)

Shiue, C. -J. and Pauley, S. S. [1961], Some considerations on the statistical design for provenance and progeny tests in tree improvement programs. Forest Sci. 7: 116-122. A14;C (7876)

Shrikhande, V. J. [1957], Some considerations in designing experiments on coconut trees. JISAS 9:82-99. A14;C (7877)

Shubik, M. [1960], Bibliography on simulation, gaming, artificial intellegence and allied topics. JASA 55:736-751. B (7878)

Shukla, G. K. [1966], An alternative multivariate ratio estimate for finite population. Calcutta Stat. Assoc. Bull. 15:127-134. A7 (7879)

Sibuya, M. [1960], Bivariate extreme statistics, I. Ann. Inst. Stat. Math. 11:195-210. A6,7 (7880)

Sibuya, M. and Haga, T. [1959], Orthogonal polynomials without constant term. Ann. Inst. Stat. Math. 10:209-222. A7 (7881)

Sidak, Z. [1967], Rectangular confidence regions for the means of multivariate normal distributions. JASA 62:626-633. A3,7 (7882)

Sieben, J. W. [1962], Experimenten. (Experiments.) Stat. Neerlandica 16:1-16. C (7883)

Siegel, S. [1957], Nonparametric statistics. Amer. Stat. 11(3):13-19. A12 (7884)

Siegel, S. and Tukey, J. W. [1960], A nonparametric sum of ranks procedure for relative spread in unpaired samples. JASA 55:429-445. A12 (7885)

Simon, K. H. and Thiele, H. [1961], Die Auswertung eines Kiefernkulturversuches mit Hilfe der mehrfachen Varianzanalyse bei ungleicher Besetzung der Untergruppen mit Beobachtungen. Biom. Zeit. 3:92-112. A10 (7886)

Simon, M. and Roussel, N. [1960], L'emploi des méthodes biométriques dans l'étude et l'expérimentation des variéties de betterave sucrière en Belgique. Biométrie-Praximétrie 1(3-4):167-179. C (7887)

Simon, M. and Roussel, N. [1960], L'emploi des méthodes biométriques dans l'étude et l'expérimentation des variétés de betterave sucrière en Belgique. (The use of biometry in the experimentation of sugar-beet varieties in Belgium.) Bull. Inst. Agronomique Sta. Recherches Gembloux (Hors Série) 1:305-317. C (7888)

Simone, M. and Pangborn, R. M. [1957], Consumer acceptance methodology: One vs. two samples. Food Tech. 11:25-29. A14;C (7889)

Singh, D., Krishnan, K. S., and Bhargava, P. N. [1965], Recent study on plot sizes for estimation of crop-production (abstract). JISAS 17:123. A14 (7890)

Singh, J. [1955], A suggested application of Wald's sequential analysis to railway operation. Sankhyā 14:389-392. S2 (7891)

Singh, J. [1963], Statistical theory of selection (selection problems). Ph. D. Thesis, Univ. Toronto. A2;S2 (7892)

Siotani, M. [1955], The significance of the discordant variance estimates. Ann. Inst. Stat. Math. 7:39-55. A11 (7893)

Siotani, M. [1956], Order statistics for discrete case with a numerical application to the binomial distribution. Ann. Inst. Stat. Math. 8:95-104. A12 (7894)

Siotani, M. [1960], Notes on multivariate confidence bounds. Ann. Inst. Stat. Math. 11:167-182. A7 (7895)

Siotani, M. [1960], A note on the interval estimation related to the regression matrix. Ann. Inst. Stat, Math. 12:147-149. A7 (7896)

Sirota, M. [1965], Minimum sample sizes for superiority comparisons of prior tested items. IQC 21:603-605. S1 (7897)

Siskind, V. [1964], On certain suggested formulae applied to the sequential t-test. Biometrika 51:97-106. S2 (7898)

Sisson, D. V. [1963], Combining data over years (abstract). Biometrics 19:663-664. A13;C (7899)

Sisson, D. V., Brindley, T. A., and Bancroft, T. A. [1965], Combining biological data from European corn borer experiments over years. Iowa State J. Sci. 39: 403-415. A13 (7900)

Sitgreaves, R. [1964], Classification into one of k populations (abstract). Technometrics 6:124. A2,7 (7901)

Sittig, J. [1955], De statistische basis van de planning in een machinefabriek. (Statistical planning in an engine factory.) Stat. Neerlandica 9:47-69. C (7902)

Sixtl, F. and Wender, K. [1964], Der Zusammenhang zwischen multidimensionalem Skalieren und Faktorenanalyse. (On the relation between multidimensional scaling and factor analysis.) Biom. Zeit. 6:251-261. A7 (7903)

Sjöström, L. B., Cairncross, S. E., and Caul, J. F. [1957], Methodology of the flavor profile. Food Tech. 11:20-25. C (7904)

Skellam, J. G., Brian, M. V., and Proctor, J. R. [1959], The simultaneous growth of interacting systems. Acta Biotheoretica 13:131-144. A7 (7905)

Skinner, J. C. [1960], Bunch planting experiments. Proc. Tenth Int. Congress Soc. Sugar Cane Tech. pp. 708-714. A14 (7906)

Skinner, J. C. [1961], Sugar cane selection experiments. 2. Competition between varieties. Bur. Sugar Expt. Sta. Queensland, Tech. Comm. 1:1-26. A14 (7907)

Smid, L. J. [1959], De symmetrietoets van Wilcoxon, indien, gelijken optreden. (Wilcoxon's set test for symmetry when ties are present.) Stat. Neerlandica 13:463-464. A12 (7908)

Smillie, K. W. [1964], Remarks on a recent paper on round-off errors in regression analysis. Amer. Stat. 18(4):26. A7 (7909)

Smith, B. B. [1951], On some difficulties encountered in the use of factorial designs and analysis of variance with psychological experiments. British J. Psychology 42:250-268. C (7910)

Smith, D. [1967], Investigation of rejection rules for outliers in small samples from the normal distribution. Ph. D. Thesis, Univ. Wisconsin. A6 (7911)

Smith, D. L. and Chamorro M. R. [1958], Óptimo tamaño de parcela para ensayos de rendimiento en maíz. (Optimum plot size for yield trials of maize.) III. Reunión Interamericana de Fitogenetistas, Fitopatólogos, Entomólogos y Edafólogos, Bogotá, D. E., Colombia, 20 de Junio a 1º de Julio de 1955, p. 411. A14 (7912)

Smith, H. F. [1950], Estimating precision of measuring instruments. JASA 45:447-451. C (7913)

Smith, H. F. [1950], Effect of fertilizers on growth of hevea: A study in combination of data from a heterogeneous group of experiments. J. Rubber Res. Inst. Malaya 12:127-166. A13 (7914)

Smith, H. F. [1951], Analysis of variance with unequal but proportionate numbers of observations in the sub-classes of a two-way classification. Biometrics 7:70-74. A10 (7915)

Smith, H. F. [1956], A multivariate analysis of covariance. Inst. Stat. Mimeo. Series No. 157, Univ. N. Carolina. A7 (7916)

Smith, H. F. [1957], Interpretation of regressions in analysis of covariance (summary). JASA 52:381. A7 (7917)

Smith, H. F. [1957], Interpretation of adjusted treatment means and regressions in analysis of covariance. Biometrics 13:282-308. A7;B (7918)

Smith, H. F. [1958], A multivariate analysis of covariance. Biometrics 14:107-127. A7 (7919)

Smith, K. L. [1961], Sequential analysis applied to biological control tests for pharmacopoeial substances (with discussion). Quantitative Methods in Pharmacology pp. 49-55. S2 (7920)

Smith, N. M. [1963], The role of intuition in the scientific method. Proc. Eighth. Conf Design Expt. Army Res. Dev. Testing. pp 439-451. C (7921)

Snedecor, G. W. [1950], The statistical part of the scientific method. Ann. New York Academy Sci. 52:792-799. C (7922)

Snedecor, G. W. [1953], Query 98. Biometrics 9:107-108. A12 (7923)

Snedecor, G. W. [1953], Query 100. Biometrics 9:253-255. A10 (7924)

Snedecor, G. W. [1954], Biometry, its makers and concepts. Statistics and Mathematics in Biology pp. 3-10. C (7925)

Snell, E. S. and Armitage, P. [1957], Clinical comparison of diamorphine and pholcodine as cough suppressants by a new method of sequential analysis. Lancet 272:860-862. S2 (7926)

Sobel, M. [1966], Selecting the t populations with the largest α- percentiles (abstract). AMS 37:546. A2 (7927)

Sobel, M. [1967], Nonparametric procedures for selection of the t populations with the α-quantiles. AMS 38:1804-1816. A2, 12 (7928)

Sobel, M. and Huyett, M. J. [1958], Nonparametric definition of the representativeness of a sample--with tables. Bell System Tech. J. 37:135-161. A12;S1 (7929)

Sobel, M. and Rizvi, M. H. [1966], Selecting a subset containing the population with the largest α th-percentile (abstract). AMS 37:315. A2 (7930)

Somberg, S. I. [1967], A simplified method of solving simple and multiple regression equations. Amer. Stat. 21(1):31-32. A7 (7931)

Somermeyer, W. H. [1953], Kwantificering in de sociale wetenschappen. (Quantification in the social sciences.) Statistica 7:209-222. C (7932)

Somerville, P. N. [1953], Optimum sample sizes for choosing the largest of (k + 1) means using minimax methods (abstract). AMS 24:490. A2 (7933)

Somerville, P. N. [1954], Some problems of optimum sampling. Biometrika 41:420-429. A2 (7934)

Specht, G. [1957], Zu einigen Problemen bei der Verrechnung und Auswertung von Feldversuchen. Z. Landw. Vers. Untersuch Wes. 2:107-116. C (7935)

Specht, G. and Müller, K.-H. [1960], Probleme bei der Durchführung von Grossflächenstreuversuchen. Z. Landw. Vers. Untersuch Wes. 6:148-165. A13;C (7936)

Spicer, C. C. [1961], Problems in the analysis of a large scale clinical trial. Biometrics 17:332. C (7937)

Spicer, C. C. [1962], Some new closed sequential designs for clinical trials. Biometrics 18:203-211. A2;S2 (7938)

Sprent, P. [1961], Some hypothesis concerning two phase regression lines. Biometrics 17:634-645. A7 (7939)

Sprent, P. [1965], Fitting a polynomial to correlated equally spaced observations. Biometrika 52:275-276. A7 (7940)

Sprent, P. [1966], A generalized least-squares approach to linear functional relationships (with discussion). JRSSB 28:278-297. A7 (7941)

Spurrell, D. J. [1967], Letter to the Editor. Technometrics 9:495. A7 (7942)

Spurway, H. [1956], Is biometry a separate discipline? Sankhyā 16:215-220. C (7943)

Sreenath, P. R. and Sardana, M. G. [1967], Fitting of Mitscherlich's curve to unequally spaced levels. JISAS 19(2):53-66. A7 (7944)

Srikantan, K. S. [1961], Testing for the single outlier in a regression model. Sankhyā A 23:251-260. A6,7 (7945)

Srivastava, M. S. [1964], On a multivariate slippage problem (abstract). AMS 35:1848. A2,7 (7946)

Srivastava, M. S. [1966], On the asymptotic theory of sequential confidence intervals (abstract). AMS 37:313. A3;S2 (7947)

Srivastava, M. S. [1966], On a multivariate slippage problem. Ann. Inst. Stat. Math. 18:299-305. A2,7 (7948)

Srivastava, M. S. [1966], Some asymptotically efficient sequential procedures for ranking and slippage problems. JRSSB 28:370-380. A2;S2 (7949)

Srivastava, R. S. [1963], Distribution of sugarcane clumps with regard to tiller number and its transformation for analysis of variance. JISAS 15:185-193. A9 (7950)

Srivastava, S. R. and Gupta, V. P. [1965], Estimation after preliminary testing in ANOVA Model I. Calcutta Stat. Assoc. Bull. 14:151-159. A11 (7951)

Stanley, J. C. [1957], Partially annotated bibliography to supplement Julian C. Stanley's chapter "Research methodology: Experimental design", which appears in the December 1957 (vol.27, No. 5) issue of the Review of Educational Research and lists 83 references. Dept. Education Univ. Wisconsin (mimeo.), 16 pp. B (7952)

Stanley, J. C. [1957], Research Methods: Experimental design. Rev. Educational Res. 37:449-459. C (7953)

Stanley, J. C. [1957], Additional "post-mortem" tests of experimental comparisons. Psychological Bull. 54:128-130. A3 (7954)

Stanley, J. C. [1962], Analysis-of-variance principles applied to the grading of essay tests. J. Expt. Education 30:279-283. C (7955)

Stanley, W. C. [1954], Note on score transformations and nonparametric statistics. Psychological Bull. 51:517. A12 (7956)

Starr, N. [1966], The performance of a sequential procedure for the fixed-width interval estimation of the mean. AMS 37:36-50. S2 (7957)

Steel, R. G. D. [1955], An analysis of perennial crop data. Biometrics 11:201-212. A7 (7958)

Steel, R. G. D. [1959], A multiple comparison rank sum test: treatments versus control. Biometrics 15:560-572 (abstract 15:495-496; Math. Res. Center Tech. Summary Report No. 82, U.S. Army and Univ. Wisconsin). A3,12 (7959)

Steel, R. G. D. [1959], A multiple comparison sign test: treatments versus control (abstract). Biometrics 15:496. A3,12 (7960)

Steel, R. G. D. [1959], A multiple comparison sign test: Treatments versus control. JASA 54:767-775 (Math. Res. Center Tech. Summary Report No. 83, U.S. Army and Univ. Wisconsin). A3,12 (7961)

Steel, R. G. D. [1959], A rank sum test for comparing all pairs of treatments (abstract). AMS 30:1278. A3,12 (7962)

Steel, R. G. D. [1960], A rank sum test for comparing all pairs of treatments. Technometrics 2:197-207 (Math. Res. Center Tech. Summary Report No. 108, U.S. Army and Univ. Wisconsin. A3,12 (7963)

Steel, R. G. D. [1961], Query 163. Error rates in multiple comparisons. Biometrics 17:326-328. A1 (7964)

Steel, R. G. D. [1961], A variable rank sum multiple comparisons test (abstract). Biometrics 17:176. A3 (7965)

Steel, R. G. D. [1961], Some rank sum multiple comparisons tests. Biometrics 17:539-552. A3,12 (7966)

Steel, R. G. D. [1962], Multiple comparisons sign tests (abstract). Biometrics 18:265. A3,12 (7967)

Stein, C. [1960], Multiple regression. <u>Contributions to Probability and Statistics</u> pp. 424-443. A7 (7968)

Stern, K. [1954], Zur Entwicklung eines forstlichen Sortenversuchswesens. Z. Forstgenetik Forstpflanzenzüchtung 3:91-98. C (7969)

Stern, K. [1954], Die Tielstückgrösse im Sortenversuch (abstract). Z. Forstgenetik Forstpflanzenzüchtung 3:141. A14 (7970)

Stern, K. [1954], Ein Model für die Wechselwirkungen des Wachstums. Der Züchter 24:216-220. C (7971)

Stern, W. R. [1965], The effect of density on the performance of individual plants in subterranean clover swards. Australian J. Agri. Res. 16:541-555. A14 (7972)

Stevens, M. E. and Alexander, S. N. [1957], Comparative characteristics of medium-priced fully automatic computers for statistical applications. Proc. First Conf. Design Expt. Army Res. Dev. Testing. pp. 137-144. A15 (7973)

Stevens, W. L. [1951], Asymptotic regression. Biometrics 7:247-267. A7 (7974)

Stevens, W. L. [1953], Tables of the angular transformation. Biometrika 40:70-73. A9 (7975)

Stewart, G. F. [1954], Analytical sensory tests for food products. Statistics and Mathematics in Biology pp. 371-381. C (7976)

Stewart, H. L. and Herrold, K. M. [1962], A critique of experiments on attempts to induce cancer with tobacco derivatives. Bull. ISI 39(3):457-477. B;C (7977)

Stickler, F. C. [1960], Estimates of optimum plot size from grain sorghum uniformity trial data. Kansas State Univ. Agri. Expt. Sta. Tech. Bull. 109:1-21. A14 (7978)

Stickler, F. C. [1964], Row width and plant population studies with corn. Agron. J. 56:438-441. A14 (7979)

Stickler, F. C. and Younis, M. A. [1966], Plant height as a factor affecting responses of sorghum to row width and stand density. Agron. J. 58:371-373. A14 (7980)

Stivers, R. K. [1956], Influence of interplanting of corn and ladino clover on the yields of corn. Agron. J. 48:97-98. A14 (7981)

Stolurow, L. M. and Frincke, G. [1966], A study of sample size in making decisions about instructional materials. Educational Psycological Measurement 26:643-659. S1 (7982)

Stone, M. [1967], Extreme tail probabilities for the null distribution of the two-sample Wilcoxon statistic. Biometrika 54:629-640. A12 (7983)

Storch, J. M. [1953], Sequente analyse bij keuring van goederen. (Sequential sampling Statistica 7:3-14. S2 (7984)

Stout, L. E. [1957], Application of sequential analysis to catapult testing. Proc. First Conf. Design Expt. Army Res. Dev. Testing. pp. 217-227. S2 (7985)

St-Pierre, J. and Zinger, A. [1956], The null distribution of the difference between the two largest sample values. AMS 27:849-851. A2 (7986)

Strand, L. [1955], Plot sizes in field trails. Z. Forstgenetik Forstpflanzenzüchtung 4:157-162. A14 (7987)

Strand, L. [1958], Die statistische Auswertung von vergluchenden Versuchen. Silvae Genetica 7:9-12. A3 (7988)

Strydom, G. J. [1966], Studies on the planning of field experiments with vegetable crops. Suid-Afrikaanse Tyskrif Landbouwetenskap 9:183-194. A14 (7989)

Stuart, A. [1951], An application of the distribution of the ranking concordance coefficient. Biometrika 38:33-42. A12 (7990)

Studden, W. J. [1967], On selecting a subset of k populations containing the best. AMS 38:1072-1078. A2 (7991)

Sugimura, M. and Asano, C. [1967], Optimum designs for selecting one of two medical treatments sequential plan 1. Bull. Math. Stat. 12(3&4):1-9. S2 (7992)

Suits, D. B. [1957], Use of dummy variables in regression equations. JASA 52: 548-551. A7 (7993)

Sukhatme, B. V. [1958], A two sample distribution free test for comparing variances. Biometrika 45:544-548. A12 (7994)

Sule, A. G. [1967], Genesis of experimental designs (abstract). JISAS 19(1):154-155. C (7995)

Sullivan, J. W. W. [1955], Nonrandomness. ASQC, Nat. Conv. Trans. 9:5-7. C (7996)

Sung, C.-H. [1957], A study of sampling technique in rice field trials (in Chinese). Nung-yeh-hsueh Pao/Act. Agri. Sinica 8:178-184. A14 (7997)

Sunseon, C. A. and Ramage, R. T. [1962], Competition between near-isogenic genotypes. Crop Sci. 2:249-250. A14 (7998)

Sutherland, W. H. [1965], Problems in the design of statistics-generating war games. Proc. Tenth Conf. Design Expt. Army Res. Dev. Testing. pp. 685-690. C (7999)

Suzuki, Y. [1961], On sequential decision procedures. Bull. ISI 38(4):201-205. S2 (8000)

Suzuki, Y. [1964], On the use of some extraneous information in the estimation of the coefficients of regression. Ann. Inst. Stat. Math. 16:161-173. A7 (8001)

Sváb, J. [1956], A fajta és tenyészterület kölcsönhatása kukoricakísérletekben. (The interaction between variety and planting distance in maize trials.) Kísérl-Ügy. Közl. 50:89-101. A14 (8002)

Sváb, J. [1957], A nagyüzemi fajtakísérletek módszerei. (Methods for large-scale varietal trials.) Agrártudomány 9(9):1-7. C (8003)

Sváb, J. [1957], Überlegungen zu methodischen Fragen der Grossflächenversuche. Z. Landw. Vers.- Untersuch Wes. 3:268-278. C (8004)

Sváb, J. [1959], Szubjektív fajtabírálat módszere nagyüzemi fajtakísérletekben. (A method of subjective evaluation of varieties in large-scale varietal trials.) Nemesített Növényfajtakkal Végzett országos fajtakísérletek eredményei 1957. (Results of the national varietal trials of improved crop varieties, 1957), Budapest, pp. 50-61. A13 (8005)

Sváb, J. [1960], A termésmennyiség, minöség és idényszerüség együttes gazdasági értékének meghatározása kertészeti kísérletekben. (The determination of the combined economic value of crop yield, quality and seasonal value in horticultural experiments.) Kísérl-Ügy. Közl. 53(C)(3):3-18. C (8006)

Sváb, J. [1960], A nagyüzemi kísérletek pontossága. (The accuracy of large-scale trials.) Kísérl-Ügy. Közl. 53(s):119-129. A13;C (8007)

Sváb, J. [1960], Zur Methode der Grossflächen-Streuversuche: Meinungsforschung. Z. Landw. Vers.- Untersuch Wes. 6:182-191. C (8008)

Sváb, J. [1961-2], Über die Realisierung der Ergebnisse aus Exaktversuchen unter den Bedingungen des Grossanbaues. Acta Agronomica, Budapest 11:321-328. A13;C (8009)

Swain, A. K. P. C. [1964], On Stein's two sample theory to test the difference between means of normal populations with common unknown variance and unequal amount of sampling from the populations. JISA 2:165-172. S2 (8010)

Swain, A. K. P. C. and Ghosh, M. N. [1964], Mixed model analysis of covariance in agricultural experiments (abstract). JISAS 16:171. A7 (8011)

Swaroop, S. [1961], Statistical techniques in medical research. Bull. ISI 38(3): 153-167. C (8012)

Takahashi, K. [1961], Factor analysis of autonomic regulating system. Bull. ISI 38(3):169-195. A7 (8013)

Talacko, J. [1962], Professor Ronald A. Fisher: su vida, trabajos y publicaciones. Trabajos Estadica 13:155-172. B (8014)

Tamura, R. [1963], On a modification of certain rank tests. AMS 34:1101-1103. A12 (8015)

Tarver, M. G. and Ellis, B. H. [1960], Selection of flavor panels for complex flavor difference. ASQC, Nat. Conv. Trans. 14:277-284. C (8016)

Tarver, M. G. and Ellis, B. H. [1961], Selection of flavor panels for complex flavor differences. IQC 17(12):22-26. C (8017)

Tarver, M. G. and Schenck, A. M. [1957], The SQC approach to multiple correlation and regression in food packaging research. Food Tech. 11:558-562 (erratum 11:692). A7 (8018)

Tattersfield, J. R. [1960], Field experiments: their design, management and interpretation. Rhodesia Agri. J. 57:362-371. C (8019)

Taylor, H. L. [1951], An examination of the effect of plot shape on experimental error. M. S. Thesis, Iowa State Univ. A14 (8020)

Taylor, J. [1951], Statistical studies on strawberry crop and vigour measurements. Annual Report, East Malling Res. Sta., 1950, pp. 100-107. A14 (8021)

Taylor, J. [1956], Exact linear sequential tests for the mean of a normal distribution. Biometrika 43:452-455. S2 (8022)

Taylor, L. H. [1964], Methods of planting orchardgrass for evaluating yield and disease resistance (abstract). Agron. Abstracts p. 81. A14 (8023)

Taylor, R. J. and David, H. A. [1961], Sequential allocation of patients in clinical trials (abstract). Biometrics 17:504-505. S2 (8024)

Tedin, O. and Gösta, J. [1953], Ett fall, där variansanalysen lämnar felaktig uppfattning om den statistiska säkerheten hos ett försöksresultat. (A case, where the analysis of variance gives an erroneous estimate of the statistical significance of an observed difference.) Sveriges Utsädesförenings Tidskrift (Sweden) 63(5): 469-474. A11 (8025)

Tedin, O. and Julén, G. [1953], Ett fall, där variansanalysen lämnar felaktig uppfattning om den statistiska säkerheten hos ett försöksresultat. (A case in which the analysis of variance gives a false impression regarding the statistical reliability of the result of an experiment.) Sveriges Utsädesförenings Tidskrift, Malmö 63:469-474. A3 (8026)

Teghem, J. [1954], Sur la regression polynomiale, dans le cas ou les valeurs de la variable independante sont en progression arithmetique lacunaire. Bull. ISI 34(2):109-121. A7 (8027)

Telser, L. G. [1964], Iterative estimation of a set of linear regression equations. JASA 59:845-862. A7 (8028)

Terpstra, T. J. [1954], A non-parametric test for the problem of k samples. Koninklijke Nederlandse Akademie van Wetenschappen 57(A):505-512. A12 (8029)

Terry, M. E. [1955], On the analysis of planned experiments. ASQC, Nat. Conv. Trans. 9:553-556. A6 (8030)

Terry, M. E. [1959], Polynomial curve fitting on electronic computers. ASQC, Nat. Conv. Trans. 13:565-573. A7,15 (8031)

Terry, M. E. [1960], An optimum replicated two-sample test using ranks. Contributions to Probability and Statistics pp. 444-447. A12 (8032)

Tetreault, F. G. [1966], Statistical outlier methodology for observed points and lines. Ph. D. Thesis, Iowa State Univ. A6 (8033)

Theil, H. [1951], Verdelingsvrije methoden in de regressieanalyse van twee variabelen. (Distribution-free methods in the regression analysis of two variables.) Statistica 5:97-117. A7,12 (8034)

Theil, H. [1963], On the use of incomplete prior information in regression analysis. JASA 58:401-414. A7 (8035)

Theil, H. and Schweitzer, A. [1961], The best quadratic estimator of the residual variance in regression analysis. Stat. Neerlandica 15:19-23. A7 (8036)

Thiers, R. E. [1951], A punch card system for the bibliography of analytical chemistry. Virginia J. Sci. 2:28-45. B (8037)

Thionet, P. [1966], Sur certains tests non parametriques bien connus. Rev. ISI 34:13-26. A12 (8038)

Thoele, H. W. and Hervey, M. C. [1952], Growth uniformity trials with identical twin dairy heifers--estimates of heritability and twin efficiency (abstract). J. Dairy Sci. 35:494-495. A14 (8039)

Thomas, W. [1956], Effect of plant population and rates of fertilizer nitrogen on average weight of ears and yield of corn in the South. Agron. J. 48:228-230. A14 (8040)

Thompson, D. J. and Hutchcroft, C. D. [1951], Accuracy of estimating the mean percentage of nondetasseled plants in double cross corn seed production fields. Agron. J. 43:609-616. S1,2 (8041)

Thompson, N. R. and Terry, M. E. [1951], A Latin square grazing plan for dairy cattle (abstract). Virginia J. Sci. 2:375. A13 (8042)

Thompson, W. A. [1956], A note on ranking means (abstract). AMS 27:548. A2,3 (8043)

Thompson, W. A. [1962], Simultaneous confidence intervals for the dispersion parameters (preliminary report) (abstract). AMS 33:824. A3 (8044)

Thompson, W. A. [1963], Precision of simultaneous measurement procedures. Proc. Eighth Conf. Design Expt. Army Res. Dev. Testing. pp. 175-185. A11;C (8045)

Thompson, W. A. and Endriss, J. [1961], The required sample size when estimating variances. American Stat. 15(3):22-23. S1 (8046)

Thompson, W. A. and Willke, T. A. [1963], On an extreme rank sum test for outliers. Biometrika 50:375-383. A6,12 (8047)

Thompson, W. A. and Willke, T. A. [1963], On a rank sum test for outliers (summary). JASA 58:563. A6,12 (8048)

Thomson, K. F. [1957], An example of experimental design in personnel research. Proc. First Conf. Design Expt. Army Res. Dev. Testing. pp. 29-33. C (8049)

Thöni, H. [1967], Transformations of variables used in the analysis of experimental and observational data. A review. Stat. Lab. Tech. Report No.7, Iowa State Univ. A9;B (8050)

Thrall, R. M. [1967], Best fitting linear varieties. Proc. Twelfth Conf. Design Expt. Army Res. Dev. Testing. pp. 311-318. A7 (8051)

Tiao, G. C. [1966], Bayesian analysis of hierarchical design model (abstract). AMS 37:767-768. A11 (8052)

Tiao, G. C. and Guttman, I. [1967], Analysis of outliers with adjusted residuals. Technometrics 8:541-559. A6 (8053)

Tiao, G. C. and Zellner, A. [1964], On the Bayesian estimation of multivariate regression. JRSSB 26:277-285. A7 (8054)

Tiao, G. C. and Zellner, A. [1964], Bayes's theorem and the use of prior knowledge in regression analysis. Biometrika 51:219-230. A7 (8055)

Tiedemann, H. J. [1964], Statistical testing techniques used in the development of the Pratt and Whitney Aircraft RL10 rocket engine for the National Aeronautics and Space Administration. Proc. Ninth Conf. Design Expt. Army Res. Dev. Testing. pp. 191-203. C (8056)

Tingey, D. C. [1952], Effect of spacing, irrigation, and fertilization on rubber production in guayule sown directly in the field. Agron. J. 44:298-302. Al4 (8057)

Tintner, G. [1950], A test for linear relations between weighted regression coefficients. JRSSB 12:273-277. A7 (8058)

Tippett, H. C. [1963], Quelques plans d'expérience réalisés dans l'industrie textile. Rev. Stat. Appl. 11(1):77-86. C (8059)

Tippett, L. H. C. [1959], Statistics in the textile industry. Proc. Fourth Conf. Design Expt. Army Res. Dev. Testing. pp. 273-289. C (8060)

Tobach, E., Smith, M., Rose, G., and Richter, D. [1967], A table for rank sum multiple paired comparisons. Technometrics 9:561-567. A3,12 (8061)

Tong, Y. L. [1967], On partitioning a set of normal populations into those worse and those better than a standard by a single stage, two stage and sequential procedure (abstract). AMS 38:958-959. A3;S2 (8062)

Tormann, H. [1958], Blindversuche in Feldbeständen von Futterrüben und Winterweizen als Beitrag zur Felderversuchsmethodik. Albrecht-Thaer-Arch. 2:316-317. Al4 (8063)

Torrie, J. H., Schmidt, D. R., and Tenpas, G. H. [1963], Estimates of optimum plot size and shape and replicate number for forage yield of alfalfa-bromegrass mixtures. Agron. J. 55:258-260. Al4 (8064)

Truax, D. [1954], An optimum slippage test for the variance of K normal populations (summary). JASA 49:359. A2 (8065)

Tucker, L. R. [1958], Determination of parameters of a functional relation by factor analysis. Psychometrika 23:19-23. A7 (8066)

Tucker, L. R. [1958], An inter-battery method of factor analysis. Psychometrika 23:111-136. A7 (8067)

Tukey, J. W. [1950], Some sampling simplified. JASA 45:501-519. Al4 (8068)

Tukey, J. W. [1951], Components in regression. Biometrics 7:33-69. A7 (8069)

Tukey, J. W. [1951], Quick and dirty methods in statistics, Part II, Simple analyses for standard designs. ASQC, Ann. Conv. Trans. 5:189-197. Al2 (8070)

Tukey, J. W. [1953], The problem of multiple comparisons. Unpublished dittoed notes, Princeton Univ., 396 pp. Al,3 (8071)

Tukey, J. [1953], Multiple comparisons (summary). JASA 48:624-625. A3 (8072)

Tukey, J. W. [1954], Causation, regression, and path analysis. Statistics and Mathematics in Biology pp. 35-66. A7 (8073)

Tukey, J. W. [1954], Unsolved problems of experimental statistics. JASA 49:706-731 (summary 49:343). A14;C (8074)

Tukey, J. W. [1954], Query 111. Biometrics 10:412. A4 (8075)

Tukey, J. W. [1955], Interpolations and approximations related to the normal range. Biometrika 42:480-485. A3 (8076)

Tukey, J. W. [1955], Mathematical consultants, computational mathematics and mathematical engineering. American Math. Monthly 62:565-571. C (8077)

Tukey, J. W. [1957]. On the comparative anatomy of transformations. AMS 28: 602-632. A9 (8078)

Tukey, J. W. [1958], Bias and confidence in not-quite large samples (preliminary report) (abstract). AMS 29:614. A14 (8079)

Tukey, J. W. [1959], A quick, compact, two-sample test to Duckworth's specifications. Technometrics 1:31-48. A12 (8080)

Tukey, J. W. [1959], A survey of sampling from contaminated distributions. Contribution to Probability and Statistics (Stat. Tech. Res. Group Tech. Report No. 33 Princeton Univ). A6 (8081)

Tukey, J. W. [1960], Where do we go from here? JASA 55:80-93. C (8082)

Tukey, J. W. [1960], Conclusions vs decisions. Technometrics. 2:423-433. C (8083)

Tukey, J. W. [1962], The future of data analysis. AMS 33:1-67 (correction 33:812). C (8084)

Tukey, J. W. [1963], A citation index for statistics and probability (with discussion). Bull. ISI 40:747-756. B (8085)

Tukey, J. W. [1965], The technical tools of statistics. Amer. Stat. 19(2):23-28. C (8086)

Tukey, J. W. [1965], The inevitable collision between computation and data analysis. Proc. IBM Scientific Computing Symp. Stat. pp. 141-152. C (8087)

Tukey, J. W. [1965], The future of processes of data analysis. Proc. Tenth Conf. Design Expt. Army Res. Dev. Testing. pp. 691-729. A14 (8088)

Tukey, J. W. et al. [1957], Panel discussion on how and where do statisticians fit in. Proc. First Conf. Design Expt. Army Res. Dev. Testing. pp. 255-278. C (8089)

Tung, S. S. Y., Koontz, R. S. and Chasteen, J. R. [1966], Multiple correlation applications in design analysis. ASQC Ann. Tech. Conf. Trans. pp. 169-178. A7 (8090)

Tünnerhoff, K. [1962], Der individuelle therapeutische Vergleich bei chronischen Formen der Lungentuberkulose. Methods Information Medicine 1:120-129. C (8091)

Turnbull, K. J. and Pienaar, L. V. [1966], A non-linear mathematical model for analysis and prediction of growth in non-normal and thinned forest stands. Int. Advisory Group Forest Stat., Second Conf. Institutionen Skoglig Matematisk Stat. pp. 319-333. A7 (8092)

Turner, M. E. [1960], Straight line regression through the origin. Biometrics 16:483-485. A7 (8093)

Turner, M. E., Monroe, R. J., and Lucas, H. L. [1961], Generalized asymptotic regression and non-linear path analysis. Biometrics 17:120-143. A7 (8094)

Uematu, T. [1959], Note on the numerical computation in the discrimination problem. Inst. Stat. Math. 10:131-135. A7 (8095)

Uematu, T. [1964], On a multidimensional linear discriminant function. Ann. Inst. Stat, Math. 16:431-437. A7 (8096)

Uhlmann, W. [1959], Zu einem nichtparametrischen Test von E. L. Lehmann. Metrika 2:169-185. A12 (8097)

Ulmo, J. [1958], Applications de la technique des plans d'expériences à des essais industriels exécutés à l'échelon international. Bull. ISI 36(3):563-572. C (8098)

Ulmo, J. [1963], Contribution a l'étude de la régression linéare et des plans d'expériences. Colloques Internationaux du Centre National de la Recherche Scientifique No. 110 Le Plan d'Expériences pp. 215-232. A7 (8099)

Ungria, A. D., Camacho, A., and Rios, S. [1955], Analisis discriminante de dos muestras de indios venezolanos. Trabajos Estadistica 6:237-242. A7 (8100)

Unterstenhöfer, G. [1957], Die Grundlagen des Pflanzenschutz-Freilandversuches. Höfchen-Briefe 10(4):169-232. C (8101)

Ura, S. [1960], Selection of judges by ranking method. Reports Stat. Appl. Res. 7:120-130. C (8102)

Ury, H. K. [1966], A note on taking a covariable into account. JASA 61:490-495. A7,12 (8103)

Ury, H. K. [1966], The behavior of some tests for ordered alternatives under interior slippage (abstract). AMS 37:1866. A2 (8104)

Utida, S. [1953], Interspecific competition between two species of bean weevil. Ecology 34:301-307. A14 (8105)

Vágnerová, V. [1961], Příspěvek k mežnostem uspořádání zkoušky hromadným křížením při šlechtění trav (trojúhelníkový systém). (A contribution to the possibilities of setting up polycross tests in grass breeding (the triangular system).) Sborn. Vysoké Skol. Zeměd. les. Brně Řada A. No:1-2:113-123. A14 (8106)

van Dalfsen, J. W. [1955], Statistische beschouwingen in verband met textiel. (Statistics in textile industry.) Stat. Neerlandica 9:37-42. C (8107)

van Dantzig, D. [1953], De verantwoordelijkheden van de statisticus. (Responsibilities of the statistician.) Statistica 7:199-208. C (8108)

van Dantzig, D. [1957], Statistical priesthood. (Savage on personal probabilities [1]). Stat. Neerlandica 11:1-16. C (8109)

van Dantzig, D. [1957], Statistical priesthood II. Sir Ronald on scientific inference. Stat. Neerlandica 11:185-200. C (8110)

van Dantzig, D. and Hemelrijk, J. [1954], Statistical methods based on few assumptions. Bull. ISI 34(2):239-265. A12 (8111)

van der Burg, A. R. [1958], Statistiek of experiment. (Statistics or experiment.) Stat. Neerlandica 12:131-134. C (8112)

van der Knaap, W. P. [1955], Blancoproeven met cacao en koffie. (Uniformity trails with cacao and coffee.) Archief voor Koffiecultuur 17:187-239. A14 (8113)

van der Laan, P. [1966], A sequential distribution-free two-sample grouped test with three possible decisions. Stat. Neerlandica 20:31-41. A12;S2 (8114)

van der Mooren, A. L. [1965], Dimensionering van homogeniseringsinstallaties. (Designing of homogenizing equipment.) Stat. Neerlandica 19:107-127. A14 (8115)

van der Paauw, F. [1952], Critical remarks concerning the validity of the Mitscherlich effect law. Plant Soil 4:97-106. A7 (8116)

van der Reyden, D. [1952], A simple statistical significance test. Rhodesia Agri. J. 49:96-104. A12 (8117)

van der Heiden, J. A. [1952], Methode van de kleinste kwadraten, toegepast op een stel simultane vergelijkingen die een of meer constanten gemeen hebben. (The method of least squares, applied to a set of simultaneous equations which have one or more constants in common.) Statistica 6:107-112. A7 (8118)

van der Vaart, H. R. [1962], The role of mathematical models in biological research. Bull. ISI 39(2):31-59. C (8119)

van der Vaart, H. R. and van Dantzig, D. [1951], De toets van Wilcoxon. (Two lectures on Wilcoxon's two sample test.) Statistica 5:185-190.
A12 (8120)

van der Waerden, B. L. [1953], Ein neuer Test für das Problem der zwei Stichproben. Mathematische Annalen 126:93-107.
A11 (8121)

van Dyke, J. [1964], Fitting y = βx when the variance depends on x. J. Res. NBS 68B:67-72.
A7 (8122)

van Dyne, G. M. [1960], A method for random location of sample units in range investigations. J. Range Management 13:152-153.
A14 (8123)

van Dyne, G. M., Vogel, W. G., and Fisser, H. G. [1963], Influence of small plot size and shape on range herbage production estimates. Ecology 44:746-759.
A14 (8124)

van Eeden, C. [1955], A sequential test with three possible decisions for comparing two unknown probabilities, based on groups of observations. Rev. ISI 23:20-28.
S2 (8125)

van Eeden, C. [1956], Verdelingsvrije toetsen voor twee steekproeven en de methode der 2 × 2-tabel. (Distributionfree two sample tests and the method of the 2 × 2-table.) Stat Neerlandica 10:157-162.
A12 (8126)

van Eeden, C. and Benard, A. [1957], A general class of distributionfree tests for symmetry containing the test of Wilcoxon and Fisher. Koninklijke Nederlandse Akademie van Wetenschappen, Proc. Ser. A, 60:381-408.
A12 (8127)

van Eeden, C. and Rümke, C. L. [1958], Wilcoxon's two sample test. Stat. Neerlandica 12:275-280.
A12 (8128)

van Eijk, C. J. and Moerman, H. [1957], Opmerkingen over de standaardafwijking van de constante term bij regressie-analyse. (Some remarks on the standard error of the constant term in regression analysis.) Stat. Neerlandica 11:23-30.
A7 (8129)

van Elteren, P. [1960], On the combination of independent two sample tests of Wilcoxon. Bull. ISI 37(3):351-361.
A12 (8130)

van Elteren, P. [1961], Two restricted sequential sign tests. Stat. Neerlandica 15:91-96.
A12;S2 (8131)

van Ijzeren, J. [1954], De theoretische zijde van de methode der kleinste kwadraten. (The theoretical aspect of least squares.) Statistica 8:21-45.
A7 (8132)

van Ijzeren, J. [1957], Practische berekening van de standaardafwijking van de constante term bij regressie-analyse. (Practical calculation of the standard error of the constant term in regression analysis.) Stat. Neerlandica 11:161-167.
A7 (8133)

van Noordwijk, J. [1964], Communication between the experimental animal and the pharmacologist. Stat. Neerlandica 18:403-416. C (8134)

van Strik, R. [1958], Statistiek en experiment rondom het poliomyelitis vaccin. (Statistics and experiment about the poliomyelitis vaccine.) Stat. Neerlandica 12:159-169. C (8135)

Varangot, V. [1951], Grafische bepaling van de tweevoudige correlatiecoëfficiënt. (A nomograph for the multiple correlation coefficient.) Statistica 5:145-147. A7 (8136)

Vengris, J. [1966], Competition between barnyardgrass and alfalfa. Agron. J. 58:478-479. A14 (8137)

Verdade, F. da C., Venturini, W. R., do Amaral, A. Z. and Wutke, A. C. P. [1966], Níveis de fertilidade do solo para a cultura algodoeira. II - Correlação entre a produção e o teor de fósforo no solo. (Level of soil fertility for cotton crops. II - Correlation between yield and phosphorus content in the soil.) Bragantia 25:41-55. A13 (8138)

Verdooren, L. R. [1963], Extended tables of critical values for Wilcoxon's test statistic. Biometrika 50:177-186 (corrigendum 51:527). A12 (8139)

Verdooren, L. R. [1964], Rassenproeven als voorbeeld van een proeftechnisch compromis. (Varietal trials as an example of compromise in trial technique.) Landbouwkundig Tijdschrift, Wageningen 76:554-561. C (8140)

Verhagen, A. M. W. [1960], Growth curves and their functional form. Australian J. Stat. 2:122-127. A7 (8141)

Verhagen, A. M. W. [1961], The estimation of regression and error-scale parameters, when the joint distribution of the errors is of any continuous form and known apart from a scale parameter. Biometrika 48:125-132. A7 (8142)

Verhagen, A. M. W. [1963], The "caution-level" in multiple tests of significance. Australian J. Stat. 5:41-48. A3 (8143)

Verma, M. C. and Ghosh, M. N. [1963], Simultaneous tests of linear hypotheses and confidence interval estimation. JISAS 15:194-212. A3 (8144)

Vermetten, J. B. and Lisman, J. H. C. [1964], Het quotiënt van Von Neumann bij regressie-analyse. Stat. Neerlandica 18:129-138. A7 (8145)

Vervelde, G. J. [1960], Enkele gedachten over de kosten van proefvelden. (Some thoughts on the cost of experimental plots.) Landbouwkundig Tijdschrift, Wageningen 72:830-837. C (8146)

Vincent, S. E. [1961], A test of homogeneity for ordered variances. JRSSB 23:195-206. A11 (8147)

Vogel, W. [1960], A sequential design for the two armed bandit. AMS 31:430-443.
S2 (8148)

Vogel, W. [1961], Sequentielle Versuchs-Pläne. Metrika 4:140-157.
S2 (8149)

Voss, R. and Pesek, J. [1955], Geometrical determination of uncontrolled-controlled factor relationships affecting crop yield. Agron. J. 57:460-463.
A7 (8150)

Wadley, F. M. [1958], Evaluation of infective virus preparations as to potency. Proc. Third Conf. Design Expt. Army Res. Dev. Testing. pp. 166-171. C (8151)

Wagner, F. [1959], Die technische Durchführung von Feldversuchen. Verlag Paul Parey, Berlin and Hamburg pp. 174. C (8152)

Wagner, G. [1964], Versuchsplanung in der Fehlerforschung. Methods Information Medicine 3:117-127. C (8153)

Wald, A. [1951], Sobre los principios de la inferencia estadistica. Trabajos Estadistica 2:113-148. C (8154)

Walker, A. M. [1950], Note on sequential sampling formulae for a binomial population. JRSSB 12:301-307. S2 (8155)

Wallace, A. T., and Chapman, W. H. [1956], Studies in plot technique for oat clipping experiments. Agron. J. 48:32-35. A14 (8156)

Wallace, A., Mueller, R. T. and Squier, M. G. [1952], Variability in orange leaves of the same age and collected from a single tree. Proc. American Soc. Hort. Sci. 60:51-54. S1 (8157)

Wallace, D. L. [1954], Multiple tests and intersection region procedures (abstract). AMS 25:175-176. A3 (8158)

Wallace, D. L. [1957], Multiple comparisons in the analysis of variance. ASQC, Ann. Conv. Trans. 11:279-285. A3 (8159)

Wallis, W. A. [1951], Tolerance intervals for linear regression. Proc. Second Berkeley Symp. Math. Stat. Prob. pp. 43-51. A7 (8160)

Wallis, W. A. [1952], Rough-and-ready statistical tests. IQC 8(5):35-40. A12 (8161)

Walsh, J. E. [1950], Some nonparametric tests of whether the largest observations of a set are too large or too small. AMS 21:583-592. A6,12 (8162)

Walsh, J. E. [1950], Correction to: On the best choice of sample sizes for a t-test when the ratio of variances in known, JASA 44:554-558. JASA 45:111. S1 (8163)

Walsh, J. E. [1951], Some nonparametric results for experimental designs (abstract). AMS 22:608. A12 (8164)

Walsh, J. E. [1952], Some nonparametric tests for Student's hypothesis in experimental studies. JASA 47:401-415. A12 (8165)

Walsh, J.E. [1953], Large sample confidence intervals for density function values at percentage points. Sankhyā 12:265-276. S1 (8166)

Walsh, J. E. [1956], Nonparametric mean estimation of percentage points and density function values. Ann. Inst. Stat. Math 8:167-180. A12 (8167)

Walsh, J. E. [1958], Efficient small sample nonparametric median tests with bounded significance levels. Ann. Inst. Stat. Math, 9:185-199. A12 (8168)

Walsh, J. E. [1959], Large sample nonparametric rejection of outlying observations. Ann. Inst. Stat. Math. 10:223-232. A6,12 (8169)

Walsh, J. E. [1959], Comments on "The simplest signed-rank tests". JASA 54: 213-224. A12 (8170)

Walsh, J. E. [1960], Nonparametric tests for median by interpolation from sign tests. Ann. Inst. Stat. Math. 11:183-188. A12 (8171)

Walsh, J. E. [1963], Simultaneous confidence intervals for differences of classification probabilities. Biom. Zeit. 5:231-234. A3 (8172)

Wang, C.-M. [1955], New methods of testing significance of mean difference between two samples (in Chinese). Taiwan Tahsueh Nunghsuehyuan Yenchiu Paokao/Memoirs College Agri. Nat. Taiwan Univ. 4(10):8-22. A11 (8173)

Wang, C.-M. [1956], The past and present of tests for the significance of differences between means (in Chinese). Chunghua Nunghsueh Hui Pao/J. Agri. Assoc. China New Series No. 13:28-34. A3 (8174)

Wang, C. -M. [1956], The problem of comparing two samples (in Chinese). Taiwan Ta-hsueh Nung-hsueh-yuan Chuan-Kan/Spec. Pub. Taiwan Univ. College Agri. No. 2:1-30. A11 (8175)

Wang, C. -M. [1958], Theoretical bases for the statistical treatment of percentages (in Chinese). Chunghua Nung-hsueh Hui Pao/J. Agri. Assoc. China 24:1-9. A9 (8176)

Wang, C. -M. [1959], Some problems of statistical analysis of fractions (abstract). Biometrics 15:155-156. A9 (8177)

Wang, C.-M. and Yeh, S.-F. [1963], Preliminary studies on sampling units for the principal agricultural crops of Taiwan (in Chinese). Chung-hua Nung-hsueh Hui Pao/J. Agri. Assoc. China 44:9-17; 46:47-90. A14 (8178)

Ward, A. C. and Boggs, M. M. [1951], Comparison of scoring results for two and four samples of corn per taste session. Food Tech. 5:219-220. C (8179)

Ward, D. H. [1959], Elimination of bias in the preliminary examination of data. Appl. Stat. 8:21-25. A6 (8180)

Warren, J. A. [1963], Use of empirical equations to describe the effects of plant density on the yield of corn and the application of such equations to variety evaluation. Crop Sci. 3:197-201. A7 (8181)

Wassom, C. E. and Kalton, R. R. [1953], Estimations of optimum plot size using data from bromegrass uniformity trials. Iowa State Agri. Expt. Sta. Bull. 396: 295-320. A14 (8182)

Watson, F. R. [1964], A new method for solving simultaneous linear equations associated with multivariate analysis. Psychometrika 29:75-86. A7 (8183)

Watson, G. S. [1964], Smooth regression analysis. Sankhyā A 26:359-372. A7 (8184)

Wearden, S. [1959], The use of the power function to determine an adequate number of progeny per sire in a genetic experiment involving half-sibs. Biometrics 15: 417-423. S1 (8185)

Webb, W. B. and Lemmon, V. W. [1950], A qualification in the use of analysis of variance. Psychological Bull. 47:130-136. A3 (8186)

Weber, C. R. and Horner, T. W. [1957], Estimate of cost and optimum plot size and shape for measuring yield and chemical characters in soybeans. Agron. J. 49: 444-449. A14 (8187)

Weber, C. R., Shibles, R. M., and Byth, D. E. [1966], Effect of plant population and row spacing on soybean development and production. Agron. J. 58:99-102. A14 (8188)

Weber, C. R. and Staniforth, D. W. [1957], Competitive relationships in variable weed and soybean stands. Agron. J. 49:440-444. A14 (8189)

Weber, E. [1957], Betrachtungen zur Diskriminanzanalyse. Z. Pflanzenzüchtung 38:1-36. A7 (8190)

Weber, E. [1964], Partielle Bestimmtheitsmasse bei Einbezug von Polynomen in multiple lineare Regressionsanalysen. (Partial coefficients of determination in anlaysis of multiple linear regression containing polynomials.) Biom. Zeit. 6:262-269. A7 (8191)

Weber, E. [1967], I. Statistical Theorie. Biometrische Bearbeitung multipler Regressionen unter besonderer Berücksichtigung der Auswahl, der Transformation und der Linearkombination von Variablen. Statistische Heft. Int. Z. Theorie Praxis 8:228-251. A7,9 (8192)

Weber, E. and Brott, C. [1963], Ein Linearitätstest mit Hilfe elektronischer Datenverarbeitungsanlagen. Biom. Zeit. 5:188-205. A7,15 (8193)

Webster, J. T. [1965], On the use of a biased estimator in linear regression. JISA 3:82-90. A7 (8194)

Weetman, L. M. and Hundertmark, B. W. [1960], Differences in varietal responses on inside and outside rows of small sugarcane plots. Proc. Tenth Congress Int. Soc. Sugar Cane Tech. pp. 743-745. A14 (8195)

Weiler, H., Chapman, R. E., and Doney, J. M. [1960], The total number of fibres on sheep. II. Estimation by greasy fleece and fibre weights. Australian J. Agri. Res. 11:628-635. A14 (8196)

Weiling, F. [1959], Die Bedeutung der Transformation für die Versuchsanalyse insbesondere bei Bonituren mit beliebiger Breite der Boniturskala. Der Züchter 29:281-284. A4,9 (8197)

Weiling, F. [1960], Vereinfachte Prüfung der Additivität bei Streuungszerlegungen (Varianzanalysen). Der Züchter 30:269-272. A4 (8198)

Weiling, F. [1961], Über Möglichkeit und Interpretation der Zerlegung von Reststreuungen (Restvarianzen). Biom. Zeit. 3:178-185. A11 (8199)

Weiling, F. [1963], Weitere Hinweise zur Prüfung der Additivität bei Streuungszerlegungen (Varianzanalysen). Der Züchter 33:74-77. A4 (8200)

Weiling, F. [1964], Über Bedeutung und Handhabung der multiplen Regressionanalyse bei der Untersuchung von Zusammenhängen in biologischen Bereich. Biom. Zeit. 6:24-36. A7 (8201)

Weinberg, G. H. and Tripp, C. A. [1957], A simplification of the sign test. Psychological Bull. 54:79-80. A12 (8202)

Weir, J. B. De V. [1960], Significance of the difference between two means when the population variances may be unequal. Nature 187:438. A11 (8203)

Weiss, G. H. and Zelen, M. [1963], A stochastic model for the interpretation of clinical trials. Proc. Nat. Academy Sci. 50:988-994. C (8204)

Weiss, L. [1962], On sequential tests which minimize the maximum expected sample size. JASA 57:551-566. S2 (8205)

Weiss, L. [1962], A sequential test of fit for multivariate distributions. Sankhyā A 24:377-384. A7;S2 (8206)

Weiss, L. [1963], Sequential Bayes procedures which never observe more than a bounded number of observations. Ann. Inst. Stat. Math. 15(3):177-185. S2 (8207)

Welch, B. L. [1951], On the comparison of several mean values: an alternative approach. Biometrika 38:330-336. A11 (8208)

Welch, B. L. [1956], Note on some criticisms made by Sir Ronald Fisher. JRSSB 18:297-302. A11;C (8209)

Welch, N. H., Burnett, E., and Eck, H. V. [1966], Effect of row spacing, plant population, and nitrogen fertilization on dryland grain sorghum production. Agron. J. 58:160-163. A14 (8210)

Wellisch, P. [1962], Nagyüzemi kísérletek statisztikai értékelése, ha a gazdaságok száma kezelésenként változó. (The statistical evalutaion of farm-scale trials where the number of farms per treatment varies.) Nemesített növényfajtákkal végzett orzágos Fajtakisérletek eredményei 1960. Budapest, pp. 89-104. A13;C (8211)

Welte, E. [1954], Design of experiments in clinical medicine (abstract). Biometrics 10:312-313. C (8212)

Wermke, M. [1956], Zusammenfassende Auswertung von Feldversuchen nach der Varianzanalyse durch Anwendung des Lochkartenverfahrens. Ruhr-Stickstoff, Aktiengesellschaft, Wittener Strasse 45, Bochum, pp. 1-22. A15 (8213)

Wermke, M. [1958], Zur Frage der Streuversuche. Z. Landw. Vers.- Untersuch Wes. 4:50-64. A13 (8214)

Wermke, M. [1963], Die Anlage und Auswertung von Weiderversuch en nach mathematisch-statistichen Gesichtspunkten. Pflanzenbau 117:32-54. A14;C (8215)

Wernimont, G. [1951], Design and interpretation of interlaboratory studies of test methods. Analytical Chemistry 23:1572-1576. C (8216)

Wetherill, G. B. [1961], Bayesian sequential analysis. Biometrika 48:281-292. S2 (8217)

Wette, R. [1959], Regressions- und Kausalanalyse in der Biologie. Metrika 2:131-137. A7 (8218)

Wheeler, J. L. [1960], Field experiments on systems of management for mesophytic pastures. Australian C.S.I.R.O. Div. Plant Industry, Div. Report 20:1-51. C (8219)

Wheeler, S. and Watson, G. S. [1964], A distribution-free two-sample test on a circle. Biometrika 51:256-257. A12 (8220)

White, C. [1952], The use of ranks in a test of significance for comparing two treatments. Biometrics 8:33-41. A12 (8221)

White, W. C. and Black, C. A. [1954], Willcox's agrobiology: III. The inverse yield-nitrogen law. Agron. J. 46:310-315. A7 (8222)

Whitfield, J. W. [1950], Uses of the ranking method in psychology (with discussion on pages 182-191). JRSSB 12:163-170. A12 (8223)

Whitlock, J. H. [1956], The problem of animal disease reporting and vital statistics. J. American Veterinary Medical Assoc. 128(2):63-66. C (8224)

Whitney, D. R. [1951], A bivariate extension of the U statistic. AMS 22:274-282. A7,12 (8225)

Whittle, P. [1956], On the variation of yield variance with plot size. Biometrika 43:337-343. A14 (8226)

Whittle, P. [1964], Some general results in sequential analysis. Biometrika 51: 123-141. S2 (8227)

Whittle, P. [1965], Some general results in sequential design (with discussion). JRSSB 27:371-394. S2 (8228)

Whittle, P. and Lane, R. O. D. [1967], A class of situations in which a sequential estimation procedure is non-sequential. Biometrika 54:229-234. S2 (8229)

Wiebe, G. A., Petr, F. C., and Stevens, H. [1963], Interplant competition between barley genotypes. Statistical Genetics and Plant Breeding. pp. 546-557. A14 (8230)

Wiedemann, A. M. and Leininger, L. N. [1963], Estimation of optimum plot size and shape for safflower yield trials. Agron. J. 55:222-225. A14 (8231)

Wiegert, R. G. [1962], The selection of an optimum quadrat size for sampling the standing crop of grasses and forbs. Ecology 43:125-129. A14 (8232)

Wiegmann, F. H. [1954], Use of regressions and test procedures. J. Farm Economics 36:633-640. A7 (8233)

Wienhues, F. [1957], Anlage und Auswertung von Mikroprüfungen in Moderne Methoden der Pflanzenzüchtung 44:82-96. C (8234)

Wiggers, B. G. [1965], Statistiek, operationele research en computer. Stat. Neerlandica 19:173-198. C (8235)

Wilcoxon, F. [1950], Some rapid approximate statistical procedures. Ann. New York Academy Sci. 52:808-814. A12 (8236)

Wilcoxon, F. and Bradley, R. A. [1964], A note on the paper two sequential two-sample grouped rank tests with application to screening experiments. Biometrics 20:892-895. A12;S2 (8237)

Wilcoxon, F., Katti, S. K., and Wilcox, R. A. [1963], Critical values and probability levels for the Wilcoxon rank sum test and the Wilcoxon signed rank test. Dept. Stat., American Cyanamid Co. Florida State Univ. A12 (8238)

Wilcoxon, F., Rhodes, L. J., and Bradley, R. A. [1963], Two sequential two-sample grouped rank tests with application to screening experiments. Biometrics 19:58-84. A2,12;S2 (8239)

Wiley, R. C., Briant, A. M., Fagerson, I. S., Murphy, E. F., and Sabry, J. H. [1957], The Northeast regional approach to collaborative panel testing. Food Tech. 11(9):43-48. C (8240)

Wilk, M. B. [1958], Linear models in the analysis of variance. Proc. Second Conf. Design Expt. Army Res. Dev. Testing. pp. 243-257. C (8241)

Wilk, M. B. and Gnanadesikan, R. [1961], Some remarks on plotting procedures in the analysis of experiments. Biometrics 17:178. C (8242)

Wilk, M. B. and Gnanadesikan, R. [1962], Analysis of multi-research experiments (summary). JASA 57:501. C (8243)

Wilk, M. B. and Gnanadesikan, R. [1964], Graphical analysis of variance. Technometrics 6:125. A9;C (8244)

Wilkinson, G. N. [1957], The analysis of covariance with incomplete data. Biometrics 13:363-372. A7 (8245)

Wilkinson, G. N. [1962], Query 178. Biometrics 18:407-409. A14 (8246)

Wilks, S. S. [1955], Statistical aspects of the design of experiments. Proc. American Philosophical Soc. 99:169-173. C (8247)

Wilks, S. S. [1963], Statistical inference in geology. The Earth Sciences - Problems and Progress in Current Research pp. 105-136. C (8248)

Wilks, S. S. [1963], Multivariate statistical outliers. Sankhyā A 25:407-426. A6,7 (8249)

Willcox, O. W. [1954], Quantitative agrobiology: I. The inverse yield-nitrogen law. Agron J. 46:315-320. A7 (8250)

Willcox, O. W. [1954], Quantitative agrobiology: II. The nitrogen constant 318. Agron. J. 46:320-322. A7 (8251)

Willcox, O. W. [1954], Quantitative agrobiology: III. The Mitscherlich equation and its constants. Agron. J. 46:323-326. A7 (8252)

Willcox, O. W. [1954], Quantitative agrobiology: IV: Apparent exceptions to the Mitscherlich law. Agron. J. 46:326-328. A7 (8253)

Willcox, O. W. [1955], Quantitative agrobiology: V: Further comments on Black, Kempthorne and White's criticism of "Wilcox's agrobiology". Agron. J. 47: 499-502. A7 (8254)

Willett, E. L., Ohms, J. I. and Torrie, J. H. [1955], Factors influencing experimental error in field trials in artificial insemination. J. Dairy Sci. 38:1375-1384. A14;C (8255)

Williams, E. J. [1953], A method of analysis for double classifications. Australian J. Appl. Sci. 4:357-370. A10 (8256)

Williams, E. J. [1953], Tests of significance for concurrent regression lines. Biometrika 40:297-305. A7 (8257)

Williams, E. J. [1955], Significance tests for discriminant functions and linear functional relationships. Biometrika 42:360-381. A7 (8258)

Williams, E. J. [1958], Simultaneous regression equations in experimentation. Biometrika 45:96-110. A7 (8259)

Williams, E. J. [1959], The comparison of regression variables. JRSSB 20:396-399. A7 (8260)

Williams, E. J. [1961], Tests for discriminant functions. J. Australian Math. Soc. 2:243-252. A7 (8261)

Williams, E. J. [1962], Exact fiducial limits in non-linear estimation. JRSSB 24: 125-139. A7 (8262)

Williams, E. J. [1962], The analysis of competition experiments. Australian J. Biological Sci. 15:509-525. A14 (8263)

Williams, E. J. [1967], The analysis of association among many variates (with discussion). JRSSB 29:199-242. A7 (8264)

Willke, T. A. [1964], General application of Youden's rank sum test for outliers and tables of one-sided percentage points. J. Res. NBS 68B:55-58. A6,12 (8265)

Willke, T. A. [1966], A note on contaminated samples of size three. J. Res. NBS 70B:149-151. A6 (8266)

Willoughby, W. F. [1967], Estimation of time fuze characteristics by non-linear regression methods. Proc. Twelfth Conf. Design Expt. Army Res. Dev. Testing. pp. 365-393. A7 (8267)

Wilm, H. G. [1952], A pattern of scientific inquiry for applied research. J. Forestry 50:120-125. C (8268)

Wilson, W. [1962], A note on the inconsistency inherent in the necessity to perform multiple comparisons. Psychological Bull. 59:296-300. A3 (8269)

Wine, R. L. [1956], On the power of multiple range tests (summary). JASA 51:527. A3 (8270)

Wine, R. L. [1956], On the comparison of multiple test procedures (abstract). Biometrics 12:535. A3 (8271)

Wine, R. L. and Freund, J. E. [1957], On the enumeration of decision patterns involving n means, AMS 28:256-259. A3 (8272)

Winne, D. [1965], Die sequentiellen statistischen Verfahren in der Medizin. Arzneimittel-Forschung 15:1088-1091. S2 (8273)

Winsor, C. P. [1950], Query 78. Biometrics 6:167-168. S1 (8274)

Wishart, J. [1955], Multivariate analysis. Appl. Stat. 4:103-116. A7 (8275)

Wishart, J. [1955], Influence of agronomic investigation on the science of statistics. JISAS 7:73-80. C (8276)

Wishart, J. [1956], 'Significant difference' in the analysis of field trials. Agri. Rev. May, pp. 44-46. C (8277)

Wishart, J. and Metakides, T. [1953], Orthogonal polynomial fitting. Biometrika 40:361-369. A7 (8278)

Wohlzogen, F. X. and Wohlzogen-Bukovics, E. [1966], Sequentielle Parameterschätzung bei biologischen Alles-order-Nichts-Reaktionen. (Sequential estimation of parameters in biological all-or-nothing-reactions.) Biom. Zeit. 8:84-120. S2 (8279)

Wold, H. [1956], Casual inference from observational data. A review of ends and means (discussion). JRSSA 119:28-60. C (8280)

Wolfowitz, J. [1953], Estimation by the minimum distance method. Ann. Inst. Stat. Math. 5:9-23. A7 (8281)

Wolfowitz, J, Menger, K., and Tintner, G. [1952], Abraham Wald, 1902-1950. The formative years of Abraham Wald and his work in geometry. Abraham Wald's contributions to econometrics. The publications of Abraham Wald. AMS 23:1-33. B (8282)

Woods, J. E., Hafenrichter, A. L., Schwendiman, J. L., and Law, A. G. [1953], The effect of grasses on yield of forage and production of roots by alfalfa-grass mixtures with special reference to soil conservation. Agron. J. 45:590-595. A14 (8283)

Woolf, B. [1951], Computation and interpretation of multiple regressions. JRSSB 13:100-119. A7 (8284)

Wormleighton, R. [1960], A useful generalization of the Stein two-sample procedure. AMS 31:217-221. S2 (8285)

Wright, J. W. [1963], The design of field tests. Proc. World Consultation Forest Genetics Tree Improvement, Stockholm, Sweden, 23-30 August 3/4:iv+6. C (8286)

Wright, J. W. and Freeland, F. D. [1958], Plot size in forest genetic research. Papers Mich. Academy Sci. 44:177-182. A14 (8287)

Wright, J. W. and Freeland, F. D. [1960], Plot size and experimental efficiency in forest genetic research. Michigan State Univ. Agri. Expt. Sta. Tech. Bull. 280. A14 (8288)

Wright, S. [1954], The interpretation of multivariate systems. Statistics and Mathematics in Biology, pp. 11-33. A7 (8289)

Wrigley, C. [1957], The distinction between common and specific variance in factor theory. British J. Stat. Psychology 10:81-97. A7;C (8290)

Yamada, T. [1953], Studies on the occurrence and mechanism of non-genetic variation due to interplant competition. III. Enlargement of non-genetic variation due to intensified competition in a red clover population. Japanese J. Breeding 3(1):17-22. A14 (8291)

Yamada, T. and Horiuchi, S. [1953], Studies on the occurrence and mechanism of non-genetic variation due to competition among different types of plant. I. On the warp of normal phenotypes due to the competition among different barley varieties. Japanese J. Breeding 2:159-172. A14 (8292)

Yamada, T. and Horiuchi, S. [1953], Studies on the occurrence and mechanism of non-genetic variation due to competition among different types of plant. II. On the warp of normal phenotypes due to the interplant competition between different soybean varieties. Japanese J. Breeding 3(1):9-16. A14 (8293)

Yang, C. S., Yang, T.C., and Li, H. W. [1959], Varietal competition in bunch planting in sugarcane breeding. Taiwan Sugar 6(4):13-21. A14 (8294)

Yang, T.-C. and Chu, T.-L. [1952], Studies on the technique of field trials of sugar cane. II. Block lay-out (in Chinese). Taiwan Tangyeh Shihyenso Yenchiu Huipao/ Reports Taiwan Sugar Expt. Sta. 8:1-14. A14 (8295)

Yao, Y. [1965], An approximate degrees of freedom solution to the multivariate Behrens-Fisher problem. Biometrika 52:139-147. A7,11 (8296)

Yates, F. [1950], Recent applications of biometrical methods in genetics. (1) Experimental techniques in plant improvement. Biometrics 6:200-207 (abstract 6:226). C (8297)

Yates, F. [1951], Bases logiuqes de la planification des expériences. Ann. Inst. Henri Poincaré 12:97-112. C (8298)

Yates, F. [1951], Quelques développements modernes dans la planification des expériences. Ann. Inst. Henri Poincaré 12:113-130. C (8299)

Yates, F. [1952], Principles governing the amount of experimentation in developmental work. Nature 170.138-140. S1,2 (8300)

Yates, F. [1954], The place of simple experiments on cultivators' fields in agricultural development (abstract). Biometrics 10:174. C (8301)

Yates, F. [1955], The combination of data from a set of 2 × 2 tables (abstract). Biometrics 11:249. A11 (8302)

Yates, F. [1961], Marginal percentages in multiway tables of quantal data with disproportionate frequencies. Biometrics 17:1-9. A9,10 (8303)

Yates, F. [1964], Sir Ronald Fisher and the design of experiments. Biometrics 20: 307-321. C (8304)

Yates, F. [1964], Statistics Department. Report Rothamsted Expt. Sta. pp. 203-211. A15 (8305)

Yates, F. [1966], Computers, the second revolution in statistics. Biometrics 22: 233-251. A15;C (8306)

Yates, F., Healy, M. J. R., and Lipton, S. [1957], Routine analysis of replicated experiments on an electronic computer (discussion). JRSSB 19:234-254. A15 (8307)

Yates, F., Healy, M. J. R., and Lipton, S. [1959], L'analyse des expériences répétees sur calculatrice électronique (discussion). Rev. Stat. Appl. 7(2):47-75. A15 (8308)

Yates, F. and Lipton, S. [1957], An automatic programming routine for the Elliott 401. J. Assoc. Computing Machinery 4:151-156. A15 (8309)

Yates, F. and Mather, K. [1963], Ronald Aylmer Fisher 1890-1962. Biographical Memoirs Fellows Royal Soc. 9:89-129. B;C (8310)

Yates, F. and Simpson, H. R. [1960], The analysis of surveys: Processing and printing the basic tables. Computer J. 4(1):1-5. A15 (8311)

Yates, F., Vernon, A. J., and Nelson, S. W. [1964], An example of the analysis of uniformity trial data on an electronic computer. Empire J. Expt. Agri. 32:25-30. A14,15 (8312)

Yawalkar, K. S. and Schmid, A. R. [1954], Performance of birdsfoot trefoil alone and in competition with other species in pastures. Agron. J. 46:407-411. A14 (8313)

Yeh, S.-F. [1953], Studies on the technique of field experiments with flax (in Chinese). Taiwan Tahsueh Nunghsuehyuan Yenchiu Paokao/Memoirs College Agri. Nat. Taiwan Univ. 2(5):35-64. A14 (8314)

Yen, E. H. [1964], On two-stage non-parametric estimation. AMS 35:1099-1114. A12;S2 (8315)

Yen, E. H. [1964], Generalization of Wilcoxon statistic for the case of k samples. Stat. Neerlandica 18:293-302. A12 (8316)

Yllö, L. [1965], Ruudun koko perunan lajikekokeissa. (Plot size in potato variety trials.) Maataloustieteellinen Aikakauskioja, Helsinki 37:7-12. A14 (8317)

Yntema, L. [1955], Statistische aspecten de muziek. (Statistical aspects of music.) Stat. Neerlandica 9:161-171. C (8318)

Yoda, K., Kira, T., and Hozumi, K. [1957], Intraspecific competition among higher plants, IX. Further analysis of the competitive interaction between adjacent individuals. J. Inst. Polytechnics, Osaka City Univ. D, 8:161-178. A14 (8319)

Yoshimura, I. [1962], On a method of the analysis of ranking data. Reports Stat. Appl. Res. 9:1-3. A12 (8320)

Youden, W. J. [1950], Statistics in analytic chemistry. Ann. New York Academy Sci. 52:815-819. C (8321)

Youden, W. J. [1951], The Fisherian revolution in methods of experimentation. JASA 46:47-50. C (8322)

Youden, W. J. [1951], Locating sources of variability in a process. Ind. Eng. Chemistry 43:2059-2062. C (8323)

Youden, W. J. [1952], Statistical aspects of analytical determinations (discussion). Analyst 77:874-878; 889-891. C (8324)

Youden, W. J. [1954], Modern statistics is still a new subject with a long list of unsolved statistical problems. Ind. Eng. Chemistry 46(2):107A-109A. C (8325)

Youden, W. J. [1954], Statistical techniques are most productive when incorporated into projects at the planning stage. Ind. Eng. Chemistry 46(4):111A-112A, 114A. C (8326)

Youden, W. J. [1954], Modern statistical design is based on techniques traditionally employed in good experimentation. Ind. Eng. Chemistry 46(6):115A-116A. C (8327)

Youden, W. J. [1954], Instrumental drift. Science 120:627-631. C (8328)

Youden, W. J. [1955], Industrial application of statistical design requires more active support from management. Ind. Eng. Chemistry 47(4):107A-109A. C (8329)

Youden, W. J. [1955], Industrial research problems are stimulating the development of new statistical design techniques. Ind. Eng. Chemistry 47(6):111A-113A. C (8330)

Youden, W. J. [1955], Gun problem illustrates diverse classes of statistical designs available in research work. Ind. Eng. Chemistry 47(10):99A-100A. C (8331)

Youden, W. J. [1955], Design of experiments in the physical sciences (abstract). Biometrics 11:541-542. C (8332)

Youden W. J. [1956], Recent advances in statistical design provide useful guides for planning research programs. Ind. Eng. Chemistry 48(2):98A-100A. C (8333)

Youden, W. J. [1956], Method for fitting lines to data illustrates statistical procedure for testing hypotheses. Ind. Eng. Chemistry 48(4)107A-109A. A7 (8334)

Youden, W. J. [1956], Data on physical properties of rubber exemplify problems of data evaluation and estimation of error. Ind. Eng. Chemistry 48(6)104A-105A. C (8335)

Youden, W. J. [1956], Who makes designs? (Summary). JASA 51:527. C (8336)

Youden, W. J. [1957], Testing uniformity of sheets and plates. Ind. Eng. Chemistry 49(8):71A-72A (correction 50(2):90A). A14 (8337)

Youden, W. J. [1957], Random numbers aren't nonsense. Ind. Eng. Chemistry 49(10):89A-90A. C (8338)

Youden, W. J. [1957], Design of experiments in industrial research and development. Proc. First Conf. Design Expt. Army Res. Dev. Testing. pp. 9-14. C (8339)

Youden, W. J. [1958], Problems of the experimenter. Ind. Eng. Chemistry 50(2): 89A-90A. C (8340)

Youden, W. J. [1958], Problems in testing materials. Ind. Eng. Chemistry 50(4): 81A-82A. C (8341)

Youden, W. J. [1958], How to pick a winner. Ind. Eng. Chemistry 50(6):81A-82A. A2 (8342)

Youden, W. J. [1959], Evaluation of chemical analyses on two rocks. Technometrics 1:409-417. A13 (8343)

Youden, W. J. [1959], What is a measurement? Ind. Eng. Chemistry 51(2):81A-82A. C (8344)

Youden, W. J. [1959], A sampling study. Ind. Eng. Chemistry 51(4):81A-82A. A14 (8345)

Youden, W. J. [1959], Experiments in experimentation. Ind. Eng. Chemistry 51(8): 65A-66A. C (8346)

Youden, W. J. [1959], Problems of the experimenter. ASQC, Nat. Conv. Trans. 13:41-47. C (8347)

Youden, W. J. [1959], Graphical diagnosis of interlaboratory tests results. IQC 15(11):1-5. A2 (8348)

Youden, W. J. [1960], The sample, the procedure, and the laboratory. Analytical Chemistry 32:23A-37A. C (8349)

Youden, W. J. [1961], How to evaluate accuracy. Materials Res. Standards 1:268-271. C (8350)

Youden, W. J. [1962], Experimentation and measurement. National Sci. Teachers Assoc. Vistas Sci. 2, (Scholastic Book Services, New York 36, N.Y.) 127 pp. C (8351)

Youden, W. J. [1962], Accuracy of analytical procedures. J. Assoc. Official Agri. Chemists 45:169-173. A14 (8352)

Youden, W. J. [1962], Systematic errors in physical constants. Technometrics 4:111-123. C (8353)

Youden, W. J. [1963], Physical measurements and experiment design (discussion). Colloques Internationaux du Centre National de la Recherche Scientifique No. 110, Le Plan d'Expériences pp. 115-128. C (8354)

Youden, W. J. [1965], The evaluation of designed experiments (with discussion). Proc. IBM Scientific Computing Symp. Stat. pp. 59-67. C (8355)

Youden, W. J. [1967], Statistical techniques for collaborative tests. Manual Assoc. Official Analytical Chemists, Inc. Washington, D.C. A13;C (8356)

Youden, W. J. and Cameron, J. M. [1950], Use of statistics to determine precision of test methods. Symp. Application Stat., (American Soc. Testing Materials) Special Tech. Publication No. 103, pp. 27-34. C (8357)

Youden, W. J. and Connor, W. S. [1953], Making one measurement do the work of two. Chemical Eng. Progress 49:549-552. C (8358)

Yudowitch, K. L. [1958], Problems in a particular military field experiment. Proc. Second Conf. Design Expt. Army Res. Dev. Testing. pp. 51-63. C (8359)

Zaalberg, J. [1958], Auxiliary tables for Wilcoxon's two sample test. Stat. Neerlandica 12:265-273. A12 (8360)

Zaat, J. C. A. [1964], Poisson-proces met interrupties, een waarschijnlijkheids-model voor het machinaal dunnen van bieten. Stat. Neerlandica 18:311-323. A14 (8361)

Žák, V. [1957], Příspěvek k problematice odrůdových pokusů s cukrovkou. (A contribution to the problem of varietal trials of sugar beet.) Listy Cukr. 73:265-273. A14 (8362)

Zeegers, J. J. B. [1965], Vectoren en lineaire regressie. Biométrie-Praximétrie 6:93-106. A7 (8363)

Zelen. M. [1955], Query 115. Biometrics 11:238-239. C (8364)

Zelen, M. [1962], The role of constraints in the theory of least squares. Math. Res. Center Tech. Summary Report No. 314, U.S. Army Univ. Wisconsin. A7 (8365)

Zelen, M. [1966], The interpretation of clinical trials (abstract). Biometrics 22:209. C (8366)

Zeller, A. [1950], Zur Terminologie einінger Begriffe der naturwissenschaftlichen und der Stichproben-Statistik. Stat. Vjschr. 3:141-145. C (8367)

Zeller, A. [1951], Die neuen Verfahren zur Berechnung und Anlage landwirtschaftlicher Versuche. Veröff. Bundesanst. Alp. Landw. Admont 4:43-68. C (8368)

Zellner, A. [1962], An efficient method of estimating seemingly unrelated regressions and tests for aggregation bias. JASA 57:348-368. A7 (8369)

Zimmermann, K. [1954], Feldversuchen, Probleme und Versuche. Der Züchter 24: 116-127. C (8370)

Zimmermann, K. [1955], Technik des Versuchswesens und der Pflanzenzüchtung. S. Hirzel Verlag, Leipzig, pp. viii + 414. C (8371)

Zinn, M. H. [1958], Problems in analysis of electron tube experiments. Proc. Third Conf. Design. Expt. Army Res. Dev. Testing. pp. 283-287. C (8372)

Zinn, M. H. [1959], Analysis of cathode interface resistance experiment. Proc. Fourth Conf. Design Expt. Army Res. Testing. pp. 49-74. C (8373)

Zlobik, E. T. [1961], Subjective product evaluation methods. ASQC, Ann. Conv. Trans. 15:515-520. C (8374)

Zoellner, J. A. and Gunther, P. [1956], Decision theory aspects in the design and analysis of engineering experiments (summary). JASA 51:527. C (8375)

Zweigbaum, H. and Donaldson, D. [1958], Human engineering experiment on tube tester TV-2/U. Proc. Second Conf. Design Expt. Army Res. Dev. Testing. pp. 73-79. C (8376)

Zyskind, G. [1962], On structure, relation, Σ, and expectation of mean squares. Sankhyā A 24:115-148. A4,5 (8377)

Zyskind, G. [1963], A note on residual analysis. JASA 58:1125-1132. A6 (8378)

Co-authors list for category "Other"

Name	Ref. No.
Abrams, D.	6003
Aitchison, J.	7564
Aitken, J.R.	7319
Alexander, S.N.	7973
Allen, D.D.	7326
Amerine, M.A.	5719
	5720
	5723
	7425
	7467
	7689
Antle, C.E.	6439
Armitage, J.V.	7085
Armitage, P.	7399
	7683
	7819
	7820
	7926
Asahira, T.	6855
Asano, C.	7992
Atkins, R.E.	7152
Axtell, L.A.	7398
Baer, R.M.	6099
Baker, G.A.	5725
	6718
	6852
	6853
	6854
	7425
	7467
	7689
Baker, R.E.	5721
Bancroft, T.A.	5939
	6436
	7122
	7900
Bandyopadhyay, K.S.	6999
	7000
Bannink, R.	7028
Barber, J.C.	6375
Bargmann, R.E.	7717
Barnes, T.W.	7247
Barnett, B.N.	7173
Barron, B.A.	5676
Bartlett, B.O.	7788
Baten, W.D.	6465
	7523
Beall, G.	6224
Benard, A.	8127
Berg, H. W.	6812
	6813
Berry, M.O.	7310
Betz, E.H.	7321
Bhargava, P.N.	7890
Bickle, A.	7365
Bielorai, H.	7575
Biswas, A. K.	7845
Black, C.A.	6344
	8222
Blackmon, G.H.	7868
Bliss, C.I.	6605
Boggs, M.M.	8179
Booker, A.	6756
Bonner, R.G.	6303
Borojević, K.	5918
Bose, P.K.	7628
Bose, R.C.	7718
	7719
	7720
Bosson, R.P.	6980
Bradley, R.A.	8237
	8239
Brady, N.C.	5885
Bray, D.F.	7017
Bray, R.H.	5731
	7105
Brian, M.V.	7905
Briant, A.M.	8240
Brim, C.A.	6721
	7830
Brindley, T.A.	7900
Broster, W.H.	5710
Brott, C.	8193
Brown, A.H.F.	7520
Brown, B.M.	7149

Nelson, W.L.	7665
Newton, B.S.	6779
Neyman, J.	6099
Nigam, L.N.	7000
Nissen, Ø.	6410
	7439
Nissim, J.A.	5727
Nogueira, I.R.	6562
Nomachi, Y.	7027
Nottingham, R.B.	7145
Nylund, R.E.	7424
O'Flaherty, C.A.	6194
Ogawa, H.	7019
	7020
	7021
Ogle, W.L.	7062
Ohms, J.I.	8255
Olkin, I.	6210
O'Regan, W.G.	7474
Orlando, P.A.	6487
Oxnard, C.E.	5700
Oyler, J.R.	5986
Page, E.S.	5677
Palca, H.	6411
Palmer, F.H.	7116
Panettiere, B.R.	7167
Pangborn, R.M.	7889
Patel, I.R.	7866
Pauley, S.S.	7876
Pauls, J.F.	6837
Paulson, A.S.	6193
Pearson, E.S.	6758
Peirce, L.C.	6834
Pendleton, J.W.	7114
Penfound, W.T.	7658
Penny, L.H.	6329
Peryam, D.	6345
Peryam, D.R.	6540
Pesek, J.	8150
Petersen, H.D.	5964
Petersen, R.G.	6624
	7604
Petersen, W.E.	7464
Peterson, L.C.	6268
Petr, F.C.	8230
Pienaar, L.V.	8092
Pigatti, A.	6487
Pilgrim, F.J.	7535

Pinkham, R.S.	6829
Pitman, E.N.T.	7641
Plaisted, R.L.	6268
Plotnikov, N.I.	7053
	7054
Popa, T.	7796
Potgieter, M.	6605
Powell, R.W.	6772
Prins, H.J.	6272
	6273
	6274
Proctor, J.R.	7905
Puri, M.L.	7580
Puri, P.S.	7578
Purser, A.F.	6267
Quandt, R.E.	6555
Quarles, D.A.	5796
Rădoi, A.	7796
Ram, B.M.	7862
Ramachandran, K.V.	7867
Ramage, R.T.	7998
Rao, J.S.	7859
Rao, M.M.	6061
Rastogi, S.C.	6702
Reed, J.W.	6760
Rees, D.H.	6626
	6627
Reinker, K.A.	5898
Rhodes, L.J.	8239
Richardson, B.A.	6207
Richardson, H.B.	7425
Richter, D.	8061
Rigney, J.A.	7040
Rios, S.	8100
Rizvi, M.H.	5641
	5642
	5750
	7930
Roberts, P.	6943
Robbins, H.	6078
	6530
Robson, D.S.	5706
Rodgers, P.	6466
Roessler, E.B.	5651
	5720
Rogers, T.H.	6436

X. BOOKS ON EXPERIMENT AND TREATMENT DESIGN

The books listed in this part of the bibliography are those which may be considered as texts in a course on the design of experiments. The two exceptions are the books on tables by Fisher and Yates [1963] and Kitagawa and Mitome [1953], which were included because they contain material of considerable usefulness to researchers on designs. It should be pointed out that all these books lag considerably behind the published literature on experiment and treatment designs.

The table of contents for each book is given. The reader may determine the material covered in a particular book but he will not, in general, be able to determine the level of the book from the table of contents. In a few instances, the table of contents was in a language the authors were unable to translate and hence none is given.

Chakrabarti, M. C. [1962], MATHEMATICS OF DESIGN AND ANALYSIS OF EXPERIMENTS. Asia Publishing House, Bombay, Calcutta, New Delhi, Madras, London, New York, pp. 1-120.

Chapter I - THEORY OF LINEAR ESTIMATION
Notations -- Estimable Parametric Functions and Condition of Estimability -- Method of Least Squares and Markoff's Theorem -- Tests of Hypothesis regarding Parameters -- Test Involving Several Linear Functions of Parameters -- Tests of Subhypotheses -- A Particular Test Involving Equality of Some of the Parameters -- Linear Estimation with Correlated Variables -- Exercises.

Chapter II - GENERAL STRUCTURE OF ANALYSIS OF DESIGNS
General Theory of Analysis of Experimental Designs -- Two-way Elimination of Heterogeneity -- Analysis with Recovery of Interblock Information -- Expectations of Sums of Squares in Intrablock Analysis with the Model of Section 3 -- Total and Partial Confounding -- Exercises.

Chapter III - STANDARD DESIGNS
Randomised Block Design -- Latin Square Design -- Greaco-Latin and Higher Order Squares -- Cross-over Design -- Balanced Incomplete Block Design -- Youden Square -- Lattice Design -- Partially Balanced Incomplete Block Design -- Exercises.

Chapter IV - STANDARD DESIGNS (continued)
Factorial Designs -- Split-plot Experiments -- Split-plot in a Latin Square -- Repeated Subdivision -- Split-plot with Subunits in Strips -- Confounded Arrangement in Split-plot Designs -- Exercises.

Chapter V - APPLICATIONS OF GALOIS FIELDS AND FINITE GEOMETRY IN THE CONSTRUCTION OF DESIGNS
Galois Fields -- Finite Geometry -- Orthogonal Latin Squares -- Construction of BIBD -- Construction of PBIBD -- Construction of Confounded Designs -- Hypercubes of Strength d -- Balancing -- Exercises.

Chapter VI - SOME SELECTED TOPICS IN DESIGN OF EXPERIMENTS
Missing-Plot Technique -- Fractional Replication -- Confounded Asymmetrical Factorial Designs -- Weighing Designs -- Construction of Youden Squares -- Exercises.

LIST OF REFERENCES
INDEX

Chapin, F. S. [1955], EXPERIMENTAL DESIGNS IN SOCIOLOGICAL RESEARCH. Harper & Brothers Publishers, New York, pp. xii+297. First printing by Harper & Brothers Publishers, 1947.

INTRODUCTION TO THE FIRST EDITION

FOREWARD TO THE REVISED EDITION

Chapter I - NATURAL SOCIAL EXPERIMENTS BY TRIAL AND ERROR
The Theory and Practice of the Experimental Method -- Natural Experiments -- Experimentation upon Human Beings -- The Utopian Community Experiments -- Social Legislation Is Social Experimentation -- The Difficulties of Experimentation in Sociology -- Trial and Error, Planning and Prediction.

Chapter II - THREE EXPERIMENTAL DESIGNS FOR CONTROLLED OBSERVATION

Chapter III - CROSS-SECTIONAL EXPERIMENTAL DESIGN
The Personal Adjustment of Boy Scouts -- The Personal Adjustment of Work Relief Clients.

Chapter IV - PROJECTED EXPERIMENTAL DESIGN. THE CLASSICAL PATTERN OF "BEFORE" AND "AFTER" EXPERIMENTS THAT OPERATE FROM THE PRESENT TO THE FUTURE
The Effects of Rural Hygiene in Syria -- The Social Effects of Public Housing in Minneapolis -- An Experimental Study of Staff Stimulation to Social Participation and Social Adjustment -- Delinquency Treatment in the Controlled Activity Group.

Chapter V - EX POST FACTO EXPERIMENTAL DESIGN: FROM PRESENT TO PAST
Delinquency Before and After Admission to a New Haven Housing Development -- The Effects of Length of High School Education on Economic Adjustment in the Community of St. Paul -- Rentals and Tuberculosis Death Rates in New York City -- Summary Comments.

Chapter VI - SOCIOMETRIC SCALES AVAILABLE FOR CONTROL AND THE MEASUREMENT OF EFFECTS
Sociometry or Social Measurement -- An Illustration of an Application of Semantics and Syntactics to the Improvement of Sociometric Tools.

Chapter VII - SOME FUNDAMENTAL PROBLEMS AND LIMITATIONS TO STUDY BY EXPERIMENTAL DESIGNS
Practical Obstacles -- Theoretical Limitations -- Obstacles to Replication.

Chapter VIII - ANALYSIS OF VARIANCE AND THE t-STATISTIC: UNDERLYING ASSUMPTIONS

Chapter IX - NON-PARAMETRIC OR DISTRIBUTION-FREE STATISTICAL METHODS

Chapin (Cont'd.)

Chapter X - THE EX POST FACTO DESIGN: REPLICATION AND EXTENSION
Medical Care Costs Among the Aged -- Religious Activity and Personal Adjustment Among the Aged -- The Logic of Ex Post Facto Design.

Chapter XI - SOME PROBLEMS IN PSYCHO-SOCIAL MEASUREMENT
Extreme Individuals -- Social Obstacles to the Acceptance of Existing Social Science Knowledge -- Social Attitudes and Opinions -- Some Variations in Response Behavior.

APPENDIX A TENTATIVE NORMS OF SOCIAL STATUS SCALE

APPENDIX B TENTATIVE NORMS OF SOCIAL PARTICIPATION SCALE

APPENDIX C COMMENT ON Experimental Designs in Sociological Research

APPENDIX D HABERMAN'S TABLE OF PROBABILITY VALUES FOR RANDOM SAMPLES

APPENDIX E LIST OF PERIODICAL ABBREVIATIONS

INDEX OF NAMES

INDEX OF SUBJECTS

Cochran, W. G. and Cox, G. M. [1957], EXPERIMENTAL DESIGNS. John Wiley and Sons, Inc., New York, London, Sydney, pp. xiv+611. First Edition by John Wiley and Sons, Inc., 1950. Seventh printing by John Wiley and Sons, Inc., December, 1966. Spanish Edition, 1965 by Editorial F. Trillas, S.A., México.

Chapter I - INTRODUCTION
The Contribution of Statistics to Experimentation -- Initial Steps in the Planning of Experiments -- References.

Chapter 2 - METHODS FOR INCREASING THE ACCURACY OF EXPERIMENTS
Introduction -- Number of Replications -- Other Methods for Increasing Accuracy -- The Grouping of Experimental Units -- References.

Chapter 3 - NOTES ON THE STATISTICAL ANALYSIS OF THE RESULTS
Introduction -- The General Method of Analysis -- Accuracy in Computations -- Subdivision of the Sum of Squares for Treatments -- Calculation of Standard Errors for Comparisons among Treatment Means -- Subdivision of the Sum of Squares for Error -- Missing Data -- The Analysis of Covariance -- Effects of Errors in the Assumptions Underlying the Analysis of Variance -- References.

Chapter 4 - COMPLETELY RANDOMIZED, RANDOMIZED BLOCK, AND LATIN SQUARE DESIGNS
Completely Randomized Designs -- Single Grouping: Randomized Blocks -- Double Grouping: Latin Squares -- Cross-over Designs -- Triple Grouping: Graeco-Latin Squares -- Designs for Estimating Residual Effects When Treatments are Applied in Sequence -- References -- Plans.

Chapter 5 - FACTORIAL EXPERIMENTS
Description -- Calculation of Main Effects and Interactions -- Designs for Factorial Experiments -- References.

Chapter 6 - CONFOUNDING
The Principle of Confounding -- The Use of Confounded Designs -- Notes on the Plans and Statistical Analysis -- References -- Plans.

Chapter 6A - FACTORIAL EXPERIMENTS IN FRACTIONAL REPLICATION
Construction and Properties of Fractionally Replicated Designs -- The Use of Fractional Factorial Designs in Practice -- Designs with Factors at More Than Two Levels -- References -- Plans.

Chapter 7 - FACTORIAL EXPERIMENTS WITH MAIN EFFECTS CONFOUNDED: SPLIT-PLOT DESIGNS
The Simple Split-plot Design -- Repeated Subdivision -- Some Variants of the Split-plot Design - References.

Cochran and Cox (cont'd.)

Chapter 8 - FACTORIAL EXPERIMENTS CONFOUNDED IN QUASI-LATIN SQUARES
Introduction -- Randomization of Quasi-latin Squares -- Notes on the Plans and Statistical Analysis -- Other Quasi-latin Squares -- Estimation of the Efficiency of Quasi-latin Squares -- Treatments Applied to Complete Rows of a Latin Square -- Treatments Applied to Complete Rows and Columns of a Latin Square -- References -- Plans.

Chapter 8A - SOME METHODS FOR THE STUDY OF RESPONSE SURFACES
First Order Designs -- Second Order Designs -- Methods for Determining the Optimum Combination of Factor Levels -- The Single-factor Method -- The Method of Steepest Ascent -- Summary Comments -- References -- Plans.

Chapter 9 - INCOMPLETE BLOCK DESIGNS
Balanced Designs -- Partially Balanced Designs -- Basis of the Statistical Analysis -- Comparison of Incomplete Block and Randomized Block Designs -- Comparisons with Other Designs -- Choice of Incomplete Block Design -- References.

Chapter 10 - LATTICE DESIGNS
Balanced Lattices -- Partially Balanced Lattices -- Rectangular Lattices -- Cubic Lattices -- References -- Plans.

Chapter 11 - BALANCED AND PARTIALLY BALANCED INCOMPLETE BLOCK DESIGNS
Balanced Incomplete Blocks -- Balanced Incomplete Blocks in Taste and Preference Testing -- Comparisons with Other Designs -- Arrangement of Experimental Material -- Randomization -- Statistical Analysis -- Partially Balanced Incomplete Block Designs -- Chain Block Designs -- References -- Plans.

Chapter 12 - LATTICE SQUARES
Description -- Statistical Analysis -- References -- Plans.

Chapter 13 - INCOMPLETE LATIN SQUARES
Description -- Statistical Analysis -- Other Designs for Small Numbers of Treatments -- Partially Balanced Designs -- References -- Plans.

Chapter 14 - ANALYSIS OF THE RESULTS OF A SERIES OF EXPERIMENTS
Initial Steps in the Analysis -- Criticisms of the Preliminary Analysis -- Experiments of Unequal Size -- A Test of the Treatments x Places Interactions -- Repetitions in Both Space and Time -- References.

Chapter 15 - RANDOM PERMUTATIONS OF 9 and 16 NUMBERS
Use of the Random Permutations -- Construction of the Random Permutations -- Randomization of More than 16 Numbers -- Tests of Randomness -- References -- Tables of Random Permutations - Permutations of 9; Permutations of 16.

SELECTED BIBLIOGRAPHY
LIST OF AUTHOR REFERENCES
INDEX
TABLES OF t and F.

Cox, D. R. [1958], PLANNING OF EXPERIMENTS. John Wiley and Sons, Inc., New York, N.Y., U.S.A., pp. vii+308. Fifth printing by John Wiley and Sons, Inc., December, 1966.

Chapter 1 - PRELIMINARIES
Comparative Experiments -- Requirements for a Good Experiment - Absence of Systematic Error; Precision; Range of Validity; Simplicity; The Calculation of Uncertainty -- Summary.

Chapter 2 - SOME KEY ASSUMPTIONS
Introduction -- Additivity -- Constancy of Treatment Effects -- Interference Between Different Units -- Summary.

Chapter 3 - DESIGNS FOR THE REDUCTION OF ERROR
Introduction -- Paired Comparisons -- Randomized Blocks - Introduction and Example; Missing Values; Further Examples -- Elimination of Error by Several Groupings of the Units - Latin Squares; Combined Latin Squares; Graeco-Latin Squares -- The Need for More Complicated Arrangements -- Summary.

Chapter 4 - USE OF SUPPLEMENTARY OBSERVATIONS TO REDUCE ERROR
Introduction -- Nature of Concomitant Observations -- The Use of a Concomitant Observation as an Alternative to Blocking -- Alternative Procedures -- The Use of a Concomitant Observation in Addition to Blocking -- Some General Points -- Several Concomitant Variables -- Summary.

Chapter 5 - RANDOMIZATION
Introduction -- The Mechanics of Randomization -- Nature of Random Numbers and Randomness -- Justification of Randomization - Introduction; Systematic Arrangements; Subjective Assignment; Randomization as a Device for Concealment; Summing Up -- Errors Arising in Several Stages -- Statistical Discussion of Randomization -- Some Further Points -- Summary.

Chapter 6 - BASIC IDEAS ABOUT FACTORIAL EXPERIMENTS
Introduction -- General Definitions and Discussion -- Types of Factors - Specific Qualitative Factors; Quantitative Factors; Ranked Qualitative Factors; Sampled Qualitative Factors -- Main Effects and Interactions in a Two-Factor Experiment -- The Interpretation of Interactions -- Experiments with More than Two Factors -- Main Effects and Interactions in Experiments with All Factors at Two Levels -- A Single Quantitative Factor - Types of Response Curve; Statistical Analysis of Response Curves -- Several Quantitative Factors -- Further Discussion of Main Effects and Interactions - Two Specific Qualitative Treatment Factors; Two Specific Qualitative Factors, One a Treatment Factor and One a Classification Factor; One Specific Qualitative Treatment Factor and a Sampled Qualitative Classification Factor; Two Quantitative Treatment Factors; Two Treatment Factors, One Quantitative and One Specific Qualitative; Quantitative Treatment Factor and a Specific Qualitative Classification Factor; Quantitative Treatment Factor and a Sampled Qualitative Classification Factor; General Discussion -- The Estimation of Error in Factorial Experiments -- Summary.

Cox (cont'd.)
Chapter 7 - DESIGN OF SIMPLE FACTORIAL EXPERIMENTS
Introduction -- The Choice of Factors -- The Choice of Levels - Qualitative Factors; Quantitative Factors: Choice of Extreme Levels; Choice of Number and Position of Levels -- The Control of Error and the Split Unit Principle -- The Final Choice of a Design -- Summary.

Chapter 8 - THE CHOICE OF THE NUMBER OF OBSERVATIONS
Introduction -- The Measurement of Precision -- The Estimation of Precision - Use of Observed Variation between Experimental Units; Use of High-Order Interactions in a Factorial Experiment; From Theoretical Consideration; From Within-Unit Sampling Variation; From the Results of Previous Similar Experiments; Summing Up -- Some Standard Formulas -- Some Sequential Techniques -- Summary.

Chapter 9 - CHOICE OF UNITS, TREATMENTS, AND OBSERVATIONS
Introduction -- Choice of Experimental Units - Size of Units; Representative Nature of Units; Independence of Different Units -- Choice of Treatments -- Choice of Observations - Primary Observations; Substitute Primary Observations; Explanatory Observations; Supplementary Observations for Increasing Precision; Supplementary Observations for Detecting Interactions; Observations for Checking the Application of the Treatments; Observations to Check on External Conditions -- Summary.

Chapter 10 - MORE ABOUT LATIN SQUARES
Introduction -- A Table of Latin Squares -- A Table of Graeco-Latin Squares -- Orthogonal Partitions of Latin Squares -- Squares with More Than Two Alphabets -- Miscellaneous Examples -- Summary.

Chapter 11 - INCOMPLETE NONFACTORIAL DESIGNS
Introduction -- Balanced Incomplete Block Designs - General; Existence and Form of Designs; Analysis of Observations -- Incomplete Designs for the Two-Way Elimination of Error - General; Youden Squares; Lattice Squares -- Further Incomplete Block Designs - General; Designs for Differential Precision; More Units per Block than there are Treatments; Designs for a Large Number of Treatments; Designs with Two Units per Block; Designs with a Small Number of Replicates; Absence of a Suitable Balanced Design; Summing Up -- Further Designs for Two-Way Elimination of Error -- Summary.

Chapter 12 - FRACTIONAL REPLICATION AND CONFOUNDING
Introduction -- Fractional Replication - General Idea; A Simple Case; The Two-Level Factorial System; Factors at More than Two Levels; Latin Squares as Fractional Factorials; General Discussion -- Confounding - General; A Special Case; Confounding in the 2^n System; Factors at More than Two Levels; Confounding in Split Plot Experiments; Confounding in Fractionally Replicated Experiments; Double Confounding; Summing Up -- Summary.

Chapter 13 - CROSS-OVER DESIGNS
Introduction -- Experiments without Carry-Over Effects -- Cross-Over Designs

Cox (cont'd.)
with a Limited Number of Periods per Individual -- Designs with a Large Number of Observations per Subject -- Some Other Possibilities -- Summary.

Chapter 14 - SOME SPECIAL PROBLEMS
Introduction -- Trend-Free Systematic Designs -- Optimum Allocation -- Search for Optimum Conditions -- Assays -- Summary.

GENERAL BIBLIOGRAPHY
APPENDIX - TABLES OF RANDOM PERMUTATIONS AND RANDOM DIGITS
AUTHOR INDEX
SUBJECT INDEX

Davies, O. L. (Editor) [1956], THE DESIGN AND ANALYSIS OF INDUSTRIAL EXPERIMENTS. Second Edition, Oliver and Boyd, Edinburgh, pp. xiii+637. First Edition by Oliver and Boyd, 1954.

LIST OF AUTHORS

AUTHORS' PREFACE

LIST OF EXAMPLES

Chapter 1 - INTRODUCTION
Object of the Handbook -- Nature and Value of Experimental Design -- Deciding the Number of Observations Required -- Sequential Tests -- Investigations of Sampling and Testing Methods -- Randomised Blocks and Latin Squares -- Factorial Designs -- Sequential Methods in Experimental Design.

Chapter 2 - THE PLANNING OF SIMPLE COMPARATIVE EXPERIMENTS
Principles of Planning Illustrated by a Simple Experiment -- Reasoning behind the Significance Test -- Need for Randomisation -- Taking the Specimens in Pairs -- Choice of Experimental Error -- Advantages and Disadvantages of Making the Experiments Self-contained -- Reliability of Mean Difference -- Single- and Double-sided Tests -- Comparison of Means, σ known (Normal Curve test), σ estimated from Sample (t-test) -- Comparison of Proportions -- Comparison of Variances (χ^2- and F-tests) -- Confidence Limits for Variances and Ratio of Variances -- Controlling the Risks of Errors of the Second Kind -- Number of Observations Required -- Operating Characteristic Curve -- The Number of Observations Required in the Comparison of Means, Proportions, and Variances -- Assumptions made in Statistical Tests -- Transformation of Data to Obtain Approximate Normality -- Tests based on the Randomisation Distribution.

APPENDIX: 2A - The Assumption that the Errors are Normally Distributed.

Chapter 3 - SEQUENTIAL TESTS OF SIGNIFICANCE
The Sequential Test -- Testing for a Difference in Mean Value, σ known -- Single-sided Alternative Hypothesis -- Operating Characteristic Curve -- Number of Observations -- Double-sided Alternative Hypotheses -- Testing for a Difference in Mean Value, σ estimated from Sample -- Barnard's Sequential t-test -- Testing for a Difference in Proportions, in Frequencies and in Variances -- Summary of Sequential Tests.

APPENDICES: 3A - Sequential Tests of Simple Hypotheses -- 3B - The Operating Characteristic Curve, or Power Curve -- 3C - The Average Sample Number.

Chapter 4 - INVESTIGATION OF SAMPLING AND TESTING METHODS
Objects of Experiments -- Interaction -- Replicate Analyses -- Errors of Chemical Analysis -- Analysis of Experiment with b Batches and k Samples, with n Tests on Each Sample -- Sampling and Testing Errors -- Economics of Sampling and Testing -- Precision of Estimates of Variance -- Data on Testing Bulk Density of Chalk -- Testing Dyestuffs -- Sampling and Analysis of a Fertiliser.

Davies (cont'd.)

APPENDICES: 4A - Expectation of Mean Squares with Unequal Numbers of Observations in the Groups -- 4B - Confidence Limits for the Ratio of Two Variances Estimated from an Analysis of Variance -- 4C - Computation of Sampling and Testing Errors for an Organic Chemical -- 4D - Computation of Errors in Testing the Bulk Density of Chalk -- 4E - Computation of Errors in the Testing of Dyestuffs -- 4F - Computation of the Sampling and Analytical Errors of a Fertiliser.

Chapter 5 - RANDOMISED BLOCKS AND LATIN SQUARES
Randomised Blocks: Effect of Interactions -- Effect of Time Trends -- Analysis -- Conditions for Use -- Errors of the Second Kind -- Confidence Limits -- Summary -- Missing Values. Latin Squares: Conditions for Use -- Randomisation -- Analysis -- Missing Values -- Orthogonal Squares -- Graeco-Latin Squares -- The 3 × 3 Latin Square -- A Design Using Several 3 × 3 Graeco-Latin Squares -- Latin Cubes.

APPENDICES: 5A - Expectations of Mean Squares -- 5B - Computations for a Randomised Block Experiment -- 5C - Computations for a 4 × 4 Latin Square -- 5D - Computations for a 7 × 7 Latin Square -- 5E - Computations for an 8 × 8 Graeco-Latin Square -- 5F - Computations for Four 3 × 3 Graeco-Latin Squares.

Chapter 6 - INCOMPLETE RANDOMISED BLOCK DESIGNS
Use of the Designs -- Example -- Analysis of Balanced Incomplete Block Designs -- Comparison of the Efficiency of Balanced Incomplete Blocks with Designs Using Controls in Each Block -- Symmetrical Balanced Incomplete Blocks -- Youden Square -- Non-balanced Incomplete Blocks -- Lattice Square -- Quasi-factorial and Quasi-Latin Square Designs.

APPENDICES: 6A - Tables of Balanced Incomplete Blocks and Youden Squares -- 6B - General Treatment of Two-way Tables with Missing Values -- 6C - Unbalanced Incomplete Blocks.

Chapter 7 - FACTORIAL EXPERIMENTS: ELEMENTARY PRINCIPLES
Simple Factorial Design -- Definitions -- The Advantages of Factorial Design -- Interpretation of Main Effects and Interactions -- Expectation of Mean Squares in Factorial Experiments -- The Assumption of Normality -- Designs with Factors at Two Levels Only -- Examples -- Symbolic Expressions for Effects -- Standard Order of Treatment Combinations and Effects -- Calculation of Effects and Analysis of Variance -- Analysis of General 2^n Factorial Design -- Orthogonality -- Randomisation.

APPENDICES: 7A - Expectations of Mean Squares in Factorial Experiments -- 7B - Systematic Checking in Yates' Method for the Analysis of 2^n Design -- 7C - Significance of One and Several Mean Squares -- 7D - The Combination of Interaction Mean Squares to form an Estimate of Error Variance.

Davies (cont'd.)

Chapter 8 - FACTORIAL EXPERIMENTS WITH FACTORS AT MORE THAN TWO LEVELS
Qualitative and Quantitative Factors -- Analysis of Multi-factor Experiments with All Factors Qualitative -- Quantitative Factors, Polynomial Representation -- Analysis of Factorial Designs of Two Factors, One Quantitative and the Other Qualitative -- Analysis of n x m Factorial Designs with Both Factors Quantitative -- Factorial Designs with All Factors at Three Levels -- Notation -- Analysis.

APPENDICES: 8A - General Method of Analysis of Factorial Designs for Qualitative Factors -- 8B - Details of Analysis of a 4 x 4 x 9 Factorial Experiment, All Factors Qualitative -- 8C - Partitioning of Degrees of Freedom for Quantitative Factors by means of Orthogonal Polynomials -- 8D - Divisors for Sums of Squares in an Analysis of Variance -- 8E - Details of Analysis of a 5 x 4 Factorial Experiment, One Factor Quantitative, the other Qualitative -- 8F - Statistical Analysis of a 3^3 Factorial Experiment -- 8G - Systematic Method for the Analysis of a 3^n Design when All Factors are Quantitative and the Levels of Each Factor are Equally Spaced.

Chapter 9 - CONFOUNDING IN FACTORIAL DESIGNS. FACTORIAL EXPERIMENTATION WHEN UNIFORM CONDITIONS CANNOT BE MAINTAINED THROUGHOUT THE EXPERIMENT
Experimental Conditions -- Analysis of Confounded Designs -- Situations requiring the Use of Confounding -- Partial Confounding -- General Principles of confounding 2^n Designs in Two Blocks -- Higher Degrees of Confounding in 2^n Factorial Designs -- Rules for confounding Given Interactions -- Generalised Interaction -- Confounding 2^n Factorial Designs in Four Blocks -- Defining Contrasts -- Principal Block -- Double Confounding -- Confounding in 3^n Designs -- The I-, J-, W-, X-, Y-, and Z-Interactions -- Confounding 3^n Designs in Three and in Nine Blocks.

APPENDICES: 9A - Details of the Statistical Analysis of a 2^3 Factorial Design replicated twice and Confounded between Four Blocks -- 9B - Partial Confounding in a 2^3 Factorial Design -- 9C - Details of Statistical Analysis of a 2^5 Experiment confounded between Two Blocks -- 9D - Confounding from the Standpoint of Finite Groups -- 9E - Systems of Confounding in 2^n Factorial Designs -- 9F - Example of a 3^n Design Counfounded in Three Blocks and Details of Analysis -- 9G - Confounded 3^n Factorial Designs.

Chapter 10 - FRACTIONAL FACTORIAL EXPERIMENTS
Basic Principles -- The Weighing Problem -- Designs of Eight Observations, All Factors at Two Levels -- Relation between Fractional Factorial Designs and Confounding -- Confusion of Effects -- Aliases. General Methods of Construction -- Analysis -- Designs of Sixteen Observations, All Factors at Two Levels -- List of 2^n Designs -- Confounding -- Discussion on Application of Fractional Factorial Designs -- Experimental Error -- Use of Fractional Factorial Designs in Sequence -- Separation of Aliases -- Fractional Replication in 3^n Factorial Designs -- List of 3^n Designs -- Designs of Nine Observations -- Designs of 27 Observations.

Davies (cont'd.)

APPENDICES: 10A - Fractional Replicates up to Eight Factors in Sixteen Observations -- 10B - Aliases in Fractional Replicates of 3^n Designs -- 10C - Systematic Analysis of Results of Fractional Factorial Designs.

Chapter 11 - THE DETERMINATION OF OPTIMUM CONDITIONS
Notation and Terminology -- Representation of Surfaces by Contours -- Polynomial Representation of a Surface -- Methods of Finding Maximum Response -- Exploration and Location of Stationary Point -- Scales of Measurement -- Designs to Determine First-order Effects -- Bias -- An Example of Steepest Ascent -- Exploration of Yield Surface in Near-stationary Region -- Fitting a Second-degree Equation -- Analysis of a Fitted Surface -- Canonical Form -- Ridges -- Design to Determine Second-order Effects, Composite Designs -- Orthogonal and Non-orthogonal Composite Designs -- Analysis of a Fitted Second-degree Equation -- Three-dimensional Contour Systems -- Point, Line and Plane Maxima. Stationary Ridges -- Types of Surfaces -- Canonical Analysis -- General Conclusions.

APPENDICES: 11A - Solutions of Linear Equations. Inversion of Matrices. Calculation of Determinants -- 11B - The Use of the Method of Least Squares in the Fitting of Response Surfaces.

GLOSSARY OF TECHNICAL TERMS

TABLES OF STATISTICAL FUNCTIONS

GENERAL INDEX

INDEX OF PROPER NAMES

Edwards, A. L. [1950], EXPERIMENTAL DESIGN IN PSYCHOLOGICAL RESEARCH. Rinehart and Company, Inc., Publishers: New York, pp. xiv+446. Second printing by Rinehart and Company, Inc., January, 1951.

Chapter 1 - THE NATURE OF PSYCHOLOGICAL RESEARCH
Introduction -- Observations -- Stimulus Variables -- Response Variables -- Organismic Variables -- Research Problems -- Dependent and Independent Variables -- Examples.

Chapter 2 - PRINCIPLES OF EXPERIMENTAL DESIGN
Samples in Research -- Sampling Distributions -- Randomization and Experimental Design -- Tables of Random Numbers -- The Difference between 2 Means -- The Test of Hypotheses -- Two Kinds of Errors -- Practical Versus Statistical Significance -- Examples.

Chapter 3 - PROBABILITY AND EXPERIMENTAL DESIGN
The Farmer's Divining Rod -- A Simple Experimental Design -- Permutations and Combinations -- Experimental Controls -- A Limitation in the Design -- Increasing the Sensitivity of the Experiment -- The Binomial Expansion -- An Experiment on Taste -- Examples.

Chapter 4 - THE NORMAL AND χ^2 APPROXIMATIONS OF THE BINOMIAL PROBABILITIES
Introduction -- The Normal Distribution -- Relation of the Binomial to the Normal Distribution -- Parameters of the Binomial Distribution -- Approximations of the Binomial Probabilities from the Table of the Normal Curve -- Evaluation of the Experiment on Taste -- A Problem in Opinion Polling -- The χ^2 Distribution -- The Relation between z and χ^2 for 1 Degree of Freedom -- Summary -- Examples.

Chapter 5 - EXPERIMENTS INVOLVING A COMPARISON OF THE DIFFERENCE BETWEEN 2 FREQUENCIES OR PROPORTIONS
Introduction -- The Null Hypothesis -- Standard Error of the Difference between 2 Uncorrelated Proportions -- One-tailed and Two-tailed Tests of Significance -- The χ^2 Test for the Difference between Uncorrelated Proportions -- The Correction for Continuity -- Another Method for Calculating χ^2 -- Correlated Proportions -- Standard Error of the Difference between Correlated Proportions -- The χ^2 Test for Correlated Proportions -- Correcting for Continuity: Correlated Proportions -- The Influence of Correlation on the Test of Significance -- Examples.

Chapter 6 - THE APPLICATION OF THE χ^2 DISTRIBUTION TO RESEARCH PROBLEMS INVOLVING MORE THAN 1 DEGREE OF FREEDOM
Introduction -- A Study of Preferences -- A Study of Industrial Accidents -- A Study of Vocational Advisement -- The j x 2 Table -- Planning the Comparisons to be Made -- An Experiment Involving a Test of Technique -- χ^2 for More than 30 Degrees of Freedom -- Examples.

Edwards (cont'd.)

Chapter 7 - TESTING HYPOTHESES ABOUT CORRELATION COEFFICIENTS
Introduction -- The Sampling Distribution of the Correlation Coefficient -- The Normal Curve Test of the Hypothesis of Zero Correlation -- The t Test of the Hypothesis of Zero Correlation -- Table of Significant Values of the Correlation Coefficient -- The z' Transformation for the Correlation Coefficient -- Establishing the Fiducial Limits -- Testing the Significance of the Difference between 2 Correlation Coefficients --Finding an Average Value of the Correlation Coefficient for 2 Samples -- An Average Value of the Correlation Coefficient Based upon Several Samples -- Testing the Hypothesis that Several Samples are from a Common Population -- The Fiducial Limits for an Average Value of the Correlation Coefficient -- A Correction for a Systematic Bias in Averaging z' Values -- Examples.

Chapter 8 - THE t TEST AND THE SIGNIFICANCE OF MEANS AND DIFFERENCES BETWEEN MEANS
The Sampling Distribution of the Mean -- The t Distribution -- The Fiducial Limits for the Mean -- Increasing the Size of the Sample -- Fiducial Probability -- An Experiment on Retention -- The Standard Error of the Difference between 2 Means -- Testing the Null Hypothesis -- The Fiducial Limits for the Difference between 2 Means -- Estimating the Size of the Sample for a Repetition of the Experiment -- The Influence of Changes in C^2 on the Subsequent Value of t -- Examples.

Chapter 9 - HETEROGENEITY OF VARIANCE AND THE t TEST
Introduction -- The F Distribution -- Testing for Homogeneity of Variance -- The Effects of Nonnormality -- Testing the Hypothesis of a Common Population Mean When n_1 and n_2 Differ -- Obtaining the Value of t Which Will be Regarded as Significant -- The Influence of Heterogeneity of Variance upon the t Test -- Testing the Hypothesis of a Common Population Mean When n_1 Equals n_2 -- Examples.

Chapter 10 - AN INTRODUCTION TO THE ANALYSIS OF VARIANCE
Introduction -- The Partitioning of the Total Sum of Squares for r Random Samples of n Cases -- The Mean Square within Groups -- The Mean Square between Groups -- Independence of the Mean Squares -- The Test of Significance -- An Analysis of Variance for 2 Groups -- Some Problems to Which the Analysis of Variance Might be Applied -- An Experiment Involving 5 Experimental Conditions -- Summary of the Calculations -- Examples.

Chapter 11 - HETEROGENEITY OF VARIANCE AND TRANSFORMATIONS OF THE SCALE
Introduction -- The Test for Homogeneity of Variance: Equal n's -- The Test for Homogeneity of Variance: Unequal n's -- Transformations -- The Square Root Transformation -- Two Examples of the Application of the Square Root Transformation -- The Logarithmic and Angular Transformations -- The Reciprocal Transformation -- The Analysis of Variance without Transformation -- Examples.

Chapter 12 - THE 2^n FACTORIAL DESIGN FOR EXPERIMENTS IN WHICH VARIABLES ARE VARIED IN ONLY 2 WAYS
Introduction -- The 2 × 2 Factorial Design -- Testing for Homogeneity of Variance

Edwards (cont'd.)

-- Partitioning the Sum of Squares between Groups -- Calculation of the Interaction Sum of Squares Based on 1 Degree of Freedom -- Allocation of the Degrees of Freedom -- Interpretation of the Experiment -- The 2 × 2 × 2 Factorial Design -- Sums of Squares for the Main Experimental Variables -- The Interaction Sums of Squares -- The Interpretation of the Experiment -- Testing Whether One Mean Square is Significantly Smaller than Another -- A Method of Showing the Comparisons in the 2^n Factorial Design -- Orthogonal Comparisons -- Some Advantages of the Factorial Design -- Examples.

Chapter 13 - COMPLEX FACTORIAL DESIGNS

Introduction -- A 4 × 3 × 2 Factorial Design -- Calculation of the Sum of Squares -- Direct Calculation of a Second- or Higher-order Interaction -- Summary of the Analysis -- The Use of Interactions as Error Terms Instead of the Usual Estimates of Error -- Factorial Designs without Replication -- The Assumptions involved in Pooling Higher-order Interactions -- An Example of an Analysis without Replication -- Testing the Interaction Mean Squares for Homogeneity of Variance -- Summary -- Examples.

Chapter 14 - EXPERIMENTAL DESIGNS INVOLVING MATCHED GROUPS

An Experiment Described -- A Change in the Design -- Analysis of the Data -- The Nature of the Residual Sum of Squares -- The Efficiency of Design Involving Matched Groups -- The Matching Variables -- An Analysis of Several Matched Groups -- The t Test Applied to 2 Matched Groups -- The Importance of Considering Possible Interactions -- Examples.

Chapter 15 - EXPERIMENTAL DESIGNS INVOLVING REPEATED MEASUREMENTS OF THE SAME SUBJECTS

A Problem in Experimental Design -- Testing the Significance of Practice Effects for a Single Group Given 2 Trials -- The Significance of Practice Effects for a Series of Trials -- The Analysis of Repeated Measurements on Several Independent Groups -- Calculation of the Sums of Squares -- The Tests of Significance -- Relating the Analysis to Previous Methods -- Examples.

Chapter 16 - THE LATIN SQUARE DESIGN IN PSYCHOLOGICAL RESEARCH

An Experiment in Color Recognition -- The Bliss and Rose Experiment -- A Method for Assigning Treatments in the Latin Square -- The Analysis of Variance for the Latin Square -- The Direct Calculation of the Residual Sum of Squares -- Applications of the Latin Square Design -- Combining the Factorial Design with the Latin Square -- The Nature of the Row Mean Square -- An Experimental Design with Replication of the Same Latin Square -- Analysis of the Independent Observations -- Analysis of the Correlated Observations -- Summary of the Analysis -- Examples.

Chapter 17 - APPLICATIONS OF THE ANALYSIS OF COVARIANCE

Introduction -- The Analysis of Covariance -- Correlation and Regression -- Partitioning the Total Sum of Cross Products -- The Sums of Squares of Errors of Estimate -- An Application of the Analysis of Covariance -- Interpretation of the Analysis -- Another Application of Covariance Analysis -- Interpretation of the Analysis -- The Use of the Analysis of Covariance in Research -- Examples.

Edwards (cont'd.)

BIBLIOGRAPHY

LIST OF FORMULAS

APPENDIX

TABLE I. TABLE OF RANDOM NUMBERS

TABLE II. SQUARES, SQUARE ROOTS, AND RECIPROCALS OF NUMBERS FROM 1 TO 1,000

TABLE III. AREAS AND ORDINATES OF THE NORMAL CURVE IN TERMS OF x/σ

TABLE IV. TABLE OF χ^2

TABLE V. TABLE OF t

TABLE VI. VALUES OF r AT THE 5 AND 1 PER CENT LEVELS OF SIGNIFICANCE

TABLE VII. TABLE OF z' VALUES FOR r

TABLE VIII. THE 5 AND 1 PER CENT POINTS FOR THE DISTRIBUTION OF F

ANSWERS TO EXAMPLES

INDEX OF NAMES

INDEX OF SUBJECTS

Federer, W. T. [1955], EXPERIMENTAL DESIGN - Theory and Application. The Macmillan Company, New York, N.Y., U.S.A., pp. xix+544+47. Second printing by the Macmillan Company, 1963. Indian Edition, 1967, Oxford & IBH Publishing Company, Calcutta, India.

Chapter I - INTRODUCTION
The Principles of Scientific Experimentation - Formulation of questions to be answered and hypotheses to be tested; A critical and logical analysis of the problem or problems raised; Selection of a procedure for research; Selection of suitable measuring instruments and control of the personal equation; A complete analysis of the data and the interpretation of results in light of experimental conditions and hypotheses tested; Preparation of a complete, correct, and readable report of the experiment; Statistics in relation to the principles of scientific experimentation -- Classification of Experimental Designs - Systematic designs; Randomized designs -- Selection of an Experimental Design -- Validity and Choice of an Experimental Error -- Symbolism.

Chapter II - SOME USEFUL STATISTICAL TOOLS AND CONCEPTS
Tests of Significance for a Group of Ranked Means - Multiple range tests (The lsd test or the multiple t test; Student-Newman-Keuls test; Duncan's multiple range test; Tukey's test based on allowances; Short cut to allowances; Comments on multiple range tests); Multiple F tests (Fisher's least significant difference test; Duncan's multiple comparisons test; Scheffé's test; Comments on multiple F tests); Tukey's gap, straggler, and variance test; Selection of largest n means from a set of v means -- Transformation of Data - General comments; Square root transformation; The arcsine or angular transformation; The logarithmic transformation; Other transformations -- Test for Homogeneity of Variances -- Tests for Additivity of Data -- Nonparametric Tests in the Analysis of Variance -- Probability Levels for t, F, and χ^2.

Chapter III - PLOT OR PEN TECHNIQUE
Competition - Intra-experimental unit competition; Inter-experimental unit competition -- Size of Experimental Unit - Factors affecting the size of the experimental unit; Methods of determining the size of the experimental unit (Maximum curvature method; Fairfield Smith's Variance Law; Some mathematical developments related to "Fairfield Smith's Variance Law"; Some additional results) -- Shape of Experimental Unit -- Replication - Shape of the replicate; Size of replicate; Number of replications (The Harris-Horvitz-Mood method; Tang's method; Tukey's method) -- Sampling the Experimental Unit - Loss in information due to sampling; Determination of the optimal sampling rate; Sampling for one character when a related character of the experimental unit is known.

Chapter IV - THE COMPLETELY RANDOMIZED DESIGN
Application of the Completely Randomized Design - Introduction; Advantages and disadvantages; Layout of the design; Analysis for equal replication of the treatments; Analysis for unequal replication of the treatments; Hierarchial classifications -- Least Squares Estimates and Expectation of Mean Squares - Samples of n observations (Least squares estimate of the mean; Variance of the sample

Federer (cont'd.)

mean); Completely randomized design with an equal number of replicates per treatment (Least squares estimates of effects; Expectation of mean squares); Completely randomized design with unequal numbers of replicates per treatment (Least squares estimates of effects; Expectation of mean squares); Expectation of mean squares in hierarchial classifications (Three categories of variation; Five categories of variation); Regression coefficients and the intercept (Least squares estimate of the regression coefficient and of the intercept; Variances and covariances of the estimates).

Chapter V - RANDOMIZED COMPLETE BLOCK DESIGN

Applications of the Randomized Complete Block Design - Introduction; Advantages and disadvantages; Layout and analysis; Analysis for one observation per experimental unit; Analysis for more than one observation per experimental unit; Unequal numbers of observations per experimental unit and unequal replication per treatment (Missing experimental units; Disproportionate numbers per experimental unit; Other situations) -- Least Squares Estimates and Expectation of Mean Squares - One unit per experimental unit; k units per experimental unit -- Development of Formulae for Missing or Deleted Values.

Chapter VI - THE LATIN SQUARE DESIGN

Applications of the Latin Square Design - Introduction; Advantages and disadvantages; Construction and arrangements; Randomization; Experimental layout; Statistical analysis for one observation per experimental unit; Statistical analysis for a group of latin squares with a single determination per plot; Analysis for more than one observation per experimental unit; Missing data (Missing experimental units; Disproportionate numbers in the experimental unit; Other situations) -- Least Squares Estimates and Expectation of Mean Squares - One unit per experimental unit (Least squares estimates; Expectation of mean squares (Model II)); s squares of k×k latin squares (Least squares estimates; Expectation of mean squares (Model II)); k×k latin square with p items per cell and d determinations on each item -- Development of Formulae for Missing Experimental Units.

Chapter VII - THE CHOICE OF TREATMENTS AND THE FACTORIAL EXPERIMENT – p^n SERIES

Selection of Treatments and Treatment Combinations -- The Factorial Experiment - Definitions of main effects and interactions; Advantages and disadvantages of factorial arrangements; Choice of levels -- Procedures to Investigate Specific Objectives -- The Factorial Experiment – 2^n Series - The 2×2 factorial; The 2^3 factorial; The 2^4 factorial; Generalized interaction in the 2^n series -- The Factorial Experiment - 3^n Series - The 3^2 factorial; The 3^3 factorial; Generalized interaction in the 3^n series -- Other Factorials of the p^n Series - The 4×4 factorial; The 5×5 factorial; General case -- Experimental Designs for Factorial Experiments - Experiments with repetition of the p^n treatments; Experiments without repetition of the p^n treatments.

Chapter VIII - OTHER FACTORIAL EXPERIMENTS

The p×q×k... Series of Factorials - The $2^n \times 3^s$ series; Other factorials --

Federer (cont'd.)

Additional Treatments in Factorial Experiments - Factorial arrangement of treatments plus other treatments; Additional combinations included in the factorial experiment; Proportional factorials; Comments on the inclusion of additional treatments -- Designs for p×q×k... Factorials -- Missing Data -- Least Squares Estimates of Effects and Expectation of Mean Squares - The p×q factorial in a randomized complete block design; The p×q×k factorial in a randomized complete block design; Analysis for all possible single crosses among k lines in a randomized complete block design; Analysis for $k(k-1)/2$ possible crosses among k lines in a randomized complete block design.

Chapter IX - CONFOUNDING IN FACTORIAL EXPERIMENTS

Confounding – Use and Types -- Advantages and Disadvantages of Confounding -- Complete Confounding - Randomization; Complete confounding in the 2^n series; Complete confounding in the 3^n series; Complete confounding in other factorials; Applying treatments to strata which are designed to control heterogeneity -- Partial Confounding - Partial confounding in the 2^n series; Partial confounding in the 3^n series; Partial confounding in other factorials; General methods of analysis; Recovery of interblock information in partial confounding -- Fractional Replication -- Missing Data.

Chapter X - FACTORIAL EXPERIMENTS WITH MAIN EFFECTS CONFOUNDED – SPLIT PLOT AND SPLIT BLOCK DESIGNS WITH VARIATIONS

The Split Plot Design - Introduction; Advantages and disadvantages; Randomization; Analysis (Whole plots in a randomized complete block design; Whole plots in a latin square design; Whole plots and split plots arranged in latin square designs; Repetitions of experimental results); Missing data -- The Split Block Design - Introduction; Advantages and disadvantages; Randomization and experimental layouts; Analysis; Missing data -- The Split Split Plot Design and Further Subdivisions -- Some Variations -- Expectation of Mean Squares - Split plot design; Randomized complete block experiment repeated at several locations; Expectation of mean squares from an experiment on perennial crops.

Chapter XI - INCOMPLETE BLOCK DESIGNS – GENERAL CONSIDERATIONS AND THE ONE-RESTRICTIONAL LATTICES WITH TREATMENTS IN COMPLETE REPLICATES

Introduction - Advantages and disadvantages; Randomization procedure for lattice designs with one restriction; Analysis for lattice designs with one restriction; Size and shape of incomplete block and layout of blocks -- Classification of Lattice Designs -- Two-Dimensional One-Restrictional Lattice Designs - Double (simple) lattice; Triple lattice; Balanced lattice; Rectangular lattice -- Three-Dimensional One-Restrictional Lattices - The three-dimensional triple lattice (cubic lattice); Three-dimensional lattices with more than three arrangements (k = a prime number or power of a prime number) -- n-Dimensional One-Restrictional Designs - k^4 treatments in blocks of k^s $(s<4)$ (k^4 treatments in blocks of k; k^4 treatments in blocks of k^3); k^5 treatments in blocks of k^s $(s<5)$ (k^5 treatments in blocks of k; k^5 treatments in blocks of k^2; k^5 treatments in blocks of k^3) -- Missing Data -- Tests of Significance -- Least Squares Estimates for a Double Lattice and Expectation of Mean Squares - Least squares estimates of effects with recovery of interblock information; Expectation of mean squares --

Federer (cont'd.)

Derivation of Standard Errors - The double lattice; The three-dimensional triple lattice (the cubic lattice)

Chapter XII - LATTICE DESIGNS WITH MORE THAN ONE RESTRICTION ON THE ALLOCATION OF TREATMENTS IN THE COMPLETE BLOCK

Introduction -- Lattice Square Designs - Semi-balanced lattice square; Balanced lattice square; Other lattice squares (Analysis for 2, 3, ..., (k-1)/2 arrangements; Analysis for more than (k+1)/2 arrangements; Lattice square designs for k = 6, 10, and 12 in 3 replicates; Incomplete lattice square in 3q replicates) -- k^3 Treatments in k Whole Plots, k Split Plots, and k Split Split Plots, a Two-Restrictional Design -- Three-Dimensional Three-Restrictional Designs -- Four-Dimensional Lattices with Two or More Restrictions -- Missing Data -- Tests of Significance.

Chapter XIII - OTHER INCOMPLETE BLOCK DESIGNS

Introduction -- Incomplete Block Designs for v Treatments in b Incomplete Blocks of Size k - Balanced incomplete block designs (One set; More than one basic set); Partially balanced incomplete block designs -- Incomplete Block Designs for the Two-Way Elimination of Heterogeneity - One or more missing rows or columns in a latin square; The semi-latin square and similar designs; Youden squares; Other designs -- Split Plot Designs for Nonfactorial Experiments - Groups of similar treatments in each whole plot; Inclusion of random controls in the split plots -- Linked Block Designs.

Chapter XIV - BALANCED DESIGNS

Introduction -- Change-Over Designs - The simple change-over design; Double-reversal designs with extensions — no residual effect from previous treatment; Change-over designs to measure residual effects; Other designs -- Designs for Long-Term Experiments - Same treatment applied on a single crop every year; Experiments on a single rotation of crops; Comparison of several rotations.

Chapter XV - SOME ADDITIONAL DESIGNS

Three-Way and Higher-Way Grouping of the Treatments - Graeco-latin squares; Hyper-graeco-latin squares; Latin and hyper-graeco-latin cubes and hypercubes -- Quasi-Latin Squares - Confounding in latin squares; Half-plaid latin squares; Plaid latin squares -- Magic Latin Squares and Super Magic Latin Squares -- Weighing Designs.

Chapter XVI - COVARIANCE

Introduction -- Completely Randomized Design -- Randomized Complete Block Design -- Latin Square Design -- Split Plot Design -- Split Block Design -- One-Restrictional Lattice Designs - Covariance analysis without recovery of interblock information; Covariance analysis with recovery of interblock information -- Two- and Higher-Restrictional Lattice Designs - Two-restrictional lattice designs; Lattice designs with more than two restrictions -- Change-Over Designs -- Covariance and Unequal Numbers of Observations per Treatment - Covariance on dummy variates for missing or mixed-up plot yields; Covariance on dummy variates for non-orthogonal two-way classifications; Covariance

Federer (cont'd.)

analysis of n-way classifications with unequal numbers in the subclasses - The randomized complete block design with one pair of values for each experimental unit; The randomized complete block design with n_{ij} pairs of observations per experimental unit; n-Way Classifications and multiple covariance -- Covariance Versus Stratification -- Expectation of Mean Squares - Completely randomized design; Randomized complete block design; Double lattice design.

Finney, D. J. [1960], AN INTRODUCTION TO THE THEORY OF EXPERIMENTAL DESIGN. The University of Chicago Press, Chicago and London, pp. xii+223. Third printing, 1963.

Chapter 1 - EXPERIMENTATION AND THE MATHEMATICIAN
"Of Making Many Books ..." -- Comparative Experiments -- Experimenter and Statistician -- Nomenclature -- The Meaning of Design -- Randomization -- Economic Considerations -- The Interpretation of Experiments.

Chapter 2 - ANALYSIS OF VARIANCE
The Partition of a Sum of Squares -- Normality -- Completely Randomized Design -- Models -- Variance Components.

Chapter 3 - RANDOMIZED BLOCKS AND ORTHOGONAL SQUARES
The Randomized Block Design -- Use of Randomized Blocks -- Latin Squares -- Analysis of a Latin Square Experiment -- Greco-Latin Squares -- Orthogonal Squares of Higher Order -- Orthogonal Partitions of Latin Squares.

Chapter 4 - FACTORIAL DESIGN AND FRACTIONAL REPLICATION
The Factorial Principle -- The 2^n Designs -- Factorial Experiments -- Single Replication -- Fractional Replication -- Different Fractions of One Design -- Prime Power Designs -- Single Replication of 3^n Designs -- Fractional Replication of 3^n Designs -- Fisher's Theorem -- Number of Levels Equal to Power of a Prime -- Mixed Levels -- Weighing Designs -- Other Designs for Zero Interactions.

Chapter 5 - CONFOUNDING
Blocks -- A Simple Confounded Design -- Confounding as a Form of Fractional Replication -- General Confounding of 2^n Designs -- Partial Confounding -- Confounding in Single Replication -- Confounding of Fractional Replicates -- Confounding of Prime Power Designs -- Fractional Replication of Prime Power Designs -- Fisher's Theorem on Minimal Confounding -- Mixed Levels -- Double Confounding and Quasi-Latin Squares -- Split-Plot Designs -- Fractional Replication and Orthogonal Squares -- Inclusion of Factors -- The Merits of Factorial Design.

Chapter 6 - INCOMPLETE BLOCK DESIGNS
The Need for Incomplete Blocks -- Balanced Incomplete Blocks -- Existence of Designs -- Fisher's Theorem -- Schützenberger's Theorem -- Analysis of a Balanced Incomplete Block Experiment -- Doubly Balanced Incomplete Blocks -- Youden Squares -- Lattice Designs -- Rectangular Lattices -- Multidimensional Lattices -- Lattice Squares -- Balanced Lattice Squares -- Randomized Block Analysis -- Partially Balanced Incomplete Blocks -- Partially Balanced Youden Squares -- Choice of Design.

Chapter 7 - EXPERIMENTS INVOLVING CHANGES OF TREATMENT
Time as an Experimental Factor -- Agricultural Rotations -- Fruit and Tree Crops -- Cross-over Designs -- Balanced Sequences -- Sequential Use of Fractional Replication.

Finney (cont'd.)

Chapter 8 - SEQUENTIAL EXPERIMENTATION
Sequential Sampling -- Sequential Experiments -- Sequential Estimation -- Estimation of Optimal Conditions -- Evolutionary Operation in Industry.

Chapter 9 - EFFICIENCY OF EXPERIMENTATION
Analysis, Design, and Planning -- Inclusion of Controls -- Number of Treatments -- Levels of the Factors -- Another Approach to the Choice of Levels -- Replication -- Balance and Covariance -- Choice of Design -- Selection -- The Screening of Drugs.

Chapter 10 - ECONOMICS OF EXPERIMENTATION
Internal and External Economics -- Estimation of an Optimal Rate -- Optimal Choice between Two Alternatives -- External Economy of Selection -- "The Little Black Box".

REFERENCES
SUBJECT INDEX
AUTHOR INDEX

Finney, D. J. [1955], EXPERIMENTAL DESIGN AND ITS STATISTICAL BASIS. The University of Chicago Press, pp. xi+169.

Chapter I - STATISTICAL SCIENCE
Why Statistics? -- Experimental Design -- Books on Statistical Method -- Books on Experimental Design -- This Book.

Chapter II - COUNTS
Records of Frequencies -- Deviations from a Theoretical Proportion -- The χ^2 Distribution -- Counts in More than Two Classes -- Disproof, Proof, and Estimation -- The Planning of Genetical Experiments -- The Comparison of Proportions -- Interpretation of a Significance Test -- Experimental Design -- Design of Clinical Experiments.

Chapter III - MEASUREMENTS
Measurements -- Design for a Simple Experiment -- Results and Statistical Analysis -- The Normal Distribution -- Homogeneity of Variance - An Improvement in Design -- Estimation -- Precision and Efficiency -- A Further Complication.

Chapter IV - RANDOMIZED BLOCKS AND LATIN SQUARES
Agricultural Research and Experimental Design -- Experimental Units -- Experiments on Several Treatments -- Replication -- Randomization -- Completely Randomized Design -- Blocks -- Randomized Blocks -- Counts and Measurements -- Latin Squares -- Statistical Analysis of a Latin Square -- Orthogonality -- Graeco-Latin Squares -- Sets of Latin Squares -- Latin Cubes.

Chapter V - INCOMPLETE BLOCK DESIGNS
Limitations on Block Size -- An Experiment on Self-Administered Analgesia -- Balanced Incomplete Blocks -- Youden Squares -- Lattice Designs -- Partially Balanced Incomplete Blocks -- Analysis of Incomplete Block Designs -- Use of Incomplete Block Designs -- Other Designs.

Chapter VI - FACTORIAL EXPERIMENTS
Factorial Design in Agricultural Research -- Factorial Design in Other Sciences -- Example of a Factorial Experiment -- Specification of Factorial Designs -- A 2^3 Experiment -- Notation -- Analysis of Variance -- Replication -- Fractional Replication -- Confounding -- Partial Confounding -- Split-plot Designs -- Double Confounding -- Designs for Zero Interactions -- Information from Factorial Experiments.

Chapter VII - SEQUENTIAL EXPERIMENTS
Sequential Nature of Research -- Factorial Experiments -- Experimental Search for Optimal Conditions -- Sequential Rules for Terminating Experiments -- Staircase Methods.

Finney (cont'd.)

Chapter VIII - BIOLOGICAL ASSAY

Types of Biological Assay -- The Standard Response Curve -- The Planning of Assays -- Parallel Line Assays -- Choice of Doses for Parallel Line Assays -- Slope Ratio Assays -- Choice of Doses for Slope Ratio Assays -- Quantal Responses -- The Choice of Subjects and of Responses.

Chapter IX - THE SELECTION OF A DESIGN

Design, Analysis, and Interpretation -- The Number of Factors -- The Choice of Levels -- Controls -- Number of Replications -- Allocation of Treatments to Plots -- The Number of Experiments -- The Measurements -- Concomitant Measurements -- Statistical Analysis.

REFERENCES
INDEX

Fisher, R. A. [1966], THE DESIGN OF EXPERIMENTS. (Eighth Edition). Oliver and Boyd, Edinburgh, pp. xiv+248. First Edition by Oliver and Boyd, 1935.

Chapter I - INTRODUCTION
The Grounds on which Evidence is Disputed -- The Mathematical Attitude towards Induction -- The Rejection of Inverse Probability -- The Logic of the Laboratory.

Chapter II - THE PRINCIPLES OF EXPERIMENTATION, ILLUSTRATED BY A PSYCHO-PHYSICAL EXPERIMENT
Statement of Experiment -- Interpretation and its Reasoned Basis -- The Test of Significance -- The Null Hypothesis -- Randomisation; the Physical Basis of the Validity of the Test -- The Effectiveness of Randomisation -- The Sensitiveness of an Experiment. Effects of Enlargement and Repetition -- Qualitative Methods of Increasing Sensitiveness - Scientific Inference and Acceptance Procedures.

Chapter III - A HISTORICAL EXPERIMENT ON GROWTH RATE
Darwin's Discussion of the Data -- Galton's Method of Interpretation -- Pairing and Grouping -- "Student's" t Test -- Fallacious Use of Statistics -- Manipulation of the Data -- Validity and Randomisation -- Test of a Wider Hypothesis - 'Non-parametric' tests.

Chapter IV - AN AGRICULTURAL EXPERIMENT IN RANDOMISED BLOCKS
Description of the Experiment -- Statistical Analysis of the Observations -- Precision of the Comparisons -- The Purposes of Replication -- Validity of the Estimation of Error -- Bias of Systematic Arrangements -- Partial Elimination of Error -- Shape of Blocks and Plots -- Practical Example.

Chapter V - THE LATIN SQUARE
Randomisation subject to Double Restriction -- The Estimation of Error -- Faulty Treatment of Square Designs -- Systematic Squares -- Graeco-Latin and Higher Squares - Configurations in three or more dimensions; an exception design -- Practical Exercises.

Chapter VI - THE FACTORIAL DESIGN IN EXPERIMENTATION
The Single Factor -- A Simple Factorial Scheme -- The Basis of Inductive Inference -- Inclusion of Subsidiary Factors -- Experiments without Replication.

Chapter VII - CONFOUNDING
The Problem of Controlling Heterogeneity -- Example with 8 Treatments, Notation -- Design Suited to Confounding the Triple Interaction -- Effect on Analysis of Variance - General system of confounding in powers of 2; Double confounding -- Example with 27 Treatments -- Partial Confounding - Practical exercises.

Fisher (cont'd.)

Chapter VIII - SPECIAL CASES OF PARTIAL CONFOUNDING
Dummy Comparisons -- Interaction of Quantity and Quality -- Resolution of Three Comparisons among Four Materials -- An Early Example -- Interpretation of Results -- An Experiment with 81 Plots.

Chapter IX - THE INCREASE OF PRECISION BY CONCOMITANT MEASUREMENTS. STATISTICAL CONTROL
Occasions Suitable for Concomitant Measurements -- Arbitrary Corrections -- Calculation of the Adjustment -- The Test of Significance - Missing values -- Practical Examples.

Chapter X - THE GENERALISATION OF NULL HYPOTHESES. FIDUCIAL PROBABILITY
Precision regarded as Amount of Information -- Multiplicity of Tests of the same Hypothesis -- Extension of the t Test - Fiducial limits of a ratio -- The χ^2 Test -- Wider Tests based on the Analysis of Variance -- Comparisons with Interactions.

Chapter XI - THE MEASUREMENT OF AMOUNT OF INFORMATION IN GENERAL
Estimation in General -- Frequencies of Two Alternatives -- Functional Relationships among Parameters -- The Frequency Ratio in Biological Assay -- Linkage Values inferred from Frequency Ratios -- Linkage Values inferred from the Progeny of Self-fertilised or Intercrossed Heterozygotes -- Information as to Linkage derived from Human Families -- The Information elicited by Different Methods of Estimation -- The Information Lost in the Estimation of Error.

INDEX

Fisher, R. A. and Yates, F. [1963], STATISTICAL TABLES FOR BIOLOGICAL, AGRICULTURAL AND MEDICAL RESEARCH. Sixth Edition. Oliver and Boyd Ltd., Edinburgh, Great Britain and Hafner Publishing Company Inc., New York, pp. x+146. First printing by Oliver and Boyd Ltd., 1938. Second Edition, 1943. Third Edition, 1948; reprinted in 1949. Fourth Edition, 1953. Fifth Edition, 1957.

The Table numbers of the earlier editions, when different from those of the fifth and sixth editions, are given in brackets.

INTRODUCTION

TABLE I. The Normal Distribution

II. Ordinates of the Normal Distribution

II_1. The Normal Probability Integral (VIII_4)

III. Distribution of t

IV. Distribution of χ^2

V. Distribution of z and the Variance Ratio

V_1. Fiducial Limits for a Variance Component

VI. Significance of Difference between Two Means (V_1)

VI_1. Significance of Difference between Two Means – Behrens' Test: Odd Degrees of Freedom

VI_2. Significance of Difference between Two Means – One Component of Error distributed normally, the Other in Student's Distribution (V_2)

VII. The Correlation Coefficient – Values for Different Levels of Significance (VI)

VII_1. The Correlation Coefficient – Transformation of r to z (VII)

VIII. Tests of Significance for 2×2 Contingency Tables

VIII_1. Binomial and Poisson Distributions: Limits of the Expectation

VIII_2. Densities of Organisms estimated by the Dilution Method

VIII_3. Significance of Leading Periodic Components

IX. Probits – Transformation of the Sigmoid Dosage Mortality Curve to a Straight Line

Fisher and Yates (Cont'd.)

TABLE IX_1. Probits – Simple Quantiles of the Normal Distribution (X)

IX_2. Probits – Weighting Coefficients and Probit Values to be used for Final Adjustments (XI)

IX_3. Probits – Weighting Coefficients for Use when there is Natural Mortality (XI_1)

X. The Angular Transformation – Transformation of Percentages to Degrees (XII)

X_1. The Angular Transformation – Transformation of Proper Fractions to Degrees (XIII)

X_2. The Angular Transformation – Angular Values for Final Adjustments (XIV)

XI. The Logit or r, z Transformation

XI_1. Logits – Weighting Coefficients and Logit Values to be used for Final Adjustments

XII. The Complementary Loglog Transformation

XII_1. The Complementary Loglog Transformation – Working Values

XIII. Scores for Linkage Data from Intercrosses (XIV_1)

XIII_1. Product Ratios for Different Recombination Fractions

XIV. Segmental Functions (XIV_2)

XV. Latin Squares

XVI. Complete Sets of Orthogonal Latin Squares

XVII. Balanced Incomplete Blocks – Combinatorial Solutions

XVIII. Balanced Incomplete Blocks – Index by Number of Replications

XIX. Balanced Incomplete Blocks – Index by Number of Units in a Block

XIX_1. Balanced Incomplete Blocks – Cyclic Solutions, r = 11-15

XIX_2. Balanced Incomplete Blocks – Index by Number of Replications, r = 11-15

XX. Scores for Ordinal (or Ranked) Data

Fisher and Yates (Cont'd.)

TABLE XXI. Scores for Ordinal Data — Sums of Squares of Mean Deviations Tabulated

XXII. Initial Differences of Powers of Natural Numbers

XXIII. Orthogonal Polynomials

XXIV. Calculation of Integrals from equally spaced Ordinates

XXV. Logarithms

XXVI. Natural Logarithms

XXVII. Squares

XXVIII. Square Roots

XXIX. Reciprocals

XXX. Factorials

XXXI. Natural Sines

XXXII. Natural Tangents

XXXIII. Random Numbers

XXXIII$_1$. Random Permutations of 10 Numbers

XXXIII$_2$. Random Permutations of 20 Numbers

XXXIV. Constants, Weights and Measures, etc.

Hicks, C. R. [1964], FUNDAMENTAL CONCEPTS IN THE DESIGN OF EXPERIMENTS. Holt, Rinehart and Winston, New York, Chicago, San Francisco, Toronto, London, pp. x+293.

Chapter 1 - THE EXPERIMENT, THE DESIGN, AND THE ANALYSIS
Introduction -- The Experiment -- The Design -- The Analysis -- Summary in Outline -- Example.

Chapter 2 - REVIEW OF STATISTICAL INFERENCE
Introduction -- Estimation -- Tests of Hypotheses -- Power of a Test -- How Large a Sample? -- Problems.

Chapter 3 - SINGLE-FACTOR EXPERIMENTS WITH NO RESTRICTIONS ON RANDOMIZATION
Introduction -- Analysis of Variance Rationale -- After ANOVA — Tests on Means -- Confidence Limits on Means -- Components of Variance -- General Regression Significance Test -- Summary -- Problems.

Chapter 4 - SINGLE-FACTOR EXPERIMENTS — RANDOMIZED BLOCK DESIGN
Introduction -- Randomized Complete Block Design -- ANOVA Rationale -- Interpretations -- General Regression Significance Test Approach -- Missing Values -- Randomized Incomplete Blocks — Restriction on Experimentation -- Summary -- Problems.

Chapter 5 - SINGLE-FACTOR EXPERIMENTS — LATIN AND OTHER SQUARES
Introduction -- Latin Squares -- Graeco-Latin Squares -- Youden Squares -- Summary -- Problems.

Chapter 6 - FACTORIAL EXPERIMENTS
Introduction -- ANOVA Rationale -- Remarks -- Summary -- Problems.

Chapter 7 - 2^n FACTORIAL EXPERIMENTS
Introduction -- 2^2 Factorial -- 2^3 Factorial -- 2^n — Remarks -- Summary -- Problems.

Chapter 8 - QUALITATIVE AND QUANTITATIVE FACTORS
Introduction -- Single Factor — Quantitative Levels -- Two Factors — One Qualitative, One Quantitative -- Two Factors — Both Quantitative -- Summary -- Problems.

Chapter 9 - 3^n FACTORIAL EXPERIMENTS
Introduction -- 3^2 Factorial -- 3^3 Factorial -- Summary -- Problems.

Chapter 10 - FIXED, RANDOM AND MIXED MODELS
Introduction -- Single-factor Models -- Two-factor Models -- EMS Rules -- EMS Derivations -- Remarks -- Problems.

Hicks (cont'd.)

Chapter 11- NESTED AND NESTED-FACTORIAL EXPERIMENTS
Introduction -- Nested Experiments -- ANOVA Rationale -- Nested-factorial Experiments -- Summary -- Problems.

Chapter 12 - EXPERIMENTS OF TWO OR MORE FACTORS – RESTRICTIONS ON RANDOMIZATION
Introduction -- Factorial Experiment in a Randomized Block Design -- Factorial Experiment in a Latin Square Design -- Remarks -- Summary -- Problems.

Chapter 13 - FACTORIAL EXPERIMENT – SPLIT-PLOT DESIGN
Introduction -- A Pseudo F Test -- Summary -- Problems.

Chapter 14 - FACTORIAL EXPERIMENT – CONFOUNDING IN BLOCKS
Introduction -- Confounding Systems -- Block Confounding with Replication -- Block Confounding – No Replication -- Summary -- Problems.

Chapter 15 - FRACTIONAL REPLICATION
Introduction -- Aliases -- Fractional Replications -- Summary -- Problems.

Chapter 16 - MISCELLANEOUS TOPICS
Introduction -- Response Surface Experimentation -- Evolutionary Operations (EVOP).

Chapter 17 - SUMMARY

GLOSSARY OF TERMS

REFERENCES

STATISTICAL TABLES

ANSWERS TO ODD-NUMBERED PROBLEMS

INDEX

Jeffers, J. N. R. [1960], EXPERIMENTAL DESIGN AND ANALYSIS IN FOREST RESEARCH. Almqvist & Wiksell, Stockholm, pp. 1-172.

INTRODUCTION

Chapter 1 - EXPERIMENTAL DESIGN
Randomised Blocks -- Latin Squares -- Factorial Experiments -- Split-plot Designs -- Balanced Incomplete Block Designs -- Miscellaneous Aspects of the Design of Experiments.

Chapter 2 - ANALYSIS OF EXPERIMENTS
Randomised Block Experiments - General; Preliminary analysis; Detailed analysis; Tests of significance -- Latin Square Experiments - Description of experiment; Preliminary analysis; Detailed analysis; Tests of significance; Subdivision of treatment sum of squares -- Analysis of Factorial Experiments - Characteristics of factorial designs; Description of typical experiment; Preliminary analysis; Detailed analysis; Analysis of confounded factorial experiments -- Split-plot Experiments - Characteristics of split-plot experiments; Description of a typical experiment; Preliminary analysis; Detailed analysis -- Analysis of Balanced Incomplete Block Experiments - Preliminary analysis; Detailed analysis -- Tests of Significance -- Combination of Experiments Carried out in Different Sites and Different Years -- Analysis of Experiments with Missing Plots - Replacement of missing values by use of formulae; Replacement of missing values by calculation from a trial analysis.

Chapter 3 - TRANSFORMATIONS FROM DISTRIBUTIONS OTHER THAN THE NORMAL
Assumptions Made in the Analysis of Variance -- Subdivision of Error Sum of Squares -- Use of Transformations -- Angular Transformation -- Logarithmic Transformation -- Square Root Transformation -- Other Transformations -- Tests to Determine the Appropriateness of Transformations -- Presentation of Transformed Data -- Interactions between Blocks and Treatments.

Chapter 4 - CHI-SQUARE AND THE ANALYSIS OF ATTRIBUTES
Definition of Chi-square -- Calculation of Chi-square -- Sampling Distribution of Chi-square -- Application of Chi-square Tests - Tests of hypotheses; Test of independence -- Other Applications of Chi-square.

Chapter 5 - METHODS OF SAMPLING
Methods of Sampling Available - Subjective sampling; Unrestricted random sampling; Stratified random sampling; Systematic sampling; Sequential sampling -- Shape and Size of Sampling Units -- Selection of Sampling Units -- Estimation of Numbers of Sample Units Required -- Uniformity Trials and Pilot Surveys.

Chapter 6 - REGRESSION AND CORRELATION
Introduction -- Calculation of Linear Regression Equation -- Sampling Errors and Tests of Significance for Linear Regression -- Calculation of Linear Regression from Weighted Data -- Calculation of Correlation Coefficients -- Curvilinear

Jeffers (cont'd.)

Regression -- Calculation by Use of Orthogonal Polynomials -- Calculation of Multiple Regression -- Discriminant Functions -- Approximate Methods of Calculating Regression and Correlation -- Approximate Method of Calculating Linear Regression Equations -- Calculation of Correlation by Use of Ranking.

Chapter 7 - ANALYSIS OF COVARIANCE

Introduction -- Purpose of Covariance Analysis -- Assumptions Made in the Analysis of Covariance -- Application of Analysis of Covariance to a Randomised Block Experiment -- Evaluating the Gain in Using Covariance -- Application of the Analysis of Covariance to other Designs -- Uses of Covariance Analysis.

Chapter 8 - PRESENTATION OF RESULTS OF EXPERIMENTS

The Objects of the Experiment -- Design of the Experiment -- List of Variables Assessed -- Practical Considerations and Difficulties -- Result of the Experiment in Detail -- Criticisms of Experiment and Suggestions for Improvements -- Summary of Results and Conclusions -- Summary of Data -- Methods Used in Analysis.

APPENDICES

I. Working Sheets for the Analysis of Experiments

II. References

III. Shortened Forms of Statistical Tables

IV. Glossary of Statistical Terms

Kempthorne, O. [1952], THE DESIGN AND ANALYSIS OF EXPERIMENTS. John Wiley and Sons, Inc., New York, London, Sydney, pp. xix+631. Fourth printing by John Wiley and Sons, Inc., January, 1965.

Chapter 1 - INTRODUCTION
The Scientific Method -- The Formulation and Testing of Hypotheses -- The Role of Statistics -- The Design of Experiments -- The Use of Prior Information -- The Decision Function Approach.

Chapter 2 - THE PRINCIPLES OF EXPERIMENTAL DESIGN
Introduction -- An Illustrative Example -- An Example of Flavor Discrimination.

Chapter 3 - ELEMENTARY STATISTICAL NOTIONS
Populations, Distributions, Parameters -- The Normal Distribution and Derived Distributions -- Linear Functions of Normally Distributed Variates -- Orthogonality -- Other Distributions -- Estimation -- The Testing of Hypotheses -- Interval Estimation.

Chapter 4 - AN INTRODUCTION TO THE THEORY OF LEAST SQUARES
Introduction -- The Markoff Theorem -- A Non-linear Example.

Chapter 5 - THE GENERAL LINEAR HYPOTHESIS OR MULTIPLE REGRESSION AND THE ANALYSIS OF VARIANCE
Description of Statistical Procedures -- Extension -- The Likelihood Ratio Test -- Reduction of Other Cases of Regression -- Orthogonality -- Quantity of information -- Proof of the Results of This Chapter -- The Testing of a Subhypothesis -- The Canonical Form of the General Linear Hypothesis.

Chapter 6 - THE ANALYSIS OF MULTIPLE CLASSIFICATIONS
The 2-way Classification with One Observation per Cell -- Alternative Approach to Hypotheses Not of Full Rank -- The 2-way Classification with Unequal Numbers and No Interaction - A Numerical example -- The Case of Proportional Frequencies -- More Complex Classifications - The 2-way classification with interaction; The general p-way classification without interaction; The case of missing observations in planned n-way classifications; The alternative method of analyzing incomplete experiments; Proof of alternative procedure for missing plots -- The Analysis of Covariance -- Components of Variance - The hierarchal classification; The n-fold hierarchal classification; The case of n-way classifications -- An Example of the Estimation of Components of Variance in a Genetic Problem -- Conclusion.

Chapter 7 - RANDOMIZATION
The Principle of Randomization -- Randomization in the Case of Continuous Variables -- Sampling from Finite Populations -- Randomization Tests - An example of a randomization test -- Other Formulations of the Problem.

Kempthorne (cont'd.)

Chapter 8 - THE VALIDITY OF ANALYSES OF RANDOMIZED EXPERIMENTS
Introduction -- The Analysis of Randomized Blocks When Additivity Holds -- The Analysis of Randomized Blocks with Non-additivity -- Method of Analysis Used in Subsequent Chapters -- Transformations -- An Example for the Reader -- The Analysis of Covariance -- A Test for Additivity.

Chapter 9 - RANDOMIZED BLOCKS
Introduction -- The Analysis of Randomized Blocks -- Breakdown of the Treatment Sum of Squares -- Randomization Test -- The Treatment of Randomized Block Experiments with Missing Data -- The Variance of Treatment Comparisons with Missing Plots -- Difficulties of Randomized Blocks -- The Purposes of Replication -- The Use of Concomitant Information -- The Efficiency of Randomized Blocks.

Chapter 10 - LATIN SQUARES
Introduction -- Combinatorial Properties -- Graeco-Latin Squares -- The Completely Orthogonalized Square -- The Analysis of the Latin Square Design -- Analysis of the Latin Square with the Infinite Model -- Missing Data - The case of a missing row, column, or treatment -- Efficiency of Latin Squares - Some results on efficiencies -- Designs Based on Graeco-Latin Squares -- The Use of Systematic Latin Squares -- Further Remarks on Latin Squares.

Chapter 11 - PLOT TECHNIQUE
Introduction -- Field Experiments - Size and shape of plots; Size and shape of blocks -- Animal Experiments -- The Use of Sampling in Experiments.

Chapter 12 - THE SENSITIVITY OF RANDOMIZED BLOCK AND LATIN SQUARE EXPERIMENTS
Introduction -- The Power of the Analysis of Variance Test -- Randomized Blocks - Randomized Blocks for 2 treatments; The general case; Single degree of freedom contrasts in randomized blocks -- Latin Squares -- Comparison of Randomized Blocks with Latin Squares -- The Validity of the Infinite Model Approach.

Chapter 13 - EXPERIMENTS INVOLVING SEVERAL FACTORS
Introduction -- Effects and Interactions - The case of 2 factors; The case of 3 factors; The general case -- The Interpretation of Effects and Interactions -- A Simple Example of a Factorial Experiment -- The General Case of Several Factors Each at 2 Levels - The analysis of 2^n factorial experiments -- The information Given by Factorial Experiments -- The Sensitivity of Factorial Experiments.

Chapter 14 - CONFOUNDING IN 2^n FACTORIAL EXPERIMENTS
Introduction -- An Example -- Systems of Confounding for 2^n Experiments -- The Composition of Blocks for a Particular System of Confounding -- The Effects of Differential Treatment Effects in Blocks -- The Use of Only One Replicate -- An Example.

Chapter 15 - PARTIAL CONFOUNDING IN 2^n FACTORIAL EXPERIMENTS
A Simple Case -- Efficiency of Partial Confounding -- Partial Confounding of 2^3

Kempthorne (cont'd.)

Experiments --Confounding with 4 Factors in Blocks of 4 Plots -- Partial Confounding of a 2^4 Experiment in Blocks of 2 -- Confounding in Latin Squares - The 2^3 design in two 4 × 4 Latin squares; The 2^3 design in three 4 × 4 Latin squares; 2^4 design in 8 × 8 Latin squares; Arrangements for 5 and 6 factors in an 8 × 8 square -- Double Confounding -- Missing Values in Factorial Experiments.

Chapter 16 - EXPERIMENTS INVOLVING FACTORS WITH 3 LEVELS

Introduction -- The Formal Method of Defining Effects and Interactions - Yields of treatment combinations in terms of effects and interactions -- Confounding -- Useful Systems of Confounding for 3^n Experiments - 2 factors; 3 factors; 4 factors; 5 factors -- An Example -- The Use of Only One Replicate -- Confounding 3^n Experiments in Latin Squares.

Chapter 17 - THE GENERAL p^n FACTORIAL SYSTEM

The Representation of Effects and Interactions -- Confounding with the p^n System -- Yields of Treatment Combinations in Terms of Effects and Interactions -- Analysis of p^n Factorial Systems -- The Interpretation of Effects and Interactions -- The Knut Vik Square -- The Arrangement of n Factors Each with p^m levels in Blocks of $(p^m)^s$ -- Finite Geometries.

Chapter 18 - OTHER FACTORIAL EXPERIMENTS

Introduction -- The Number of Levels Being the Same Prime Power -- Number of Levels Equal to Different Powers of Same Prime -- Number of Levels Equal to Different Prime Numbers - The 3 × 2 × 2 experiment; The 3 × 3 × 2 experiment; The 3 × 3 × 3 × 2 experiment -- Number of Levels a Product of Different Primes -- The General Method of Analyzing Partially Confounded Designs -- Other Mixed Factorial Designs -- Partially Factorial Experiments.

Chapter 19 - SPLIT-PLOT EXPERIMENTS

The Simple Split-plot Experiment -- Arrangement of Whole-plot Treatments in a Latin Square -- Extension of the Split-plot Principle -- The Efficiency of Split-plot Designs Relative to Randomized Blocks -- The 2-factor Experiment with Both Factors in Strips -- Split-plot Confounding -- The Analysis of Covariance in Split-plot Designs -- Missing Data.

Chapter 20 - FRACTIONAL REPLICATION

Introduction -- A Simple Example of Fractional Replication -- 1/2 Replication of a 2^6 Experiment -- A Simple Example of 1/4 Replication -- 1/4 Replication of a 2^8 Experiment -- One-in-2^p Replication of the 2^n Factorial System -- The Value of Fractional Designs.

Chapter 21 - THE GENERAL CASE OF FRACTIONAL REPLICATION

Factors at the Same Prime Number (p) of Levels -- The Formal Equivalence of Fractional Replication and Confounding -- The Effect of Block-treatment interactions -- Fractional Replication of Mixed Factorial Systems -- Other Fractionally Replicated Designs -- Weighing Designs -- An Example of the Use of Fractional Replication.

Kempthorne (cont'd.)

Chapter 22 - QUASIFACTORIAL OR LATTICE AND INCOMPLETE BLOCK DESIGNS
Introduction -- The Simplest Quasifactorial Design - The estimation of the weights W and W' -- Types of Lattice Design -- Types of Incomplete Block Design.

Chapter 23 - LATTICE DESIGNS
Types of Lattice Designs - 2-dimensional lattices; 3-dimensional lattices; 4-dimensional lattices; 5-dimensional lattices -- The General Method of Analysis -- The Analysis of 2-dimensional lattices - Repetition of the simple lattice -- The Analysis of Other Lattice Designs - The triple lattice; The 3-dimensional lattice; The 5-dimensional lattice -- The Variance of Treatment Comparisons -- The Effects of Inaccuracies in the Weights -- The Analysis of Lattice Designs as Complete Randomized Block Designs -- The Efficiency of Lattice Designs Relative to Complete Randomized Blocks -- The Comparison of p^n Treatments in Blocks of p^s Plots -- The Non-prime Case - The simple lattice; The triple lattice; The 3-dimensional lattice; A worked example.

Chapter 24 - LATTICE DESIGNS WITH TWO RESTRICTIONS
The Completely Balanced Lattice Square - The estimation of the weights; The variance of treatment comparisons; The effects of inaccuracies in the weights -- The Design and Analysis of Other Lattice Squares -- The Non-prime Case -- Designs with 2-restrictions which are not Square -- Further Designs.

Chapter 25 - RECTANGULAR LATTICES
Introduction -- The Analysis of a Simple Rectangular Lattice for k(k-1) Treatments - The analysis of variance; The variance of treatment comparisons -- The Analysis of the Triple Rectangular Lattice -- The Variance of Treatment Differences for a Triple Rectangular Lattice -- An Example of a Triple Rectangular Lattice -- Further Notes.

Chapter 26 - BALANCED INCOMPLETE BLOCK DESIGNS
Introduction -- Examples of Incomplete Block Designs -- The General Case -- The Analysis of Balanced Incomplete Block Experiments - The estimation of the weights; Special cases -- Youden Squares.

Chapter 27 - PARTIALLY BALANCED INCOMPLETE BLOCK DESIGNS
Introduction -- The Specifications of Partially Balanced Incomplete Block Designs -- Designs for a Small Number of Treatments -- The Analysis of Partially Balanced Incomplete Block Designs -- A Worked Example -- The Enumeration of Partially Balanced Incomplete Block Designs -- Conclusion.

Chapter 28 - EXPERIMENTS ON INFINITE POPULATIONS AND GROUPS OF EXPERIMENTS
Introduction -- The 1-factor Experiment -- The 2-factor Experiment -- The Design and Analysis of a Single Series of Experiments -- A Variety Trial at a Random Sample of Places for a Number of Years -- Difficulties in the Analysis of a Group of Experiments.

Kempthorne (cont'd.)

Chapter 29 - TREATMENTS APPLIED IN SEQUENCE
The Comparison of 3 Treatments -- The Case of 4 Treatments -- Residual Effects -- Long-term Agricultural Experiments.

APPENDIX - Table I 10%, 5%, and 1% Points of F or e^{2z}
Table II The Power of the Analysis of Variance Test

INDEX

Kitagawa, T. and Mitome, M. [1953], TABLES FOR THE DESIGN OF FACTORIAL EXPERIMENTS. Baifukan Co., Ltd. Tokyo, Japan, pp. 292+38+136+25+7+15 +13+15+4. (in Japanese)

(Table of contents in Japanese.)

The following tables are appended to the Japanese text.

Table I. LATIN SQUARES AND CUBES
$2\times2 \sim 12\times12$, 20×20 -- Complete Sets of Orthogonal Latin Squares -- Orthogonal Partitions in Latin Squares -- Orthogonal Latin Cubes.

Table II. FACTORIAL DESIGNS
$2^2 - 2^3$; 2^4; 2^5; 2^6; 2^7; 2^8; 2^9 -- $3^2 - 3^3$; 3^4 -- $4^2 - 4^3$ -- $5^2 - 5^3$ -- $7^2 - 7^3$ -- $2^2\times3 - 2^3\times3$; $2^4\times3$; $2^2\times4$; $2^3\times4$; $2^2\times5$; $2^3\times5$; $3^2\times2$; $3^3\times2$; $3^2\times4$; $4^2\times2$; $4^2\times3$; $3^2\times2^2$; $4\times3\times2$.

Table III. FRACTIONAL REPLICATION IN FACTORIAL DESIGNS
2^n-1/2 -- 2^n-1/4 -- 2^n-1/8 -- 2^n-1/16 -- 3^n-1/3.

Table IV. FACTORIAL DESIGNS WITH SPLIT-PLOT CONFOUNDING
In Randomized Blocks -- In Latin Squares.

Table V. FACTORIAL DESIGNS CONFOUNDED IN QUASI-LATIN SQUARES
Without Subsidiary Factors -- With One Subsidiary Factor: Non-plaid squares - With one subsidiary factor: Half-plaid squares -- With Two Subsidiary Factors: Half-plaid Squares - With two subsidiary factors: Plaid squares.

Table VI. LATTICE DESIGNS
2-Dimensions -- 3-Dimensions -- Lattice Square $k\times k$.

Table VII. BALANCED INCOMPLETE BLOCK DESIGNS

Table VIII. YOUDEN SQUARES

Kitagawa, T. [1955], LECTURES ON THE DESIGN OF EXPERIMENTS. Vol. 1 and 2. Baifukan Co., Ltd. Tokyo, pp. xi+378.

(Table of contents in Japanese)

LeClerg, E. L., Leonard, W. H., and Clark, A. G. [1962], FIELD PLOT TECHNIQUE. Burgess Publishing Company Minneapolis, Minnesota, U.S.A., pp. iii+373. First Edition by Leonard, W. H. and Clark, A. G., 1939.

Chapter 1 - INTRODUCTION
Field Experimentation -- Establishment of Experiment Stations - First Agricultural Experiment Station; Rothamsted Experimental Station; American Agricultural Experiment Stations; United States Department of Agriculture -- Agronomic Research Projects -- Value of Early Agronomic Experiments -- Present Trends in Agronomic Research - Design of experiments; Long-time projects; Regional cooperation in research; Basic research -- References.

Chapter 2 - METHODS OF SCIENTIFIC INQUIRY
Some Preliminary Concepts -- Inference - Deduction; Induction -- Hypotheses - Formulation of hypotheses; Deductive development of hypotheses; Formal conditions for hypotheses; Null hypothesis; Hypotheses as statements of casual connections; Crucial tests -- Observation - Bare observation; Controlled observation -- Analogy -- Pitfalls in Experimentation - Faulty design or inferior techniques; Improper interpretation of results; The personal equation -- Summary: Essentials of Scientific Method -- References.

Chapter 3 - GENERAL TYPES OF EXPERIMENTS
Classification of Field Experiments - Variety tests; Cultural studies; Fertilizer experiments; Pesticide experiments; Crop rotation tests; Pasture experiments; Perennial crop experiments; Other types of field experiments -- Sources of Variation in Field Experiments - Variation related to the plant; Variations due to seasons; Soil heterogeneity -- Greenhouse Tests - Role of greenhouse experiments; Sources of variation in greenhouse experiments -- Other Types of Experiments - Controlled environment rooms; Lysimeters -- References.

Chapter 4 - MEASUREMENT OF SAMPLE DATA
Some Typical Statistical Terms - Statistics; Statistical variable; Sample -- The Frequency Distribution - Grouping of Data into Classes; Frequency table; Graphical representation of frequency table -- Measures of Central Value - Arithmetic mean (Direct computation; Computation from coded data); Other measures of central value -- Measures of Variability or Dispersion - Standard deviation (Direct computation; Computation from coded data); Coefficient of variability -- References -- Problems.

Chapter 5 - TESTS OF SIGNIFICANCE
Statistical Inference - Population; Sample -- Elementary Probability Theory - Definition of probability (Mathematical probability; Empirical probability); Several probabilities (Additiȯn theorem; Multiplication theorem) -- Types of Frequency Distributions - Normal distribution; Other types of distributions -- Application of Probability to the Normal Curve - Normal curve; Areas under normal curve -- Standard Error of the Mean -- Statistical Hypotheses - Experimental error; Test of statistical hypotheses; Types of inferential errors --

LeClerg, Leonard, and Clark (cont'd.)

Degrees of Freedom -- Tests of Significance with "t" Distribution - Means of paired samples; Means of two non-paired samples (Samples with same number of observations; Samples with different numbers of observations) -- Standard Error of a Difference -- Confidence Interval - Confidence interval of a mean; Confidence interval of mean differences -- General Applicability of Statistical Methods - Mathematical basis for application; Value of the statistical method; Reliability of the statistical constant; Some misconceptions of the statistical method -- References -- Problems.

Chapter 6 - CHI SQUARE (χ^2) TESTS

Chi-square as Measure of Dispersion -- Chi-square Criterion - Formula for chi-square; Sampling distribution for chi-square; Grouping of data; Probability tables for chi-square -- Goodness of Fit Tests - General method of computation; Special method for two classes; Fit of observed data to theoretical distributions -- Heterogeneity Chi-square Tests -- Chi-square Test for Independence - Contingency tables; Computation for a manifold contingency table; Computation for a 2 x 2 or 4 - fold table; Correction for continuity -- The Null Hypothesis and Chi-square -- References -- Problems.

Chapter 7 - REGRESSION AND CORRELATION

Linear Regression - Description of linear regression; Computation of regression coefficient; Significance of the regression coefficient; Significance of the difference between two regression coefficients -- Correlation - Description of correlation; Measurement of correlation; Computation of correlation coefficient from ungrouped data; Test of significance of correlation coefficient (Tabular test; t-test); Significance of difference between two correlation coefficients; Average of correlation coefficients; Interpretation of the correlation coefficient -- References -- Problems.

Chapter 8 - BASIS FOR THE ANALYSIS OF VARIANCE

Introduction -- Generalized Standard Error Method -- Assumptions in the Analysis of Variance - Independence of experimental errors; Normal distribution of experimental errors; Homogeneity of experimental errors; Additivity of variances; General applicability -- Sources of Variation -- One Criterion of Classification - Theory of completely randomized design; Computation for the completely randomized design -- Two or More Criteria of Classification - Theory for two criteria of classification; Computation for two criteria of classification -- Components of Variance - Introduction; General relationships; Fixed and random effects (Fixed effects; Random effects); Determination of sample mean squares (Sources of variation listed in hierarchal order; Identification of sources of variation; Determination of fixed and random effects); Population models for the analysis of variance; The F-test and choice of error term; Estimation of magnitude of component variability -- References -- Problems.

Chapter 9 - THE FIELD PLOT

Introduction -- Measurement of Soil Heterogeneity - Uniformity trial data; Correlation as a measure; Soil fertility maps -- Role of Soil Heterogeneity in Field

LeClerg, Leonard, and Clark (cont'd.)

Experimentation - Amount of soil heterogeneity; Causes of soil heterogeneity; Corrections for soil heterogeneity; Selection of the experimental field -- History of Field Plots - Plot sizes in early experiments; Pioneer use of replication -- Size of Field Plots - Factors that influence plot size; Size classification of experimental plots; Relation of plot size to accuracy -- Plot Shape - Data on plot shape; Some practical considerations in plot shape -- Plot Sizes and Shapes for Different Crops -- Plot Replication - Reduction of error by replication; Number of replications -- Calculation of Plot Efficiency -- Inter-plot Competition or Border Effect - Inter-plot competition; Alley effect on border rows; Control of inter-plot competition -- Intra-plot Competition or Stand Relationships - Uneven plant distribution in plots; Differences in stands in plots; Correction for uneven stands -- References.

Chapter 10 - PRINCIPLES OF EXPERIMENTAL DESIGN

Basic Principles of Experimental Design - Randomization; Replication; Local control -- Types of Experiments - Preliminary tests; Simple experiments; Factorial experiments -- Relation of Type of Experiment to Design - General types of yield tests; Cultural and fertilizer experiments; Pasture experiments; Crop rotation experiments; Tree experiments -- References.

Chapter 11 - COMPLETE BLOCK DESIGNS

Randomized Complete Block Design - General applicability; Method of analysis of data (Arrangement of pots; Computations; The F-test); Functional analysis of data; Comparison of individual means (Least significant difference test (LSD); Duncan range test); Missing values (One missing value; Two or more missing values) -- Latin Square Design - General applicability; Comparison with randomized complete block design; Randomization; Method of analysis of data (Field plot arrangement; Computations); Comparison of individual means; Missing values (One missing value; Two or more missing values) -- References -- Problems.

Chapter 12 - COMPLETE AND CONFOUNDED FACTORIAL EXPERIMENTS

Role of Factorial Experiments -- Interactions -- Complete Factorial Experiments - Comparative applicability; Example of a complete factorial experiment; Components of variance -- Confounded Factorial Experiments - The principle of confounding; Analysis of a completely confounded factorial experiment; Analysis of a partially confounded experiment; Other types of complete and confounded factorial experiments; Missing values -- References -- Problems.

Chapter 13 - SPLIT PLOT AND SPLIT BLOCK DESIGNS

Introduction -- Split-plot Design - Applications of split-plot designs; Randomization; Statistical analysis; Errors for tests of significance; Relative efficiency; Missing values (Split-plot of a randomized complete block experiment; Split-split plot of a randomized complete block experiment; Split-plot of a Latin square experiment; Split-split plot of a Latin square experiment); Standard errors; Some variations of the split-plot design (Repeated sub-division; Systematic arrangement of whole plots; Dummy plots); Randomized-block vs. split-plot experiments --

LeClerg, Leonard, and Clark (cont'd.)

Split-block Design - Advantages and disadvantages; Randomization and types of plot arrangements (Main factor in a randomized complete block design; Main factor in a randomized complete block design with sub-plots arranged in a Latin square); Statistical analysis; Missing values (No subdivision of intersecting plots; Subdivision of intersecting plots) -- Comparison of Two Designs -- Components of Variance - Fixed effect (Model I); Random effects (Model II) (Main treatment effects; Sub-treatment effects); Mixed effects (Model III) (Main treatments fixed and sub-treatments random; Main treatments random and sub-treatments fixed) -- Extension of the Split-plot Principle -- References -- Problems.

Chapter 14 - COMBINED EXPERIMENTS

Introduction -- Application to a Barley Variety Trial - Analysis of tests into components (Computation of the sums of squares; Comparison of sums of squares in simple vs. combined experiments); Interpretation of data (Significance of interactions; Components of variance and tests of significance; Variety test at one location; Evaluation of treatments x stations interaction) -- Homogeneity of Experimental Error Variances - Computations -- References -- Problems.

Chapter 15 - LATTICE DESIGNS

Introduction -- Classification of Incomplete Block Designs -- Simple Lattice Design - Arrangement of plots; Analysis of data; Test of significance; Adjustment of variety means; Standard error of the difference; Missing values -- Triple Lattice Design - Arrangement of plots; Analysis of data; Test of significance; Adjustment of variety means; Standard errors of the difference; Missing values; Covariance analysis -- Balanced Lattice Designs - Arrangement of plots; Analysis of data (Calculations for analysis of variance; Adjustments of variety totals; Standard errors of the difference); Missing values -- Other Types of Lattice Designs - Quadruple design; Quintuple design -- Lattice Designs for Different Numbers of Varieties -- References -- Problems.

Chapter 16 - RECTANGULAR LATTICES

Introduction -- Simple Rectangular Lattice - Arrangement of plots; Analysis of data; Sum of squares for component (a); Sum of squares for component (b); Test of significance; Adjustment of variety means; Standard errors of the difference; Covariance analysis; Missing data -- Triple Rectangular Lattice - Arrangement of plots; Analysis of data; Test of significance; Adjustment of variety means; Standard errors of the difference; Missing values; Covariance analysis -- References -- Problems.

Chapter 17 - LATTICE SQUARES WITH MODIFICATIONS

Lattice Square Designs - Balanced lattice squares ((k+1)/2 balanced lattice square; (k+1) balanced lattice square); Partially balanced lattice squares; Unbalanced lattice squares; Missing values; Lattice squares with split-plots -- Youden Squares -- References -- Problems.

LeClerg, Leonard, and Clark (cont'd.)

Chapter 18 - ANALYSIS OF COVARIANCE

Introduction -- Analysis of Variance Applied to Linear Regression -- Theoretical Considerations -- Computation of Covariance Analysis -- Calculation of Regression Coefficients -- Application of Covariance Technique -- Adjustment of Yields by Covariance - Adjustment of yield on the basis of stand; Preliminary data for error reduction (Variance and covariance for preliminary and experimental yields; Calculation of the regression coefficient; The adjusted treatment means; Standard error of a difference between adjusted yields) -- References -- Problems.

Chapter 19 - PLOT SAMPLING METHODS

Introduction -- Principles of Sampling -- Terminology in Sampling Problems -- Choice of Sampling Units -- Sources of Variability -- Analysis of Sampling Data - Computational procedure for analysis; Computation of number of sampling units or replications (Estimated number of replications; Estimated number of sampling units; Relation of number of sampling units and replications to precision) -- Loss of Information Due to Sampling -- Determination of Cost of Sampling -- Importance of Plot and Laboratory Errors - Chemical analyses of crop material; Soil sampling -- References -- Problems.

Chapter 20 - TRANSFORMATION OF EXPERIMENTAL DATA

Introduction -- Square Root Transformations - Example of square root transformation -- Logarithmic Transformations - Example of logarithmic transformation -- Angular Transformation - Classification of types of percentages; Example of angular transformation -- Test for Additivity - Computation of test -- Summary -- References -- Problems.

APPENDIX

Table 1. Area Under the Normal Curve
2. Values of F and t
3. Table of Chi Square (X^2)
4. Neparian or Hyperbolic Logarithms
5. Values of Percentages Transformed into Degrees of an Angle
6. Significant Studentized Ranges for a 5 Percent Level New Multiple Range Test
7. Significant Studentized Ranges for a 1 Percent Level New Multiple Range Test
8. Sets of Random Sample Numbers to 60

INDEX

Linder, A. [1953], PLANEN UND AUSWERTEN VON VERSUCHEN - Eine Einführung für Naturwissenschafter, Mediziner und Ingenieure. Verlag Birkhäuser Basel/ Stuttgart, pp. 1-182.

Kapitel 0 - GRUNDSÄTZE FUR DAS PLANEN VON VERSUCHEN
Versuche und statistische Verfahren -- Drei Grundsätze - Wiederholen; Zufällig zuordnen; Blöcke mit möglichst gleichartigen Versuchseinheiten -- Statistische Auswertung - Zweck der Auswertung; Durchschnitt und Streuung; Einfache Streuungszerlegung; Prüfverfahren (Prüfen von Durchschnitten; Prüfen von Streuungen; Anwendung der Prüfverfahren in der einfachen Streuungszerlegung).

Kapitel 1 - VERSUCHE IN BLÖCKEN MIT ZUFALLIGER ANORDNUNG
Grundsätze und Beispiel -- Doppelte Streuungszerlegung -- Vergleich mit Versuchen in völlig zufälliger Anordnung -- Fehlende Angaben -- Mehrfache Streuungszerlegung -- Orthogonale Vergleiche -- Zwei Verfahren: Anordnen in Paaren -- Gehaltsbestimmung durch Vergleich mit Standard.

Kapitel 2 - VERSUCHE IN LATEINISCHEN QUADRATEN
Grundsätze und Auswertung -- Regelmässige und zufällige Anordnung -- Fehlende. Angaben.

Kapitel 3 - ZWEI VERGLEICHSREIHEN MIT VERSCHIEDENER PRÄZISION: VERSUCHE IN TEILPARZELLEN
Grundsätze und Auswertung -- Wirkungsgrad des Versuchsplanes.

Kapitel 4 - VERSUCHE MIT MEHREREN FAKTOREN
Grundsätze -- Auswerten von Versuchen mit je zwei Stufen -- Auswerten von Versuchen mit mehr als zwei Stufen -- Vermengen von Wechselwirkungen mit Unterschieden zwischen Blöcken -- Vollständiges Vermengen -- Teilweises Vermengen -- Teilweise wiederholte Versuche.

Kapitel 5 - VERSUCHE IN UNVOLLSTÄNDIGEN BLÖCKEN
Grundsätze -- Ausgewogene Versuche in unvollständigen Blöcken -- Teilweise ausgewogener Versuch in unvollständigen Blöcken.

Kapitel 6 - MATHEMATISCHE ANMERKUNGEN
Wirkung der zufälligen Anordnung -- Orthogonale Vergleiche -- Streuungszerlegung - Doppelte Streuungszerlegung im allgemeinen; Versuche in vollständigen Blöcken (Fehlender Wert); Ausgewogene Versuche in unvollständigen Blöcken.

LITERATUR

TAFELN
I Verteilung von t und von χ^2
II Verteilung von F
III Zufällig angeordnete Zahlen

SACHREGISTER

Lindquist, E. F. [1953], DESIGN AND ANALYSIS OF EXPERIMENTS IN PSYCHOLOGY AND EDUCATION. Houghton Mifflin Company, Boston, pp. xix+393. 2nd printing, 1956.

Chapter 1 - INTRODUCTION: FUNDAMENTAL CONCEPTS AND BASIC DESIGNS
The Nature and Purpose of Educational and Psychological Experiments in General -- The Importance of Measures of Precision -- Testing Hypotheses -- The Essential Characteristics of a Good Experimental Design -- Basic Experimental Designs -- Basic Types of Error -- The Principle of Randomization -- Illustrative Applications of Basic Designs.

Chapter 2 - THE CHI-SQUARE, t, AND F DISTRIBUTIONS
Introduction -- The Chi-square Distribution -- Proof of the Independence of the Mean and Variance of Random Samples Drawn from a Normal Population -- Degrees of Freedom -- The t-Distribution -- The F-Distribution.

Chapter 3 - THE SIMPLE-RANDOMIZED DESIGN
The Importance of the Simple-Randomized Design -- The Hypothesis to be Tested -- Limitations of the Simple t-Test -- The Test of the Over-all Null Hypothesis -- The Measure of Discrepancy as a "Mean Square" Ratio -- Type I and Type II Errors -- The Importance of the Assumptions Underlying the F-Test -- The Norton Study of the Effects of Non-normality and Heterogeneity of Variance -- Testing the Significance of the Difference in Means for Individual Pairs of Treatments -- The Significance of the Difference between Two Sample Means When the Population Variances Differ -- Types of Applications of the Simple-Randomized Design to Experimental Data -- Applications of the Simple-Randomized Design to Observational Data.

Chapter 4 - ANALYSIS OF VARIANCE IN DOUBLE-ENTRY TABLES
Introduction -- Analysis of Total Sum of Squares into Four Components (Method of Arithmetic Corrections) -- Analysis of Total Sum of Squares into Four Components (Algebric Method) -- The Case of One Observation per Cell -- Computational Procedure -- A Worked Example -- The Generalized Meaning of Interaction.

Chapter 5 - TREATMENTS × LEVELS DESIGN
Generalized Case of the Treatments × Levels Designs -- The Analysis of the Total Sum of Squares -- The Meaning of Interaction -- Constituting the "Levels" in an Experiment -- Selection of the Control Variable -- Testing the Significance of the Main Effect Treatments -- Test of the Significance of the Interaction -- The Meaning of ms_A/ms_{AL} -- Treatments × Levels Designs with One Observation per Cell -- Tests of Significance Applied to Individual Differences -- Possibilities of Confounding Extraneous Factors with Levels -- Limitations and Advantages of the Treatments × Levels Design -- What to Do About Missing Cases -- The Use of Transformations.

Lindquist (cont'd.)

Chapter 6 - THE TREATMENTS x SUBJECTS DESIGN
The Generalized Case of the Treatments x Subjects Design -- Analysis of the Total Sum of Squares -- Testing the Significance of the Treatments Effects -- Limitations and Advantages of the Treatments x Subjects Design -- Randomizing or Counter-Balancing Sequence and Order Effects -- Confounding Extraneous Factors with "Subjects" -- Testing Differences in Individual Pairs of Treatment Means -- Establishing a Confidence Interval for the True Mean for Any Treatment.

Chapter 7 - THE GROUPS-WITHIN-TREATMENTS DESIGN
The Generalized Case of the Groups-Within-Treatments Design -- The Analysis of Variance in Groups-Within-Treatments Designs (The Subject as the Unit of Analysis) -- Computational Procedures (Subject the Unit of Analysis) -- Analysis of Unweighted Group Means (The Group as the Unit of Analysis) -- Test of the Hypothesis of Equal Treatment Means (Unweighted) -- The Groups Considered as Random Samples from Corresponding (Hypothetical) Subpopulations -- The Expected Values of ms_A and ms_{GwA} -- Meaning of $F = ms_{GwA}/ms_{wG}$ -- Precision of Individual Means and of Differences in Pairs of Means -- General Advantages and Limitations of the Groups-Within-Treatments Design.

Chapter 8 - THE RANDOM REPLICATIONS DESIGN
The Generalized Case of the Random Replications Design -- The Random Replications (A x R) Design. When the Population Consists of Finite Groups -- Replications of the Simple-Randomized Design with Subgroups of the Same Size (Random Sampling from Randomly Selected Subpopulations) -- The Special Case of "Simple" Replications -- Testing the AR Interaction in Random Replications of Treatments x Levels or Treatments x Subjects Designs -- Testing Differences in Individual Pairs of Treatment Means -- Establishing a Confidence Interval for the True Mean for a Given Treatment -- Important Precautions in the Planning and Administration of a Random Replications (A x R) Experiment -- Advantages and Limitations of the Random Replications Design -- The Possibilities of Simple Random Replication -- The Use of ms_{AL} as an Error Term in Treatments x Levels Designs.

Chapter 9 - FACTORIAL DESIGNS (TWO FACTORS)
The Generalized Case of the Two-Factor (A x B) Design -- The Test of the AB Interaction and Its Interpretation -- Testing the Significance of the Main Effect of Either Factor -- Testing the Simple Effects of Either Factor -- Individual Comparisons of Row, Column, or Cell Means -- The Meaning of $ms_{A\,(or\,B)}/ms_{AB}$ When the Interaction Is Significant -- The Conditions Under Which $ms_{A(or\,B)}/ms_{AB}$ Is Distributed as F -- How to Make Comprehensive Tests of Significance When the Interaction Is in Part Intrinsic and in Part Due to Randomized Type G Errors -- The Use of Transformations.

Chapter 10 - THREE-DIMENSIONAL DESIGNS
Introduction -- Analysis of Total Sum of Squares -- Computational Procedures --

Lindquist (cont'd.)

Meaning of Triple Interaction -- Applications of Three-Dimensional Designs -- Random Replications of a Two-Factor Experiment (A × B × R Designs) -- Treatments × Treatments × Subjects (A × B × S) Designs -- Random Replications of Treatments × Levels Designs (A × L × R Designs) -- Two-Factor Experiments with Matched Groups (A × B × L Designs) -- Three-Factor (A × B × C) Designs -- Testing Differences in Individual Pairs of Means in Three-Dimensional Designs.

Chapter 11 - HIGHER-DIMENSIONAL DESIGNS

Analysis in Higher-Dimensional Designs -- Computational Procedures -- Interpretation of Higher-Order Interactions -- A Notation for Factorial Designs -- Practical Limitations of Higher-Dimensional Designs -- "Complete" and "Incomplete" Factorial Designs.

Chapter 12 - LATIN SQUARE AND GRAECO-LATIN SQUARE DESIGNS

Introduction -- Analysis in Simple Latin Square Designs -- Confounding in Latin Square Designs -- Graeco-Latin Squares.

Chapter 13 - CONTROLLING INDIVIDUAL DIFFERENCES IN FACTORIAL EXPERIMENTS THROUGH THE USE OF "MIXED" DESIGNS

Introduction -- Type I Designs -- Type II Designs -- Type III Designs -- Type IV Designs -- Type V Designs -- Type VI Designs -- Tyoe VII Designs -- Summary of Two- and Three-Factor Designs -- Additional Designs -- Partial Confounding -- Mixed Higher-Order Experiments.

Chapter 14 - ANALYSIS OF COVARIANCE

Nature and Purposes of Analysis of Covariance -- Basic Formulas -- An Illustrative Example -- Importance of the Assumptions Underlying the Test of Significance of the Treatments Effect -- Test of Homogeneity of Regression -- Generalized Procedure -- Analysis of Covariance vs. the Treatments × Levels Design as a Means of Increasing the Precision of an Experiment -- Analysis of Covariance as a Means of Introducing an Additional Factor into a Factorial Experiment -- Statistical Control of More than One Concomitant Variable.

Chapter 15 - TESTS CONCERNED WITH TRENDS

Introduction -- Tests of Trend in the Simple-Randomized Design -- Tests for Trend in Treatments × Levels Designs -- Tests for Trend in Treatments × Subjects Designs -- Tests of Trend in Type II (Confounded) Designs -- Comparisons of Trends -- Designs Appropriate for Trend Comparisons.

Chapter 16 - ESTIMATION OF VARIANCE COMPONENTS IN RELIABILITY STUDIES

Introduction -- The One-Dimensional Design -- The Two-Dimensional Design -- The Three-Dimensional Design -- Groups (of Observations) Within Subjects.

Lindquist (cont'd.)

TABLES

1. Table of X^2
2. Table of t
3. Percent Points in the Distribution of F
4. Phase 1 of the Norton Study. Percents of Mean Square Ratios in Empirical Distributions Exceeding Given Percent Points in the Normal Theory F-Distribution
5. Phases 2, 3, and 4 of the Norton Study. Percents of Mean Square Ratios in Empirical Distributions. Exceeding Given Percent Points in the Normal Theory F-Distribution
6. Distribution of Intelligence Test Scores for an Experimental Sample and for the Population Represented by the Sample
7. Analysis of Total Sum of Squares in Three-Way Table into Eight Components by Successive Applications of the Method of Arithmetic Corrections
8. Computational Formulas for a Three-Classification (R × C × S) Design
9. Complete Sets of Orthogonal Latin Squares
10. Analysis of Type I Designs
11. Analysis of Type II Designs
12. Analysis of Type III Designs
13. Analysis of Type IV Designs
14. Analysis in Type V Designs
15. Analysis in Type VI Designs
16. Analysis in Type VII Designs
17. Summary of Two- and Three-Factor Mixed Designs

Appendix: Table of Random Numbers

Mann, H. B. [1949], ANALYSIS AND DESIGN OF EXPERIMENTS. Dover Publications, Inc., New York, N.Y., U.S A., pp. x+195.

Introduction

Chapter I - CHI-SQUARE DISTRIBUTION AND ANALYSIS OF VARIANCE DISTRIBUTION

Chapter II - MATRICES, QUADRATIC FORMS AND THE MULTIVARIATE NORMAL DISTRIBUTION

Chapter III - ANALYSIS OF VARIANCE IN A ONE WAY CLASSIFICATION

Chapter IV - LIKELIHOOD RATIO TESTS AND TESTS OF LINEAR HYPOTHESES

Chapter V - ANALYSIS OF VARIANCE IN AN r-WAY CLASSIFICATION DESIGN

Chapter VI - THE POWER OF ANALYSIS OF VARIANCE TESTS

Chapter VII - LATIN SQUARES AND INCOMPLETE BALANCED BLOCK DESIGNS

Chapter VIII - GALOIS FIELDS AND ORTHOGONAL LATIN SQUARES

Chapter IX - THE CONSTRUCTION OF INCOMPLETE BALANCED BLOCK DESIGNS

Chapter X - NON-ORTHOGONAL DATA

Chapter XI - FACTORIAL EXPERIMENTS

Chapter XII - RANDOMIZED DESIGNS, RANDOMIZED BLOCKS AND QUASIFACTORIAL DESIGNS

Chapter XIII - ANALYSIS OF COVARIANCE

Chapter XIV - INTERBLOCK ESTIMATES AND INTERBLOCK VARIANCE

TABLES

Myers, J. L. [1966], FUNDAMENTALS OF EXPERIMENTAL DESIGN. Allyn and Bacon, Boston, U.S.A., pp. x+407. Second printing by Allyn and Bacon, June, 1967.

Chapter 1 - PLANNING THE EXPERIMENT
Introduction -- The Independent Variable -- Irrelevant Variables -- Factors In Selecting Experimental Designs -- The Dependent Variable -- Subjects -- Concluding Remarks -- Supplementary Readings.

Chapter 2 - STATISTICAL INFERENCE
Introduction -- Point Estimation -- Interval Estimation -- Hypothesis Testing -- Confidence Intervals and Significance Tests -- Exercises -- Supplementary Readings.

Chapter 3 - NOTATION
A Single Group of Scores -- Several Groups of Scores -- Exercises -- Answers to Exercises.

Chapter 4 - COMPLETELY RANDOMIZED ONE-FACTOR DESIGNS
Introduction -- A Model for the Completely Randomized One-factor Design -- The Analysis of Variance -- Power of the F Test -- Concluding Remarks -- Exercises -- Supplementary Readings.

Chapter 5 - COMPLETELY RANDOMIZED MULTI-FACTOR DESIGNS
Introduction -- A Model for the Completely Randomized Two-factor Design -- The Analysis of Variance for the Completely Randomized Two-factor Design -- A Model for the Completely Randomized Three-factor Design -- The Analysis of Variance for the Completely Randomized Three-factor Design -- Interaction Effects in the Three-factor Design -- A Numerical Example for the Three-factor Design -- More than Three Independent Variables -- Computations for Single df Effects -- Concluding Remarks -- Exercises.

Chapter 6 - DESIGNS USING A CONCOMITANT VARIABLE
Treatments × Blocks: Introduction -- Relative Efficiency -- Selecting the Optimal Number of Blocks -- Extensions of the Treatments × Blocks Design -- The Extreme Groups Design -- Exercises.

Chapter 7 - REPEATED MEASUREMENTS DESIGNS
Introduction -- Models for the One-factor Repeated Measurements Design -- More than One Treatment Variable -- Concluding Remarks -- Exercises.

Chapter 8 - MIXED DESIGNS: BETWEEN- AND WITHIN-SUBJECTS VARIABILITY
Introduction -- One between- and One within-subjects Variable -- Additional Mixed Designs -- Concluding Remarks -- Exercises.

Myers (cont'd.)

Chapter 9 - HIERARCHICAL DESIGNS
Introduction -- Groups within Treatments -- A within-groups Variable -- Repeated Measurements in a Hierarchical Design -- Concluding Remarks -- Exercises.

Chapter 10 - LATIN SQUARE DESIGNS
Introduction -- The Latin Square -- A Simple Latin Square Design -- The Greco-Latin Square -- Replicating a Latin Square -- Between-squares Treatment Variables -- Concluding Remarks -- Exercises -- Supplementary Readings.

Chapter 11 - EXPECTED MEAN SQUARES
Introduction -- Significance Tests -- Estimation of Components of Variance -- Reliability and the Analysis of Variance -- Exercises -- Supplementary Readings.

Chapter 12 - ANALYSIS OF COVARIANCE
Introduction -- The Completely Randomized One-factor Design -- The Analysis of Covariance for Multi-factor Designs -- Interpretation of Covariance Tests -- Comparison with the Treatments × Blocks Design -- Exercises -- Supplementary Readings.

Chapter 13 - FURTHER DATA ANALYSES: QUALITATIVE INDEPENDENT VARIABLES
Introduction -- Sums of Squares for Multiple Comparisons -- Null Hypothesis Tests and Confidence Intervals for Multiple Comparisons -- The Analysis of Interaction -- Concluding Remarks -- Exercises -- Supplementary Readings.

Chapter 14 - FURTHER DATA ANALYSES: QUANTITATIVE INDEPENDENT VARIABLES
Introduction -- The Completely Randomized One-factor Design -- Multi-factor Designs -- Unequal Intervals -- Concluding Remarks -- Exercises.

APPENDIX TABLES

Peng, K. C. [1967], THE DESIGN AND ANALYSIS OF SCIENTIFIC EXPERIMENTS. Addison-Wesley Publishing Company, Reading, Massachusetts, Palo Alto, London, Don Mills, Ontario, pp. ix+252.

PRELIMINARIES
The Design of Experiments -- The Concept of Error Variance -- Estimation -- Testing Hypotheses -- Noncentral χ^2, F, the Power of the F-test, and the Determination of Sample Sizes -- Methods of Multiple Comparisons.

Chapter 1 - TWO-WAY ARRANGEMENTS
Experimental Situation and Some Terminology -- Combinatorial Configuration -- Randomization -- Descriptive Information -- Statistical Analysis -- Computations -- A Numerical Example.

Chapter 2 - THREE-WAY AND MULTIWAY ARRANGEMENTS
Three-way Arrangements -- Operator Calculus and Mapping Schemes for Programming Multifactor Analysis of Variance.

Chapter 3 - METHODS OF PARTITIONING A SUM OF SQUARES
Orthogonal Contrasts -- Orthogonal Polynomials -- Orthogonal Polynomials for Unequal Intervals -- A Numerical Example.

Chapter 4 - NESTED EXPERIMENTS
Nested Two-factor Experiment -- Completely Nested Experiment -- An Experiment with Both Nesting and Crossing.

Chapter 5 - FIXED, RANDOM, AND MIXED MODELS
Introduction -- Models for the Two-way Arrangement with r Replications per Cell -- Expected Mean Squares for the Three-way Arrangement with r Replications per Cell -- General Rules for Expected Mean Squares -- General Rules for Calculating Sum of Squares.

Chapter 6 - RANDOMIZED-BLOCKS, LATIN-SQUARE, AND SPLIT-PLOT DESIGNS
Introduction -- Randomized Blocks -- Latin Square -- Graeco-Latin Square -- Split-plot Design -- Missing Value Technique.

Chapter 7 - FRACTIONAL FACTORIAL DESIGNS AND CONFOUNDING
Introduction -- Factorial Designs with Factors at Two Levels (the 2^K Systems) -- Fractional Factorial Designs with Factors at Two Levels -- Finite Groups and Fractional Factorial Designs with Factors at Two Levels -- The Analysis of Fractional Factorial Designs with Factors at Two Levels -- Factorial Designs with Factors at 3 Levels (the 3^K Systems) -- Fractional Factorial Designs with Factors at Three Levels -- Finite Groups and Fractional Factorial Designs with Factors at Three Levels -- The Analysis of Fractional Factorial Designs with Factors at Three Levels -- Some Other Fractional Factorial Designs -- Confounding in Factorial Experiments.

Peng (cont'd.)

Chapter 8 - RESPONSE SURFACE DESIGNS
Response Surfaces -- Least Squares and Quadratic Surfaces -- Response Surface Designs -- The Method of Steepest Ascent.

Chapter 9 - SPECIAL TOPICS
General Linear Hypothesis Model -- Analysis of Variance in Experimental Design by the General Linear Hypothesis Approach -- Unbalanced Experiments -- The Assumptions Underlying the Analysis of Variance and Some Consequences of Departures from Them.

Chapter 10 - ANALYSIS OF COVARIANCE
Introduction -- Basic Theory of the Technique -- Multiple Covariance on a Multi-factor Structure -- Computational Procedure for Multiple Covariance Analysis.

Chapter 11 - NONFACTORIAL EXPERIMENTS
Introduction -- Designs for One-way Elimination of Error -- Designs for Two-way Elimination of Error -- Designs for Special Purposes.

GENERAL REFERENCES

APPENDIX I A Computer Program for the Analysis of Common Factorial Experiments

APPENDIX II A Computer Program for the Analysis of Latin Square and Graeco-Latin Square Experiments

APPENDIX III A Computer Program for the Analysis of Fractional Factorial Experiments with Factors at Two Levels

INDEX

QUENOUILLE, M. H. [1953], THE DESIGN AND ANALYSIS OF EXPERIMENT. Charles Griffin & Co. Ltd. London and Hafner Publishing Co., New York, U.S.A., pp.xiii+356.

Chapter 1 - THE DESIGN AND ANALYSIS OF EXPERIMENTS
Introduction -- Principles Involved in Experimentation -- Statistical Methods Used in the Analysis of Experiments -- Example of the Interpretation of an Experiment -- Example of the Use of Analysis of Variance -- Assumptions involved in the Analysis of Experiments -- First Steps in Planning an Experiment -- Methods of Improving the Accuracy of an Experiment -- Choosing the Design -- Randomising the Design -- Carrying out the Analysis.

Chapter 2 - RANDOMISED BLOCKS AND LATIN SQUARES
Randomised Block Design -- Example of the Analysis of a Randomised Block Experiment -- Testing Specific Comparisons -- Testing a Series of Treatments -- Latin Square Design -- Example of the Analysis of a Latin Square Experiment -- Graeco-Latin Square Design -- Efficiency of Randomised Blocks and Latin Squares -- Covariance Analysis to the Adjustment of Treatment Means -- Standard Errors of Adjusted Means -- Significance Tests in Covariance Analysis.

Chapter 3 - SIMPLE FACTORIAL AND SPLIT PLOT DESIGNS
Purposes of Factorial Experiments -- Example of a Simple Factorial Design -- The 2^3 Factorial Design -- Example of the Analysis of a 2^3 Factorial Design -- The 2^m Factorial Design -- Designs Involving Factors at Two Levels -- Halved Plot Designs -- Example of a Halved-plot Design.

Chapter 4 - GENERAL FACTORIAL AND SPLIT-PLOT DESIGNS
Factorial Designs with Two Sets of Factors -- Factorial Designs with Several Sets of Factors -- Experiments with Factors at Many Levels -- Components of Introduction -- Interactions of Quality and Quantity -- Dummy Treatments -- The General Split-plot Design -- Standard Errors in Split-plot Experiments -- Example of a Split-plot Experiment -- Covariance Analysis in Split-plot Experiments.

Chapter 5 - FACTORIAL DESIGNS INVOLVING FACTORS AT TWO LEVELS
Confounding of a Single Comparison -- Example of an Experiment with Confounding -- Confounding of Several Comparisons -- Determination of Confounded Designs -- Estimation of Error from High-order Interactions -- Analysis of a 2^m Experiment with Confounding -- Partial Confounding -- Designs Involving Factors at Four Levels.

Chapter 6 - FACTORIAL DESIGNS INVOLVING FACTORS AT THREE LEVELS
The I and J Components of Interaction -- Experiments with Three Factors at Three Levels -- Example of a 3^3 Experiment with Confounding -- Experiments with More than Three Factors at Three Levels -- Confounding with Factors at Two and Three Levels -- Dummy Treatments with Confounding.

Quenouille (cont'd.)

Chapter 7 - COMPLEX FACTORIAL DESIGNS
Modifications to the Factorial Design -- Quasi-Latin Squares for 2^m Experiments -- Quasi-Latin Squares for 3^m Experiments -- Split-plot Confounding -- A Complex Split-plot Experiment -- Fractional Replication.

Chapter 8 - INCOMPLETE BLOCK DESIGNS FOR A SINGLE SET OF TREATMENTS
Types of Design -- Split-plot Arrangements -- The Lattice -- Three-dimensional and Triple Lattices -- Balanced Incomplete Blocks -- Example of a Balanced Incomplete Block Analysis -- Youden Squares and Lattice Squares -- Relative Merits of Incomplete Block Designs.

Chapter 9 - LONG-TERM EXPERIMENTS
Problems of Long-term Policy -- Short-term Designs Involving Time as a Factor -- Adjustment for Residual or Carry-over Effects -- Some Designs for Long-term Experiments with Stable Conditions -- Example of the Estimation of Residual Effects in Stable Experiments -- Further Designs for Long-term Experiments with Stable Conditions -- Some Designs for Serial Long-term Experiments -- Further Designs for Serial Long-term Experiments -- Elimination of Site Differences in Serial Experiments.

Chapter 10 - PLANNING OF GROUPS OF EXPERIMENTS
General Considerations -- Size and Number of Experiments -- Locating the Experiments -- Choosing the Number of Treatments -- Series of Experiments -- Choosing the Designs -- Degrees of Freedom of the Residual Mean Square -- Sampling the Experiments -- Grouping of Experimental Results -- Analysing the Experiments.

Chapter 11 - COMBINATION OF EXPERIMENTAL RESULTS
General Considerations -- Methods of Combinations -- Tests of Homogeneity at Variance and Consistency of Treatment Effects -- Combination of Estimates of a Single Treatment Effect -- Series of Similar Experiments of Comparable Precision -- Example of the Analysis of a Series of Similar Experiments -- Series of Experiments of Differing Precision -- Combination of Estimates Obtained at Different Times and Places -- Combination of Results in Serial Long-term Experiments.

Chapter 12 - SPECIAL DESIGNS AND ANALYSES
Need for Special Designs -- Factorial Designs with Zero Interactions -- $p \times p$ Factorial Designs with Missing Diagonals -- Experiments on Mixtures -- Experiments Estimating Changes in Variability -- Stratification in Experiments Estimating Changes in Variability -- Sequential Experiments -- Partially-controlled and Disarranged Experiments -- Crop Rotation Experiments -- Paired Comparisons.

Quenouille (cont'd.)

Chapter 13 - MISSING OBSERVATIONS
Missing Observations -- The Estimation of Missing Observations -- Example of Analysis with Missing Observations -- Missing Split-plots -- Missing Treatments -- Example of a Missing Treatment Analysis -- Missing Blocks -- Interchanged and Misapplied Treatments -- Missing Observations Affected by Treatments -- Rejection of Extreme Observations.

Chapter 14 - SCALING OF OBSERVATIONS
Reasons for Scaling Observations -- Scaling for Additivity -- Testing for Additivity -- The Scaling of Percentages — the Probit Transformation -- Example of the Use of Probits -- Scaling to Attain Variance-homogeneity -- Scaling for Non-normality -- Analytical Scaling -- Correlated Measurements -- Presentation of Scaled Observations.

BIBLIOGRAPHY
TABLES
INDEX

Vajda, S. [1967], THE MATHEMATICS OF EXPERIMENTAL DESIGN - Incomplete Block Designs and Latin Squares. No. 23 of Griffin's Statistical Monographs and Courses. Hafner Publishing Company, New York, N.Y., U.S.A., and Charles Griffin and Co., London, pp. viii+110.

Chapter I - INTRODUCTION
Finite Groups -- Galois Fields -- Finite Projective Geometries -- Finite Euclidean Spaces -- Difference Sets, and Systems of Difference Sets.

Chapter II - BALANCED INCOMPLETE BLOCK DESIGNS
Finite Spaces -- Conics -- Incidence Matrix -- Existence Theorems -- Structure -- Exercises.

Chapter III - LATIN SQUARES AND ORTHOGONAL ARRAYS
Latin Squares -- Latin Rectangles -- Orthogonal Sets of Latin Squares -- Graeco-Latin Squares Based on Groups -- Existence Theorems -- Orthogonal Arrays -- Construction -- Maximal Strength -- MacNeish's Theorem -- Orthogonal Latin Squares -- The Falsity of Euler's Conjecture -- The Complete Theorem on Graeco-Latin Squares -- Exercises.

Chapter IV - PARTIALLY BALANCED INCOMPLETE BLOCK DESIGNS
Definition -- Construction -- Various Association Schemes -- Association Matrices -- Self-dual Designs -- Partially Balanced Incomplete Block Designs Through Inversion -- Exercises.

Chapter V - PARTIALLY BALANCED INCOMPLETE BLOCK DESIGNS WITH TWO ASSOCIATE CLASSES
Introduction -- Classification - Group-divisible designs (Types of group-divisible designs; Structure); Triangular association scheme; Latin square type designs -- Exercises.

SOLUTIONS TO EXERCISES
BIBLIOGRAPHY
INDEX

Villars, D. S. [1951], STATISTICAL DESIGN AND ANALYSIS OF EXPERIMENTS FOR DEVELOPMENT RESEARCH. Wm. C. Brown Company, Dubuque, Iowa, pp. xvii+455.

Chapter 1 - INTRODUCTION
Duality of Experimentation -- Small-Sample Statistics -- Examples.

Chapter 2 - USE OF THE CALCULATING MACHINE
Features of Calculating Machine -- Number of Significant Figures to Retain -- Formation of Totals -- Summing Squares and Cross Products of Deviations from Mean -- Coding -- Weighting -- Simultaneous Summation of Squares and Cross Products of Two Variables -- Extraction of Square Root.

Chapter 3 - STUDENT'S t-TEST
Significance of Difference between Means -- Least Significant Difference, (LSD) -- Number of Replications Required to Establish a Definite Sample Difference as Significant -- Fiducial Limits -- Confidence Limits -- Problems.

Chapter 4 - DISTRIBUTIONS AND STATISTICAL SIGNIFICANCE
Normal Distribution -- Binomial and Multinomial Distribution -- Poisson Distribution -- Parameters and Estimates -- Standard Deviation and Variance -- Degrees of Freedom -- Chi-square Distribution -- z- and F-Distribution -- Null Hypothesis and Statistical Significance -- Statistical Odds -- A Posteriori Tests -- Exercises.

Chapter 5 - ANALYSIS OF VARIANCE
Variability -- "Between and Within Treatments" Variance Analysis -- "Interactions" Variance Analysis -- Factorial Variance Analysis -- Replication Degeneracy -- Exercises and Problems.

Chapter 6 - PRINCIPLES OF DESIGN
Introduction -- Replication -- Randomization -- Symmetry.

Chapter 7 - DESIGNS AVAILABLE
Introduction -- "Between and Within" - One Treatment Classification -- Factorial Design - Several Treatment Classifications -- Medium Design -- Randomized Block -- Replication Degeneracy -- Balanced Incomplete Block -- Latin Square -- Confounding -- Exercises and Problems.

Chapter 8 - ANALYSIS OF COVARIANCE
Introduction -- Linear Regression -- Error in Both Variables -- Covariance Analysis -- Multilinear Regression -- Insertion or Removal of Independent Variable -- Multiple Covariance Analysis -- Polynomial Regression - General Case -- Polynomial Regression - Equally Spaced Independent Variable -- Exponential Regression -- Exercises and Problems.

Villars (Cont'd.)

Chapter 9 - DISCONTINUOUS DATA

Introduction -- Binomial Distribution -- Multinomial Distribution - Case of zeros -- Poisson Distribution -- Chi-square Test -- Null Hypothesis and Degrees of Freedom -- Meaning of Chi-square Probability and Comparison with Binomial Distribution Probability -- Contingency Tables -- Enumeration Data by Analysis of Variance -- Exercises and Problems.

Chapter 10 - THE CONTROL CHART

Introduction -- Statistical Control -- Estimation of Standard Deviation from Range -- Systematic Procedure for Assignment of Limit Lines for Individual Fluctuations -- Control Chart for Ranges -- Other Criteria for Lack of Statistical Control -- Criterion for Change in Trend.

Chapter 11 - SEQUENTIAL ANALYSIS

Introduction -- Type of Problem -- Risks -- Computations -- Sequential Test for Change in Trend.

Chapter 12 - MISCELLANEOUS REFINEMENTS

Sheppard's Correction -- Missing Values -- Homogeneity of Variances -- Fisher and Behrens' Test for Significance of Difference between Means of Two Samples with Different Variances -- Fiducial Limits of True Variance -- Combining Significance Tests -- Normalization of Distributions -- Linearization of Functions -- Weighting -- Statistical Comparison of Two Machines -- Coefficient of Resolution -- Use of Variance Components for Redesigning Experiment -- Rational System for Determining Correct Method of Analyzing Variance Problems.

Chapter 13 - EFFICIENCY STUDIES

Information -- Efficiency -- Treatment Phase of Design -- Optimum Ratio of Tests per Mix -- Testing Medium Phase of Design.

Chapter 14 - FUNDAMENTAL THEORY

Propagation of Errors -- Transformation from Normal to Chi-square Distribution -- Development of Dependent Variable as Linear Combination of Orthogonal Functions -- Assignment of Orthogonal Functions. Multilinear Regression -- Assignment of Orthogonal Functions. Polynomial Regression -- Mathematical Validity of Variance and Covariance Analysis -- Derivation of Student t-Distribution -- Derivation of Fisher's z-Distribution -- Components of Variance -- Variance of Subclass Means -- Exercises.

APPENDIX

1 - Table of Student t-distribution
2 - Table of Chi-square distribution
3 - Tables of F-distribution
4 - Charts of Statistical Odds
5 - Sukhatme's table of d and Fisher and Behrens' test
6 - Villars and Anderson F' significance test

Villars (Cont'd.)

7 - Latin squares
8 - Variance components
9 - Balanced incomplete blocks
10 - Control Chart tables
11 - Answers to problems
12 - Glossary

INDEX

Authors
Subjects

Winer, B. J. [1962], STATISTICAL PRINCIPLES IN EXPERIMENTAL DESIGN. McGraw-Hill Book Company, New York, San Francisco, Toronto, London, pp. x+672.

INTRODUCTION

Chapter 1 - BASIC CONCEPTS IN STATISTICAL INFERENCE
Basic Terminology in Sampling -- Basic Terminology in Statistical Estimation -- Basic Terminology in Testing Statistical Hypotheses.

Chapter 2 - TESTING HYPOTHESES ABOUT MEANS AND VARIANCES
Testing Hypotheses on Means — σ Assumed Known -- Tests of Hypotheses on Means — σ Estimated from Sample Data -- Testing Hypotheses about the Difference between Two Means — Assuming Homogeneity of Variance -- Computational Formulas for the t Statistic -- Test for Homogeneity of Variance -- Testing Hypotheses about the Difference between Two Means — Assuming that Population Variances are not Equal -- Testing Hypotheses about the Difference between Two Means — Correlated Observations -- Combining Several Independent Tests on the Same Hypothesis.

Chapter 3 - DESIGN AND ANALYSIS OF SINGLE-FACTOR EXPERIMENTS
Introduction -- Definitions and Numerical Example -- Structural Model for Single-factor Experiment — Model I -- Structural Model for Single-factor Experiment — Model II (Variance Component Model) -- Methods for Deriving Estimates and Their Expected Values -- Comparisons among Treatment Means -- Use of Orthogonal Components in Tests for Trend -- Use of the Studentized Range Statistic -- Alternative Procedures for Making a Posteriori Tests -- Comparing all Means with a Control -- Tests for Homogeneity of Variance -- Unequal Sample Sizes -- Determination of Sample Size.

Chapter 4 - SINGLE-FACTOR EXPERIMENTS HAVING REPEATED MEASURES ON THE SAME ELEMENTS
Purpose -- Notation and Computational Procedures -- Numerical Example -- Statistical Basis for the Analysis -- Use of Analysis of Variance to Estimate Reliability of Measurements -- Tests for Trend -- Analysis of Variance for Ranked Data -- Dichotomous Data.

Chapter 5 - DESIGN AND ANALYSIS OF FACTORIAL EXPERIMENTS
General Purpose -- Terminology and Notation -- Main Effects -- Interaction Effects -- Experimental Error and Its Estimation -- Estimation of Mean Squares Due to Main Effects and Interaction Effects -- Principles for Constructing F Ratios -- Higher-order Factorial Experiments -- Estimation and Tests of Significance for Three-factor Experiments -- Simple Effects and Their Tests -- Geometric Interpretation of Higher-order Interactions -- Nested Factors (hierarchal designs) -- Split-plot Designs -- Rules for Deriving the Expected Values of Mean Squares -- Quasi F Ratios -- Preliminary Tests on the Model and Pooling Procedures -- Individual Comparisons -- Partition of Main Effects and Interaction into Trend Components -- Replicated Experiments -- The Case n = 1 and a Test for Nonadditivity

Winer (cont'd.)

The Choice of a Scale of Measurement and Transformations -- Unequal Cell Frequencies -- Unequal Cell Frequencies — Least-squares Solution.

Chapter 6 - FACTORIAL EXPERIMENTS — COMPUTATIONAL PROCEDURES AND NUMERICAL EXAMPLES

General Purpose -- p × q Factorial Experiment Having n Observations per Cell -- p × q Factorial Experiment — Unequal Cell Frequencies -- Effect of Scale of Measurement on Interaction -- p × q × r Factorial Experiment Having n Observations per Cell -- Computational Procedures for Nested Factors -- Factorial Experiment with a Single Control Group -- Test for Nonadditivity -- Computation of Trend Components -- General Computational Formulas for Main Effects and Interactions -- Missing Data -- Special Computational Procedures When all Factors Have Two Levels -- Illustrative Applications -- Unequal Cell Frequencies — Least-squares Solution.

Chapter 7 - MULTIFACTOR EXPERIMENTS HAVING REPEATED MEASURES ON THE SAME ELEMENTS

General Purpose -- Two-factor Experiment with Repeated Measures on One Factor -- Three-factor Experiment with Repeated Measures (case I) -- Three-factor Experiment with Repeated Measures (case II) -- Other Multifactor Repeated-Measure Plans -- Tests on Trends -- Testing Equality and Symmetry of Covariance Matrices -- Unequal Group Size.

Chapter 8 - FACTORIAL EXPERIMENTS IN WHICH SOME OF THE INTERACTIONS ARE CONFOUNDED

General Purpose -- Modular Arithmetic -- Revised Notation for Factorial Experiments -- Method for Obtaining the Components of Interactions -- Designs for 2 × 2 × 2 Factorial Experiments in Blocks of Size 4 -- Simplified Computational Procedures for 2^k Factorial Experiments -- Numerical Example of 2 × 2 × 2 Factorial Experiment in Blocks of Size 4 -- Numerical Example of 2 × 2 × 2 Factorial Experiment in Blocks of Size 4 (repeated measures) -- Designs for 3 × 3 Factorial Experiments -- Numerical Example of 3 × 3 Factorial Experiment in Blocks of Size 3 -- Designs for 3 × 3 × 3 Factorial Experiments -- Balanced 3 × 2 × 2 Factorial Experiment in Blocks of Size 6 -- Numerical Example of 3 × 2 × 2 Factorial Experiment in Blocks of Size 6 -- 3 × 3 × 3 × 2 Factorial Experiment in Blocks of Size 6 -- Fractional Replication.

Chapter 9 - BALANCED LATTICE DESIGNS AND OTHER BALANCED INCOMPLETE-BLOCK DESIGNS

General Purpose -- Balanced Simple Lattice -- Numerical Example of Balanced Simple Lattice -- Balanced Lattice-square Designs -- Balanced Incomplete-block Designs -- Numerical Example of Balanced Incomplete-block Design -- Youden Squares -- Numerical Example of Youden Square -- Partially Balanced Designs -- Numerical Example of Partially Balanced Design -- Linked Paired-comparison Designs.

Winer (cont'd.)

Chapter 10 - LATIN SQUARES AND RELATED DESIGNS
Definition of Latin Square -- Enumeration of Latin Squares -- Structural Relation between Latin Squares and Three-factor Factorial Experiments -- Uses of Latin Squares -- Analysis of Latin-square Designs — No Repeated Measures -- Analysis of Greco-Latin Squares -- Analysis of Latin Squares — Repeated Measures.

Chapter 11 - ANALYSIS OF COVARIANCE
General Purpose -- Single-factor Experiments -- Numerical Example of Single-factor Experiment -- Factorial Experiment -- Computational Procedures for Factorial Experiment -- Factorial Experiment — Repeated Measures -- Multiple Covariates.

APPENDIX A - TOPICS CLOSELY RELATED TO THE ANALYSIS OF VARIANCE
Kruskal-Wallis H Test -- Contingency Table with Repeated Measures -- Comparing Treatment Effects with a Control -- General Partition of Degrees of Freedom in a Contingency Table -- Hotelling's T^2 Test for the Equality of k Means -- Least-squares Estimators — General Principles.

APPENDIX B - TABLES
Unit Normal Distribution -- Student's t Distribution -- F Distribution -- Distribution of the Studentized Range Statistic -- Arcsin Transformation -- Distribution of t Statistic in Comparing Treatment Means with a Control -- Distribution of F_{max} Statistic -- Critical Values for Cochran's Test for Homogeneity of Variance -- Chi-square Distribution -- Coefficients of Orthogonal Polynomials -- Curves of Constant Power for the Test on Main Effects -- Random Permutations of 16 Numbers.

CONTENT REFERENCES
REFERENCES TO EXPERIMENTS
INDEX

Yates, F. [1937], THE DESIGN AND ANALYSIS OF FACTORIAL EXPERIMENTS. Technical Communication No.35, Imperial Bureau of Soil Science, Harpenden, England, pp.1-95.

Chapter 1 - INTRODUCTION
Principles Underlying Factorial Design -- Criticisms of Factorial Design -- Scope of the Present Paper -- New Material -- Notation, etc.

Chapter 2 - A SIMPLE FACTORIAL EXPERIMENT ON POTATOES
Yields of the Different Combinations of Treatments -- Main Effects -- Interactions -- Calculation of the Main Effects and Interactions from the Experimental Yields -- Interpretation of Main Effects and Interactions -- General Remarks.

Chapter 3 - STATISTICAL ANALYSIS OF A 2 x 2 x 2 EXPERIMENT

Chapter 4 - CONFOUNDING
Example to Illustrate Confounding -- Statistical Analysis -- Presentation of Results -- Example of Partial Confounding -- Statistical Analysis -- Presentation of Results.

Chapter 5 - SYSTEMS OF CONFOUNDING FOR 2 x 2 x 2 x ... DESIGNS
Confounding with Five Factors -- Confounding with Six Factors -- Confounding with Four Factors in Blocks of 4 Plots -- General Remarks.

Chapter 6 - ESTIMATION OF ERROR FROM HIGH-ORDER INTERACTIONS

Chapter 7 - AN EXPLORATORY EXPERIMENT ON BEANS
Analysis -- Gain in Precision Due to Confounding.

Chapter 8 - CONFOUNDING IN LATIN SQUARE DESIGNS WITH FACTORS AT TWO LEVELS
2 x 2 x 2 Design in Two 4 x 4 Latin Squares -- Numerical Example -- Arrangements for Five and Six Factors in an 8 x 8 Square.

Chapter 9 - FACTORS AT MORE THAN TWO LEVELS
Two Factors -- Three or More Factors -- Simplification When One of the Factors is at Two Levels Only -- Procedure When Two or More Factors are at Two Levels Only -- Two Factors at Three Levels: Formal Sub-division of Interactions in a 3 x 3 Table -- Example.

Chapter 10 - CONFOUNDING WITH THREE AND FOUR FACTORS EACH AT THREE LEVELS
3 x 3 x 3 Designs in Blocks of 9 Plots -- Example of a 3 x 3 x 3 Design -- Adjusted Yields of Three-factor Combinations -- 3 x 3 x 3 x 3 Designs in Blocks of 9 Plots -- 3^3 and 3^4 Designs in Quasi-Latin Squares -- Extension to 3^n in Blocks of 3^{n-1} or 3^{n-2}.

Yates (cont'd.)

Chapter 11 - THE SUBDIVISION OF SETS OF DEGREES OF FREEDOM
Subdivision of Main Effects -- Subdivision of Interactions -- Example -- General Remarks.

Chapter 12 - THE 3 x 3 x 3 DESIGN: SINGLE REPLICATION
Systematic Method of Analysis -- Alternative Method -- The Linear Component of the Three-factor Interaction.

Chapter 13 - CONFOUNDING WITH SOME FACTORS AT TWO AND SOME AT THREE LEVELS
3 x 2 x 2 Design in Blocks of 6 Plots -- Statistical Analysis of 3 x 2 x 2 Design -- Example -- 3 x 2 x 2 x 2 Design in Blocks of 6 Plots -- Extension to 3×2^n Design in Blocks of $3 \times 2^{n-1}$ and $3 \times 2^{n-2}$ Plots -- 3 x 3 x 2 Design in Blocks of 6 Plots -- 3 x 3 x 3 x 2 Design in Blocks of 6 Plots -- Extension to $3^n \times 2$ Designs in Blocks of $3^{n-1} \times 2$ and $3^{n-2} \times 2$ Plots -- 3 x 3 x 2 Design in a 6 x 6 Quasi-Latin Square.

Chapter 14 - CONFOUNDING WITH ONE OR MORE FACTORS AT FOUR LEVELS OR EIGHT LEVELS
General Method -- Example: 4 x 4 Designs -- Combined Varietal and Manurial Trials in Latin Squares.

Chapter 15 - DUMMY TREATMENTS
Application of Fertilizer at Two Different Times -- Alternative Designs -- 3 x 3 x 3 Designs Including Quality Differences.

Chapter 16 - ARRANGEMENTS WITH SPLIT PLOTS
Structure and Analysis of Split-plot Designs -- Example: A Varietal and Manurial Trial on Oats -- Calculation of Standard Errors -- Efficiency -- Confounding of Interactions in Split-plot Designs -- Half-plaid Latin Squares -- Plaid Squares -- Use of Latin Squares with Split Plots in Varietal Trials -- The Graeco-Latin Square -- The Hyper-Graeco-Latin Square.

Chapter 17 - VARIETAL TRIALS — QUASI FACTORIAL DESIGNS
The Lattice -- Triple and Balanced Lattices -- Lattice Squares -- Three-dimensional Lattices -- Non-factorial Designs: Balanced Incomplete Blocks -- The Introduction of Additional Treatments in Quasi-factorial Designs.

NOTES
Number of Figures Required in the Computations and Results -- Numerical Divisors in the Analysis of Variance, etc. -- Orthogonal Functions -- Hints on the Use of Calculating Machines.

REFERENCES AND MATERIAL FOR FURTHER READING

XI. ACKNOWLEDGEMENTS

This work was partially supported by a U.S. Army Contract, No. DA-31-124-ARO-D-462, University of Wisconsin, and by a General Medical Sciences Research Grant, No. 5 RO1 GM05900, Cornell University.

The authors wish to acknowledge the generous and continued help from the secretarial staffs of the Biometrics Unit, Cornell University, and of the Biometrical Services, University of Sydney. In particular, the authors wish to express their sincere thanks to Mrs. Sevim Cadun for her careful and painstaking efforts in producing the final version of the manuscript.

XII. REFERENCES

1. Federer, W. T. [1964], Literature review of experimental design through 1949. Mathematics Research Center Technical Summary Report #405, University of Wisconsin, February.

2. Federer, W. T. and Balaam, L. N. [1970], Bibliography of Experiment Design 1950 - 1967. Mathematics Research Center Technical Summary Report #1080, University of Wisconsin, July.

3. Kendall, M. G. and Doig, A. G. [1962], Bibliography of Statistical Literature, 1950 - 1958. Oliver and Boyd, Edinburgh and London.

4. Kendall, M. G. and Doig, A. G. [1965], Bibliography of Statistical Literature, 1940 - 1949. Oliver and Boyd, Edinburgh and London and Hafner Publishing Company, New York.

5. Kendall, M. G. and Doig, A. G. [1968], Bibliography of Statistical Literature, Pre-1940. Oliver and Boyd, Edinburgh and London.